Statistical Analysis of Ecotoxicity Studies

Statistical Analysis of Ecotoxicity Studies

John W. Green, Ph.D. (Mathematics), Ph.D. (Statistics)
Principal Consultant: Biostatstics
DuPont Data Science and Informatics

Timothy A. Springer, Ph.D. (Wildlife and Fisheries Science)
Director of Special Projects and IT Operations
EAG/Wildlife International

Henrik Holbech, Ph.D. (Ecotoxicology)
Associate Professor
Institute of Biology, University of Southern Denmark

WILEY

This edition first published 2018

Registered Office
John Wiley & Sons, Inc., 111 River Street, Hoboken, NJ 07030, USA

Editorial Office
111 River Street, Hoboken, NJ 07030, USA

For details of our global editorial offices, customer services, and more information about Wiley products visit us at www.wiley.com.

Wiley also publishes its books in a variety of electronic formats and by print-on-demand. Some content that appears in standard print versions of this book may not be available in other formats.

Library of Congress Cataloging-in-Publication Data

Names: Green, John William, 1943– author. | Springer, Timothy A., 1948 March 3– author. | Holbech, Henrik, 1969– author.
Title: Statistical analysis of ecotoxicity studies / by John W. Green, Ph.D., Timothy A. Springer, Ph.D., Henrik Holbech, Ph.D.
Description: First edition. | Hoboken, NJ : John Wiley & Sons, 2018. | Identifiers: LCCN 2018008775 (print) | LCCN 2018014558 (ebook) | ISBN 9781119488828 (pdf) | ISBN 9781119488811 (epub) | ISBN 9781119088349 (cloth)
Subjects: LCSH: Environmental toxicology–Statistical methods. | Toxicity testing–Statistical methods.
Classification: LCC QH541.15.T68 (ebook) | LCC QH541.15.T68 G74 2018 (print) | DDC 615.9/07–dc23
LC record available at https://lccn.loc.gov/2018008775

Cover design by Wiley
Cover image: © Ross Collier/Alamy Stock Photo

Set in 10.25/12pt Times by SPi Global, Pondicherry, India

Printed in the United States of America

V10002140_071318

Contents

Preface ix
Acknowledgments xi
About the Companion Website xiii

1. An Introduction to Toxicity Experiments 1

1.1 Nature and Purpose of Toxicity Experiments 1
1.2 Regulatory Context for Toxicity Experiments 7
1.3 Experimental Design Basics 8
1.4 Hierarchy of Models for Simple Toxicity Experiments 12
1.5 Biological vs. Statistical Significance 13
1.6 Historical Control Information 15
1.7 Sources of Variation and Uncertainty 15
1.8 Models with More Complex Structure 16
1.9 Multiple Tools to Meet a Variety of Needs or Simple Approaches to Capture Broad Strokes? 16

2. Statistical Analysis Basics 19

2.1 Introduction 19
2.2 NOEC/LOEC 19
2.3 Probability Distributions 24
2.4 Assessing Data for Meeting Model Requirements 29
2.5 Bayesian Methodology 30
2.6 Visual Examination of Data 30
2.7 Regression Models 32
2.8 Biology-Based Models 34
2.9 Discrete Responses 35
2.10 Time-to-Event Data 37
2.11 Experiments with Multiple Controls 38
Exercises 41

3. Analysis of Continuous Data: NOECs 47

3.1 Introduction 47
3.2 Pairwise Tests 47
3.3 Preliminary Assessment of the Data to Select the Proper Method of Analysis 53
3.4 Pairwise Tests When Data do not Meet Normality or Variance Homogeneity Requirements 62
3.5 Trend Tests 67
3.6 Protocol for NOEC Determination of Continuous Response 75
3.7 Inclusion of Random Effects 75
3.8 Alternative Error Structures 76
3.9 Power Analyses of Models 77
Exercises 81

4. Analysis of Continuous Data: Regression 89

4.1 Introduction 89
4.2 Models in Common Use to Describe Ecotoxicity Dose–Response Data 92
4.3 Model Fitting and Estimation of Parameters 95
4.4 Examples 104
4.5 Summary of Model Assessment Tools for Continuous Responses 112
Exercises 114

5. Analysis of Continuous Data with Additional Factors 123

5.1 Introduction 123
5.2 Analysis of Covariance 123
5.3 Experiments with Multiple Factors 135
Exercises 154

6. Analysis of Quantal Data: NOECs 157

6.1 Introduction 157
6.2 Pairwise Tests 157
6.3 Model Assessment for Quantal Data 160
6.4 Pairwise Models that Accommodate Overdispersion 162
6.5 Trend Tests for Quantal Response 165

6.6 Power Comparisons of Tests for Quantal Responses 168
6.7 Zero-Inflated Binomial Responses 172
6.8 Survival- or Age-Adjusted Incidence Rates 175
Exercises 179

7. Analysis of Quantal Data: Regression Models 181

7.1 Introduction 181
7.2 Probit Model 181
7.3 Weibull Model 188
7.4 Logistic Model 188
7.5 Abbott's Formula and Normalization to the Control 190
7.6 Proportions Treated as Continuous Responses 197
7.7 Comparison of Models 198
7.8 Including Time-Varying Responses in Models 199
7.9 Up-and-Down Methods to Estimate LC50 204
7.10 Methods for EC*x* Estimation When there is Little or no Partial Mortality 206
Exercises 215

8. Analysis of Count Data: NOEC and Regression 219

8.1 Reproduction and Other Nonquantal Count Data 219
8.2 Transformations to Continuous 219
8.3 GLMM and NLME Models 223
8.4 Analysis of Other Types of Count Data 228
Exercises 237

9. Analysis of Ordinal Data 243

9.1 Introduction 243
9.2 Pathology Severity Scores 243
9.3 Developmental Stage 249
Exercises 255

10. Time-to-Event Data 259

10.1 Introduction 259
10.2 Kaplan–Meier Product-Limit Estimator 261
10.3 Cox Regression Proportional Hazards Estimator 266
10.4 Survival Analysis of Grouped Data 268
Exercises 271

11. Regulatory Issues 275

11.1 Introduction 275
11.2 Regulatory Tests 275
11.3 Development of International Standardized Test Guidelines 276
11.4 Strategic Approach to International Chemicals Management (SAICM) 279
11.5 The United Nations Globally Harmonized System of Classification and Labelling of Chemicals (GHS) 279
11.6 Statistical Methods in OECD Ecotoxicity Test Guidelines 279
11.7 Regulatory Testing: Structures and Approaches 279
11.8 Testing Strategies 287
11.9 Nonguideline Studies 291

12. Species Sensitivity Distributions 293

12.1 Introduction 293
12.2 Number, Choice, and Type of Species Endpoints to Include 294
12.3 Choice and Evaluation of Distribution to Fit 294
12.4 Variability and Uncertainty 300
12.5 Incorporating Censored Data in an SSD 302
Exercises 307

13. Studies with Greater Complexity 309

13.1 Introduction 309
13.2 Mesocosm and Microcosm Experiments 310
13.3 Microplate Experiments 316
13.4 Errors-in-Variables Regression 321
13.5 Analysis of Mixtures of Chemicals 323
13.6 Benchmark Dose Models 326
13.7 Limit Tests 327
13.8 Minimum Safe Dose and Maximum Unsafe Dose 329
13.9 Toxicokinetics and Toxicodynamics 331
Exercises 343

Appendix 1 Dataset 345

Appendix 2 Mathematical Framework 347

A2.1 Basic Probability Concepts 347
A2.2 Distribution Functions 348
A2.3 Method of Maximum Likelihood 350
A2.4 Bayesian Methodology 352
A2.5 Analysis of Toxicity Experiments 354

A2.6 Newton's Optimization Method 358
A2.7 The Delta Method 359
A2.8 Variance Components 360

Appendix 3 Tables 363

Table A3.1 Studentized Maximum Distribution 364
Table A3.2 Studentized Maximum Modulus Distribution 365
Table A3.3 Linear and Quadratic Contrast Coefficients 366
Table A3.4 Williams' Test $\bar{t}_{\alpha,k}$ for $\alpha = 0.05$ 367

References 371
Author Index 385
Subject Index 389

Preface

John Green and Tim Springer developed a one-day training course, Design and Analysis of Ecotox Experiments, for the Society for Environmental Toxicology and Chemistry (SETAC) and delivered it for the first time at the SETAC Europe 13th Annual Meeting in Hamburg, Germany, in 2003. Since then, in many years we have taught this course at the annual SETAC conferences in Europe and North America, updating it each time to stay abreast of the evolving regulatory requirements. In 2011, Henrik Holbech joined us and has made valuable contributions ever since. In 2014, Michael Leventhal of Wiley approached us with the idea of turning the training course into a textbook. The result is the current book, and we appreciate the opportunity to reach a wider audience.

This book covers the statistical methods in all current OECD test guidelines related to ecotoxicity. Most of these have counterparts in the United States Environmental Protection Agency (USEPA) guidelines. In addition, statistical methods in several WHO and UN guidelines are also covered, as are guidelines in development or that have been proposed. Chapter 11 provides a good coverage of all the test guidelines covered in this book with reference to the chapters in which guideline-specific statistical methods are developed. With very few exceptions, the data used in the examples and exercises are from studies done for product submissions or in developing some regulatory test guideline. The authors have been members for a combined total of more than 30 years of the OECD validation management group for ecotoxicity (VMG-eco) responsible for development and update of significant portions of numerous current test guidelines including OECD TG 210, 229, 230, 234, 236, 240, 241, 242, and 243. We have also been actively involved in designing and analyzing ecotoxicity studies for more than a combined total of 60 years. One or more of us were also members of the expert groups that developed (i) the European Framework for Probabilistic Risk Assessment (Chapman et al., 2007), (ii) OECD *Fish Toxicity Testing Framework* (OECD, 2014c), (iii) *Current Approaches in the Statistical Analysis of Ecotoxicity Data: A Guidance to Application* (OECD, 2014a, 2006a), (iv) OECD test guideline 223 that describes a sequential test designed to measure mortality in avian acute tests, (v) OECD *Guidance Document on Standardised Test Guidelines for Evaluating Chemicals for Endocrine Disruption* (OECD, 2012a) and (vi) OECD test guideline 305 for assessing bioaccumulation in fish.

Our intent is to provide an understanding of the statistical methods used in the regulatory context of ecotoxicity. However, the coverage and treatment of the topics should appeal to a much wider audience. A mathematical appendix is included to provide technical issues, but the focus is on the practical aspects of model fitting and hypothesis tests. There are numerous exercises based on real studies to help the reader enhance his or her understanding of the topics. Ample references are provided to allow the interested reader to pursue topics in greater depth. We have not shied away from controversies in the field. We think it important that the reader understand that statistics is not free of controversy and should be well-informed on these issues. Nonetheless, while we have points of view on these topics and express them, we have tried to take an even-handed approach in describing the different points of view and provide references to allow the reader to more fully appreciate the arguments on these issues.

A frequent question from participants in the training course was where one could find software to carry out the methods of analysis we taught and were required or at least recommended in regulatory test guidelines. While we have developed in-house proprietary SAS-based software for this purpose, it has not been possible to share it. One of the benefits of this textbook is the availability of a website created by Wiley where we are providing SAS and R programs for almost all methods presented. In some instances, rather than present programs, we provide a link to free online software that has been developed for specific guidelines or for a more general use. In some cases, we have been unable to find R programs to carry out the recommended methods. For those cases especially, we invite the readers of this book to develop and send such programs to us. In a few cases, no SAS program is provided. In all cases, a program or link is provided for all analyses discussed. After we test programs supplied by readers, we will put them on the website with appropriate acknowledgments. Also, if any shortcomings are found in the initially provided programs, we encourage the readers to bring them to our attention and we will post corrections or improvements. As regulatory requirements change or methods improve, we will update the website.

We have had support from numerous people over the years in developing the training material and the material for this book. Colleagues too numerous to name from DuPont, Wildlife International/EAG, USEPA, OECD, and other companies, universities, and CROs have contributed ideas and data that have been very helpful in improving our understanding of ecotoxicology. Two instructors joined us, Michael Newman of Virginia Institute of Marine Science, School of Marine Science, The College of William and Mary, and Chen Teel of DuPont, each for one offering of the course and both added value. In addition, we have SAS expertise, but more limited experience with R. As a consequence, while we developed some R programs ourselves, several very capable people were engaged to develop most R programs for the website. Several deserve special acknowledgment. We have modified their programs in minor ways to fit the needs of the website and accept responsibility for any errors.

Joe Swintek is a statistician working with the Duluth office of the USEPA. He was a contributor to one of our publications (Green et al., 2014) and turned the SAS version of the StatCHARRMS software John and Amy Saulnier developed under contract for the USEPA into an R package. The SAS version is provided in Appendix 1 (the website) and the R version is now in the CRAN library. A link is provided in the references (Swintek, 2016). In addition to the RSCABS program for histopathology severity scores (Chapter 9), StatCHARRMS contains the Dunnett and Dunn tests, the step-down trend tests Jonckheere–Terpstra (Chapter 3), Cochran–Armitage and Fisher's exact tests (Chapter 6), Shapiro–Wilk and Levene tests for normality and variance homogeneity (Chapter 3), and repeated measures ANOVA for multi-generation medaka reproduction studies (Chapter 5). Several of these tests are provided in Appendix 1 in stand-alone versions, as well as in the full CRAN version. In addition, Joe developed a versatile R program for the important Williams' test, and that is in Appendix 1 and has been added to the StatCHARRMS package. We were surprised to find that this test had not previously been released in an R package, so far as we are aware. There is an R package, multcomp, that refers to Williams' type contrasts within the function mcp, but the results deviate substantially from Williams' test. We have verified with the developer, Ludwig Hothorn, that package mcp does not provide Williams' test. More discussion on this is provided in Chapter 3. Joe also provided numerous other R programs for several chapters as well as pointing out a simple R function based on the package sas7bdat for reading a SAS dataset into R without the need to have SAS installed or converting the dataset to excel or text first. We are very grateful for his contributions.

Chapter 13 leans heavily on discussions of the expert group that developed guidance on implementation of OECD test guideline 305 on bioaccumulation in fish. In particular, Tom Aldenberg of RIVM has provided invaluable communications to us concerning the R program, bcmfR, that he has provided to OECD for analysis of bioconcentration and biomagnification studies.

Georgette Asherman also deserves special mention, primarily for her R programming work for Chapter 5. Among her notable contributions were versatile and robust versions of the Shapiro–Wilk and Levene tests, the Shirley nonparametric ANCOVA program, two parametric ANCOVA programs, programs to add confidence bounds to the graphic output for nonlinear regression, and zero-inflated binomial and beta-binomial models.

Erand Smakaj provided training in the use of R-Studio and contributed programs for survival analysis and for several topics in Chapter 13 and was very accommodating throughout the text and code development.

Xiaopei Jin made important contributions to the R programs for Chapter 8 and demonstrated useful capabilities of R that can be applied to programs in all chapters.

Finally, we would be remiss not to acknowledge the many contributions Amy Saulnier has made to SAS programming used in this book and elsewhere. John has worked with Amy over the entire 29+ years of his DuPont career. In addition to turning his SAS programs into the user-friendly StatCHARRMS program, she has done the same for two other heavily used SAS-based in-house software packages routinely used for our toxicology and ecotoxicology analyses for regulatory submissions. She has maintained these programs, updated them as needed to stay current with regulatory requirements and changes in the computing environment, and has been an essential contributor to DuPont's work for over three decades.

The term GLMM is used for generalized linear models regardless of whether there is a random term. This encompasses both generalized linear mixed models and fixed effects models. The term GLM is reserved to the classic general linear model with normal errors.

Acknowledgments

In addition to the people mentioned above for programming and other professional support, John would like to acknowledge his wife Marianne, without whose unwavering support and understanding, this book would not have been possible. He would also like to acknowledge the support he received from his daughters and step-daughter, M'Lissa, Janel, and Lauren, who encouraged him throughout. Finally, he would like to thank his companions Sam, Max, Ben, and of course Jack for their warmth and comfort through the countless hours devoted to this work. Henrik would like to acknowledge his wife Bente, always supporting the work on the book.

About the Companion Website

This book is accompanied by a companion website:

www.wiley.com/go/Green/StatAnalysEcotoxicStudy

The companion website contains programs in SAS and R to carry out the analyses that are described in the text. These programs will be updated as improvements are identified or regulations change. Readers are invited to send corrections or improvements to the authors through Wiley. Once these are verified and judged appropriate, they will be added to the website with appropriate acknowledgment. Also on the website are datasets referenced in the text but too large to include there. These are in the form of excel files or SAS datasets. An R program is provided to convert SAS datasets to R without the need to have access to SAS. In a few instances noted in the text, links are given to specialized programs developed specifically for some regulatory test guideline when there seemed no purpose in creating a new program.

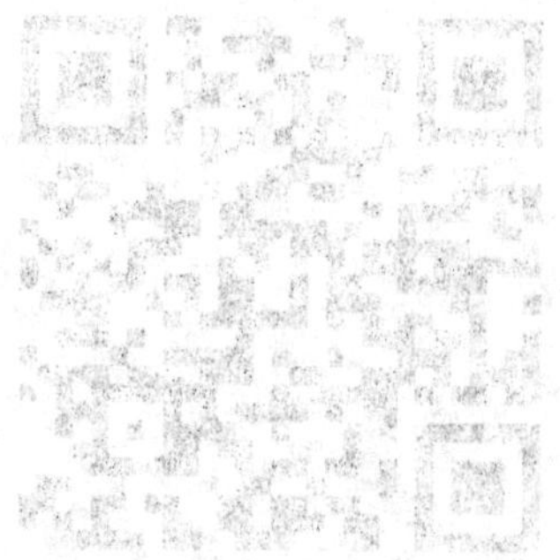

Chapter 1

An Introduction to Toxicity Experiments

This chapter introduces some basic concepts that apply to all chapters. It begins with a discussion of the nature of toxicology or ecotoxicology studies that distinguish them from experiments more generally. Then some basic experimental design issues are discussed, such as types of control groups, replicates and pseudo-replicates, and units of analysis. The various types of responses that occur are introduced, with pointers to chapters in which methods of statistical analysis of the various types of response are developed. An introduction is given to the use of historical controls and how these studies relate to regulatory risk assessment of chemicals in the environment. Then a hierarchy of statistical models is provided that, in broad terms, defines the statistics used in this field of study and, specifically, in this text. Finally, a topic is introduced that is the cause of considerable tension in ecotoxicology and biological analysis of data in general, namely the difference between biological and statistical significance.

1.1 NATURE AND PURPOSE OF TOXICITY EXPERIMENTS

The purpose of a toxicity experiment is to obtain a quantifiable measure of how toxic a given substance is to a group of organisms or community of organisms. The primary purpose of this book is to describe the design and statistical analysis of laboratory experiments on groups of organisms of a single species exposed to controlled levels of a substance thought to have the potential to produce an adverse effect on the test organisms. Such experiments have the goal of quantifying the level of exposure to the substance that has an adverse effect of biological concern. Some consideration is also given to how information from multiple toxicity experiments on different species can be combined to assess the adverse effect of the test substance on an ecological community. This chapter is intended to provide a general overview of toxicity studies and an introduction to the topics covered in this book.

1.1.1 Designed Experiments Compared to Observational Studies

Historically, the toxicity of chemicals has been studied using experiments performed under carefully controlled conditions in the laboratory and by observation of responses in uncontrolled settings such as the environment. Observational studies that gather information by survey or monitoring have the advantage of providing insight into toxicological responses under real-world conditions. Such studies are valuable in alerting researchers to potential problems resulting from chemical exposure. However, in surveys and monitoring studies, many uncontrolled factors can affect responses, and exposure of organisms to a chemical of interest (e.g. dose and concentration) usually cannot be estimated accurately. As a result, conclusions concerning the relationship between possible toxicological responses and exposure to the chemical are difficult to establish with certainty.

On the other hand, designed experiments typically control most of the factors that affect response, and dose or exposure concentration can be accurately measured. Designed experiments performed in a laboratory are usually performed at constant temperature with constant exposure to a test substance. Control of test substance exposure and other experimental factors allow the relationship between exposure and response to be modeled.

Exposure to the test substance in these experiments may be: via food or water ingested, air breathed, from

Statistical Analysis of Ecotoxicity Studies, First Edition. John W. Green, Timothy A. Springer, and Henrik Holbech.

Companion website: www.wiley.com/go/Green/StatAnalysEcotoxicStudy

contact with the soil or sediment or contact with spray application or spray drift on plants, through gavage or intravenous injection, or by direct application to the skin or eyes. The measure of exposure can be the concentration in the food or water or air, the quantity of chemical per unit of body weight, the quantity of chemical per unit of land area, or the concentration of the chemical in the blood.

Toxicity experiments are generally classified as acute, if the exposure is of short duration relative to the life span of the organism; or subchronic, if the exposure is of medium duration relative to a full life time; or chronic, if the exposure is for approximately a normal life span of the test substance.

Toxicity is measured in many ways. In its simplest form, it refers to the exposure level that kills the whole organism (e.g. laboratory rat or fish or tomato plant). Many sublethal responses are measured and the types of measurements are varied. The types of response encountered in toxicology fall broadly into one of the following categories: Continuous, quantal, count, and ordinal. Below is an introduction to each of these types of responses together with an indication of some of the challenges and methods associated with each type. Later chapters will discuss in detail all the points mentioned here.

1.1.1.1 Continuous Response

This class includes measurements such as plant yield, growth rate, weight and length of a plant or animal, the amount of some hormone in the blood, egg shell thickness, and bioconcentration of some chemical in the flesh, blood, or feathers. Typical continuous response data are shown in Tables 7.6 and 7.7 and Figures 7.2 and 7.3.

Continuous responses also include responses that exist in theory on a continuous scale, but are measured very crudely, such as days to first or last hatch or swim-up or reproduction, or time to tumor development or death, which are observed (i.e. "measured" only once per day). Hypothesis testing methods of analyzing continuous data are presented in Chapter 3 and regression models are presented in Chapters 4 and 5.

Example 1.1 Daphnia magna reproduction

The experimental design is seven daphnid individually housed in beakers in each of six test concentrations and a water control. Once each day, it is recorded whether or not each daphnid has reproduced. Ties in first day of reproduction are very common. In this typical dataset, there were a total of six distinct values across the study. While in theory, time to reproduction is continuous, the measurement is very crude and, as will be seen in Chapters 3 and 4, analysis will be different from that for responses measured on a continuous scale.

See Figure 1.1. The solid curve connects the mean responses in the treatment groups with line segments. Recall that there are seven beakers per treatment, but many beakers have the same first day of reproduction, so each diamond can represent from 1 to 6 observations. See Table 1.1 for the actual data.

1.1.1.2 Quantal Response

Quantal measures are binary (0–1 or yes/no) measurements. A subject is classified as having or not having some characteristic. For each subject, the possible values of the response can be recorded as 0 (does not have the characteristic of interest) or 1 (has the characteristic of interest). The quintessential example is mortality. Outside Hollywood films about zombies and vampires, each subject at a given point in time is either alive (value 0) or dead (value 1). Other quantal responses include immobility, the presence

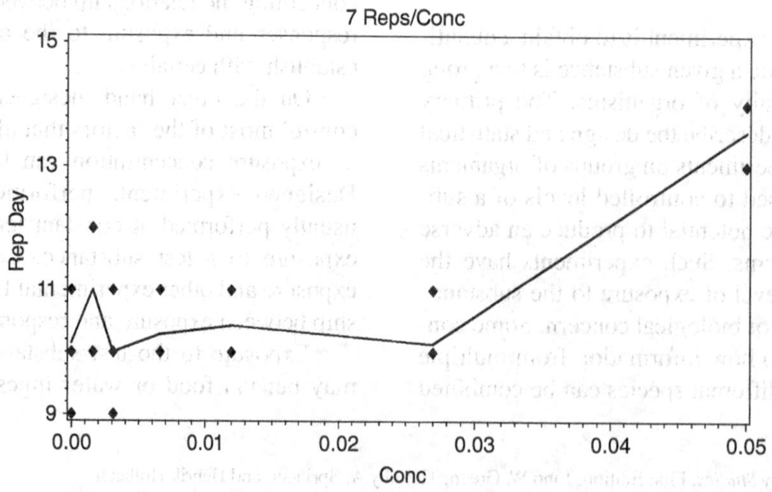

Figure 1.1 First day of daphnid reproduction. Diamonds, replicate means; solid line, joins treatment means.

Table 1.1 Daphnid First Day of Reproduction Data for Example 1.1

Conc	Rep	RepDay	Conc	Rep	RepDay	Conc	Rep	RepDay
−1	1	9	0.0016	6	10	0.012	4	10
−1	2	9	0.0016	7	12	0.012	5	11
−1	3	11	0.0031	1	10	0.012	6	10
−1	4	9	0.0031	2	11	0.012	7	10
−1	5	10	0.0031	3	10	0.027	1	10
−1	6	9	0.0031	4	10	0.027	2	10
−1	7	9	0.0031	5	10	0.027	3	10
0	1	9	0.0031	6	9	0.027	4	10
0	2	11	0.0031	7	10	0.027	5	11
0	3	10	0.0066	1	10	0.027	6	10
0	4	11	0.0066	2	10	0.027	7	10
0	5	10	0.0066	3	11	0.05	1	14
0	6	10	0.0066	4	11	0.05	2	13
0	7	10	0.0066	5	10	0.05	3	13
0.0016	1	10	0.0066	6	10	0.05	4	13
0.0016	2	12	0.0066	7	10	0.05	5	14
0.0016	3	10	0.012	1	10	0.05	6	14
0.0016	4	12	0.012	2	11	0.05	7	14
0.0016	5		0.012	3	11			

Conc = −1 is water control. Conc = 0 is solvent control. Controls should be combined (with Rep numbers altered to distinguish replicates in the two controls) prior to further analysis, or else one control should be discarded (see Sections 1.3.1 and 1.3.2). RepDay, first day of reproduction of daphnid in the beaker.

Table 1.2 Mite Survival Data

Conc	Unit	Risk	Alive	Conc	Unit	Risk	Alive
0	1	5	3	75	1	5	4
0	2	5	5	75	2	6	2
0	3	5	5	75	3	5	3
0	4	5	5	75	4	5	2
18.75	1	5	5	150	1	5	1
18.75	2	5	5	150	2	5	1
18.75	3	5	5	150	3	5	0
18.75	4	5	3	150	4	5	0
37.5	1	5	5	300	1	5	0
37.5	2	6	6	300	2	5	0
37.5	3	5	5	300	3	5	0
37.5	4	5	2	300	4	5	0

Unit, replicate vessel; Risk, number of mites placed in vessel at study start; Alive, number of mites alive at the end of the study period.

of matted fur, pregnant, lethargic, and the presence of liver tumor. Hypothesis testing methods of analyzing quantal data are presented in Chapter 6 and regression models are presented in Chapter 7. See Table 1.2 for an example of survival data for mites.

The data in Table 1.2 are from an experiment on mites. Mites were exposed to varying levels of a pesticide as part of a risk assessment for product registration. Each housing unit consists of a frame with a glass plate at the top and bottom of the frame. Pesticide residue is sprayed on the inner side of each glass plate. For the control, water is sprayed on the inner plate surface. After the plates dry, mite protonymphs are placed between the plates. Fresh air is circulated within the frame by an air pump. The mites are examined 7 days after exposure begins. Risk is the number of mites in each housing unit. Alive is the number alive at the end of the experimental period. The concentrations were in ppm. There were nominally five mites per unit, including control. Due to initial counting problems two units actually included six mites. Chapters 6 and 7 will discuss how to analyze such data.

1.1.1.3 Count Response

While quantal responses involve counts of the number of animals with the characteristic of interest, as we use the term, counts are the number of occurrences in a single subject or housing unit of some property. These include the number of eggs laid or hatched, the number of cracked eggs, the number of fetuses in a litter, the number of kidney adenomas, and the number of micronucleated cells. See Table 1.3 for an example dataset from Hackett et al. (1987) showing variable litter sizes and sex ratios.

Methods for analyzing count data will be presented in Chapter 8. As will be discussed there, count data can sometimes be analyzed as though it were continuous (usually after a transformation). Count data can also be analyzed through specialized distributions, such as Poisson

Table 1.3 Mouse Litter Size and Sex Ratio

Conc	Dam	Males	Litter	Conc	Dam	Males	Litter	Conc	Dam	Males	Litter
0	1	6	9	40	27	8	16	200	53	7	12
0	2	6	10	40	28	6	11	200	54	7	15
0	3	2	3	40	29	6	12	200	55	0	2
0	4	5	13	40	30	3	6	200	56	8	16
0	5	7	12	40	31	7	13	200	57	7	15
0	6	5	13	40	32	6	11	200	58	9	11
0	7	8	15	40	33	4	10	1000	59	4	9
0	8	7	13	40	34	8	15	1000	60	5	14
0	9	4	13	40	35	7	14	1000	61	5	11
0	10	6	13	40	36	3	14	1000	62	7	12
0	11	11	12	40	37	7	13	1000	63	6	14
0	12	6	13	200	38	7	13	1000	64	9	14
0	13	3	9	200	39	8	12	1000	65	7	15
0	14	8	13	200	40	7	12	1000	66	7	12
0	15	6	9	200	41	4	13	1000	67	9	16
0	16	3	13	200	42	6	14	1000	68	10	14
0	17	7	14	200	43	9	15	1000	69	9	14
0	18	6	14	200	44	5	11	1000	70	7	11
40	19	6	11	200	45	7	14	1000	71	4	10
40	20	3	11	200	46	6	11	1000	72	7	9
40	21	3	13	200	47	6	12	1000	73	5	12
40	22	8	14	200	48	7	11	1000	74	3	8
40	23	8	13	200	49	8	11	1000	75	6	10
40	23	11	15	200	50	5	13	1000	76	9	13
40	25	6	13	200	51	6	14	1000	77	2	11
40	26	6	12	200	52	8	12	1000	78	6	15

Dam is an ID for the pregnant female mouse. Litter is the number of fetuses for that dam. Males = number of males in the litter and conc is the exposure parentage dosage of 1,3-butadiene in ppm. Questions of interest include whether the chemical affects the litter size or sex ratio and whether there is an association between litter size and sex ratio. Fetal, placenta, and dam body weights were also included in the original dataset and other questions were also addressed.

or zero-inflated Poisson, in the context of what are called generalized linear models (GLMM). We will present and compare these methods in Chapter 8.

1.1.1.4 Ordinal Response

Ordinal responses indicate relative severity or level but not magnitude. Examples include amphibian developmental stage and histopathology severity scores. Amphibian developmental stages are represented by numbers 1–66 (as derived from Nieuwkoop and Faber, 1994), but the difference between stage 55 and 56 is not comparable to the difference between 56 and 62. The larger number indicates a more advanced development, but this development is defined by the presence or absence of specific physical characteristics, not otherwise quantifiable. Consider the following stages as examples:

1. Stage 56 typically occurs on day 38 post hatch. Forelimbs of stage 56 animals are visible beneath the skin of the tadpoles. The tadpoles are filter-feeding.
2. Stage 57 typically occurs on day 41 post hatch. Stage 57 animals lack emerged forelimbs, and metamorphosis in the alimentary canal is just beginning.
3. Stage 58 typically occurs on day 44 post hatch. Stage 58 animals have emerged forelimbs and there is significant histolysis of the duodenum (animals can no longer digest food).
4. Stage 59 typically occurs on day 45 post hatch. Stage 59 animal forelimbs now reach to the base of the hindlimb and there is now histolysis of the non-pyloric part of the stomach (animals still can no longer digest food).
5. Stage 60 typically occurs on day 46 post hatch.

In terms of development rates, a stage 57 animal is 3 days behind a stage 58 animal, whereas a stage 58 animal is only 1 day behind a stage 59 animal. Also, in terms of development rate, a stage 56 animal is 6 days behind a stage 58 animal, whereas a stage 58 animal is only 2 days behind a stage 60 animal.

The biological significance of moving between two stages might vary greatly depending on which stages are being considered. For example, a stage 56 animal can filter-feed. None of the animals in the other stages listed above can.

Developmental stage is a key endpoint in the OECD TG 231 Amphibian Metamorphosis Assay (AMA). The experimental design in the test guideline is for four tanks per test concentration, 20 tadpoles per tank, and three test concentrations plus a water control. In developing the test guideline, other designs were explored, including designs with five test concentrations plus control, two tanks per concentration, and 20 tadpoles per tank. See Table 1.4 for an example with this latter design.

In Table 1.4, there was an apparent shift right in group 5 and perhaps in group 4, but groups 2 and 3 have increased frequencies of smaller stages. It is not clear what a 10% effects concentration would mean for this response. Averaging stages in a group is meaningless (i.e. Stage 57.2 is meaningless), as stage is an ordinal, not a quantitative, variable. The response measure should not be based on simply considering the proportion of tadpoles above some stage (e.g. >stage 58), since calculation of the concentration causing a 10% increase in the percent of tadpoles with stage greater than 58 ignores the effects on the distribution of stages above and below 58. Analysis based on median stages in tanks ignores too much within-tank information. Chapter 9 will describe the analysis of such data.

Clearly, the analysis of the stage data requires care, and it is important not to think of the stages as representing equal increments of development. It should be clear that a shift in the stage of metamorphosis of a single stage *might* be, but need not be, biologically meaningful. The analyses of developmental stage data will be discussed in detail in Chapter 9.

Table 1.4 Example Developmental Stage Data from AMA Study

Stage		56	57	58	59	60	61	62
Group	Tank							
1	1	2[a]	3	7	4	2	2	
	2			9	8	2	1	
2	1			6	9	3		2
	2		7	8	4	1		
3	1		2	9	6		3	
	2		1	6	7	2	2	2
4	1			1	13	3		3
	2			1	3	8	3	5
5	1			5	6	6	1	2
	2				5	7	3	5

Stage, developmental stage reached by some tadpole in the indicated tank; Group, treatment group, with control = group 1; Tank, replicate vessel.

[a] Number in cell is the number of tadpoles in the tank at the indicated developmental stage.

Histopathology severity scores are similar to developmental stages in terms of being ordinal, not numeric, but differ in another way that requires a different type of analysis. Here, pathologist-grade organ slides on a scale 0–4, with score 0 meaning no abnormality was observed, score 1 meaning only a minimal abnormality, score 2 meaning mild abnormality, and scores 4 and 5 meaning moderate and severe abnormalities, respectively. It would be more accurate to describe score 0 as meaning there was nothing remarkable, rather than no abnormality. A severity score is assigned to a tissue sample by a trained pathologist. These scores depend on the type of tissue damage found and an assessment of its importance to the health of the animal. See Figure 1.2 for an example tissue slide. Assigning severity scores to such slides is not a simple exercise. More discussion of this and a more detailed example are provided in Chapter 10.

With most toxicology severity scores, there is no uniform change in severity between scores, that is, the difference between minimal and mild is not the same as the difference between mild and moderate or between moderate and severe. See Figure 1.3 for a simple illustration that may help keep these scores in mind.

Stated this way, the nature of severity scores is straightforward. Few people would suggest that if half of the tissue samples have a minimal finding and half have a moderate finding, then on average, the finding is mild.

Confusion arises from the common practice of labeling a finding of none as 0, minimal as 1, mild as 2, moderate as 3,

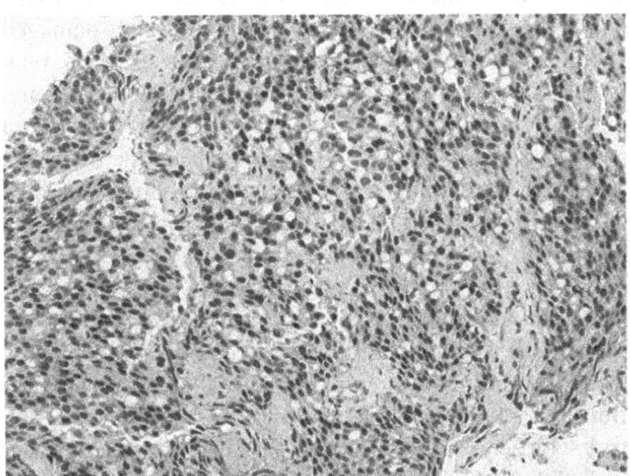

Figure 1.2 Example tissue slide for histopathology grading. Expert judgment is used to score tissue slides such as these. Image from Google images altered to black and white and cropped using Photoshop. https://image.slidesharecdn.com/cpc-4-4-2-ren-bph-pathlec-view-091013211114-phpapp02/95/pathology-of-prostate-53-728.jpg?cb=1255468480.

and severe as 4. These labels are numbers and a simple-minded statistical approach is to treat them as though these labels behave as numbers do rather than recognizing them merely as labels: that is, one can average them, compute the standard deviation, and employ all the simple statistical tools one learned in an introductory course, such as the T-test. However, moving from a score of 1 to 2 does not indicate a doubling of severity, and moving from 3 to 4 may not indicate a change in severity equal to that in moving from 1 to 2.

It should be emphasized that these scores are just labels. To average scores 1 and 2 is the same as averaging minimal and mild. What is the average of minimal and mild or of mild and severe? These scores are arbitrary except for order. We could just as well use the numbers 1, 2, 5, 7, and 12 as scores (see Figure 1.3) to emphasize that the difference between "adjacent" scores is not the same as a subject progresses from no effect to severe effect. So the average of minimal and severe could be $(1+4)/2=2.5$ or $(2+12)/2=7$. Neither average makes sense.

Such numerical approach is nonsensical, but it does highlight a real concern. If the tank in an aquatic experiment is the unit of analysis, what value do we give to the tank? Leaving aside for now how to analyze severity scores, if there are five fish in a tank with severity scores 0, 0, 3, 3, and 4, what single value do we assign to the tank for statistical analysis? Note that the arithmetic mean of these numerical labels is 2. Does 2 capture something meaningful about this set of scores?

While the mean score is objectionable, what about the median score? The median inherently treats the labels as equally spaced across the spectrum of severities. Think about where in the wide range of moderate (score 3) tissue damage assessments in Figure 1.3 the moderately damaged slide lies. As shall be discussed in Chapters 3 and 5, rank ordering is a basic idea in most nonparametric testing, and the set of all values from treatment and control are ranked as a whole and then the sum of the ranks in the treatment and control are compared. Such nonparametric tests take the spread of severity scores into account, not just the median.

One of the two approaches is typically taken in rodent histopathology analysis. (i) Apply a nonparametric test such as the Mann–Whitney (Chapter 3), which compares the median scores in treatment tanks to those in the control. But the tank median ignores the spread of the data. The tank with scores 0, 0, 3, 3, and 4 has the same median as a tank with scores 3, 3, 3, 3, and 3, but the first is much more dispersed than the second and this may signal a difference of biological importance missed by the comparison of medians. The need for a summary measure for each tank limits the appropriateness of traditional nonparametric procedures for severity score analysis. (ii) Some scientists simply do not perform a statistical analysis, either because they recognize the shortcomings of the above approach or because they have little value for statistics altogether.

Given the restricted number of possible severity scores and the small sample sizes typical in histopathology, at least in ecotoxicology studies, analysis methods for severity scores are different from those for developmental stage. See Table 1.5 for an example from a medaka multi-generation test.

In the dataset in Table 1.5, there were no score 0 fish. The empty tanks (A in treatment 1 or control and C in treatment 2) do not represent mortality. Rather, medaka could not be sexed at the initiation of the study and by chance, these tanks contained no females. This inability to know the sex at study initiation leads to highly imbalanced experimental designs. The tank is the unit of analysis, not the individual fish, it is thus important to retain tank identification and not lose the distribution of scores within the tank. Also, because fish cannot be sexed at the beginning of the study and must be analyzed by sex at the end of the

Table 1.5 Severity Scores for Liver Basophilia in Female F2 Medaka at 8 Weeks

		Frequency of score per tank						
Trt	Score	A[a]	B	C	D	E	F	Total
1	1		1[b]	1	2	2	3	9
	2		3		1			4
	3					1		1
2	1				2	1	1	4
	2	4	2		1	1		8
3	1		1			2	4	7
	2	2	2	1	1			6
4	1	1	1			1		3
	2	1		1	3	3	2	10
	3						1	1
5	1	1	2	1	1	1		6
	2			1	1	2		4
	3	1			2		1	4

Trt, treatment group, with 1=control; Total, number of fish in all tanks in the indicated treatment group with the indicated score.

[a] Tanks are labeled A, B, and F.

[b] Numbers in cells indicate the number of fish (ignoring tanks) in the treatment group with that score.

None	Minimal	Mild	Moderate	Severe

Figure 1.3 Example severity scale. Varying widths for different scores indicate possible differences in the range of severities given the same score.

study, tank sizes are highly variable and this complicates the analysis. For that reason and others, analysis of tank medians, for example, would discard important information.

Appropriate methods for analysis of ordinal data are discussed in detail in Chapter 9.

1.1.2 Analysis of Laboratory Toxicity Experiments

The variety of sublethal endpoints measured suggests the need for multiple statistical tools by which to analyze toxicity data. It is the objective of this book to discuss many of the statistical methods that have been used for this purpose and to indicate what additional tools could be brought to bear. Science is not static and advances in statistical methods and computer power and software have made available techniques that were impossible only a few years ago. It is fully expected that additional advances will be made in the time to come that cannot be foreseen today. The authors will attempt to present the main statistical methods in use now, and to the extent possible, those likely to be included in the near future.

In its simplest form, a toxicity experiment is conducted on a single species for a fixed amount of time. Different groups of subjects are exposed to difference levels of the test substance. More complex experiments include other factors, such as measurements of lethal and sublethal effects over time, differences among the sexes of the subjects, different ambient conditions, and mixtures of chemicals. The object of the statistical analysis is to identify the level of exposure that causes a biologically meaningful adverse effect under each set of conditions in the experiment. Ideally, subject matter experts (e.g. toxicologists or biologists) will determine what level of effect is biologically meaningful. Criteria for making that determination can be on the basis of the health of the individual animal or on the ability of the population as a whole to thrive. For example, it may be the scientific judgment of biologists that a 10% change in body weight of a Sprague-Dawley rat, a 3% change in the length of a Daphnia magna, and only a 300% or greater increase in vitellogenin (VTG) are of biological importance. This is not a statistical question but it is very important to the statistician in designing or interpreting a toxicity study to know what size effect it is important to find. Without the information on what size effect it is important to detect, the statistician or data analyst can only determine what is statistically significant or estimate an arbitrary percent effect that may have no inherent value. The result is unsatisfying to the statistician, biologists, and risk assessor.

Ethical concerns about the use of animals in toxicity experiments are increasingly important and the authors share this concern. There is a very active worldwide effort underway to reduce or eliminate the number of animals for various species (mice, fish, birds, etc.) used in toxicity experiments. We will not pursue the question of the desirability of animal testing. Our purpose is to provide scientifically sound methods for analyzing the range of responses that arise from toxicity experiments. Most of these methods apply whether the test subject is a fathead minnow, tomato plant, cell, or bacterium. In all cases, experiments should be designed to use the minimum number of test subjects needed to provide scientifically sound conclusions. This is an instance where ethical and cost considerations coincide.

1.2 REGULATORY CONTEXT FOR TOXICITY EXPERIMENTS

Many toxicity studies are done to meet a regulatory requirement needed to obtain permission to use a chemical that may lead to an environmental exposure. Such toxicity experiments are used by regulatory authorities, such as the United States Department of Agriculture (USDA), Animal and Plant Health Inspection Service (APHIS), United States Environmental Protection Agency (USEPA), Office of Pesticide Programs (OPP), European Food Safety Association (EFSA), the European Chemicals Bureau (ECHA), The Institute for Health and Consumer Protection (IHCP), or one of the European country environmental agencies, including the Danish Environmental Protection Agency (DK-EPA) and Umweltbundesamt (UBA) following standardized test guidelines issued by the Organization for Economic Co-operation and Development (OECD) or the USEPA to assess the likelihood of adverse impacts on populations and communities of organisms in the environment.

To minimize data requirements and avoid unnecessary tests, regulatory risk assessments in the US have a tiered structure. Tier I studies estimate hazard and exposure under "worst-case" conditions. If no adverse effects are found under these conditions, there may be no need for further data. In its simplest form, a so-called limit test may be done with a single very high concentration of the test chemical and a control. In other instances, there may be several exposure levels. In either case, except for determining lethal exposure levels, the emphasis is on testing hypotheses regarding whether an adverse effect exists, but there is no need for a precise quantification of the size effect at each exposure level. If a higher tier test is needed, the focus of such tests is usually on sublethal effects, so it is important for the tier I tests to establish exposure levels that are lethal to a substantial portion of the exposed subjects. Early tier tests tend to be simple in design and may indicate that there is no need for the more detailed information that can come from higher tiered tests. Higher tier tests are designed either to assess risk under more realistic conditions or to obtain more precise quantification of the exposure–effect relationship.

In the European Union (EU) chemicals expected to enter the environment are mainly regulated by three regulations: (i) REACH (Registration, Evaluation, Authorization, and Restriction of Chemicals) covering industrial chemicals, (2) PPPR (Plant Protection Products Regulation) covering pesticides, and (3) BPR (Biocidal Products Regulation) covering biocides. The test information requirements in REACH are driven by tonnage, i.e. the yearly volume produced in or imported to the EU. Test requirements start when more than 1 ton of a chemical is produced or imported yearly. The most test requirements are applied to chemicals exceeding 1000 ton $year^{-1}$.

Chapters 2–10 and 13 will develop methods appropriate for all levels of this tiered process. Much more information on the regulatory process will be provided in Chapter 11. Chapter 12 will develop an important tool for combining the information from individual studies into a single summary distribution useful for risk assessment. References that can be explored now and returned to throughout a course based on this text include http://www.epa.gov/pesticides/biopesticides/pips/non-target-arthropods.pdf, http://www.epa.gov/oppefed1/ecorisk_ders/toera_analysis_eco.htm, http://www.epa.gov/pesticides/health/reducing.htm, and http://www.eea.europa.eu/publications/GH-07-97-595-EN-C2/riskindex.html.

1.3 EXPERIMENTAL DESIGN BASICS

While observational studies of animals or plants captured in the wild are valuable to environmental impact studies, such studies can be quite frustrating in that routes and conditions of exposure are often unknown, sample sizes are often inadequate, and measurements are all too often non-standardized, so that comparisons among studies are very difficult. This book is not concerned with observational studies, even though one of the authors has been very actively involved in several such studies, including one major study lasting for more than 12 years. We will restrict ourselves to designed experiments.

Considerations of study objectives should include what and how measurements will be taken to address the objectives. For a study of fish, for example, how is death to be determined? It may be difficult to know with certainty whether a fish floating upside down at the top of the tank is dead or just immobile. How long should it be allowed to float before deciding it is dead or near death and should be euthanized to prevent suffering? If a fish or plant is weighed, is it weighed wet or first blotted dry or desiccated? Specific protocols should be provided to address such questions.

Experiments intended for regulatory submissions of new pharmaceuticals or crop protection products or food stuffs will receive special attention in this book. In studies done to meet regulatory requirements, objectives are generally very detailed in test guidelines that must be followed. What is often unclear in test guidelines is the size of effect it is important to detect or estimate. Guidelines, especially older guidelines, simply refer to effects that are statistically significant. As a result, it has often been argued, with some merit, that such guidelines reward poor experimentation, since the more variable the data, the less likely an observed effect will be found statistically significant. A good study should state explicitly what size effect is important to detect or estimate for each measured response and the power to detect that size effect or the maximum acceptable uncertainty for that estimate in the proposed study. Detailed discussion of statistical power is introduced in Chapter 2 and discussed in detail in Chapters 3, 5, 6, 8, and 9 in the context of specific tests. There has been increasing interest in the last 15 years or so in replacing the use of hypothesis tests to determine a NOEC by regression models to estimate a specific percent effects concentration, EC*x*. One goal of the regression approach is to replace the ill-defined connection between biological and statistical significance with an estimate of the exposure level that produces an effect of a specific size. Such methods are introduced in Chapter 2 and explored in depth in Chapters 4, 6, 7, and 8. A hypothesis testing method with the same goal is discussed in Chapter 13.

The basic toxicity experiment has a negative control, where subjects are not exposed to the test substance, and one or more treatment groups. Treatment groups differ only in the amount of the test substance to which the subjects are exposed, with all other conditions as nearly equal as possible. For example, treatment groups might be tanks of fish exposed to different concentrations of the test substance, or pots or rows of plants exposed to different application rates of the test chemical, or cages of mice with different amounts of the test substance administered by gavage. Apart from the amount of chemical exposure, the same species, strain, age, sex, ambient conditions, and diets should be the same in all treatment groups and control.

1.3.1 Multiple Controls

It is common in aquatic and certain other types of experiments that the chemical under investigation cannot be administered successfully without the addition of a solvent or vehicle. In such experiments, it is customary to include two control groups. One of these control groups receives only what is in the natural laboratory environment (e.g. dilution water in an aquatic experiment, a water spray in a pesticide application experiment, and unadulterated food in a feeding study), while the other group receives the dilution water with added solvent but no test chemical, a spray with surfactant but no test chemical, or an oral gavage with corn oil but no test substance. In ecotoxicity experiments, these are often termed negative or dilution water (non-solvent) and solvent controls. OECD recommends limiting the use of solvents (OECD, 2000); however, appropriate use of solvents should be evaluated on a case-by-case basis. Details regarding the use of solvents (e.g. recommended chemicals and maximum concentrations) are discussed in the relevant guideline documents for a specific ecotoxicity test. In addition, regulatory guidelines must be followed by both controls with regard to the range of acceptable values (e.g. minimum acceptable percent survival or mean oyster shell deposition rate). Multiple control groups can be utilized regardless of whether the experiment was intended for hypothesis testing or regression analysis.

In rodent studies where the chemical is administered by oral gavage using a corn oil vehicle (or some other vehicle), one control group should be given just the corn oil by gavage. The intention is to rule out a gavage effect or separate it from any effect from the test chemical. Not all such rodent experiments include a control group that is simply fed a standard diet with no gavage administered. The statistical treatment of multiple controls will be addressed in Chapter 2 and in specific types of analyses in later chapters.

In some experiments, a positive control group is also used. Here a different compound known to have an effect is given to one group of subjects. The purpose is to demonstrate that the experimental design and statistical test method are adequate to find an effect if one is present. If the positive control is not found to be significantly different from the control, the experiment will generally have to be repeated. More information on how to analyze experiments with a positive control group will be given in subsequent chapters. There are other ways to demonstrate the sensitivity of the design and analysis method, including power analysis and computer modeling. These topics will also be addressed later.

1.3.2 Replication

In almost all toxicity experiments, each treatment group and control is replicated, so that there are multiple subjects exposed to each treatment. The need for replication arises from the inherent variability in measurements on living creatures. Two animals or plants exposed to the same chemical need not have the same sensitivity to that chemical, so replication is needed to separate the inherent variability among subjects from the effects, if any, of the test substance. The number of replicates and the number of subjects per replicate influence the power in hypothesis testing and the confidence limits of parameter estimates and other model evaluation measures in regression models and will be discussed in depth in later chapters.

It is important to understand what constitutes a replicate and the requirements of statistical methods that will be used to analyze the data from an experiment. A replicate, or experimental unit, is the basic unit of organization of test subjects that have the same ambient conditions and exposure to the test substance. To paraphrase Hurlbert (1984), different replicates are capable of receiving different treatments and the assignment of treatments to replicates can be randomized. The ideal is that each replicate should capture all the sources of variability in the experiment other than the level of chemical exposure. Two plants in the same pot will not be considered replicates, since they will receive the same application of the test chemical and water and sunlight and other ambient conditions at the same time and in the same manner. Different pots of plants in different locations in the greenhouse will generally be considered replicates if they receive water, test compound, and the like through different means, for example, by moving the applicator and water hose. If 25 fish are housed together in a single tank and the chemical exposure is through the concentration in the water in that tank and the ambient conditions and chemical exposure in that tank are set up uniquely for that tank, then the tank constitutes one replicate, not 25. Furthermore, if two tanks sit in the same bath and receive chemical from a simple splitter attached to a single reservoir of the test substance so that the chemical exposure levels in the two tanks are the same and do not capture all the sources of variability in setting up an exposure scenario, then the two tanks are not true replicates.

Hurlbert (1984) describes at some length the notion of pseudoreplication, "defined as the use of inferential statistics to test for treatment effects with data from experiments where either treatments are not replicated (though samples may be) or replicates are not statistically independent. In ANOVA terminology, it is the testing for treatment effects with an error term inappropriate to the hypothesis being considered." Hurlbert defines the rather colorful term nondemonic intrusion as "the impingement of chance events on an experiment in progress" and considers interspersion of treatments as an essential ingredient in good experimental design. Oksanen (2004) extends the idea of spatial interspersion to interspersion along all potentially relevant environmental axes so that nondemonic intrusions cannot contribute to the apparent treatment effects. The primary requirements of good experimental design, according to Hurlbert, are replication, randomization, interspersion of treatments, and concomitant observations. Many designed experiments fail to meet these ideals to some degree. For example, in an aquatic experiment, tanks of subjects in the same nominal treatment group may receive their chemical concentrations from a common source through a physical splitter arrangement. Rodents may be housed throughout a chronic study in the same rack. The latter is usually compensated for by the rack frame that rotates positions of the racks to equalize air flow, light, room temperature variations, and other ambient conditions across the experiment as a whole. Furthermore, it is sometimes impossible to make concomitant measurements on all subjects in a large experiment, so that a staggered experimental design may be necessary in which subjects are measured at equivalent times relative to their exposure. For Oksanen (2004), "the proper interpretation of an experiment of a demonstrated contrast between two statistical populations hinges on the opinion of scientists concerning the plausibility of different putative causes." Oksanen (2001, 2004) would accept the results of an experiment if the scientific judgment was that the observed treatment effects could not be plausibly explained by the shortcomings of the experimental design, even if it was possible to *imagine* some form of nondemonic intrusion (Hurlbert, 2004) that could account for the observed effect. However, it must be stated that true replication, randomization, concomitant observation, and interspersion of treatments is the goal.

In some toxicity experiments, subjects are individually housed, such as one bird per cage, one daphnid per beaker, or one plant per pot. In these experiments, the replicate is usually the test vessel, which is the same as the subject, unless there are larger restrictions on clusters of vessels, such as the position in the lab. In other experiments, multiple subjects are housed together in the same cage or vessel and there are also multiple vessels per treatment. In these latter experiments, the replicate or experimental unit is the test vessel, not the individual subject.

In a well-designed study, one should investigate the trade-off between the number of replicates per treatment and the number of subjects per replicate. Decisions on the number of subjects per subgroup and number of subgroups per group should be based on power calculations, or in the case of regression modeling, sensitivity analyses, using historical control data to estimate the relative magnitude of within- and among-subgroup variation and correlation. If there are no subgroups (i.e. replicates), then there is no way to distinguish housing effects from concentration effects and neither between- and within-group variances nor correlations can be estimated, nor is it possible to apply any of the statistical tests to be described to subgroup means. Thus, a minimum of two subgroups per concentration is recommended; three subgroups are much better than two; and four subgroups are better than three. The improvement in modeling falls off substantially as the number of subgroups increases beyond four. (This can be understood on the following grounds. The modeling is improved if we get better estimates of both among- and within-subgroup variances. The quality of a variance estimate improves as the number of observations on which it is based increases. Either sample variance will have, at least approximately, a chi-squared distribution. The quality of a variance estimate can be measured by the width of its confidence interval and a look at a chi-squared table will verify the statements made.)

The number of subgroups per concentration and subjects per subgroup should be chosen to provide adequate power to detect an effect of magnitude judged important to detect or to yield a slope or ECx estimate with acceptably tight confidence bounds. These determinations should be based on historical control data for the species and endpoint being studied. There are two areas of general guidance. If the variance between subjects greatly exceeds the variance between replicates, then greater power or sensitivity is usually gained by increasing the number of subjects per replicate, even at the expense of reducing the number of replicates, but almost never less than two per treatment. Otherwise, greater power or sensitivity generally comes from increasing the number of replicates and reducing the number of subjects per replicate. This claim will be developed more fully in the context of specific types of data in Chapter 3. The second generality is that for hypothesis testing (NOEC determination), generally there need to be more replicates per treatment and fewer treatments, whereas with regression analysis, it is generally better to have more treatments, and there is less need for replicates. As will be illustrated in Chapter 4, the quality of regression estimates is affected by the number of replicates unless there are a large number of treatments.

Since the control group is used in every comparison of treatment to control, it is advisable to consider allocating more subjects to the control group than to the treatment groups in order to optimize power for a given total number of subjects and thoroughly base the control against which all estimates or comparisons are to be made. The optimum allocation depends on the statistical method to be used. A widely used allocation rule for hypothesis testing was given by Dunnett (1955), which states that for a total of N subjects and k treatments to be compared to a common control, if the same number, n, of subjects are allocated to every positive treatment group, then the number, n_0, to allocate to the control to optimize power is determined by the so-called square-root rule. By this rule, the value of n is (the integer part of) the solution of the equation $N = kn + n\sqrt{k}$, and $n_0 = N - kn$. (It is almost equivalent to say $n_0 = n\sqrt{k}$.) Dunnett showed this to optimize power of his test. It is used, often without formal justification, for other pairwise tests, such as the Mann–Whitney and Fisher exact test. Williams (1972) showed that the square-root rule may be somewhat suboptimal for his test and optimum power is achieved when $\sqrt{k}$ in the above equation is replaced by something between $1.1\sqrt{k}$ and $1.4\sqrt{k}$. The square-root allocation rule will be explored in more detail in Chapter 2 and in subsequent chapters in the context of specific tests or regression models.

1.3.3 Choice and Spacing of Test Concentrations/Doses

Factors that must be considered when developing experimental designs include the number and spacing of doses or exposure levels, the number of subjects per dose group, and the nature and number of subgroups within dose groups. Decisions concerning these factors are made so as to provide adequate power to detect effects that are of a magnitude deemed biologically important.

The choice of test substance concentrations or doses or rates is one aspect of experimental design that must be evaluated for each individual study. The goal is to bracket the concentration/dose/rate[1] at which biologically important effects appear and to space the levels of the test compound as closely as practical. If limited information on the toxicity of a test material is available, exposure levels can be selected to cover a range somewhat greater than the range of exposure levels expected to be encountered in the field and should include at least one concentration expected not to have a biologically important effect. If more information is available this range may be reduced, so that doses can be more closely spaced. Effects are usually expected to increase approximately in proportion to the log of concentration, so concentrations are generally approximately equally spaced on a log scale. Three to seven concentrations plus concomitant controls are suggested, with the smaller experiment size typical for acute tests and larger experiment sizes most appropriate when preliminary dose-finding information is limited.

Of course, the idea of bracketing the concentration/dose/rate at which biologically important effects appear is much simpler to state than to execute, for if we knew what that concentration was, there would no longer be a need to conduct an experiment to determine what it is. To that end, it is common to do experiments in stages. Conceptually, a small range-finding study is done to give an idea of the exposure levels likely to produce effects of interest. Based on that, a larger definitive study is done. Experience indicates that this process is not fail proof, so exposure levels generally start well below the expected level and extend well beyond. There are practical issues as well. If concentration levels are too small, analytical chemistry methods may not be sufficiently sensitive to measure these levels and it sometimes happens that there is an inversion, where some mean measured concentrations are in reverse order to the planned nominal concentrations. This complicates the interpretation of results and brings into question the entire experiment. Another issue is involved in the intended use of the test substance. For example, a pesticide or pharmaceutical product has to be administered at a high-enough level to be effective. Testing much below the effective level may only make sense if the concern is of environmental exposure that might arise from dilution in a stream or from rainfall.

At the other extreme, in aquatic experiments, chemicals have a solubility limit that cannot be exceeded and this obviously restricts the range of exposure levels that can be included. In all types of studies aimed at determining sublethal effects, the exposure levels must be below the level that produces high mortality. Generally, separate studies are done to determine lethality and that information is used in both the range finder and definitive tests for sublethal effects.

1.3.4 Randomization

Variability (often called noise) is inherent in any biological dataset. The following factors affect the level of noise in an experiment:

1. the variation between the individual animals, due to genetic differences,
2. the differences in the conditions under which the animals grew up prior to the experiment, resulting in epigenetic differences between animals,
3. the heterogeneity of the experimental conditions among the animals during the experiment,
4. variation within subjects (i.e. fluctuations in time, such as female hormones, which may be substantial for some endpoints), and
5. measurement errors.

Randomization is used in designing experiments to eliminate bias in estimates of treatment effects and to ensure independence of error terms in statistical models. Ideally, randomization should be used at every stage of the experimental process, from selection of experimental material and application of treatments to measurement of responses. To minimize the effects of the first two factors, animals need to be randomly distributed into concentration groups. To minimize the effects of the third factor (both intended and unintended, such as location in the room), application of treatments should be randomized as much as possible. To minimize the effects of the fourth factor, the measurement of responses should be randomized in time (e.g. although all responses will be recorded at 24 h, the order in which the experimental units are measured should be randomized). With good scientific methods, measurement errors can be minimized.

If any experimental processes are carried out in a non-random way, then statistical analysis of the experimental data should include a phase in which the potential effect of

1 To avoid repeated awkward phrases such as concentration/dose/rate, the text will frequently use only one of these terms, usually concentration when the context clearly requires an aquatic environment, but commonly dose regardless of context. The terms will be used interchangeably in this text except in rare instances that are clear from context.

not randomizing on the experimental results is examined and modifications are made to the model to account for this restriction on randomization.

1.3.5 Species Used for Experiments

Many species are used in ecotoxicity experiments. In aquatic toxicology, rainbow trout, fathead minnow, zebrafish, Japanese medaka, sheepshead minnow, and silverside are common fish species tested. In addition, Daphnia magna, various species of algae, macrophytes, lemna, and sediment dwelling chironomid and endobenthic species round out the common aquatic species used in laboratory experiments. To the aquatic species must be added earthworms, honeybees, mites, numerous avian species, and many non-target crop plants and wild plant species. Mammalian toxicity studies are most often done on some rat and mouse species, plus rabbits and guinea pigs. Other species are also used. With rare exceptions, humans are not test subjects for toxicity experiments. Humans are, of course, used to test pharmaceuticals in clinical trials and sublethal toxic effects may be observed in these trials. Clinical trials are not considered in this text.

Some statistical methods apply across species. There is no widely accepted specific fish statistical test or model used in toxicity studies. There is, on the other hand, much variety in the types of responses that arise and while these are sometimes linked to specific species, it is the nature of the response that determines the statistical method to be used, not the species per se.

1.3.6 Extrapolation to Human Toxicity

Given that humans are not subjects for experiments, but are exposed to various chemicals in the course of work or in food consumption, wearing apparel, and use of home products, the risk assessor needs some mechanism for extrapolating from animal studies to human exposure. It is not the purpose of this text to explore the ways in which such extrapolations are done, other than to indicate that generally this involves some uncertainty factor to apply to the animal studies. For example, the lowest level found to have a toxic effect on a rodent may be divided by 100 or 1000 in assessing the safe level of human exposure. Further discussion of such extrapolations and human risk assessment can be found in Brock et al. (2014), Vose (2000), Warren-Hicks and Moore (1998), and Hubert (1996, pp. 401ff).

1.4 HIERARCHY OF MODELS FOR SIMPLE TOXICITY EXPERIMENTS

There is a model underlying every statistical test used to derive a NOEC or estimate an EC*x*. A basic experimental design in ecotoxicity is one in which independent groups of subjects of common species, age, and sex are exposed to varying concentrations of a single test chemical for the same length of time, so that the only non-random source of difference among these subjects is the level of chemical exposure. It is expected for most species, chemicals, and responses to be analyzed that if there is an effect of the chemical it will tend to increase as the chemical concentration increases. The basic statistical model for this simple toxicity experiment is given by

$$Y_{ij} = \mu_i + e_{ij}, \qquad (1.1)$$

where μ_i is the expected mean response in the *i*th concentration, and e_{ij} are independent identically distributed random errors, often assumed to be normally distributed with homogeneous variances, though that is not by any means an absolute requirement. What distinguishes one model from another is what additional restrictions or assumptions are placed on the treatment means, μ_i.

The simplest model that is used for hypothesis testing is usually stated in terms of null and alternative hypotheses as

$$H_{02}: \mu_0 = \mu_1 = \mu_2 = \cdots = \mu_k \text{ vs. } H_{a2}: \mu_0 \neq \mu_i \text{ for some } i, \qquad (1.2)$$

where μ_0 is the control mean.

This model implies no relationship among the treatment means and one merely tests each treatment against the control. In the contest of toxicology, Model (1.2) ignores the expected relationship between the exposure concentration and the response. A more appropriate model is given by

$$H_{03}: \mu_0 \geq \mu_1 \geq \mu_2 \geq \cdots \geq \mu_k \text{ vs. } H_{a3}: \mu_0 > \mu_i \text{ for some } i. \qquad (1.3)$$

This model assumes a non-increasing concentration–response, which is what is expected biologically for most responses from ecotoxicity experiments. It should also be evident that if $\mu_0 > \mu_i$ for some *i*, then $\mu_0 > \mu_j$ for all $j > i$. Of course, the inequalities can be reversed where an increased response is expected with increasing concentrations. A two-sided version of this model is also possible for those situations where the researcher is confident that the subjects will respond to the chemical insult in a monotone dose–response fashion, but they are unsure for the compound and endpoint in question whether the direction of effect will be an increase or a decrease. For either Model (1.2) or (1.3), one would estimate the μ_i from the data.

Do we expect a monotone dose–response? If not, then tests based on monotonicity should not be used and, of course, regression models are likewise inappropriate. An exception exists for the case of hormesis where a specialized model can be employed to capture that phenomenon. Hormesis, or more generally, low-dose stimulation, is the presence of an apparently beneficial effect at low exposure levels followed by an adverse effect at higher exposure levels. Such effects are observed in many testing laboratories throughout the world and can arise from several causes. Pharmaceuticals are used because they are beneficial at low dosages but they can be toxic at high dosages. Duke et al. (2014) reported that allellochemicals, which are phytotoxins released from plants, are known to induce hormesis and some chemicals stimulate the production of allellochemicals. They speculate that some mechanisms producing hormetic responses could represent physiological attempts to compensate for chemical stress. For example, plants could produce more seeds, thereby increasing the chance of the next generation to germinate under more favorable conditions. In rodent studies, animals are often fed *ad libitum*. Since there is little opportunity for exercise or other physical stimulation, control animals sometimes get grossly overweight, which has health implications for those species as it does for humans. At low doses, the chemical can make the animals slightly ill so their appetite is reduced and they maintain a more healthy body weight and do not suffer the adverse weight-related health effects of the control animals. At higher dose levels, the adverse effects of the test compound overwhelm any benefit from reduced weight. In an aquatic experiment, the low dose of the test compound may stimulate the growth of algae because of a mild increase in nutrients in the chemical, but higher concentrations inhibit growth because the toxic effect overwhelms the nutrient effect. In a pesticide spray application, a low application rate may inhibit pests without damaging the plants, but at higher applications, plant damage may occur.

Calabrese and Baldwin (2002) have a very interesting discussion of hormesis in which they interpret it an "adaptive response with distinguishing dose-response characteristics that is induced by either direct acting or overcompensation-induced stimulatory processes at low doses. In biological terms, hormesis represents an organismal strategy for optimal resource allocation that ensures homeostasis is maintained." van der Woude et al. (2005) and Carelli and Iavicoli (2002) contain further discussion of hormesis. Lloyd (1987) discusses this in connection with mixtures of chemicals. Dixon and Sprague (1981) and Stebbing (1982) discuss this for a variety of phyla.

Even if we expect a monotone dose–response, it is important to assess the data for consistency with that judgment. There are several reasons for this. Bauer (1997) has shown that certain tests based on a monotone dose–response can have poor power properties or error rates when the monotone assumption is wrong. While our experience does not substantiate the idea, Davis and Svendsgaard (1990) and others have suggested that departures from monotonicity may be more common than previously thought. These concerns suggest that a need for caution exists. We advocate both formal tests and visual inspection to determine whether there is significant monotonicity or significant departure from monotonicity. Further discussion of this issue will be presented in Chapter 3.

Regression models assume a specific mathematical form for the relationship between treatment mean and concentration, for example, when modeling length or weight of fish from an aquatic experiment or shoot height from a non-target plant study, one might hypothesize

$$\mu_i = ae^{bx_i} \tag{1.4}$$

where x_i is the concentration in the ith treatment, and a and b are positive parameters to be estimated from the data. The essential difference between Model (1.3) and models of the type illustrated by Model (1.4) is the specific mathematical relationship assumed between concentration and response. It is rare in ecotoxicology to have a priori models based on biological principles, so the regression approach becomes an exercise in curve fitting and evaluation. An advantage of Model (1.3) is that there is no need to specify an exact form. An advantage of Model (1.4) is the ability to predict the mean response at a range of chemical concentrations by interpolating between the x_i. The typical goal in a regulatory setting is to estimate the concentration or rate that produces a specific percent change (increase or decrease, as appropriate) in the measured response compared to the control mean response. The label ECx is used for the concentration that produces an x% change from the control and is referred to as the x-percent effects concentration. There are other differences that will be addressed later.

Biologically-based models are mathematical models having the form of regression models, such as Model (1.4), but derived from, or based on, concepts observed in biology that encompass much more than just the concentration of test chemical. The Dynamic Energy Budget (DEB) theory developed by S.A.L.M. Kooijman and colleagues (e.g. Kooijman and Bedaux, 1996) is built on the idea that the hazard rate and the parameters that quantify the energy budget of the individual are proportional to the concentration of the test substance in the animal. DEB theory "specifies the rules that organisms use for the energy uptake of resources (food) and the ensuing allocations to maintenance, growth, development and propagation."

1.5 BIOLOGICAL VS. STATISTICAL SIGNIFICANCE

Statistical analysis of toxicity data, especially the hypothesis testing or NOEC approach, is often criticized for finding treatment means statistically significantly different from the control mean that are not biologically important. On the other hand, in some experiments, differences are found that are thought to be biologically important but are not found statistically significant. The difference between biological and statistical significance is real and should be appreciated. Moreover, an ideal study should be designed to have high power to find biologically important treatment effects and low likelihood of finding significant a treatment "effect" that is not biologically important. In terms of estimating an x% effects concentration, ECx, from a regression model, the percent effect, x, to be used should be biologically determined, not an arbitrary value selected purely to satisfy some legislation or convenience.

There are several factors that make it difficult or impossible to design such an ideal experiment. First, it is unusual for an experiment to have a single endpoint or response of interest. This means that while it may be possible to design the experiment to be optimal in some sense for one response, the experiment may be decidedly suboptimal for another response. Second, it has proved very difficult to find agreement within the ecotoxicity scientific community to reach agreement on the size effect for a given response that is biologically important to detect, or even the basis for making such a determination. Is it more important to judge biological importance on the health of the individual subject, as is usually the case in assessing the carcinogenic risk in rodent experiments (and by inference, to humans), or on the ability of population as a whole to thrive, as suggested by Iwasaki and Hanson (2014) and Mebane (2012)? Third, when there is, say, 90% power to detect, i.e. found to be significant, a 10% change in some response, there may also be a 50% chance of detecting a 5% change and 25% power to detect a 3% change. In plain language, then, sometimes very small effects will be found statistically significant. If a regression model can be fit to the data to try to eliminate this problem, then it might turn out that the 95% confidence interval for EC10 spans the entire range of tested concentrations including the control. (Examples will be given in Chapter 4.) This hardly seems informative. Data may be highly variable, given practical limitations on the size experiment than can be run and the inherent variability of the subjects and sensitivity of the measuring equipment, and will mean a 50% effect cannot be detected or no model can be fit or the aforementioned wide confidence bounds on EC10 estimates encountered.

One should also question the idea that 10% is some universally applicable size effect. The measurement of VTG in fish to evaluate possible endocrine effects is extremely variable and effects of 1000% or higher increases are observed. There also frequently very high inter-lab and even intra-lab variability in this measurement. The data are continuous but by no means normally distributed or homogeneous in variance. For hypothesis testing purposes, a log-transform or even a rank-order transform might be used to deal with the huge spread in the data. Regression models, even on log-transformed responses, are often very poor and generate very wide confidence bounds. It is totally pointless to estimate EC10 with such data, so what size effect should be estimated? Furthermore, do we estimate an x% effect based on the untransformed control mean or on the log-transform? If the former, the model-fitting algorithm will usually not converge. If the latter, the meaning of ECx will vary from experiment to experiment in a much bigger way than with more well-behaved data. For example, if the mean control response is 10, 100, 1000, or 10000, then a 10% increase in the logarithm corresponds to a 26, 58, 100, or 151% effect in the untransformed values.

Isnard et al. (2001) note the problem of determining an appropriate choice of x, noting "Biological arguments are scarce to help in defining a negligible level of effect x for the ECx." An insightful presentation by C. Mebane of the US Geological Survey/NOAA Fisheries Liaison at the SETAC conference in Long Beach, California in 2012 (Mebane, 2012) reported on a review of early-life stage toxicity testing with aquatic organisms (reduced growth, fecundity, and survival) in the context of responses of wild fish populations to disturbances associated with changes in mortality or growth rates. The review suggested that different ECx values would be appropriate for different endpoints and for species with different life histories. Under some conditions and for some species, differences in length of as little as 5% can disproportionately determine survival. Growth reductions of the magnitude of 20% could predict extremely high indirect mortalities for juvenile fish. For some other populations, where juveniles have to compete for limited shelter to survive their first winter, much greater than 20% loss of the young-of-year is routinely absorbed. These observations suggested to Mebane that an EC20 for fecundity or first-year survival in density-dependent fish populations would conceptually be sustainable, yet for reduced growth (as length of juvenile fish) an EC5 would be a more appropriate endpoint.

These findings and others suggest a real need to identify the size effects of biological importance across a wide spectrum of measured endpoints. In any event, to design properly an experiment, it is critical to know what size effects are of interest for each response to be analyzed. This must be tempered by the understanding mentioned above that an experiment well designed for one response may be over- or under-sensitive for other responses. One practical workaround is to design around the most important responses. It is also important to maintain a working historical control database within the testing laboratory for each type of study, species, and response, and then interpret the results of each new study in light of the distribution of control responses. This can be done informally or through Bayesian methodology applied to incorporate such information formally in the analysis.

Before leaving this topic, it is appropriate to discuss how statistical significance is decided and to appreciate a controversy on this issue. In regulatory risk assessment, a finding of statistical significance of a treatment compared to the control is almost always judged by whether the statistical test used in deciding the question is significant at the 0.05 significance level, that is, whether the test statistic has a *p*-value less than 0.05. We will not dwell on the history of the use of this particular value, but content ourselves with pointing out that it is entirely arbitrary and results in scientifically indefensible decisions that a test statistic with *p*-value 0.0499 is significant while one with a *p*-value of 0.0500 is not. Nonetheless, it is the commonly accepted value used in regulatory risk assessment. The American Statistical Association issued a statement in 2016 on the use and misuse of *p*-values (ASA, 2016). There have been numerous articles critical of *p*-values (e.g. Newman, 2008; Hothorn, 2014; Greenland et al., 2016; Schmidt et al., 2016) and the reader is urged to review those for further discussion. In the present risk assessment paradigm, some "bright line" is probably necessary to enable a risk assessor to reach a decision. Unless and until that changes, the use of $p=0.05$ in regulatory risk assessment is a fact of life and will be followed throughout this text.

1.6 HISTORICAL CONTROL INFORMATION

The use of historical control data has long been recognized as valuable. For example, Gart et al. (1979), Tarone et al. (1981), Haseman (1983), and Gart et al. (1986) discussed this in depth. Fears et al. (1977) and Tarone et al. (1981) demonstrated value in informal use of such information, while Tarone (1982) introduced the (Bayesian) beta-binomial model to formalize the use of historical control information in the analysis of tumor data. Ryan (1993) and later Ibrahim et al. (1998) extended the beta-binomial model to incorporate correlated observations, such as among litter mates in a developmental study. Much information is given in Berry and Stangl (1996), though this will not include more recent developments, such as Zhang et al. (2012), who applied Bayesian methods to incorporate historical control information in aquatic tests of reproductive inhibition and Mylchreest and Harris (2013), Ema et al. (2012), and Petterino and Argentino-Storino (2006). More discussion on the incorporation of historical control information is given in Chapters 3–5.

1.7 SOURCES OF VARIATION AND UNCERTAINTY

Biological data are generally much more variable than data from manufacturing processes. This is to be expected since even when breeding special strains of laboratory animals, the health and sensitivities of individual animals is subject to genetic and other factors outside the control of the experimenter. There is even more uncertainty when dealing with aquatic and avian species that do not have a long history of husbandry and with terrestrial plants, especially wild species.

Thus, there is often considerable subject-to-subject variation. In addition, mice, birds, fish, and plants are usually grouped into multiple replicate vessels within each treatment group and between-replicate variability must be considered. The flip side of between-replicate variability is within-replicate correlations, which should not be ignored. Water or air temperature differences, variations in water chemistry, light intensity, and noise factors in the lab, weather, and soil, water conditions in field studies, and the physical impossibility to collect all measurements at the same time, all play a part. Much of this is addressed by random allocation of subjects to replicate vessels, and replicate vessels to treatments. Factors that cannot be randomized are controlled in some other way. For many species and responses, it is important to do separate analyses for each sex or to specify a variance–covariance structure that allows for different means and variances for different sexes. With some endocrine disruption studies, sex itself varies with treatment. In other studies, sex cannot be determined at the beginning of the study, but only at the end when the subjects are more mature and are sacrificed and examined histologically. Where sex differences exist, these realities increase the uncertainty in statistical analyses.

Statistical methods are based on assumptions or requirements, such as having a normal distribution with homogeneous variances. Real data do not always meet these requirements. Determining the proper transformation to bring the data closer to meeting the requirements can be challenging. Alternatively, determining an acceptable alternative distribution can be difficult. Sample sizes in many studies are too small to decide definitively among competing distributions. Since conclusions can vary, even dramatically, according to the statistical method, distribution, or transformation used, caution is needed in analysis and the data analyst needs to be alert to indications of potential problems in the data. Power and sensitivity analyses should be used at the design stage to make sure there is adequate protection against these uncertainties and the experimenter is well advised to allow for more variation and uncertainty than a small database might suggest.

When regression models are fit to a dataset without an underlying biological basis, there is model uncertainty in addition to the uncertainties inherent in the data. Perhaps less obviously, model uncertainty also affects hypothesis testing because as discussed in Section 1.5, there are different models underlying pairwise- and trend-based analyses and within each type, there are uncertainties about distributional assumptions. Serious errors can enter data interpretation when model uncertainty is ignored. Chapters 4 and 7–9 will consider ways to assess model uncertainty. We will also explore ways to incorporate certain types of uncertainty, such as overdispersion, into the model. This can be done through including a Bayesian or a non-Bayesian variance–covariance structure. These approaches will be pursued in later chapters.

We do not claim to have listed all sources of uncertainty in toxicity studies. Rather, the hope is to raise sensitivity to these concerns so they are taken into account both at the design stage and in analysis.

1.8 MODELS WITH MORE COMPLEX STRUCTURE

Not all toxicity experiments have the simple structure that formed most of the discussion in this chapter so far. Several more complex experiments are common in toxicity studies and more are being introduced. Perhaps the most common added complexity occurs when the same subject is measured for the same response at several different times over the course of a single study. Growth and growth rate, food consumption, egg production, and mortality are often measured or recorded daily or weekly or on some other schedule. Data of this sort are referred to as repeated measures. Some experiments have two or more factors, both of which vary in a controlled way over a study. A multigeneration reproduction study may have treatment, generation, and week as factors. In this context, there may be a 3–5-week period of prime egg laying. Week would be a repeated measures variable, as the same subjects (e.g. birds or fish) lay eggs daily and these eggs are counted daily or weekly. Depending on the experimental design, generation may be a repeated measures variable. This might require tracking from parent to child to grandchild and keeping replicates (if any) fixed across generations. In an experiment of this sort, it is entirely possible that the effect of the treatment may vary across generations or across weeks within generation or vary across generations within the same nominal week. This leads to a study of interactions of two or more factors. Another common complexity is analysis of covariance (ANCOVA), where some independent concomitant variable unaffected by the treatment of interest may have an effect on the response of interest. By accounting for the variance in the response that arises solely from differences in the covariate, the variability due to the treatment, and hence the treatment effects, becomes clearer. These three topics are developed in Chapter 5. Another type of study where time plays a special role is time-to-event data, discussed in Chapter 10.

In a statistically similar neurotoxicity experiment, animals exposed to various dosages of a chemical may be tested for learning effects. This can have the form of rapid repeat measurements of several minutes on the same day, with a week or more between days. So there are two repeated measures variables (weeks and minutes) with interactions of treatment and the time variables.

A study that explores both a chemical and a method of administration (e.g. mixed in the food or water or by oral gavage or by injection) does not involve time as a factor, but there may well be different treatment effects across methods of administration. Multifactor studies with interactions will be considered in Chapter 5.

1.9 MULTIPLE TOOLS TO MEET A VARIETY OF NEEDS OR SIMPLE APPROACHES TO CAPTURE BROAD STROKES?

An issue on which there is no consensus of opinion among statisticians or among biologists or toxicologists is whether to analyze the same response in always the same way for a given type of study or to let the data affect the analysis. For example, some scientists argue that, based on observing many studies, rainbow trout weight is always log-normally distributed and should always be analyzed as such. Even when data from a particular study are found to depart significantly from log-normality, that should be ignored (indeed, no test for normality should be performed). The reasoning is that log-normality is the true state of nature and should not be overturned by the peculiarities of a particular dataset. This allows uniformity of approach and simplifies the comparison of conclusions across studies. If a hypothesis testing approach is followed, the Dunnett's test is always used (or Williams' or whatever is thought appropriate for all studies of this type). Some go even further and say that if the data do not agree with the agreed model, then the data are wrong and the experiment should be repeated until data are obtained that are in agreement with the model.

If a regression approach is used, then, according to this school of thought, the same regression model is always used (apart from the specific parameters estimated from the data) regardless of how well that model fits the current data.

Additional argument for this approach is that ANOVA and regression methods are robust against mild violations of the assumptions on which they are based. Slob and Setzer (2014) argue that an exponential or a Hill model with four parameters adequately describes toxicological dose–responses for continuous responses in toxicity studies. They found no exception to this in a large historical database. The natural conclusion is that one of these two models can be picked and always used. This would greatly simplify the regulatory process and eliminate contentious arguments between companies submitting product registration packages and the regulators who must decide whether to reject the product or approve it subject to certain restrictions on use. They conclude "more realistic ranges of parameter values could be derived from re-analyses of large numbers of historical dose–response datasets in the same endpoint and study type, which could be used as parameter constraints in future individual datasets. This approach will be particularly useful for weak datasets (e.g. few doses, much scatter). In addition, this approach may open the way to use fewer animals in future studies. In the discussion, we argue that distinctions between linear, sub/supralinear or thresholded dose–response shapes, based on visual inspection of plots, are not biologically meaningful nor useful for risk assessment."

We disagree. Generally, a statistical regression model itself does not have any meaning and the choice of the model is somewhat arbitrary. It is the data, not the model, that determines the dose–response. However, the purpose of regression modeling in toxicology is not merely to obtain a curve that fits the data well. We want to use this curve to estimate something of inherent interest in evaluating the toxicity of the compound under study. A generally agreed measure of interest is the concentration that induces a specified percent inhibition (or stimulation) in the mean of the exposed population compared to the control population mean. The concentration that induces an $x\%$ change in the population mean is called the ECx. It should be obvious that an improper choice of the model can lead to a poor estimate of the ECx and we will provide numerous examples to illustrate this claim. The choice of the model should be governed by the data and we find compelling reason to allow flexibility in model selection. Similarly, not all hypothesis testing methods are equally valid and we find, and will present, what we think are compelling reasons for having multiple techniques to fit the needs of a large variety of data. We do look for commonalities and see no reason to make statistical analysis more complicated than it needs to be. We support the parsimony principle by which the simplest model that adequately explains the data should be used. We apply this idea in both the regression and hypothesis testing realms.

Chapter 2

Statistical Analysis Basics

2.1 INTRODUCTION

A basic principle in selecting statistical methods is to attempt to use underlying statistical models that are consistent with the actual experimental design and underlying biology. This principle has historically been tempered by widely-adopted conventions. For example, it is traditional in ecotoxicological studies to analyze the same response measured at different time points separately by time point, although in many cases unified analysis methods may be available. In this chapter, we will not pursue the appropriateness of this convention in depth. Instead, the present discussion will be restricted to the most appropriate analysis of a response at a single time point and to situations where all other relevant factors, such as sex, are fixed at a single value. It is entirely possible to use a statistical model that incorporates and accounts for differences in sex, time of measurement, and other pertinent factors and we will consider such models in Chapter 10.

In this chapter, we will introduce basic issues related to hypothesis testing methods to determine a so-called No Observed Effects Concentration (NOEC), regression methods to estimate an $x\%$ effects concentration (ECx), Bayesian adaptations of both, and what are called biologically-based models to determine a toxic threshold. There is controversy within the ecotoxicity community regarding the relative merits of NOEC and ECx approaches to statistical analysis. The NOEC approach has been in place for many years but there is a strong movement away from that toward regression. We will address that controversy primarily in the context of numerous examples and developments throughout this text. In general, if a regression model is available that is appropriate for the data and which provides scientifically sound estimates, then that would be the preferred approach. However, as we shall demonstrate, there are numerous datasets and whole classes of responses for which there is no currently acceptable regression model. References to pursue the controversy include Green (2014b, 2015a, b) and Green et al. (2012). In addition, the wide variety of types of measurements will be presented that make clear the need for a variety of statistical models and methods to analyze toxicity data. The discussion begins with the NOEC, as the underlying model is conceptually simpler and many of the underlying principles apply to the other methods.

2.2 NOEC/LOEC

The NOEC is defined in terms of LOEC, or Lowest Observed Effects Concentration. The LOEC is the lowest tested concentration that results in a statistically significant adverse effect. The NOEC is defined as the highest test concentration below the LOEC. Thus, the LOEC and NOEC are tested concentrations and experiments with different sets of tested concentrations will produce different NOEC and LOEC values. The NOEC has often been defined as the highest tested concentration at which there is no statistically significant effect compared to the control. This is especially true in older test guidelines (e.g. OECD, 1984a, b). Newer test guidelines (e.g. OECD, 2006a, b, c) sometimes still use this definition but include a phrase indicating that the NOEC is smaller than the LOEC, which does not follow from the definition used. Other more recent test guidelines use our definition (e.g. OECD, 2013). While the two definitions of NOEC often produce the same value, they are not equivalent and different values can occur. The difference arises when there is an exposure level in the experiment at which a statistically significant effect is

Statistical Analysis of Ecotoxicity Studies, First Edition. John W. Green, Timothy A. Springer, and Henrik Holbech.

Companion website: www.wiley.com/go/Green/StatAnalysEcotoxicStudy

Table 2.1 Rainbow Trout Weight

Conc[a]	Mean[b]	StdErr[c]	Conc	Mean	StdErr
Water control	0.692	0.017	0.62	0.650	0.017
DMF control	0.638	0.017	1.4	0.610	0.017
0.11	0.669	0.017	4.1	0.655	0.019
0.24	0.670	0.017			

[a] Concentration in $mg\,l^{-1}$.

[b] Mean wet weight of two replicate tanks of 15 fish each.

[c] Standard error of the replicate mean responses.

observed and a higher exposure level at which no statistically significant effect is observed. As will be shown, whether this can occur depends on the statistical method used to determine the NOEC. See Table 2.1 for a dataset that will illustrate this, depending on how the data are analyzed.

It should be observed in Table 2.1 that the mean weight of the $1.4\,mg\,l^{-1}$ treatment group is notably lower than the two controls and of the high-treatment group. If Dunnett's test (Chapter 3) is used to analyze these data, the mean response is significantly lower than the combined or solvent only control only at $1.4\,mg\,l^{-1}$. In particular, the mean response at $4.1\,mg\,l^{-1}$ is not significantly lower than the control mean response. Thus, by our definition, the NOEC (by Dunnett's test) is $0.62\,mg\,l^{-1}$, but by the alternative definition, NOEC $= 4.1\,mg\,l^{-1}$.

2.2.1 Maximum Allowable Toxic Concentration

We will explore how statistical significance is established and the implications for experimental design. A related concept that is used in risk assessment is the MATC or Maximum Allowable Toxic Concentration, which is the geometric mean of the NOEC and LOEC. That is,

$$\text{MATC} = \sqrt{\text{NOEC*LOEC}}$$

Equivalently, the MATC is the antilog of the mean of the logarithms of the NOEC and LOEC. This concept can probably best be understood from the fact that traditionally the effect of a toxic chemical was expected to increase linearly with the logarithm of concentration. On a log scale, the MATC is midway between a NOEC and a LOEC and MATC is a compromise between those wanting to base risk assessment on the LOEC and those wanting to base risk assessment on the highest concentration in which no adverse effect was observed. The idea is that since the NOEC and LOEC are tested concentrations subject to the "whim" of experimental design, there may be concentrations below the LOEC that produce adverse effects and there may be concentrations above the NOEC that do not. The midpoint (on the log scale) is a "best guess" at where the true dividing point is between the adverse effect and non-adverse effect. Clearly, this is not a true measure of the exposure level dividing adverse from non-adverse. Indeed, it is not at all clear that such a dividing line exists. It is well known that if an experiment is done twice under conditions as identical as can be done, the NOECs derived can be different tested concentrations. As will be shown in great depth in Chapters 4, 7, and 8, using regression models to estimate EC10 (or ECx for some other value of x) does not change this, as the uncertainty in ECx estimates can be quite large. Biological variation is a very real issue in toxicity studies and uncertainty in results needs to be well understood to avoid drawing erroneous conclusions.

2.2.2 Biological Meaning of NOEC

As to the NOEC, note that no claim can be made that the condition of organisms exposed to toxicants at the NOEC is the same as the condition of organisms in the control group, or that the NOEC is an estimate of the threshold of toxicity (if such exists) or of any ECx. Rather, no effect could be distinguished from the control response in this particular experiment at that concentration. The detectability of an effect depends on the quality and the size of the experiment and the statistical procedure used. The fact that a statistically significant effect was not found does not mean that the effect was zero.

The relationship between the detectability of effects and the quality of the experiment can be quantified by the concept of statistical power. For a given null and alternative hypothesis, statistical power is the probability that a particular magnitude of effect will result in a significant test outcome. In large experiments (i.e. many replicates) smaller sized effects are detectable as compared to small experiments. Thus, one may consider the detectable effect size of a particular experiment as an analogue of the detection limit of a particular chemical analysis. The detectable effect size can be increased not only by using larger sample sizes but also by taking measures to make the experimental (residual) error smaller and by selecting more powerful statistical tests.

How accurate is the NOEC? It must be emphasized that the NOEC is not an estimate of the threshold of toxicity, so accuracy is not an appropriate term to use. We cannot provide a confidence interval for the NOEC. We can provide a confidence interval for the absolute or percent effect at the NOEC. A better term than accuracy might be utility. What is relevant for the NOEC is how likely is the process of determining the NOEC to identify an exposure level of biological importance and how likely is it to declare an exposure level as having biological importance when in fact that level is not of real concern. In other words, the power to detect an important effect and the false-positive rate (FPR) are what determine the utility of the NOEC. We will pursue the concept of statistical power of selected tests in some detail.

2.2.3 Statistical Models for NOEC/LOEC

The statistical test that is used depends on the hypothesis to be tested, the associated statistical model, and the distribution of the values (e.g. are data normally distributed?). Thus, it is necessary to determine the appropriate statistical model and specify the hypotheses to be tested before selecting the test procedure.

All methods of statistical assessment (regression and hypothesis testing alike) assume a statistical model. The hypothesis testing approach to evaluation of toxicity data is based in part on keeping to a reasonable number the untestable or difficult-to-test assumptions, particularly those regarding the statistical model, which will be used in reaching conclusions. The models used in regression-based methods use stronger assumptions (such as the dose–response shape has a specific parametric form) than the models used in the hypothesis testing approach but result in more quantitative conclusions.

2.2.3.1 Pairwise Models

The simplest statistical model generally used in hypothesis testing assumes only that the distributions of responses within these populations are identical except for a location parameter (e.g. the mean or median of the distribution of values from each group). The basic hypothesis (in one-sided form) can be stated as follows:

$$H_0\colon \mu_0 = \mu_1 = \mu_2 = \cdots = \mu_k \text{ vs } H_1\colon \mu_0 > \mu_i \text{ for at least one } i, \quad \text{(Model 1)}$$

where μ_i, $i=0, 1, 2, 3, \ldots, k$ denotes the means (or medians) of the control and test *populations*, respectively. Thus, one tests the null hypothesis of no differences among the population means against the alternative that at least one population mean is smaller than the control mean. There is no investigation of differences among the treatment means, only whether treatment means differ from the control mean.

In this formulation, an adverse effect is one that reduces the population mean response. This will generally be appropriate for responses such as size, biomass, and reproduction. Where an adverse effect is indicated by an increase in the population mean response, the inequality Model (1) is reversed.

In some situations, any change, i.e. either an increase or decrease, in the mean response is indicative of an adverse effect. In these situations, a two-sided alternative hypothesis is appropriate. Then the alternative is

$$H_1 : \mu_0 \neq \mu_i \text{ for at least one } i.$$

The test statistics for the above hypotheses are based on comparing each treatment to the control, independent of the other treatments. Many tests have been developed for this approach, some of which will be discussed in this text. Most such tests were developed for experiments in which treatments are qualitatively different, as, for example, in comparing various new therapies or drug formulations to a standard.

2.2.3.2 Trend Models

In toxicology, the treatment groups generally differ only in the exposure concentration (or dose) of a single chemical. It is further often true that biology and experience suggest that if the chemical is toxic, then as the level of exposure is increased, the magnitude of effect will tend to increase. Depending on what response is measured, the effect of increasing exposure may show up as an increase or as a decrease in the measured response, but not both. The statistical model underlying this biological expectation is what will be called as a *trend model* or a model assuming monotonicity of the population means:

$$\mu_0 \geq \mu_1 \geq \mu_2 \geq \mu_3 \geq \cdots \geq \mu_k \text{ (or with inequalities reversed)}. \quad \text{(Model 2)}$$

The null and alternative hypotheses can then be stated as

$$H_{02}\colon \mu_0 = \mu_1 = \mu_2 = \cdots = \mu_k \text{ vs } H_{12}\colon \mu_0 \geq \mu_1 \geq \mu_2 \geq \mu_3 \geq \cdots \geq \mu_k, \text{ with } \mu_0 > \mu_k.$$

Note that $\mu_0 > \mu_k$ is equivalent, under the specified alternative, to $\mu_0 > \mu_i$ for at least one i. If this monotone model is accepted as representing the true responses of test organisms to exposure to toxicants, it is not possible for, say, μ_3 to be smaller than μ_0 and μ_6 not to be smaller.

Under the trend model and tests designed for that model, if tests of hypotheses H_{02} vs H_{12} reveal that μ_3 is different from μ_0, but μ_2 is not, the NOEC has been determined (i.e. it is the test concentration associated with μ_2) and there is no need to test whether μ_1 differs from μ_0. Also, finding that μ_3 differs from μ_0 implies that a significant trend exists across the span of doses including μ_0 and μ_3, the span including μ_0 and μ_4, and so on. For the majority of toxicological studies, a test of the trend hypothesis based on Model (2) is consistent with the basic expectations for a model for dose–response. In addition, statistical tests for trend tend to be more powerful than alternative non-trend tests and should be the preferred tests if they are applicable. Thus, a necessary early step in the analysis of results from a study is to consider each endpoint, decide whether a trend model is appropriate, and then choose the initial statistical test based on that decision. Only after it is concluded that the trend is not appropriate do specific pairwise comparisons make sense to illuminate sources of variability.

Toxicologists sometimes do not know whether a compound will cause measurements of continuous variables such as growth or weight to increase or decrease, but they are confident it will act in only one direction. For such endpoints, the two-sided trend test is appropriate. One difference between implementing step-down procedures for quantal data and continuous data is that two-sided tests are much more likely to be of interest for continuous variables. Such a model is rarely appropriate for quantal data, as only increased incidence rates above background (control) incidence are of interest in toxicology. The two-sided version of the step-down procedure is based on the underlying model:

$$\mu_0 \geq \mu_1 \geq \mu_2 \geq \mu_3 \geq \cdots \geq \mu_k$$

or

$$\mu_0 \leq \mu_1 \leq \mu_2 \leq \mu_3 \leq \cdots \leq \mu_k.$$

Under this model, in testing the hypothesis that all population means are equal against the alternative that at least one inequality is strict, one first tests separately each one-sided alternative at the 0.025-level of significance with all doses present. If neither of these tests is significant, the NOEC is higher than the highest concentration. If both of these tests are significant, a trend-based procedure should not be used, as the direction of the trend is unclear. (Note: for several tests, such as the Jonckheere-Terpstra and Cochran-Armitage, it is not possible for tests in both directions to be significant.) If exactly one of these tests with all the data is significant, then the direction of all further tests is in the direction of the significant test with all groups. Thereafter, the procedure is as in the one-sided test, except all tests are at the 0.025 significance level to maintain the overall 0.05 FPR

The use of $p=0.05$ as a decision rule is arbitrary. The latter point has been made by authors too numerous to list. While we recognize that there is no real scientific basis for this use of 0.05, the regulatory context probably requires such fixed rules to avoid whimsical or arbitrary decisions or endless controversy. In any case, we will follow this convention in this book, except where explicitly stated otherwise.

Where it is biologically sensible, it is preferable to test the one-sided hypothesis, because then random variation in one direction can be ignored, and as a result, statistical tests of the one-sided hypothesis are more powerful than tests of the two-sided hypothesis. Note that a hypothesis test based on Model (2) assumes only a monotone dose–response rather than a precise mathematical form, such as is required for regression methods.

2.2.3.3 Comparison of Single-step (Pairwise Comparisons) to Step-down Trend Tests for NOEC

In general, determining the NOEC for a study involves multiple tests of hypotheses (i.e. a family of hypotheses is tested), either pairwise comparisons of treatment groups or a sequence of tests of the significance of trend. Statisticians have developed tests to control the family-wise error (FWE) rate (the probability that one or more of the null hypotheses in the family will be rejected incorrectly) in the multiple comparisons performed to identify the NOEC. The method of controlling the FWE rate has important implications for the power of the test. There are two general approaches that will be discussed: Single-step procedures and step-down procedures.

Single-step procedures amount to performing all possible comparisons of treatment groups to the control. Multiple comparisons to the control may be made but there is no ordered set of hypotheses to test and no use of the sequence of outcomes in deciding which comparisons to make. Examples of the single-step approach include the use of the Fisher's exact test for quantal responses, and the Mann–Whitney, Dunnett, and Dunn tests for continuous responses. Since many comparisons to the control are made, some adjustment must be made for the number of such comparisons to keep the FWE rate at a fixed level, generally 0.05.

There has been considerable discussion over the years about whether to make adjustments for multiple comparisons, and if so, what adjustments should be made. While no attempt will be made here to cover that topic, the article by Gelman et al. (2012) and its references provide some background and potential alternatives. One of the advantages of step-down trend tests (discussed below) is that there is no need to make any adjustment within a single response. The larger issue, discussed in Gelman et al. (2012), is whether adjustments should be made when, as is typical in toxicology and ecotoxicology, many responses are measured on the same subjects in a single study. In these situations, responses are not independent, so the question arises as to whether and how to take account of these dependencies. By far the most common tactic is to ignore correlations among responses in formal analyses but to make an informal assessment.

2.2.3.4 Adjustments for Number of Comparisons

With tests that are inherently single comparison tests, such as the t-test, Fisher's exact, and Mann–Whitney, a Bonferroni adjustment can be made: a study with k treatment levels would be analyzed by performing the pair-wise comparisons of each of the treatment groups to the control group, each performed at a significance level of α/k instead of α. (This is the Bonferroni adjustment.) The Bonferroni adjustment is generally overly conservative, especially for large k. Modifications reduce the conservatism while preserving the FWE at 0.05 or less.

For the Holm modification (Holm, 1979) of the Bonferroni adjustment, arrange the k unadjusted p-values for all comparisons of treatments to control in rank order, i.e. $p_{(1)} \le p_{(2)} \le p_{(3)} \le \cdots \le p_{(k)}$. Beginning with $p_{(1)}$, compare $p_{(i)}$ with $\alpha/(k-i+1)$, stopping at the first nonsignificant comparison. If the smallest i for which $p_{(i)}$ exceeds $\alpha/(k-i+1)$ is $i=j$, then all comparisons with $i>j$ are judged nonsignificant without further comparisons. It is helpful (Wright, 1992) to report adjusted p-values rather than the above comparisons. Thus, report $p^*_{(i)}=p_{(i)} * (k-i+1)$ and then compare each adjusted p-value to α. In the case of ties among the unadjusted p-values, all adjustments are the same as for the smallest rank of those p-values. See Table 2.2 for an illustration of the advantage of the Bonferroni–Holm method. In this hypothetical example, only the comparison of treatment 4 with the control would be significant if the Bonferroni adjustment is used, whereas all comparisons except the comparison of the control with treatment 1 would be significant if the Bonferroni–Holm adjustment is used.

Alternatives based on the Sidak inequality (each comparison at level $1-(1-\alpha)^k$) are also available. The Bonferroni and Bonferroni–Holm adjustments guarantee that the FWE rate is less than α, but they are conservative.

Other tests, such as Dunnett's, have a "built-in" adjustment for the number of comparisons made and are less conservative (hence, more powerful). There is no additional adjustment for the number of comparisons.

Step-down procedures are generally preferred where they are applicable. All step-down procedures discussed are based on a sequential process consisting of testing an ordered set of hypotheses concerning means, ranks, or trend. A step-down procedure based on trend works as follows: First, the hypothesis that there is no trend in response with increasing dose is tested when the control and all dose groups are included in the test. Then, if the test for trend is significant, the high-dose group is dropped from the dataset, and the hypothesis that there is no trend in the reduced dataset is tested. This process of dropping the highest treatment group and retesting the reduced dataset is continued until the first time the trend test is not significant. The highest dose in the reduced dataset at that stage is then declared to be the NOEC. Distinguishing features of step-down procedures are that the tests of hypothesis must be performed in a given order, and that the outcome of each hypothesis test is evaluated before deciding whether to test the next hypothesis in the ordered sequence of hypotheses. It is these two aspects of these procedures that account for controlling the FWE rate. One of the earliest treatments of step-down tests was presented in Marcus et al. (1976).

A step-down method typically uses a critical level larger than that used in single-step procedures, and seeks to limit the number of comparisons that need to be made. Indeed, the special class of "fixed-sequence" tests that we will describe fixes the critical level at 0.05 for each comparison but still bound the FWE rate at 0.05. Thus, step-down methods are generally preferable to the single-step methods as long as the population response means are monotonic.

Tests based on trend are logically consistent with the anticipated monotone pattern of responses in toxicity tests. Step-down procedures make use of this ordered alternative by ordering the tests of hypotheses. This minimizes the number of comparisons that need to be made, and in all the methods discussed here, a trend model is explicitly assumed (and tested) as a part of the procedure.

Procedures that employ step-down trend tests have more power than procedures that rely on multiple pairwise comparisons when there is a monotone dose–response because they make more use of the biology and experimental design being analyzed. When there is a monotone dose–response, procedures that compare single treatment means or medians against the control, independent of the results in other treatments (i.e. single-step procedures), ignore important and relevant information and suffer power loss as a result.

The trend models used in the step-down procedures do not assume a particular precise mathematical relationship between dose and response but rather use only monotonicity of the dose–response relationship. The underlying statistical model assumes a monotone dose–response in the *population* means, not the *observed* means.

Table 2.2 Comparison of Adjusted and Unadjusted p-Values

Comparison	Unadjusted p value	Bonferroni–Holm adjusted p value $p^*_{(i)}$	α/k (Bonferroni adjustment of α)
Control–Treatment 4	$p_{(1)}=0.002$	0.002 * 4=0.008	0.05/4=0.0125
Control–Treatment 2	$p_{(2)}=0.013$	0.013 * 3=0.039	0.05/4=0.0125
Control–Treatment 3	$p_{(3)}=0.020$	0.020 * 2=0.040	0.05/4=0.0125
Control–Treatment 1	$p_{(4)}=0.310$	0.310 * 1=0.310	0.05/4=0.0125

Rejection of the null hypothesis (i.e. rejecting the hypothesis that all group means, or medians, or distributions are equal) in favor of the stated alternative implies that the high dose is significantly different from the control. The same logic applies at each stage in the step-down application of the test to imply, whenever the test is significant, that the high dose remaining at that stage is significantly different from the control. These tests are all applied in a one-sided manner with the direction of the alternative hypothesis always the same. Moreover, this methodology is general and applies to any legitimate test of the stated hypotheses under the stated model. That is, one can use this fixed-sequence approach with the Cochran-Armitage test on quantal data and the Jonckheere-Terpstra or Williams or Brown-Forsythe tests of trend on continuous data. Other tests of trend can also be used in this manner.

2.2.3.5 Deciding Between the Two Approaches

The first consideration is biological. Do we expect a monotone dose–response? If not, then tests based on monotonicity should not be used. Even if we expect a monotone dose–response, it is important to assess the data for consistency with that judgment. There are several reasons for this. Bauer (1997) has shown that certain tests based on a monotone dose–response can have poor power properties or error rates when the monotone assumption is wrong. Davis and Svendsgaard (1990) and others have suggested that departures from monotonicity may be more common than previously thought. These results suggest that a need for caution exists.

We advocate both formal tests and visual inspection to determine whether there is significant monotonicity or significant departure from monotonicity. One can use a "positive" test to affirm monotonicity before employing a test based on monotonicity, or one can use a "negative" test for departure from monotonicity and proceed unless there is evidence of non-monotonicity. Details of these procedures are given later. The second approach (and our preference) is grounded in the idea that monotonicity is the rule and that it should take strong evidence to depart from this rule.

One simple procedure that can be used for continuous responses is to construct linear and quadratic contrasts of normalized rank statistics (to avoid the complications that can arise from non-normal or heterogeneous data). If the linear contrast is not significant and the quadratic contrast is significant, there is evidence of possible non-monotonicity that calls for closer examination of the data or the use of pairwise comparison methods. Otherwise, a trend-based analysis is used. As we shall see, for quantal data using the Cochran-Armitage test, there is a built-in test for lack of monotonicity.

2.3 PROBABILITY DISTRIBUTIONS

Before we discuss methods for assessing whether a dataset meets the requirements of a statistical model, it is necessary to introduce the concepts of probability distribution functions and probability density functions. The distribution function for a random variable X is a non-negative, real-valued function F defined by

$$F(x) = p[X \le x] \text{ for all real numbers } x,$$

where *p[E]* is the probability that E occurs. Discussion of probability functions and events, as well as much more on distribution and density functions, is provided in Appendix 2.

2.3.1 Distribution Functions

A *distribution function* F has the following characteristic properties:

$$0 \le F(x) \le 1 \text{ for all real numbers } x.$$

$$\text{If } x_1 < x_2, \text{ then } F(x_1) \le F(x_2).$$

$$\lim_{x\to\infty} F(x) = 1.$$

$$\lim_{x\to-\infty} F(x) = 0.$$

$$\log_{x\to a+} F(x) = F(a) \text{ for each real number a.}$$

In Chapter 1, four types of responses were defined: Continuous, quantal, count, and ordinal. These correspond to random variables of the same names. The simplest is perhaps the quantal random variable. The distribution for a quantal random variable is a step function. That is, it has the form

$$\begin{aligned} F(x) &= 0 \quad \text{if } x < 0 \\ &= p_0 \quad \text{if } 0 \le x < 1 \\ &= p_0 + p_1 \quad \text{if } 1 \le x < 2 \\ &= p_0 + p_1 + p_2 \quad \text{if } 2 \le x < 3 \\ &\vdots \end{aligned}$$

See Figure 2.1 for an illustration of such a step function.

2.3.2 Probability Density Functions

For quantal random variables, it is usually more convenient to define the *probability density function* (pdf). The pdf for a quantal random variable is defined as follows:

$$\begin{aligned} f(x) &= 0 \quad \text{if } x < 0 \\ &= p_0 \quad \text{if } 0 \le x < 1 \\ &= p_1 \quad \text{if } 1 \le x < 2 \\ &= p_2 \quad \text{if } 2 \le x < 3 \\ &\vdots \end{aligned}$$

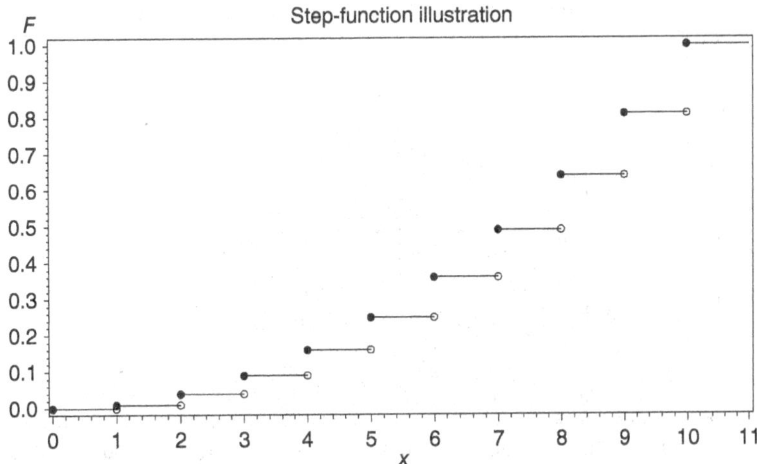

Figure 2.1 Step function. The endpoints are exaggerated for effect, with solid dots indicating that the point is on the plot and the open circles indicate that the point is not on the plot.

The relationship between the distribution function and the density function for a quantal random variable is given by

$$F(x) = \sum_{t \le x} f(t). \tag{2.1}$$

The most important quantal distribution in toxicity studies is the binomial. The Poisson distribution is sometimes preferred for quantal data and is also very useful for count data that are not quantal, such as the number of eggs or young produced. These two distributions are easily defined by their density functions.

2.3.3 Binomial Probability Density Function with Parameters *n* and *p*

$$f(x) = \frac{n!}{x!(n-x)!} p^x (1-p)^{n-x} \text{ if } 0 \le x \le n \text{ and } x \text{ is an integer}$$
$$f(x) = 0 \quad \text{otherwise} \tag{2.2}$$

To put this in a toxicity context, if n subjects are exposed to some substance and the probability of death for each individual subject is p, then $f(x)$ is the probability that exactly x subjects from the group of n die. The mean and variance of a binomial distribution are np and $np(1-p)$, respectively.

2.3.4 Poisson Probability Density Function with Parameters *n* and *p*

$$f(x) = e^{-np\frac{(np)^x}{x!}} \text{ if } 0 \le x \le n \text{ and } x \text{ is an integer} \tag{2.3}$$
$$f(x) = 0 \quad \text{otherwise}$$

Usually, the term np in the Poisson density function is replaced by a single symbol, such as λ, and that is the value of both the mean and variance of this distribution. The Poisson is often used when n is very large and p is very small, so that np is modest in size. Both density functions apply to situations where individual subjects all have the same probability, p, exhibiting the random phenomenon of interest, e.g. death. These density functions would not be appropriate if different subjects have different probabilities. As we shall see, Bayesian methods provide a mechanism for modeling variable p. This is especially important when different replicate vessels at the same nominal test concentration vary in the probability of the phenomenon of interest. We will examine that and other quantal density functions later in context of specific types of toxicity experiments.

In the above two examples, the density function depends on the values of *parameters*, n and p, and it will often be useful to include those parameters in the functional form by writing $f(x; n, p)$ for the density function and $F(x; n, p)$ for the corresponding distribution function.

The discussion for quantal distributions and density functions can also be extended to include the discrete case. It is also possible to define probability distribution and density functions for continuous random variables, but the relationship between them is different from that for quantal or discrete variables. One critical difference is that for a continuous random variable, the probability that it have a specific value is generally zero. For example, if we are describing body weight of an adult male Sprague-Dawley rat, it makes little sense to try to calculate the probability that it weighs exactly 475 g. It is much more reasonable to calculate the probability that its weight falls in the interval 470–480. See Figure 2.2 where a histogram is constructed with weight intervals $340 \le x < 360$, $360 \le x < 380$, and $380 \le x < 400$, where the heights of the histogram bars

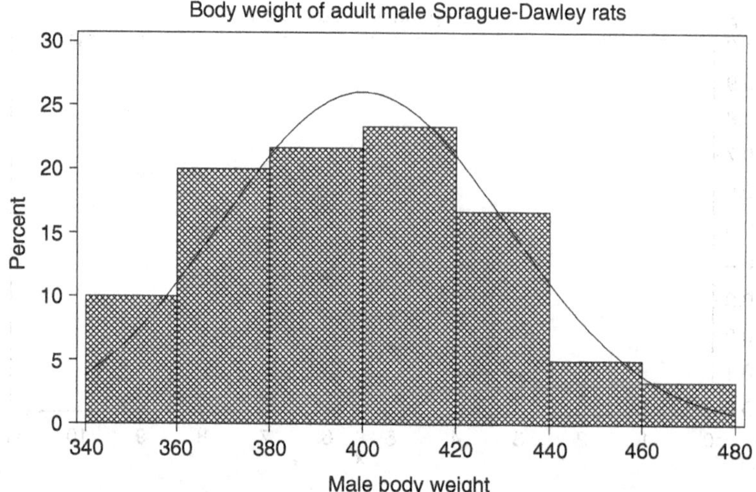

Figure 2.2 Histogram of Sprague-Dawley adult male body weights probabilities (vertical axis values) that have been multiplied by 100. Overlaid on the histogram is a theoretical curve that attempts to capture the general shape of the histogram. The curve displayed here is a normal density function, as described in Eq. (2.6). We will defer to for the moment how one obtains the values of the parameters θ and σ in that equation. Appendix 2 contains a description of methods for estimating the parameters for a model from the data collected.

represent the probability of the weight falling in the indicated interval.

If there were sufficient data, we could draw histograms with shorter intervals that might do a better job of capturing the shape. This would lead to approximate slopes to the curve on each interval, i.e.

$$\frac{F(x_{i+1})-F(x_i)}{x_{i+1}-x_i}.$$

In the limit (and in the abstract, not for a simple example of real data), as x_{i+1} approaches x_i, we get the instantaneous slope of the distribution function at $x=x_i$. So, provided the distribution function enjoys nice mathematical properties, the pdf for a continuous distribution is given by

$$f(x)=F'(x)=\frac{dF}{dx}. \tag{2.4}$$

That is, the pdf is the derivative of the distribution function. We can also describe the relationship between density function and distribution function in general by writing

$$F(x)=\int_{-\infty}^{x} f(t)\,dt. \tag{2.5}$$

2.3.5 Normal Distribution

Familiar examples of continuous density functions used in toxicity include the normal or bell-shaped curve defined (for mean $=\mu$, variance $=\sigma^2$) by

$$f(x;\mu,\sigma)=\frac{1}{\sqrt{2\pi}\sigma}e^{-\frac{(x-\mu)^2}{2\sigma^2}}. \tag{2.6}$$

The overlaid density curve in Figure 2.2 is a normal density function with mean and variance estimated from the data. It will be convenient to have a standard notation for this widely used density function. The standard normal density function is the normal with mean 0 and variance 1 and is denoted by

$$\varphi(x)=\frac{1}{\sqrt{2\pi}}e^{-\frac{x^2}{2}}, \tag{2.7}$$

and the corresponding standard normal distribution function is given by

$$\Phi(x)=\frac{1}{\sqrt{2\pi}}\int_{-\infty}^{x} e^{-\frac{t^2}{2}}\,dt. \tag{2.8}$$

Then, a normal distribution function with mean μ and variance σ^2 is given by

$$F(x;\mu,\sigma)=\Phi\left(\frac{x-\mu}{\sigma}\right), \tag{2.9}$$

and the density function is

$$f(x;\mu,\sigma)=\frac{1}{\sigma}\varphi\left(\frac{x-\mu}{\sigma}\right). \tag{2.10}$$

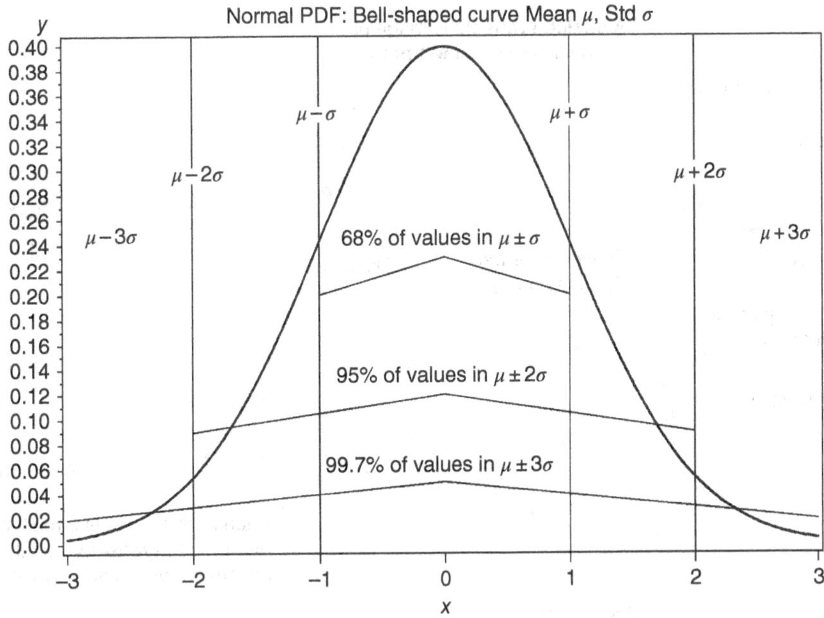

Figure 2.3 Normal density function with probability intervals scale on horizontal axis is in multiples of standard deviation above or below the zero mean.

See Figure 2.3 for an illustration of the normal pdf and a few of its characteristics. For example, 68% of the observations from a normal distribution are within 1 standard deviation of the mean, 95% are within 1.96 standard deviations of the mean, and 99.7% are within 3 standard deviations of the mean.

Undoubtedly, the normal distribution is a fundamentally important distribution in statistics in general, and in ecotoxicity studies in particular. There are several reasons for this. On a purely statistical level, the Central Limit Theorem gives a strong reason for expecting normality to play an important role.

2.3.5.1 Central Limit Theorem

This says roughly that if we take a random sample from any population, regardless of its distribution, the sum and mean of the sample values are approximately normally distributed, with the approximation improved as the sample size increases. A more mathematical formulation is given in Appendix 1.

Many biological processes can be thought of as *sums* of many independent small subprocesses on a cellular level, so that the observed responses of these processes tend to follow a normal distribution. The tiny contraction of thousands of sarcomeres in muscle myofibrils involved in movement through muscle action in animals is an example. One might then expect, e.g. the heart rate and blood pressure to follow a normal distribution. See Figure 2.4 for a relevant illustration. This figure is reproduced from Google Search (2015). Anyone interested in exploring the biology in this concept is encouraged to consult Nowicki (2004). Exercise 2.1 illustrates the statistical properties of the Central Limit Theorem.

In might be more realistic to think of biological responses as *products* of many independent actions and, indeed, Figure 2.4 can be used in support of this idea. Since the logarithm of a product is the sum of its logarithms, this leads to log-normal responses. Many, but not all, biological processes are log-normal and there are other continuous distributions of importance in toxicology that will be introduced in the context of specific types of studies.

Statistics deals primarily with random samples from populations. By a random sample from a population, we mean a subset of the subjects from that population with the property that any two subjects from that population have the same chance, i.e. probability, of being in the sample. We are not generally interested in the subjects per se but rather in some measurement made on those subjects, such as body weight, whether alive or dead at the end of the study, or when it reproduced. The measurement of interest will have some distribution, F, over the population. We record the value of that measurement on the jth subject as X_j. Each such measurement is the value of a random variable with distribution F, so that our random sample (of measurements, not subjects) is a sequence $X_1, X_2, X_3, \ldots, X_n$ of random variables with distribution F. Initially, we will be concerned with independent observations, so X_j are *iid* random variables with distribution F. We will consider correlated observations later but will keep the discussion simpler for now.

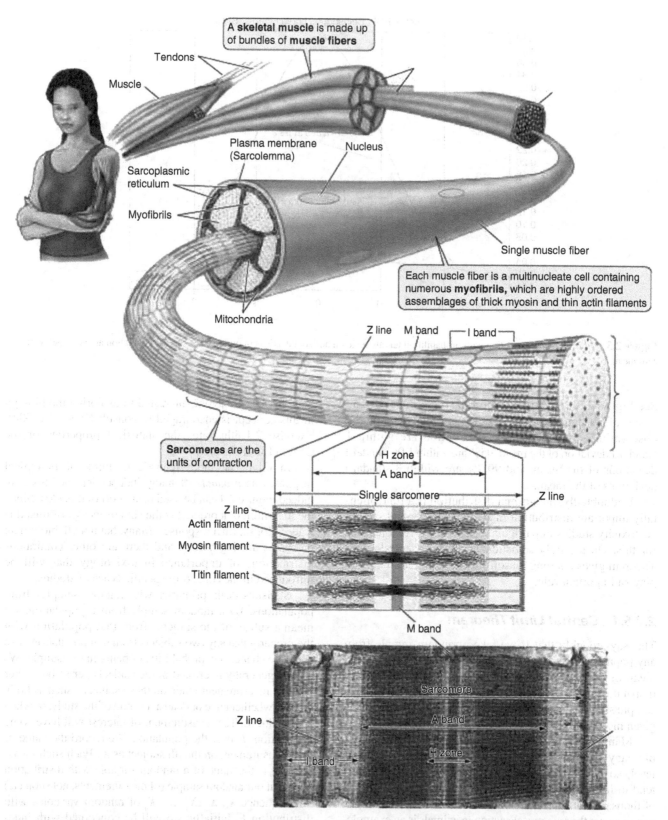

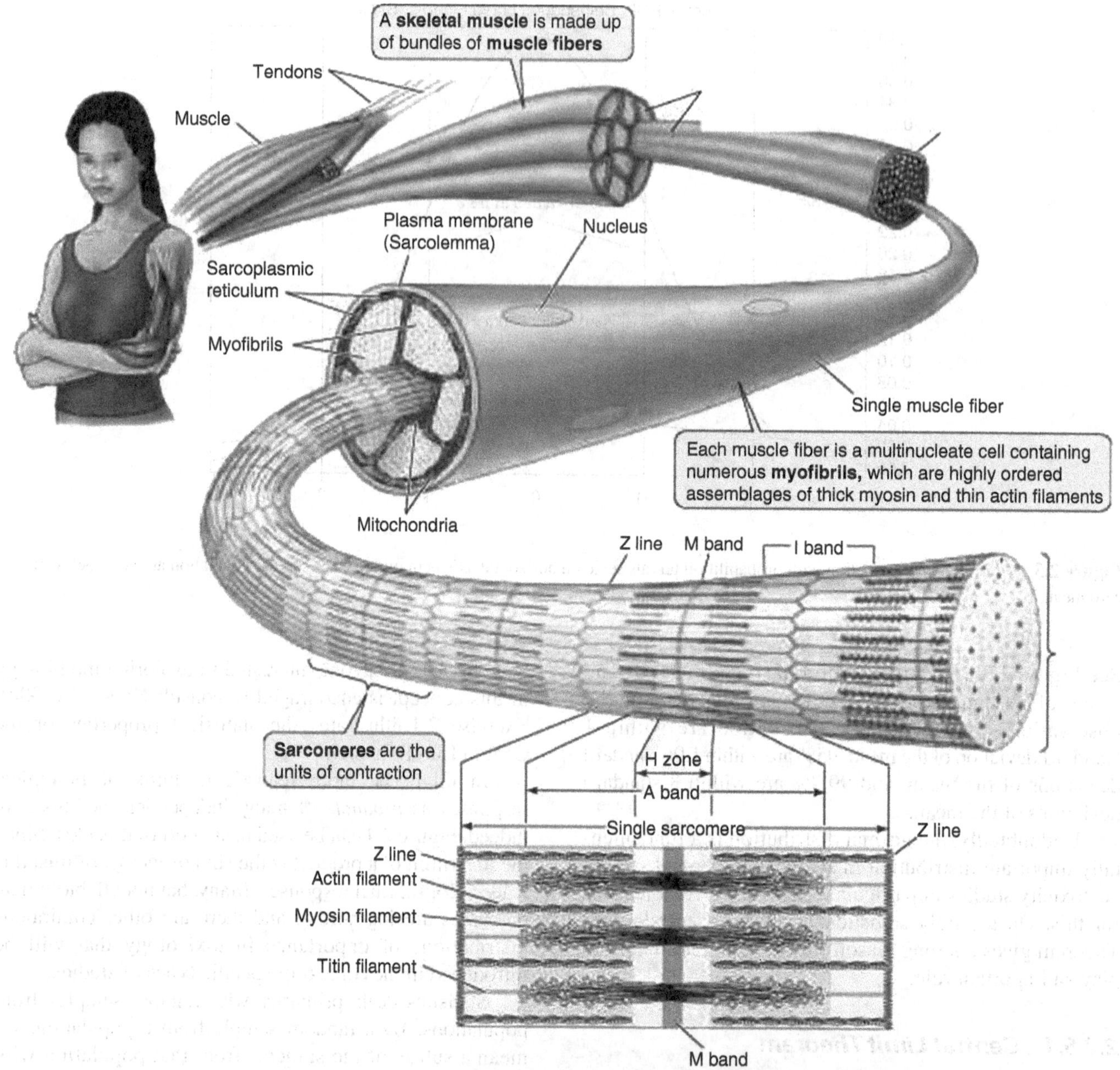

Figure 2.4 Muscle myrofibrils. Reproduced from Google Search (2015) after transforming to black and white using Photoshop. https://www.78stepshealth.us/plasma-membrane/images/3342_1291_1759-myofibrils-micrograph.jpg

2.4 ASSESSING DATA FOR MEETING MODEL REQUIREMENTS

All models have requirements that must be satisfied, at least to a reasonable approximation, in order for the results from the model to be considered valid. The application of an incorrect model to a dataset can lead to erroneous results, so it is important to understand the requirements of each model considered for evaluating a given dataset and whether those requirements are met.

Ideally, a professional statistician is actively involved in all data analysis. Where such expert judgment is available, formal tests for some model assumptions, such as monotonicity or its lack, may be replaced by visual inspection of the data, especially of the mean or median responses. The same concept applies to assessing normality and variance homogeneity. However, it is a rare testing facility that has sufficient qualified personnel to do this on every study and response, and it is not entirely clear that employment of such personnel in all routine analyses is good use of such resources. In the absence of sufficient personnel for this purpose, some automation of routine statistical tests is necessary.

Some object to the testing of model assumptions before applying a model on the grounds that such preliminary tests affect the FPR of subsequent hypothesis tests in an unknown way (Rasch et al., 2011). Rochon et al. (2012) explored this concern and found that the *overall* FPR and power are not much affected by preliminary tests of assumptions. The alternative to preliminary tests of assumptions is to ignore the actual nature of the data in favor of a preconceived theory of what should be true but that decision can impact the validity of subsequent hypothesis tests. It is commonly asserted that models are robust against violations of their assumptions and this justifies ignoring the issue of whether a dataset meets those assumptions. Box (1953) famously stated that "…to make the preliminary test on variances is rather like putting to sea in a rowing boat to find out whether conditions are sufficiently calm for an ocean linear to leave port!" Khan and Rayner (2003) and its references discuss normality but primarily in comparing the power of the ANOVA *F*-test and Kruskal–Wallis test rather than the multiple comparison tests more relevant for toxicity studies. They conclude that kurtosis is a much more serious issue than skewness.

Continuous distributions with a sharp peak but heavy tails have a significant impact on results when a normality-based test is used. Lorenzen and Anderson (1993) provide a valuable discussion of variance heterogeneity in the context of linear regression. More relevant to hypothesis testing in toxicology studies is Dunnett's test. Tamhane (1979), Dunnett (1980), Dunnett and Tamhane (1991), and Tamhane and Dunnett (1996) explored the use of Dunnett's classical test and alternatives in the presence of variance heterogeneity and developed a robust version of Dunnett's test, referred to below as the Tamhane–Dunnett test or T3-test (Hochberg and Tamhane, 2009). Williams (2010) discusses variance heterogeneity in logit and probit models. Anderson and Walsh (2013) showed that heterogeneity could have a major impact on some multivariate tests. The corresponding term for quantal and count data is overdispersion. Green et al. (2014) have demonstrated the impact of overdispersion on certain tests of ordinal responses. Cameron and Trivedi (2013) and Hilbe (2007) discuss the impact of overdispersion on models for count data. Further discussion of the impact of variance heterogeneity on specific tests will be given in later chapters.

While mild violations of model assumptions generally have little impact on results, serious violations assuredly can affect the validity of conclusions. We will illustrate this point numerous times in later chapters. The real issue is how serious the violation of assumptions must be before an alternative analysis is needed and what alternative is preferred. While there is ample room for disagreement on precise rules on when to use alternative methods, the need for some assessment of the assumptions is strong.

A different basis for criticism of assessing model requirements, at least for routine studies, is that for such studies, there is an ample history of the nature of the data from such studies and an assessment of normality, for example, is to be made on the basis of that history and not from each individual study. If the data are "typically" normal (however that is defined), then normality is assumed for all such studies. This begs the following questions: (i) How big and varied a database is required to make this assessment? (ii) What assurance is there that a new chemical being tested will produce the same sort of data? (iii) Unless 100% of the data in this database meet the model requirement, what proportion must meet the requirement before it is considered to apply universally? (iv) Is this process not itself in fact a preliminary test of model requirements with unknown implications for subsequent hypothesis tests and model fitting?

Our view is that the data drive the modeling process and less harm is done by checking requirements than by ignoring them. Furthermore, some assessment of model requirements is unavoidable, whether done formally or informally. For example, one assesses, possibly only visually, whether the dose–response tends to increase or decrease before attempting to fit an increasing or decreasing regression model. In the same vein, if there is a clear U- or inverted U-shape to the dose–response, then fitting either an increasing or a decreasing model or applying a trend test is inappropriate and counterproductive. If there is a gross outlier or influential observation in the data, ignoring that is unwise. Finally, it is undeniable that one assesses the data for whether it is quantal, continuous, discrete, ordinal, or categorical and that (informal) assessment obviously affects the subsequent analysis and its conclusions. It is not possible to analyze data without doing some assessment of the data and making a judgment of how it should be analyzed. All of that said, we do not advocate blind adherence to any strict set of rules in deciding whether model requirements are met. Expert judgment is best. Formal and informal assessments help guide that judgment.

2.5 BAYESIAN METHODOLOGY

Bayesian methodology attempts, in part, to formalize preliminary assessment of the nature of the data with the actual observations. This is done by relating the "posterior" distribution of a model parameter to its prior distribution. The joint pdf of the data, y, and the unknown parameter, θ, can be written as the product of a *prior density function*, $p(\theta)$, which represents our understanding of θ prior to collecting data, and the sampling density function, $p(y|\theta)$, which represents our understanding of the data independently of θ, as follows:

$$p(\theta, y) = p(\theta)p(y|\theta). \tag{2.11}$$

Using Bayes' rule, this can be rewritten as

$$p(\theta|y) = \frac{p(\theta, y)}{p(y)} = \frac{p(\theta)p(y|\theta)}{p(y)}, \tag{2.12}$$

where

$$p(y) = \sum_{\theta} p(\theta)p(y|\theta), \quad \text{for a discrete density, and}$$

$$p(y) = \int_{\theta} p(\theta)p(y|\theta)d(\theta), \quad \text{for a continuous density.} \tag{2.13}$$

The density function $p(\theta|y)$ is called the posterior density function of θ and will be recognized as the likelihood function. We can think of this as evaluating our prior understanding of the parameter in light of the data. A natural question to ask is how to obtain the prior density function $p(\theta)$. There are several basic approaches to answering this question. Essentially, every scientist uses his or her expert opinion in deciding how to design and interpret experiments. One of the rationales for Bayesian methodology is that expert opinion can be incorporated into the decision-making process in a formal way. One way of doing this is called eliciting expert opinion. In the words of Choy et al. (2009) "This use of the Bayesian framework reflects the scientific 'learning cycle,' where prior or initial estimates are updated when new data become available." Choy et al. (2009) is an excellent introduction to eliciting expert opinion and have an extensive list of references. Other references for this approach include Warren-Hicks and Moore (1998), Hora and Iman (1989), Ortiz et al. (1991), Cooke (1991), Meyer and Booker (2001), Morgan and Henrion (1992), and Hora (1993). In practice, it can be very challenging to build prior density functions from elicited expert judgment.

Either as an alternative to eliciting expert judgment or as a supplement, prior information can be obtained from experiments or other studies. Indeed, some authors think that the only objective prior information comes from previous studies (Clyde, 1999; Hobbs and Hilborn, 2006). In the absence of other information, sometimes a class of "conjugate" priors can be identified. If the posterior distribution, $p(\theta|x)$, is in the same family as the prior probability distribution, $p(\theta)$, the prior and posterior are then called conjugate distributions, and the prior is called a conjugate prior for the likelihood function. It can be inferred from Eq. (2.13) that the integration required can be daunting. If the prior and posterior are conjugates, the integration will often lead to a simple algebraic, closed form. Thus, the use of conjugate priors is often a mathematical convenience rather than the result of insight. Gelman et al. (2004) contains a list of conjugate priors for many familiar (and some unfamiliar) likelihood functions.

Finally, so-called uninformative priors can be used. An uninformative prior is one that is not subjectively elicited. The uniform prior is commonly used for this purpose. The use of an uninformative prior often yields results similar to non-Bayesian statistical analysis, since the likelihood function often yields more information than the uninformative prior. One uses whatever prior information is available to understand the data. Gelman et al. (2004) is among several comprehensive texts describing Bayesian methods in much greater detail than is possible here. We will illustrate the Bayesian approach below after introducing quantal responses.

We will provide methods by which to assess whether the data from a given study satisfies the requirements for a particular modeling approach and will provide examples that demonstrate the fallacies that arise from applying a model to data not meeting the requirements of that model.

Bayesian methodology will be illustrated in Chapter 6 to develop the beta-binomial model where it will be applied to overdispersed quantal data that arise in many studies. It will be explored further in Section 2.9.1 to develop the gamma-Poisson model, which will be applied to snail reproduction in Chapter 8. Bayesian methods will be used in Chapter 12 for species sensitivity distributions with censored data.

2.6 VISUAL EXAMINATION OF DATA

It will sometimes be convenient to refer to the control as treatment number 0 or treatment level 0, and refer to the lowest test concentration or application rate or dosage as treatment level 1, the next lowest as treatment level 2, and so forth. Thus, treatment levels refer to test concentrations in *order* of application rate but not magnitude.

For a continuous response, a simple scatter plot shows the treatment number on the horizontal axis and response value on the vertical axis. Figure 2.5 is a simple example where the response is the length in mm of rainbow trout in a fish early life stage experiment. The treatment mean responses were plotted as stars and joined by line segments to help reveal any treatment–response trend in the data. The spread of the data within treatments can be compared visually to look for evidence of unequal within-treatment variance. The variance in treatment level 7 is certainly less than in other groups, but as Table 2.3 shows, this is an artifact of the reduced sample size in this group. The spread in group 6 also appears smaller than in groups 1–5. Whether it is sufficiently smaller than the others to indicate a problem for analysis is unclear and it is for such situations that a formal test for variance homogeneity might be useful.

The reason for plotting the data with equal spacing of treatment levels rather than using actual concentrations is that often most of the test concentrations are clustered near zero and it is hard to discern information using the natural scale. Given that most toxicity experiments are approximately equally spaced on a log-scale, a similar plot can be obtained using log-spacing, provided the controls are treated appropriately. Rather than developing special rules for handling controls and also for accommodating the less common experiments where the concentrations are not geometrically spaced, it is simpler to use the equal spacing scheme exhibited in Figure 2.5.

The plot, especially with the joined means, can be assessed easily as consistent with a monotone dose–response, the slight upward ticks in levels 2 and 5 notwithstanding. See Figure 2.5 for an example to demonstrate this point. The data and SAS program, *RBT_Size_Scatterplot.sas* and R program *scatter lines2.R*, to generate Figure 2.5 are given in Appendix 1. An alternative plot displaying box and whisker plots in each treatment group instead of individual points is given by *scatter boxes.R*. More general programs for generating scatterplots are *Scatterplot MACRO.sas* and *Scatter.R*. For more details on graphics and basic statistics in R, Kabacoff (2015), Kleinman and Horton (2014), and Crawley (2013) are valuable references. More advanced statistical methods are given by Hothorn (2016). See Table 2.3 for the mean treatment length and weight for the trout data in Figure 2.5, where as usual, the two controls were combined.

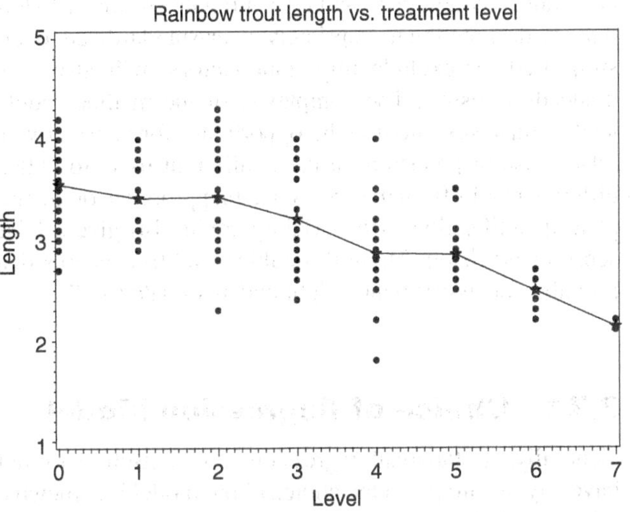

Figure 2.5 Scatterplot with Trend Line Dots = observations. Stars = mean of observations for the indicated treatment level. Solid line merely joins the means.

Table 2.3 Mean Sizes of Rainbow Trout for Figure 2.5

Level	0	1	2	3	4	5	6	7
Weight	0.53	0.43	0.44	0.37	0.26	0.24	0.14	0.08
Length	3.55	3.41	3.43	3.20	2.86	2.85	2.49	2.13
N	56	29	27	29	28	23	18	3

Dots are individual fish lengths, not replicate values. Solid line joins treatment means. Level = treatment group, with 0 = control. *N* = number of fish in treatment group. Length, weight = treatment mean responses.

A scatter plot can also be useful to identify "outliers," i.e. observations that seem to fall outside the usual spread of the data within a treatment group. No observations stand out in Figure 2.5. Other observations that can deviate from the usual pattern of data are what are called influential observations. See Figure 2.6 for an example.

An influential observation is one that has a major influence on an analysis. While an outlier can be influential in this sense, the term is generally restricted to unusually large or small value of an explanatory variable rather than an unusual response value. Figure 2.6 illustrates such an observation.

The dose values are all in the range 1–6 except for a single observation at dose = 30. A linear regression model fit to the data has an R^2 value of 0.86 but the fitted line in no way describes the data. Meaningless high correlations such as this are unlikely to occur in a designed experiment, especially those following regulatory guidelines, but are not uncommon in observational studies, such as that might occur in an environmental sampling study. Several examples from such a study are given in exercises for this chapter. One should guard against the dangers of inferring causal relationships in such situations. Many other visual and formal tools for assessing continuous responses will be presented in Chapters 3 and 4.

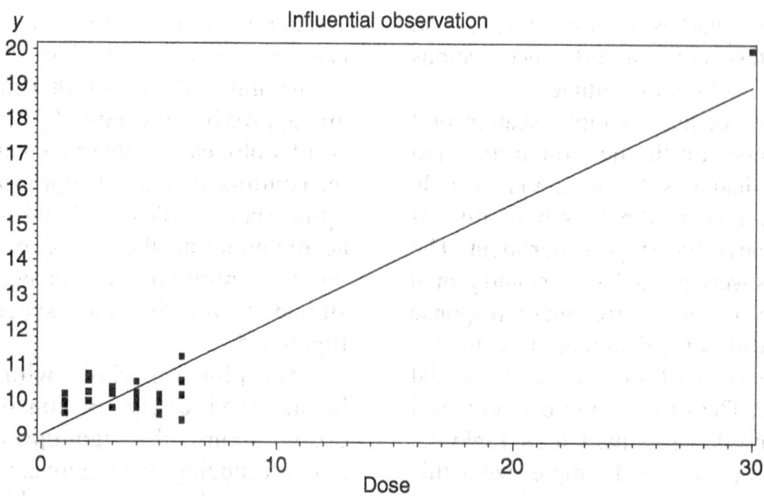

Figure 2.6 Example of influential observation. Single observation at dose = 30 is an influential observation. The fitted regression line shown is highly influenced by this observation. Without this observation, there is little apparent relationship between dose and response.

2.7 REGRESSION MODELS

Regression modeling attempts to describe the dose–response data as a whole, by means of a precise mathematical model. In general terms, it is assumed that the response, y, can be described as a function of concentration (or dose), δ:

$$y = f(\delta), \tag{2.14}$$

where f can be any function that is potentially suitable for describing a particular dataset. For example, the linear function

$$y = a + b\delta \tag{2.15}$$

indicates the response changes linearly with the concentration δ. Here, a and b are called the model parameters. By changing parameter a one may shift the line upwards or downwards, while by changing the parameter b one may rotate the line. Fitting a line to a dataset is the process of finding those values of a and b that result in "the best fit." Similarly, for any other dose–response model, or function f, the best fit may be achieved by adjusting the model parameters according to some criterion of best fit. There are competing criteria for determining which line, or some other type of curve, provides the best fit to a given dataset. Some of the most commonly used of these criteria are discussed in Appendix 2 and Sections 4.3.3 and 4.3.4. Beyond what is found in there, the reader is referred to standard texts, such as Draper and Smith (2012) or Montgomery et al. (2013) for linear models, Bates and Watts (2007) for nonlinear models, and Kennedy and Gentle (1980) and Motulsky and Christopoulus (2004) for both linear and nonlinear models. These methods have been implemented in numerous widely available software packages, such as SAS and R. The focus of this text is on selecting and evaluating the models to be fit to a given dataset.

Hart (2001) discussed the "risk of misleading results, which could arise in several ways. Inappropriate or imprecise data or assumptions will give unreliable results. Models that are designed inappropriately, over-standardized, over-simplified, or exclude important factors, will also give misleading results. The complexity of the methods could lead to mistakes, provide the opportunity for abuse, give a false sense of precision, or cause different users to obtain different results from the same data. Inappropriate or incomplete quantification of uncertainty could also give a false sense of precision. These difficulties lead to concerns that probabilistic assessments might underestimate risk."

2.7.1 Choice of Regression Model

Generally, a statistical regression model itself does not have any meaning and the choice of the model is somewhat arbitrary. It is the data, not the model, that determines the dose–response. However, the purpose of regression modeling in toxicology is not merely to obtain a curve that fits the data well. We want to use this curve to estimate something of inherent interest in evaluating the toxicity of the compound under study. A generally agreed measure of interest is the concentration that induces a specified percent inhibition (or stimulation) in the mean of the exposed population compared to the control population mean. The concentration that induces an x% change in the population mean is called the *ECx* or *x% effects concentration.* Obviously, an improper choice of the model can lead to a poor estimate of the ECx and the choice of the model should be governed primarily by the data. An alternative approach is to model a "threshold" or no effect concentration (NEC), not to be confused with an NOEC, below which no effect is assumed to exist. The existence of a toxic threshold is not universally accepted and, in any event, is highly model dependent. Models to estimate an NEC or toxic threshold will be presented in Chapter 4. Controversy over the existence of a toxic threshold dates back at least to Mantel (1963), Chen and Selleck (1969), and Epstein (1973). Gad (2006) contains an interesting, if somewhat dated, discussion of the threshold concept. Kooijman and Bedaux (1996) developed models for ecotoxicity studies that allow estimation of a threshold value and contain extensive references relevant to ecotoxicology studies. OECD (2014a) contains an updated and abbreviated assessment of Kooijman and Bedaux and is further extracted and compressed in Section 2.10. We will consider such models in greater detail in Chapter 4.

Numerous dose–response models are theoretically possible. Indeed, there are many models that have been seriously proposed for continuous responses and numerous (though less so) in use for quantal responses. Quantal regression models have long been used to estimate EC50 and a few other ECx values, and historically, only a few of these models have shown much promise. Only recently has interest in modeling continuous responses been given significant attention. This is not to imply that regression applied to continuous responses is of recent origin. Indeed, regression has been used extensively in some fields for over two centuries. Its application to toxicology has been limited, so there has not been a history from which to judge whether only a few models (and which ones) are appropriate. Only a few useful models will be discussed.

For continuous data, unlike quantal, there is no inherent restriction on the range of response values. Furthermore, it is not at all uncommon for a dose–response curve to exhibit what is called low-dose stimulation. That is, for some species and chemicals, there is an apparent beneficial response at low concentrations, with an adverse effect at higher concentrations, so that the dose–response is not monotone. This is a well-known, widely observed, repeatable phenomenon. Such a thing is extremely rare for quantal responses, but not for continuous responses, so that the class of models that must be considered is considerably larger than for quantal responses. Further complicating things is that there is a much wider variety of dose–response shapes that are observed. Assessing models for goodness of fit to the data is more important and, in many ways, more difficult for continuous responses than for quantal.

It is important to keep in mind that in the model y represents the true response, which may be thought of as the absolute or percent inhibition from the control population mean corresponding to the given concentration level. For a continuous response, the function $f(\delta)$ *must* include a term corresponding to the control population mean (or background response level) since the ECx is defined as the concentration that causes an x% change from the control population mean *as estimated from the model.* This might be accomplished by writing f in the form

$$y = f(\delta) = a + (1-a)g(\delta), \tag{2.16}$$

where $g(0)=0$, so that the parameter a is the control population mean, which must be estimated from the model fit to the whole dataset.

It may be more appropriate or convenient to write the function f in the form

$$y = f(\delta) = ag(\delta), \tag{2.17}$$

where $g(0)=1$ and a is the again the control population mean to be estimated from the model. In either case, it is not appropriate to use the control sample mean alone to estimate a.

In regression, the role of the confidence interval for ECx is analogous to the notion of power in hypothesis testing. Small experiments (few doses and few subjects per dose) generally produce wide confidence intervals. In addition, one must recognize that the confidence interval does not capture model uncertainty, so that, especially with few doses, a narrow confidence interval is not a sure indication of a good estimate. Similarly, in hypothesis testing, power may depend on the unknown shape of the dose–response curve, at least for some tests. Both power in NOEC determination and confidence intervals in ECx estimation are important but one should not over-interpret them.

2.7.2 Regression Models and Percent Effect

As discussed in Chapter 1, a regression model is fit to the data from a simple dose–response experiment to capture the overall shape of the dose–response relationship. For risk assessment, the primary objective is to estimate for some value of x, ECx, the concentration at which there is an x% effect relative to the control. It is important to understand the meaning of an x% effect. For a continuous response, such as size or yield, the meaning is straightforward. There is a mean response in the control and we seek the concentration, ECx, at which the mean response is x% less than the control mean, so that at ECx, the mean response $=(1-x/100) * Y_0$, where Y_0 is the estimated control mean response. Alternatively, if the effect of interest is an increase compared to the control, then ECx is the concentration at which the mean response is $(1+x/100) * Y_0$.

It should be evident that ECx refers to the concentration at which there is an x% effect *relative to the model-estimated control mean.* It is very important that the model estimates the control mean well, since errors in that estimate translate into errors in ECx. Thus, a key model assessment tool is an examination of how well the model estimates the control response.

For a quantal response, the meaning of ECx is less clear. For example, if the response is mortality, then the control mean response will often be zero. In this case, it makes no sense to estimate the concentration at which the mortality rate is an x% increase of the control mortality since x% of zero is still zero. If there is control mortality, it will usually be small. For example, if there is 10% control mortality, then we are not really interested in the concentration that increases the mortality rate by x%, for EC10 would then refer to the concentration at which the mortality rate is 11%. There are three possible meanings of an x% effect for quantal data:

1. Absolute risk: x% of the population is affected.
2. Added risk: x% in excess of the background is affected (total = $100c\% + x\%$).
3. Relative (or extra) risk: x% of the population above background is affected (total = $100c\% + x^*(1-c)\%$).

In ecotoxicity, we almost always use relative risk. Of course, mortality is not the only quantal response of interest and a more general term is incidence, such as the incidence of liver sarcoma. Control incidence is often referred to as background incidence. An example will illustrate the three different meanings of risk. Suppose the background incidence rate is $C = 0.2$, i.e. there is 20% background incidence. Then EC50 can have the following three meanings:

1. Absolute risk: At EC50, 50% of the population is affected (only 30% above background).
2. Added risk: At EC50, 70% is affected (total = 20% + 50%).
3. Extra (or relative) risk: At EC50, 60% is affected since total = 20% + 0.5 * (100 − 20)% = 20% + 40% = 60%.

In this text and in virtually all ecotoxicity studies of quantal data, we will use relative risk. It should be noted that this is consistent with the meaning of ECx for continuous responses if we model survival rather than mortality. If the background mortality is 100C%, then the background survival is $Y_0 = 100(1-C)\%$ and an x% decrease in survival means the survival of $(1 - x/100) * 100(1 - C) = (1 - x/100) * Y_0$.

2.8 BIOLOGY-BASED MODELS

So-called biology-based methods aim to model observed effects in terms of underlying processes such as toxicokinetics, mortality, feeding, growth, and reproduction. One distinction between biologically-based models and common regression models is that exposure time is treated explicitly. This approach makes use of prior knowledge about the chemistry and biology that at least partially accounts for the observed effects. The model is a function of the test chemical concentration and the time during which the test organism is exposed to the test compound. A key objective of such a model is the estimation of the NEC together with its confidence interval. As indicated previously, the NEC is not to be confused with the NOEC. The models explicitly assume that ECx decreases as exposure time increases, provided the experimental conditions do not change (e.g. concentration levels). Also, these models do not allow for hormesis or low-dose stimulation, which is unfortunate, since such effects are widely observed in laboratories across the world. These models allow the use of several datasets simultaneously, such as survival data, sublethal effect data, and data on the concentration of test compound inside the bodies of the test organisms during accumulation/depuration experiments. Nonetheless, if survival, reproduction, body size data, and time-varying concentration datasets are combined, the stochastic deviations from the mean are assumed to be independent for the different datasets defining these processes, so maximum likelihood methods can be applied in terms of separate functions. Otherwise, the mathematics becomes unwieldy.

The concept behind biology-based models is one of full balance, or as Kooijman put it in OECD (2014a, chapter 7): What goes in must come out. Offspring are produced indirectly from food, so reproduction is linked to feeding. Large animals eat more than small ones, so feeding is linked to growth. Maintenance of the organism implies a drain of resources over and above that from reproduction. Consequently, if maintenance is increased, then there is a reduction of growth or reproduction unless feeding is increased (i.e. replenishing resources). If food conditions are good, maintenance comprises only a small fraction of the daily food budget of individuals. Small effects of a toxicant on maintenance result in small effects on the growth rate. If food conditions are poor, however, maintenance comprises a large fraction of the daily food budget and small effects on maintenance can now have substantial effects on the population size. Mode of action is particularly important for predicting the effects of chemical exposure at the population level.

Each model parameter is assumed to be linearly related to *internal* concentration to characterize small effects. In turn, "internal concentration boils down to the independent action at the molecular level. Each molecule that exceeds individual's capacity to repress effects acts independent of the other molecules." It should be noted that a linear relationship between internal concentration and effect does not imply a linear relationship between external concentration and effect. Rather, that relationship is generally sigmoid in nature. The mathematical and statistical specifications of these models must await developments to be presented later in this text.

Kooijman and Bedaux support the use of the Michaelis-Menten and Hill models for receptor-mediated effects and the von Bertalanffy growth curve is used for growth. All of these will be described in detail in Chapter 4. Standard survival models are used for mortality and are covered in Chapter 11. Models for reproduction are given in Kooijman and Bedaux (1996) and are covered in Chapter 8.

2.9 DISCRETE RESPONSES

Discrete data are in some sense between quantal and continuous. Discrete data in ecotoxicity studies typically arise as counts. Unlike survival data, there is no fixed number of subjects at risk, some of which die or are otherwise affected. Instead, each subject gives rise to a count of some biological process, such as the number of offspring produced or the number of breeding tubercles (an outward sign of sexual development in males). If these counts vary greatly in value, it will often be acceptable to model them, possibly after a square-root transform, as a continuous response. Since analysis is typically on replicate mean or total count and there are multiple organisms in each replicate vessel, the Central Limit Theorem serves as a partial justification for treating them as normally distributed. Alternatively, these counts may follow a Poisson distribution, in which case a generalized linear mixed model (GLMM) is conceptually more appealing and is both tractable and flexible. See Table 2.4 for a typical example.

The data in Table 2.4 present some challenges. The untransformed total live young on day 21 are found consistent with normality by the Shapiro–Wilk test and consistent with variance homogeneity by Levene's test. Dunnett's test on TLY21 find no treatment group mean significantly less than the control mean at $p=0.05$ (but the p-value for a decrease in the highest test concentration is 0.0501). Similarly, a GLMM analysis using a Poisson error distribution finds the same. When a square-root transform is done, the resulting data are still consistent with normality and variance homogeneity. See Figure 2.7 for scatter plots of the transformed and untransformed TLY2. There is very little difference except for the scale on the vertical axis and no clear heterogeneity evident in either plot. The use of a log-transform does not change anything of importance in the interpretation of these data. There is also a downward trend evident in both plots, so these formal tests not finding any significant decrease from the control is troubling. It is of interest that the trend tests Jonckheere-Terpstra and Williams both find a significant decrease in the high-treatment group and the Jonckheere-Terpstra test also finds a significant decrease at the second highest test concentration. These results will be verified in Chapter 3. This example illustrates several points: (i) The use of the 0.05 significance level in regulatory risk assessment can seem arbitrary. (ii) The selection of the statistical test used to evaluate the

Table 2.4 Daphnia Reproduction Data

Rep	Conc	RepDay	TLY21	Immob21	Rep	Conc	RepDay	TLY21	Immob21
A	0	10	87	3	A	0.19	10	91	0
B	0	10	79	2	B	0.19	10	84	3
C	0	10	97	5	C	0.19	10	77	3
D	0	13	57	1	D	0.19	10	60	3
E	0	10	93	3	E	0.19	10	70	4
F	0	10	63	7	F	0.19			
G	0	10	116	3	G	0.19	10	86	1
A	0.015	10	91	2	A	0.67	10	62	10
B	0.015	10	73	2	B	0.67	10	58	13
C	0.015	10	102	0	C	0.67	10	85	6
D	0.015	10	79	0	D	0.67	10	73	3
E	0.015	10	73	2	E	0.67	11	81	1
F	0.015	10	99	2	F	0.67	11	70	4
G	0.015	10	89	1	G	0.67	10	71	9
A	0.053	10	85	15	A	2.35			
B	0.053	10	98	1	B	2.35	13	39	5
C	0.053	10	87	1	C	2.35	10	67	13
D	0.053	10	84	5	D	2.35	11	53	12
E	0.053	10	60	5	E	2.35	10	85	9
F	0.053	12	45	8	F	2.35	10	64	2
G	0.053	14	47	5	G	2.35	13	76	3

Conc is test concentration in $\mu g\, l^{-1}$. RepDay is first day of reproduction. TLY21 is the total number of live young produced by test day 21. Immob21 is the number of those young that are immobilized.

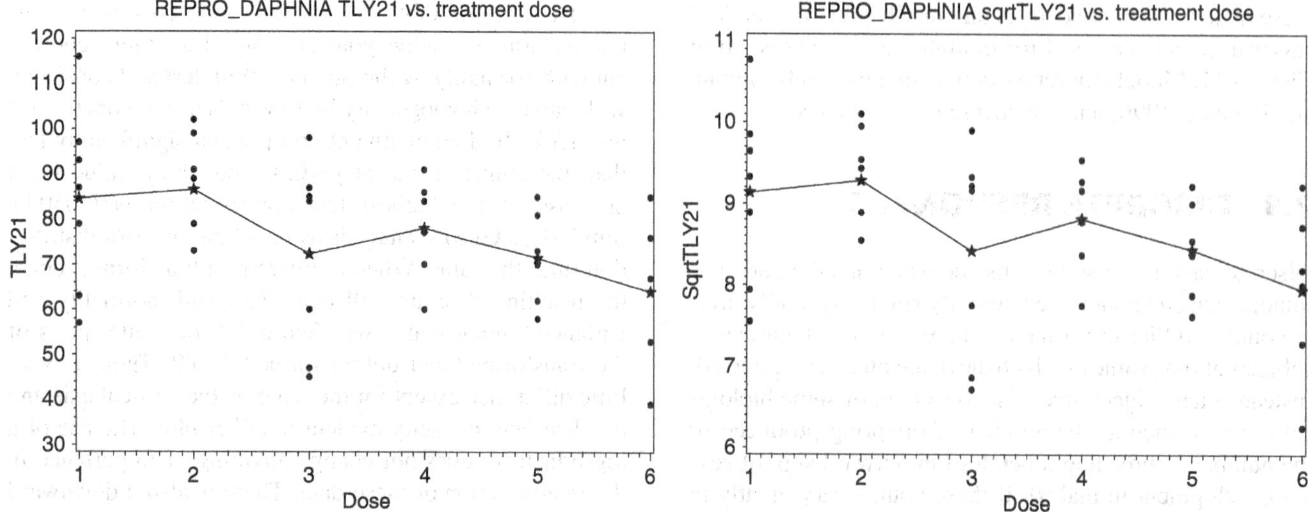

Figure 2.7 Scatter plots of TLY21 and sqrt(TLY21). The transform had little impact on the trend or spread in the response.

data is important. (iii) Plotting the data can be helpful in understanding the data and can point the way to a more careful analysis when a formal analysis might seem counterintuitive.

With regard to the arbitrariness of the 0.05 significance level as a hard rule in risk assessment, one should realize that it is necessary to assess risk in order to approve or disallow a chemical to be used in the environment and no matter what criteria are used there will be examples that make these criteria seem arbitrary. This is also the reason that many species, life-stages, experiment durations, and responses are used in regulatory risk assessment. No assessment rests solely on a single dataset or analysis.

Programs to analyze the Daphnia reproduction data will be discussed in Chapters 3 and 8 when the appropriate tests are developed and in Chapters 4 and 8 when the relevant models are introduced.

Given the significant heterogeneity indicated by Levene's test (Chapter 3), it is appropriate to consider the possibility that a Poisson model for the count data model is not appropriate. It is necessary to have an alternative count data model that accommodates overdispersion and a test to determine whether to accept the hypothesis of Poisson distribution against an appropriate alternative distribution.

2.9.1 The Gamma-Poisson or Negative Binomial Distribution

The beta binomial distribution is derived from a binomial distribution by postulating a beta distribution for the population parameters in a Bayesian framework. The beta is a conjugate prior distribution to the binomial. The gamma distribution is the conjugate distribution to the Poisson and so it is natural to consider the gamma-Poisson distribution for count data to accommodate overdispersion. Recall the Poisson density function from Eq. (2.3).

$$f(x|\mu) = e^{-\mu}\frac{(\mu)^x}{x!} \quad \text{if } x \text{ is an integer}$$

$$f(x|\mu) = 0 \quad \text{otherwise.}$$

Here the term np from Eq. (2.3) has been replaced by term μ. The gamma density function with parameters α and β is given by

$$g(\mu|\alpha,\beta) = \frac{\beta^\alpha}{\Gamma(\alpha)}\mu^{\alpha-1}e^{-\beta\mu} \tag{2.18}$$

Analogous to the development of the beta-binomial given in Appendix A2.12, the posterior density function of X is thus given by

$$h(x) = \int_0^\infty f(x|\mu)g(\mu|\alpha,\beta)d\mu = \frac{\beta^\alpha}{\Gamma(\alpha)}\int_0^\infty e^{-\mu}\frac{(\mu)^x}{x!}\mu^{\alpha-1}e^{-\beta\mu}d\mu,$$

which can be rewritten as

$$\begin{aligned} h(x|\alpha,\beta) &= \frac{\beta^\alpha}{x!\Gamma(\alpha)}\int_0^\infty \mu^{\alpha+x-1}e^{-\mu(1+\beta)}d\mu \\ &= \frac{\beta^\alpha\Gamma(\alpha+x)}{\Gamma(x+1)\Gamma(\alpha)}\left(\frac{1}{\beta+1}\right)^{\alpha+x}, \\ &= \binom{\alpha+x-1}{x}\left(\frac{\beta}{\beta+1}\right)^\alpha\left(1-\frac{\beta}{\beta+1}\right)^x, \end{aligned} \tag{2.19}$$

which is the negative binomial pdf. The mean and variance of the negative binomial distribution are

$$E[X|\alpha,\beta]=\frac{1}{\beta}$$
$$\mathrm{Var}[X|\alpha,\beta]=\frac{1}{\beta}+\left(\frac{1}{\alpha}\right)\left(\frac{1}{\beta}\right)^2. \quad (2.20)$$

It is convenient to reparameterize the negative binomial by replacing $1/\beta$ by μ and $1/\alpha$ by θ so that

$$E[X|\mu,\theta]=\mu$$
$$\mathrm{Var}[X|\mu,\theta]=\mu+\theta\mu^2. \quad (2.21)$$

It is then clear that overdispersion occurs when $\theta>0$, underdispersion occurs when $\theta<0$, and an estimate of θ is needed or else a test of whether $\theta>0$. There are several estimators of θ and tests for overdispersion in a nominally Poisson distribution. From Eq. (2.21), it can be seen that $E[(y-\mu)^2-\mu]=\theta\mu^2$. This leads to the simple estimator of θ given by

$$\hat{\theta}=\frac{1}{n.-k}\sum_{i=1}^{n}\frac{(y_{ij}-\hat{\mu}_i)^2-\hat{\mu}_{ij}}{\hat{\mu}_i^2\sqrt{2}}, \quad (2.22)$$

where $\hat{\mu}_i$ is the Poisson model prediction and y_{ij} is the jth observed value from the ith treatment group, $j=1, 2, 3, n_{i\cdot}$, $n.$ is the total number of observations, and k is the number of independent model terms. The estimator $\hat{\theta}$ has a t-distribution with $df=n.-k$. Cameron and Trivedi (2013) do not include the square-root term $\sqrt{2}$ in the denominator while Hilbe (2007) includes it and treats $\hat{\theta}$ as a standard normal rather than a t-distribution, justifying that by the Central Limit Theorem. If Eq. (2.22) is applied to the daphnia reproduction data from Table 2.4, the result is $\hat{\theta}=-0.19$. This clearly does not support overdispersion and so the results from a Poisson GLMM model can be accepted.

An alternative method developed in Cameron and Trivedi (2013) is to do a regression of the Z-scores test defined by Eq. (2.23) on the predicted values, $\hat{\mu}$ with no intercept. The estimated slope from this regression is an estimate of θ. When that is done for the daphnia reproduction data, the resulting estimate is −0.16, which again indicates *under*dispersion.

$$Z_{ij}=\frac{(y_{ij}-\hat{\mu}_i)^2-y_{ij}}{\hat{\mu}_i\sqrt{2}}. \quad (2.23)$$

2.10 TIME-TO-EVENT DATA

Time-to-event data has been studied for many years (e.g. Lee and Wang, 2013; Collett, 2014) but has received little attention in ecotoxicity studies until recently. Test guidelines issued in recent years by the Organization for Economic Cooperation and Development (OECD) or the United States Environmental Protection Agency (USEPA) that specifically require collection and assessment of time-to-event data include the Japanese quail toxicity test (JQTT: USEPA 890-2100), the larval amphibian growth and development assay (LAGDA: OECD TG 241), and the Medaka extended one-generation reproduction test (MEOGRT: OECD 240), which are discussed in Chapter 12. One of the characteristics of time-to-event data that require careful attention is censoring. Depending on the specific response and experimental design, the event of interest may not occur for one or more subjects by the time the study ends. For example, in the LAGDA test, a response of interest is the time to reach developmental stage 62. Some tadpoles may not reach stage 62 by the time the experiment ends 70 days after exposure. In that situation, all one can report for that tadpole is that the time to reach stage 62 exceeds 70 days. Such an observation is called censored, more specifically right-censored. The statistical analysis of censored data for time-to-event data will be developed in Chapter 11 using survival models specially adopted to handle the replicate structure common in ecotoxicity studies. Another type of censored data will be discussed in Chapter 13.

For the present discussion, attention will be confined to time-to-event data from test guidelines that have been in place for many years but are not analyzed by survival models. The reason is that the nature of the data does not lend itself to traditional survival analysis methods. Consider a chronic Daphnia magna reproduction test with seven daphnia per treatment group in individual beakers, where the time for a daphnid to reproduce is measured in two ways: the first and last days of reproduction. The measurement is crude, since the daphnia are checked only once per day to determine whether reproduction has occurred. Also, these studies typically use only seven daphnia per test concentration. As a result, there is often very little variation in these measures. See Table 2.5 for data from a typical study,

Table 2.5 Daphnia First Day of Reproduction

	Rep						
Conc	1	2	3	4	5	6	7
0	9	11	10	11	10	10	10
0.0016	10	12	10	12		10	12
0.0031	10	11	10	10	10	9	10
0.0066	10	10	11	11	10	10	10
0.012	10	11	11	10	11	10	10
0.027	10	10	10	10	11	10	10
0.05	14	13	13	13	14	14	14

The number in each cell in a column in Table 2.5 labeled reps 1–7 is the first day of reproduction of the daphnia in that replicate beaker. Concentrations were measured in mg l^{-1}.

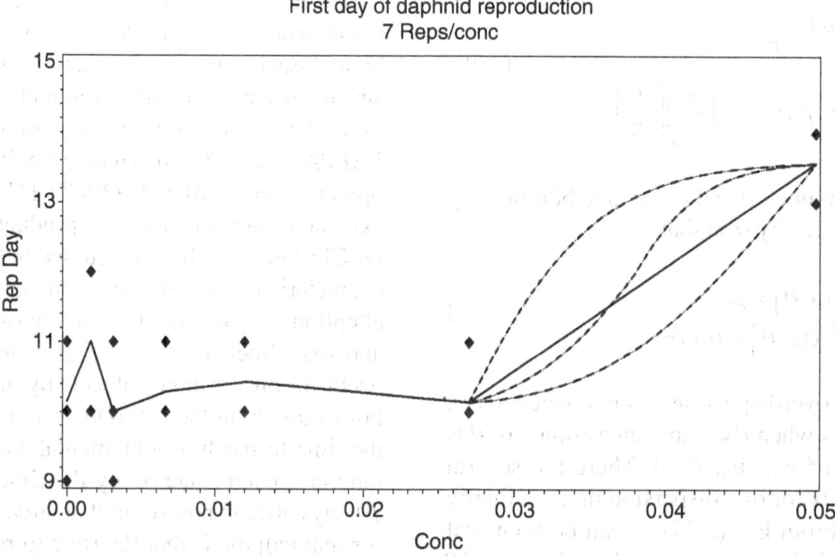

Figure 2.8 Daphnid first day of reproduction. The response is flat until the highest test concentration. There is no information from which to determine an appropriate curve to describe the data. Four possible dose–response shapes connecting the two highest test concentrations are shown to illustrate the range of possibilities.

where it will be observed that in the entire study, there were only six distinct values of first day of reproduction. As will be discussed in depth in Chapter 11, such data are poorly analyzed by the available survival models for either regression or NOEC. See Figure 2.8 where the problem with regression modeling of such data is illustrated. As will be shown in Chapter 11, these data are easily analyzed by the Jonckheere-Terpstra test to determine a NOEC = 0.027 mg l^{-1}.

Other time-to-event data that arise in ecotoxicity studies include first and last day of swim-up and first and last day of egg hatch.

2.11 EXPERIMENTS WITH MULTIPLE CONTROLS

It is common in aquatic and certain other types of experiments that the chemical under investigation cannot be administered successfully without the addition of a solvent or vehicle. In such experiments, it is customary to include two control groups. One of these control groups receives only what is in the natural laboratory environment (e.g. dilution water), while the other group receives the dilution water with added solvent but no test chemical. In ecotoxicity experiments, these are often termed negative or dilution water (non-solvent) and solvent controls. OECD recommends limiting the use of solvents (OECD, 2000); however, appropriate use of solvents should be evaluated on a case-by-case basis for difficult-to-test substances. Details regarding the use of solvents (e.g. recommended chemicals and maximum concentrations) are discussed in the relevant guideline documents for specific ecotoxicity tests. In addition, regulatory guidelines must be followed by both controls with regard to the range of acceptable values (e.g. minimum acceptable percent survival or mean oyster shell deposition rate). Multiple control groups can be utilized regardless of whether the experiment was intended for hypothesis testing or regression analysis.

For each measured response, data from the two negative control groups are analyzed to determine whether the two controls were statistically different for that response. The challenge is what to do with the outcome of these comparisons. It must be understood that in comparing the controls, the test used has a FPR, usually 5%. However, if there is a statistically significant difference between the negative and solvent control groups, there is no way from the single study to know whether the difference is a chance result (i.e. false positive) or a real solvent effect. In guideline studies, only a small number of solvents are permitted and these are well understood and documented to have no meaningful effect on the test organisms used in laboratory studies. Hutchinson et al. (2006) investigated whether commonly used solvents have an effect on common laboratory organisms. They found that the toxic levels of these solvents are well above the levels currently recommended in OECD test guidelines, especially for acute studies, but also by a smaller margin, for chronic studies. Those data and others were discussed in Green and Wheeler (2013). See Figure 2.9 for a summary of those results in relation to regulatory guidelines for acute studies and see Figure 2.10 for the corresponding plot for chronic studies. These figures contain the solvents that are commonly used in guideline studies.

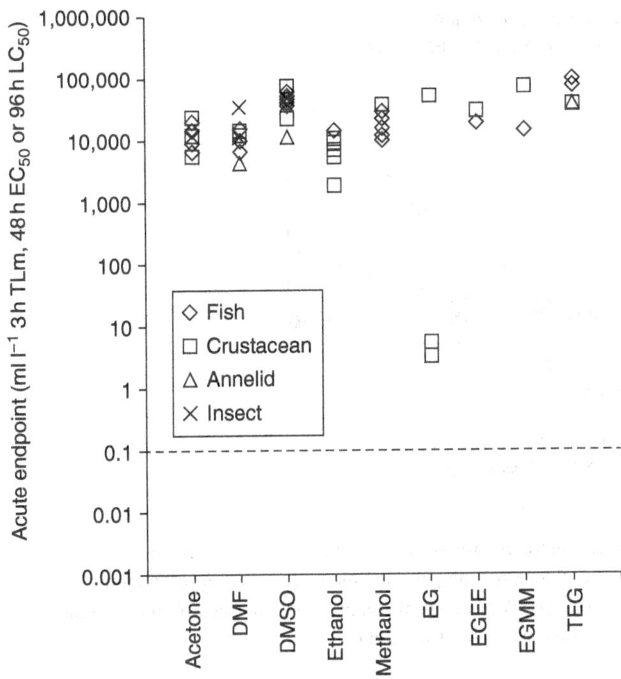

Figure 2.9 Solvent effects in aquatic studies: Acute toxicity endpoints far exceed the regulatory solvent limit of 0.1 mg l^{-1} (dotted line). Figure and related results are discussed further in Green and Wheeler (2013), which in turn is based on data in Hutchinson et al. (2006).

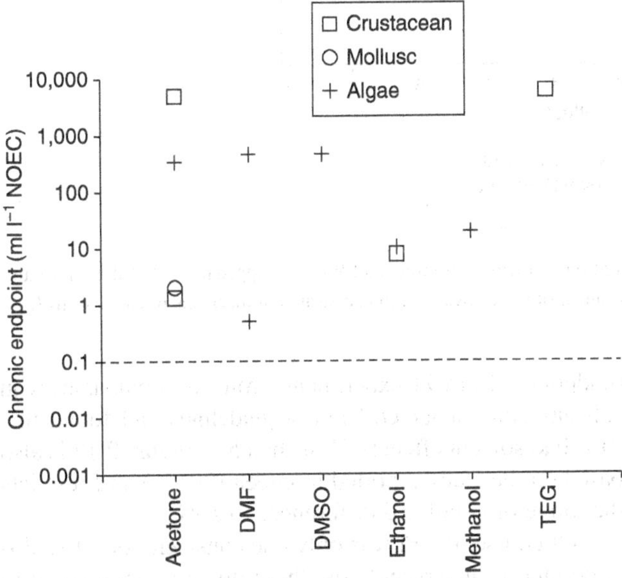

Figure 2.10 Solvent effects in aquatic studies: Chronic studies. Chronic toxicity endpoints exceed the regulatory solvent limit of 0.1 mg l^{-1} (dotted line). Figure and related results are discussed further in Green and Wheeler (2013), which in turn is based on data in Hutchinson et al. (2006).

2.11.1 Statistical Issues Related to Testing or Modeling with Multiple Negative Controls

Nevertheless, it could be that despite our expectations, the solvent might have an effect on the organisms under study due to some impurity or hitherto unsuspected biochemical phenomenon. For that reason, some regulatory authorities have expressed concern that any conclusions and inferences based on a study where the two controls differ significantly for one or more responses could be impacted due to the presence of a solvent effect. It has at times been proposed to reject a study for regulatory purposes when a significant difference is found between the water and solvent controls for even one test guideline required response. This is problematic in part because with few exceptions, multiple responses are measured and analyzed statistically, so that if the FPR for each comparison is 5%, the chance of at least one false positive among numerous comparisons can be quite large. See Table 2.6 for the FPR associated with this idea for various numbers of tested responses.

In testing for an effect in a treatment group, either for a NOEC determination or for ECx estimation, the issue is what control to use for studies with both a water control and a solvent control. There are five schemes that have been used by various regulatory bodies at one time or another.

1. Always combine the controls.
2. Use solvent control only.
3. Use water control only.
4. Compare treatments to each control separately.
5. Protocol (or *Sequential*): Combine controls unless statistically significantly different, in which case use solvent control only.

Green (2014a) did a computer simulation study to investigate this question. The simulations dealt only with a single treatment group, but they indicate important differences among these five approaches. See Figure 2.11 for an illustration of FPRs over a wide range of possible

Table 2.6 False Positive Rate Associated with Number of Tested Responses

K	*N*	Prob	*K*	*N*	Prob
1	1	0.05	2	1	0
1	2	0.0975	2	2	0.0025
1	3	0.1426	2	3	0.0072
1	4	0.1855	2	4	0.0140
1	5	0.2262	2	5	0.0226

K, number of significant tests required to invalidate study; *N*, number of responses tested at 0.05 significance level; Prob, probability of at least *K* false positive significant tests.

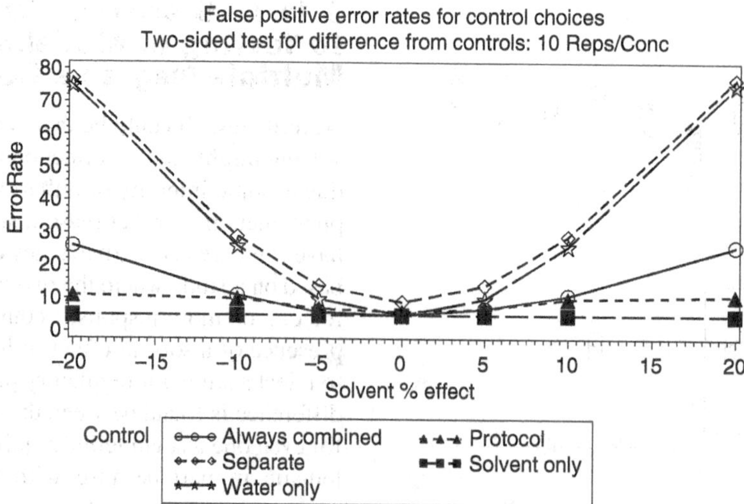

Figure 2.11 False-positive rates for five schemes for using two controls test for treatment effects: Related material is given in Green (2014a). Protocol, combine if controls not significantly different, use solvent only otherwise. Separate, compare treatments to each control: Claim effect if either is significant. Solvent % effect, 100 * (Solvent control mean – Water control mean)/Water control mean, where the true difference in mean response is indicated. ErrorRate = False – Positive rate, the likelihood of claiming a treatment effect when there is none.

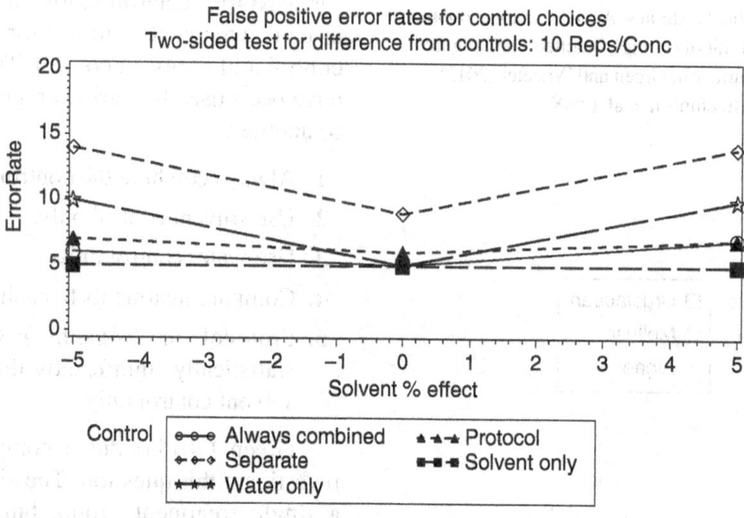

Figure 2.12 Little difference is observed in false-positive rates among solvent only, always combine, and Protocol approaches, but there remain much higher false-positive rates for the approaches of always using the water control only or always doing separate comparisons to each control.

solvent effects. See Figure 2.12 for the FPRs associated over a more restricted range. In each of the five schemes illustrated, the experimental design has 10 replicates per control and treatment.

It follows from the results summarized in Figure 2.11 that the solvent only FPR is always 5%, while the water only and separate comparisons FPR can be as high as 75%, the always combine controls approach FPR can be as high as 20%, and the protocol FPR can be as high as 9%. These FPRs should not be overinterpreted, as the high rates are associated with differences between the two controls that are rarely observed. Green and Wheeler (2013) studied 141 responses (endpoints) from 71 experiments from five companies with solvents done under OECD test guidelines and found that 91% had solvent effects of less than 5%. Figure 2.12 is also based on the study reported in Green (2014a) and restricts the range of simulated differences to ±5%.

Of course, the FPR is only one consideration. It is also important to understand how these different approaches to using the two controls affect the power to detect a real effect. See Figure 2.13 for power curves for one set of conditions.

Green and Wheeler (2013) found a distinct power advantage when using the pooled controls. A strategy of pooling the controls for a given response unless the controls

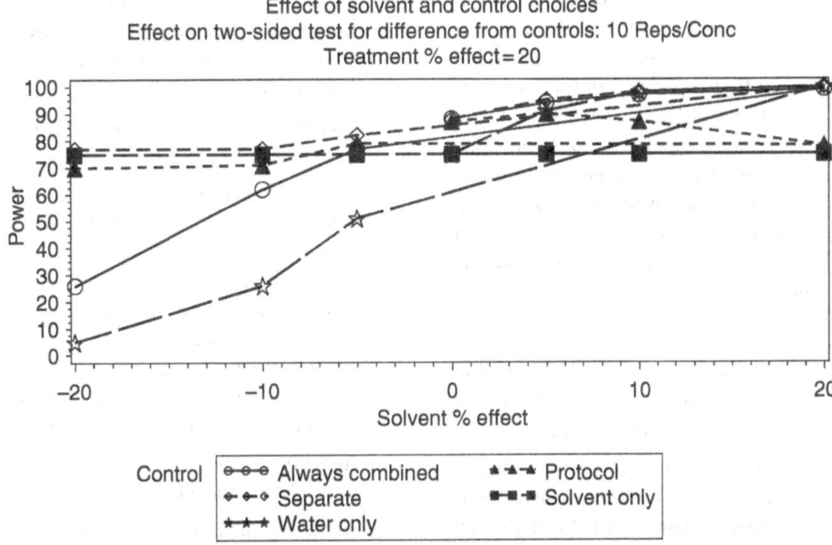

Figure 2.13 Power to detect a real effect. The water only and always combine approaches have very poor power when the solvent "effect" is opposite to treatment effect. The power for the protocol approach ranges from 6% below to 9% above the power for the solvent-only approach over this entire range.

are significantly different for that response and using the solvent control alone otherwise provides almost as much power and greater protection against reporting false-negative treatment effects. No evidence is found of a systematic tendency for the dilution water control mean response to be greater or less than the solvent control mean response. The differences between the control mean responses tend to be small, infrequently exceeding 10% and simple statistical tests to compare the controls have acceptable power to detect differences of 10% or more and often find significant differences of 5% or less. Neither the data nor the scientific literature supports the theory of an interaction between solvents and the chemical treatments under investigation at the low solvent concentrations used in these experiments.

The conclusion reached by Green (2014a) and Green and Wheeler (2013) is to prefer the so-called protocol method as the best combination of low FPR and high power, as well as making use of all the data, rather than discarding the water control except for the punitive purpose of possibly rejecting the entire study. As shown in Figures 2.9 and 2.10, at the levels used, solvents have little or no real effect on toxicity measures and apparent solvent effects, i.e. differences in control mean responses, can result in higher or lower means compared to water controls. Combining controls, under these conditions, bases the estimated natural condition (control mean) on all of the available data, which increases the control degree of freedom, thus enhancing the power to detect treatment effects. But see also van der Hoeven (2010), who recommends using only the solvent control regardless of how the controls compare.

2.11.2 Treatment of Positive Control Data

Some toxicity studies include a positive control group. This is a group that usually is identical in design to other treatment groups in the study but using a different test substance known to have a biologically important effect on the test organisms with respect to one or more responses to be analyzed statistically. The usual purpose for including a positive control is to demonstrate that the experimental design and statistical test protocol have the power to detect biologically important effects if they exist. If a proper power analysis is done for the experimental design and statistical test protocol, the utility of including a positive control is limited. In any case, a separate analysis is done to compare the positive control to the negative control (following one of the procedures described in later chapters). The positive control is not included in the analysis of the test substance of interest. There are two reasons for that approach. First, a trend type test, such as Williams or Jonckheere-Terpstra, is not applicable when the treatments differ in kind instead of, or in addition to, differences in concentration or level. Second, even if a pairwise method is used, such as Dunnett's test, the need to adjust for another treatment of no importance to the central question of the safety of the test substance lowers the power to detect a real effect.

EXERCISES

In all datasets in the text and exercises, if a concentration is given with the value −1 means that a solvent used in the study and the water control is recorded as −1 and the solvent control is listed as 0.

The reader should determine whether to combine the two controls for NOEC determination or ECx estimation. If the two controls are to be combined, then the water control should be changed to 0 and the replicates renamed so that replicates in the two controls can be distinguished. If only the solvent control group is to be used, then the water control should be discarded. If only the water control should be used, then the solvent control should be discarded and the concentration for the water control should be changed to 0. Under no circumstances should the concentration value −1 be used for NOEC or ECx purposes. Guidance on this is provided in Chapters 2 and 3.

2.1 Central Limit Theorem

(a) Use a random number generator to generate a random sample of size 5 from a uniform (0,100) distribution. Compute the sample mean and record the result.

(b) Repeat that process 500 times. This generates a new population of 500 sample means. Calculate the mean and standard deviation of this population and construct a histogram of the data. Overlay the histogram with a normal density function with the mean and standard deviation computed above. Programs to draw a histogram are given by *Histogram.sas* and *Histogram.R*. A more general R program that creates the residuals from a raw dataset and plots a histogram and QQ-plot of the residuals is given by *Univar.R*. QQ-plots will be discussed in Chapter 3.

(c) Repeat this for sample size 10, 50, and 100. This will generate three more populations of sample means. Plot histograms of these four populations so generated.

(d) Compute the mean and standard deviation of the uniform distribution in part (a) and for $n=5$, 10, 50, 100, compute the mean and standard deviations of the populations generated in steps (b) and (c). Compare them to μ and $\sigma/\sqrt{n}$, where μ and σ are the mean and standard deviation of the uniform distribution from (a).

The next three exercises refer to spreadsheets in the excel file *Poisson Data.xlsx*. These same data can be revisited in Chapters 6 and 7.

2.2 The data in spreadsheet Daphnia 1 consisted of five treatment groups and a control in which 10 daphnid in individual beakers were exposed to a test chemical. The responses recorded were total live young at 14 and 21 days. Nominal concentrations were in mg l^{-1}.

(a) Assess each of these responses for agreement with a Poisson distribution in each group.

(b) Assess the square-root transform of these data in each group for agreement with a normal distribution.

(c) Subtract the group mean from each observation and assess the set of residuals from the entire study for agreement with a normal distribution.

(d) Repeat (c) for the square-root-transformed counts.

2.3 The data in spreadsheet Daphnia 2 consisted of data from a 21-day daphnia chronic study in which there were seven treatment groups and a control. In each group there were two replicate vessels of 10 daphnid each exposed to the indicated concentration of a test chemical. The responses were survival of adult daphnid at 14 and 21 days, first day of reproduction, and total live young produced by days 14 and 21. Also both the nominal and mean measured concentrations of the test chemical are given in mg l^{-1}.

(a)–(d) Follow the instructions of Exercise 2.2 for TLY14 and TLY21.

2.4 The data are in spreadsheet Daphnia 3 and come from a 21-day daphnia chronic study in which there were seven concentrations of a chemical and a water control. In each group, there were 10 replicate vessels of four daphnia each. The responses are the number of survival adults at 7, 14, and 21 days; total live young produced by days 14 and 21; the number of immobilized young on day 21; and two measures of stress. These stress responses were the number of winter eggs and number of eggs on the bottom of the vessel at day 21.

(a) Follow the instructions of the previous two exercises for TLY14 and TLY21.

2.5 A 21-day daphnia chronic study was done in which there were seven test concentrations of a chemical, a water control, and a DMF solvent control. There were seven replicate vessels of one daphnid each and three replicate vessels of five daphnid each in every group. The responses were survival at 7, 14, and 21 days; total live young at 14 and 21 days; first day of reproduction; immobilized young on day 21; and length of adult daphnia at 21 days. Reproduction responses were recorded only for the seven replicate vessels of one daphnid each. The water control is recorded as −1 and the DMF control is recorded as zero. The data are given in Appendix 1 as three SAS datasets named Study131survive, study131repro, and study131length. The same names are given as spreadsheets in the excel file study131.xlsx. These data can be revisited in later chapters for NOEC and regression analyses.

(a) Compare the survival counts for the two controls. Think carefully about whether the seven singleton vessels and the three vessels of five daphnid should be compared together or separately.

(b) Compare Immob21, TLY14, and TLY21 for the two controls.

(c) Compare length measurements for the two controls.

2.6 A 21-day daphnia chronic study was done with seven concentrations of the test chemical, a water control, and a ph-7 adjusted water control. The data are given in three SAS datasets in Appendix 1 with names Study132survive, study132repro, and study132length and in the excel file study132 with spreadsheets of the same names as the SAS datasets. In this study, there were 10 replicate vessels of four daphnid in each group. Follow the instructions of Exercise 2.5 without the complication of different size replicates.

2.7 Data from the development of the Fish Sexual Development Test guideline (OECD TG 234) are given in the SAS dataset VTGMNS in Appendix 2 and in the excel file of the same name. The response of interest is vitellogenin or VTG, which is a measure of reproductive health. The production of VTG, especially in male fish, is a sensitive indicator that a chemical is an endocrine disrupting chemical or EDC and as such is important in assessing chemicals for environmental safety.

There are eleven datasets identified by measure (e.g. VTG1 and VTG11) and chemical (e.g. trenbolone and flutimide), concentration (conc) and rep.

(a) For each dataset calculate the treatment means and standard errors across replicates of VTG, Log(VTG), and Log(1 + VTG) and the ratio of standard error of Log(VTG) over that of Log(1 + VTG) for each treatment. The point of this exercise is to indicate the need for a log-transform for some responses to reduce variability, and the importance of using Log(1 + x) rather than Log(x) when x can be a very small positive number or zero. Also, these data illustrate that there is no harm in using the transform Log(1 + x) when there are no very small values of x that might cause problems for analysis. These data can be revisited in Chapter 3 to determine the effect of transforms on hypothesis tests. Some of these data will also present some challenges for the regression approach in Chapter 4.

(b) To make sure one understands the importance of the 1 + x adjustment rather that s + x for some small positive value, repeat the exercise using the transform Log(0.001 + VTG).

2.8 A study was done of surface water in a river system polluted by mercury. As part of an effort to track the source of the pollution and determine its effect on the environment, water samples were collected over a period of years and these samples were measured for the amounts of various chemicals in addition to mercury. The data are in Appendix 1 in a SAS dataset named SW (for surface water) and spreadsheet of the same name in the excel file river.xlsx. Figures 2.14–2.17 compare the correlations of Ln(Mercury) and the log-transform of various analytes when all the data were used and when some influential observations were omitted. The purpose of this exercise is to illustrate the risks of naively using correlations or linear regression to determine the relationship between two variables. A SAS program, River plots2.sas, is given in Appendix 2 to carry out the correlations and plots. The reader can explore whether river reach or month or year of sample might help account for influential observations or outliers among the variables.

(a) Verify the correlation and regression for mercury vs cadmium. See Figure 2.14.

(b) Verify the correlation and regression for mercury vs chromium. See Figure 2.15.

(c) Verify the correlation and regression for mercury vs copper. See Figure 2.16.

(d) Verify the correlation and regression for mercury vs lead. See Figure 2.17.

2.9 The study referenced in Exercise 2.8 also included sediment samples. Those data are contained in the SAS dataset SED in Appendix 1 and spreadsheet of the same name in the excel file river.xlsx. The SAS program, River plots2.sas, given in Appendix 2, can also be used to carry out correlation and regression plots similar to those in Exercise 2.8. Explore mercury vs. copper and lead to illustrate that influential observations can hide relationships as well as exaggerate them. Explore other relationships of your choice.

2.10 Normality

The purpose of this exercise is to illustrate visual and formal checks of normality of various datasets. In each case requested, first calculate the residuals of each treatment group obtained by subtracting the group mean from each observation. This can be done using a simple ANOVA program or manually. The program *Univar.R* will be useful for this purpose. Then construct a histogram and a QQ-plot of all residuals and assess them for visual evidence to support or reject normality. In Chapter 3, revisit this example and do a Shapiro–Wilk test for normality of these residuals. Alternatively, do the Anderson–Darling test.

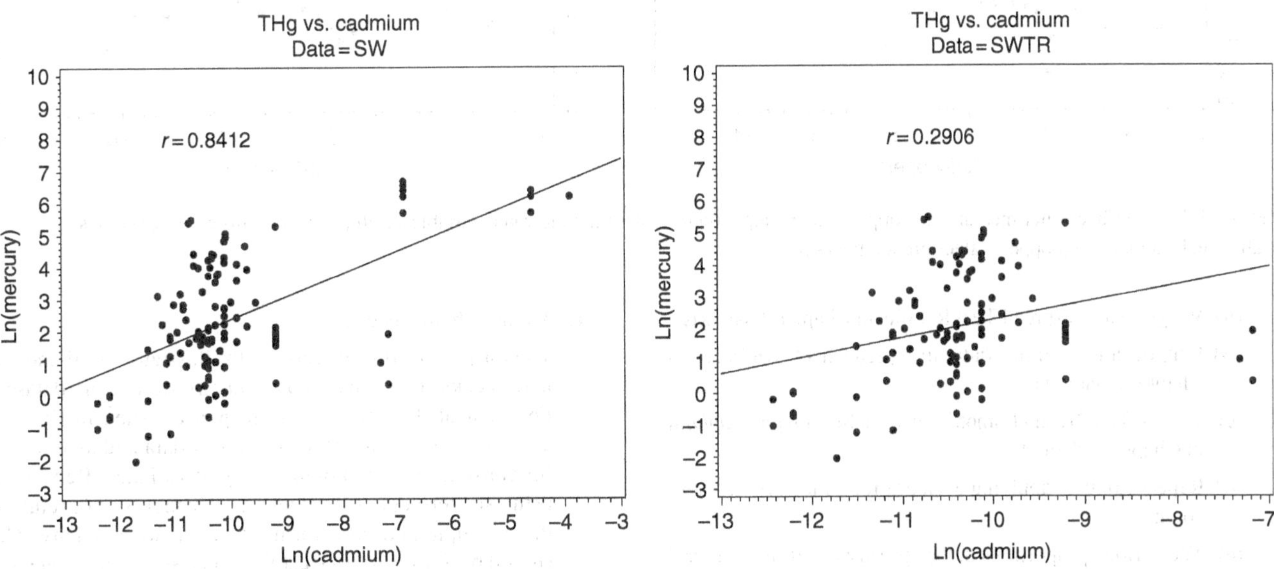

Figure 2.14 Surface water mercury vs. cadmium. Line = simple regression of the indicated variables. Left panel contains all observations. Right panel excludes Ln(cadmium) values greater than −7.

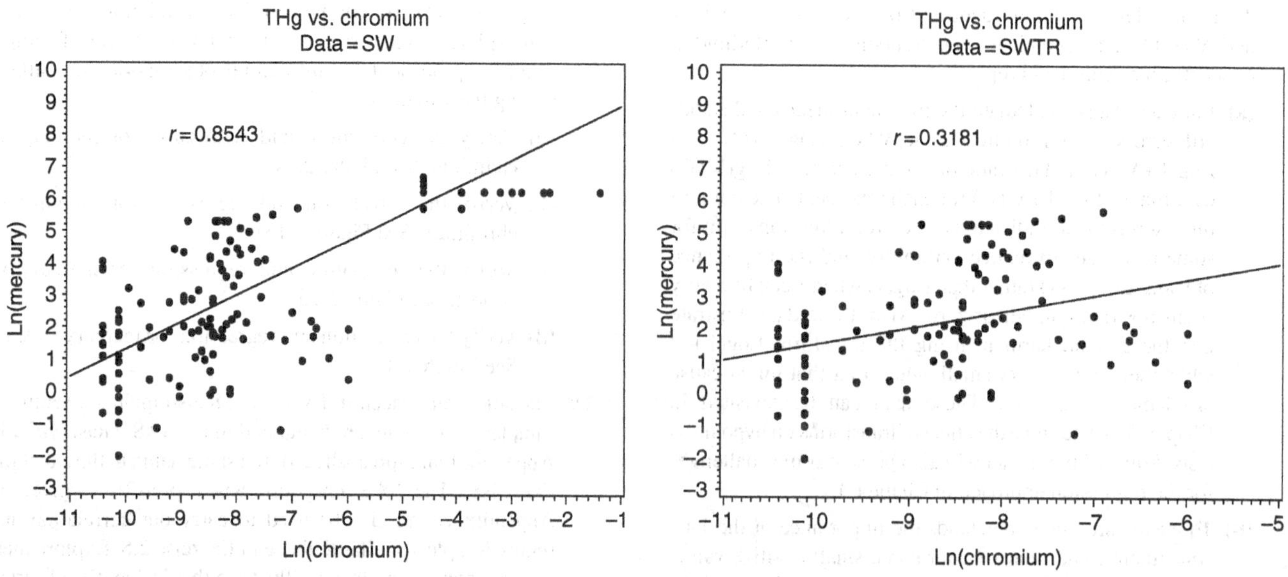

Figure 2.15 Surface water mercury vs. chromium. Line = simple regression of the indicated variables. Left panel contains all observations. Right panel excludes Ln(chromium) values greater than –5.

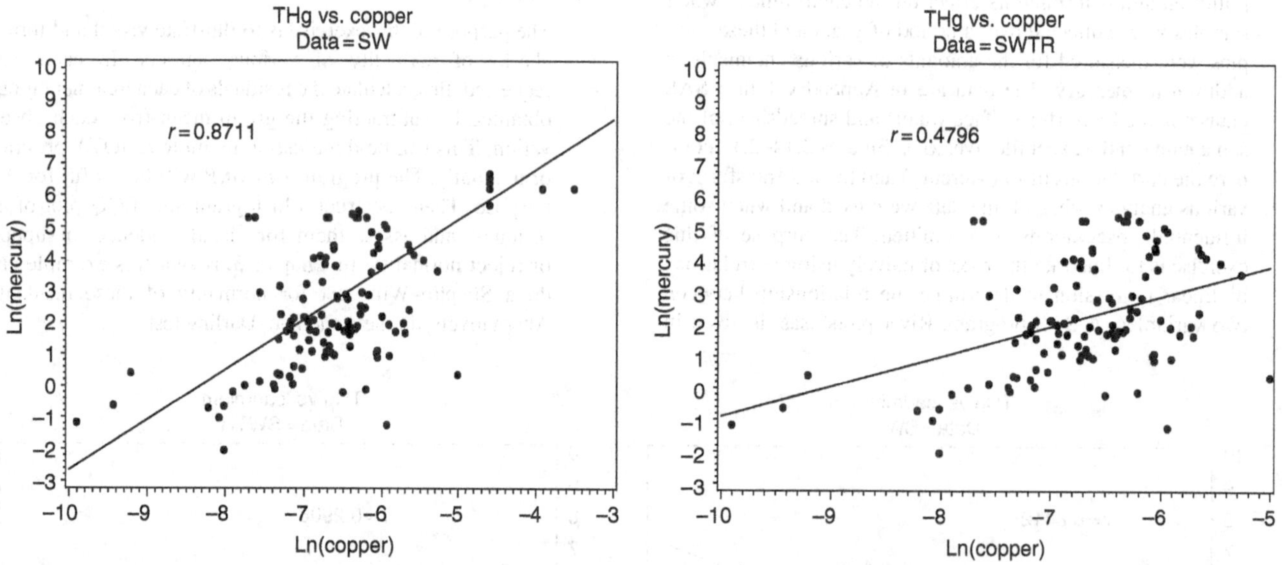

Figure 2.16 Surface water mercury vs. copper. Line = simple regression of the indicated variables. Left panel contains all observations. Right panel excludes Ln(copper) values greater than –5.

(a) Weight data in spreadsheet Rabbits in Chapter 2 data.xls.

(b) Length and weight data in spreadsheet *RBTsize* in Chapter 2 data.xls.

(c) TLY14, TLY21, and Immob21 in spreadsheet *repro_daphnia* in Chapter 2 data.xls.

(d) Repeat (c) after applying a square-root transform to these counts.

(e) The various proportions in the spreadsheet zf in Chapter 2 data.xls.

(f) Repeat (e) after applying an arc-sine square-root transform to the proportions.

2.11 Variance homogeneity

The purpose of this exercise is to illustrate visual and formal checks of variance homogeneity of various datasets. For each of the datasets and responses listed in Exercise 2.5, construct scatter plots of the raw data and assess them for consistency with homogeneity of variance. Repeat this using scatterplots of residuals. In Chapter 3 or 6, return to this example and do a formal test for homogeneity. This should be Tarone's $C(\alpha)$ test for overdispersion (Chapter 6) for proportions and Levene's test for continuous responses (Chapter 3).

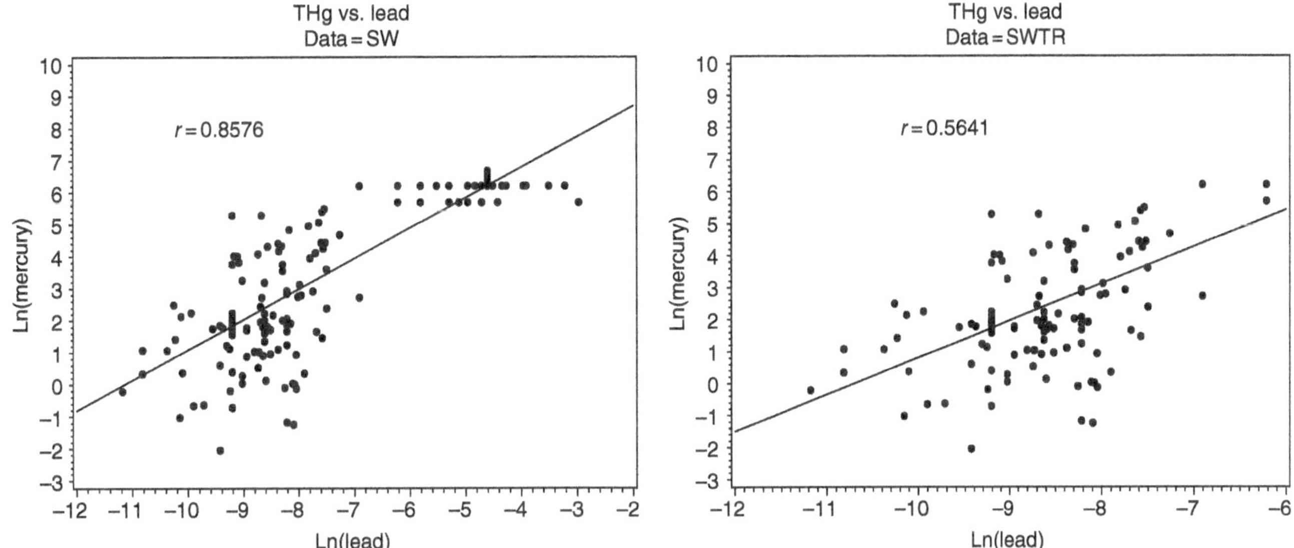

Figure 2.17 Surface water mercury vs. lead. Line = simple regression of the indicated variables. Left panel contains all observations. Right panel excludes Ln(lead) values greater than −6.

2.12 Monotonicity

For the scatter plots in Exercise 2.6, join the treatment means or medians with straight-line segments and assess the means for consistency with a monotone concentration–response. Trend tests and regression models (other than hormetic models) require monotonicity. Recall that some deviation from monotonicity is expected from random variation but "serious" deviations are inconsistent with monotonicity. If a small number of outliers are present, remove them and reassess. Programs *Scatterplot MACRO.sas* and *Scatter.R* can be used to draw scatterplots.

2.13 Outliers

Use a scatterplot, such as provided by the program *scatterplot macro.sas* or *scatter.R*, to do a visual check for outliers in the following datasets and responses:

(a) data = daphnia_repro, response = TLY21

(b) data = daphnia_repro, response = TLY14

(c) data = daphnia_repro, response = Immob21.

2.14 Data from nontarget terrestrial plant studies (OECD TG 208) is given in SAS datasets named NTTP*x*, where *x* is a study number, e.g. 2, 5, and 47. In these studies, various plant species are exposed to a chemical intended to control insect pests, but could also have an adverse effect on plant growth. These data contain information only on plant dry weight. The purpose of this exercise is to compare the Bonferroni and Bonferroni–Holm adjustments to significance levels. For convenience, since the appropriate test statistics will not be developed until Chapter 3, tables named Study*x* for the same choices of *x* are provided that give unadjusted *p*-values, labeled PROBT, that came from comparing each treatment to the control using a simple T-Test. These are in the excel file NTTP.xlsx.

(a) Verify the Bonferroni adjusted *p*-values, labeled BONP, and the Bonferroni–Holm adjusted *p*-values, labeled PBHadj. Of possible interest are the Dunnett *p*-values labeled ADJP. The raw data are provided for use in Chapters 3 and 4. Programs to calculate the Bonferroni–Holm adjusted *p*-values are *bholm.sas* and *bholm.R*, both given in Appendix 1.

(b) Compare the BONP and PBHadj values and find those for which the comparison is not significant by the Bonferroni adjustment (BONP) but is significant for the Bonferroni–Holm adjustment (PBHadj).

(c) Make the same sort of comparison of the Dunnett adjusted *p*-values with the Bonferroni–Holm and Bonferroni. Another adjustment to the comparisons is given by the trend tests discussed in Chapter 3.

Chapter 3

Analysis of Continuous Data: NOECs

3.1 INTRODUCTION

Some regulatory test guidelines call for the determination of an NOEC for one or more responses, and it is important to understand the statistical models that are available for this purpose. This requires an understanding of the experimental designs that support these tests, the properties, such as power, sensitivities, and limitations that determine their utility. This chapter will address tests that are appropriate for analysis of continuous responses that arise from toxicity studies and other types of studies as well. Later chapters will deal with quantal, discrete, and ordinal data.

As discussed in Sections 1.4, 2.2.3.1, and 2.2.3.2, there are two broad categories of hypothesis tests used for NOEC determination. These are pairwise and trend tests. Within each of these categories, there are two types of tests that will be developed. These are parametric and nonparametric tests. All tests are based on models that have requirements, and use of a test for data that do not meet the requirements of that test can lead to erroneous conclusions, sometimes dramatic. With that in mind, this chapter will explore both formal and informal tools to help determine whether a given dataset meets model requirements such as normality and variance homogeneity and whether the data are consistent with a monotone dose–response, which is a requirement of the trend tests.

A point to be emphasized through this and subsequent chapters, except as specifically noted otherwise, is that the basic unit of analysis is the replicate, not the individual animal or plant. Many toxicity studies house multiple subjects together in a small number of test vessels (tanks, pots, etc.), and organisms in the same vessel generally cannot be considered independent. The reader is referred to Section 1.3.2 for a detailed discussion of replicates and the unit of analysis.

3.2 PAIRWISE TESTS

While trend tests are the preferred type of test for most NOEC determinations, there are situations and datasets for which only pairwise tests are appropriate. When the data do not follow a monotone dose–response, use of a trend test based on an assumed monotone dose–response can lead to a nonsensical NOEC. For example, in a pesticide spray application on a crop, a decrease from control in dry weight of the crop is considered adverse, but it is not uncommon for there to be increase in dry weight at one or more low application rates, a phenomenon known as hormesis or low dose stimulation. Use of a trend test for such data can lead to an LOEC at which the mean dry weight is higher than the control mean, which is patently absurd. This can occur by forcing a decreasing trend model on data that do not follow such a trend. A similar problem arises when a decreasing regression model is force fit to non-monotone data. An example of hormesis and its effect on analysis will be given in Sections 3.5, 3.6, and 4.2.2. Pairwise methods are better suited for hormetic data than trend models. Calabrese and Baldwin (2001) discuss hormesis, as do Calabrese (2005a, b), Calabrese and Blain (2005), Cedergreen et al. (2006). Davis and Svendsgaard (1990) provide a discussion of non-monotone dose–responses in toxicology and an extensive set of references. Almstrup et al. (2002) reported on phytoestrogens that were aromatase inhibitors at low concentrations but estrogenic at higher concentrations, giving rise to U-shaped dose–response curves. Monte-Silva et al. (2009) reported on inverted U-shaped dose–response curves. Another reason for developing pairwise tests is that such tests are frequently reported in the toxicity literature even when trend tests should be used.

Statistical Analysis of Ecotoxicity Studies, First Edition. John W. Green, Timothy A. Springer, and Henrik Holbech.

Companion website: www.wiley.com/go/Green/StatAnalysEcotoxicStudy

A question that must be addressed is whether to use a one- or two-sided test, i.e. which of the following three sets of hypotheses to test:

$$H_{0i}: \mu_i = \mu_0 \text{ vs. } H_{1i}: \mu_i \neq \mu_0, i = 1,\ldots,k,$$

or

$$H_{0i}: \mu_i = \mu_0 \text{ vs. } H_{1i}: \mu_i < \mu_0, i = 1,\ldots,k,$$

or

$$H_{0i}: \mu_i = \mu_0 \text{ vs. } H_{1i}: \mu_i > \mu_0, i = 1,\ldots,k.$$

Since the focus of most toxicity experiments is the determination of chemical exposures that have an adverse effect, there must be an understanding of what is an adverse effect. For proper statistical analysis, one must know whether an adverse effect is a *change* from control or only a decrease from control or only an increase from control. Usually, one-sided tests are appropriate in toxicology. For example, there is concern that animals may gain less weight after exposure to a chemical, so a decrease in body weight or weight gain compared with the control is an adverse effect, or the concern is that there is a decrease in yield of a crop after treatment by a pesticide. While crop yield might increase after application of a pesticide because the pests have been reduced and therefore crop damage is reduced, that is not an adverse effect and the purpose of the test is to determine whether there is evidence of an adverse effect. In this event, a one-sided test against a decreased alternative hypothesis is appropriate. On the other hand, the concern might be whether the time to reproduction is either delayed or accelerated, either of which is an adverse effect with possible population survival or health implications. In that event, any change is a possibly adverse effect and a two-sided test is appropriate.

Of course, the identification of adverse effects is not the only question that might be of interest where pairwise tests are used. In a pharmaceutical experiment, it is important to determine dose regimens that are efficacious, so some measure of benefit would be important. Except for the language used to describe the statistical application, the methods discussed here all apply in either situation.

3.2.1 Adjustments for the Number of Comparisons Made

When there are multiple doses or concentrations of the test chemical used in most toxicity experiments, another important question is whether to use the nominal false positive rate for every comparison of treatment to control, i.e. use the so-called comparison-wise error (CWE) rate or to use the family-wise error (FWE) rate, i.e. adjusting the critical value of the test statistic to account for the number of comparisons made to control. Generally, some adjustment for the number of comparisons to control is made to guard against an inflated false positive rate (Wright, 1992). Two common adjustments for pairwise comparisons will be discussed. References for other approaches include Hochberg and Tamhane (2009), Hsu (1996), Miller (1981), and Wright (1992).

The Bonferroni adjustment for multiple comparisons is a simple, but conservative, adjustment that makes the family-wise false positive error rate at most the stated nominal level α, usually specified as 5% or $\alpha=0.05$. This adjustment is based on Boole's inequality that states that if A and B are events, then

$$P(A \cup B) \leq P(A) + P(B), \tag{3.1}$$

where $P(E)$ is the probability that event E occurs and $A \cup B$ is the event that A occurs or B occurs and the logical operator "or" is taken as inclusive, i.e. it includes the possibility that both A and B occur. Equation (3.1) is easily generalized to include k events, E_i, to wit:

$$P\left(\bigcup_{i=1}^{k} E_i\right) \leq \sum_{i=1}^{k} P(E_i). \tag{3.2}$$

Bonferroni's inequality states that in a study with k treatment levels plus control, if a pairwise comparison is made of each treatment groups to the control group at the significance level of α/k, then the probability of at least one false positive occurring is α. Alternatively (Wright, 1992), the significance level of each comparison could be multiplied by k and then compared with α. The Bonferroni adjustment can be overly conservative, especially for large k.

Modifications exist that reduce the conservatism while preserving the FWE at 0.05 or less. One such modification is the Holm adjustment (Holm, 1979) to Bonferroni, which states the following, taken[1] with modifications from OECD (2014a). Arrange the k unadjusted p-values for all comparisons of treatments with control in rank order, i.e. $p_{(1)} \leq p_{(2)} \leq p_{(3)} \leq \cdots \leq p_{(k)}$. Beginning with $p_{(1)}$, calculate $p^*_{(i)} = p_{(i)} * (k - i + 1)$ and compare with α, stopping at the first comparison not significant at level α. If the smallest i for which $p^*_{(i)}$ exceeds α is $i = j$, then all comparisons with $i > j$ are judged nonsignificant without further comparisons. An example given in Section 3.4 shows how this handles two or more identical p-values. See Table 3.1 for a contrived example that compares these two adjustments. In this example, only the comparison of treatment 4 with the control would be significant

1 Green and Springer were major contributors to OECD 2014 and developed the discussion of Bonferroni–Holm in that reference, which is adapted here in modified form.

Table 3.1 Comparison of Bonferroni and Bonferroni–Holm Adjusted *p*-Values

Comparison	Unadjusted *p* value	Bonferroni–Holm adjusted *p* value	Bonferroni adjusted *p* value
Control–treatment 4	$p_{(1)}=0.002$	0.002 * 4=0.008	0.002 * 4=0.008
Control–treatment 2	$p_{(2)}=0.013$	0.013 * 3=0.039	0.013 * 4=0.052
Control–treatment 3	$p_{(3)}=0.020$	0.020 * 2=0.040	0.020 * 4=0.080
Control–treatment 1	$p_{(4)}=0.310$	0.310 * 1=0.310	0.310 * 4=1.000[a]

[a] Probability cannot exceed 1, so any product exceeding 1 is replaced by 1.

Table 3.2 Total Protein (TP) Levels in Blood from Rats After 90 Days on Various Diets

Diet	TP	Std	*n*	Contrast	TValue	ProbT	PRank	Bon_p	BHolm_p	Signif
Basel	6.77	0.177	10	CTRL 2 vs. CTRL 1	3.350	0.002	3	0.014	0.008	*
Fish oil	7.10	0.226	10	LOW DOSE vs. CTRL 1	0.812	0.421	5	1	1	
5000ł	6.69	0.218	10	INT DOSE vs. CTRL 1	0.200	0.84	7	1	0.84	
15000	6.75	0.268	10	HI DOSE vs. CTRL 1	0.624	0.535	6	1	1	
50000	6.83	0.200	9	LOW DOSE vs. CTRL 2	4.162	0	1	0	0.001	*
				INT DOSE vs. CTRL 2	3.552	0.001	2	0.007	0.006	*
				HI DOSE vs. CTRL 2	2.634	0.012	4	0.084	0.046	*

ł test diet in ppm TP is the mean TP value for the indicated diet, Std is the standard deviation of the individual animal TP values, and *n* is the number of animals that survived to the end of the study. CTRL 1, basel diet; CTRL 2, Fish oil diet. Signif = * indicates significance with the Bonferroni–Holm adjusted *p*-values.

by the Bonferroni adjustment, whereas all comparisons except that of comparison of treatment 1 versus the control would be significant by the Bonferroni–Holm adjustment. An alternative application of these adjustments to an atypical study is given in Section 3.2.2.

There are other adjustments that can be made to control the FWE rate. References include Hochberg and Tamhane (2009), Hsu (1996), Hoppe (1993), and especially Wright (1992). Related to this is whether to make adjustments to take the number of different responses that are tested into account. While there have been some attempts to do this, it is not common in toxicology and will not be pursued here.

For "well-behaved" continuous data, i.e. data that follow a normal distribution with homogeneous variances, there are several tests that are very widely used in toxicity studies and beyond. Among these are the Dunnett and *t*-test with or without adjustments for the number of comparisons, for pairwise comparisons of treatments to control that ignore any dose–response trend. Other tests that could be used include Steel, Duncan, Tukey honest significant difference (HSD), and Scheffe, but there are numerous other tests. No attempt will be made to develop all of these tests.

When data do not meet the normality and variance homogeneity requirements for the use of some tests, a normalizing, variance-stabilizing transformation may be sought. Otherwise, a nonparametric analysis can be done. A discussion of the Mann–Whitney (or Wilcoxon) and Dunn tests will be given.

3.2.2 Student T-Test

This is familiar from elementary statistics courses and requires the response to be normally distributed with homogeneous variances. The test statistic is given by

$$T_i = \frac{\bar{x}_i - \bar{x}_0}{s\sqrt{\dfrac{1}{n_i} + \dfrac{1}{n_0}}}, \tag{3.3}$$

where *s* is the square root of the usual pooled within-group sample variance from Eq. (A2.33) based on the control and all treatment groups, not just the two groups indicated in Eq. (3.1). This test statistic is calculated for each treatment group and for each such group, T_i is compared with the *one-* or *two*-sided critical value of the $t_{n.-k-1}$ distribution, where *k* is the number of treatment groups *excluding* the control and *n*. is the total number of replicates in control and treatment groups. See Table 3.2 for a summary of the results of a t-test in the context of an atypical study where neither Dunnett's test nor one of the trend tests is applicable.

3.2.3 Application to Multiple Controls

A useful application of the *t*-test and adjustments for multiple controls arises in some dietary studies. In a dietary study in rats, there were two reference groups or controls.

These were a basel diet and a fish oil diet. Three concentrations of an experimental dietary supplement were to be compared with each of the reference diets, and the two reference diets were compared with each other. Unlike the water and solvent control experimental design discussed in Section 2.11, differences between the two reference diets are expected, and comparisons of the test diets with each reference diet are of interest, as is the comparison of the two reference diets. In this study, rats were individually caged, so the subject and replicate are the same. Within-treatment variability was assessed across all treatments, and these variances were found homogeneous by Levene's test and the data were furthermore found consistent with a normal distribution by the Shapiro–Wilk test. Presentation of those tests is deferred to Section 3.3. An analysis was needed that used the pooled variance and used *p*-values adjusted only for the comparisons of interest. A standard ANOVA was done by specifying the desired contrasts. This produced *t*-tests with corresponding *p*-values to which the Bonferroni and Bonferroni–Holm adjustments were applied. A summary table of total protein (TP) values and comparisons is given in Table 3.2. Note that statistical significance following the two adjustments agrees except for the comparison of the high dose against the fish oil diet, where the Bonferroni-adjusted *p*-value exceeded 0.05, while the Bonferroni–Holm adjusted *p*-value was less than 0.05. In this type of study, an NOEC is not appropriate, but the same methodology is used to compare diets as used in some NOEC determinations.

3.2.4 Dunnett's Test

Dunnett's test assumes a normally distributed response with homogeneous within-group variances. The statistic is the same as that for the Student *t*-test given in Eq. (3.3), except this time, as established by Dunnett (1955, 1964), T_i is compared with the *one-* or *two-*sided upper α equicoordinate point of a *k*-variate *t*-distribution with $df=n_0+n_1+n_2+\cdots+n_{k-1}-k$ and correlation between the *i*th and *j*th comparisons given by

$$\rho_{ij}=\left[\frac{n_i n_j}{(n_i+n_0)(n_j+n_0)}\right]^{1/2}. \tag{3.4}$$

If all treatment sample sizes are equal, but the control sample size is possibly different, then Eq. (3.4) reduces to $1/(1+r)$, where $r=n_0/n$ and n is the common treatment sample size. Note that if $n_0=n$, then $\rho_{ij}=0.5$. Tables for various values of r are available in many places and the calculations based on the general correlation structure are computed automatically by several commercial software packages (e.g. SAS, R, ToxRat (2016), CETIS (2016)).

3.2.4.1 Dunnett's Test Example with Rainbow Trout Length

To illustrate Dunnett's test, see Table 3.3 for the data from an early life stage study (OECD TG 210) with rainbow trout. There were two reps of 15 fish in both dilution water and solvent controls and in each of five test concentrations. The rep is the unit of analysis, and Table 3.3 only provides replicate information. The pooled within-group variance estimate of 0.0056 had 7 degrees of freedom. The water and solvent controls were found statistically indistinguishable by a *t*-test and subsequently combined for determination of treatment effects. The replicate and treatment means are displayed in Figure 3.1, which makes clear the small rise in

Table 3.3 Rainbow Trout Size Data from Early Life-Stage Study

Dose	Rep	Conc (mg l^{-1})	Control	Length (cm)	Weight (g)	N
1	A	0	Water	4.51	1.31	15
1	B	0	Water	4.57	1.33	15
2	A	0	Solvent	4.47	1.30	15
2	B	0	Solvent	4.37	1.33	15
3	A	0.63		4.37	1.21	15
3	B	0.63		4.31	1.22	15
4	A	1.3		4.27	1.17	15
4	B	1.3		4.25	1.18	15
5	A	2.5		4.35	1.19	15
5	B	2.5		4.18	1.17	15
6	A	5		4.37	1.25	15
6	B	5		4.35	1.23	15
7	A	10		4.25	1.13	15
7	B	10		4.08	1.14	15

N, number of trout per replicate tank.

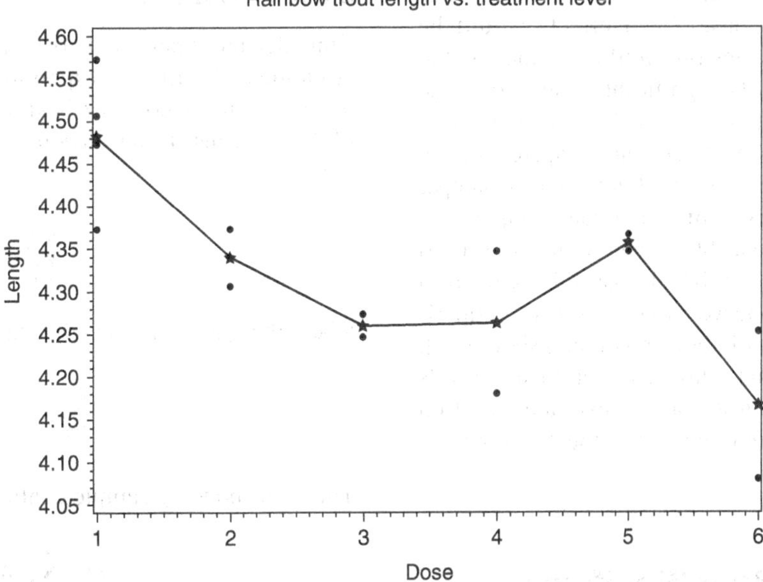

Figure 3.1 Rainbow trout length. Replicate mean lengths of rainbow trout are plotted as dots. Treatment means are plotted as stars and joined by line segments. The slight rise at dose 4 compared to dose 3, and the larger rise at dose 5 are evident, but overall, there is a downward trend.

Table 3.4 Dunnett' Test Results for Rainbow Trout Length

Estimate	Signif	*p*-Value
DOSE 2 − 1		0.14
DOSE 3 − 1	*	0.03
DOSE 4 − 1	*	0.03
DOSE 5 − 1		0.19
DOSE 6 − 1	**	0

Signif: * means significant at the 0.05 level; ** means significant at the 0.01 level.

observed lengths at 5 mg l^{-1} (dose 5). See Table 3.4 for the results from Dunnett's test on length, where it will be evident that the mean lengths at concentrations 1.3, 2.5, and 10 mg l^{-1} were significantly different from the control by Dunnett's test, but the mean at 5 mg l^{-1} was not. This is a somewhat confusing, but not uncommon, result in that there appears to be an adverse effect at two low concentrations but not at one of the higher concentrations. If one defines the NOEC as the highest test concentration at which there was no statistically significant effect, then the NOEC would be 5 mg l^{-1}. If one defines the LOEC as the lowest test concentration at which there is a statistically significant effect and the NOEC as the highest test concentration less than the LOEC as we do, then the NOEC = 0.63 mg l^{-1}. Gaps like this can create confusion in interpretation. We will return to this issue in the discussion of trend tests.

Variations on this and computational formulas are given in Hsu (1996) and in Hochberg and Tamhane (2009). If there are subgroups in the treatment groups, the sample variances in the above formula can be replaced by a Satterthwaite-type expression such as that used for the Tamhane–Dunnett procedure to be described in Section 3.4.1.

Dunnett's test is readily available in many books and statistical software packages, is familiar to most toxicologists as well as statisticians, and is simple to calculate. *Unfortunately* and more importantly, it ignores monotonicity in the dose–response that is present and thereby ignores an important component of the experimental design and biology. Consequently, it thereby loses power in relationship to step-down trend methods. The test was developed for qualitatively different treatments, such as different formulations of a drug or different drugs. As with all pairwise tests, this can lead to a low dose being found statistically significant and a higher dose not statistically different from the control.

It is also important to understand regarding pairwise tests, including *t*-tests, Dunnett, and Tamhane–Dunnett, that even though these tests might be done in an ANOVA setting or using ANOVA software programs, they stand alone. They do not require a significant *F*-test to justify their application. Indeed, requiring a significant *F*-test in order to use one of these tests changes the properties of these tests. As pointed out in Green et al. (2012), "the *F*-test (and similar statements apply to the Kruskal-Wallis test for nonparametric analysis) guards against many comparisons that are typically of no interest in toxicology, such as comparisons of one treatment group against another. Consequently, the *F*-test may not be significant when there are real treatment effects that show up in the form of trends in the concentration–response. Such effects would be missed, because no further testing would be done. On the other hand, the F-test may be significant because of some difference among the treatments that does not indicate a significant difference of any treatment compared to the control. In either case, the use of the F-test as a gatekeeper distorts the significance levels of subsequent comparisons of treatments to control. Hochberg and Tamhane (2009) and Hsu (1996, esp. pp. 177ff) discuss this issue in a broader context. Even if one were to do pairwise comparisons of treatments to control, they do not, and should not, depend on a prior significant F-test. The significance levels associated with Dunnett's test, for example, are based on an assumption of its use independent of the F-test or any other prior test."

3.2.5 Use of a Prior Significant F-test Before Using Dunnett's Test

Since the point made in the previous paragraph is so important and so often misunderstood and abused in the field of ecotoxicology, even by some statisticians, some additional discussion will be made. Toxicity experiments differ from many other types of experiments in some important ways. In a general single factor experiment, treatments are not related in a quantitative fashion. Instead, different formulations of drugs are to be compared or different brands of milk or tires or cars. In multifactor experiments, the responses under different combinations of the factors are of interest. In these types of studies, one sometimes does not have a specific hypothesis in mind at the beginning of the experiment to test other than whether there are differences among the various treatments or combinations of treatments. The statistical analysis is then done in stages. If the requirements of ANOVA are satisfied, then the ANOVA F-test is examined for significance at, say, the 0.05 level. If that test is not significant, testing stops and one concludes that there are no differences among the treatments. If the F-test for overall differences is significant, then further testing is done to elucidate where those differences lie. These further tests are generally described as post hoc test because the questions they answer only arose after the fact, i.e. after the experiment was done.

Toxicity experiments are different. The treatments differ only quantitatively in the level of exposure to a single test compound. It is known before the experiment begins that one intends to compare the treatments with the control to determine at what level a toxic effect is observed. These are planned comparisons and, indeed, the purpose of the experiment. The ANOVA F-test is irrelevant and can be misleading. To understand why this is so, it will be helpful to define the notion of a contrast of treatment means.

3.2.5.1 Contrasts

Consider an experiment to investigate k populations or "treatments" with corresponding population means μ_1, μ_2, μ_3, …, μ_k. A contrast is simply a linear combination of those means. That is, a contrast is any expression of the form

$$C = \sum_{i=1}^{k} c_i \mu_i,$$

where the c_i are constants that sum to 0, i.e.

$$\sum_{i=1}^{k} c_i = 0.$$

The corresponding sample contrast is given by

$$\hat{C} = \sum_{i=1}^{k} c_i \hat{\mu}_{i,},$$

where $\hat{\mu}_i$ is the sample mean of the ith treatment. For example, $\mu_k - \mu_1$ is a contrast comparing the kth treatment with the first defined by $c_1 = -1$, $c_k = 1$, and $c_i = 0$ for i different from 1 and k, as is $3\mu_1 - \mu_2 - \mu_3 - \mu_4$, where $c_1 = 3$, $c_2 = -1$, $c_3 = -1$, $c_4 = -1$, and $c_i = 0$ for $i > 4$. Contrasts are very important in ANOVA settings and allow one to explore many hypotheses of possible interest through appropriately defined combinations of treatment means.

In toxicology experiments, generally only a few contrasts are of interest. Typically, these are contrasts of the form $\mu_j - \mu_1$, for some j, $2 \le j \le k$, where μ_1 is the control mean and the question of interest is whether there is a significant "effect" in treatment j. The hypothesis of interest can be stated in the form of a question as follows: Is there a difference between the mean response of the jth treatment and the control mean response? We may also be interested in whether there is a trend in the treatment means. For example, if $k = 4$, a linear trend in the treatment means can be tested using the contrast $-3\mu_1 - \mu_2 + \mu_3 + 3\mu_4$.

The ANOVA F-test guards against many comparisons of no interest in toxicology. In fact, the ANOVA F-test is significant at the 0.05 significance level if and only if some contrast of the treatment means is significantly different from zero by the Scheffe test (see, for example, Scheffe (1959, p. 67) at the 0.05 significance level. To understand the implications of this, consider a simple one-way ANOVA with k treatment levels plus a control, with s^2 the sample variance obtained from the ANOVA residuals. The F-test is significant at the 0.05 level if and only if some contrast is significant by Scheffe's test at that level, i.e. if and only if there is some contrast C for which

$$\hat{C}^2 > s^2 F_{0.05}(k, n-k-1),$$

where n is the total sample size and $F_{0.05}(k,n-k-1)$ is the value of the F-statistic with k and n-$k-1$ degrees of freedom. A t-test for a comparison of a treatment with the control is a special type of contrast of the form $C=\mu_j-\mu_1$ and the t-test is significant at the 0.05 level if and only if

$$\hat{C}^2 > s^2 F_{0.05}\left(1,n-k-1\right),$$

where $T_{0.05}(n-k-1)=[F_{0.05}(1,n-k-1)]^{1/2}$ is the value of the T-statistic with $n-k-1$ degrees of freedom. So, the t-test (expressed in terms of the F-statistic) is smaller by virtue of replacing the factor k by 1 and larger by using the critical value of the F-distribution with 1 degree of freedom in the numerator rather than k.

There is thus no direct correspondence between the t-test and the ANOVA F-test. That is, there can be a contrast that is significantly different from zero by the Scheffe test without any simple t-test of interest in toxicology being significant, and there can be a simple contrast of interest in toxicology that is significant by a t-test without that or any other contrast being significant by the Scheffe test. Stated more simply, a significant F-test can happen without any treatment effect being significant and a significant treatment effect can exist without the F-test being significant. A simple situation where the former can happen is that some two exposed groups will differ significantly (e.g. through a hormetic effect at a low concentration and a small but apparently adverse effect at a higher concentration), and this will trigger the F-test, when there is really no exposed group that differs significantly from the control. Thus, requiring a significant F-test prior to testing for treatment effects is unjustified and misleading. There are additional reasons why the F-test is inappropriate that will be explored below.

In a basic toxicity test, the treatments form an ordered set of exposure concentrations of a single chemical. As pointed out in Green et al. (2012), the F-test (and similar statements apply to the Kruskal–Wallis test for nonparametric analysis) guards against many comparisons that are typically of no interest in toxicology, such as comparisons of one treatment group against another. Consequently, the F-test may not be significant when there are real treatment effects that show up in the form of trends in the concentration–response or simple differences from the control. Such effects would be missed, because no further testing would be done. On the other hand, the F-test may be significant because of some contrast of the treatments that does not indicate a significant difference of any treatment compared with the control. In either case, the use of the F-test as a gatekeeper distorts the significance levels of subsequent comparisons of treatments to control. Additional discussion of this issue in a broader context is given in Hochberg and Tamhane (2009) and Hsu (1996, esp. pp. 177ff). Additional references on this point include Dunnett (1955, 1964), Williams (1971, 1972, 1975, 1977), Robertson et al. (1988), Jonckheere (1954), and Dunn (1964).

To reiterate, if one were to do pairwise comparisons of treatments to control, they do not, and should not, depend on a prior significant F-test. The significance levels associated with Dunnett's test, for example, are based on an assumption of its use independent of the F-test or any other prior test. However, as we shall see, Dunnett's test is not generally appropriate for toxicity experiments, and there are much better ways to analyze toxicity data that do take the concentration–response trends into account.

While it may seem that this point has been given a great deal of attention, it is altogether too clear that not all authors in this field seem to be aware of the above, and incorrect analyses are not uncommon in the toxicology literature. For example, Fox (2008) and Landis and Chapman (2011) both indicate a need for a significant F-test prior to comparing treatments with control. It is unfortunate that there are numerous other papers in ecotoxicology and toxicology where the same mistake is made. As will be described in detail in this and several following chapters, there are other ways to analyze toxicity experiments where the F-test is clearly seen as irrelevant.

3.3 PRELIMINARY ASSESSMENT OF THE DATA TO SELECT THE PROPER METHOD OF ANALYSIS

In Section 3.2.4.1, we argued against using a preliminary F-test or Kruskal–Wallis test prior to comparing treatment groups with the control to determine an NOEC. Indeed, rejecting such a preliminary test was demonstrated not just to be a matter of judgment but integral to the understanding of toxicology and the statistical properties of the tests to be employed. In this section, we explore a concept that is at least conceptually related. It is this: All statistical tests and procedures are based on assumptions about the nature of the data to be analyzed. Of these, two key assumptions are generally considered of great importance. These are assumptions about the distribution of the data within and between treatment groups. In this chapter, this comes down to determining whether the data in each treatment group are consistent with a normal distribution and, if so, whether the data in different treatment groups are consistent with an assumption of homogeneity of variance across treatments. Later chapters will explore quantal or count data where the concerns are consistency with a binomial or Poisson distribution and whether there is evidence of overdispersion, a term analogous to variance heterogeneity. As will be clear from the following discussion, we think it is imperative to assess data for consistency with the assumptions underlying the tests that will be used to analyze the data.

3.3.1 Residuals and the Assessment of Normality

Some statistical tests assume that the continuous data being modeled are normally distributed within treatment groups. With sample sizes as large as shown in Table 2.3, it is feasible to assess normality within each treatment, except for level 7. However, sample sizes are often smaller, especially when only replicate means are analyzed, and it is generally preferable in any event to assess normality from the study as a whole. This does not mean simply investigating whether the data as recorded follow a normal distribution, for we expect a treatment effect and that would affect the distribution of responses. Instead, we first remove the treatment effects and then assess normality from what remains. For the present purpose, this is very simple.

If y_{ij} is the jth observation from treatment i and $\bar{y}_{i.}$ is the sample mean response from treatment i, then the "pure error" residual is $y_{ij}-\bar{y}_{i.}$, the observed value minus the predicted (i.e. estimated) mean. See Figure 3.2 for an illustration of this. It will be observed that some residuals are positive and others are negative, and it is a simple matter to show that the sum of the pure error residuals from a single treatment is zero.

The pure error residuals collected from all treatment groups can be used to assess whether the data are consistent with a proposed statistical model. There are several visual and formal tools available from which to make this assessment.

3.3.2 Histogram of Pure Error Residuals

One simple, but useful, tool for assessing data whether it comes from a specified distribution is a plot of the residuals from all treatment groups in a histogram, often overlaid by the probability density function from the supposed distribution. To construct a histogram for a residual dataset, one divides the data into nonoverlapping intervals that collectively span the range of observations. Generally, these intervals are of equal length, though that is not required and sometimes unequal length intervals are replaced by intervals of equal frequency or some other scheme. To be more precise, each interval includes its left endpoint but not its right endpoint, as was done in Figure 2.1 for step functions used to describe a quantal distribution function. One then constructs a box for each interval whose height is the proportion or percentage of the data observed in that interval. See Figure 3.3 for an example from the rainbow

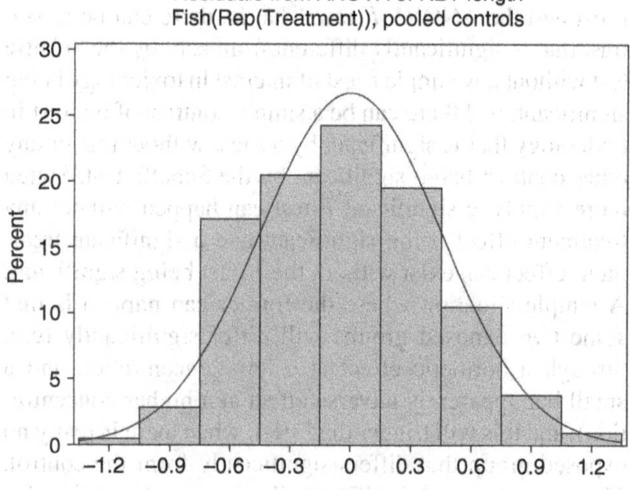

Figure 3.3 Histogram of rainbow trout length. Solid line is overlaid normal density function. A slight skew toward lower values is evident, but overall, plot is consistent with a normal distribution.

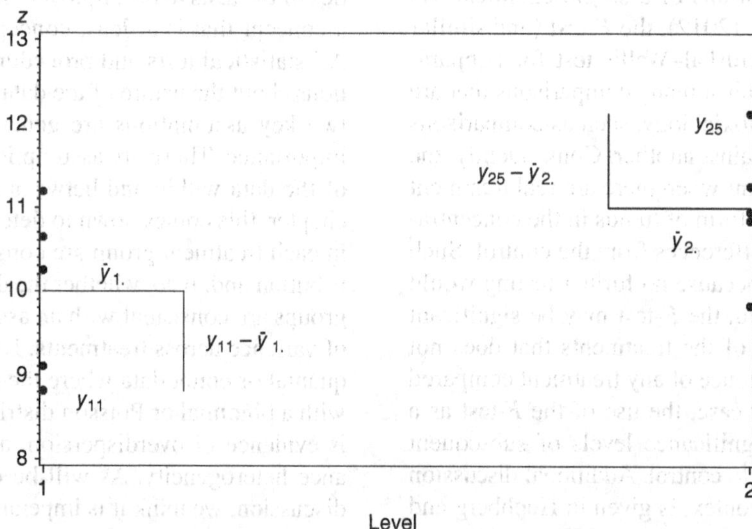

Figure 3.2 Pure Error Residuals Dots are observations. Horizontal segments are plotted at the arithmetic mean, $\bar{y}_{i.}$, of observations for indicated treatment level. Vertical segment indicates the residual.

trout length dataset underlying Table 3.1. It should be obvious that the shape of the histogram depends heavily on the number and choice of intervals used in its display. We will not attempt to discuss the various rules that have been suggested for the construction of histograms and the closely related concept of stem-and-leaf plots. There are many introductory statistics text books that provide such discussion (e.g. Thisted, 1988; Tamhane and Dunlop, 1999; Hoaglin et al., 2000). Figures 3.3 and 3.4 can be generated by the programs *Make histo.sas*, *histogram.R*, *Univar.r*, and *Table 55.r* that, with one exception, were first presented in Chapter 2 and are given in Appendix 1.

3.3.3 Quantile Plots

A related plot that avoids some of the problems associated with histograms is the quantile–quantile or QQ-plot. To do this, first rank-order the residuals, $r_{(1)} \le r_{(2)} \le r_{(3)} \le \cdots \le r_{(N)}$, where there are N residuals in all (including any repeat values). Next, calculate $z_i = i/(N+1)$ and then $q_i = F^{-1}(z_i)$, the quantile corresponding to the ith ordered observation from a standard distribution of the type being investigated. For example, F might be Φ, the standard normal distribution. Finally, plot $r_{(i)}$ against q_i. If the residuals come from the expected distribution, the plot should be linear in appearance. If not, the data may not be from that distribution. Consider QQ-plots (see Figure 3.4) from the trout length data also used to generate Figures 3.1 and 3.3, where the distribution being assessed is the normal.

The left plot in Figure 3.4 shows pure error residuals from the trout data and appears consistent with the expected normal distribution. There is a slight deviation at each extreme, suggesting more extreme values than expected, but the deviation is so slight as not to indicate a serious problem. A straight reference line from a normal distribution is shown to make the assessment easier. A slight rise of observations above the reference line toward the lower end is an indication of the slight skewness mentioned in Figure 3.3. The right plot is presented as an example of what a QQ-plot might look like when the data are not consistent with the expected (normal) distribution. The data for the right plot are not residuals from an ANOVA. We will consider formal tests of goodness of fit to a specified distribution in Section 3.3.5.

3.3.4 Scatterplot

In addition to assessing the distribution, it is also often important to assess variance homogeneity. Many simple models for hypothesis testing or regression assume that the

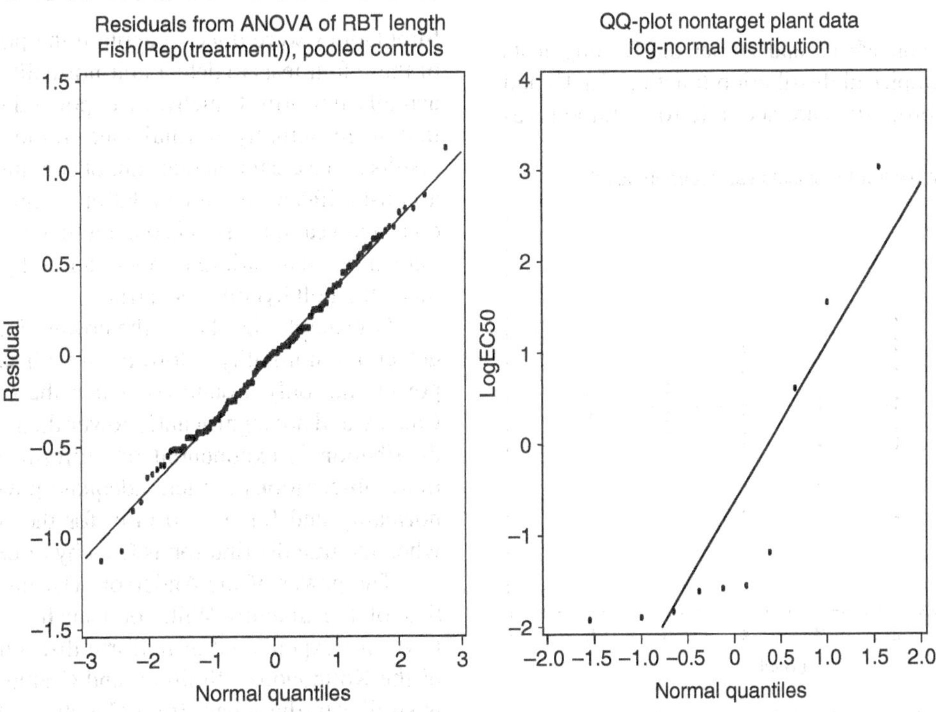

Figure 3.4 QQ-plots. The left plot shows pure error residuals from the trout data and appears consistent with the expected normal distribution. There is a slight deviation at each extreme, suggesting more extreme values than expected, but the deviation is so slight as not to indicate a serious problem. A straight reference line from a normal distribution is shown to make the assessment easier. A slight rise of observations above the reference line toward the lower end is an indication of the slight skewness mentioned for Figure 3.3. The right plot is presented as an example of what a QQ-plot might look like when the data are not consistent with the expected (normal) distribution. The data for the right plot are not residuals from an ANOVA. We will consider formal tests of goodness of fit to a specified distribution in Section 3.3.5.

data are equally variable across treatments. A scatterplot of residuals against treatment level is often helpful in assessing variance homogeneity, i.e. whether the variances in different treatment groups (or different values of the explanatory variable) are the same. See Figure 3.5 for an example for the same trout length data used to construct Figures 2.4, 3.3, and 3.4. In fact, Figure 3.5 is the same as Figure 2.4 except the trend has been removed.

While the spreads in Figure 3.5 are not all equal in width, the differences do not appear sufficient to bring variance homogeneity into question when the smaller sample size in the highest treatment group is taken into account. There are several formal tests of variance homogeneity that will be discussed below.

3.3.5 Formal Tests for Normality

In addition to these visual assessments, there are formal tests of variance heterogeneity and normality that can be especially important in a small- or high-volume testing facility where no statistician reviews every analysis. D'Agostino and Stephens (1986) describe many tests for normality and explore their properties. We will confine our treatment to Cramer–von Mises, Kolmogorov–Smirnov, Anderson–Darling, and Shapiro–Wilk, which are probably the most commonly cited and are widely available in commercial software.

The Cramer–von Mises and Anderson–Darling tests are based on the empirical distribution function (EDF) and their asymptotic properties are accurate for sample sizes of 20 or more (D'Agostino and Stephens, 1986, p. 124). Of those listed here, D'Agostino and Stephens (1986, p. 407) considered the Shapiro–Wilk best for judging normality from residuals from regression.

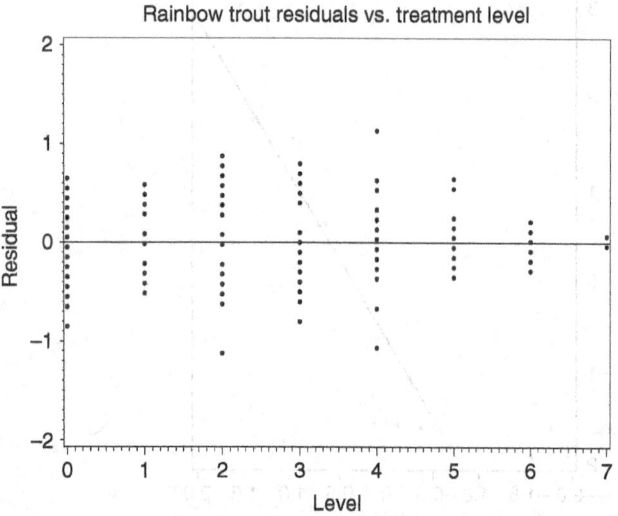

Figure 3.5 Scatterplot of Residuals Dots is individual fish lengths. The horizontal reference line at 0 is provided as a visual aid. One expects roughly the same number of points above the zero-line as below, both randomly distributed across treatments, and the width of the vertical data spreads across treatment groups should be roughly the same.

D'Agostino and Stephens (1986, p. 395) state that the Anderson–Darling test requires at least eight observations while the Shapiro–Wilk can be used with as few as three, though as shown below, the power properties are poor for very small samples. D'Agostino and Stephens (1986, p. 404) do not give minimum sample sizes for the Kolmogorov–Smirnov and Cramer–von Mises tests but assert that these tests should not be used. The power properties are shown below to be decidedly inferior to the Shapiro–Wilk and Anderson–Darling in some situations and are at best comparable in the other situations that have been investigated by the authors of the present text. D'Agostino and Stephens (1986, p. 215) also indicate that the Shapiro–Wilk is best for small samples, which makes it appropriate for many ecotoxicity studies.

It should be understood that the power of a test for normality depends not only on the sample size but also on the true distribution. For example, the power of a test is also affected by skewness to the right or left, the presence of thick tails, and multimodality.

3.3.6 Power to Detect Non-Normality

Simulations were done to explore the power or likelihood of these four tests to detect non-normality of data that were actually uniform, Cauchy, or exponential, as well as data that were actually normal (see Tables 3.5–3.8). These results make clear that one cannot conclude normality from a nonsignificant test for violation of normality performed on a small sample. This is but another example of the well-known fact that failure to reject a null hypothesis does not mean the null hypothesis is true.

For sample size 4 or 5, the power of all of these tests to detect non-normality is low. Even with sample size 10, the powers are only around 0.6 when the true distribution is Cauchy and are significantly lower than that when the true distribution is exponential or uniform. Only with 20 or more observations is there adequate power to detect non-normality and for $n = 20$ only for the Shapiro–Wilk test when the true distribution is Cauchy or exponential.

The power of the Anderson–Darling test is similar to that of the Shapiro–Wilk for Cauchy, though somewhat lower for exponential or uniform distributions. The power of the Kolmogorov–Smirnov and Cramer–von Mises tests is similar to the others for a Cauchy distribution but significantly lowers for exponential distributions. For exponential distributions, the power of the Cramer–von Mises test is only slightly lower than that for the Shapiro–Wilk and Anderson–Darling, but the power of the Kolmogorov–Smirnov is much worse.

Table 3.5 Power of Tests to Detect Non-normality in Cauchy-Distributed Samples of Sizes 4–30

Signif_sw	Signif_ks	Signif_ad	Signif_cm	Size	Count
2127	2248	2092	1871	4	10000
2888	2935	2935	2827	5	10000
3625	3687	3762	3710	6	10000
4858	4784	5002	5011	8	10000
5866	5750	6096	6076	10	10000
7600	7437	7818	7787	15	10000
8642	8397	8735	8721	20	10000
9602	9428	9658	9641	30	10000

sw, Shapiro–Wilk; ks, Kolmogorov–Smirnov; ad, Anderson–Darling; cm, Cramer–von Mises. Power is cell entry divided by count.

Table 3.6 Power of Tests to Detect Non-normality in Exponentially Distributed Samples of Sizes 4–30

Signif_sw	Signif_ks	Signif_ad	Signif_cm	Size	Count
1284	1302	1195	969	4	10000
1659	1425	1612	1422	5	10000
2163	1762	2096	1949	6	10000
3360	2345	3127	2871	8	10000
4449	2965	4093	3790	10	10000
6790	4456	6255	5772	15	10000
8353	5879	7759	7294	20	10000
9683	7749	9360	8988	30	10000

sw, Shapiro–Wilk; ks, Kolmogorov–Smirnov; ad, Anderson–Darling; cm, Cramer–von Mises. Power is cell entry divided by count.

Table 3.7 Power of Tests to Detect Non-normality in Uniformly Distributed Samples of Sizes 4–30

Signif_sw	Signif_ks	Signif_ad	Signif_cm	Size	Count
653	653	571	464	4	10000
602	510	579	457	5	10000
648	558	620	538	6	10000
707	580	651	580	8	10000
793	628	788	708	10	10000
1264	765	1164	1017	15	10000
1971	949	1651	1349	20	10000
3861	1476	2981	2334	30	10000

sw, Shapiro–Wilk; ks, Kolmogorov–Smirnov; ad, Anderson–Darling; cm, Cramer–von Mises. Power is cell entry divided by count.

Table 3.8 False Positive Rate of Tests for Normally Distributed Samples of Sizes 4–30

Signif_sw	Signif_ks	Signif_ad	Signif_cm	Size	Count
519	573	477	374	4	10000
449	462	447	369	5	10000
444	474	441	410	6	10000
534	505	520	500	8	10000
471	508	483	458	10	10000
505	500	516	509	15	10000
495	492	506	504	20	10000
516	476	497	497	30	10000

sw, Shapiro–Wilk; ks, Kolmogorov–Smirnov; ad, Anderson–Darling; cm, Cramer–von Mises. Power is cell entry divided by count.

In Tables 3.5–3.8, count is the number of simulated datasets of the indicated size from the indicated distribution, size refers to the sample size simulated in each of the *count* samples, and signif_xx is the number of significant tests (p=0.05) for the test xx. Here xx=sw for Shapiro–Wilk, ad for Anderson–Darling, cm for Cramer–von Mises, and ks for Kolmogorov–Smirnov. For each sample size shown, 10000 random samples from the indicated, distributed were simulated. Thus, signif_sw is the number of significant Shapiro–Wilk tests out of the *count* tests done. For example, the power of the Shapiro–Wilk test to detect the non-normality of a sample of size 4 from a Cauchy distribution is 2127/10000 or 0.2127. Normality should be judged from residuals from a simple ANOVA of replicate means with the treatment group as the sole explanatory variable. (Alternatively, if a nested variance structure is used, then normality is assessed from residuals from an ANOVA of individual subject values using dose as the sole explanatory variable and subject(rep(dose)) and Rep(Dose) as the random effects.) Thus, if there are five treatment groups plus control and four replicates per treatment, that generates an n=24, and the Shapiro–Wilk, Anderson–Darling, and Cramer–von Mises tests have good power to detect non-normality if the true distribution is Cauchy or exponential but poor power if the true distribution is uniform. If there are only two reps per treatment, then n=12, and the tests all have low power. Since these alternative distributions are only three possible alternatives among many, there is little reason to be confident of detecting non-normality. The false positive rate is close to 5% in all cases (see Table 3.8), so that if a test indicates non-normality, one can have some confidence that the true distribution is not normal.

3.3.7 Power to Detect Lack of Agreement with Other Distributions

Testing consistency with lognormality is simple: one merely needs to do a log-transform of the data and then test for normality. While the normal distribution is important and many natural processes lead to normally or lognormally distributed data, not all natural processes follow one of these distributions. The normal distribution has been much more thoroughly investigated than the others.

There are alternative approaches to assess goodness of fit for distributions other than normal. For some specific distributions, such as the exponential and logistic, there are specific tests, a situation similar to that for the normal. Fortunately, it is not necessary to develop new tests for agreement with every distribution of possible interest in ecotoxicity studies. Indeed, if X is a random variable with distribution $F(x)$, then the random variable $F(X)$ has a uniform distribution. It is thus possible to investigate the appropriateness of the distribution F for a given sample by investigating whether $F(X)$ is uniformly distributed. Furthermore, the data can be normalized by applying the inverse normal distribution to the uniform just indicated and then test that for normality. This is discussed in detail in D'Agostino and Stephens (1986), chapter 6, where the authors recommend the Shapiro–Wilk or Anderson–Darling tests on the transformed normal values.

Now the transformation approach requires estimating the parameters of the assumed distribution, usually by maximum likelihood (ML) methods, to obtain the presumed uniform $U(0,1)$ sample. If a standard normal inverse function is applied to the $U(0,1)$ sample so obtained, there is no need to estimate the mean and standard deviation of the sample from the normal transform. It will be a standard normal and one would test for that, rather than for normality in general. However, as noted in D'Agostino and Stephens (1986), chapter 9, knowledge of the mean and standard deviation of this normal have little bearing on the power of these tests. Thus, one can apply the Shapiro–Wilk or Anderson–Darling test in the usual way, just as if one did not know these parameters.

3.3.8 Assessing Variance Homogeneity

Residual plots (plot of the residuals vs. the predicted values) and formal tests of heterogeneity of variance can be used to find heterogeneity or dismiss it as a problem. Levene's test (Box, 1953; Brown and Forsythe, 1974) is recommended for toxicity studies where there is only one factor (dose), since it is not as sensitive to outliers and other data issues such as non-normality (Box, 1954; Miller, 1968) as some alternatives such as Bartlett's test (Snedecor and Cochran, 1989) or Hartley's F-max test (Hartley, 1950). Levene's test can have high false positive rates for small samples (Miller, 1968) but is reliable for sample sizes of 10 or more (Brown and Forsythe, 1974). Other references on this topic include Box and Cox (1964), Box and Hill (1974), Box and Tidwell (1962), Draper and Cox (1969), Hartley (1950), and O'Brien (1981). The basic idea is to compute the residuals from a one-way ANOVA, calculate the absolute deviation of each residual from the corresponding group mean, and then do an ANOVA on those deviations. More precisely, given k treatment groups and a control with variances $\sigma_0^2, \sigma_1^2, \sigma_2^2, \sigma_3^2, \ldots, \sigma_k^2$, and random samples y_{ij}, $i=0$ to k, $j=1$ to n_i, from those $k+1$ populations, to test the hypothesis H_0: $\sigma_0^2 = \sigma_1^2 = \sigma_2^2 = \sigma_3^2 = \cdots = \sigma_k^2$ vs. H_1: $\sigma_i^2 \neq \sigma_j^2$ for some i and j, Levene's test statistic is given by

$$F = \frac{n.-k-1}{k} \frac{\sum_{i=0}^{k} n_i \left(L_i - L\right)^2}{\sum_{i=0}^{k}\sum_{j=1}^{n_i}\left(L_{ij} - L_i\right)^2}, \tag{3.5}$$

where

$$L_{ij} = \left| y_{ij} - \tilde{y}_{i.} \right|$$

and $\tilde{y}_{i.}$ is the median value of y in the ith sample.

Alternatives define $\tilde{y}_{i.}$ to be the mean or trimmed mean of the ith sample, but the median provides a test statistic robust against many distributions with reasonable power properties. If the underlying populations are normally distributed, then H_0 is rejected in favor of H_1 with $100(1-\alpha)\%$ confidence if $F > F_{(1-\alpha, k, n.-k-1)}$, the upper α-tail of the F-distribution with degrees of freedom k and $n.-k-1$.

When heterogeneity of variances appears to be due to real biological heterogeneity in responses among individual organisms, then more care must be taken. For example, when the organisms (or experimental units) consist of two distinct subpopulations, one responding, and the other not, an estimated change in mean response has no useful meaning unless the subpopulations can be distinguished from known causes, such as sex. In that event, the appropriate way to proceed is to analyze the subpopulations separately, or using the subpopulation identity or value as a covariate or use a multifactor ANOVA taking the subpopulations into account.

Heterogeneity of variances may be a matter of scaling that can be removed by the right transformation. Another approach is to omit the transformation and use methods that are directly based on the assumed distribution (i.e. generalized linear models). If no such distribution or transformation can be found or is readily available, then nonparametric procedures can be used.

3.3.9 Transformation of Data

Many common parametric methods assume normality and variance homogeneity, and these assumptions are not always met. When these requirements are not satisfied, results from these common methods can be unreliable. If a normalizing, variance-stabilizing transformation can be found, then the statistical analysis can be done on the transformed data. This is not problematic for the NOEC approach but requires great care when regression analysis is done. That will be discussed further in Chapter 4.

In some cases, there is a theoretical distribution of the response variable, and this can guide the choice of a transformation. For example, if the underlying distribution is Poisson, the square-root transformation,

$$y_i' = \left(y_i^{1/2}\right) \text{ or } y_i' = \left[\left(y_i+1\right)^{1/2}\right] \text{ or } y_i' = \left(y_i+1\right)^{1/2} + \left(y_i^{1/2}\right),$$

is used. Indeed, the square-root transform is often useful for count data even if a Poisson distribution is not assumed. When the underlying data are lognormal, the log-transformation, $y_i' = \log(y_i)$, is useful. However, if the data contains very small numbers or zero as well as larger numbers, it is preferable to use $y_i' = \log(1+y_i)$ to avoid skewing the transformed data to exaggerate the contribution of the small numbers: Consider log(0.001) for example. For proportions, the underlying distribution is generally binomial and an arc-sin square-root transformation $[y_i' = \arcsin(y_i^{1/2})]$ or Freeman–Tukey transformation (Eq. 3.6) is often used:

$$y_i' = \arcsin\left(\sqrt{\frac{n_i y_i + 1}{n_i}}\right) + \arcsin\left(\sqrt{\frac{n_i y_i + 1}{n_i + 1}}\right). \tag{3.6}$$

If the underlying theoretical distribution is unknown, a data-based procedure (Box–Cox transformation) can be used (Box and Cox, 1964), though this can present challenges to the nonstatistical scientists trying to understand and interpret the results.

The use of transformations will often simplify the analysis since more familiar and traditional data analysis methods can be used, but care must be taken in interpreting the results. Also, there have been many advances in the treatment of non-normal data in recent years and generalized linear mixed models (GLMM) that offer alternative error structures can provide conceptually more appealing approaches to analysis that work well with regression as well as for NOEC/LOEC. See, for example, Bolker et al. (2008) and also Section 3.8.

If a transformation is used, then means and confidence bounds should be back-transformed when reporting results. Back-transformed means generally differ from the arithmetic means of the original data. Back-transformed means should be interpreted as estimates of the median of the untransformed data distribution if the transformed data are symmetrically distributed. In particular, the back-transformed mean after a log-transformation is the geometric mean of the untransformed data, and this value estimates the median of the underlying lognormal distribution.

When a transformation is not used in the data analysis, the difference in the means is a measure for the size of an effect. The back-transformed difference in means (of the transformed data) is often more challenging to interpret. In the case of a log-transformation, the difference estimates the ratio of the median responses.

Transformations may not maintain additivity of effects in more complex experiments, for example, interactions among factors, such as test substance, sex, and age. Transformation of the response especially complicates regression methods to estimate an ECx, since an x% change in the untransformed response does not correspond to an x% change in the transformed response. Great care must be taken to estimate a meaningful quantity after a transform. This is less of a problem with hypothesis testing, since we are only inquiring whether a response mean or median (or other measure of location) differs significantly from the control. It is not necessary to quantify the amount of change to determine the NOEC. Additional discussion of transformations is given in OECD (2006b) to which the present authors contributed and on which some of this discussion is based.

Not all data problems can be fixed by transformation of the response. For example, if a large percentage of the responses have the same measured value (ties), no transformation will address that issue. In the case of very small samples, it may not be possible to assess in a meaningful way the distribution of responses (see Section 3.3.6). One can either use an assumed distribution based on some natural philosophy or past experience or treat this situation in the same way as data with massive ties. Exact permutation methodology may be an appropriate tool to use in these cases. (See Good, 2005, and references therein.)

3.3.10 Outliers and Influential Observations

An observation is labeled an outlier if it deviates in a noteworthy way from other observations made under similar circumstances. Some outliers arise from an experimental condition that is not accounted for in the statistical model that is used to describe the data. For instance, several organisms might be housed in the same vessel but one of them contained an infection. Alternatively, a male might be mistakenly housed in a vessel that was intended for female subject or a sterile female might be housed undetected and unexpectedly with fertile females in a reproduction study. Perhaps the subjects in one replicate vessel were fed or sacrificed at a different time than those in other replicates. Another possibility is simple data entry error. In one amusing case dealing with rat organs, the brain weight of one subject was recorded as zero. Describing this to the study director as an outlier was a true "no-brainer." Detection of this type of outliers typically cannot be enforced by any formal statistical method, and one has to rely on careful planning, visual inspection, judgment, and experience.

Obviously, experimental unit or treatment group outliers are highly undesirable, since they directly interfere with the effect that one wishes to measure, thereby increasing the probability of both false positive and false negative results. For example, an NOEC may be assessed at a level where substantial effects do occur or an LOEC may be assessed at a level without real effects from the treatment under investigation. The best way to reduce the chance of an outlier at the group or replicate level is a design that is well randomized with respect to all experimental actions that may potentially influence the (observed response of) biological system. In practical biological studies, however, perfect randomization is hard to realize. Therefore, it is important to make the study design relatively insensitive to potential outliers, i.e. by randomized replicated dose groups or by increasing the number of treatments, replicates, or subjects per replicate.

Outliers at the individual level can be detected in continuous data. When a particular distribution is assumed for the scatter in the data, the judgment of outliers may be based on a specific, small probability that any single data point could occur. This implies that the judgment of outliers can depend quite strongly on the assumed distribution. For example, values that appear to be extremely high when assuming a normal distribution may be judged as non-extreme when assuming a lognormal distribution. On the other hand, low values may be judged as extremes when assuming a lognormal distribution, but not so when a normal distribution is assumed.

The statistical analysis of the data is sensitive to individual outliers, although often less dramatically than to group outliers. On the one hand, individual outliers may result in biased estimates of the effect (either too small or too large). On the other hand, the estimate of the residual variance (the "noise") will be increased, implying that statistical tests tend to be less powerful, and estimated parameters (e.g. EC*x*) less precise. Therefore, if biological reasons can be found explaining the outliers, it is favorable to delete them from the analysis. As a reality check, one can also repeat the analysis with the outliers omitted. If this does not change the conclusions, one may have greater confidence in the result if no biological justification for omitting the outliers is found. This is a helpful step before expending great effort to find a reason for omitting an observation.

In addition to plots and visual inspection, there are formal outlier rules that are useful, particularly when a large volume of data must be processed in a short period of time. The Tukey (1977) outlier rule is practical, useful, and available in numerous statistical software packages. Furthermore, it is not itself sensitive to the effects outliers, which make it a decided improvement over methods based on standard deviations, which are quite sensitive to the effects of outliers. Tukey's outlier rule can be used as a formal test with outliers being assessed from residuals (results of subtracting treatment means from individual values) to avoid confounding outliers and treatment effects. This rule will be described in Section 3.3.11.

3.3.11 Definition of Percentile and Quantile

For any number x in the interval $0 \le x \le 100$, the xth percentile, Px, of a finite dataset, is defined as a number that equals or exceeds at least $x\%$ of the data and is equal to or exceeded by at least $(100-x)\%$ of the data. More generally, for any distribution F, the Qth percentile of the F-distribution is a number Q satisfying $x/100 \le F(Q) \le (1-x/100)$. For example, suppose the data consists of the 10 numbers 0, 1, 3, 5, 10, 12, 12, 15, 20, and 50. Now 25% of 10 is 2.5, so $P25$ must equal or exceed at least three observations and equal or be exceeded by at least eight observations. The number 3 meets these conditions since 3 equals or exceeds 1, 2, and 3, which is more than 25% of the data and 3 equals or is exceeded by 3, 5, 10, 12, 12, 15, 20, and 50, which is more than 75% of the data. $P25$ is unique for this dataset. Similarly, $P75=15$. Note that every number in the interval $10<x<12$ satisfies the definition of $P50$, so it is not unique. Where there are multiple values that satisfy the conditions of Px (and this is often true), we usually report the midpoint of all such values. We make no distinction between the meanings of the terms percentile and quantile. $P25$ and $P75$ are also referred to as the first and third quartiles, labeled $Q1$ and $Q3$, respectively, and the difference $Q3-Q1$ is called the interquartile range (IQR).

3.3.11.1 Tukey Outlier Rule (Named After John W. Tukey. See Tukey, 1977)

Any response more than 1.5 times the IQR above the third quartile (75th percentile) or below the first quartile (25th percentile) is considered an outlier by Tukey's rule.

The IQR of a dataset is a measure of the spread or variability of the data. Finally, Tukey constructs a "fence" around the middle 50% of the data by computing $Q1-1.5\,\text{IQR}$ and $Q3+1.5\,\text{IQR}$. Any observation outside this interval or "fence" is considered an outlier by the Tukey test. This is preferable to outlier rules that depend on standard deviations (such as any observation more than 2.5 standard deviations away from the mean), since the standard deviation is itself highly influenced by outliers, whereas the fence in the Tukey rule is not much affected by outliers. This rule is motivated by the properties of a normal distribution, though it applies to any distribution. In a standard normal distribution, $Q1=-0.67449$, $Q3=0.67449$, and $\text{IQR}=1.34898$, and the probability of a random observation from a standard normal being less than $Q1-1.5\,\text{IQR}$ is 0.00349. Thus, in a normal distribution, less than 0.7% of randomly selected data will be classified as outliers. Such outliers should be reported with the results of the analysis.

3.3.12 Purpose of Outlier Identification

The identification of outliers can be helpful in determining how to interpret data. As noted above, in a normally distributed dataset, only a tiny fraction of the data is expected to fall outside the "Tukey fence." If a substantial portion of a given dataset is outside this range, the dataset may not be normally distributed or there may be as yet unexplained factors that should be included in the model that account for the unusual data. Even one or a small number of outliers can affect the conclusions drawn from the data. The question naturally arises, as to what to do once an outlier has been identified.

Should we also eliminate the observation identified as an outlier? Unless there is sound biological or physical reason for omitting an observation, it is highly inadvisable to do that and it is also not generally permissible under regulatory guidelines. We strongly advise against any automatic elimination of data from analysis. Outliers may contain important information that a thorough examination might illuminate. In any case, these are the data collected and they should not be discarded without reason. It may be nonetheless useful and interesting to reanalyze the data with the outlier(s) omitted. If the conclusions are unchanged (e.g. the same NOEC is obtained), or changed in only a modest way (e.g. EC10 and its confidence bounds changes <10%), then one might feel comfortable with the analysis of the full dataset. Alternatively, if a dataset is found not to be normally distributed or to have heterogeneous variances, but the omission of one or two outliers makes the data normally distributed with homogeneous variances and the conclusions are otherwise little affected, then the use of a model on the full dataset that assumes normality and variance homogeneity may be justified. Conversely, if the omission of outliers changes the conclusions in a nontrivial way (e.g. different NOEC or greatly altered ECx value), then great care should be taken in interpreting the data. If there is no convincing biological or physical explanation that justifies omission of the outlier, then an alternative method of analysis should be considered, such as a nonparametric model or a model with a different error structure. While it is true that nonparametric analyses are less sensitive to outliers than parametric analyses, omission of outliers can still change conclusions, especially when sample sizes are small or outliers are numerous.

Outliers can often be found from scatterplots of the data, including scatterplots of residuals. However, it must be recognized that residual plots can mask outliers. This can happen when a single observation is so different from the others that the mean for the corresponding treatment group is pulled toward the outlier, generating apparent outliers in the residuals of the other observations from that group. See Table 3.9 for an example dataset.

It is useful to plot residuals from a simple one-way ANOVA with a horizontal reference line at height 0 (see Figure 3.6). Tukey's outlier rule identified all observations in the high dose as outliers. Examination of the plot suggests that there is only one outlier, the observation

20 for rep 5 in dose 6. It is left as an exercise to show that the full data are found non-normal by the Shapiro–Wilk test and no treatment group mean response is significantly different from the control mean by the Dunn, Dunnett, Williams, or Jonckheere–Terpstra test. If rep 5 of dose 6 is omitted, then the two trend tests (Willams and Jonckheere–Terpstra) set the NOEC to be dose 4. There is no effect except in the high dose group, so regression modeling is challenging.

Table 3.9 Outlier Example Data

Dose	Rep	y	Dose	Rep	y
1	1	9.95	4	1	10.03
1	2	9.85	4	2	9.96
1	3	9.61	4	3	10.52
1	4	9.85	4	4	10.20
1	5	10.16	4	5	9.76
2	1	10.64	5	1	9.92
2	2	10.54	5	2	10.14
2	3	9.96	5	3	9.63
2	4	10.22	5	4	9.58
2	5	10.73	5	5	9.97
3	1	10.36	6	1	8.73
3	2	10.20	6	2	9.81
3	3	10.26	6	3	9.86
3	4	9.77	6	4	9.43
3	5	10.17	6	5	20.00

Rep 5 in dose 6 is an outlier.

3.4 PAIRWISE TESTS WHEN DATA DO NOT MEET NORMALITY OR VARIANCE HOMOGENEITY REQUIREMENTS

Sometimes data do not meet the normality and variance homogeneity requirements for the t-test or Dunnett test and no normalizing, variance-stabilizing transform can be found. The statistician is then forced to ignore these concerns or seek an alternative method of analysis. Ignoring the distribution of the data can lead to incorrect evaluation and, in any case, puts conclusions on shaky statistical grounds. Numerous examples are given in this book to illustrate this. Fortunately, there are alternatives in such situations.

In the context of continuous data, a rank-order transform is sometimes helpful when all else fails. The Mann–Whitney, Dunn, and Jonckheere–Terpstra tests, discussed below, are all based in essence on a rank-order transform, but these tests do not assume that normality is achieved, except asymptotically. An alternative is to apply Dunnett's or Williams' or some ANOVA test to the rank-order transformed data. This may not be optimal and caution should be employed.

Several publications (Akritas, 1990; Akritas et al., 1997; Bathke and Lankowski, 2005) present simulation studies to show that the rank-order transform can inflate the false positive rate for repeated measures and some other multifactor experimental designs, though others reach different conclusions (e.g. Danbaba, 2012, and references therein). This is not an issue for a one-factor experiment analyzed at a single point in time, which is by far the most common experimental design in laboratory ecotoxicology

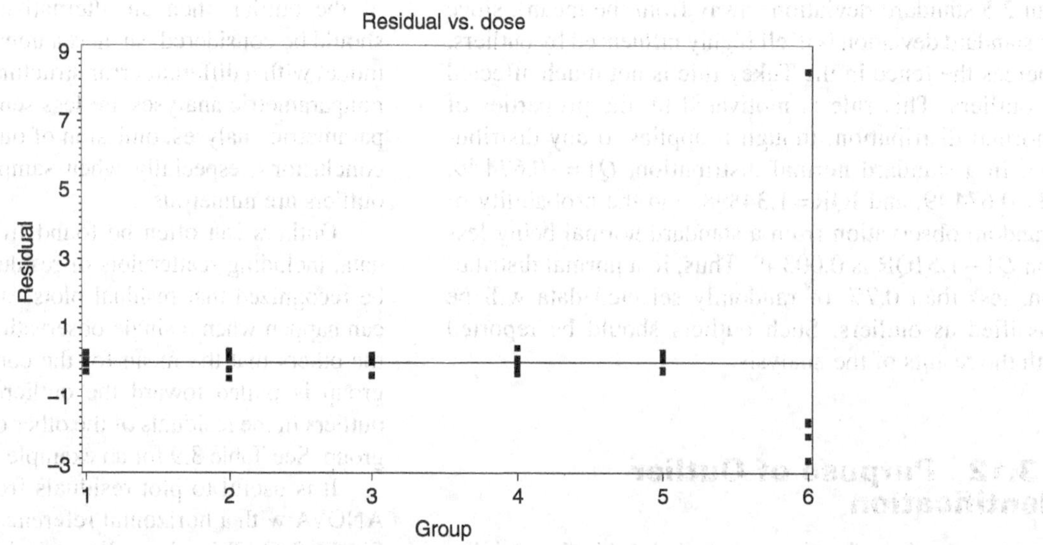

Figure 3.6 Scatterplot of residuals from outlier data Dots is observations. Solid line is a mean residual, which is always zero. High residual in treatment 6 is an apparent outlier.

studies. The rank transform is well validated and remains appropriate for those studies.

It is also advisable to consider models that do not require normality or variance homogeneity. Three such pairwise tests will now be described. Later sections and chapters will describe additional methods of analysis of such data.

3.4.1 Tamhane–Dunnett Test

When the data have heterogeneous variances but are normally distributed within each treatment group, then the Tamhane–Dunnett test is a valuable tool for analysis. This test assumes normally distributed data but is optimized for variance heterogeneity. There is some power loss for homoscedastic data, so it is not recommended as a universal replacement for Dunnett's test. It was developed in a series of papers: Dunnett (1980), Tamhane (1979), Dunnett and Tamhane (1991, 1992, 1995, 1998), Tamhane and Dunnett (1996) and is described in Hochberg and Tamhane (2009) but labeled there as the T3 test. The basic statistic is the same as in Welch's unequal variance t-test, but it uses a different critical value that accounts for the multiple correlated comparisons in a manner similar to the way Dunnett's critical values replace those from a t-distribution in evaluating a simple t-test:

$$T_i = \frac{\bar{x}_i - \bar{x}_0}{\sqrt{\dfrac{s_i^2}{n_i} + \dfrac{s_0^2}{n_0}}}. \tag{3.7}$$

Here, n_i is the number of replicates in the ith group. For a two-sided test, or to construct confidence intervals for the difference in means, critical values come from the Studentized maximum modulus distribution for k comparisons and degrees of freedom calculated using the Satterthwaite formula (3.8), where fractions are replaced by the least integer exceeding the computed value. For a one-sided test, T_i is compared with the Studentized maximum distribution for the same k and df. These distributions are given in Appendix 3.

$$df(\text{Satterthwaite}) = \frac{\left(\dfrac{s_i^2}{n_i} + \dfrac{s_0^2}{n_0}\right)^2}{\dfrac{s_i^4}{n_i^2(n_i - 1)} + \dfrac{s_0^4}{n_0^2(n_0 - 1)}}. \tag{3.8}$$

If there are subgroups in the treatment groups, the sample variances in the above formula are replaced by Satterthwaite-type expressions. This can be used to account for different sample sizes in the replicates, for example, from differential mortality or from some other experimental design feature:

$$\text{Let } C_g = \sum_j \frac{n_{gj}^2}{n_g^2}, \tag{3.9}$$

where n_{gj} is the number of subjects in the jth subgroup of treatment group g, $g=0$ or i, and n_g is the total number of subjects in group g. Let

$$U_g = \frac{\text{VWR}_g}{n_g} + C_g{}^*\text{VBR}_g, \tag{3.10}$$

where VWR_g and VBR_g are the within- and between-rep variances, respectively, of treatment group g. In the expression for T_i in Eq. (3.7), substitute U_i, U_0 for s_i^2, s_0^2. Then the approximate degrees of freedom for this variance estimate is

$$df_i = \frac{(U_0 + U_i)^2}{\dfrac{U_0^2}{n_0 - 1} + \dfrac{U_i^2}{n_i - 1}}. \tag{3.11}$$

Tables for the Studentized maximum and maximum modulus distributions are provided in Appendix 3 for convenience and are also found in Tamhane (2009), Hochberg and Tamhane (2009), and Stoline and Ury (1979). Alternatively, tables for these distributions can be computed using the following formulas, which were used to construct Tables A3.1 and A3.2:

Maximum Modulus Distribution

$$\Pr\{\max |T_i| \le t\} = \int_0^\infty \left[2\Phi(tx) - 1\right]^k \mathrm{d}F_\nu(x). \tag{3.12}$$

Studentized Maximum Distribution

$$\Pr\{\max T_i \le t\} = \int_0^\infty \left[\Phi(tx)\right]^k \mathrm{d}F_\nu(x). \tag{3.13}$$

In Eqs. (3.12) and (3.13), k is the number of treatment groups excluding the control and ν is the degree of freedom from Eq. (3.11).

This test assumes within-group (or subgroup) normality and makes no use of any monotone dose–response that usually exists in an ecotoxicity study and, thus, loses power and some interpretability in comparison with tests that do.

Dunnett (1980) and Hochberg and Tamhane (2009) state that the power of this test is very similar to that of Dunnett's test when the assumptions underlying that test are met. However, that depends on the sample sizes used in the comparison. Given that analysis of ecotoxicology data is on replicate means or medians and there are usually no more than four replicates per treatment, a power loss from the reduced degrees of freedom in this test compared with the degrees of freedom in the pooled variance estimate used in Dunnett's test, it is recommended that Dunnett's test be used when the requirements for it are met. The power for the unequal variance case will depend on the particular pattern of variances and experience suggests substantially lower power can occur compared with the Jonckheere–Terpstra trend test, which is a nonparametric test described in Section 3.5.2 that is robust against moderate heterogeneity.

3.4.1.1 Tamhane–Dunnett Example with Collembola Reproduction

A study was done following OECD TG 232 on the *Collembola Folsomia candida* exposed to a test chemical for 28 days. Ten adult *Collembola* were placed in each replicate test chamber containing soil of an approved kind, with four replicates in each test concentration and eight in the control. The objective of the study was to determine whether the test substance reduced reproductive output. The response was the number of juveniles produced. This is a count variable with a wide range of possible values, so an ANOVA-type analysis following a square-root transform is a reasonable approach. An alternative analysis using a Poisson error structure and no transform will be described in Section 3.8. Summary data are presented below (see Table 3.10). The full dataset is called *Collembola* and is given in Appendix 1.

The residuals from ANOVA are consistent with normality by the Shapiro–Wilk test, but Levene's test found significant heterogeneity. It can be seen from a plot of the *Collembola* data (see Figure 3.7) that the variability at doses 2 and 3 is less than in the other dose groups, especially dose 4. Accordingly, the Tamhane–Dunnett test was used. That test found significant decreases in mean reproductive output at 62.5, 31.3, and 7.81 mg l^{-1} but not at 15.6 mg l^{-1}, so the NOEC was 3.91 mg l^{-1}. As an example calculation, the 95% confidence interval for the difference between the mean response at 15.6 mg l^{-1} and the control is given by

$$\hat{\mu}_3 - \hat{\mu}_0 = \bar{y}_3 - \bar{y}_0 \pm T_3\sqrt{\frac{s_3^2}{n_3} + \frac{s_0^2}{n_0}},$$

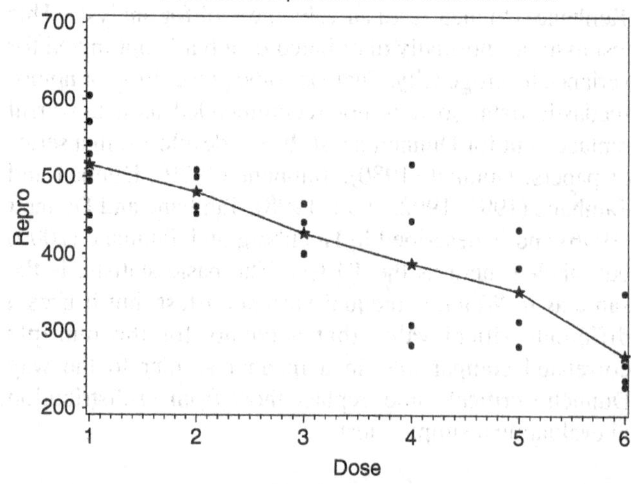

Figure 3.7 Collembola reproduction. Reproduction in individual test chambers shown as dots. Treatment means are shown as stars and connected by straight line segments. Variance in doses 2 and 3 is noticeably less than in the control and especially in dose 4. Treatment groups (doses) are plotted by order rather than concentration. Downward concentration–response trend is very clear.

Table 3.10 Collembola Summary Data

Group	Conc	Rep	Repro	Survive	Group	Conc	Rep	Repro	Survive
1	0	1	449	10	3	7.81	3	401	8
1	0	2	430	10	3	7.81	4	441	8
1	0	3	543	9	4	15.6	1	443	9
1	0	4	494	9	4	15.6	2	517	8
1	0	5	605	10	4	15.6	3	308	10
1	0	6	571	9	4	15.6	4	282	7
1	0	7	529	9	5	31.3	1	313	8
1	0	8	504	8	5	31.3	2	383	9
2	3.91	1	500	9	5	31.3	3	432	9
2	3.91	2	461	8	5	31.3	4	281	7
2	3.91	3	452	9	6	62.5	1	237	7
2	3.91	4	509	10	6	62.5	2	229	7
3	7.81	1	465	9	6	62.5	3	350	9
3	7.81	2	400	9	6	62.5	4	257	6

Repro, number of live young; Survive, number of adults that survive.

where $s_0=58.987$, $n_0=8$, $s_3=111.506$, $n_3=4$, so $df_3=4$, $T_3=4.203$, and $\hat{\mu}_3-\hat{\mu}_0=128.125\pm250.187$.

3.4.2 Dunn's Test

Dunn's test is often described in texts as a way of comparing all possible pairs of treatment groups, but her original paper (Dunn, 1964) provided a means of estimating any number of general contrasts and adjusting for the number of contrasts estimated. This is quite useful for handling data such as in the multiple control example in Section 3.2.3 for the *t*-test. For present purposes, only comparisons of treatments with a common control will be described. Begin by ranking all the data from the control and combined treatment groups by using the mean rank for tied responses. Next, from the ranking of all data, compute the mean rank, R_i, for each treatment group and the total sample size, N, and the sample sizes n_i for each group. Finally, compute the variance of R_i-R_0 using Eq. (3.14):

$$V_{i0}=\left[\frac{N(N+1)}{12}-\frac{\sum\left(t^3-t\right)}{12(N-1)}\right]\left[\frac{1}{n_i}+\frac{1}{n_0}\right], \quad (3.14)$$

where the sum is over all distinct responses and t is the number of observations tied at that response. The test statistic is given by Eq. (3.15):

$$Z_i=\frac{R_i-R_0}{\sqrt{V_{i0}}}. \quad (3.15)$$

Z is compared with a standard normal distribution at critical levels adjusted for the number of comparisons made. Under the Bonferroni adjustment, the critical level is $p=1-\alpha/k$, where k is the number of comparisons to control and α is 0.05 or 0.025, as a one- or two-sided test is used. The Bonferroni–Holm adjustment is less conservative.

Dunn's test is based on ranks and is robust against a wide variety of distributions and heteroscedasticity. It allows arbitrary contrasts to be tested, not just comparisons of treatments with control. This test does not permit modeling multiple sources of variances (e.g. within and between subgroups). There is no exact permutation counterpart to this test, so it has low power for very small samples or experiments with massive ties among the response values.

Published power properties of Dunn's test are limited. Dunn (1964) describes an extreme situation, where all but one of k population means equal the control, but that every observation in that treatment group exceeds every observation in the control. The power of her two-sided test (for large samples) to detect this difference is approximately

$$1-\Phi\left[\frac{z_{1-\alpha/2k}\sqrt{\frac{k(kn+1)}{6}}-kn/2}{\sqrt{\frac{\left[(k-1)n+1\right][k-2]}{12}}}\right], \quad (3.16)$$

where n is the common within-group sample size. For a one-sided test, the divisor $2k$ in the subscript of z is replaced by k. The power of Dunn's test depends on the specific alternative, as is the case for other procedures. While the above provides an approximate power that can be used for design assessment purposes, an SAS program, *simjtdunit.sas*, is provided in Appendix 1 as an example program to compute power for several of the tests discussed in this chapter, including Dunn's, that is specific to the sample sizes and study designs used in some regulatory guideline studies. Another test that could have been used for these data is the Steel–Dwass test (Hochberg and Tamhane, 2009). This is similar in concept to Dunn's test but requires equal sample sizes. Given that mortality in ecotoxicity tests is common, the utility of Steel–Dwass is limited and will not be pursued here.

3.4.2.1 Dunn Test Example with Algae Density

Example cell density data are available from a study with the algae *Anabaena flos-aquae* after 48 h exposure to a test compound (see Table 3.11). There were four replicate test vessels in the control and each test concentration. Cell density is a count variable, but the range of possible values is large, and analyzing the data as a continuous response, possibly after a square-root transform, is reasonable. However, the transformed data were still not consistent with a normal distribution or variance homogeneity. The nonparametric Dunn test was used to determine the NOEC. There were no ties, so Eq. (3.14) simplifies. It is straightforward to apply Eqs. (3.14) and (3.15) to verify that $V_{i0}=5.8166$ for each group, $Z_1=-0.8166$, $Z_2=-0.4722$, $Z_3=-1.805$, $Z_4=-2.4928$, $Z_5=-3.1805$, and $Z_6=-3.868$. If the Bonferroni adjustment for multiple comparisons is used, then these Z_i are compared with the critical standard normal value C_p, for $p=0.05/6=0.00833$ for a one-sided test against a decreasing alternative. So $C_p=-2.394$ and the calculated Z-values of only the three highest test concentrations exceed C_p, making the NOEC 0.26 mg kg^{-1}. When the Bonferroni–Holm adjustment is used, the critical values are −1.96, −1.645, −2.128, −2.241, −2.326, and −2.394 for the Z-values in order from Z_1 to Z_6. The NOEC does not change. The details are left as an exercise.

An alternative GLMM analysis treating density as a Poisson variable is presented in 3.8. That analysis also found the NOEC to be 0.26 mg l^{-1}. These data were also analyzed using the step-down Jonckheere–Terpstra test, and that test, reported in Table 3.15, set the NOEC as 0.1 mg l^{-1}.

Table 3.11 Algae Density 48 h After Treatment

Group	Rep	Conc	Density	Rank	Sum	Group	Rep	Conc	Density	Rank	Sum
0	A	0	742 104	28	100	3	C	0.26	284 181	14	
0	B	0	708 884	27		3	D	0.26	295 655	15	
0	C	0	549 377	26		4	A	0.64	133 218	12	42
0	D	0	453 174	19		4	B	0.64	131 304	11	
1	A	0.041	450 007	18	81	4	C	0.64	126 781	9	
1	B	0.041	481 867	22		4	D	0.64	130 290	10	
1	C	0.041	473 340	21		5	A	1.6	73 238	6	26
1	D	0.041	463 705	20		5	B	1.6	73 245	7	
2	A	0.1	528 456	25	89	5	C	1.6	68 884	5	
2	B	0.1	504 289	23		5	D	1.6	74 047	8	
2	C	0.1	509 052	24		6	A	4	44 658	4	10
2	D	0.1	442 682	17		6	B	4	44 304	2	
3	A	0.26	243 353	13	58	6	C	4	42 933	1	
3	B	0.26	297 722	16		6	D	4	44 363	3	

Rank, rank-order of replicate density value; Sum, sum of ranks in group.

3.4.3 Mann–Whitney (or Wilcoxon) Test

Similar to Dunn's test, the Mann–Whitney rank sum test compares the ranks of measurements in two independent random samples of n_1 and n_2 observations. Unlike Dunn's test, the Mann–Whitney test uses only the information from those two groups. This has the effect of reducing the power and underutilizing the full experimental dataset. However, according to Lehmann (2006), if the ranks used in the test are computed from the combined samples from all groups, then the power properties are improved.

To use the Mann–Whitney rank sum test, first rank all (n_1+n_2) observations, with tied observations assigned the average of the ranks of the tied observations. Call the sum the ranks of the observations in the ith group T_i. If the distributions of ranks in the two groups are identical, then T_1 and T_2 are identical. Otherwise, the difference between T_1 and T_2 is a measure of how different these distributions are. There are one-tailed and two-tailed versions of the test, as well as small sample and large sample (asymptotic) versions.

For a one-tailed test, the test statistic is T_1 if $n_1 < n_2$ or T_2 if $n_2 \le n_1$. The rejection region is $T_1 \ge T_U$, if T_1 is test statistic; or $T_2 \le T_L$, if T_1 is test statistic. The critical values, T_U and T_L, can be obtained from tables available in many statistics texts. SAS and R programs, MWTEST.SAS and MWtest.R, are provided in Appendix 1.

A normal approximation to the Mann–Whitney test statistic has the form

$$Z = \frac{T - \mathrm{Exp}(T)}{\sqrt{\mathrm{Var}(T)}},$$

where T is the smaller of the sums of the ranks of the two samples.

Elementary calculations show that $\mathrm{Exp}(T) = n_1(n_1 + n_2 + 1)/2$ and $\mathrm{Var}(T) = n_1 n_2(n_1 + n_2 + 1)/12$, so

$$z = \frac{T - n_1(n_1 + n_2 + 1)/2}{\sqrt{n_1 n_2 (n_1 + n_2 + 1)/12}}. \tag{3.17}$$

This normal approximation is generally good if $n_1 \ge 10$ and $n_2 \ge 10$, though it is often used for smaller sample sizes. Sheskin (2011) recommends adding a continuity correction factor of 0.5 to the numerator of Eq. (3.17) for small samples.

If more than one comparison is to be made, then Bonferroni–Holm adjustment is made to the p-values. The assumptions of the Mann–Whitney rank sum test are only that samples must be independent, and distributions of values must be symmetrical. The Mann–Whitney is nearly as efficient as the t-test when data are normally distributed with equal variances and is generally more efficient when the test data fail to meet the assumptions for the t-test.

The Mann–Whitney test is based on ranks and is robust against a wide variety of distributions and heteroscedasticity. It has low power for small sample sizes such as those that are common in ecotoxicity studies, but there is also an exact permutation version of this test for small samples, which has better power properties.

The test does not permit modeling multiple sources of variances (e.g. within and between subgroups). The Bonferroni–Holm adjustment for multiple comparisons is statistically conservative, whereas an unadjusted multiple comparison procedure has an unacceptable false positive. The previously mentioned computer program, *simjtdunit.sas*,

Table 3.12 Mann–Whitney Calculations

Variable	Dose	N	T	Exp(T)	Var(T)	Z	Probz	Adjp	Prob_BH
MWrank01	1	4	13	18	12	−1.44338	0.074457	0.14891	0.14891
MWrank02	2	4	13	18	12	−1.44338	0.074457	0.07446	0.14891
MWrank03	3	4	10	18	12	−2.3094	0.010461	0.06276	0.06276
MWrank04	4	4	10	18	12	−2.3094	0.010461	0.0523	0.06276
MWrank05	5	4	10	18	12	−2.3094	0.010461	0.04184	0.06276
MWrank06	6	4	10	18	12	−2.3094	0.010461	0.03138	0.06276

N is the number of replicates; T is the sum of ranks for the smaller of the two samples; Exp(T) and Var(T) are the expected value (mean) and variance, respectively, of T, Z is the z-statistic from Eq. (3.17); Prob$z$ is the normal probability associated with the Z-statistic; Adj$p$=min(1,Probz * (7 − Q)), where Q is the rank-order of the probz p-values; and Prob_BH is the Bonferroni–Holm adjusted p-value. Note how the tied z-values are handled for the highest four treatment group. Thus, in this example, the Mann–Whitney test gave a lower NOEC than Dunn's test. The opposite is very often true for other datasets. SAS programs, *DunnTest.sas* and *Mann Whitney Macro.sas*, and R code, DunnTest.r and Mann_Whit_Test.r, are given in Appendix 1 to carry out the Dunn and Mann–Whitney test calculations.

provided in Appendix 1 generates power properties of the Mann–Whitney test.

The Mann–Whitney test will be illustrated with the algae data in Table 3.11. Note that the ranks in that table must be recalculated to do the separate Mann–Whitney calculations. See Table 3.12 for partial calculations for the normal approximation for this test. The programs MWTEST.SAS and MWtest.R given in Appendix 1 do not use the normal approximation. It is instructive to compare the results from those programs with the normal approximations in Table 3.12. It should be recalled that $n=4$ for each treatment group, well below the threshold of 10 indicated above.

3.5 TREND TESTS

As discussed in Section 3.2 and in Section 2.2.3.2, trend tests are the preferred type of test for most NOEC determinations, since they take the expected monotone dose–response shape into account and use all the treatment groups in determining the NOEC, not just individual treatment groups separately compared with the control. As already discussed in Section 3.2, not every dataset should be analyzed by a trend model, and it is important to assess the data for consistency with a trend model. It is to be emphasized that the treatment means need not be monotonic to use a trend test, but there should be an overall monotone trend. Usually, a visual inspection of the data, such as that given in Figure 3.1, will often suffice, and it is highly recommended to plot the data in this manner as part of the process for selecting the method of analysis. However, in a high-volume testing facility, where statistically trained personnel do not do all routine statistical analyses, it may be helpful to have formal tests for lack of monotonicity to alert the study director or technician to take a closer look at the data. There are several formal procedures than can be considered. One that has been useful to the authors uses contrasts of treatment means. If $\mu_0, \mu_1, \mu_2, \ldots, \mu_k$ are the population means, a contrast of those means is a linear combination

$$C = \sum_{i=0}^{k} C_i \mu_i, \text{where} \sum_{i=0}^{k} C_i = 0. \quad (3.18)$$

The estimate $\hat{C}$ of the contrast C is obtained by replacing the population means μ_i by the sample means. The variance of this estimator is given by

$$\mathrm{Var}\left(\hat{C}\right) = \sum_{i=0}^{k} C_i^2 \frac{\sigma_i^2}{n_i}, \quad (3.19)$$

where σ_i^2 is the variance of the ith population and n_i is the sample size from that population. The sample variance is obtained by replacing the population variances by the sample variances. Assuming the samples are normally distributed, a test statistic to test whether the contrast is zero is given by

$$T = \frac{\hat{C}}{\sqrt{\mathrm{Var}(\hat{C})}}. \quad (3.20)$$

The test statistic T has a t-distribution with degrees of freedom for the pooled variance estimator in the homogeneous case or by Satterthwaite's formula in the heterogeneous case.

Then define linear and quadratic contrasts of the rank-transformed data across treatment groups, including the control, with coefficients given in Table A3.3. In an ANOVA setting, test whether contrasts are significantly different from zero. Given that we expect a trend, we reject the trend model only if both the quadratic trend is significant and the linear trend is not. This is in keeping with our approach to tests for non-normality, heterogeneity, and overdispersion, where we accept the hypothesis of normality, etc. unless the data compels us otherwise. This test is reasonably robust against non-normality, variance heterogeneity, and outliers, but is by no means foolproof, as a strong quadratic trend might well be monotone but not linear. It is strongly advised to plot the data. Other tests exist, such as Bartholomew's (Robertson et al., 1988) and Brown–Forsythe (Brown and Forsythe, 1974).

3.5.1 Step-down Williams' Test and Parametric Alternatives

For toxicology studies Williams' test is the most commonly used. Williams' test (Williams, 1971, 1972, 1977) assumes the response is normally distributed with homogeneous variances, and the population means are monotonically ordered. Unlike the Jonckheere–Terpstra test discussed in Section 3.3.1, Williams' test formally incorporates the presumed monotone dose–response in the estimated mean effects in each dose. The resulting means are called isotonic estimates and are based on ML theory, given the dose–response is monotone. Isotonic estimators were developed by Ayer et al. (1955), who called their method the pool-the-adjacent-violators algorithm (PAVA). Isotonic regression was introduced by Barlow et al. (1972). This discussion is modified and expanded from material written by Green and Springer for annexes to OECD (2014a) and in OECD (2006a, b).

If the population means are $\mu_0 \le \mu_1 \le \mu_2 \le \cdots \le \mu_k$, then the ML estimate of μ_i under the monotonicity assumption is

$$\tilde{Y}_i = \operatorname*{Max}_{1 \le u \le i} \operatorname*{Min}_{i \le v \le k} \frac{\sum_{j=u}^{v} n_j \bar{Y}_j}{\sum_{j=u}^{v} n_j}, \tag{3.21}$$

where $\bar{Y}_i$ is the arithmetic mean of the replicate values in the ith dose and n_j is the sample size, i.e. the number of replicate values. The roles of Min and Max are reversed for a nonincreasing trend. It is easier to compute isotonic estimators than to describe them. Given that a nondecreasing trend in the means is expected, we look for a violation of this expected result in two adjacent means. If $\bar{Y}_j > \bar{Y}_{j+1}$, then we pool (or amalgamate) these two means using a simple weighted average, weighted by the sample sizes. The resulting reduced set of means is again examined for violations of the expected order and additional amalgamation is done as needed. It makes no difference to the final result which adjacent means we amalgamate first. We continue this amalgamation procedure until the means are in the expected nondecreasing order. Note that the control is never amalgamated with positive dose groups.

Williams' test statistic to compare the mean response at dose i with the control mean response is

$$\bar{t}_i = \frac{\tilde{Y}_i - Y_0}{s\sqrt{\frac{1}{n_i} + \frac{1}{n_0}}}, \tag{3.22}$$

where s^2 is the usual ANOVA pooled within-group variance estimator. The statistic $\bar{t}_i$ is compared with a Williams' critical value $\bar{t}_{i,\alpha}$, which decreases as i increases. The amalgamation and tests can be carried out in some software packages, such as Probmc in SAS and a program to implement this test in given in Appendix 1 with the name DOWILL_DECR_INCR_ALT2.SAS. The corresponding R program was developed by Joe Swintek for this book and is called *WilliamsTest V2.r*. It contains two examples that demonstrate how to use the program. Swintek will add this program to the StatCHARRMS R-package that he developed, StatCHARRMS (2017), in the CRAN library and requires R version 3.3.3 or later. We are indebted to Joe for this valuable program. To use this program and all other R programs provided in Appendix 1, it is necessary to define an R library where the programs and code and the table used by the Williams program reside. Code such as the following inserted at the beginning of the program will do this, where the user has previously created the folder inside the quotation marks. This folder will serve as the referenced R library:

```
setwd("c:/Personal/ecostats book/
R-project/Rlib/")
```

It should be pointed out that the R program for Williams' test cited above uses a table lookup functionality, as there is no built-in computation of the critical values in R so far as we are aware at the time of writing this book. This program agrees well with the SAS and other implementations of the test (CETIS and ToxRat) and with Williams' original papers (Williams, 1972, 1977). To avoid possible confusion, the reader should be aware that the R package *multcomp* offers a function *mcp* with possible value Dose = "Williams." This might suggest that the Williams' test can be done with this function. The issue is not made entirely clear in the valuable text by Hothorn (2016) or the documentation for *multcomp* (Cran, 2016), where "Williams-type" contrasts are defined. While the overall Williams' test from this R function agrees roughly with the actual Williams' test, the step-down calculations are far from Williams' test results. In a series of papers (e.g. Hothorn, 2014) and in the text (Hothorn, 2016), the author makes clear his criticism of the Williams' test and offers the *mcp* functionality to implement an alternative. We have not found this alternative to be useful in ecotoxicology applications, but the reader is advised to read the references cited for an understanding of these criticisms.

3.5.1.1 Williams' Test Example Applied to Rainbow Trout

It is useful to present a plot showing the effect of the PAVA amalgamations done to implement Williams' test. (See Figure 3.8 for an example plot and Table 3.13 for PAVA means.).

The experiment had five doses of 10, 5, 2.5, 1.25, and 0.0625 ppm, plus a zero-dose control, with the following means and sample sizes. The response is trout length in cm. Table 3.13 contains the arithmetic and amalgamated means, sample sizes, and amalgamated sample sizes.

The null hypothesis is H_0: $\mu_0=\mu_1=\mu_2=\cdots=\mu_5$ and the alternative is H_1: $\mu_0 \geq \mu_1 \geq \mu_2 \geq \cdots \geq \mu_5$, a nonincreasing trend. The first violation of the expected trend is from dose 3 to dose 4. We amalgamate these to get $\tilde{Y}_3=\tilde{Y}_4=(2*4.26+2*4.36)/(2+2)=4.31$. The means are now 4.48, 4.34, 4.26, 4.31, and 4.17 with sample sizes 4, 2, 2, 4, and 2. The next violation is from doses 2 to 3. We amalgamate these to obtain $\tilde{Y}_2=\tilde{Y}_3=(2*4.26+4*4.31)/(2+4)=4.29$ and the amalgamation is complete.

To complete the example, the pooled within-group standard deviation, $s=0.07456$, obtained from Appendix 2, Section 2.4, (Eq. A2.50), is used along with the critical value from the Williams distribution. These critical values can be obtained from Table A3.4, which expands on table 1 of Williams (1972) and was created using the Probmc function in SAS software (SAS 9.4) using a program presented in Appendix 1. To get the critical value when all five groups are present, we use for DF=8, $k=5$, and w=#control reps/#treatment reps=2, the critical values are $\bar{t}_{\alpha,5}=Q-10^{-2}*B(1/1/w)=2.024-10^{-2}*5(1-1/2)=1.999$, $\bar{t}_{\alpha,4}=2.014-10^{-2}*5(1-1/2)=1.989$, $\bar{t}_{\alpha,3}=1.997-10^{-2}*4(1-1/2)=1.977$, and $\bar{t}_{\alpha,2}=1.962-10^{-2}*2(1-1/2)=1.952$. Table A3.4 does not contain the value of $\bar{t}_{\alpha,1}$, but that is just the Student t-distribution critical value, which is 1.859.

The resulting calculations are as follows (see Table 3.14), where WILL is the value of Williams' statistic obtained from Eq. (3.22), CRIT is the associated critical values of Williams' test, YTILDE is the isotonic mean, Y0 is the mean of the control group, and DF is the degrees of freedom for Williams' test. In the rainbow trout example,

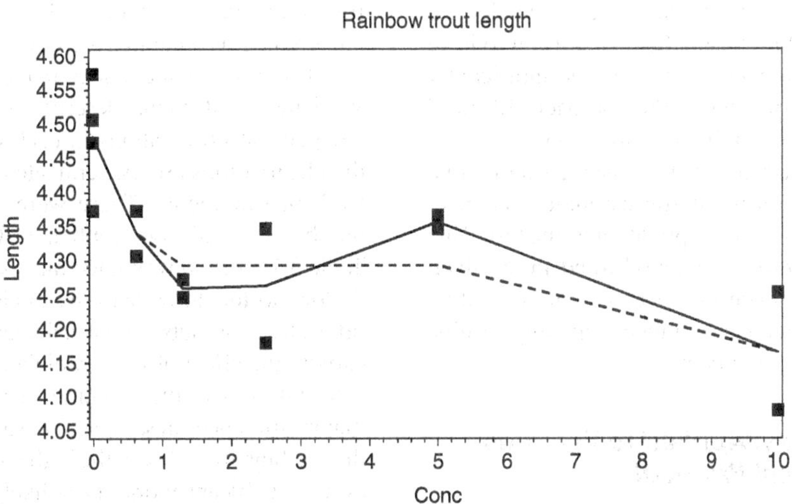

Figure 3.8 Rainbow trout length data with PAVA means. Dashed line is for PAVA means. Solid line for arithmetic means.

Table 3.13 Rainbow Trout Summary Statistics for Williams' Test

Dose	0	1	2	3	4	5
$\bar{Y}_i$	4.48	4.34	4.26	4.26	4.36	4.17
% Change	0	−3	−5	−5	−3	−7
N_i	4	2	2	2	2	2
$\tilde{Y}$	4.48	4.34	4.29	4.29	4.29	4.17
$\tilde{N}_i$	4	2	6			2

$\bar{Y}_i$ and N_i, treatment mean and sample size; $\tilde{Y}$ and $\tilde{N}_i$, amalgamated treatment mean and sample size; % change, % change of treatment mean compared to control mean.

Table 3.14 Williams' Test Calculations for Rainbow Trout Length

Conc	YTilde	Y0	DF	Will	Crit	Signif
10	4.16667	4.48167	8	4.48904	1.999	*
5	4.29333	4.48167	8	2.68392	1.989	*
2.5	4.29333	4.48167	8	2.68392	1.977	*
1.3	4.29333	4.48167	8	2.68392	1.952	*
0.63	4.34	4.48167	8	2.01888	1.859	*

YTilde, amalgamated treatment mean; Y0, control mean; DF, degrees of freedom for Williams' test; Will, value of Williams' test statistic; Crit, critical value of Williams distribution; Signif, * means treatment mean significantly different from control mean at $p=0.05$.

DF = 8 if the controls are pooled before calculating the control variance and $k = 5$. It will be seen that the NOEC < 0.63 mg l^{-1} as the reduction in mean length is significant ($p < 0.05$) at every test concentration. Recall that Dunnett's test did not find a significant effect in the lowest test concentration, which illustrates the increased power that comes from incorporating the expected monotone dose–response shape in the test statistic. However, there was only a 3% decrease in mean length at 0.63 mg l^{-1}, so Williams' test may be overly sensitive. Dunnett's test found that the 5% decrease at 1.3 mg l^{-1} to be statistically significant, but not the 3% decrease at 0.63. By interpolation in Table A3.4, it is possible to approximate the p-value for each comparison. That is left as an exercise.

With unequal sample sizes, it is possible that doses i and $i+1$ are amalgamated and $\tilde{Y}_i$ is significantly different from the control but $\tilde{Y}_{i+1}$ is not. There are several ways to avoid this undesirable and confusing situation. Williams suggests recomputing the isotonic means at each stage of the step-down procedure, using only those means remaining at each stage. In the balanced case, this is equivalent to the procedure already described. Another alternative is to use only the reduced set of amalgamated means and declare all dose groups involved in the amalgamated mean to have significant effects if the amalgamated mean is significantly different from the control. This is the approach followed here. Williams also suggests other modifications.

This test makes direct use of the assumed monotone dose–response both in terms of the estimated mean at each dose and in the step-down application of the test. This works well if in fact there is a general trend in the data. However, when the presence of a trend is questionable, forcing a trend on the data can produce confusing results. Consider the following experiment.

3.5.1.2 *Williams' Test Applied to* Xenopus laevis *with Statistical Protocol*

A study was done with the African clawed frog *Xenopus laevis* in the embryonic life stage. The design was three runs of four 96-well microplates. Each well is either empty, or a blank, or control medium, or a concentration of the test chemical plus medium. Except for empty and blank wells, each well also contains a single Xenopus embryo. There were nominally five wells per group in each microplate for a total of 60 wells in each group. Given the manual loading and measuring of the wells, some wells labeled as blank contain an embryo and some labeled as containing an embryo are actually empty except for the medium, and a well labeled blank may actually be empty, that is, have no medium. Thus, the actual number of embryos analyzed was sometimes less than 5 per microplate. Because of the way wells were filled with test material and embryos were placed, the scientific judgment was to treat wells as independent. The response was a measure of GFP protein fluorescence in the transgenic THbZIP tadpoles by way of a robotic imaging system that transforms the fluorescence signal to a numerical format. See Table 3.15 and Figure 3.9 for summary information. The full dataset is given in Appendix 1 as Xenopus_eco.sas7bdat.

Table 3.15 Meanlog and STD Are the Mean and Natural Logarithms of the Fluorescence Values and PAVA Are the Amalgamated Meanlog Values

Treatment	MeanLog	STD	PAVA	Mean	STDLog
Control	8.197	529.32	8.197	3664.66	0.141
1 mg l^{-1}	8.235	737.85	8.235	3839.48	0.189
3 mg l^{-1}	8.181	496.94	8.200	3606.05	0.137
10 mg l^{-1}	8.184	579.94	8.200	3626.79	0.156
30 mg l^{-1}	8.135	517.40	8.200	3450.57	0.148
100 mg l^{-1}	8.300	601.04	8.200	4069.34	0.148

A decreasing trend has been assumed and Williams' test finds no significant decrease at all, so the NOEC >100. It should be noted that the Shapiro–Wilk test found these data not to be normally distributed and no normalizing transform was found, so Williams' test is really not appropriate. These data were also analyzed by the nonparametric Jonckheere–Terpstra test in Section 3.5.

There is an issue hidden in the above description and in Williams' test itself that deserves discussion. The test assumes independent observations in each treatment group. That is, the observations are the replicate means, not the individual well measurements. There were minor differences in the number of wells per replicate. The treatment means in Table 3.15 were unweighted means of these replicate means. So long as the differences in sample sizes are small, this has little effect on subsequent calculations. Exercise 3.13 will explore the effect of using weighted means for the Xenopus data. Of course, this becomes a serious issue if there are major differences in sample size, such as that might happen if there is high mortality in the high test concentration(s). In that event, significant mortality is likely to be found. Williams' test is generally applied only to sublethal effects, so the analysis in this situation is probably best restricted to those concentrations in which there was not significant mortality.

Had an increasing trend been assumed, Williams' test would find a significant upward trend, due to the 11% increase in the high test concentration, and set the NOEC at 30 mg l^{-1}, where there was a 6% decrease from the control. This illustrates a need for care in what trend is requested when Williams' test is used. It should be observed that the highest four treatment groups were amalgamated. The data analyst should investigate if at any time large portions of the treatment groups are combined by the PAVA algorithm. Ideally, plots such as in Figure 3.4 will be done and examined as a routine part of data interpretation. See Figure 3.10 for a possible flow chart for use of Williams' test that was developed specifically for *Xenopus laevis* fluorescence data.

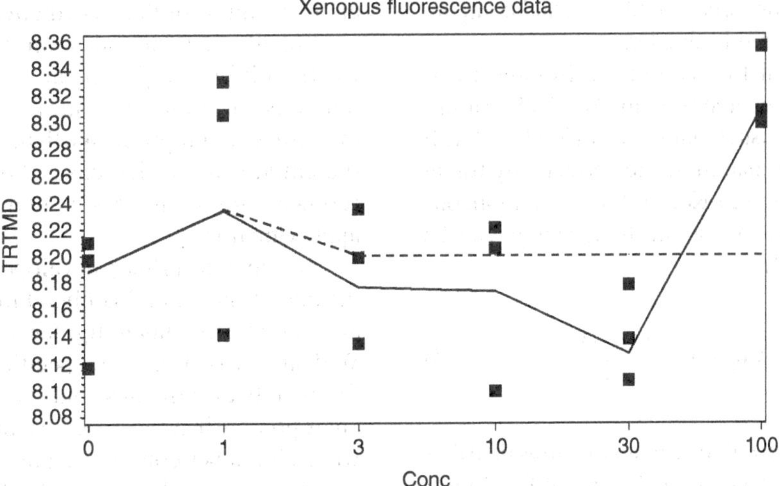

Figure 3.9 *Xenopus laevis* fluorescence replicate means. Run means of log-transformed *Xenopus* fluorescence data are plotted as solid boxes, solid line segments join treatment means, and dashed line segments join amalgamated means. Concentrations are plotted by order rather than value. While there seems to be a general downward trend in treatment means through 30 mg l^{-1}, there is a sharp rise at 100 mg l^{-1} that the amalgamated means discount.

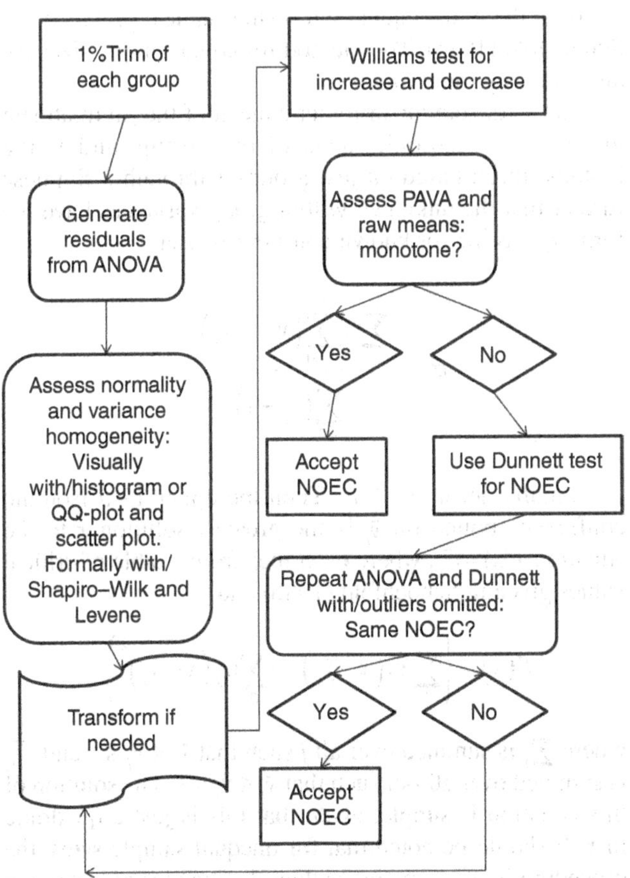

Figure 3.10 Possible flowchart for use of Williams' test. The 1% trim is arbitrary. While in general, we do not support automatic outlier elimination, for data that are prone to strange aberrations, possible from empty wells read by automated machines, a trim of the 1–5% most extreme values at the low and high ends of the distribution may be useful.

Under the recommended modifications discussed in the previous paragraph, it is not possible for a low dose to show an effect and a higher dose not. Additionally, the method used in the Tamhane–Dunnett test in Section 3.4.1 can be used with Williams' test to take multiple sources of variation into account.

3.5.1.3 Further Properties of Williams' Test

Williams' test loses power when additional higher dose groups are added to the design unless the effects at the new doses are substantially greater than at the lower dose levels. There can be a loss in power if there is a change in direction at the high dose, which can happen when there is high mortality in the highest one or two test concentrations. If the data are not normally distributed or not homogeneous, then the Jonckheere–Terpstra test (Section 3.5.2) or Shirley's (1979) nonparametric alternative is advised. References for further discussion of the effect on this test of non-normality, variance heterogeneity, or change in direction are given in Bretz (1999), Bretz and Hothorn (2000), and Williams (1972).

Limited power simulations have been published by Marcus (1976), Poon (1980), Potter and Sturm (1981), Shirley (1979), Williams (1971, 1972), and Puri (1965). These suggest power characteristics for Williams' test similar to those for the Jonckheere–Terpstra test, when the data meet the requirements for Williams' test and there is no change in direction at the high dose. Power simulations done by one of the present authors (Green) for the smaller sample sizes common in ecotoxicity studies also found comparable power for Williams and Jonckheere–Terpstra tests, but lower power than these large sample results indicate. Annex 1 provides computer code that can generate the power of Williams' test for certain conditions and compare it with that of other tests in this chapter.

A confidence interval for each of the isotonic treatment means (i.e. those estimated from the PAVA algorithm) can be constructed in the usual manner from Table A3.4. It should be noted, however, that these are correct only for an individual mean. According to Korn (1982), simultaneous $1-\alpha$ confidence intervals for the all the means μ_i can be calculated from Eq. (3.23):

$$\bar{Y}_j - \frac{m_{k,N-k}s}{\sqrt{n_j}} < \mu_j < \bar{Y}_j + \frac{m_{k,N-k}s}{\sqrt{n_j}}, \tag{3.23}$$

where $m_{k,N-k}$ is the Studentized maximum modulus distribution for k treatment groups and $N-k$ degrees of freedom, s is square root of the pooled within-group variance estimate, N is the total sample size, and $\bar{Y}_j$ is the arithmetic mean of the jth group. However, as Korn and others observe, these confidence intervals are wider than necessary for a monotone dose–response, in part because they make no use of the monotonicity. OECD (2014a, b, c) describes several ways to improve simultaneous confidence bounds for isotonic means and their differences generated by the Williams' test procedure. Some results from OECD (2014a, b, c) are provided here in modified form for convenience.

Three methods for constructing more appropriate simultaneous confidence bounds are presented, all of which assume normality of the response within groups. In addition, methods are presented for the case where confidence bounds for the mean or difference from the control are wanted only at the NOEC.

Isotonic regression yields ML estimates of the treatment means based on a monotone dose–response model. Korn suggests the following modification of Eq. (3.23):

$$\max_{j\le i}\left(\bar{Y}_i - \frac{m_{k,N-k}s}{\sqrt{n_i}}\right) < \mu_j < \min_{i\le j}\left(\bar{Y}_i + \frac{m_{k,N-k}s}{\sqrt{n_i}}\right). \tag{3.24}$$

These intervals are not derived from the isotonic estimates of means but from the sample means and the monotone (i.e. isotonic) assumption on the means. Korn suggests doing linear interpolation to obtain approximate confidence bounds for the response between tested doses.

Williams (1977) uses the isotonic estimators of the mean responses to show that a lower $1-\alpha$ confidence bound on the true mean for the jth treatment is given by $\tilde{\mu}_j - A_{\nu,\alpha}s$, where s is the square root of the usual pooled variance estimate, $\tilde{\mu}_j$ is the isotonic estimate of the jth treatment mean, and $A_{\nu,\alpha}$ is the critical value from the distribution of $(\tilde{\mu}_j - \mu)/s$ under the assumption $\mu_{j-1} = \infty$ and $\mu_{j+1} = \mu_{j+2} = \cdots = \mu_k$. Williams (1977) contains tabulated critical values for this distribution. He also gives a modification of this to provide confidence bounds or intervals for the difference $\tilde{\mu}_0 - \tilde{\mu}_j$ and simultaneous confidence intervals for arbitrary contrasts of the means under the monotone assumption. Williams (1977) considered only the equal sample size case, though if sample size differences are small, his bounds should be reasonable approximations.

Schoenfeld (1986) developed isotonic estimators of the bounds based on likelihood ratio methods. His bounds are sometimes much tighter than those proposed by Williams (1977), and he specifically recommended them for toxicity experiments, and states that they are especially appropriate when the dose–response curve is shallow and we desire upper confidence bounds on the effects at concentrations at or below the NOEC. For unequal sample sizes, the critical values used in constructing Schoenfeld's upper bounds must be estimated by simulation. This makes the procedure more computer intensive than desirable for routine use. If the differences in sample sizes are not large, then approximate confidence bounds can be obtained by treating them as equal and using tables provided by Schoenfeld (1986). The method for equal sample sizes for an increasing trend follows.

Let $\bar{y}_j$ denote the arithmetic mean of the jth treatment group, $\tilde{y}_j$ the isotonic mean of that group, and $\hat{y}_j$ the isotonic mean based on just groups 1 through j. Suppose further that the unknown within-group variances have the form $\sigma_i^2 = \sigma^2/a_i$ for known constants a_i. Let

$$\hat{\sigma}^2 = \frac{\sum_{j=1}^{k} a_j \sum_{m=1}^{n_j}\left(y_{jm} - \bar{y}_j\right)^2}{\sum_{j=1}^{k}\left(n_j - 1\right)}.$$

Finally, let $w_j = n_j * a_j$. Then the upper $1-\alpha$ isotonic confidence bound on $\tilde{y}_i$ is the greatest solution x to the equation $T(x) = C_\alpha$, where C_α comes from a table of critical values given in Schoenfeld (1986) and

$$T(x) = \left\{\sum_1 w_j\left(x - \tilde{y}_j\right)^2 - \sum_2 w_j\left(x - \hat{y}_j\right)^2\right\},$$

where $\sum_1$ is summed over all j such that $\tilde{y}_i \le \tilde{y}_j \le x$ and $\sum_2$ is summed over all $j<i$ such that $\hat{y}_i \le \hat{y}_j \le x$. The solution of this equation is simple, given that this is just a quadratic in x. It should be noted that for unequal sample sizes, the procedure is the same other than the manner of obtaining the critical value C_α. It should be noted that this is an upper confidence bound for a single isotonic mean. Schoenfeld (1986) also gives a modification of this procedure that gives simultaneous confidence intervals for multiple isotonic means.

3.5.2 Step-down Jonckheere–Terpstra Test and Nonparametric Alternatives

A nonparametric trend test that can be used to analyze proportion data is the Jonckheere–Terpstra trend test, which is intended for use when the underlying response on each subject is continuous and the measurement scale is at least ordinal. An example would be first day of reproduction of a daphnid, where the underlying response, time to first reproduction, is continuous, but the measurement is taken only once per day. Of course, any truly continuous response can be analyzed by this test, provided the means or medians are consistent with monotonicity. One of the appealing properties of this test is the requirement of monotonicity rather than linearity or any other specific mathematical shape in the dose–response. Another advantage is that an exact permutation version of this test is available to meet special needs such as very small samples or massive ties.

The Jonckheere–Terpstra test statistic is based on Mann–Whitney counts. The idea of the test is very simple. First, a decision is made on biological or physical grounds, what direction (increasing or decreasing) the dose–response has. Suppose an increasing trend is expected. Then the observations from all groups are combined and ordered or ranked from smallest to largest. For each two dose groups $d_i < d_j$, with $i<j$, count the number, O_{ij}, of pairs (x_i, x_j) of observations that can be made, with x_i from group i and x_j from group j, which follow the expected order, $x_i<x_j$. Add all of the O_{ij} and compare that sum with what would be expected if the dose–response were flat. A large positive difference is evidence of a significant increasing dose–response. The only change needed for a decreasing trend is that O_{ij} should count the pairs

$$(x_i, x_j) \text{ for which } x_i > x_j.$$

The test statistic is given by

$$J = 2W - \sum_{i<j} n_i n_j, \tag{3.25}$$

where $W=\Sigma_{i<j}O_{ij}$.

For an *exact* test with equal sample sizes, tables of critical values are given in several sources, such as Hollander and Wolfe (1973), Lehmann (2006), and Sheskin (2011). The mean and variance of J are readily obtained from

$$E[W] = \frac{1}{2}\sum_{i<j} n_i n_j = \frac{n^2 - \sum_i n_i^2}{4},$$

$$\mathrm{Var}(W) = \frac{n^2(2n+3) - \sum_i n_i^2(2n_i+3)}{72}.$$

Equation (3.25) assumes there are no ties among the observations. When there are ties, then tie corrections should be made. The tie-corrected expected value is obtained from Eq. (3.25) by replacing O_{ij} by the tie-corrected Mann–Whitney sum O_{ij}^* obtained by adding half the number of pairs (x_i, x_j) for which $x_i=x_j$. The tie-corrected variance for W is rather more tedious to calculate. According to Lehmann (2006),

$$\mathrm{Var}(W^*) = \frac{1}{72}\left[n(n-1)(2n+5) - \sum_{i=1}^{k} n_i(n_i-1)(2n_i+5) - \sum_{j=1}^{e} d_j(d_j-1)(2d_j+5)\right] + \frac{1}{36n(n-1)(n-2)}\left[\sum_{i=1}^{k} n_i(n_i-1)(n_i-2)\right]\left[\sum_{j=1}^{e} d_j(d_j-1)(d_j-2)\right] + \frac{1}{8n(n-1)}\left[\sum_{i=1}^{k} n_i(n_i-1)\right]\left[\sum_{j=1}^{e} d_j(d_j-1)\right], \tag{3.26}$$

where d_i is the number of instances of the jth distinct value and e is the number of distinct values.

For large samples, an *approximate* test is given by

$$Z = \frac{J - E[J]}{\mathrm{Var}(j)}, \tag{3.27}$$

which has a standard normal distribution.

Fortunately, there are software packages that carry out the calculations for both the normal approximation and the exact test. The SAS program for this test is *Jonckheere macro.SAS* provided in Appendix 1. The R program is the following:

```
#JT test requires library StatCharrms
 require(clinfun)
  monocityTest(Algae ,'dose','Density48')
  #Reference StatCharrms R package
  jonckheereTerpstraTest(Algae ,'dose',
  'Density48','Decreasing',0.05)
#Reference StatCharrms (2017). This may
take a moment.
```

The Jonckheere–Terpstra test is robust against both mild and major violations of normality and homoscedasticity. It is based on a presumed monotone dose–response and is powerful against ordered alternatives for a wide variety of dose–response patterns and distributions, but unlike Williams' test, no amalgamation of adjacent treatment groups is done. There is no problem with unequal numbers of replicates, but if the replicate means or medians are based on unequal sample sizes, there is no weighting that can account for that. This should not be a problem if the replicate sample sizes are not too different, and simulations suggest it is not a problem in general. However, Brünning and Kössler (1999) have developed what they call a Jonckheere-type test that provides a way to include replicate sample sizes in the test statistic.

There is no way to take into account multiple sources of variance, such as within and between replicates. There is no built-in procedure for deciding whether the data are consistent with a monotone dose–response model. Asymptotic power properties of this test have been published (Bartholomew, 1961; Puri, 1965; Odeh, 1971, 1972; Poon, 1980), but computer simulations demonstrate that they overestimate the power, sometimes grossly, when there are fewer than 30 replicates, and in any case they are not computed for the step-down application described in this document. Since the number of replicates in a guideline ecotoxicity study is generally much smaller than 30, the asymptotic power properties are not very useful. A sample program for simulating the power characteristics is given in Appendix 1. There is no true confidence interval associated with the mean response at the NOEC or any other treatment group, but a normal approximation could be applied. This test will be illustrated with the rainbow trout data in Table 3.3 and the algae data in Table 3.11. See Table 3.16 for summary results for the trout data.

The Jonckheere–Terpstra sets the NOEC for the rainbow trout data at 0.63 mg l^{-1}, the same as that found by Dunnett's test in Table 3.4, and one level higher than found by Williams' test in Table 3.14. Recall from the discussion of Williams' test for these data that there was only a 3% decrease in mean length at 0.63 mg l^{-1}, so the Jonckheere–Terpstra and Dunnett tests are probably more reasonable in terms of what is biologically relevant. Indeed, the question of the size effect that is biologically important to detect or estimate is increasingly important for regulatory assessment but has received far too little attention. Some discussion of this is given in the challenge posed by Green (2015b) and the responses to that, especially that of Mebane (2015). Note that there is no gap in doses found significantly different from zero, making interpretation simpler than with Dunnett's test.

It is not surprising that not all of these tests find the same NOEC. The natural question to ask is what should be done with these different results. We advise against doing multiple tests of the same data, except as here for illustration. There is no statistically defensible way to select one result over another so long as the requirements of all methods used are satisfied, though biology would suggest a trend test over a pairwise approach. We strongly recommend establishing a statistical protocol before the data are analyzed and following that protocol unless there are compelling reasons to do otherwise. This is to avoid being suspected (or guilty!) of picking a test to give the desired answer. A possible protocol for continuous responses is given in Section 3.6. See Table 3.17 for Jonckheere–Terpstra test summary results for the algal data.

Table 3.16 Rainbow Trout Length Data Analyzed by Jonckheere–Terpstra Test

Highest dose	%Change	_JT_	Z_JT	XPL_JT	XPR_JT	Signif	Dose
≤10 mg l^{-1}	−7	19.5	−2.803	0.002	.	**	6
≤15 mg l^{-1}	−3	17	−2.130	0.018	.	**	5
≤2.5 mg l^{-1}	−5	7.5	−2.734	0.002	.	**	4
≤1.3 mg l^{-1}	−5	4	−2.587	0.005	.	**	3
≤0.6 mg l^{-1}	−3	3.5	−1.644	0.133	.		2

_JT_ is the value of the Jonckheere–Terpstra test statistic for RBT reproduction. Z_JT is the normalized test statistic. XPL_JT and XPR_JT are exact 1-sided Bonferroni-Holm adjusted p-value for decreasing and increasing trends, respectively. Dose is the highest dose or group in the step-down process. % change is the percent change of treatment mean length from control. Signif = "**" means significant at the 0.01 level.

Table 3.17 Algal Data Analyzed by Jonckheere–Terpstra Test

Highest conc	_JT_	Z_JT	PL_JT	PR_JT	Signif	Dose
Conc ≤ 4	39	−5.998 401	0.000 1	.	**	7
Conc ≤ 1.6	36	−5.143 039	0.000 1	.	**	6
Conc ≤ 0.64	33	−4.118 109	0.000 1	.	**	5
Conc ≤ 0.26	30	−2.801 578	0.002 5	.	**	4
Conc ≤ 0.1	27	−0.878 31	0.215 7	.		3

_JT_ is the value of the Jonckheere–Terpstra test statistic. Z_JT is the normalized test statistic. PL_JT (PR_JT) is the Bonferroni–Holm adjusted p-value for decrease (increase) from normal approximation of the test, appropriate when there are four or more replicates per treatment. Dose is the highest dose or group in the step-down process. Signif = "**" means significant at the 0.01 level.

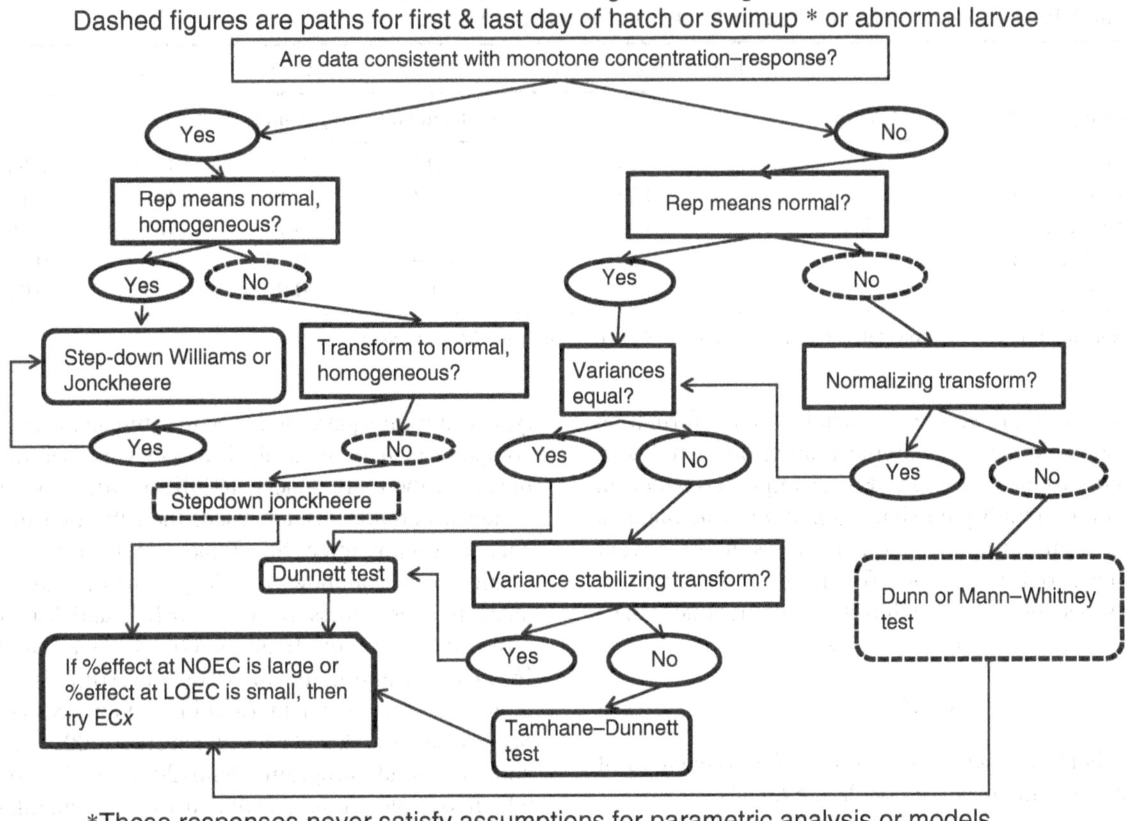

Figure 3.11 Possible statistics flow chart for continuous responses.

The Jonckheere–Terpstra test sets the NOEC for the algal data to be 0.1 mg kg^{-1}, where there was an observed 28% decrease in density, compared with 0.26 determined by the Tamhane–Dunnett test discussed in connection with Table 3.11, where there was a 54% decrease. In Chapter 4, regression models will be fit to these data to estimate EC*x* for several choices of *x*.

No significant trend in either direction was found in the Xenopus data when it was analyzed by the Jonckheere–Terpstra test. Given the striking non-monotonicity in the concentration–response, this should not be alarming. Indeed, the formal contrast-based test discussed in Section 3.5 found significant non-monotonicity, since the linear contrast was not significant ($p=0.2203$) and the quadratic contrast was significant ($p=0.0007$).

3.6 PROTOCOL FOR NOEC DETERMINATION OF CONTINUOUS RESPONSE

As stated previously, e.g. in Section 3.3.2, it is advisable to have a statistical protocol in place before a study is done that indicates how the data will be analyzed. Such a protocol should cover all types of responses to be analyzed and all types of analysis. In this section, a possible protocol for NOEC determination for a continuous response is presented. In Chapter 4, a protocol for regression modeling to estimate EC*x* will be given, and in Chapters 5 and 6 analogous protocols will be given for quantal responses. Special protocols may also be required, such as the one presented in Figure 3.5 for the Xenopus embryonic toxicity assay. See Figure 3.11 for an example flow chart for NOEC determination developed by the authors for OECD Fish Early Life Stage Test Guideline 210.

Figure 3.11 is intended only as an illustration. Other statistically sound protocols exist. This addresses several responses in fish ELS studies. Discussion of transformations of data called for in flow chart is given in Section 3.5 as well. Discussion of time-to-event data such as first and last day of hatch or swim-up will be covered in Chapter 10.

3.7 INCLUSION OF RANDOM EFFECTS

In all the examples to this point, the unit of analysis was the replicate, and we did not explore the observations used to calculate the replicate mean or median that formed the data.

Table 3.18 Comparison of Dunnett' Test Results for RBT Length Under Type 3, Maximum Likelihood, and REML Estimation Methods with Analysis of Replicate Means

		Rep means analysis	Type 3	REML	ML	R
Comparison	% Change	Dunnett adjusted *p*-values				
DOSE 2 – 1	–3	0.14	0.14	0.32	0.31	0.58
DOSE 3 – 1	–5	0.03	0.03	0.11	0.11	0.15
DOSE 4 – 1	–5	0.03	0.03	0.12	0.11	0.16
DOSE 5 – 1	–3	0.19	0.19	0.38	0.37	0.7
DOSE 6 – 1	–7	0	0.00	0.03	0.03	0.01

Rep means results are from Table 3.4. Other results are from program *study24.sas* or *Study24.r*.

Sometimes, it is preferable to include such information in the model. Perhaps the simplest situation of this sort is a nested variance structure. For example, consider an experiment with multiple fish in each of multiple tanks in each of several nominal treatment groups, and we want to examine variability among fish in the same tank and variability between tanks within the same treatment. The simplest model to capture such a study is

$$Y_{ijk} = \mu_i + e_{ijk},$$

where Y_{ijk} is the response on the subject *k* in replicate *j* of treatment *i*, μ_i is the mean of population *I*, and

$$\begin{aligned} \operatorname{cov}\left(Y_{ijk.}\ Y_{i^*j^*k^*}\right) &= \operatorname{cov}\left(e_{ijk.}\ e_{i^*j^*k^*}\right) = 0 \quad \text{if } i \neq i^*, \\ &= \varphi_1 \quad \text{if } i = i^*,\ j \neq j^*, \\ &= \varphi_0 + \varphi_1 \quad \text{if } i, j = i^*, j^*. \end{aligned} \tag{3.28}$$

Observations in the same tank are correlated and their common covariance is φ_1, which can be positive or negative. Appendix 2 provides more detail on this parameterization and references. An alternative parameterization is obtained by writing the model as

$$Y_{ijk} = \mu_i + \rho_{j(i)} + e_{ijk}, \tag{3.29}$$

where $\rho_{j(i)}$ are zero-mean *iid* random effects of replicates within treatment with variance σ_r^2 and e_{ijk} are zero-mean *iid* random effects of subjects *k* within replicate *j* within treatment *i* with variance σ_0^2. In this form, both variance components, σ_r^2 and σ_0^2, are restricted to be nonnegative, and the covariance between observations in the same replicate can be worked out easily from Eq. (3.29). In either parameterization, Dunnett's test, for example, could be done taking the nested variance structure into account. For unbalanced designs, a weighted analysis of replicate means, weighted by replicate sample size, is needed to capture the nested effect. Whether this yields the same results as obtained from an analysis of replicate means depends on the method of estimation. If what are called type 3 sums of squares are used to estimate the two variance components, then the analysis agrees with that for replicate means. If the more general REML or ML estimates of the variance components are used, then the two methods of analysis do not agree. See Table 3.18 for a comparison of results for the rainbow trout length data in Table 3.3. The analysis in R agrees with the REML and ML results in terms of what is significant, but the *p*-values are different. Further discussion of methods of estimating variance components is given in Hocking (1985), Searle (1987), Searle et al. (2007), and McCulloch et al. (2008). Appendix 1 contains simple programs, *Study24.sas* and *Study24.r*, by which to explore this type of analysis. In general, we think the analysis of replicate means is a practical procedure unless there are major differences in replicate sample sizes. In that case, either a weighted analysis of replicate means using sample size as weight or the type of nested variance structure discussed in this section is preferable.

Other restrictions on randomization, such as repeated measures of the same subject over time, or experiments with multiple generations or mesocosm experiments with multiple species interactions will be discussed in later chapters.

3.8 ALTERNATIVE ERROR STRUCTURES

Generalized linear models (GLMM) and generalized linear mixed models (GLMM) are increasingly used as a more natural way to analyze data from ecotoxicity studies than to use transforms to mimic normality. For example, fecundity data, such as the number of eggs produced, is a count variable that could be analyzed assuming a Poisson error structure rather than using a square-root transform to approximate a normal distribution. Variance heterogeneity would then be overdispersion and a negative binomial error structure could be used rather than an attempt to find a variance-stabilizing transform. These models were introduced in Section 2.11, and a GLMM model with Poisson error

Table 3.19 Summary Statistics for Daphnia Reproduction Data

Dose	Conc (mg l^{-1})	Mean	Median	StdErr	%Change
1	0	84.6	87	7.7	0
2	0.015	86.6	89	4.5	2
3	0.053	72.3	84	8.0	−15
4	0.19	78.0	80.5	4.7	−8
5	0.67	71.4	71	3.6	−16
6	2.35	64.0	65.5	6.7	−24

A positive % change indicates an increase in mean response compared with the control, while a negative value indicates a decrease. StdErr is the standard error of the estimated mean.

structure will be used to analyze the daphnia reproduction data in Table 3.19, the algae density data in Table 3.17, and the Xenopus fluorescence data in Table 3.15 using programs provided in Appendix 1.

For the daphnia reproduction data, it was shown in Section 2.11.1 that there was no evidence of overdispersion, so a Poisson error model is appropriate. This test was also analyzed by Dunnett's test assuming a normal error structure and by the Williams and Jonckheere–Terpstra tests. For the normal approximation, since total live young at day 21 is a count variable, a square-root transform is recommended. Neither under the square-root transform nor untransformed was there evidence of non-normality or variance heterogeneity, nor was there significant indication of non-monotonicity by the test of Section 3.5, though clearly from the table, there was some departure from monotonicity. See Table 3.19 for summary statistics for this experiment. Again, not all methods agree, with Dunnett's test under the GLMM or normal model both finding no significant decrease in any test concentration (p-values = 0.053 for both), but both the Jonckheere–Terpstra and Williams' test found a significant effect in the high test concentration (p-values = 0.0017 and 0.0029, respectively). This illustrates three points: (i) There is a need to follow a set protocol established before the experiment is run. (ii) Trend tests tend to have greater sensitivity of over pairwise tests when a monotone dose–response is expected. (iii) The use of $p = 0.05$ as a decision rule is arbitrary. The latter point has been made by authors too numerous to list. While we recognize that there is no real scientific basis for this use of 0.05, the regulatory context probably requires such fixed rules to avoid whimsical or arbitrary decisions or endless controversy. It is left as an exercise to verify the results of these four tests. In Chapter 4, regression models will be fit to these data, and it will be interesting to find whether such models can do better than an NOEC at which there was a 16% effect.

See Table 3.20 for summary statistics for the algae density example first presented in Section 3.4.2.1. To put the percent change at the NOEC for algae density in perspective, it should be noted that the standard error of the control is 11% of the control mean, so a 19% change is less than two standard errors of the control mean. While 19% may seem large, the data are highly variable and that restricts the size effect that can be detected. This issue will be revisited in Chapter 8, where we analyze these same data with a Poisson error regression model and evaluate the confidence intervals for ECx estimates. The Williams and Dunnett tests with normal error structure were not done because the data were found not to be normally distributed by the Shapiro–Wilk test.

Table 3.20 Summary Statistics for Algae Density at 48 h

Dose	Conc	Mean	%Change	StdErr
1	0	613385	0	67979
2	0.041	467230	−24	6835
3	0.1	496120	−19	18563
4	0.26	280228	−54	12647
5	0.64	130398	−79	1350
6	1.6	72354	−88	1172
7	4	44065	−93	385

Dunn's test found the NOEC to be 0.26 mg l^{-1}, where there was a 54% decrease in density. The Jonckheere–Terpstra test set the NOEC = 0.1 mg l^{-1}, where there was a 19% decrease in density. The GLMM version of Dunnett's test with a Poisson error structure found the NOEC < 0.041.

There was no evidence for overdispersion in the Xenopus data by using the tests developed in Section 2.11.1. Thus, a Poisson error structure was deemed appropriate. Dunnett's test with Poisson error structure found no significant decrease in fluorescence values in any treatment group. This is consistent with the Jonckheere–Terpstra analysis done in Section 3.5.2. Details are left as an exercise.

3.9 POWER ANALYSES OF MODELS

In designing an experiment, it is important to know in advance how the data will be analyzed. There are several reasons for that. First, it is important to know how likely the proposed method of analysis will find biologically important effects if they are present. An important part of that evaluation is to determine the power of the proposed statistical procedure, that is, the probability of finding a statistically significant effect if the true effect size is Δ. A computer program is provided in Appendix 1 to generate the power as a function of Δ for all of the tests described in Sections 3.2 and 3.4 and display those powers as overlaid plots with Δ along the horizontal axis and power or probability of a significant finding along the vertical axis. See Figure 3.13 for the power to detect effects on length of rainbow trout at the end of studies done under OECD TG 210, the fish early life-stage study. In such studies, there are four replicate tanks of 20 fish in the control and each treatment group, five treatment groups plus a dilution water control, and a solvent control if a solvent is used. A large control database was available when this test guideline was revised in 2012, and the distribution of within- and between-tank variances was constructed. From those distributions the 90th percentiles of each of these variances was used in developing the power curves.

The power of trend tests depends not only on the variability, replication, and number and spacing of test concentrations but also on the shape of the concentration–response curve. See Figure 3.12 for three dose–response curves corresponding to an initially rather flat dose–response,

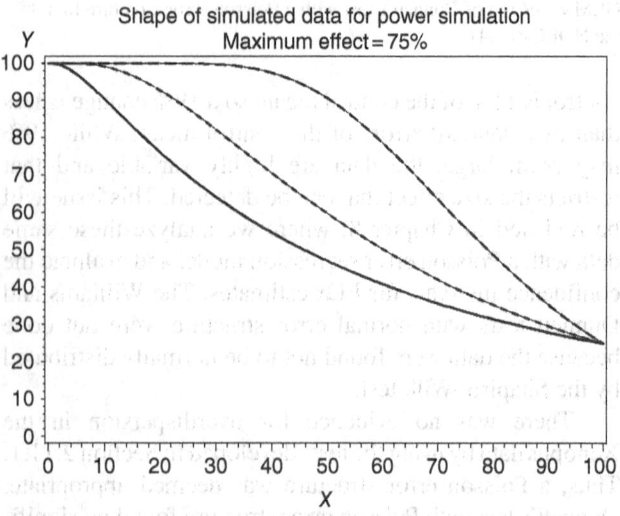

Figure 3.12 Three dose–response shapes illustrate a range of dose–response relationships used for power simulations. The maximum simulated effect at the highest concentration is varied from 1 to 99%.

a curve that bends down moderately from the control, and one that bends downward very shapely just below the control. The curves shown in Figure 3.13 all have maximum effect of 20% at the highest tested concentration. To obtain the power curves, the maximum effect was simulated to range from 5 to 50%. The general shape simulated was a Bruce–Versteeg model that has a shape parameter that also affects the shape. That model will be described in Chapter 4. For now, we will focus on the visual patterns to get a feel for how shape influences power for the Williams and Jonckheere–Terpstra tests. The shape will not affect the power of the pairwise tests, Dunnett, Dunn, and Mann–Whitney, as they depend only on the size effect and variability simulated.

A large database was available at the time OECD test guideline 210 for early life-stage studies was revised. Roughly a third of the rainbow trout length datasets followed a relatively steep Bruce–Versteeg-shaped concentration–response curve. The 90th percentiles of the between- and within-rep variances were used to simulate data from such experiments in order to compare power properties of several tests discussed in this chapter for this concentration–response shape. Figure 3.13 shows the resulting power comparisons. Other concentration–response shapes will exhibit different power comparisons. For non-normal or heterogeneous data, Williams and Dunnett tests are not appropriate, but Jonckheere–Terpstra is and has reasonable power properties. Program to generate these powers is given in Appendix 1.

It is recommended to use a historical database for each type of response, species, and study to generate data typical for that situation. The within- and between-replicate variances and dose–response shapes do vary considerably as

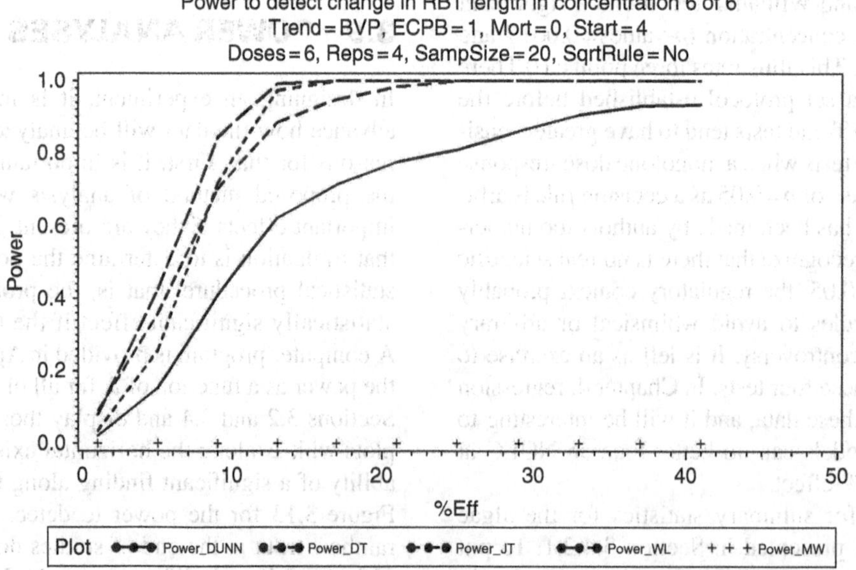

Figure 3.13 Power comparison of several tests for effects on rainbow trout in ELS studies.

these factors change and the implications for experimental design and test selection vary accordingly. The same issues affect selection and performance of regression models and simulations to explore the distribution of ECx estimates and their confidence bounds, as will be explored in Chapters 4 and 6.

Williams' test has 80+% power to detect a 9% decrease in length, whereas the Jonckheere–Terpstra test achieves that power only for an 11% decrease. Dunnett power is in between those two and the power of Dunn's test is markedly lower. Mann–Whitney with a Bonferroni–Holm adjustment has zero power to detect a change regardless of the magnitude.

3.9.1 The Concept of Power

It is important to understand the power of any test used to determine an NOEC, or more generally, a statistical finding. Recall that power is the probability of getting a significant result for a particular test concentration, given the true mean response at that concentration differs from the control by a specified amount, Δ. This power can be approximated using the distribution functions for Dunnett's test and the t-test. Programs for computing the power of Dunnett's test are given in Appendix 1. References to power computations for Dunnett's test include Dunnett and Tamhane (1992), Dunnett et al. (2001), and Horn and Dunnett (2004).

3.9.2 Power of Hypothesis Tests

For hypothesis tests, it is important to know the false positive and false negative rates; otherwise the resulting NOEC and LOEC can be misleading. Traditionally, attention has been focused on the false positive rate, but recent test guidelines have rightly emphasized the need to establish the false negative rate as well. Usually, this requirement is expressed in terms of the power to detect an effect. It may be helpful to restate the hypotheses being tested in their simplest form:

$$H_0\text{: } \mu_0 = \mu_k \text{ vs. } H_1\text{: } \mu_0 > \mu_k,$$

where $\mu_{i,}$ $i=0$ and k denote the means of the control and test *population* k, respectively. The false positive rate is conceptually very simple:it is the probability, α, that the sample treatment mean will be significantly less than the sample control mean using the specified experimental design and statistical test. There is random variation in sample results since we are dealing with samples rather than entire populations, so there is a chance that sampling will mislead us as to the true relationship between control and treatment. The false positive rate, α, quantifies one of the uncertainties. Power measures the probability that the sampling process will lead to a significant decrease in the treatment mean compared with the control mean. It is not possible to determine this probability as stated. Instead, we calculate the probability of a significant decrease as a function of the true decrease. That is, we calculate

$$\text{Power}(d) = \text{Probability}\left[\text{Reject } H_0 \text{ for } H_1 | \mu_0 - \mu_k = d\right]. \tag{3.30}$$

This probability function can be displayed in a table or a figure. (See Figure 3.14 for an example).

As a general rule, a power of 80% or greater to detect an effect of biological importance is preferred. The power curves make clear what size effect the experimental design and statistical test can be expected to find statistically significant effect. Power curves should be plotted in the planning stages of the experiment, so if there is inadequate power to detect an effect deemed important, then design or statistical test can be modified before the experiment is conducted.

3.9.3 Calculation of Power

For some simple situations, there is an easy way to calculate power. Consider the case there is just one treatment group, with mean μ_1, being compared with a control, with mean μ_0, and random samples of sizes n_0 and n_1 are taken from the control and treatment groups, respectively, and the response is normally distributed within each group with

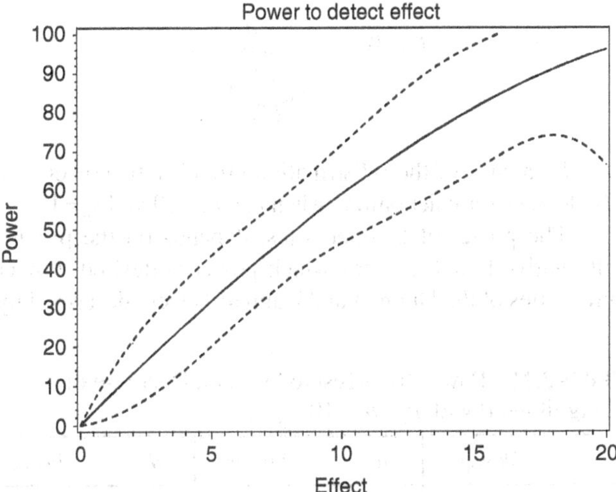

Figure 3.14 Example power plot solid line is the power (expressed as a percent rather than a probability) to detect a percent change from control of magnitude given on the horizontal axis. From the plot it can be seen that the power to detect a 15% change from the control is 80%. The dotted lines are 95% confidence bounds on the power estimated from a computer simulation study.

a common within-group variance. Then a test statistic to test H_0: $\mu_1=\mu_0$ vs. H_1: $\mu_1<\mu_0$ is

$$T^* = \frac{\bar{x}_0 - \bar{x}_1}{s_p\sqrt{\frac{1}{n_0}+\frac{1}{n_1}}}, \tag{3.31}$$

where n_0 and n_1 are the sample sizes, s_p^2 is the pooled sample variance given by

$$s_p^2 = \frac{(n_1-1)s_1^2+(n_0-1)s_0^2}{n_1-n_0-2}, \tag{3.32}$$

where s_0^2 and s_1^2 are the sample variances and $\bar{x}_0$ and $\bar{x}_1$ are the sample means.

The treatment mean $\bar{x}_1$ is declared significantly from the control mean $\bar{x}_0$ if T^* exceeds Tcrit, the 95th percentile of Student's t-distribution with $df=n_0+n_1-2$. Now

$$T = \frac{\bar{x}_0 - \bar{x}_1 - (\mu_0-\mu_1)}{s_p\sqrt{\frac{1}{n_0}+\frac{1}{n_1}}}. \tag{3.33}$$

has a Student's t-distribution with $df=n_0+n_1-2$, so

$$T^* = T - \frac{d}{s_p\sqrt{\frac{1}{n_0}+\frac{1}{n_1}}}. \tag{3.34}$$

T^* exceeds Tcrit if and only if

$$T > \text{Tcrit} - \frac{d}{s_p\sqrt{\frac{1}{n_0}+\frac{1}{n_1}}}.$$

So a table of the t-distribution provides the power. See Table 3.21 for calculations when $n_0=n_1=10$ and $s_p=1$.

The power of Dunnett's test depends on the precise alternative hypothesis for which power is desired. Power properties of the Dunnett and Dunn tests were described by the first two authors of this text in OECD (2006a, b), and the following discussion follows those descriptions. One relevant power of Dunnett's test is the probability of a significant result for a treatment group, given the true mean response for that group differs from the control by a specified amount, Δ. Such a power can be approximated using the distribution functions for Dunnett's test and the t-test. In SAS, such an approximate power can be computed in SAS in the case of equal sample sizes in all treatments as follows:

$$\text{Let } \lambda = \frac{\Delta}{\sigma\sqrt{2/n}}.$$

```
POWER1 = 1 - PROBT(PROBMC('DUNNETT1',.,1 - α,
DF,GRPS),DF,λ),
POWER2 = 1 - PROBT(PROBMC('DUNNETT2',.,1 - α,
DF,GRPS),DF,λ),
```

where α is the size of the test, Δ is the difference in means to be detected in the specified treatment, n is the number of replicates in each treatment, DF is the ANOVA degrees of freedom for error, GRPS is the number of concentrations to be compared to the control, and σ is the estimated within-group standard deviation. POWER1 and POWER2 are the approximate powers of one- and two-sided tests, respectively. The R function POWDT in the package "DunnettTests" can be used to estimate the power of Dunnett's test under some restrictions.

This and related powers can also be obtained using the %IndividualPower macro given in Westfall et al. (2011). Additional references to power computations for Dunnett's test are Dunnett and Tamhane (1992) and Dunnett et al. (2001).

Power of Dunn's Test. Dunn (1964) describes an extreme situation, where all but one of k population means equal the control, but that every observation in that treatment group exceeds every observation in the control. The power of her two-sided test to detect this difference is approximately

$$1-\Phi\left[\frac{z_{1-\alpha/2k}\sqrt{\frac{k(kn+1)}{6}}-\frac{kn}{2}}{\sqrt{\frac{[(k-1)n+1][k-2]}{12}}}\right],$$

where n is the common within-group sample size. For a one-sided test, the divisor $2k$ in the subscript of z is replaced by k. The power of Dunn's test depends on the specific alternative. Nevertheless, the above function provides an approximate power that can be used for design assessment purposes.

We do not know any published power properties of the Jonckheere–Terpstra test except those based on a large

Table 3.21 Power for t-Test to Detect a Decrease of Magnitude d with $n_0=n_1=10$, $s_p=1$

d	Power	d	Power	d	Power
0	5	0.7	43	1.4	91
0.1	7	0.8	52	1.5	94
0.2	11	0.9	61	1.6	96
0.3	15	1	69	1.7	97
0.4	21	1.1	76	1.8	98
0.5	27	1.2	82	1.9	99

sample, i.e. asymptotic power, e.g. Poon (1980) and Potter and Sturm (1981). Very extensive computer simulation studies undertaken by the first named author of this book demonstrate that these asymptotic powers exceed, often grossly, the power realized by small samples. A small number of the aforementioned power simulations are given in the Annex to OECD (2006a, b, c).

3.9.4 Minimum Detectable Difference

It is sometimes more convenient to calculate the minimum detectable difference, MDD, which is the minimum magnitude difference that can be detected with 95% confidence. This is obtained from Eq. (3.34) by replacing $\bar{x}_0 - \bar{x}_1$ by MDD and T^* by Tcrit and solving for MDD, to get

$$\text{MDD} = \text{Tcrit} * s_p \sqrt{\frac{1}{n_0} + \frac{1}{n_1}}. \quad (3.35)$$

When $n_0 = n_1 = 10$, it follows that Tcrit = 1.734 06 and if $s_p = 1$, then MDD = 0.775 5. By interpolation in Table 3.19, it can be seen that the probability of getting a significant t-test when the true difference equals MDD is 0.5.

If the variances of the two populations are not equal, then Eq. (3.34) is replaced by

$$T^* = \frac{\bar{x}_0 - \bar{x}_1}{\sqrt{\frac{s_0^2}{n_0} + \frac{s_1^2}{n_1}}}, \quad (3.36)$$

and

$$\text{MDD} = \text{Tcrit} \sqrt{\frac{s_0^2}{n_0} + \frac{s_1^2}{n_1}}. \quad (3.37)$$

If there are more treatments and some test other than the t-test is used to determine the NOEC, a proper MDD calculation becomes more challenging. It is common to report the same MDD as discussed above in these more challenging situations as a useful approximation.

When the data are proportions, then

$$s_i^2 = p_i(1 - p_i), i = 0,1. \quad (3.38)$$

So MDD depends on the proportion p_1, which is to be determined,
and

$$T^* = \frac{\hat{p}_0 - \hat{p}_1}{\sqrt{\frac{s_0^2}{n_0} + \frac{s_1^2}{n_1}}}. \quad (3.39)$$

For proportions, what can be done is to replace $\hat{p}_0 - \hat{p}_1$ in Eq. (3.39) by D and compute the probability of obtaining a smaller value than T^* for every choice of $p_1 < p_0$. The largest value of $p_1 < p_0$ that makes that probability 0.95 is the MDD, that is, the smallest change from p_0 that is statistically significant at the 0.05 level.

MDD is applied to mesocosm studies in Chapter 13, to the size effect that can be estimated by regression in Staveley et al. (2018), and to responses requiring transformation in Brock et al. (2015).

The following programs to carry out the tests in this chapter are given in Appendix 1:

- Bonferroni–Holm adjusted p-values: BHolm.SAS, BHolm.r
- Shapiro–Wilk test: Data Check.SAS, DunnettTest.r, SHAPIRO WILK NORMALITY TEST.R
  - Data Check.SAS also does Levene test for variance homogeneity and Tukey outlier test.
  - Shapiro–Wilk normality test.r also does QQ-plot and histogram of residuals and creates residual dataset from a model.
- Levene variance homogeneity test: Data Check.SAS, BFL VARIANCE HOMOGENEITY TEST.R
- Dunn test: DunnTest.SAS and DunnTest.r
- Dunnett test: dunnett test.SAS, DunnettTest.r
  - DunnettTest.r also does Shapiro–Wilk test for normality.
- Tamhane–Dunnett: Tam_Dun2.SAS, TamhaneDunnett.r
- Williams' test: DOWILL_DECR_INCR_ALT.SAS, WilliamsTest V2.r
- Jonckheere–Terpstra test: Jonckheere macro.SAS, JonckheereTest.r
- Mann–Whitney test: MANN WHITNEY MACRO.SAS, MANN_WHIT_TEST.r

EXERCISES

Note. Exercise 3.13 contains a reasonably complete description of how to use the Williams' test programs in SAS and R.

In all datasets in the text and exercises, a concentration is given with the value −1, which means a solvent used in the study and the water control is recorded as −1 and the solvent control is listed as 0. The reader should determine whether to combine the two controls for NOEC determination or ECx estimation. If the two controls are to be combined, then the water control should be changed to 0 and the replicates renamed so that replicates in the two controls can be distinguished. If only the solvent control group is to be used, then the water control should be discarded. If only the water control should be used, then the solvent control should be discarded and the concentration for the water control should be changed to 0. Under no circumstances should the concentration value −1 be used for NOEC or ECx purposes. Guidance on this is provided in this chapter and in Chapter 2.

3.1 **(a)** Verify the calculations for V_{i0}, Z_1–Z_6, used for Dunn's test applied to the algae density data in Table 3.11.

(b) Verify the Dunn NOEC stated there.

(c) Apply the Bonferroni and Bonferroni–Holm adjustments to the p-values used to derive the Dunn NOEC and determine whether those adjustments change the NOEC compared with using unadjusted or Bonferroni-adjusted p-values or each other.

3.2 Verify the Dunnett and Williams calculations and NOECs in Tables 3.4 and 3.14.

3.3 Verify the NOECs from the Jonckheere–Terpstra, Williams, Dunnett (normal), and Dunnett (GLMM) for the daphnia reproduction data summarized in Table 3.19.

3.4 **(a)** Compute the amalgamated means and sample sizes for the Xenopus data summarized in Table 3.15, and verify the Williams' test result.

(b) Verify the Jonckheere–Terpstra and Dunnett (Poisson) NOECS for the same data.

3.5 A study was done to determine the effect on reproduction of earthworms exposed to a test chemical for 4 weeks. There were eight replicate test sites in the control and four replicate sites in each of eight test concentrations. Ten adult worms were initially placed in each test site. The responses of interest were adult survival (surv28), bodyweight (wgt28), body weight gain (wgtdiff), and reproduction (juveniles) after 4 weeks exposure. The initial body weight of worms in each replicate was also provided (StartWgt). See Table 3.22 for a summary of the data for this exercise. The data full are provided in an excel spreadsheet and SAS dataset in Appendix 1, both named *Exercise35*. Body weight is measured in milligrams and concentration (conc) in $mg\,kg^{-1}$ dry soil.

(a) Test the responses Wgt28, Wgtdiff, and juveniles for normality using the Shapiro–Wilk test and for variance homogeneity using Levene's test after eliminating the highest test concentration where there were no adult survivors and create a scatterplot of the data.

(b) Find the NOEC for Wgt28 using the Tamhane–Dunnett and Jonckheere–Terpstra tests and discuss the results.

(c) Analyze Wgtdiff using the Williams and Jonckheere–Terpstra tests.

(d) Analyze juveniles using Williams and Dunnett tests and discuss which is more appropriate for these data.

Note: In Chapter 4, regression models will be fit to these data.

3.6 A rainbow trout early life-stage experiment was run to determine the effect of a test chemical. There were four replicate tanks of 20 fish in the control and each of five test concentrations. See Table 3.23 for the replicate mean values of length and weight. These data are also given in the SAS dataset *ExerciseE36* in Appendix 1.

(a) Test each response for normality using the Shapiro–Wilk test and for variance homogeneity using Levene's test, and create a scatterplot of the data.

(b) Determine the NOEC for length using the Tamhane–Dunnett and Jonckheere–Terpstra tests and discuss the differences.

(c) Determine any outliers in the data for (a) and repeat the analyses.

(d) Determine the NOEC for weight using the Dunn and Jonckheere–Terpstra test and discuss the differences.

(e) Determine any outliers in the data for (d) and repeat the analyses.

3.7 A pond mesocosm study was done to determine the effects of a test chemical. One of the species measured was an aquatic plant of the family Myriophyllum and several length measurements were taken at 14 and 21 days post exposure. See Table 3.24 for the data. With expanded response names, the same data are given as an SAS dataset *Exercise37* in Appendix 1. Lengths are measured in cm and concentration in $mg\,l^{-1}$.

(a) Test the number of side shoots at day 14 (NSS14) for normality and variance homogeneity and create a scatterplot of the data.

(b) Determine the NOEC for sideshoots14 using the Tamhane–Dunnett, Dunn, and Jonckheere–Terpstra tests.

(c) Find any outliers in the NSS14 data and analyze the data with outliers omitted.

(d) Do a square-root transform of the count data NSS14, and determine the NOEC by the Dunnett, Williams, and Jonckheere–Terpstra tests.

(e) Repeat similar tasks (a)–(d) for the lengths of side shoots at days 14 and 21 (responses TLSS14 and TLSS21), making sure to use tests justified by the data.

(f, g) Repeat similar tasks for the lengths of the main shoots at days 14 and 21 (LenMS14 and LenMS21).

(h, i) Repeat similar tasks for total length on days 14 and 21 (TotLen14 and TotLen21).

3.8 A 21-day growth and toxicity test was done on duckweed. There were four replicate test vessels in each of two controls (water and solvent) and six test concentrations. The basic responses were frond count at days 3, 5, and 7 after exposure. In addition yield and growth rate were calculated. These are shown as Fyield7, the increase in frond count from day 0 to day 7, and GRATE7, the ratio (Ln(Fcount7)-Ln(Fcount0))/(7-0), that is, the average increase per day of the natural log of frond count. See Table 3.25 for the data. Concentration was measured in $\mu g\,l^{-1}$.

(a) Compare the two controls for Fyield7, realizing that the water control is coded as −1 and the solvent as 0. Combine the controls for further analysis and recode both as concentration = 0, if justified by the formal comparison. Otherwise, omit the water control and use only the solvent control for further testing.

(b) Evaluate Fyield7 for normality and variance homogeneity and create a scatterplot of the data.

(c) Based on the results of (b), determine an NOEC for Fyield7 from appropriate pairwise and trend tests assuming normal error structure.

(d) Repeat (a–c) after doing a square-root transform of the response.

(e) Repeat (a–c) using Dunnett's test based on a generalized linear model with Poisson error structure.

Table 3.22 Earthworm Reproduction

Group	Conc	Rep	StartWgt	Wgt28	WgtDiff	AtRisk	Surv28	Juveniles	Dead28
1	0	1	410.5	475.9	65.4	10	10	142	0
1	0	2	380	460.4	80.4	10	10	134	0
1	0	3	378.9	451.1	72.2	10	10	134	0
1	0	4	392.5	472	79.5	10	10	179	0
1	0	5	412.1	462.8	50.7	10	10	145	0
1	0	6	398.1	464	65.9	10	10	146	0
1	0	7	381	474.6	93.6	10	10	193	0
1	0	8	390.4	478	87.6	10	10	173	0
2	6.5	1	407.7	491.3	83.6	10	10	148	0
2	6.5	2	416.7	491.3	74.6	10	10	150	0
2	6.5	3	374.5	459.1	84.6	10	10	214	0
2	6.5	4	375.5	428.3	52.8	10	10	155	0
3	11.8	1	388.4	470.5	82.1	10	10	189	0
3	11.8	2	355.7	482.5	126.8	10	10	195	0
3	11.8	3	367	448.8	81.8	10	10	214	0
3	11.8	4	398.1	500.6	102.5	10	10	206	0
4	21.2	1	369.1	430.6	61.5	10	10	303	0
4	21.2	2	363.4	444.7	81.3	10	10	220	0
4	21.2	3	383	473.9	90.9	10	10	222	0
4	21.2	4	378.7	435.9	57.2	10	10	237	0
5	38.1	1	367.5	479.7	112.2	10	10	244	0
5	38.1	2	343.3	438.4	95.1	10	10	238	0
5	38.1	3	361.1	427.7	66.6	10	10	254	0
5	38.1	4	365.3	467.7	102.4	10	10	263	0
6	68.6	1	366.8	484.3	117.5	10	10	225	0
6	68.6	2	323.2	451.8	128.6	10	10	219	0
6	68.6	3	357.9	480	122.1	10	10	258	0
6	68.6	4	354.6	466.9	112.3	10	10	248	0
7	123.5	1	331.7	476.8	145.1	10	10	197	0
7	123.5	2	369.5	510.3	140.8	10	10	183	0
7	123.5	3	368.3	491.4	123.1	10	10	232	0
7	123.5	4	353.5	453.8	100.3	10	10	185	0
8	222.2	1	406.6	521.3	114.7	10	3	32	7
8	222.2	2	367.7	454.6	86.9	10	9	78	1
8	222.2	3	355.5	509.6	154.1	10	8	66	2
8	222.2	4	337.1	446.5	109.4	10	8	64	2
9	400	1	383.3			10	0	0	10
9	400	2	369.6			10	0	0	10
9	400	3	395.2			10	0	0	10
9	400	4	369.7			10	0	0	10

StartWgt, weight of earthworms at the start of the study; Wgt28, weight of worms on day 28; WgtDiff, weight gain from the start of the study to day 28; AtRisk, number of adult earthworms at the start of the study; Surv28, number of adults that survived up to day 28; Juveniles, number of young produced by day 28; Dead28, number of adults dead by day 28.

(f) Calculate the growth rate GRATE7 and repeat the procedures in (a–c) for that response.

(g) Repeat the above processes for 3- and 5-day yield and, using only normal error structures, for growth rate.

These data will be revisited after regression methods are developed in Chapters 6 and 7.

3.9 A 21-day daphnia chronic study was done, with two replicate vessels of 10 daphnid each in the control and each of seven test concentrations. See Table 3.26 for the data, where concentration is measured in $\mu g\, l^{-1}$. Basic responses were the number of surviving adults, numbers of immobilized young and total young produced at 14 and 21 days. From these, the primary

Table 3.23 RBT ELS Results

Group	Rep	Length	Weight	Conc	Group	Rep	Length	Weight	Conc
1	1	19.6	0.147	0	4	1	18.9	0.137	25
1	2	19.6	0.149	0	4	2	19.0	0.138	25
1	3	19.5	0.151	0	4	3	19.0	0.143	25
1	4	19.5	0.143	0	4	4	19.3	0.137	25
2	1	19.3	0.150	6.25	5	1	18.6	0.128	50
2	2	19.3	0.143	6.25	5	2	18.4	0.126	50
2	3	19.2	0.140	6.25	5	3	18.3	0.124	50
2	4	19.3	0.144	6.25	5	4	18.3	0.128	50
3	1	19.4	0.145	12.5	6	1	17.3	0.112	100
3	2	18.7	0.128	12.5	6	2	18.2	0.113	100
3	3	18.8	0.127	12.5	6	3	18.4	0.148	100
3	4	19.5	0.153	12.5	6	4	17.5	0.110	100

Length and weight, replicate means.

Table 3.24 Myriophyllum Lengths

Group	Rep	LenMS14	NSS14	TLSS14	TotLen14	LenMS21	NSS21	TLSS21	TotLen21	Conc
1	A	13	4	10.5	23.5	18	6	22	40	0
1	B	11	3	11.5	22.5	13	4	28	41	0
1	A	12	5	21	33	17	8	54	71	0
1	B	4.5	1	2.5		6.5	2	6		0
1	A	15.5	5	27.5	43	22	6	51	73	0
1	B	9.5	3	10	19.5	18.5	3	20	38.5	0
2	A	8	1	1.5	9.5	11	1	5.5	16.5	0.0001
2	B	12	2	10	22	18	4	26	44	0.0001
2	A	12.5	5	16	28.5	18	5	30	48	0.0001
2	B	13	6	18.5	31.5	15.5	6	37	52.5	0.0001
2	A	11.5	4	9.5	21	14	4	20.5	34.5	0.0001
2	B	20	5	18.5	38.5	27	5	27.5	54.5	0.0001
3	A	8	3	9.5	17.5	15	5	21	36	0.001
3	B	12.5	3	9	21.5	12.5	4	22	34.5	0.001
3	A	8	1	4	12	15	4	25	40	0.001
3	B	12	1	7	19	12	1	5.5	17.5	0.001
3	A	12	2	11	23	14	3	17.5	31.5	0.001
3	B	13	3	7.5	20.5	13	3	26.5	39.5	0.001
4	A	5	0	0	5	5.5	0	0	5.5	0.01
4	B	3	1	3	6	3.5	1	3.5	7	0.01
4	A	6	0	0	6	5.5	0	0	5.5	0.01
4	B	4	0	0	4	6	2	1.5	7.5	0.01
4	A	6	0	0	6	8	0	0	8	0.01
4	B	4.5	2	4.5	9	5	3	6	11	0.01
5	A	6.5	0	0	6.5	8	0	0	8	0.1
5	B	8	1	1	9	9	1	1	10	0.1
5	A	4.5	0	0	4.5	6	1	1	7	0.1
5	B	8	0	0	8	9	1	1	10	0.1
5	A	9	1	3	12	9.5	1	3	12.5	0.1
5	B	7	0	0	7	7	0	0	7	0.1
6	A	5.5	0	0	5.5	5.5	0	0	5.5	1
6	B	7	1	3	10	8	1	3	11	1
6	A	6.5	0	0	6.5	8	0	0	8	1
6	B	5	0	0	5	2.5	0	0	2.5	1
6	A	6.5	0	0	6.5	7.5	0	0	7.5	1
6	B	6	0	0	6	6	1	1	7	1

LenMS*x*, length of main stem on day *x*; NSS*x*, number of side shoots on day *x*; TLSS*x*, total length of side shoots on day *x*; TotLen*x*, total length of plant on day *x*.

Table 3.25 Duckweed Growth

Conc	Group	Replicate	Fronds0	Fronds3	Fronds5	Fronds7	FYield7
−1	1	A	12	30	69	157	145
−1	1	B	12	30	73	155	143
−1	1	C	12	33	70	131	119
−1	1	D	12	36	86	177	165
0	1	A	12	28	68	129	117
0	1	B	12	39	82	169	157
0	1	C	12	30	77	159	147
0	1	D	12	36	82	174	162
0.16	2	A	12	34	69	147	135
0.16	2	B	12	30	69	160	148
0.16	2	C	12	39	90	174	162
0.16	2	D	12	32	74	153	141
0.49	3	A	12	36	89	179	167
0.49	3	B	12	33	81	156	144
0.49	3	C	12	34	75	148	136
0.49	3	D	12	29	69	149	137
1.5	4	A	12	24	47	94	82
1.5	4	B	12	21	38	80	68
1.5	4	C	12	24	42	100	88
1.5	4	D	12	23	49	119	107
4.4	5	A	12	13	15	18	6
4.4	5	B	12	16	18	23	11
4.4	5	C	12	15	20	22	10
4.4	5	D	12	14	19	22	10
13	6	A	12	14	16	16	4
13	6	B	12	13	15	16	4
13	6	C	12	14	19	20	8
13	6	D	12	13	14	15	3
40	7	A	12	12	12	13	1
40	7	B	12	13	14	14	2
40	7	C	12	14	14	14	2
40	7	D	12	14	14	15	3

Fronds*x*, number of fronds on day *x*; Fyield7, frond yield on day 7.

Table 3.26 Daphnia Magna Reproduction

Conc	Group	Rep	Imm14	Imm21	Surv14	Surv21	Young14	Young21	YperA14	YperA21
0	1	1	0	2	10	8	207	600	20.70	75.00
0	1	2	0	0	10	10	497	1193	49.70	119.30
0.05	2	1	0	0	10	10	512	1142	51.20	114.20
0.05	2	2	6	6	4	4	258	502	64.50	125.50
0.1	3	1	7	7	3	3	207	339	69.00	113.00
0.1	3	2	0	0	10	10	314	666	31.40	66.60
0.589	4	1	0	0	10	10	462	1011	46.20	101.10
0.589	4	2	0	2	10	8	449	875	44.90	109.38
4.14	5	1	0	0	10	10	482	966	48.20	96.60
4.14	5	2	1	1	9	9	339	536	37.67	59.56
20	6	1	0	2	10	8	426	705	42.60	88.13
20	6	2	1	1	9	9	353	593	39.22	65.89
105	7	1	1	1	9	9	275	609	30.56	67.67
105	7	2	1	4	9	6	372	557	41.33	92.83
403	8	1	11	11	0	0	24			
403	8	2	11	11	0	0	20			

Imm*x*, number of immobilized young on day *x*; Surv*x*, number of adults surviving on day *x*; Young*x*, number of live young on day *x*; YperA*x*, number of live young per surviving adult on day *x*.

responses of interest were young per surviving adult at 14 and 21 days.

(a) Check the responses YperA14 and YperA21 for normality and variance homogeneity using the Shapiro–Wilk and Levene tests, respectively.

(b) For YperA21, carry out the Tamhane–Dunnett and Jonckheere–Terpstra tests and determine the NOEC from each test. Discuss the differences.

(c) Repeat (a) and (b) for YperA14.

(d) Analyze Young21 using a GLMM with Poisson error structure and Dunnett's test.

(e) Analyze Young21 using a normal error structure after a square-root transform of the response.

These data will be revisited in Chapters 4, 6, and 7.

3.10 A rainbow trout study was done, and, among other responses, the first day of hatching was recorded for each replicate. There were four replicate tanks of 15 trout embryos at the start of the study. The first day of hatching was recorded for each embryo. See Table 3.27 for a summary of the data. Repsize is the number of trout that hatched by the end of the hatching period. Concentrations were in μg l^{-1}.

(a) Check the data for normality and variance homogeneity using the Shapiro–Wilk and Levene tests, respectively, and create a scatterplot of the data.

(b) Plot a QQ-plot and histogram of residuals and do a visual check of normality.

(c) Carry out the Tamhane–Dunnett and Jonckheere–Terpstra tests to determine an NOEC and discuss any difference in results.

3.11 A daphnia chronic study was done with 10 replicate vessels of one daphnid each in the control and five treatment groups. The test concentrations were measured in μg l^{-1}. The length of each daphnid was measured at the end of the 21-day study. (See Table 3.28.)

(a) Check the data for normality and variance homogeneity using the Shapiro–Wilk and Levene tests, respectively, and create a scatterplot of the data.

(b) Plot a QQ-plot and histogram of residuals and do a visual check of normality.

(c) Carry out the Jonckheere–Terpstra, Dunn, and Mann–Whitney tests on these data. For the latter two tests, do a Bonferroni–Holm adjustment on the p-values. Note that the maximum decrease in length is just under 10%.

3.12 See Table 3.29 for length data from another 21-day daphnia chronic study. There were four replicate vessels of 10 daphnia in each treatment group and control. A solvent was used. Concentrations were measured in μg l^{-1} and length was measured in mm.

Table 3.28 Daphnia Chronic Length

Rep	Length	Conc	Rep	Length	Conc	Rep	Length	Conc
A	3.9	0	A	3.5	5.7	A	3.8	23
B	4	0	B	3.9	5.7	B		23
C	3.9	0	C	3.9	5.7	C	3.7	23
D	4	0	D	3.8	5.7	D	3.8	23
E		0	E	3.9	5.7	E		23
F	3.9	0	F	3.9	5.7	F	3.7	23
G	4	0	G		5.7	G	3.5	23
H	3.9	0	H	3.9	5.7	H	3.6	23
I	3.8	0	I		5.7	I		23
J	3.9	0	J	3.9	5.7	J	3.9	23
A	3.9	2.8	A	3.8	11	A	3.6	45
B	3.9	2.8	B		11	B	3.4	45
C	3.8	2.8	C	3.9	11	C	3.6	45
D	3.7	2.8	D	3.8	11	D	3.3	45
E	3.9	2.8	E	3.8	11	E	3.5	45
F	3.8	2.8	F	3.9	11	F	3.7	45
G	3.7	2.8	G	3.9	11	G	3.5	45
H	3.8	2.8	H		11	H	3.6	45
I	3.9	2.8	I		11	I	3.6	45
J	3.8	2.8	J	3.7	11	J	3.7	45

Length, length of individual daphnid at the end of the study. One daphnid per beaker.

Table 3.27 Rainbow Trout First Day of Hatch

Rep	Group	FirstHat	RepSize	Conc	Rep	Group	FirstHat	RepSize	Conc
A	1	33	15	0	A	4	32	15	25
B	1	33	15	0	B	4	32	14	25
C	1	32	14	0	C	4	31	15	25
D	1	33	15	0	D	4	33	15	25
A	2	33	13	6.25	A	5	31	14	50
B	2	32	15	6.25	B	5	31	15	50
C	2	32	15	6.25	C	5	32	15	50
D	2	31	13	6.25	D	5	32	14	50
A	3	32	14	12.5	A	6	32	14	100
B	3	32	15	12.5	B	6	32	15	100
C	3	32	15	12.5	C	6	32	14	100
D	3	31	15	12.5	D	6	31	15	100

FirstHat, first day of hatching.

Table 3.29 Daphnia Length Data

Conc	Rep	Length	RepSize	Conc	Rep	Length	RepSize
−1	1	3.35	8	0.4	3	3.85	6
−1	2	3.53	9	0.4	4	3.83	8
−1	3	3.38	8	1.39	1	3.90	7
−1	4	3.37	10	1.39	2	3.96	5
0	1	3.66	7	1.39	3	4.00	7
0	2	3.77	6	1.39	4	3.90	9
0	3	3.74	9	4.86	1	3.90	7
0	4	3.70	7	4.86	2	3.88	8
0.113	1	3.75	10	4.86	3	4.07	6
0.113	2	3.80	10	4.86	4	3.83	6
0.113	3	3.87	10	17	1	3.53	7
0.113	4	3.82	9	17	2	3.40	3
0.4	1	3.65	8	17	3	3.60	1
0.4	2	3.71	9	17	4	3.28	5

Conc=−1 is water control; Conc=0 is solvent control. Rep, replicate identity. Length, replicate mean length. RepSize, number of daphnid in the replicate vessel.

(a) Compare the two controls, where the water control is coded −1 and the solvent control is coded 0. Combine them and recode as all concentration 0 if justified by the test used to compare them and otherwise use only the solvent control for further analysis.

(b) Determine whether the data are consistent with normality and variance homogeneity using the Shapiro–Wilk and Levene tests.

(c) Create scatter and QQ-plots and a histogram of the residuals from ANOVA.

(d) Do an appropriate pairwise test to determine an NOEC. Do not overlook the unequal sample sizes. The complete data with individual lengths is given in Appendix 1.

(e) Are these data consistent with a monotone concentration–response? Should a one-sided test for decrease be used or a two-sided test that examines increases as well as decreases?

(f) Do a trend test to determine the NOEC and justify not using such a test.

We will revisit these data in Chapter 4.

3.13 Williams' test with individual subjects

The full Xenopus dataset is given in worksheet Data_for_exercise_E3_13 in the Excel file Xenopus.xls and also in the SAS dataset *Xenopusr*. The values for Well in those data have no physical meaning as to actual location on the titerplate. They are there merely distinguish wells within replicates. The replicate means are given in the Excel file Xenopus_repmns.xls and the SAS dataset of the same name. An SAS program to implement Williams' test is DOWILL_DECR_INCR_ALT2.SAS, and the corresponding R program is *Will.r*. Both are given in Appendix A. Both programs require a variable named Dose with values 1 for control, 2 for the lowest test concentration, 3 for the next, and so on, unless there are both a water and a solvent control. If there is a solvent control, the water control is designated by conc=−1 and dose=1, the solvent control is designated by conc=0 and dose=2, and the program will combine the controls. If one does not want to combine controls, the control not to be used should be omitted from the dataset prior to using these programs. Also, these programs require a variable that identifies replicates. If individual subjects within replicate are in the dataset, then variable identifying subjects must be present.

(a) Use the *Xenopus_repmns* data to verify the calculations done for Table 3.15. In the SAS program, this requires leaving blank the Subject=entry in the macro.

(b) Use the full data in the SAS dataset *Xenopusr* or the Excel file *Xenopus.xls* to analyze these data. In the SAS program, it is necessary to specify Subject=Well to get the correct analysis. For both (a) and (b), test for an increasing trend and compare results.

(c) Repeat the above, except test for a decreasing trend. This will illustrate the effect of mis-specifying the direction with Williams' test.

(d) Repeat (b), except leave blank the Subject=entry in the SAS macro. Note the differences, especially in the degrees of freedom used. Degrees of freedom for Williams' test should be the same as for Dunnett's test when rep is nested in dose and subject is, by implication, nested in rep. This will illustrate the consequences of careless use of programs.

3.14 Williams' test with increasing dose–response

(a) Analyze VTG and LogVTG from Vtg_medaka_metocean2F using Williams' test. Check assumptions and note differences in results. Address the question: Does it matter whether a test is used when data requirements are not met?

(b) Analyze the same responses using other tests as appropriate and compare results.

These data will be revisited in Chapter 4.

3.15 Increasing dose–response

In each of the following 21-day fish screening experiments, check the VTG and LogVTG data for requirements for Williams' test. Do the test regardless of whether the data meet those requirements. Address the question: Does it matter whether a test is used when data requirements are not met? Also apply appropriate pairwise tests and Jonckheere–Terpstra. The dataset name applies to SAS and to spreadsheets in the Excel file Exercise315. These data will be revisited in Chapter 4.

(a) Dataset Ex315a_repmns.

(b) Dataset Ex315b_repmns.

(c) Dataset Ex315c_repmns.

(d) Dataset Ex315d_repmns.

Chapter 4

Analysis of Continuous Data: Regression

4.1 INTRODUCTION

Regression modeling of toxicity data is becoming increasingly required in the ecotoxicology community. The advantages of regression over hypothesis testing have been promoted by numerous authors. Chapter 1 introduces some of this discussion. OECD (2014a) contains much more and many references. It is not the purpose of this chapter to explore the relative merits of hypothesis testing and regression modeling to evaluate ecotoxicity data. Rather, we present a discussion of models that have been found useful in that evaluation and how to evaluate these models for appropriateness and to select the more useful ones for a particular dataset. All of these models were developed originally for continuous responses with normally distributed errors. They can all be fit to count data with Poisson or negative binomial errors. Such models are explored in detail in Chapter 8. Programs to carry out both normal error models and Poisson error models will be provided.

4.1.1 Interpolation Versus Extrapolation

Once an increasing regression model has been fit to a dose–response dataset, it is possible to estimate an $x\%$ effects concentration, ECx, for any positive value of x and if a decreasing regression model is fit, it is possible to estimate ECx for any x in the range $0<x<100$. Specialized models, e.g. the "hockey-stick model" apparently introduced by Cox (1987) and based on an *assumed* toxic threshold, will be discussed later that may also allow estimation of EC0, also known as the no effect concentration (NEC), not to be confused with the no observed effects concentration (NOEC). Threshold models have been proposed for carcinogenic studies (e.g. Purchase and Auton, 1995), teratogenesis (e.g. Gaylor et al., 1988), epidemiology (e.g. Ulm, 1991), and ecotoxicology (e.g. Klepper and Bedaux, 1997) and are an important part of the biology-based models of Kooijman (e.g. OECD, 2006a, b, chapter 7). Not all researchers accept the existence of a toxic threshold (e.g. Brenner and Sachs, 2006; Brenner and Hall, 2007), while others (e.g. Scott, 2008) regard it as critically important. This lack of universality underscores the caution that should be used in extrapolating below the lowest tested positive concentration from one of the simple models commonly used in toxicology.

A threshold model contains a parameter representing the threshold. The form of a threshold model is given by

$$\begin{aligned} y &= a && \text{if } x < c \\ y &= a + f(x) && \text{if } x \geq c \end{aligned},$$

where the parameter c is the threshold concentration and $f(x)$ may be any function that is zero at $x=c$. For a decreasing dose–response, $f(x)<0$ for $x>c$, and for an increasing function, $f(x)>0$ for $x>c$. An example is the piecewise exponential "hockey-stick" model given by

$$\begin{aligned} y &= a && \text{if } x < c \\ y &= ae^{x-c} && \text{if } x \geq c \end{aligned}.$$

This response is exponential above the threshold and the model is smooth (i.e. continuous and differentiable at $x=c$), which gives it nice mathematical properties.

It is important to understand that the ability to obtain an estimate from a model for ECx or some other parameter

Statistical Analysis of Ecotoxicity Studies, First Edition. John W. Green, Timothy A. Springer, and Henrik Holbech.

Companion website: www.wiley.com/go/Green/StatAnalysEcotoxicStudy

does not mean that all such estimates are useful or even meaningful. For example, if no more than a 10% effect is observed at any test concentration, estimates of ECx for $x \geq 50$ are probably meaningless and certainly unjustified by the data. This is an example of extrapolation beyond the range of the data. It is quite easy to construct examples showing the potential for error in such extrapolations. See Table 4.1 for data that will illustrate this issue.

Replicate mean lengths of rainbow trout from an early life stage experiment are given in Table 4.1. If a model is fit to the data excluding the high test concentration and EC50 is estimated (by extrapolation), the estimated EC50 is 237 with 95% confidence bounds (−561, 1035). See Figure 4.1 to see that the model fits the data reasonably well. See Table 4.2 to see the observed and predicted percent change from control to help assess goodness of fit.

The extremely wide confidence bounds on the EC50 estimate serve as a warning about the uncertainty in the EC50 estimate, which requires considerable extrapolation beyond the range of the data used. In this case, data is actually available for a higher test concentration, and if all the data are used, the resulting EC50 estimate is 6.5 with 95% confidence bounds (5.8, 7.2) and the curve appears quite different. See Figure 4.2 for the model fit to the full dataset, and see Table 4.3 for a comparison of the observed and predicted values from the full model.

Table 4.1 Extrapolation of Rainbow Trout Length

Rep	Conc (mg l^{-1})	Length (cm)	Species	Conc (mg l^{-1})	Rep	Length (cm)
1	−1	3.01	RBT	0.692	1	2.87
2	−1	2.99	RBT	0.692	2	2.77
1	0	2.92	RBT	1.35	1	2.83
2	0	2.91	RBT	1.35	2	2.77
1	0.162	2.91	RBT	2.63	1	2.73
2	0.162	2.86	RBT	2.63	2	2.63
1	0.334	2.80	RBT	5.5	1	1.80
2	0.334	2.82	RBT	5.5	2	.

Length, replicate mean length. As always, conc = −1 is the code for the water control when a solvent is present. Conc = 0 is solvent control. For analysis, the two controls should be pooled (with replicates relabeled to keep them distinct), or only the solvent control is used following the guidance in Chapter 3.

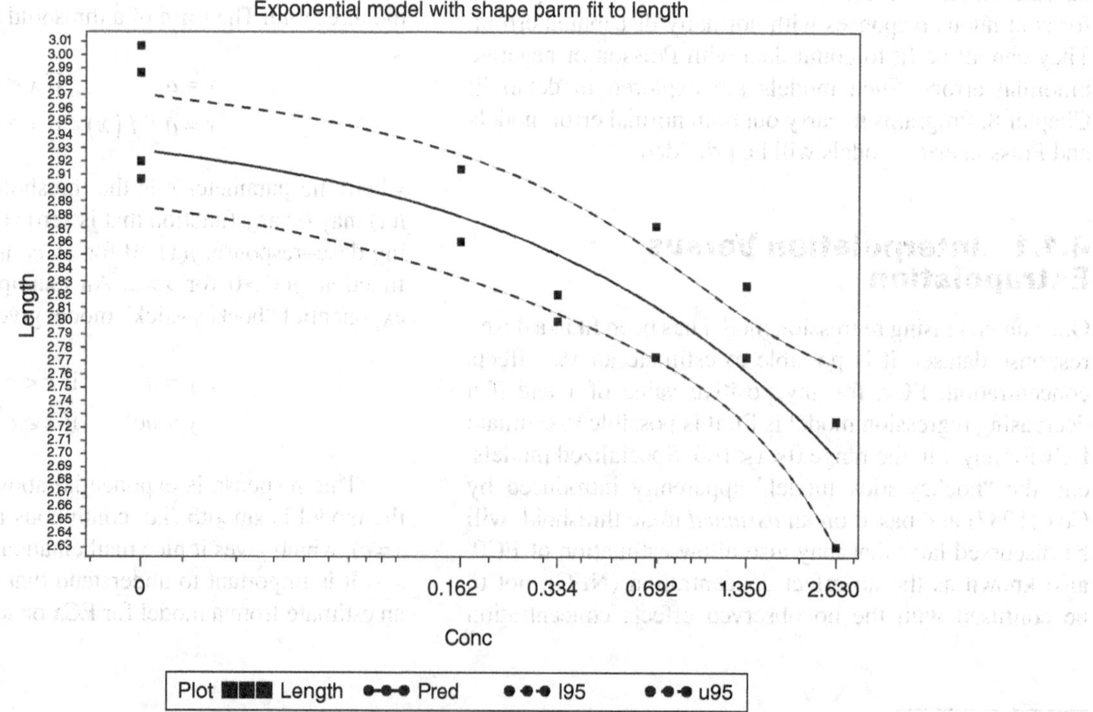

Figure 4.1 Exponential curve fit to rainbow trout length data. Solid square, replicate means; Solid line, predicted concentration–response curve; Dashed lines, 95% confidence bounds on predicted mean response.

Table 4.2 Observed and Predicted Percent Change from Control

Conc	N	Observed mean	Predicted mean	Observed % change	Predicted % change
0	4	2.955	2.953 12	0	0
0.162	2	2.887 15	2.878 69	−2	−3
0.334	2	2.81	2.850 33	−5	−3
0.692	2	2.822 38	2.811 14	−4	−5
1.35	2	2.8	2.762 61	−5	−6
2.63	2	2.677 89	2.698 41	−9	−9

Observed % change, % change in treatment mean from *observed* control mean; Predicted % change, % change in treatment mean from *predicted* control mean.

Table 4.3 Observed and Predicted Percent Change from Control from Expanded Data

Conc	N	Observed mean	Predicted mean	Observed % change	Predicted % change
0	4	2.96	2.89	0	0
0.162	2	2.89	2.89	−2	0
0.334	2	2.81	2.88	−5	0
0.692	2	2.82	2.88	−4	0
1.35	2	2.80	2.83	−5	−2
2.63	2	2.68	2.65	−9	−8
5.5	1	1.80	1.81	−39	−37

Observed % change, % change in treatment mean from *observed* control mean; Predicted % change, % change in treatment mean from *predicted* control mean.

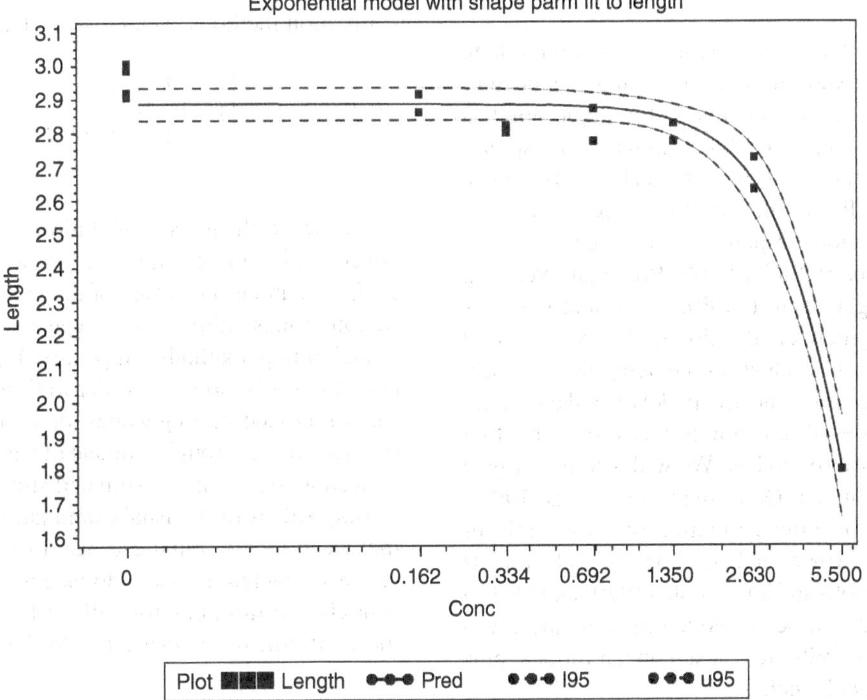

Figure 4.2 Exponential curve fit to expanded rainbow trout length data. Control mean is underestimated.

Estimation of EC50 still requires extrapolation, but extrapolating from 37% effect to 50% is much more reasonable than extrapolating from 9 to 50%. The horizontal axis on both plots is on a log-scale. It should be noted that in Figure 4.2, the control mean is somewhat underestimated. While this affects all ECx estimates, the impact on EC50 is likely to be minimal. More attention to the impact of misestimation of the control mean and other model diagnostics will be given in this chapter.

Caution should also be used in extrapolating below the lowest tested positive concentration. Examples similar to that for extrapolating beyond the highest tested concentration exist. This concern applies even though the tested concentrations include the zero concentration control. The concerns regarding low-dose extrapolation are well documented. The articles cited above for and against threshold models indicate some of the concerns. There is little reason to expect that the shape of the curve outside the range of tested concentrations follows the same pattern as within that range. Of course, in the abstract, one could make such a claim about interpolating between two tested positive concentrations. However, more information is available to address that issue. Natural phenomena do not typically behave in arbitrary or whimsical fashion with strange ups and downs. While there are exceptions and unusual laboratory results certainly occur, generally a sublethal response follows a smooth path over a short range of concentrations. It is thus reasonable to assume a given mathematical shape persists between tested concentrations. How rapidly a dose–response tends to zero below the lowest tested positive concentration is much more uncertain. It can continue on a gradual decline toward zero as the concentration approaches zero (but not necessarily along the same mathematical curve) or the response may decrease toward zero at an apparent positive threshold of toxicity.

4.2 MODELS IN COMMON USE TO DESCRIBE ECOTOXICITY DOSE–RESPONSE DATA

The focus of this section is on experiments in which there is a single chemical exposed in an increasing sequence of concentration levels or rates or doses at a single point in time to independent groups of subjects in which a response measured on a continuous scale is observed for each subject. In Chapter 5, we will consider different types of experiments where a continuous response is measured.

OECD (2014a, b, c), Slob (2002), Bruce and Versteeg (1992), Schabenberger et al. (1999), Brain and Cousens (1989), and Cedergreen et al. (2005) describe several models that are useful to evaluate ecotoxicity data. The last three of these references concern models for data where hormesis or low-dose stimulation is observed. The first three are general-purpose models. We will also investigate some biology-based models (Kooijman and Bedaux, 1996), specifically those that allow estimation of a threshold or true NEC, e.g. Cox (1987) and the hockey-stick models discussed in Yanagimoto and Yamamoto (1979) and OECD (2010a). Models that will be developed in detail are given in Eqs. (4.1)–(4.15). With five test concentrations plus control, potential models include:

- Bruce–Versteeg (BVP)
- Simple exponential (OE2)
- Exponential with shape parameter (OE3)
- Simple exponential with lower bound (OE4)
- 3-Parameter logistic (3PL)
- Linear hockey stick (HS3PL)

With six or more test concentrations plus control, models include:

- Exponential with shape parameter and lower bound (OE5)
- Michaelis–Menten
- Hill
- 4-Parameter logistic (4PL)
- Hockey stick (HS4PL, HS4PE, and HS5PE)

4.2.1 Bruce–Versteeg Model

Motivated by the probit model for quantal data, *Bruce and Versteeg* proposed the following model:

$$BVP: \quad \begin{aligned} y &= y_0\left[1-\Phi\left(a+b^* \log(U)\right)\right] && \text{for } U>0 \\ &= y_0 && \text{for } U=0 \end{aligned}, \qquad (4.1)$$

where U is the concentration, Φ is the cumulative standard normal distribution function, y_0 is the estimated control mean, and a and b are jointly estimated from the data. It can be shown that the BVP curve in Eq. (4.1) is continuous at $U=0$, and indeed, has derivative (slope) 0 there. It is possible to reparameterize this model to make ECx a model parameter, and it is advantageous to do so because both point estimate and confidence interval (CI) for ECx can be obtained directly from standard maximum likelihood parameter estimation methods. The reparameterized model is

$$BVP: \quad y = y_0\left\{1-\Phi\left[\Phi^{-1}\left(\frac{x}{100}\right)+\frac{\log(U)-\log(ECx)}{\gamma}\right]\right\}. \qquad (4.1^*)$$

Here x is the percent inhibition from control defining the concentration to be estimated (i.e. we are estimating ECx), $\gamma=1/b>0$ and is a measure of the variance in tested concentration values. Either model can be fit through nonlinear model fitting methods. Appendix 1 provides SAS and R code to fit this model. In order to fit this model, it is necessary to provide starting values for the iterative algorithm the software uses. A rough estimate of γ is EC50/EC15, and that as well as ECx can be estimated sufficiently well to obtain starting values from visual examination. The sample control mean can be the initial estimate of y_0. Bruce and Versteeg also use a weighting factor to take into account that the percent change from control will tend to be more variable for more extreme percentages than for intermediate values.

4.2.2 Hormetic Models

Brain and Cousens (1989) proposed a hormetic logistic model based on the logistic that incorporates low-dose stimulation and is useful in modeling some toxicity data. This and related models are discussed in Schabenberger et al. (1999). What is given below is a modification of their model, which eliminates one parameter used to estimate a lower horizontal asymptote for the dose–response. Over the range of concentrations generally included in laboratory experiments, this asymptote is often not approached and interest rarely lies in the extreme upper range of concentration values:

$$\text{HORM1}: \quad y = \frac{y_0+\gamma^* U}{1+w^* e^{\beta(\log U-\log ECx)}}, \qquad (4.2)$$

where

$$w = \frac{p^* y_0 + \gamma^* \text{EC}x}{y_0(1-p)},$$

$\log \text{EC}x = \log(\text{EC}x)$ is the natural logarithm of ECx, $p = x/100$, U is dose or concentration, and $\log U = \log(U)$ is the natural logarithm of U.

Thus, there are four parameters to estimate, y_0, γ, β, and ECx. Schabenberger et al. (1999) also introduce a weight function in estimating these parameters to take account of a probable relationship between the mean response and variance. An alternative hormetic model was proposed by Cedergreen et al. (2005) that requires five parameters. A modified formulation of this model is given in Eq. (4.3).

$$\text{HORM2}: \quad y = \frac{y_0 + fe^{-U^{-a}}}{1+(U/d)^b}, \tag{4.3}$$

An unfortunate aspect of this model is that there is no reparameterization that makes ECx a model parameter. This means maximum likelihood estimates (MLEs) of ECx and its confidence bounds cannot be obtained readily. A program is given in Appendix 1 that computes ECx and approximate CIs from the fiducial bounds method (described in Eq. (4.3)) and the delta method (adapted from Mi, 2010, and described in Appendix 2). We have found HORM2 fits some datasets where our modification of the Brain–Cousens model leaves some model parameter with wide confidence bounds that include zero. The reverse is also true, and examples of both are provided in this chapter. As is generally true in fitting models to data, there is a need for more than one model for hormetic data.

4.2.3 Exponential Models

Slob (2002) introduced a class of five exponential models that in his experience fit most ecotoxicity data. These are also discussed in OECD (2014a, b, c). We have found need to expand on this list to include, at a minimum, the Bruce–Versteeg and Brain–Cousens models and many in the field find 3- or 4-parameter log-logistic models to be very important. In some cases, the Hill and Michaelis–Menten models have fit data that could not be fit by any of the others, but that has been rare in our experience. In describing the exponential models (4.4)–(4.8), the parameters in all cases satisfy $a > 0$, $b > 0$, $1 > c > 0$, and $d \geq 1$, though occasionally, it is useful to relax the requirement that $d \geq 1$ to $d > 0$:

$$\text{OE1}: \quad y = a, \tag{4.4}$$

$$\text{OE2}: \quad y = a\exp\left(-\frac{\text{U}}{b}\right), \tag{4.5}$$

$$\text{OE3}: \quad y = a\exp\left(-\left(\frac{\text{U}}{b}\right)^d\right), \tag{4.6}$$

$$\text{OE4}: \quad y = a\left[c-(c-1)\exp\left(\frac{-U}{b}\right)\right], \tag{4.7}$$

$$\text{OE5}: \quad y = a\left[c-(c-1)\exp\left(-\left(\frac{U}{b}\right)^d\right)\right], \tag{4.8}$$

where y is any continuous endpoint and U denotes the dose (or concentration). In all models the parameter a represents the level of the endpoint at dose zero, and b can be considered as the parameter reflecting the efficacy of the chemical (or the sensitivity of the subject). At high doses, models OE4 and OE5 level off to the value ac, so parameter c can be interpreted as the maximum relative change. Models OE3 and OE5 have the flexibility to mimic threshold-like responses (i.e. slowly changing at low doses and more rapidly at higher doses). All these models are nested to each other, except models OE3 and OE4, which both have three parameters.

Equations (4.4)–(4.8) describe decreasing concentration–response curves, which are the most commonly needed in ecotoxicity studies, but removing the minus sign in the exponent and restricting $c > 1$ will define corresponding increasing concentration–responses.

In all of these models, the parameter a is constrained to be positive because it denotes the value of the endpoint at dose zero. Parameter d is constrained to values larger than (or equal to) one, to prevent the slope of the function at dose zero being infinite, which seems biologically implausible. Parameter b is constrained to be positive in all models to avoid confusion about whether the model defines an increasing or decreasing function.

As with other models, these (with the exception of OE1) can be reparameterized so that ECx is a model parameter:

$$\text{OE2}: \quad y = y_0 e^{-(U/\text{EC}x)\text{Log}\left(\frac{1}{1-x/100}\right)}, \tag{4.5*}$$

$$\text{OE3}: \quad y = y_0 e^{-(U/\text{EC}x)^d \text{Log}\left(\frac{1}{1-x/100}\right)}, \tag{4.6*}$$

$$\text{OE4}: \quad y = y_0\left[C-(1-C)e^{-(U/\text{EC}x)\text{Log}\left(\frac{C-1}{C-1-x/100}\right)}\right], \tag{4.7*}$$

$$\text{OE5}: \quad y = y_0\left[C-(1-C)e^{-(U/\text{EC}x)^d \text{Log}\left(\frac{C-1}{C-1-x/100}\right)}\right]. \tag{4.8*}$$

Alternatively, as shown in Slob (2002), a point estimate for ECx can be obtained for all five models, from an explicit formula derived from Eqs. (4.5) to (4.8):

$$\mathrm{EC}x = \left\{-\frac{\mathrm{Ln}\left[(\omega+1-c)/(1-c)\right]}{b}\right\}^{1/d},$$

where ω is defined as $\omega = y(\mathrm{EC}x)/a - 1$, y is the value of the model for $U=0$, $c=0$ for models OE2 and OE3, and $d=1$ for models OE2 and OE4. To obtain a CI from this expression, however, requires the delta method (Appendix 2) or some other approximation method, so the reparameterization in Eqs. (4.5*)–(4.8*) is appealing.

4.2.4 Log-logistic Models

The 3- and 4-parameter log-logistic models for a decreasing dose–response are given by the following, where

$$\text{3PL}: \quad y = \frac{y_0}{1+bU^d}, \tag{4.9}$$

$$y = \frac{y_0}{1+\left(U/\mathrm{EC}x\right)^d\left(x/100-x\right)}, \tag{4.9*}$$

$$\text{4PL}: \quad y = a + \frac{y_0 - a}{1+\left(U/b\right)^d}, \tag{4.10}$$

$$y = (1-p)y_0 + py_0\frac{1-\left(U/\mathrm{EC}x\right)^d}{1+\left(U/b\right)^d}, \tag{4.10*}$$

where $p = x/100$, EC$x>0$, $y_0>0$, $b>0$, and $d>0$.

The form of model 3PL is that given in Brown (1978) except for the y_0 term in the numerator to make it applicable to a general continuous response. The forms for 3PL and 4PL are also those given in Seefeldt et al. (1995) and Ritz et al. (2015), where the lower bound is set at zero. An alternative and closely related (but not equivalent) formulation of the log-logistic model is discussed in Chapter 8.

4.2.5 Hill Model

The Hill model is usually formulated for a function that increases from zero to a maximum (e.g. Goutelle et al., 2008). It is formulated here for a decrease from a maximum to zero:

$$\text{Hill}: \quad y = y_0\left[1-\frac{U^d}{a+U^d}\right], \tag{4.11}$$

$$y = y_0\left[1-\frac{U^d}{U^d + \frac{(\mathrm{EC}x)^d(100-x)}{x}}\right]. \tag{4.11*}$$

4.2.6 Threshold Models

Threshold models are used to make the "true" NEC or EC0 a model parameter. This class of models assumes that there is a threshold dose or concentration below which the substance has no effect on the organism. Threshold models are typically described as increasing functions, starting as some baseline value and then rising after the threshold value has been reached. This concept is useful in modeling mortality rates and some other statistically similar effects. We will consider such models in Chapter 6. In the present context, we need nonincreasing functions. The general form of a threshold model in this context is given by Eq. (4.12):

$$y = y_0 - F\left(U \mid a, \theta\right), \tag{4.12}$$

where U is dose, a is the threshold value, and F is a continuous functions satisfying $F(U|a,\theta)=0$ if $U \le a$. The most familiar threshold model in ecotoxicology is probably the linear threshold or hockey-stick model given by

$$\text{HS4PL}: \quad y = \begin{Bmatrix} y_0 & \text{if } U \le x_0 \\ y_0 + \dfrac{C-y_0}{b-x_0}\left(U-x_0\right) & \text{if } x_0 < U \le b \\ C & \text{if } b < U \end{Bmatrix}. \tag{4.13}$$

Parameter C establishes a floor or baseline below which the response cannot fall. C can be taken to be zero to yield HS3PL, but this is biologically implausible for most responses, as there is usually some minimum value below which the organism dies. There cannot be a zero weight fish, for example. Model HS4PL thus has four parameters to estimate, y_0, x_0, b, and C. If C is set to 0, the resulting model is labeled HS3PL.

Mathematically and biologically more attractive threshold functions can be obtained by replacing the linear portion of HS4PL with some other curve made "smooth" at the threshold value by requiring the derivative (slope) at that point be zero. For example, HS3PE is continuous at x_0 but not differentiable, but HS4PE is differentiable at x_0, provided $d>1$. An advantage of HS3PE over HS4PL is that there are only three parameters to estimate. Also, there is little reason to assume linearity over the range of tested concentrations. See Figure 4.3 to appreciate the nature of the three hockey-stick models, with $C=0$:

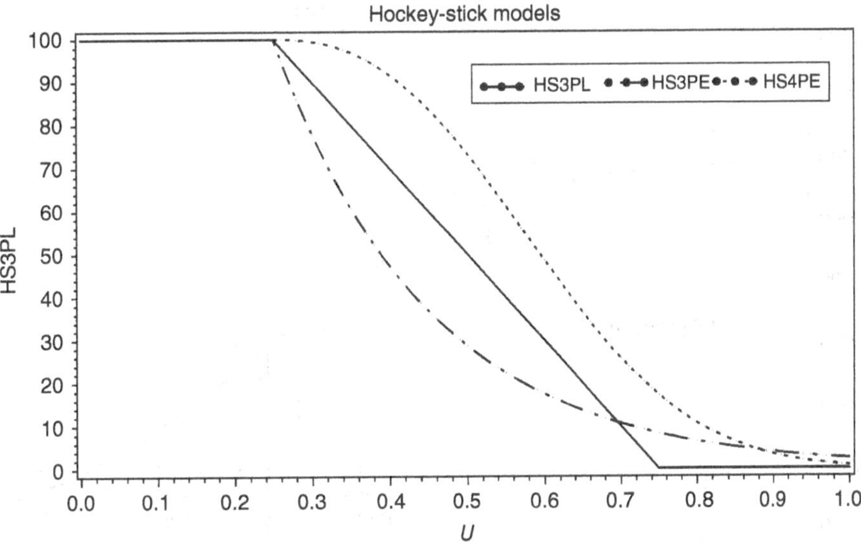

Figure 4.3 Hockey-stick models. Three example hockey-stick models that might describe the same data.

$$\text{HS3PE}: \quad y = \begin{Bmatrix} y_0 & \text{if } U \le x_0 \\ y_0 e^{-b(U-x_0)} & \text{if } x_0 < U \end{Bmatrix}. \qquad (4.14)$$

$$\text{HS4PE}: \quad y = \begin{Bmatrix} y_0 & \text{if } U \le x_0 \\ y_0 e^{-b(U-x_0)^d} & \text{if } x_0 < U \end{Bmatrix}. \qquad (4.15)$$

The exponential part of models HS3PE and HS4PE will be recognized as OE2 and OE3. To incorporate a floor in models HS3PE or HS4PE, it is merely necessary to use OE4 or OE5 in their place, but this adds another parameter to estimate and requires enough test concentrations to permit its estimation. More complex models have been investigated, for example, in Kavlock and Setzer (1996). Further discussion of threshold models is given in Cox (1987) and, specifically for hockey-stick models, in Yanagimoto and Yamamoto (1979).

These hockey-stick models can be useful when the dose–response relationship is flat from zero through one or more positive test concentrations. They are unlikely to be useful when change in the response is evident in the lowest test concentration, since there is no basis for estimating a threshold in such circumstances. An example with selenastrum cell counts is given below. It was not possible to fit any hockey-stick model to the rainbow trout length data of Table 3.3 or the collembola data of Table 3.5. An implication for experimental design if an NEC or threshold is to be estimated is that lower concentrations need to be included than are necessary to estimate EC50, EC20, or EC10. There is no guarantee that a threshold model can be fit, which can lead to wasteful additional studies. So far as we are aware, there are no international ecotoxicity test guidelines that require estimation of a threshold value. That could, of course, change in the future.

Programs in SAS and R for fitting all of the models defined above are given in Appendix 1 with names consistent with the names given in the models in this section, for example, BVPdecr.SAS and BVP Model.r. The file *Example model fits.r* provides additional datasets and model fits that can be run once all the R modules have been loaded. Each SAS program indicates additional data that to be fit.

4.3 MODEL FITTING AND ESTIMATION OF PARAMETERS

Fitting a model to dose–response data may be done by using any suitable software, e.g. SAS (www.sas.com), R (https://cran.r-project.org/), SPSS (www.spss.com), splus (www.insighful.com), and PROAST (http://www.rivm.nl/en/Documents_and_publications/Scientific/Models/PROAST). It is not necessary to understand the computational details, but some understanding of the basic principles in nonlinear regression is useful. It is also important to understand the assumptions underlying the fitting algorithm. For continuous data, it is often assumed that the data from each treatment group follow a normal or a lognormal distribution. In the latter case, a log-transformation is used to make the data (more closely) normally distributed. When a normal distribution with homogenous variances is assumed (possibly after transformation), the parameter estimates are often obtained by the method of maximum likelihood (Appendix 2).

It must be understood that if the responses are log-transformed (or some other transform is used), then after the model is fit, one must use a back-transform to return the response to its original scale before estimating ECx. This is because the back-transform of an x% change in the transformed

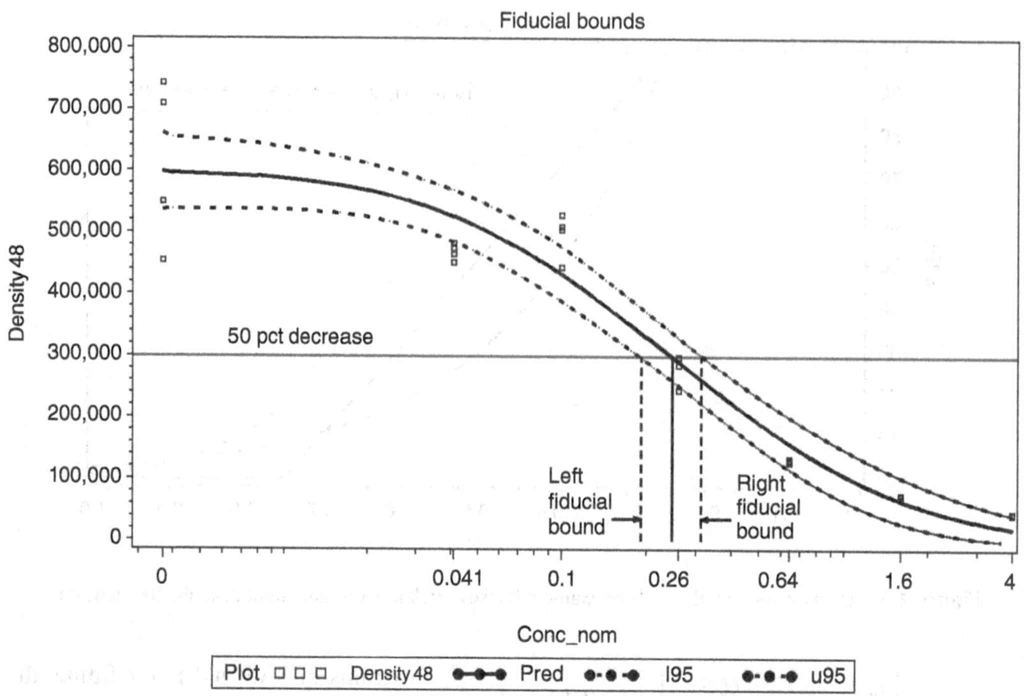

Figure 4.4 Find on the vertical scale of the untransformed model and confidence band, the point corresponding to an $x\%$ change from the estimated control mean. Draw a horizontal line through that point. Where this line intersects the curve determines the estimated ECx. Where this line intersects the confidence band determines the fiducial bounds on ECx. If the line does not intersect the upper (respectively, lower) band, then no upper (respectively, lower) fiducial bound can be obtained. In this illustration using the density data of Table 3.11, the fiducial bounds are (0.19, 0.31), while the maximum likelihood bounds are (0.17, 0.35). The solid vertical line segment locates the fiducial estimate of EC50, which is always the same as the MLE.

response need not be an $x\%$ change in the untransformed response. Obtaining point and interval estimates for ECx after a transform can be challenging. It may be very difficult to use the device discussed above for making ECx one of the model parameters. In Chapter 8, regression models on square root and log-transforms of count data are explored and compared with models treating untransformed counts as normally distributed or with alternative error structures.

A graphical technique that can be used is to obtain CIs for each theoretical concentration over a wide interval containing the experimental dose range. Then the curve and CI are plotted in back-transformed values and a graphical or analytical technique can be used to obtain approximate bounds on ECx. Such bounds are sometimes called fiducial bounds to distinguish them from true confidence bounds. See Figure 4.4 for an illustration of the technique.

4.3.1 Fitting the Model Assuming Normal Variation After Log-Transformation (Continuous Data)

When the residual variation is assumed to be lognormal, the model may be fitted after first log-transforming both the model and the data, and then either maximize the log-likelihood function based on the normal distribution, or minimize the sum of squares. Both methods will result in the same estimates of the regression parameters and the residual variance, which are MLEs in both cases. It should be noted that the resulting parameter estimates do relate to the original parameters of the (untransformed) model. Substituting the estimated regression parameters in the model results in a prediction of the *median* (or geometric mean) response as a function of dose. Therefore, in plotting the model together with the data, the back-transformed means (which are equivalent to the geometric means) should be plotted.

While the MLEs of the regression parameters relate to the model on the original scale, the MLE of the variance (s^2) relates to the log-transformed data. One may report this value, or, equivalently, the geometric standard deviation (GSD), which is the back-transformed square root of the variance of the log-data. Also, one may translate the estimated variance of the log-data in an (MLE) estimate of the coefficient of variation (CV), as a measure of the (residual) variability in the data on the original scale, by

$$CV = \sqrt{\exp\left(s^2\right) - 1},$$

when s^2 relates to the variance of the data after natural log-transformation.

4.3.2 Starting Values

Fitting any of these models by a computer program requires supplying the program with starting values to get the iterative estimation process started. Making ECx a model parameter simplifies the choice of starting values. Since y_0 is one parameter and estimates the control mean, the arithmetic mean of the control values is a good starting point. Additionally, it is usually possible to estimate visually a starting value for ECx. For models OE4 and OE5, the floor parameter C must be in the range $0<C<1-x/100$, and usually any value in that range will do to start. Parameter D, if present, is related to curvature and is difficult to describe. However, constraining $D \geq 1$ prevents infinite slope at $U=0$, which seems biologically implausible for most situations. Furthermore, larger values of d correspond to thresholds or near thresholds in toxicity. That is, if there is little or no effect evident at low concentrations, then D will tend to be large. If there is a sharp and immediate effect even at low concentrations, D will tend to be near 1 or even less than 1 in extreme cases. A starting value of $D=1$ is often reasonable.

The initial estimate of γ or gamma in the BVP model is EC50/EC15. The initial range of estimates is from the lower bound for EC50 over the upper bound on EC15 to the upper bound on EC50 over the lower bound on EC15, in increments of one-tenth that range.

For HORM1, the parameter beta, β, is related to the inflection point of the curve and a range of initial values running from the lowest concentration exhibiting a decrease from the arithmetic control mean to the highest concentration in increments of one-tenth that range usually provides good results. The remaining parameter, γ or gamma, is related to the mathematical concept of curvature and is difficult to describe. Taking 1 as a starting value often works reasonably well.

Model HORM2 is more difficult in terms of starting values. The value of B is positively associated with the magnitude of low-dose stimulation and higher values of D mean sharper drops in the response once U exceeds the hormetic high point. The value of F is inversely associated with how rapidly the initial rise starts and with the magnitude of the rise. The interaction of B and F is unclear. A is positively associated with the steepness of the initial rise.

4.3.3 Assessing Goodness of Fit for Regression Models

In addition to obtaining a good estimate of the control mean, there are many other model assessment criteria. The following comments apply to regression models for continuous responses. Typical regression models fit a curve to the mean response at each test concentration. In most ecotoxicity experiments, there are multiple replicates at each test concentration. It is thus possible to assess normality and variance homogeneity independently of the model using pure error residuals as described in Section 3.3.1. It is also helpful to calculate residuals about the fitted model. If the data have the form (x_i, y_{ij}), $x=0$ to k, $j=1$ to n_i, and $\hat{y}_{ij} = f(x_i)$ is the regression model estimate at the jth replicate of concentration x_i, then the model or regression residual or residual about the curve is $y_{ij} - \hat{y}_{ij}$. See Figure 4.5 for a visualization of the two types of residuals.

If the model passes through the mean response for some value of x, then the pure error and regression residuals are identical for that value of x. The greater the difference between the types of residuals, the worse the fit. Since residuals can be either positive or negative, a useful measure of residuals is the sum of squared residuals. The pure error and (regression) error sums of squares are given by

$$\text{SSPE} = \sum_{i=1}^{k}\sum_{j=1}^{n_i}\left(y_{ij} - \bar{y}_i\right)^2, \quad \text{with df} = \sum_{i=1}^{k}\left(n_i - 1\right) = n_. - k \tag{4.16}$$

and

$$\text{SSERR} = \sum_{i=1}^{k}\sum_{j=1}^{n_i}\left(y_{ij} - \hat{y}_i\right)^2, \quad \text{with df} = n. - p, \tag{4.17}$$

where p is the number of parameters to be estimated in the model and k is the number of groups including the control (where $i=1$).

4.3.3.1 Method of Least Squares

If the model is fit by minimizing SSERR, then as Weill (1968) and Jennrich (1969) demonstrate, a formal test for lack-of-fit is provided by

$$F = \frac{n_. - k}{k - p}\frac{\text{SSERR}}{\text{SSPE}}, \tag{4.18}$$

where F has an $F(k-p,\ n.-p)$ distribution. To be more accurate, F has a noncentral F-distribution, but the critical value to test the hypothesis of model fit uses a central F value. The model of a given form that is determined by minimizing SSERR is said to have been obtained by the method of least squares. Often maximum likelihood (ML) and restricted maximum likelihood (REML) estimates coincide with least squares estimates (e.g. for balanced designs), but this is not always true.

A minor alternative to Eq. (4.18) is a *likelihood ratio test*, given by

$$C = n.\left[\log(\text{SSERR}) - \log(\text{SSPE})\right], \tag{4.19}$$

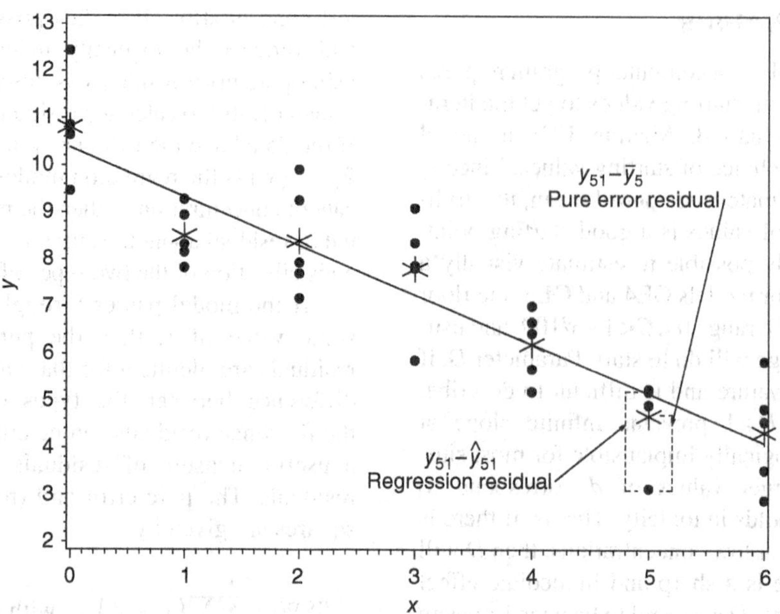

Figure 4.5 Pure error and regression residuals. The fitted model is shown as a solid line. Individual observations are dots. The mean response at each value of x is shown as a star. Here y_{51} is an observed response at $x=5$, $\hat{y}_{51}=\hat{y}_5$ is the response predicted by the model at $x=5$, and $\bar{y}_5$ is the mean of the observed responses at $x=5$.

which asymptotically has a χ^2-distribution with $k-p$ degrees of freedom (Huet et al., 2004). Note that both tests require $k>p$, which means the number of treatment groups must exceed the number of model parameters to be estimated.

Fitting a nonlinear model by minimizing SSERR is challenging, as pointed out by Dennis and Schnabel (1996), among others. The most common approach to fitting nonlinear models is through Gauss or Newton–Raphson or some other optimization method. Thus, Eqs. (4.17) and (4.18) are only asymptotically true. Neither test alone is sufficient in assessing goodness of fit, and additional model assessment tools are presented next.

For linear models, R^2 is a useful measure of goodness of fit, where

$$R^2 = 1-\frac{\text{SSERR}}{\text{SSTOT}}, \tag{4.20}$$

where

$$\text{SSTOT} = \sum_{i=1}^{k}\sum_{j=1}^{n_i}\left(y_{ij}-\bar{y}\right)^2, \tag{4.21}$$

and $\bar{y}$ is the mean of all observed response values.

For a linear model, $0 \le R^2 \le 1$, but for a nonlinear model, it is possible for SSERR to exceed SSTOT, which would make $R^2<0$. This could happen, for example, when an increasing curve is fit to a decreasing dose–response (see Motulsky and Christopoulos, 2004, p. 35).

It is important to evaluate the parameter estimates for the fitted model. The variance–covariance matrix of the parameter vector is given in Appendix 2. These estimates should be examined carefully. Ideally, each parameter should be significantly different from zero and the correlations between parameter estimates should not be very close to 1 or −1, as that would indicate a possible overparameterization. Also, the parameters should be sensible scientifically. For example, if a parameter that can only be positive for biological or physical reasons is estimated to be negative, then a different model should be selected. In addition, if the CI for a parameter, especially a key parameter such as ECx, is overly wide, then the utility of the model is questionable. As will be shown in examples in Chapter 4, it can happen that the CI for EC20 can span the entire range of test concentrations. Such an estimate has little scientific value.

It is also important to understand the difference between interpolation and extrapolation when estimating ECx for some choice of x. Extrapolation refers to estimated concentrations or predictions beyond the range of tested concentrations. If ECx exceeds the highest tested concentration, then that ECx value is an extrapolation. It is often not appreciated that if ECx is below the lowest tested positive concentration, it too is an extrapolation since the behavior of the data below the lowest tested concentration is not known. This is true even though the model allows estimation of the mean response at zero. We have fit a curve to capture the overall trend or shape, but the behavior of the data below the lowest tested concentration or above the

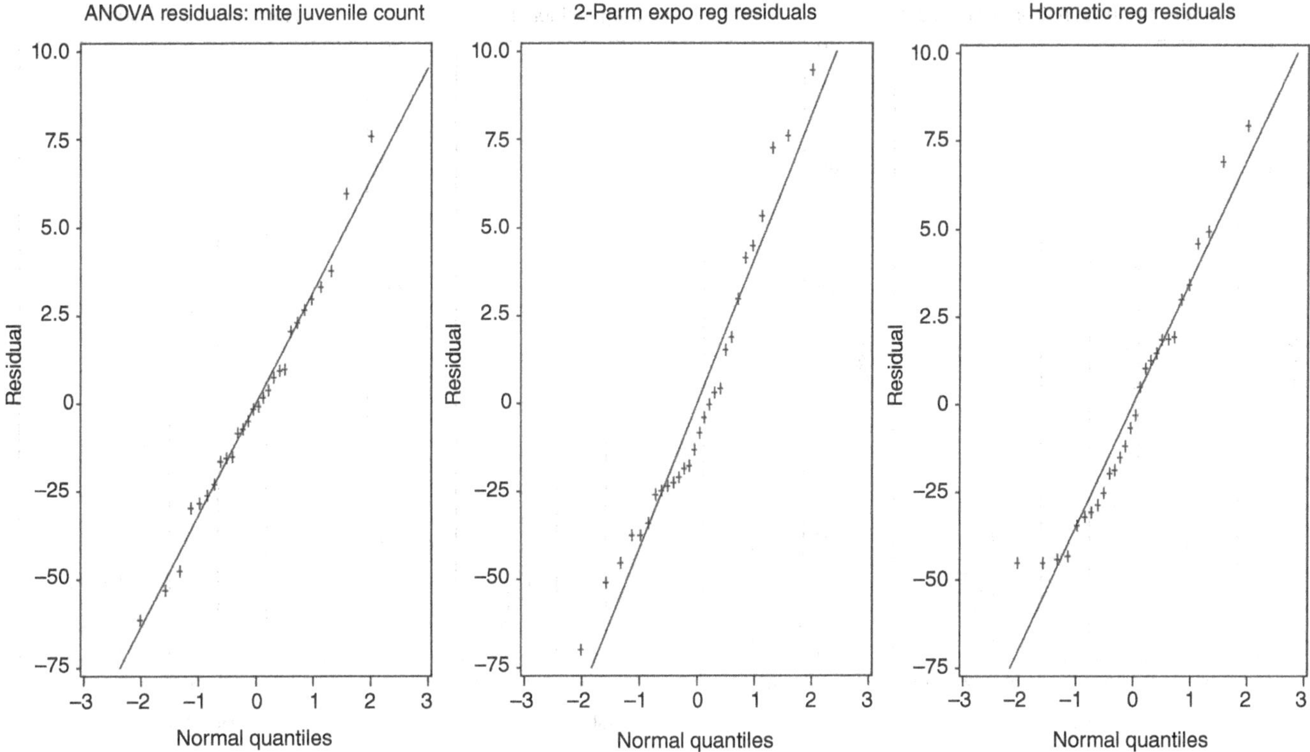

Figure 4.6 Comparison of QQ-plots. Pure error residuals in the left panel are very consistent with a normal distribution. The top two residuals deviate slightly from the reference line, indicating slightly heavy tail. The middle panel shows higher residuals at the extremes and lower residuals in the mid-range. The right panel shows more marked deviation from the reference line in the low range, indicating a very heavy tail, followed by a group of lower residuals in the lower range.

highest tested concentration may not follow that shape. While one could argue that the same concern applies to interpolation between two tested concentrations, nature generally does not behave in random or perverse ways. Below the lowest tested concentration, on other hand, the trend may continue or there may be a true no effect level or toxic threshold. That need not affect the ability of the model to estimate the control mean response, but there is little reason to trust the model to estimate points between the control and lowest positive tested concentration.

It is highly recommended to plot the data. There are several plots that are useful. A plot showing the replicate values and model predictions together with 95% confidence bounds can give a good overall impression of how well the model fits the data. Such a plot will also give visual evidence of how well the model estimates the control mean response and whether the model is in reasonable agreement with the data near the estimated EC*x* value of interest.

Plots of residuals against predicted values and against concentration or concentration order can make it easier to detect runs where the model over- or underestimates the observed mean response and also reveal variance heterogeneity if it exists. A histogram or QQ-plot of residuals may be sufficient to judge normality. See Figure 4.6 for three example QQ-plots. See Figure 4.7 for histograms for the same three datasets. Finally, see Figure 4.8 for scatterplots associated with these datasets.

4.3.3.2 Model Assessment Illustration

An experiment was done with mites (*Hypoaspis aculeifer*) exposed to a test chemical with four replicates in each of five test concentrations and eight replicates in the control. The response was the number of juvenile mites produced after 14-day exposure to the parents. While the response is a count, the variation in values justifies treating it as a continuous response. Two models were fit, a 2-parameter exponential and a hormetic model. An alternative analysis treating this as a count variable will be presented in Chapter 8. Details of the models are given later in this chapter. The purpose of this example is to illustrate ways to assess model fit. See Figure 4.9 for side-by-side comparison of these two models fit with the same mite data. See Table 4.4 for a numerical comparison of observations versus predictions from these two models.

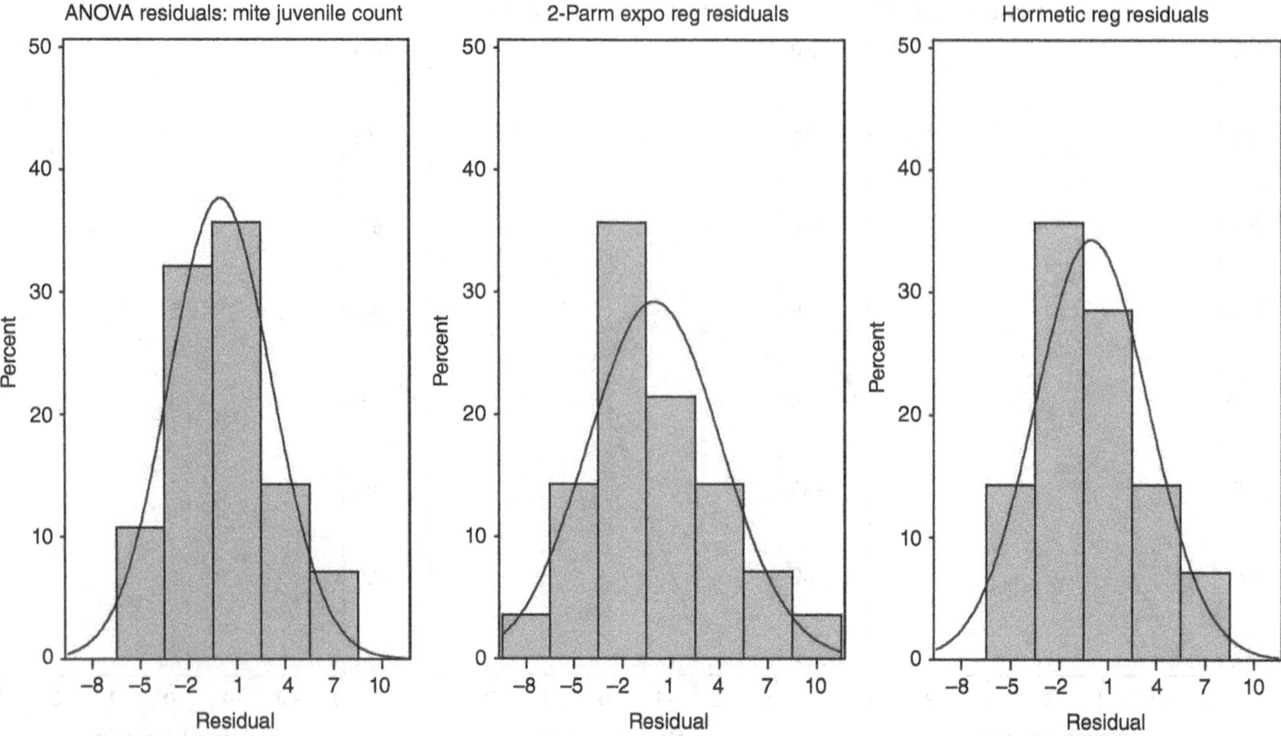

Figure 4.7 Comparison of residual histograms. Some skewness is evident in the right panel. The middle panel shows that a wider spread of residuals than in the other two panels, indicating greater variability about the two-parameter exponential fitted curve compared with the hormetic model or the pure error residuals.

Table 4.4 gives additional detail on the goodness of fit displayed in Figure 4.9. A formal lack-of-fit F-test is not significant for either model. Overall, of these two models, the hormetic model provides the better fit to the data. The EC20 estimates (and their 95% confidence bounds) were 46.3 (−1.2, 93.8) and 49 (26.8, 89.6) for the 2-parameter exponential and hormetic models, respectively. In other words, the EC20 estimate from OE2 is not significantly different from zero and its CI spans the entire range of tested concentrations including zero, while the CI from the hormetic model, which may perhaps seem wide, spans only one test concentration. It is of interest that the two-point estimates of EC20 are rather similar and that might suggest acceptance of a point estimate in spite of a very wide CI. However, it is only by its near agreement with the point estimate from another model with much tighter CI that one might have confidence in the estimate. As will be shown by additional examples in this chapter, point estimates can be very far from reasonable and only estimates with relatively short CIs from models that fit the data should be considered valid.

Figures 4.1, 4.2, 4.4, 4.9, and numerous others following those show the replicate means, fitted curve, and confidence bounds for those curves. These bounds indicate the uncertainty in each predicted value. They do not reflect the total uncertainty of the model and ignore the issue of multiple confidence bounds for multiple predictions. Nevertheless, they are valuable in giving a visual indication of the uncertainty in the model and a caution against exaggerating the precision. One of the dangers in risk assessments is the temptation to read more precision in model predictions than is justified by the data. Partly for that reason, it seems unfortunate that at present, there seems to be no function in the Cran library to generate these bounding curves for nonlinear models in a manner comparable to what is possible in SAS. Two interesting discussions of this are given in R-blogger (2017a, b). We are indebted to Georgette Asherman for developing R code to generate plots for the examples in this book using the Monte Carlo approach in R-Blogger (2017a) that while not identical to those generated by SAS are useful and informative.

It should be noted that for risk assessment, the focus is on estimating ECx for some choice of x rather than on getting the model that best fits the entire dose–response. The shortest CI for ECx does not necessarily come from the

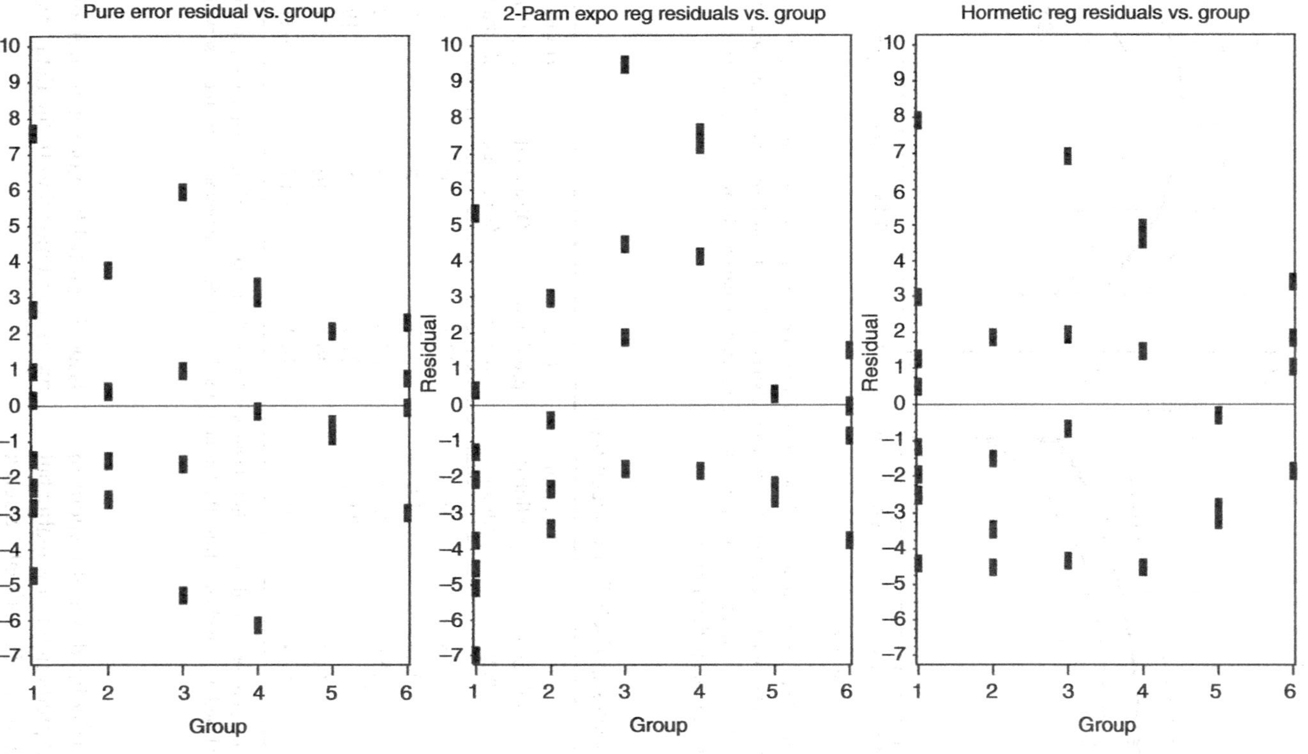

Figure 4.8 Comparison of residual scatter plots. The residuals from the hormetic model (right panel) resemble more closely the pure error residuals (left panel) and are more centered on the control mean than those from the two-parameter exponential model (middle panel). The latter model captures the middle treatment groups better than the alternative regression model. Both models tend to overestimate group 5.

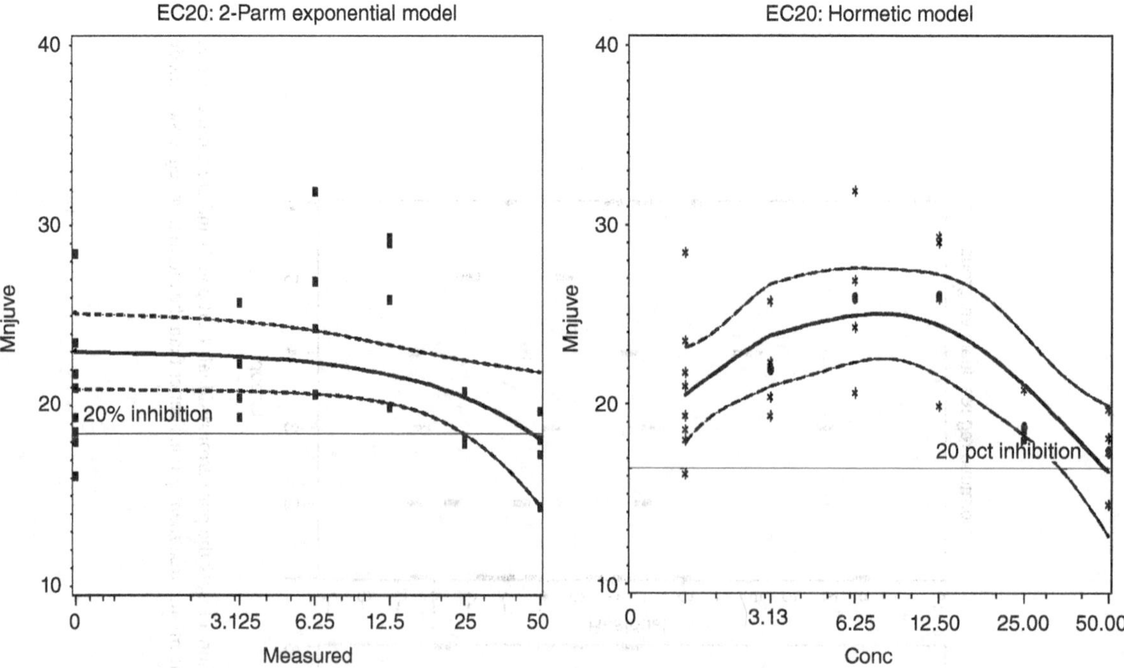

Figure 4.9 Comparison of model fits. The underestimation of the middle concentrations is evident in the left panel, as it is the better fit to those groups in the right panel. The difference in estimates of the control mean is also clear. In left panel, replicate means are indicated as solid squares. In right panel, replicate means are asterisks, and treatment means are solid dots. In both panels, the regression curve is a solid line, and dotted lines indicate 95% confidence bounds on the mean predicted value.

Table 4.4 Comparison of Estimates in Each Test Concentration

Group	Conc	Reps	Observed mean	Hormetic predicted mean	OE2 predicted mean	Observed % change	Hormetic predicted % change	OE2 predicted % change
1	0	8	20.8	20.5	23.1	0	0	0
2	3.125	4	21.9	23.8	22.7	5	16	−1
3	6.25	4	25.9	25.0	22.4	24	22	−3
4	12.5	4	26.0	24.4	21.7	25	19	−6
5	25	4	18.7	21.1	20.5	−10	3	−11
6	50	4	17.4	16.3	18.1	−17	−21	−21

Model OE2 overestimates the control mean response and the predicted percent change from that model does not agree with the observed values except in the high test concentration. The predicted percent change from the hormetic model is in better agreement with the observed values, but that model overestimates in groups 2 and 5.

best fitting model. However, as pointed out elsewhere in this volume, the CI for a model prediction is a valuable but limited measure of the uncertainty in the estimate. Such a CI assumes that the model is correct and in no way captures model uncertainty. Our thinking is that a key model prediction such as ECx cannot be considered reliable if the model does not fit the data reasonably well. It should be acknowledged that some others in the field seem to have a different perspective, for example, Hothorn (2016, p. 187), but the issue is probably more a matter of interpreting what it means for a model to fit the data "reasonably well."

4.3.4 Comparison of Models

In addition to assessing the goodness of fit of a single model to the data, it is also necessary to compare two or more different models fit with the same data in order to

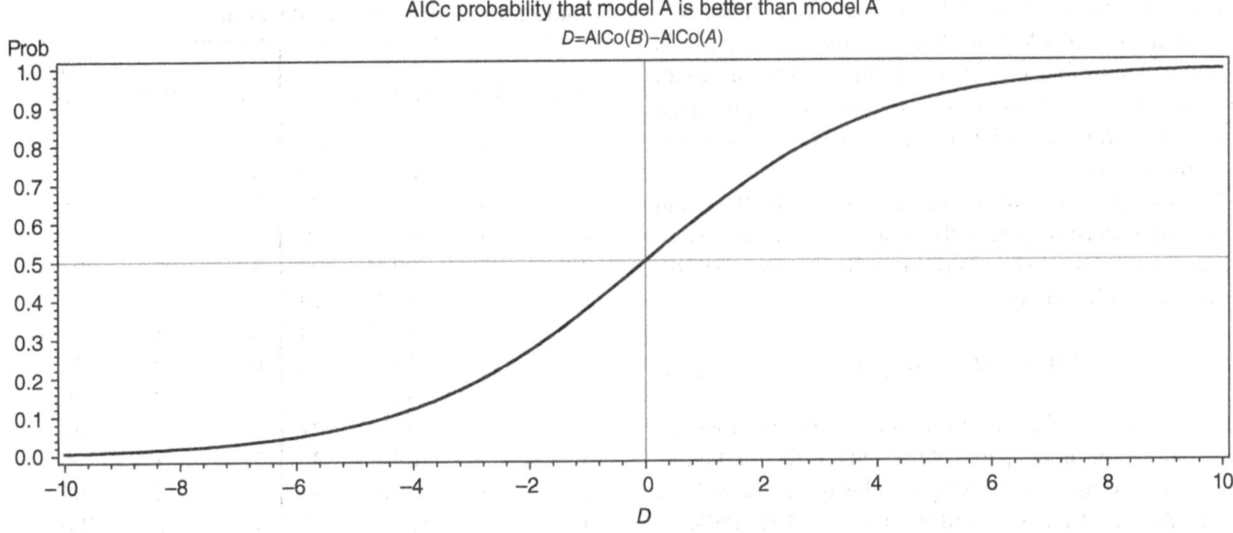

Figure 4.10 AICc probability function. Horizontal axis is unitless.

select the best model. One useful formal criterion for doing that is Akaiki's information criteria (AIC) for model selection, which uses the calculation

$$AIC = -2L + 2m, \tag{4.22}$$

where L is the log-likelihood function for the model and m is the number of parameters estimated for the model. In general, the model with the smaller AIC is preferred. When the total number, n, of observations is not large ($n/m \leq 40$), there is a finite sample correction factor, which is expressed as

$$AICc = -2L + \frac{2nm}{n-m-1}. \tag{4.23}$$

If the model is fit by the method of least square and the error distribution is normal, then Eq. (4.23) can be written as

$$AICc = nLn\left(\frac{SSERR}{n}\right) + \frac{2nm}{n-m-1}, \tag{4.24}$$

where SSERR is the residual sum of squares from the model.

There are several minor variations in formulas for AIC and AICc. We have used the calculations given in Burnham and Anderson (2004). So long as one is consistent in which formula is used, it should not matter, as only the difference in AICc values when comparing models is needed. If the values of AICc are close for two models, it is helpful to compute the probability of that model A is better than model B using the following formula:

$$Prob = \frac{e^{-D/2}}{1+e^{-D/2}},$$

where

$$D = AICc(B) - AICc(A).$$

Here AICc(B) denotes the value of AICc for model B. If the probability is high, then model A is favored. It can be shown that if AICc(B)−AICc(A) ≥ 10, the model A is almost certainly better than model B. See Figure 4.10 for the shape of the AICc probability function.

AICc values for the models summarized in Table 4.4 were 85 and 82 for OE2 and hormetic models, respectively, so while the hormetic value is lower, the difference is not significant. The AICc criterion is limited, however, where weighted fits are used, as two models should be compared using this criterion only if they use the same weights. Thus, in comparing two unweighted model fits (i.e. weight = 1), the criterion is sound. For weighted models where the weights depend on the function being estimated, the criterion is questionable and comparing an unweighted with a weighted model fit is yet more questionable. Some discussion of weights in using AICc is given in http://www.boomer.org/manual/ch05.html and in http://www.micromath.com/products.php?p=scientist&m=statistical_analysis.

Additional discussion of AICc is given in Motulsky and Christopoulos, 2004. In addition to AICc, generally, the parsimony principle should be used. In its general form, the parsimony principle states that the simplest of competing explanations is the most likely to be correct. It was proposed in the fourteenth century by William of

Ockam, who was an English Franciscan friar, scholastic philosopher, and theologian. The parsimony principle is also referred to as Occam's Razor. In the context of model fitting, the simplest model providing good fit is preferred over a more complex model. The emphasis is on good fit, not simplicity alone.

A related model comparison technique is the Bayesian information criterion (BIC), also known as the Schwartz information criterion (SIC), introduced by Schwarz (1978). The calculation formula is

$$\mathrm{BIC} = -2L + k\log(n), \tag{4.25}$$

where L is the log-likelihood function of the distribution, k is the number of model parameters, and n is the total number of observations. This is very similar to Akaiki's AIC. As with AIC and AICc, smaller values of BIC indicate better fit of the model to the data after adjusting for the number of model terms. Numerous authors have argued that the BIC tends to favor less complex models more than AIC. However, as Burnham and Anderson point out, this advantage goes away when AICc is used in place of AIC. References include Neath and Cavanaugh (2012), Posada and Buckley (2004), Schwarz (2011), Kass and Raftery (1995), and Kass and Wasserman (1995). An excellent discussion of BIC and its relation to model averaging and Bayesian factors is given in Cavanaugh (2012).

4.4 EXAMPLES

4.4.1 Hockey-Stick Model Fit to Selenastrum Cell Counts

The green algae Selenastrum were exposed to a test substance in an experiment with six test concentrations plus a zero concentration water control. There were four replicate beakers of selenastrum per concentration. Initially there were 10000 cells per beaker. Seventy-two hours after exposure, selenastrum cells were counted again. See Table 4.5 for the data.

The selenastrum data were found to be normally distributed with homogeneous variances by the Shapiro–Wilk and Levene tests, respectively, applied to the residuals from an ANOVA with dose as the sole factor. See Figure 4.11 for a plot of the model HS4PE to the data. See Table 4.6 for a summary of the fitted model. Figure 4.11 also shows an apparent lower bound on the effect even though a floor parameter was not included in the model. This is not unusual and is one reason a floor (or ceiling) parameter usually is not included in models. Another reason, of course, is that this reduces the number of parameters to be estimated making the models available for studies with fewer treatment levels.

Table 4.5 Selenastrum Cell Counts

Dose	Rep	Conc ($\mathrm{mg\,l^{-1}}$)	*y*	Dose	Rep	Conc ($\mathrm{mg\,l^{-1}}$)	*y*
1	1	0	32	4	3	12.5	28
1	2	0	28	4	4	12.5	26
1	3	0	32	5	1	25	33
1	4	0	30	5	2	25	23
2	1	3.125	27	5	3	25	27
2	2	3.125	29	5	4	25	27
2	3	3.125	34	6	1	50	11
2	4	3.125	30	6	2	50	16
3	1	6.25	32	6	3	50	12
3	2	6.25	27	6	4	50	13
3	3	6.25	30	7	1	100	1
3	4	6.25	30	7	2	100	0
4	1	12.5	28	7	3	100	3
4	2	12.5	31	7	4	100	2

y is cell count divided by 100000.

Programs to fit HS4PE, HS3PE, HS4PL, and HS3PL are provided in Appendix 1. Fitting the other three hockey-stick models to the selenastrum data is left as an exercise. It is also of interest to fit a nonthreshold model to these same data and compare the results. The Models OE3 and BVD were fit. See Table 4.7 for a summary of the model fits. For visual representations of these models, see Figures 4.12 and 4.13. OE3 provides slightly better agreement to the observed values near the 10% effects concentration, but Model BVP provides the best overall fit and the shortest CI for EC10 and the CI for the OE3 prediction at the highest test concentration is not believable. AICc makes no meaningful distinction between these two models but both are marginally better than HS4PE. Only HS4PE allows estimation of the threshold. The others are designed to estimate EC10 or ECx for another $x>0$.

4.4.2 Rainbow Trout Weight with Hormesis

Rainbow trout were exposed to a test substance in an early life stage study (OECD TG 210). There were four replicates of 20 fish each in the water control and every test concentration. See Table 4.8 for the replicate means, measured in grams. For a visual assessment of normality based on the pure error residuals, see Figure 4.14. See Table 4.9 for a detailed comparison of the fit of the Cedergreen and Brain–Cousens models, while Figures 4.15 and 4.16 give visual results of these model fits.

Both models HORM1 and HORM2 were fit to these data. Table 4.9 and Figures 4.15–4.16 summarize the models fit. By the AICc criterion, the two models are indistinguishable, though there is a very slight preference for

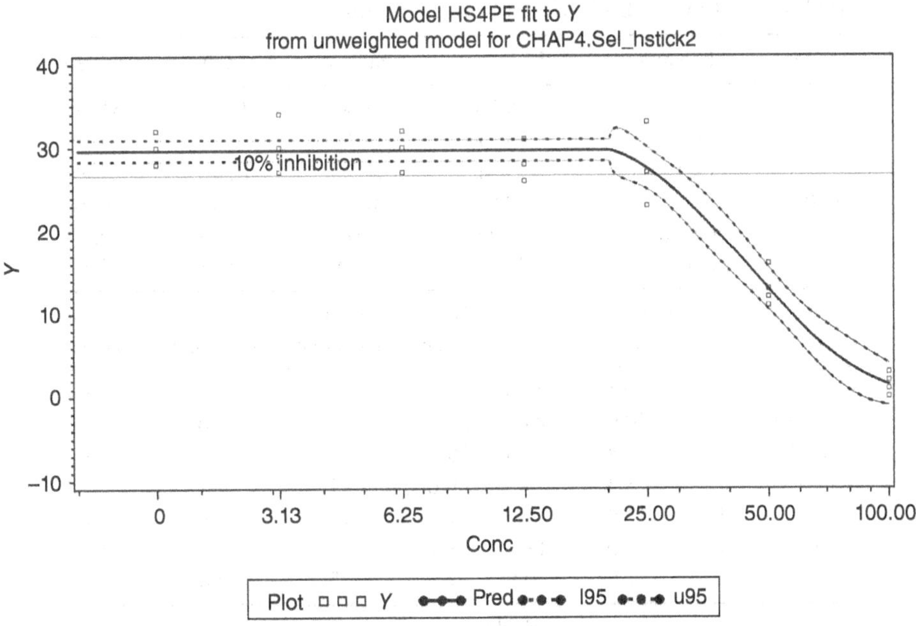

Figure 4.11 Hockey-stick model HS4PE fit to selenastrum data. The horizontal axis is in log-scale, with an adjustment to plot the control observations.

Table 4.6 HS4PE Model Fitting Summary

Source	DF	Sum of squares	Mean square	*F* value	Pr>*F*
Model	3	3075.2	1025.1	166.79	<0.0001
Error	24	147.5	6.1458		
Corrected total	27	3222.7			

Parameter	Estimate	StdErr	95% Confidence bounds	
X0	19.673	5.5911	8.1336	31.2124
Y0	29.625	0.6198	28.3459	30.9041
B	0.00737	0.0138	−0.0211	0.0358
D	1.3821	0.4699	0.4124	2.3519

Conc	Count	Observed mean	Predicted mean	Observed % change	Predicted % change
0	4	30.50	29.63	0	0
3.125	4	30.00	29.63	−2	0
6.25	4	29.75	29.63	−2	0
12.5	4	28.25	29.63	−7	0
25	4	27.50	27.50	−10	−7
50	4	13.00	13.00	−57	−56
100	4	1.25	1.25	−96	−96

Model evaluation measures

SSReg	SSErr	SSPE	FLOF	PLOF	SSTotC	RSqr	AICc
3032.11	143.75	132.5	0.95	0.55	3175.86	0.95	60.98

One model parameter, *B*, has a confidence interval containing zero.

HORM2. HORM2 agrees better with the observed data, especially the control mean. The EC10 estimates from HORM1 are similar, but that model does not fit the data quite as well. All model parameters for both models are significantly different from zero. In conclusion, HORM2 is preferred for these data.

4.4.3 Soybean Shoot Height with Hormesis

Hormesis is common in the study of the effects of crop protection chemicals on terrestrial plants. An example is shoot height in soybeans exposed at an early age to such a chemical in what is called a vegetative vigor study following OECD TG 227. See Table 4.10 for the data, where the application rate units were grams of active substance per hectare (g a.s. ha^{-1}). There was no mortality among exposed plants. See Table 4.11 for a summary of the HORM1 model fit to these data.

There is good agreement of the model to the data evident in Figure 4.17 and Table 4.11. All model parameters are significantly different from zero and confidence bounds are reasonably tight except for gamma. However, there was a significant Shapiro–Wilk test, indicating nonnormality and that can be traced to the outlier evident in Figure 4.17 at application rate 0.59 g a.s. ha^{-1}. When that observation is omitted, the Shapiro–Wilk test is no longer significant, and there are only modest effects on the model parameters. For example, EC25 = 1.72 (1.21, 2.23). We do not recommend removing observations without good reason, but it is useful to determine whether the apparent outlier is seriously affecting the model and whether that one observation

Table 4.7 OE3 and BVP Model Fitting Summary

Model	Parameter	Estimate	StdErr	LCB	UCB
OE3	Y0	29.80	0.66	28.45	31.16
	D	3.11	0.78	1.51	4.71
	EC10	25.81	4.29	16.96	34.65
BVP	Y0	29.67	0.60	28.44	30.91
	Gamma	0.45	0.07	0.30	0.59
	ec10	26.41	2.85	20.54	32.27

Conc	Reps	Observed mean	OE3 mean	BVP mean	Observed % change	OE3 % change	BVP % change
0	4	30.5	29.8	29.7	0	0	0
3.125	4	30.0	29.8	29.7	−2	0	0
6.25	4	29.8	29.8	29.7	−2	0	0
12.5	4	28.3	29.5	29.6	−7	−1	0
25	4	27.5	27.1	27.3	−10	−9	−8
50	4	13.0	13.1	13.1	−57	−56	−56
100	4	1.5	0.0	1.3	−95	−100	−96

Model Evaluation Measures

Model	SSReg	SSErr	SSPE	FLOF	PLOF	SSTotC	RSqr	AICc
OE3	17746	150.0	132.5	0.951	0.552	3 175.86	0.99	56.8
BVP	17752.4	143.6	132.5	0.910	0.593	3 175.9	0.99	55.6

ANOVA table and fit summary for two models: BVP and OE3.

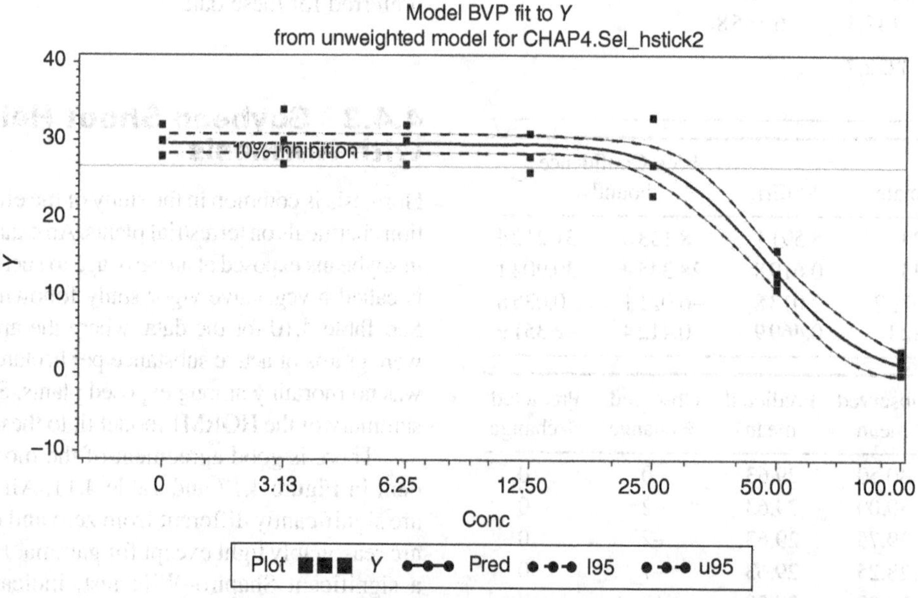

Figure 4.12 Model BVP fit to selenastrum data. There is good fit of the model to the data across the entire range of test concentrations.

accounts for the nonnormality test result. We would report the model based on the full dataset with the observations just made about the effect of that one observation.

A good question to ask is whether there really is hormesis in these soybean data. If the two lowest control values are omitted, the visual evidence for hormesis is weakened. It is instructive to fit nonhormetic models to these data. The Bruce–Versteeg model fit to these data poorly estimated the control mean and all of the observations at 0.029 g a.s. ha^{-1} lie outside the confidence bounds

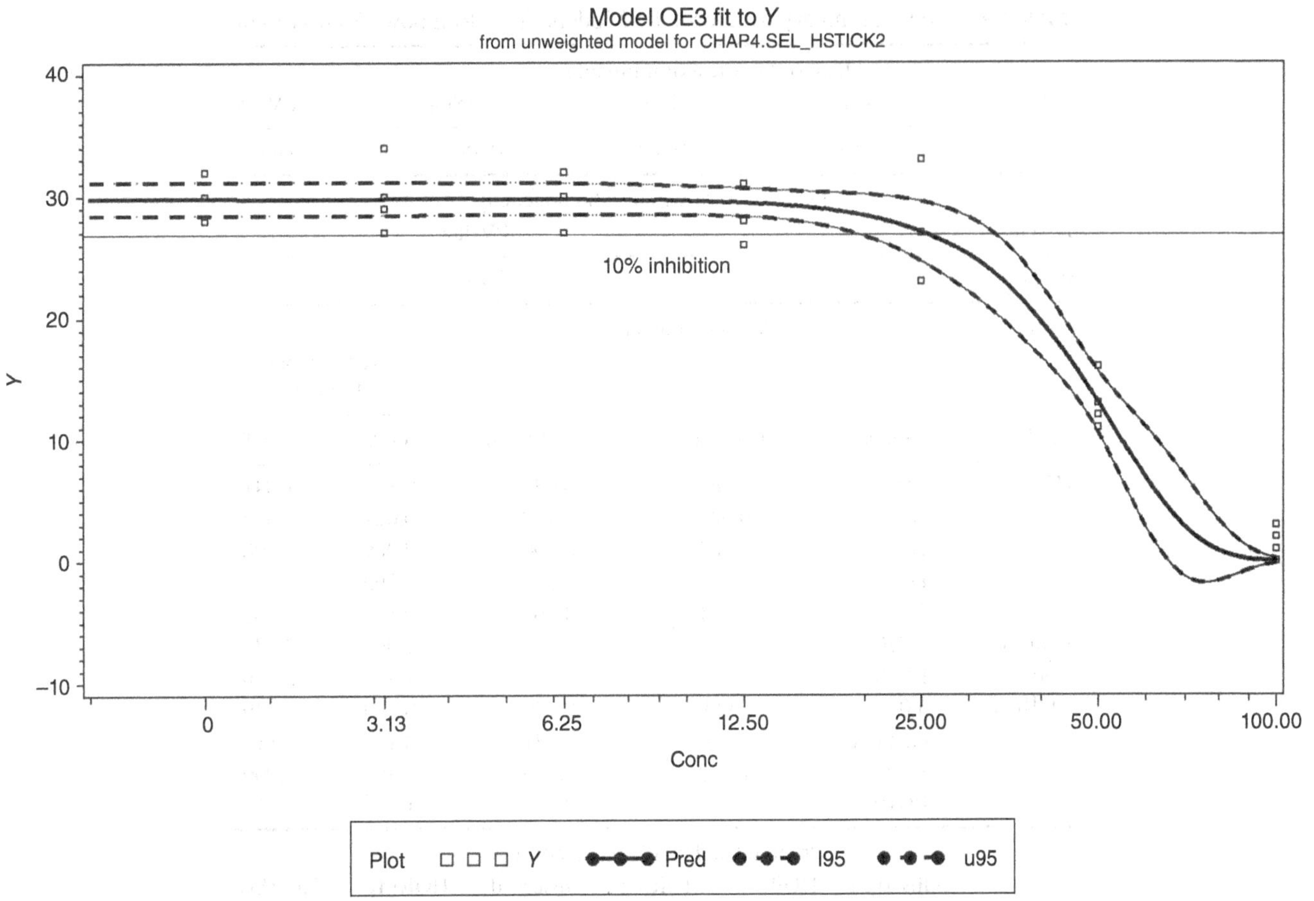

Figure 4.13 Model OE3 fit to selenastrum data. The model fits the data well except at the highest test concentration, where the model underestimates the observations and the confidence interval for the prediction at that concentration is unrealistically short.

Table 4.8 Replicate Mean Rainbow Trout Dry Weights at the End of an ELS Study Are Presented

Conc	Rep	DryWgt	Conc	Rep	DryWgt	Conc	Rep	DryWgt
0	1	0.915	0.12	4	1.120	0.96	3	0.975
0	2	0.940	0.24	1	1.008	0.96	4	1.056
0	3	0.930	0.24	2	1.048	2.4	1	0.757
0	4	1.078	0.24	3	1.018	2.4	2	0.984
0.6	1	1.200	0.24	4	1.083	2.4	3	0.878
0.6	2	1.188	0.48	1	1.170	2.4	4	0.876
0.6	3	1.084	0.48	2	1.178	6	1	0.515
0.6	4	1.045	0.48	3	1.150	6	2	0.425
0.12	1	0.947	0.48	4	1.102	6	3	0.367
0.12	2	1.126	0.96	1	1.171	6	4	0.469
0.12	3	1.119	0.96	2	1.077			

Conc is the mean measured concentration of the test substance in $mg\,l^{-1}$.

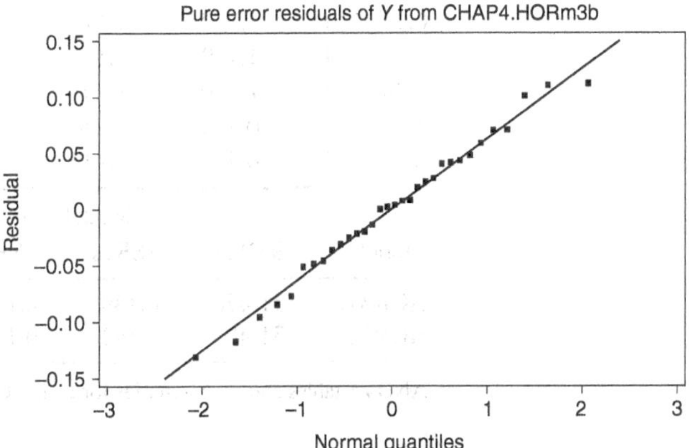

Figure 4.14 QQ-Plot of pure error residuals from ANOVA of rainbow trout weight data.

for the predicted response. EC25 was also estimated lower, at 1.3 g a.s. ha^{-1} and the AICc value was 128 versus 110 from HORM1. Model OE3 is not better and the AICc value is even higher. Model OE4 provided a better fit than OE3 or BVP and AICc = 114, only a little higher than that for HORM1. All of the monotonic models ignore the lowest two control replicate values. It should also be noted that the estimated EC10 under this model is

Table 4.9 Model Summary for Hormetic Models Fit to Rainbow Trout Weight

Shapiro–Wilk test of normality

Std	Skew	Kurt	SWStat	PValue
0.063	−0.160	−0.378	0.983	0.883

Levene test of variance homogeneity

Levene	PValue
0.321	0.937

Model parameters

Model	Parameter	Estimate	ApproxStdErr	ApproxConf Bounds LCB	ApproxConf Bounds UCB
HORM2	Y0	0.967	0.036	0.892	1.041
	A	0.328	0.148	0.024	0.632
	B	1.584	0.300	0.969	2.198
	D	3.811	0.740	2.293	5.328
	F	0.681	0.268	0.131	1.231
Fiducial	EC10	2.399		1.994	2.837
Delta	EC10	2.399		1.821	2.976
HORM1	Y0	0.989	0.029	0.929	1.049
	Gamma	0.464	0.171	0.114	0.814
	Beta	1.753	0.125	1.496	2.009
	EC10	2.200	0.203	1.783	2.616

Observed and predicted response

Conc	*N*	Observed mean	HORM2 mean	HORM1 mean	Observed % change	HORM2 % change	HORM1 % change
0	4	0.966	0.967	0.989	0	0	0
0.06	4	1.078	1.054	1.037	12	9	5
0.12	4	1.039	1.091	1.073	8	13	8
0.24	4	1.150	1.116	1.115	19	15	13
0.48	4	1.129	1.116	1.123	17	15	13
0.96	4	1.070	1.091	1.108	11	13	12
2.4	4	0.874	0.870	0.853	−10	−10	−14
6	4	0.444	0.445	0.455	−54	−54	−54

Model evaluation measures

Model	SSReg	SSRes	SSPE	PLOF	SSTotC	RSqr	AICc
HORM1	31.476	0.149	0.122	0.459	31.625	0.995	−160
HORM2	31.483	0.142	0.122	0.467	1.601	0.996	−158

ANOVA tables and fit summaries for Brain–Cousens and Cedgergreen hormetic models.

0.425, at which rate the predicted shoot height is 29.8, just 1% below the observed control mean. This seems a poor estimate of EC10 if the control replicates are to be believed. If the two low control values are to be ignored for some reason not discernable from the data alone, then model OE4 may be reasonable. Based just on the data, the understanding that these are replicate means, not individual plant values, and any compelling reason to question the control reps is absent, HORM1 is our model choice.

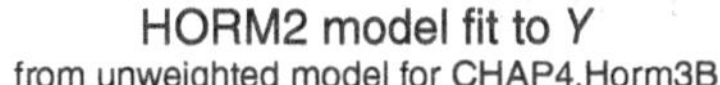

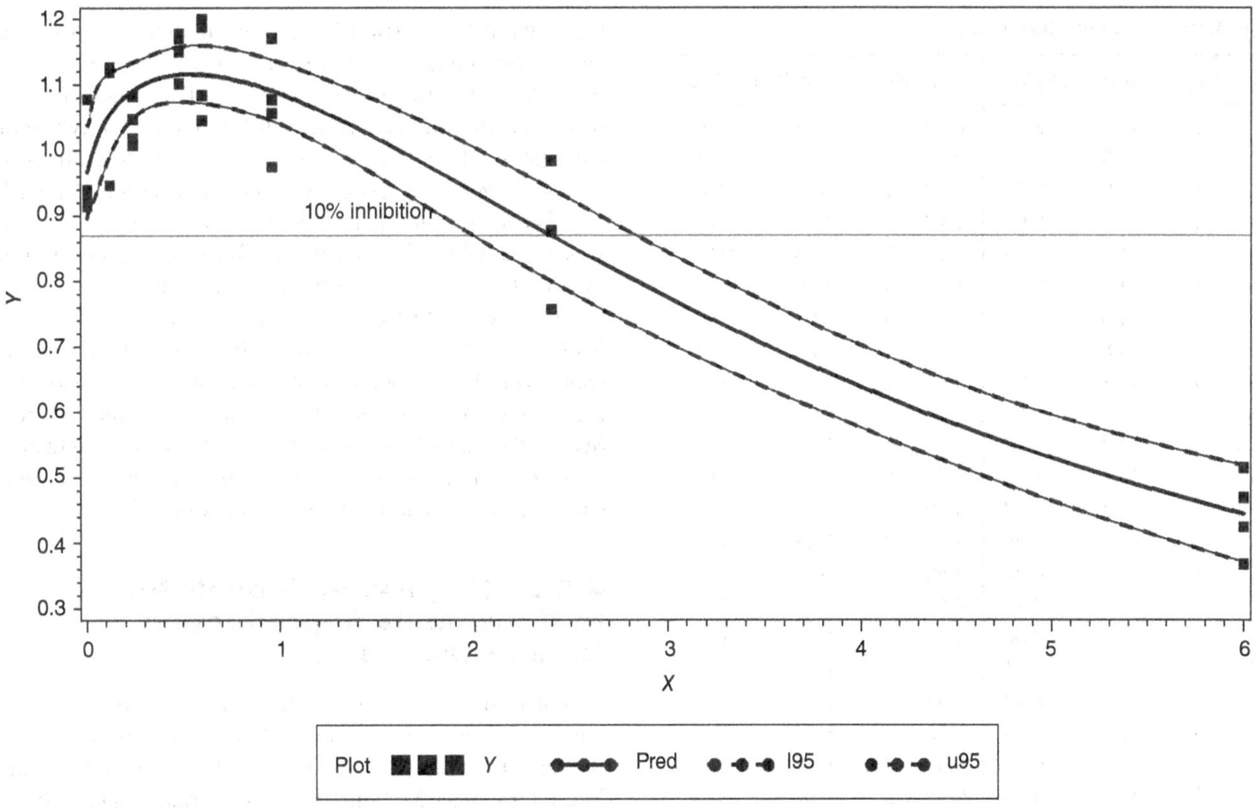

Figure 4.15 HORM2 (Cedergreen) model fit to rainbow trout weight data.

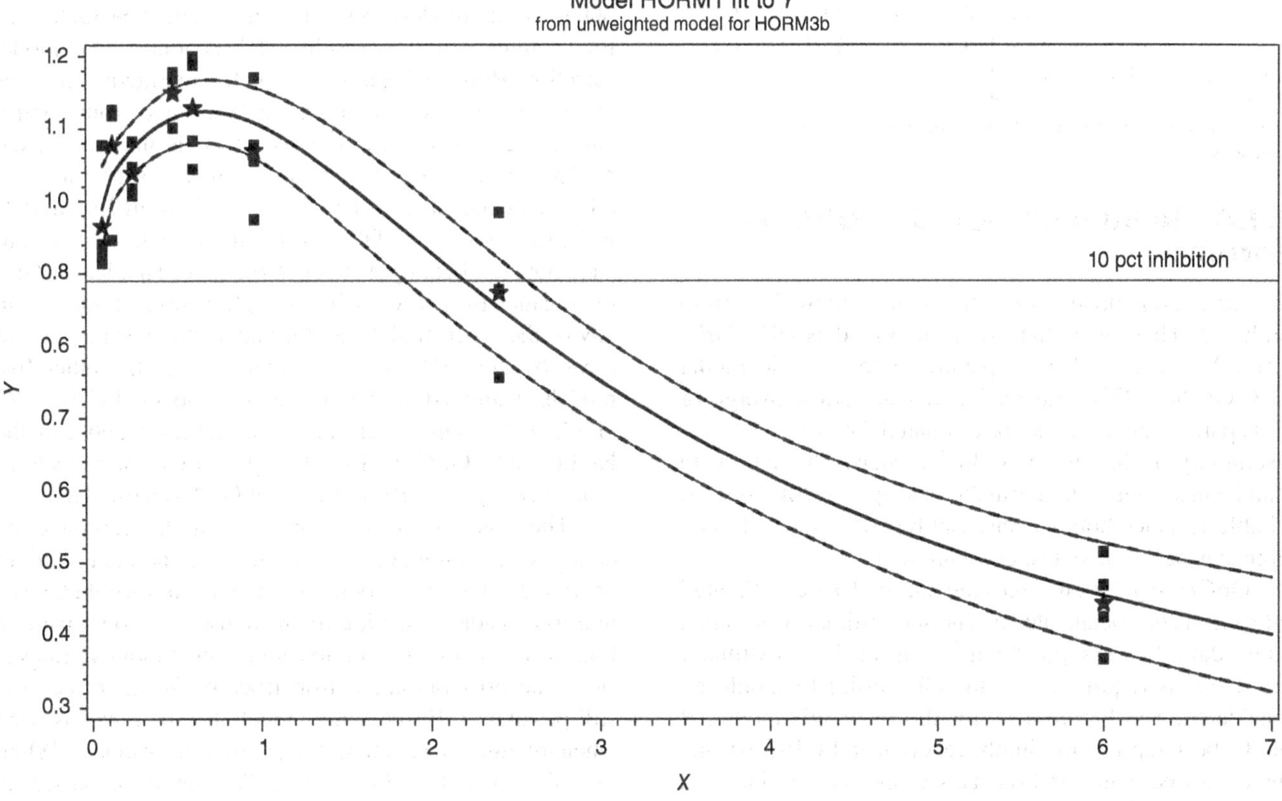

Figure 4.16 HORM1 (Brain–Cousens) model fit to rainbow trout weight data. Star, treatment mean; Filled square, replicate mean; Solid line, model; and Dotted lines, confidence curves.

Table 4.10 Soybean Shoot Height

Rate	Rep	ShootHt	DryWgt	Rate	Rep	ShootHt	DryWgt
0	1	32	8.98	0.59	4	30.8	8.99
0	2	32.8	8.77	0.59	5	29.3	8.29
0	3	26.8	9.07	1.17	1	26	8.56
0	4	24.8	8.02	1.17	2	25.5	8.07
0	5	34	8.27	1.17	3	27.5	9.8
0.037	1	34	8.92	1.17	4	26.8	8
0.037	2	34.8	8.65	1.17	5	24	7.78
0.037	3	35.8	9.1	2.34	1	20.3	6.18
0.037	4	33.8	7.87	2.34	2	22.5	7.33
0.037	5	35.3	10.6	2.34	3	23.3	6.35
0.073	1	32.8	9.95	2.34	4	17.8	4.7
0.073	2	30	8.51	2.34	5	20	6.01
0.073	3	31.3	8.53	4.69	1	18.5	4.97
0.073	4	32.3	9.49	4.69	2	16.3	3.73
0.073	5	35.3	10.46	4.69	3	16.5	4.16
0.147	1	33.5	10.24	4.69	4	17.8	4.68
0.147	2	31	9.06	4.69	5	17	4.72
0.147	3	36.3	9.51	9.38	1	13.8	3.3
0.147	4	33	8.38	9.38	2	14.8	2.79
0.147	5	31.8	7.64	9.38	3	14.3	2.48
0.29	1	32.5	9.22	9.38	4	17.3	3.38
0.29	2	32.5	9.35	9.38	5	16	2.88
0.29	3	31.5	9.58	18.75	1	15.5	1.98
0.29	4	33.5	8.17	18.75	2	15.3	1.85
0.29	5	34	8.93	18.75	3	14.5	1.77
0.59	1	26.3	8.48	18.75	4	14.3	1.49
0.59	2	17	2.15	18.75	5	14.3	1.4
0.59	3	27.3	9.1				

Shoot height and dry weight are typical responses in nontarget terrestrial plant studies.

4.4.4 Rainbow Trout Size Without Hormesis

Six regression models were fit to the length data from Table 3.3. These were the exponential models OE2–OE5, Bruce–Versteeg, and the 3-parameter log-logistic model LL3. Of these OE4 and OE5 either did not converge or some parameters could not be estimated. See Table 4.12 for a summary of the remaining four models. The data were found consistent with normality (Shapiro–Wilk) but no reliable variance homogeneity can be done with only two replicates per positive test concentration.

OECD regulations mandate an EC10 be estimated unless it can be documented that no such estimate is feasible. These data do not support a meaningful EC10 estimate, but if one is required, only the OE2 estimate should be considered. See Figure 4.18 for the fit of this model. It would be tempting to simply report that EC10 > 10, the highest tested concentration. This cannot be considered a statistically derived conclusion from what has so far been presented, however, as the CI for EC10 for all the models presented reach below 10. There is an alternative hypothesis test that could be employed to test the hypothesis that the effect at 10 mg l^{-1} is less than 10%. Such a test is described in Tamhane et al. (2001). With only two replicates per treatment group, however, this test does not establish the 10% effects concentration as exceeding 10 mg l^{-1}.

Recall that in Chapter 3, the LOEC for this response was 1.3 mg l^{-1} by the Jonckheere–Terpstra test, corresponding to a 4% decrease from control, and even the 3% decrease in the lowest tested concentration was significant by Williams' test. Such small effects found statistically significant may be hard to defend as biologically relevant. However, these regression estimates of EC10 are also of dubious value. Such is the state of regulatory testing. While one might argue for estimating ECx for a smaller value of x in this instance, such decisions should not be reached arbitrarily.

4.4.5 Regression Models Fit to Collembola Reproduction Data in Table 3.10

As noted in Section 3.4.1.1, these data exhibit heterogeneous variances, though the heterogeneity is not severe (Levene p-value = 0.034). That will be ignored initially. The NOEC was 3.91 mg l^{-1} by the Tamhane–Dunnett test, where a 7% effect was observed, making the LOEC 7.81, where a 17% effect was observed. This would seem to be a candidate where a meaningful EC10 estimate might be obtained. Six models were fit to these data. See Table 4.13 for a summary of these results. AICc is of no help in model selection. R^2 is misleading. OE2 is least attractive in terms of model agreement to observed % change from control and that is because the control mean is underestimated by 4%. Also, the EC10 estimate from OE2 seems to be at odds with the observed % decrease 7.81 mg l^{-1}. Models BVP, LL3, OE3, and OE5 are ruled out by having parameters not significantly different from zero. That leaves OE4 to consider further. OE4 has a slightly smaller AICc than any of the other models, and it underestimates the control mean by only 1%, so it is preferred over the other five models examined so far. From that model, EC10 = 5.75 (1.72, 9.78), which puts the point estimate between the NOEC and LOEC, and the CI spans two test concentrations. See Figure 4.19 for the fit of OE4 to these data.

The analysis presented above treats the reproduction data as a continuous response. As pointed out in Section 3.4.1.1, the response is a count, so it would be natural to consider modeling the response as a count using a Poisson error structure, or, to capture the minor overdispersion, a negative binomial error structure. In either case, it is still possible to fit the same models for the mean as used when treating the response as a continuous response. When the BVP model is fit using a Poisson error structure, EC10 = 4.61 (2.68, 6.54), and when OE4 is fit, EC10 = 6.08 (4.57, 7.58), so much tighter confidence bounds are

Table 4.11 Summary of the Fit of the Model HORM1 to the Soybean Data.

Shapiro–Wilk test of normality

Obsr	Std	Skew	Kurt	SWStat	PValue
55	2.191 14	−1.268 71	5.056 65	0.906 95	0.000 438

Levene test for variance heterogeneity of ShootHt

Levene	PValue
1.520 78	0.164 27

Observed and predicted % change from control mean

Dose	Rate	*N*	Observed mean	Predicted mean	Observed % change	Predicted % change
1	0	5	30.1	30.4	0	0
2	0.037	5	34.7	33.3	15	9
3	0.073	5	32.3	33.9	8	12
4	0.147	5	33.1	33.4	10	10
5	0.29	5	32.8	31.2	9	3
6	0.59	5	26.1	27.8	−13	−8
7	1.17	5	26.0	24.3	−14	−20
8	2.34	5	20.8	21.1	−31	−31
9	4.69	5	17.2	18.2	−43	−40
10	9.38	5	15.2	15.8	−49	−48
11	18.75	5	14.8	13.7	−51	−55

Estimates and 95% confidence limits for model parameter

Parameter	Estimate	LowerCL	UpperCL	Probt
Gamma	255.8	10.2127	501.4	0.041 5
Y0	30.400 5	28.135 2	32.665 9	<0.000 1
Beta	1.201 3	1.167 1	1.235 5	<0.000 1
EC25	1.606	0.950 5	2.261 4	<0.000 1

Covariance matrix and standard error of parameter estimates

NAME	EC*x*	Gamma	Y0	Beta	SE
Gamma	15.93	149 67.61	−84.67	−0.77	122.34
Y0	−0.32	−84.67	1.27	0.00	1.13
Beta	0.00	−0.77	0.00	0.00	0.02
EC*x*	0.11	15.93	−0.32	0.00	0.33

Model parameter estimates, EC25 and 95% confidence bounds

EC25	Gamma	Beta	Y0	LB	UB
1.61	255.83	1.20	30.40	0.97	2.25

Brain–Cousens hormetic model evaluation measures

SSReg	SSRes	SSPE	FLOF	PLOF	SSTotC	RSqr	AICc
39 256.6	335.6	259.3	1.1	0.4	39 592.2	1.0	110.7

All parameters are bounded away from zero. Predicted and observed responses are in good agreement. Good fit of model to data.

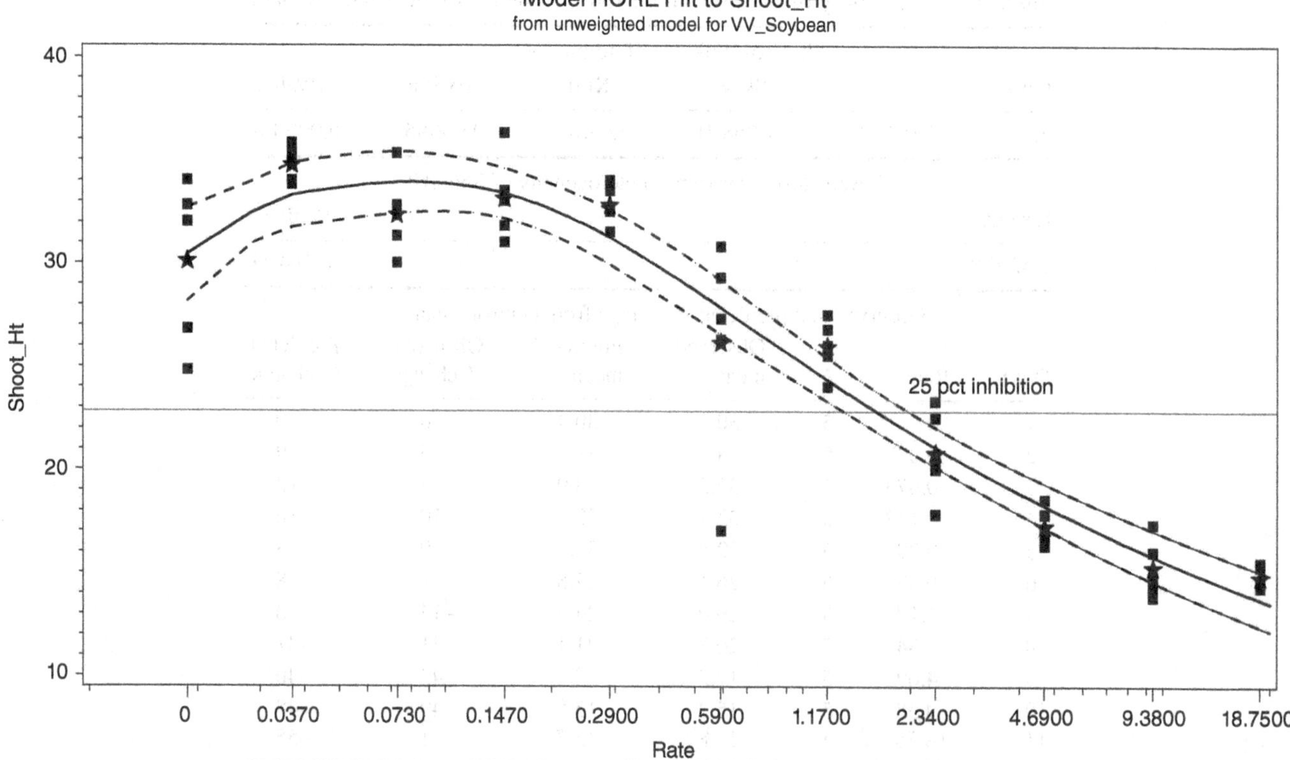

Figure 4.17 HORM1 fit to soybean shoot height data. Star, treatment mean; Filled square, replicate mean; Solid line, model; and Dashed lines, confidence curves.

obtained. Interestingly, when the BVP model is fit with Poisson error structure, EC10=4.61 (2.28, 6.54), and so the point estimate is almost identical to that obtained from the normal errors model, while the CI is much tighter.

4.4.6 Regression Models Fit to Algal Density Data in Table 3.11

The density data were found to be nonnormal, which means fitting any of the normal error models discussed in this chapter questionable. Furthermore, density is a count and so an alternative Poisson error structure would be more appropriate. We will return to this example in Chapter 8.

4.4.7 Analysis of Cucumber Dry Weight

Cucumbers were exposed to various application rates of a pesticide in a vegetative vigor study (OECD TG 227). Except in the highest application rate, 180 g a.s. ha^{-1}, there was no appreciable mortality of the plants by the end of the study. The measurement of interest in this example is dry weight at the end of the study. See Table 4.14 for the data and see Table 4.15 for a summary of two models fit to the data. The model fits are displayed in Figures 4.20 and 4.21. The primary reason for including this example is to illustrate the applicability of model OE5. The fit of OE5 is in contrast to that of OE3, which differs from OE5 only by the presence or absence of a floor. The biological necessity of a floor for these data may not be clear, but the effect on the fit is evident.

4.5 SUMMARY OF MODEL ASSESSMENT TOOLS FOR CONTINUOUS RESPONSES

Distributional issues

Assess distribution of residuals (e.g. normality) for consistency with model requirements.

Assess variance for homogeneity across treatment groups.

Goodness of fit of a single model

Visually compare observed and predicted percent decrease at each test concentration (Motulsky and Christopoulos, 2004; Draper and Smith, 2012).

Compare regression error mean square against the pure error mean square using an F-test (Draper and Smith, 2012).

Table 4.12 Summary of Models Fit to Rainbow Trout Weights

Conc	Reps	Obsr mean	BVP mean	OE2 mean	OE3 mean	LL3 mean	Obsr %Chng	BVP %Chng	OE2 %Chng	OE3 %Chng	LL3 %Chng
0	4	4.48	4.48	4.38	4.48	4.48	0	0	0	0	0
0.63	2	4.34	4.33	4.36	4.33	4.33	−3	−3	0	−3	−3
1.3	2	4.26	4.30	4.35	4.31	4.30	−5	−4	−1	−4	−4
2.5	2	4.26	4.28	4.33	4.28	4.28	−5	−4	−1	−4	−4
5	2	4.36	4.25	4.28	4.25	4.25	−3	−5	−2	−5	−5
10	2	4.17	4.22	4.18	4.22	4.22	−7	−6	−5	−6	−6

Model evaluation measures

Model	SSReg	SSErr	SSPE	FLOF	PLOF	SSTot	RSqr	AICc
BVP	223.1	0.06	0.03	1.35	0.37	223.2	1.00	−48.07
OE2	223.1	0.09	0.03	1.74	0.26	0.15	1.00	−48.97
OE3	223.1	0.06	0.03	1.34	0.37	0.15	1.00	−48.11
LL3	223.1	0.06	0.03	1.35	0.37	223.2	1.00	−48.07

Model Parameter Estimates

Modl	Parm	Estimate	StdErr	LCB	UCB
OE2	Y0	4.38	0.04	4.29	4.46
OE2	EC10	22.80	9.43	1.79	43.8
OE3	Y0	4.48	0.06	4.35	4.62
OE3	D	0.20	0.15	−0.15	0.55
OE3	EC10	170.2	485	−928	1268
BVP	Y0	4.48	0.06	4.35	4.62
BVP	Gama	10.99	8.39	−7.99	29.9
BVP	EC10	246.1	3.21	0.17	352 334
LL3	Y0	4.48	0.06	4.35	4.62
LL3	B	0.20	0.16	−0.15	0.56
LL3	EC10	180.6	525	−1008	1370

Modl, name of model fit; Obsr, observed; SSTot, corrected total sum of square. The observed and estimate means and % change from control should be compared. Only OE2 underestimates the control mean. AICc does not distinguish among models, nor does R^2. OE2 does the worst in estimating the % change from control, because of the underestimation of the control mean, though the estimated percent changes are not terrible. To decide among these models, the critical distinctions are the estimated model parameters. Only the simplest model, OE2, has all parameters significantly different from zero. Note especially the EC10 estimates and the width of the confidence intervals for EC10. It is difficult to believe an EC10 estimate so far outside the observed range of test concentrations as predicted by the models other than OE2 and the uncertainty in such extrapolations is reflected in the width of the confidence intervals.

Check that every term in the model is significantly different from zero (i.e. determine whether all model terms are important) (Motulsky and Christopoulos, 2004).

Plots of residuals from regression versus test concentration, possibly on a log(conc) scale. There should be no pattern to this plot; the points should be randomly scattered about a horizontal line at zero height.

Compare models

Use Akaiki's AICc criteria. Smaller AICc values denote better fits and if AICc(B) − AICc(A) ≥ 10, model A is almost certainly better than model B (Motulsky and Christopoulos, 2004).

Compare the two models visually by how well they meet the single model criteria mentioned above.

The parsimony principal is advised, whereby the simplest model that fits the data reasonably well is used (Ratkowsky, 1993; Lyles et al., 2008).

Quality of EC*x* estimate

The CI for EC*x* should not be too wide.

Statistical judgment is needed in deciding how wide the CI can be and EC*x* still be useful. Simulations for regression models fit to egg hatching and size data

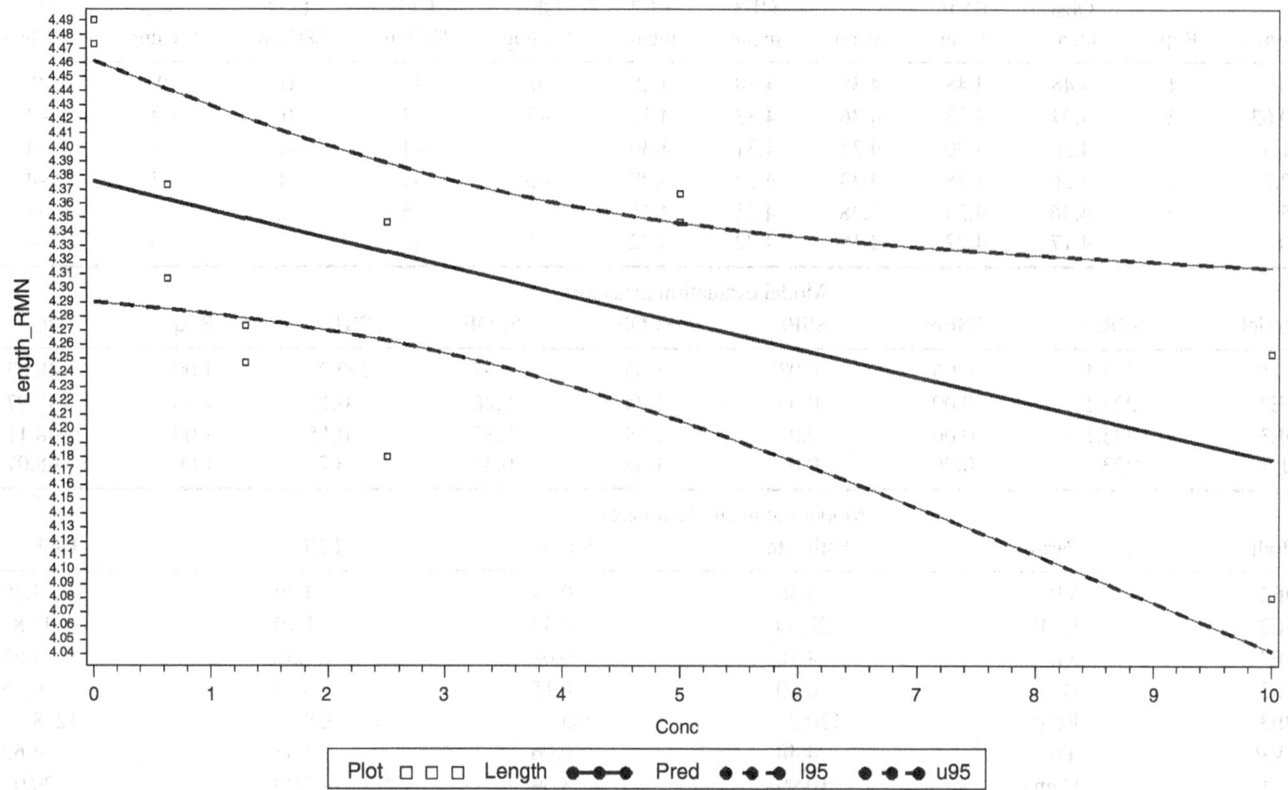

Figure 4.18 OE2 Model fit to rainbow trout length data in Table 3.3. Open squares, replicate means; Solid line, fitted curve; Dashed lines, 95% confidence bounds. Wide bounds on predicted values are apparent.

show that about 75% of CIs for ECx (x=10, 20, or 30) span no more than two test concentrations. This provides a general guide for what is acceptable and a practical guide for what is achievable.

Numerous authors assert the need to report CIs for all model parameters and that wide CIs for model parameters indicate unacceptable models (Alvord and Rossio, 1993; Bunke et al., 1999; Seber and Wild, 2003; Motulsky and Christopoulos, 2004; Lyles et al., 2008; Ott and Longnecker, 2015).

The CI for ECx (or any other model parameter) should not contain zero (Motulsky and Christopoulos, 2004).

Are the parameter estimates scientifically plausible? For example, if the CI for y_0 is ±20%, no EC10 estimate is plausible. If the model predicts a 20% effect at a concentration C and the maximum observed effect at C and lower concentrations is 10%, then the EC20 is not plausible (Wang et al., 2000; Motulsky and Christopoulos, 2004).

ECx should not require extrapolation outside the range of positive concentrations (OECD, OECD, 2006a, b, c; Draper and Smith, 2012). For example, a general guide might be for ECx to be not more than about 25% below the lowest tested concentration or above the highest tested concentration.

The CI for the mean response at ECx should not contain the estimated control mean response. This is the regression equivalent the minimum significant difference that is often cited in hypothesis testing approaches (e.g. Wang et al., 2000). It also corresponds to the requirement that the CI for the mean response at the LOEC does not contain the control mean.

EXERCISES

In all datasets in the text and exercises, if a concentration is given with the value −1 means that a solvent used in the study and the water control is recorded as −1 and the solvent control is listed as 0. The reader should determine whether to combine the two controls for NOEC determination or ECx estimation. If the two controls are to be combined, then the water control should be changed to 0 and the replicates renamed so that replicates in the two controls can be distinguished. If only the solvent control group is to be used, then the water control should be discarded. If only the water control should be used, then the solvent control should be discarded and the concentration

Table 4.13 Summary of Models Fit to Collembola Data

Conc	Obsr %Chng	BVP %Chng	LL3 %Chng	OE2 %Chng	OE3 %Chng	OE4 %Chng	OE5 %Chng
0	0	0	0	0	0	0	0
3.91	−7	−9	−9	−4	−10	−7	−10
7.81	−17	−15	−15	−8	−15	−13	−15
15.6	−25	−24	−23	−16	−23	−23	−23
31.3	−32	−35	−34	−29	−34	−35	−34
62.5	−48	−47	−47	−50	−48	−46	−48

Conc	Obsr mean	BVP mean	LL3 mean	OE2 mean	OE3 mean	OE4 mean	OE5 mean
0	515.6	516.2	515.6	496.6	517.0	511.7	516.8
3.91	480.5	470.4	480.5	475.6	464.9	475.7	465.9
7.81	426.8	438.2	426.8	455.7	437.6	444.8	437.9
15.6	387.5	393.7	387.5	418.2	397.8	395.3	397.1
31.3	352.3	337.6	352.3	351.9	342.1	330.9	340.8
62.5	268.3	274.4	268.3	249.6	270.2	276.4	271.2

Model	Parm	Est	StdErr	LCB	UCB
	EC10	4.53	2.67	−0.97	10.03
BVP	Gamma	2.18	0.55	1.05	3.32
	Y0	516.20	21.19	472.50	559.80
	EC10	4.22	2.71	−1.36	9.80
LL3	b	0.77	0.20	0.35	1.19
	Y0	516.60	21.12	473.10	560.10
OE2	EC10	9.58	1.59	6.30	12.85
	Y0	496.50	16.48	462.70	530.40
	EC10	3.87	2.71	−1.71	9.44
OE3	D	0.65	0.18	0.29	1.02
	Y0	517.00	21.17	473.40	560.60
	EC10	5.75	1.96	1.72	9.78
OE4	C	0.49	0.11	0.27	0.72
	Y0	511.70	19.44	471.70	551.80
	EC10	4.00	3.59	−3.42	11.41
OE5	C	0.17	1.93	−3.81	4.14
	D	0.69	0.56	−0.48	1.85
	Y0	516.80	21.91	471.60	562.00

Model	SSErr	SSPE	FLOF	PLOF	RSqr	AICc
BVP	92439	90336	0.900	0.603	0.982	237
LL3	92578	90336	0.902	0.601	0.982	237
OE2	101850	90336	0.954	0.550	0.981	237
OE3	92644	90336	0.902	0.601	0.982	237
OE4	94186	90336	0.918	0.585	0.679	234
OE5	92626	90336	0.940	0.561	0.684	237

Six models fit to data. AICc and PLOF are not helpful. OE4 has the lowest R^2 but the best overall fit.

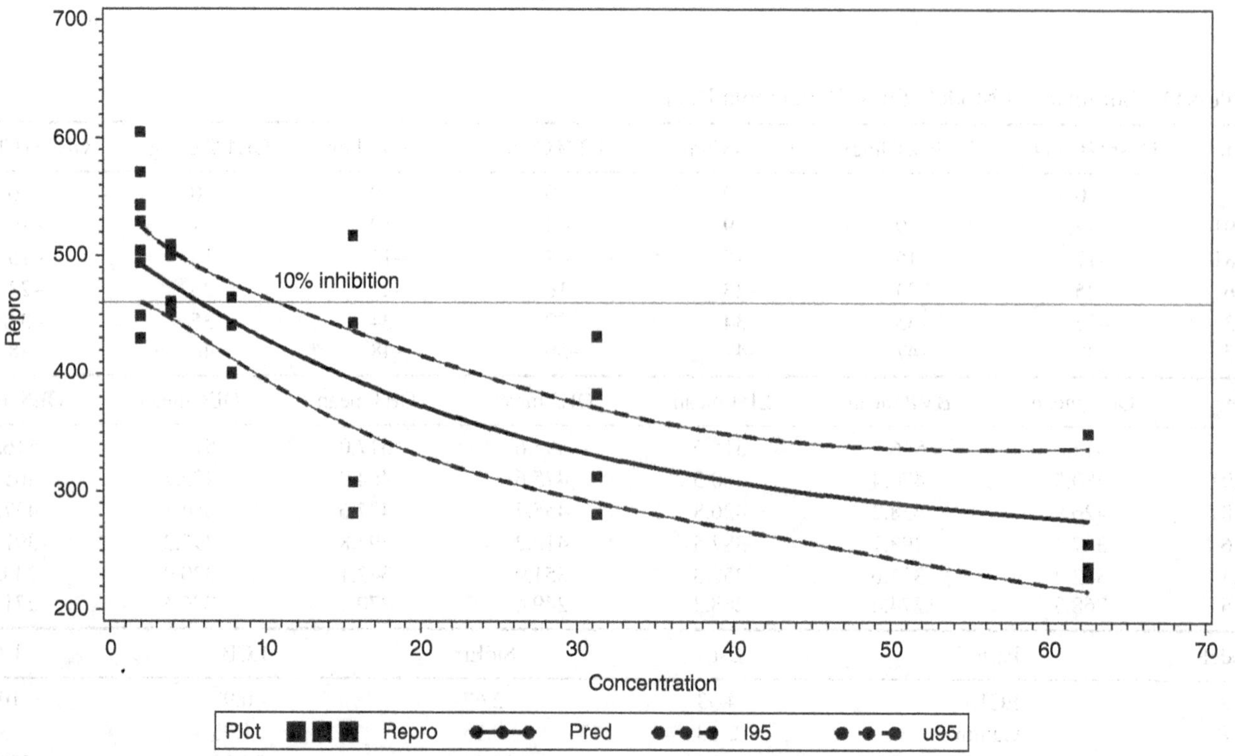

Figure 4.19 Model OE4 fit to collembola reproduction data. Filled squares, replicate means; Solid line, fitted curve; Dashed lines, 95% confidence curves.

Table 4.14 Cucumber Data from Vegetative Vigor Study

Rate	Rep	ShootHt	DryWgt	%Survive	Rate	Rep	ShootHt	DryWgt	%Survive
0	1	994	29.904	100	5.65	4	853	25.098	100
0	2	984	26.633	100	5.65	5	716	22.39	100
0	3	961	23.462	100	11.3	1	616	18.815	100
0	4	969	21.77	100	11.3	2	598	18.054	100
0	5	890	23.851	100	11.3	3	792	19.287	100
0.355	1	992	23.635	100	11.3	4	471	15.499	100
0.355	2	826	24.108	100	11.3	5	413	14.76	100
0.355	3	863	23.45	100	22.5	1	330	11.094	100
0.355	4	843	23.09	100	22.5	2	239	7.473	100
0.355	5	1073	24.941	100	22.5	3	224	7.958	100
0.71	1	1000	29.835	100	22.5	4	341	13.327	100
0.71	2	935	24.253	100	22.5	5	168	6.087	100
0.71	3	929	27.155	100	45	1	122	3.291	100
0.71	4	898	24.494	100	45	2	136	5.009	100
0.71	5	1016	28.311	100	45	3	121	3.777	100
1.42	1	1047	29.066	100	45	4	181	7.481	100
1.42	2	810	27.059	100	45	5	179	7.619	100
1.42	3	911	24.849	100	90	1	114	3.45	100
1.42	4	922	23.091	100	90	2	128	1.786	50
1.42	5	797	23.747	100	90	3	115	4.413	100
2.83	1	945	20.755	100	90	4	151	7.623	100
2.83	2	947	24.911	100	90	5	114	3.299	75
2.83	3	820	25.294	100	180	1	122	2.879	75
2.83	4	916	25.835	100	180	2	116	2.943	50
2.83	5	1034	27.77	100	180	3	112	0.96	25
5.65	1	712	28.166	100	180	4	124	2.537	75
5.65	2	668	24.982	100	180	5	125	3.831	75
5.65	3	931	25.65	100					

Replicate summary value.

Table 4.15 Model OE3 Overestimates the Effect Near the Observed 10% Effects Rate and in the Two Highest Application Rates

Model	Parameter	Estimate	StdErr	LCB	UCB
OE5	Y0	25.5	0.4	24.7	26.4
	C	0.2	0.0	0.1	0.3
	D	2.2	0.4	1.4	3.0
	EC10	6.6	1.0	4.6	8.7

Model	Parm	Estimate	StdErr	LCB	UCB
OE3	Y0	26.2	0.7	24.7	27.7
	D	1.1	0.2	0.8	1.4
	EC10	3.7	1.1	1.6	5.8

Rate	*N*	Observed mean	OE5 mean	OE3 mean	Observed % change	OE5 % change	OE3 % change
0	5	25.1	25.5	26.2	0	0	0
0.355	5	23.8	25.5	26.0	−5	0	−1
0.71	5	26.8	25.5	25.8	7	0	−2
1.42	5	25.6	25.4	25.3	2	0	−4
2.83	5	24.9	25.1	24.2	−1	−2	−8
5.65	5	25.3	23.7	22.1	1	−7	−16
11.3	5	17.3	18.5	18.1	−31	−28	−31
22.5	5	9.2	7.9	11.7	−63	−69	−55
45	5	5.4	4.4	4.5	−78	−83	−83
90	5	4.1	4.4	0.6	−84	−83	−98
180	5	2.6	4.4	0.0	−90	−83	−100

Model	SSReg	SSErr	SSPE	FLOF	PLOF	SSTotC	RSqr	AICc
OE5	37011.5	297.5	225.2	1.14	0.33	5123.6	0.99	101.7
OE3	21111.9	448.6	225.2	1.69	0.04	5123.6	0.98	124.3

Model OE3 also overestimates the control mean, albeit by only 4%, and that contributes to an apparent underestimation of EC10. There is a striking difference in AICc values that strongly favors OE5. It is also notable that the lack-of-fit test for OE3 is significant ($p=0.04$), while that for OE5 is not.

for the water control should be changed to 0. Under no circumstances should the concentration value −1 be used for NOEC or EC*x* purposes. Guidance on this is provided in Chapters 2 and 3.

4.1 The data are frond counts of Lemna gibba after 7 days of exposure to a test chemical. See Table 4.16 for the data, which are also available in the SAS dataset Lemna_Ex41 in Appendix 1. There were four replicate test vessels in the control and each treatment group. Day0 is the initial number of fronds placed in the vessel. Day7 and Day14 are the frond counts after 7 and 14 days of exposure, respectively, and biomass is the mass of the fronds after 14 days.

(a) If possible, fit regression models to Day7, Day14, and biomass assuming normal error structure to estimate EC10 and EC20.

(b) Do a square-root transform of Day7 and Day14 and obtain approximate estimates of EC10 and EC20 by fitting suitable regression models to the transformed data and back-transforming to the original scale. For this purpose, use the following methodology.

EC*p* for count can be approximated by EC*q* for square root of count in the following way.

If Y=count and Z=SQRT(Y), then at concentration U=ECp(Y), it must be that

$$Y(U)=\left(1-\frac{p}{100}\right)Y_0, \tag{4.26}$$

where Y_0 is the control mean. From this, it can be seen that

$$Z(U)=\sqrt{1-\frac{p}{100}}\sqrt{Y_0}=\sqrt{1-\frac{p}{100}}Z_0, \tag{4.27}$$

where Z_0 is the square root of the control mean. The corresponding change in Z at concentration U is given by

$$Z(U)=\left(1-\frac{q}{100}\right)Z_0. \tag{4.28}$$

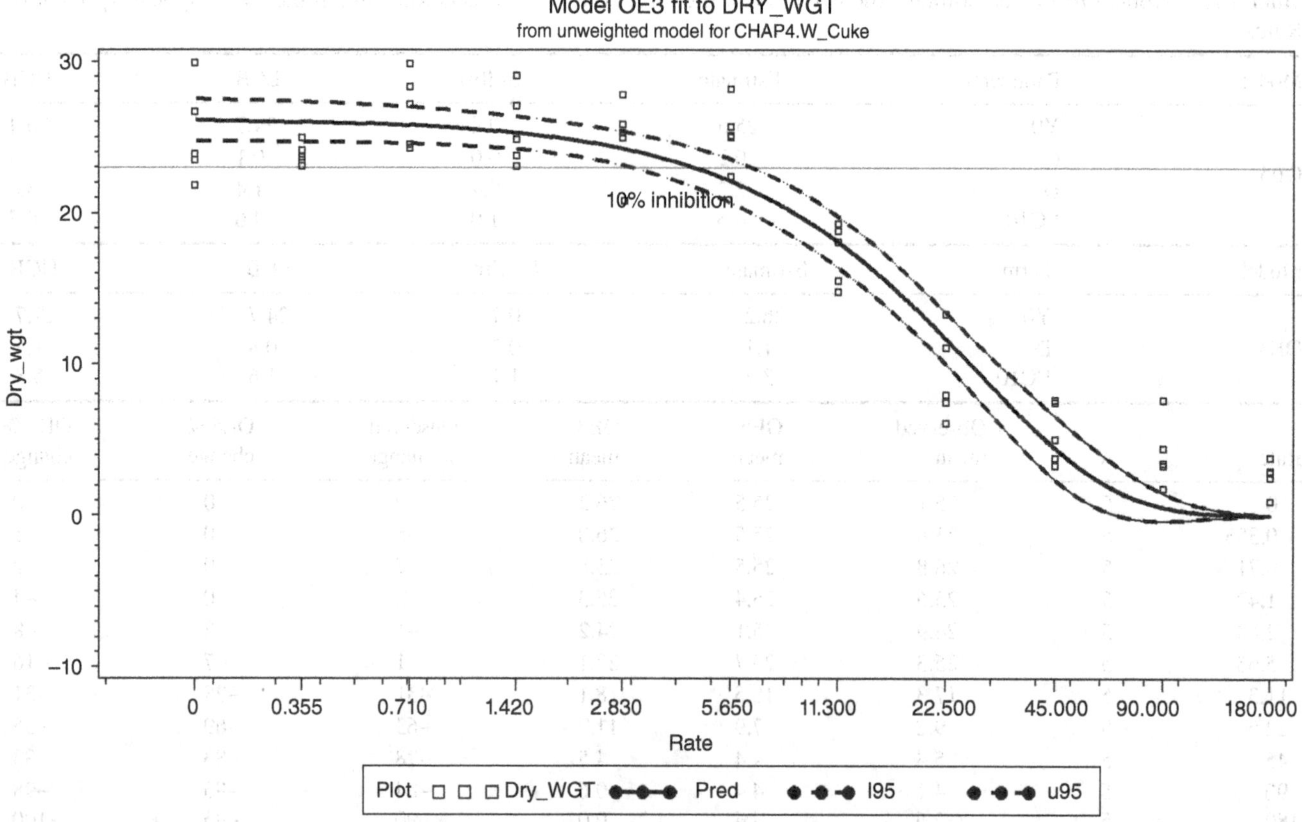

Figure 4.20 Observations in the highest two application rates are underestimated and the width of the confidence interval for the predicted mean at 180 g a.s. ha^{-1} appears unrealistically small.

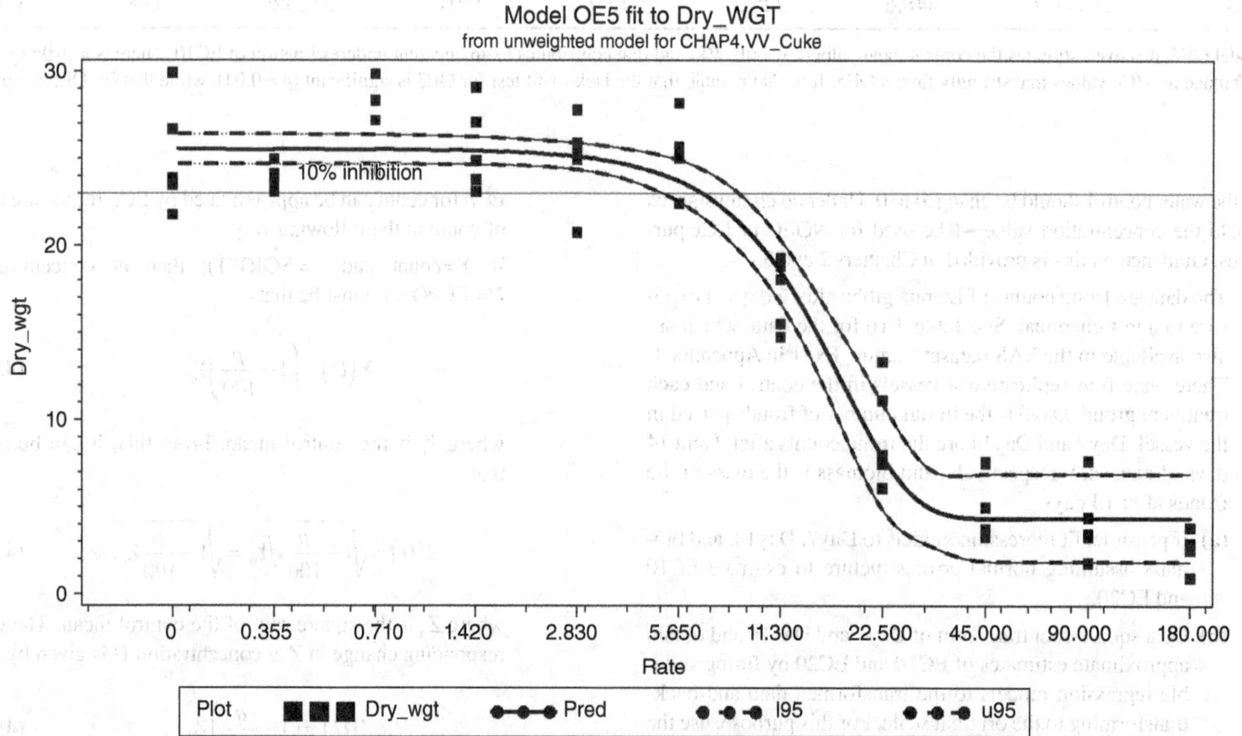

Figure 4.21 Model OE5 fit to cucumber data. Model OE5 does not capture the slight decline at the 180 g a.s. ha^{-1} application rate, but overall, the model fits the data well.

Table 4.16 Frond Counts from Lemna Gibba Study

Group	Conc	Rep	Day0	Day7	Day14	Biomass
1	0	1	15	196	450	40.6
1	0	2	15	181	440	
1	0	3	15	183	420	39.7
1	0	4	15	177	443	39.7
2	19.2	1	15	105	421	
2	19.2	2	15	114	430	36.2
2	19.2	3	15	117	424	34.9
2	19.2	4	15	84	396	34.7
3	38.4	1	15	66	265	24.5
3	38.4	2	15	82	305	27.9
3	38.4	3	15	73	289	29.2
3	38.4	4	15	64	302	
4	76.8	1	15	54	249	24.5
4	76.8	2	15	54	247	26
4	76.8	3	15	47	232	22.4
4	76.8	4	15	54	237	
5	153.6	1	15	42	239	23.4
5	153.6	2	15	33	217	
5	153.6	3	15	42	261	22.4
5	153.6	4	15	41	226	20
6	307.2	1	15	29	166	
6	307.2	2	15	22	160	12.9
6	307.2	3	15	29	189	19.7
6	307.2	4	15	28	193	19.3

Day *x*, frond count on study day *x*.

Table 4.17 Value of *q* for Selected Values of *p*

p	*q*
10	5.1
20	11.6
50	30.3

q is percent change in Sqrt(*Y*) to estimate a *p*% change in *Y*.

Equating (4.27) and (4.28) and solving for *q* yields

$$q = 100\left(1 - \sqrt{1 - \frac{p}{100}}\right). \tag{4.29}$$

See Table 4.17 for a small set of values of *p* and the associated values of *q*.

(c) Redo parts (a) and (b) after subtracting the Day0 value from the Day7 and Day14 values. Discuss which is a biologically more preferable way to analyze the effect on frond count.

4.2 A study was done under OECD TG 232 to determine whether the test chemical had a reproduction on the Collembola *Folsomia candida* over a 4-week exposure. See Table 4.18 for the data. Concentrations were in mg a.s. kg^{-1} of dry soil.

Table 4.18 Collembola Reproduction

Group	Conc	Rep	Juveniles	Group	Conc	Rep	Juveniles
1	0	1	454	3	31.25	3	462
1	0	2	374	3	31.25	4	429
1	0	3	249	4	62.5	1	429
1	0	4	406	4	62.5	2	343
1	0	5	316	4	62.5	3	380
1	0	6	347	4	62.5	4	304
1	0	7	388	5	125	1	351
1	0	8	256	5	125	2	359
2	15.625	1	409	5	125	3	235
2	15.625	2	242	5	125	4	267
2	15.625	3	374	6	250	1	184
2	15.625	4	409	6	250	2	306
3	31.25	1	424	6	250	3	204
3	31.25	2	364	6	250	4	236

Juveniles, number of young produced by the end of the study.

The response is the number of juveniles produced over 4 weeks of study from the initial 10 adults placed in each replicate vessel.

(a) Assess the data for normality, variance homogeneity, and concentration–response monotonicity.

(b) Fit the Brain–Cousens model to these data, estimate EC10 and EC20, and assess goodness of fit.

(c) Fit additional models and compare them to determine whether the Brain–Cousens models is a good choice.

(d) Determine the NOEC for these data using Williams, Jonckheere–Terpstra, and Dunnett tests. Compare the NOEC(s) with EC10.

4.3 An earthworm study was done to determine whether a test chemical had an effect on the total body weight of exposed worms during an acute toxicity test. Ten earthworms were weighed as a group and placed in each replicate test vessel at the start of the study. At the end of the exposure period, the surviving worms were again weighed as a group. The response of interest was the change in body weight. Body weight was measured in mg and concentration was in mg a.s. kg^{-1} dry soil. See Table 4.19 for the data.

Regression modeling to determine that the *x*% effects concentration, EC*x*, is usually not well modeled on weight gain (or loss) directly. The reason for this is that EC*x* is an estimate of an *x*% change from the estimate control mean response. The control mean response is 5.575 mg, so a 10% reduction from that mean would be 0.5575 mg, a very small change of no biological interest. This is not a useful measure, given that changes of up to 70 mg are observed in higher test concentrations. A more scientifically sound approach is to estimate the percent change in the ratio WGT14/WGT0, where WGT*y* is the body weight on day *y* of the test.

Fit models to this ratio, including OE4. Assess the goodness of fit and use the model selection criteria from Chapter 4 to select the model to report and estimate EC10, EC20, and EC30, or justify why an acceptable estimate of one or more of these quantities cannot be obtained from these data.

Table 4.19 Earthworm Body Weight

Conc	Group	Rep	WGT0	Alive14	WGT14
0	1	1	359.7	10	341.9
0	1	2	367.4	10	330
0	1	3	312.6	10	323.6
0	1	4	342.2	10	363.3
62.5	2	1	356.1	10	302
62.5	2	2	333.5	10	298.5
62.5	2	3	348.3	10	308.8
62.5	2	4	362.6	10	320.8
125	3	1	339.6	10	273.4
125	3	2	351.4	10	290.1
125	3	3	370.7	10	308.3
125	3	4	348	10	301.4
250	4	1	361.2	10	287.3
250	4	2	360.9	10	292.4
250	4	3	349.4	10	285.4
250	4	4	329.3	10	278.5
500	5	1	354.1	10	273
500	5	2	354.8	10	270.4
500	5	3	348.2	10	277.4
500	5	4	361.5	10	311.3
1000	6	1	320.9	10	254.2
1000	6	2	332.2	10	257.8
1000	6	3	362.6	10	278.5
1000	6	4	325.8	10	271.5

WGT*x*, weight of earthworms on day *x*; Alive14, number of adult earthworms alive on day *x*.

4.4 A vegetative vigor study was done on sugar beets following OECD TG 227. A surfactant was used so that there were both water and surfactant controls. There were 10 replicate pots per treatment and control, with one seed each at the start of the study, and 11 application rates (concentrations). The primary responses were shoot height and dry weight, root dry weight, and total dry weight. Also given was a somewhat subjective assessment of visual health, labeled visual_pct, ranging from 100 (good) to 0 (bad). The data are given in Appendix 1 in the Excel file *Exercise_E44* in the spreadsheet *sugarbeet E4.4*.

(a) Compare the two controls for each response and combine them for modeling if justified. Otherwise, use only the surfactant control.

(b) Fit models to the four continuous measurements and assess the model fits and data for meeting model requirements.

(c) Use the model selection criteria in Chapter 4 to select a model for each response and estimate EC10, EC25, and EC50 or explain why a satisfactory estimate of one or more of these endpoints is not possible from these data.

4.5 In the same study described in Exercise 4.4, wheat was also tested. The data for that nontarget terrestrial plant is in the same Excel file in the spreadsheet wheat. Follow the instructions in Exercise 4.4 for dry root weight. Since one of the concerns of the study was whether hormesis was present, one of the models fit should be HORM1. Optionally, fit models to the other responses in this spreadsheet.

4.6 Fit a regression model to the earthworm reproduction data in Exercise 3.5. Assess the data for the requirements of the model.

Rainbow Trout Early Life Stage data

In Exercises 4.7–4.12, data from an early life stage study on rainbow trout was done under an older version of OECD TG 210 that required only two replicate tanks per treatment group. The data are given in the Excel file RBT size data in a spreadsheet named study *X* and the SAS dataset of the same name in Appendix 1, where *X* is the study number (1, 6, 9, etc.). Concentrations were in mg l^{-1}.

(a) If a value of concentration of −1 is shown, then a solvent was used in the study and the water control is coded as −1 and the solvent control is coded as 0. If no concentration is shown with a value of −1, then no solvent was used. Where a solvent was used, compare the two controls and determine whether they are significantly different for either response. Either pool the two controls (changing conc=−1 to conc=0 and renumbering reps if needed so as not to duplicate rep designations) or drop the water control for further analysis. Justify your decision.

(b) Since there are only two reps per concentration, no good assessment of variance homogeneity is possible. Discuss whether a test of normality is appropriate and do one if so.

(c) Fit several regression models to each of length and weight. Assess goodness of fit of these models and use the criteria developed in this chapter to select the most appropriate model from those fit or justify a conclusion that no useful regression model could be fit to these data.

(d) Where a useful regression model could be found, plot the model and data in a single figure and estimate EC10 and EC20 where the data support that and explain why not if one or more of these EC*x* values cannot be well estimated.

(e) If there was high mortality in some treatment group(s), indicated by notable variability in the resize variable *n*, then either take steps to adjust the model fit (such as by weighting the fit by *n*) or explain why this is not necessary.

(f) Compute the NOEC and compare it with EC10 and EC20.

4.7 Use the rainbow trout ELS study in Exercise 3.6.

4.8 Use the ELS study 1 data.

4.9 Use the ELS study 6 data.

ELS study 6. No meaningful regression is possible on length, OE2 on weight w/very wide bounds.

4.10 Use the ELS study 9 data.

ELS study 9. BVP on length and weight good EC10, EC20.

4.11 Use the ELS study 11 data.

ELS study 11. BVP on length and weight. Good estimates, strong trend.

4.12 Use the ELS study 20 data.

ELS study 20. Brain–Cousens on length and weight. Good estimates.

4.13 A Daphnia Chronic study was done following OECD TG 211. The data are presented in the SAS dataset Exercise413 and the Excel file of the same name, both in Appendix 1. Concentration is measured in mg l^{-1}. There were 10 replicate vessels of one daphnid in the water control and each treatment group.

(a) Find a suitable regression model for length and estimate EC10 and EC20 or explain why no useful estimate can be obtained from these data. Comment on the maximum observed effect and the high mortality in top 2 concentrations.

(b) Compute the NOEC.

4.14 A Daphnia Chronic study was done following OECD TG 211. The data are presented in the SAS dataset Exercise414 and the Excel file of the same name, both in Appendix 1. Concentration is measured in mg l^{-1}. There were 10 replicate vessels of one daphnid in the water control and each treatment group. Find a suitable regression model for

(a) length,

(b) weight,

(c) TLY14, and

(d) TLY21,

where TLYx=total live young at day x, and estimate EC10 and EC20 or explain why no useful estimate can be obtained from these data. Think carefully about how to treat the single observation in the highest test concentration in the length and weight measurements and explain your reasoning. Check each response for outliers and repeat the modeling with outliers omitted. Comment on the comparison of results.

(e) Discuss whether any of the models in this chapter are appropriate for REPDAY, the first day of reproduction. This response will be revisited in later chapters dealing with count and time-to-event data.

4.15 A Daphnia Chronic study was done following OECD TG 211. The data are presented in the SAS dataset Exercise415 and the Excel file of the same name, both in Appendix 1. Concentration is measured in mg l^{-1}. There were 10 replicate vessels of one daphnid in the water control and each treatment group. Find a suitable regression model for

(a) length,

(b) weight, and

(c) TYS21,

where TYS21 = total live young per survival adult at day 21, and estimate EC10 and EC20 or explain why no useful estimate can be obtained from these data. Check each response for outliers and repeat the modeling with outliers omitted. Comment on the comparison of results.

Chapter 5

Analysis of Continuous Data with Additional Factors

5.1 INTRODUCTION

Not all designed experiments in toxicology follow the simple dose–response designs that have been the subject of Chapters 2–4, where the only relevant factors are dose or concentration and replicate. Other factors sometimes must be considered. Section 5.2 will deal with uncontrolled factors that should be taken into account to avoid bias. Initial body weight of subjects in a study of weight gain in animals is a common factor that cannot be ignored. While care is taken to assure the same species, strain, sex, age, and source of subjects are used in all treatment groups, there can still be some variation in initial body weight that could affect the interpretation of the data. Animals that are larger initially tend to be larger at the end of the study, especially for short-term studies, and may be more robust than smaller animals and thus not be as affected by exposure to the test substance. If organ weights are to be analyzed, it must be recognized that larger animals tend to have larger organs, so some adjustment for the total body weight may need to be considered in assessing the effect of the test substance on organ weight. One of the mechanisms to handle such issues is analysis of covariance (ANCOVA). That is the topic of Section 5.2.

Section 5.3 will consider how to analyze data when the same subjects are analyzed at multiple times. When the same subject is measured several times, the individual observations cannot be considered to be independent, and an error structure should be used that reflects the correlations among distinct observations on the same subject. Such repeated measures studies are common in neurotoxicity studies and some ecological measures, such as earthworm abundance in treated soil over time, and in blood or tissue levels of mercury in birds or fish constantly exposed to a contaminant and monitored over time.

Section 5.3 will deal with studies in which two or more factors are controlled over the course of the study. A simple example would be to analyze both male and female mice in a simple dose–response setting. In this instance, one might expect a general tendency for some measured response, such as liver weight, to decrease over time, but males and females might respond differently, and those differences might vary depending on the dose.

5.2 ANALYSIS OF COVARIANCE

The primary reason for doing ANCOVA rather than just ANOVA is to reduce extraneous sources of variation in order to improve the ability to separate true treatment effects from noise. Sometimes there is an independent concomitant variable unaffected by the treatment of interest that may have an effect on the response of interest. By accounting for the variance in the response that arises solely from differences in the covariate, the variability due to the treatment, and hence the treatment effects, becomes clearer. An example of this sort of phenomenon is growth or growth rate. If animals or plants are different in size before exposure to the treatment, they may well exhibit different growth just because of the initial differences, and

Statistical Analysis of Ecotoxicity Studies, First Edition. John W. Green, Timothy A. Springer, and Henrik Holbech.

Companion website: www.wiley.com/go/Green/StatAnalysEcotoxicStudy

this difference should not be allowed to cloud the treatment effect. CHMP (2013) emphasizes that covariates are baseline variables, not variables measured later in the study. Generally in a laboratory study with lab species, it is possible to begin with subjects that are very similar in size. In field studies, such as bovine growth studies with different types of feed, animals may differ by 100 lb or more at study start, and this should not be ignored.

In rodent studies, fetal body weight is affected by litter size, and it is tempting to use litter size as a covariate in analyzing treatment effects on fetal weight. Mead et al. (2002) suggests this in a study of piglets. However, if the treatment itself may affect litter size, it would be improper to use litter size as a covariate, because then we would be using a treatment effect to explain away a treatment effect. In this situation, litter size and fetal weight are both dependent variables to be analyzed for the effect of the treatment. Consider a treatment that tends to increase litter size. It is well known that pups in larger litters tend to be smaller than pups in small liters. In an experiment of this sort, there are two factors that may tend to reduce pup weight: larger litters and exposure to the test substance. The effects are confounded and a simple ANCOVA may be misleading. The reverse can happen, where the treatment tends to decrease litter size, which, in the absence of any other effect, would tend to increase pup weight. If the test substance also tends to reduce pup weight, the two factors acting in opposition may make it appear that there are no effects on pup weight. If litter size is first shown not to be affected, then it may be acceptable to use it as a covariate. The problem with this reasoning is that the test for litter size effect may be low power, so that failure to be significant does not rule out a treatment effect.

It should be noted that there are papers or books in the nonstatistical scientific literature that use ANCOVA even when the above assumptions are violated. Gad (2006) is an example, but even he recommends caution and cites papers that explain the issue. Shirley and Newnham (1984) advocate the use of final body weight as a covariate in analyzing organ weights only if there is a linear relationship between body weight and organ weight that passes through the origin. If the relationship is linear but does not pass through the origin, then it is also claimed that ANCOVA is still preferred since it takes into account indirect effects on body weight. The statistical justification for this is questionable, though the motivation for such an adjustment is clear. Hayes (2007) repeats the need for care in the same context. An alternative way to take body weight into account when analyzing organ weights is to use the ratio of the two, so that the response is the proportion of total body weight represented by the organ. However, Stevens (1976) showed that this can be misleading except under the same condition as noted by Shirley and Newnham. Takizawa (1978) explores this topic further and shows that ANCOVA of organ weights using body weight as a covariate when there is a treatment effect on body weight still provides an unbiased test for treatment effects on organ weights. Similar concerns would seem to apply to using concurrent maternal body weight (MBW) as a covariate in the analysis of fetal weights, though this has not been confirmed. Miller and Chapman (2001) provide a good discussion of the requirements of ANCOVA and provide examples of abuse as well as good practice. The above discussion and citations indicate that while ANCOVA is used in toxicology, its use is not free of controversy and does require care.

There is no statistical objection to the use of initial body weight as a covariate in the analysis of organ weight, but the use of final body weight as a covariate is problematic, since it ignores the possible effect of the test substance on body weight. Analysis of the ratio of organ weight to final body weight overcomes this objection. Provided final body weight is also analyzed, there is little objection to analysis of the ratio. Both approaches aim to reduce the variability in organ weight data arising from variation in body weight, but while they are sometimes thought to be equivalent, they are not. This will be explored in more detail below.

Similar concerns would seem to apply to using concurrent MBW as a covariate in the analysis of fetal weights, though this has not been confirmed. The above discussion and citations indicate that while ANCOVA is used in toxicology, its use is not free of controversy and does require care. Similar issues arise in some mesocosm studies. For example, one such study included *Myriophyllum proserpinacoides*, for which responses included main stem length, number of side shoots per main stem, and number of side shoots per meter of main stem. The last two of these, especially the last, are similar to the organ weight issue, since it is expected that the number of side shoots per plant is positively correlated with the number and length of main stems. ANCOVA was not used in this study, as the ratios were judged to be more appropriate and avoided the use of a covariate that was itself affected by the treatment. Additional discussion of mesocosm studies is provided in Chapter 13.

In addition to the usual requirements of ANOVA (briefly summarized as independence of observations, normality, and variance homogeneity), a standard ANCOVA assumes that (i) the slope of the covariate is the same across all treatment groups, (ii) the variance of the adjusted response (e.g. fetal weight adjusted for the covariate) is the same for all treatments, and (iii) the treatments have no effect on the covariate. It is the last of these assumptions that is most problematic and the most frequently abused in application. Standard applications of ANCOVA use pretreatment values of an uncontrolled variable as a covariate, such as the pretreatment maternal weight for pregnant mice in a reproduction study. There are numerous references on this issue. Miller and Chapman (2001) provide a good discussion of the requirements of ANCOVA and provide examples of abuse as well as good practice. See also Mead et al. (2002). A very readable account is given in Field (2016), with more detail in one of three similar books that discuss SPSS, SAS, or R implementation (Field and Miles, 2010; Field et al., 2012). It should be noted also that assumption 1 bears close consideration, because if the slopes of the regression relative to the covariate is not the same in all treatment groups, those differences in slope indicate a treatment effect.

At a minimum, a candidate covariate should be examined for treatment effects. There are three parts to this. First, a simple visual inspection of the means of each covariate versus treatment may indicate a potential problem. ANOVA of the covariate versus treatment adds statistical power to this visual inspection. Second, the slopes of the regression lines in each treatment group relative to the covariate should be equal. Once again, a plot might suffice, but there is a formal test for equality of slopes as well. Third, biological judgment is required to determine that a treatment effect on the candidate covariate is unlikely. This third criterion is obviously not statistical. It is included to avoid giving a risk assessor reason to question the analysis.

See Table 5.1 for data from a mouse reproductive study. The purpose of the study was to determine whether the test substance had an adverse effect on fetal body weight. The response was average fetal body weight per litter. Several possible covariates that were considered by the principal investigator are included in the table.

See Table 5.2 for a summary of the data in Table 5.1. The table contains several potential covariates that might be used to reduce extraneous variability in fetal weights to facilitate assessment of the test substance on fetal weight. The table indicates possible problems with all of the covariates except sex ratio (%Male in the table). There are downward trends in treatment means for MBW and body weight gain, food consumption, and litter size, as well as for fetal weight. ANOVA of the candidate covariate versus treatment adds statistical power to this visual inspection. See Table 5.3 for an analysis summary that shows there are significant treatment effects only for maternal food consumption (and fetal weight).

The slope of the regression on a given covariate should be the same in all treatment groups, since differences in slope indicate a treatment effect. Table 5.3 gives tests for homogeneity (equality) of slopes and shows heterogeneous slopes for MBW on day 29 and litter size. FETWGT PRELIM.SAS and FETWGT PRELIM.r are given in Appendix 1 that will generate the data in Tables 5.3–5.8. The data are in the SAS dataset *fetalwgt* and the Excel file *fetalwgt.xls*. To test for equal slopes across treatments, it is sufficient to fit the model

$$Y_{ij} = \mu + \tau_i + \beta X_{ij} + \gamma_j X_{ij} + e_{ij}, \tag{5.1}$$

and test $\gamma_1 = \gamma_2 = \gamma_3 = \cdots = \gamma_k = 0$ vs. $\gamma_j \neq 0$ for some j. Alternatively, the model can be expressed as

$$E[y] = \text{Conc} + \text{Covar} + \text{Dose*Covar}$$

where Conc is the treatment variable and the appropriate test is of whether the interaction is significant. In this alternative, no grand mean term is included, which is equivalent to replacing each $\mu + \tau_i$ in Eq. (5.1) by τ_i. It would also be equivalent to replace $\beta X_{ij} + \gamma_j X_{ij}$ by $\gamma_j X_{ij}$ and testing the hypothesis $\gamma_1 = \gamma_2 = \gamma_3 = \cdots = \gamma_k$ against the alternative $\gamma_j \neq \gamma_1$ for some j.

Thus, the formal tests indicate unequal slopes relative to MBW, maternal body weight gain (MBWG), and litter size, making these unsuitable as covariates. While food consumption and sex ratio (%Male) do not exhibit statistically significant differences in slopes across treatments by these formal tests, there is a downward trend in the food consumption evident in Table 5.1. While it is not shown in that table, there is also evidence of variance heterogeneity in the food consumption data, with the variance in the 400 ppm treatment approximately six times that in the control group. There is no indication in the data of a problem with sex ratio (%Male) as a covariate. Biological judgment was that a treatment effect on sex ratio was unlikely. Given that judgment and the absence in the data of a treatment effect, sex ratio can be used as a covariate. It should also be noted, as stated by Shirley and Newnham (1984), that lack of significance of formal tests is not sufficient justification, by itself, for inclusion of a covariate.

See Table 5.4 for the results of ANCOVA with the various candidate covariates. This will make clear the need for care, as there are major differences in conclusions from the various choices. These range from significant effects found in concentrations 200 and 400 when litter size is used as a covariate to no significant effect in any treatment when food consumption is used as a covariate. When no covariate is used or sex ratio or maternal weight on day 29 is used, there is a significant effect at 400 ppm only. With maternal weight gain, the effect at 400 ppm is only marginally significant ($p = 0.09$). Of these, only the no covariate or sex ratio options appear scientifically sound. An option not explored is preexposure MBW, as no data were available for that.

To understand how Table 5.4 was developed and to further understand ANCOVA, it will be helpful to delve into the statistical theory of ANCOVA. This is the subject of Sections 5.2.1 and 5.2.2.

Table 5.1 Mouse Reproduction Study

MBW	MBWG	FC	LS	%Male	FWGT	DAMID	Conc
3070.52	−86.38	2748.8	8	25	44.131	101	0
3277.07	−49.73	2760.6	9	44.4	39.804	102	0
2852.83	−70.77	2733.7	9	44.4	43.974	103	0
2999.3	−36.6	2752.3	8	50	39.037	104	0
3377.08	−113.12	2763	10	50	36.104	105	0
3277.11	−52.79	2759.1	10	50	37.204	106	0
3787.31	−199.49	2747.9	10	50	41.34	107	0
2771.49	−148.11	2509	9	44.4	41.096	108	0
2972.36	−48.94	2750.7	9	55.6	39.826	109	0
2932.49	−61.01	2784.3	7	71.4	49.269	110	0
2967.42	49.12	2747.8	6	50	42.162	111	0
3002.24	−71.36	2782.7	8	37.5	40.602	112	0
3196.77	24.37	2779.1	9	44.4	37.943	113	0
3008.18	−75.32	2745.3	8	62.5	37.447	114	0
3066.94	−70.16	2733.8	7	57.1	44.367	115	0
2930.83	−141.67	2762.5	9	44.4	36.579	116	0
3130.7	−25.2	2648.3	7	85.7	48.551	117	0
3323.66	−6.74	2755.9	10	60	34.875	118	0
3070.94	−113.56	2745.4	8	37.5	41.661	119	0
2772.59	−48.81	2612.5	8	62.5	42.814	120	0
3173.19	−80.11	2751	7	71.4	44.637	121	0
3290.9	−27.3	2785	9	55.6	39.988	122	0
3172.81	−12.29	2518.7	7	42.9	38.354	201	100
3119.89	28.79	2777.8	8	37.5	39.791	202	100
2970.84	−34.66	2725.7	9	66.7	39.894	203	100
3172.17	−109.93	2760.3	9	44.4	36.008	204	100
2857.17	−42.63	2378.1	7	28.6	41.459	205	100
2895.18	−330.42	2393.4	10	60	37.334	206	100
3100	53.2	2729	8	50	41.075	207	100
2808.82	−64.58	2498.6	7	28.6	41.081	208	100
3046.83	−122.57	2279	8	62.5	40.756	209	100
2937.84	−160.96	2497.6	12	41.7	39.458	210	100
3098.55	81.25	2767.4	3	33.3	45.977	211	100
3377.99	32.69	2757	7	42.9	44.079	212	100
3035.72	−28.68	2617.9	8	37.5	37.623	213	100
2773.81	−235.89	2404.5	10	50	37.693	214	100
3244.11	−287.29	2707	10	50	42.69	215	100
3157.14	74.84	2763.7	6	50	45.922	216	100
3087.73	−82.17	2767	10	50	42.674	217	100
2675.69	−132.21	2435.2	10	60	37.225	218	100
3146.29	−3.51	2736.4	8	37.5	37.82	219	100
3358.32	18.72	2673.1	8	50	44.04	220	100
3131.96	−103.84	2748.6	7	28.6	44.36	221	100
3087.81	−113.69	2720.4	10	60	35.887	301	200
3696.35	−19.55	2781.4	8	37.5	38	302	200
2729.87	−103.13	2238.6	11	54.5	24.512	303	200
3131.07	48.97	2704.5	7	57.1	38.863	304	200
2949.65	−184.35	2680.9	9	55.6	39.002	305	200
3000.56	−41.24	2338	5	60	40.786	306	200
3340.79	−273.41	2175.2	9	44.4	41.202	307	200

Table 5.1 (Continued)

MBW	MBWG	FC	LS	%Male	FWGT	DAMID	Conc
3223.84	−266.86	2450.3	10	40	36.334	308	200
2833.46	−2.54	2583.2	10	50	31.14	309	200
3074.87	31.17	2656.5	4	50	47.398	310	200
3000.36	−58.14	2790.1	3	100	49.877	311	200
3170.53	−72.97	2326.6	7	42.9	42.174	312	200
2845.83	−196.37	2297.3	9	66.7	36.734	313	200
3086.54	−85.66	2681.5	8	75	40.349	314	200
3562.48	42.08	2773.8	7	71.4	41.686	315	200
3277.83	57.43	2677.6	4	25	49.953	316	200
2932.66	−177.14	2500.8	7	71.4	45.083	317	200
3092.63	−43.47	2635.6	9	44.4	41.222	318	200
2974.57	−51.63	2350.6	6	33.3	39.64	319	200
3032.17	−31.03	2503.8	8	25	38.624	320	200
3438.54	11.14	2764.8	8	62.5	43.308	401	400
2510.24	−283.56	1767.2	7	57.1	34.064	402	400
3425.51	−101.29	2560.7	10	60	33.229	403	400
3108.62	−281.38	2209.7	7	42.9	41.736	404	400
2970.26	−101.64	2680.3	9	66.7	36.737	405	400
3041.44	−4.46	2762.6	11	27.3	36.928	406	400
2982.61	−7.79	2584.3	7	28.6	37.84	407	400
2690.45	−326.85	1547.4	8	25	32.838	408	400
2781.91	−90.29	2566.6	5	60	44.304	409	400
3385.71	103.21	2677.7	5	40	42.966	410	400
2700.35	−413.85	1611	8	75	30.615	411	400
2862.97	−69.43	2326	7	57.1	39.307	412	400
2790.81	−298.39	2455.7	10	70	35.002	413	400
2876.3	−75.6	2222.9	7	28.6	34.521	414	400
3221.57	−80.43	2707.3	6	50	44.47	415	400
2941.72	−120.48	2439.2	9	33.3	36.582	416	400
2951.37	−198.53	2015.3	8	75	35.134	417	400
3132.26	−171.44	2565.9	10	80	33.011	418	400
3250.7	−75.2	2352.3	8	50	35.779	419	400

MBW, maternal body weight on day 29; MBWG, change in maternal weight days 7–29; FC, maternal food consumption; LS, litter size; FWGT, mean fetal weight; %Male, percent of litter that was male; DAMID, maternal ID; Conc, test concentration in food (ppm).

Table 5.2 Summary of Treatment Mean Responses of Potential Covariates

Conc	MBW	MBWG	FC	LS	FWGT	%Male
0	3102	−66	2734	8.41	41.06	52
100	3056	−70	2616	8.19	40.73	45
200	3102	−77	2543	7.55	39.92	53
400	3003	−136	2359	7.89	37.28	52

Conc, concentration of test chemical in ppm; MBW, maternal body weight on day 29; MBWG, change in maternal weight days 7–29; FC, maternal food consumption; LS, litter size; FWGT, mean fetal weight; %Male, percent of litter that was male.

5.2.1 Parametric ANCOVA

The basic ANCOVA model assumes that the slope relative to the covariate, *X*, is the same in all treatment groups (indicated by the dose or concentration *U*) and in the simplest case is given by the equation

$$Y_{ij} = \mu_i + \beta X_{ij} + e_{ij}, \tag{5.2}$$

where Y_{ij} is the response on the *j*th subject in the *i*th treatment, X_{ij} is the value of the covariate associated with that subject, μ_i is the mean response in the *i*th treatment,

Table 5.3 Tests for Equality of Slope Relative to Potential Mouse Covariates

Effect	NDF	DDF	*F* value	ProbF	Effect	NDF	DDF	*F* value	ProbF
Conc (no covar)	3	78	3.1	0.0315	Conc	3	74	0.34	0.7939
					MBWG	1	74	4.22	0.0434
Conc	3	74	2.69	0.0523	MBWG*Conc	3	74	0.36	0.7805
LS	1	74	65.82	<0.0001					
LS*Conc	3	74	2.94	0.0384	Conc	3	74	0.94	0.4236
					FC	1	74	0.56	0.458
Conc	3	74	2.49	0.067	FC*Conc	3	74	0.98	0.4082
MBW	1	74	4.62	0.0348					
MBW*Conc	3	74	2.29	0.0858	Conc	3	74	0.8	0.4968
					%Male	1	74	0.17	0.677
					%Male*Conc	3	74	1.24	0.2998

Covariates are LS, MBWG, MBW, FC, and %Male. A significant interaction term indicates unequal slopes, such as LS*Conc.

Table 5.4 ANCOVA Results for Mouse Data

Covariate	Conc	StdDiff	StdErr	Adj*P*	Covariate	Conc	StdDiff	StdErr	Adj*P*
None	100	−0.335	1.320	0.989	MBW	100	−0.139	1.299	0.999
	200	−1.141	1.336	0.732		200	−1.140	1.312	0.722
	400	−3.782	1.355	0.018		400	−3.366	1.346	0.039
Sexratio	100	−0.205	1.345	0.998	MBWG	100	−0.286	1.253	0.992
	200	−1.155	1.342	0.729		200	−0.988	1.270	0.783
	400	−3.774	1.361	0.019		400	−2.808	1.324	0.096
Littersize	100	−0.708	0.954	0.803	FC	100	0.504	1.258	0.958
	200	−2.606	0.980	0.026		200	0.213	1.310	0.997
	400	−4.659	0.984	<0.0001		400	−1.120	1.482	0.788

StdDiff, standardized difference of mean fetal weight in given test concentration and control; StdErr, standard error of StdDiff; Adj*P*, *p*-value of StdDiff from Dunnett–Hsu test.

Table 5.5 ANOVA table for ANCOVA

Factor	SS	DF
Total (corrected)	T_{YY}	$n_{\cdot}-1$
Covariate	$\tilde{\beta} T_{xy}$	1
Treatment (adjusted for covariate)	$\sum_i n_i(\bar{Y}_{i.}-\bar{Y}_{..})^2+\hat{\beta}S_{xy}-\tilde{\beta}T_{xy}$	$k-1$
Residual	$S_{yy}-\hat{\beta}S_{xy}$	$n_{\cdot}-k-1$

Computational formulas to create traditional summary table for ANCOVA.

Table 5.6 Unadjusted and Adjusted Fetal Weight Treatment Means

Conc	FetalWt	AdjFWmn	SexRatio	*N*	Std
0	41.06	40.10	0.52	22	3.77
100	40.73	39.90	0.45	21	3.00
200	39.92	38.95	0.53	20	5.88
400	37.28	36.33	0.52	19	4.24

FetalWt, mean fetal weight; AdjFWmn, mean fetal weight adjusted for sexRatio; SexRatio, mean proportion males; *N*, number of adult females; Std, standard deviation of unadjusted mean fetal weight.

β is the common slope in every treatment group relative to the covariate, and e_{ij} are independent, identically normally distributed (iid $N(0, \sigma^2)$) random errors. The form given by Eq. (5.1) or Eq. (5.2) assumes a single covariate but can easily be modified in matrix terms to include two or more covariates. The single treatment mean can also be replaced by a more complex model, but the present section will focus on this simple model. It should be noted that Eq. (5.2) can be written in an overparameterized form as

$$Y_{ij} = \mu + \tau_i + \beta\left(X_{ij} - \bar{X}_{..}\right) + e_{ij}, \tag{5.3}$$

where $\bar{X}_{..}$ is the mean value of the covariate. Use of the full rank model given by Eq. (5.1) leads to simpler expressions,

Table 5.7 Calculations for Fetal Weight ANCOVA Table

S_{YY}	S_{XX}	S_{XY}	$\hat{\beta}$
1459.21	1.81907	3.33545	1.8336
T_{YY}	T_{XY}	T_{XX}	$\tilde{\beta}$
1633.16	2.19204	1.9016	1.15273
$N.$	$\bar{Y}_{..}$	$\bar{X}_{..}$	B_{YY}
82	39.82	0.51	173.946

$B_{YY} = \sum_i n_i \left(\bar{Y}_{i.} - \bar{Y}_{..}\right)^2$, the between-group sum of squares. The sums of squares and cross products S_{uv}, T_{uv}, etc. are from Eqs. (5.11), (5.15), (5.24), and (5.25). $N.$, total number of adult females. $\bar{Y}_{..}$ and $\bar{X}_{..}$ are the grand means of fetal weight and sex ratio.

Table 5.8 ANCOVA for Fetal Weight Adjusted for Sex Ratio

Source	SS	DF	MS	F	P_F
SS$\tilde{\beta}$	2.527	1	2.527	0.134	0.715
SS$\hat{\beta}$	6.116	1	6.116	0.324	0.571
Trt(adj)	177.535	3	59.178	3.136	0.030
Error	1453.090	77	18.871		
Total(Cor)	1633.160	81			

SS$\tilde{\beta} = \hat{\beta} S_{xy}$ and SS$\tilde{\beta} = \tilde{\beta} T_{xy}$.

but the two models are equivalent and yield the same tests, estimates, and conclusions.

The idea is to adjust the response, Y_{ij}, for the covariate prior to testing for any difference among the treatment groups. It will be helpful to rewrite Eq. (5.1) in matrix form:

$$Y = Z\gamma + e, \tag{5.4}$$

where

$$Z = \left[\text{Diag}_k\left(J_{n_i}\right)\ X\right],$$

and

$$\gamma = \begin{bmatrix} \mu \\ \beta \end{bmatrix}.$$

It will be evident that the left block of the matrix Z is a block diagonal matrix of dimensions $kn.\times k$, the ith diagonal element being a column vector of n_i 1s, and X is a $kn.\times 1$ column vector. From Appendix 2 (A2.62), it can be seen that

$$\hat{\gamma} = (Z'Z)^{-1} Z'Y. \tag{5.5}$$

It is necessary to find the inverse of the matrix $Z'Z$ and compute the indicated product. An outline of the necessary calculations is given below. It is left as an exercise to verify the development:

$$Z'Z = \begin{bmatrix} \text{Diag}_k(n_i) & Col_k(X_{i.}) \\ Row_k(X_{i.}) & X'X \end{bmatrix}. \tag{5.6}$$

The inverse of $Z'Z$ has the form

$$(Z'Z)^{-1} = \begin{bmatrix} A & B \\ B' & c \end{bmatrix} \tag{5.7}$$

for some $k\times k$ matrix A, k-component column vector B, and scalar c. The matrices A and B and the number c can be found by solving the equation

$$(Z'Z)^{-1}(Z'Z) = \begin{bmatrix} \mathbf{I}_k & 0 \\ 0' & 1 \end{bmatrix}. \tag{5.8}$$

Here 0 is a k-element column vector and $\mathbf{I}_k$ is the $k\times k$ identity matrix with 1s on the diagonal and 0 in every off-diagonal position. Matrix multiplication yields the following:

$$\begin{bmatrix} A\text{Diag}_k(n_i) + B\text{Row}_k(X_{i.}) & A\text{Col}_k(X_{i.}) + BX'X \\ B'\text{Diag}_k(n_i) + c\text{Row}_k(X_{i.}) & B'\text{Col}_k(X_{i.}) + c(X'X) \end{bmatrix} = \begin{bmatrix} \mathbf{I}_k & 0 \\ 0' & 1 \end{bmatrix}. \tag{5.9}$$

From the bottom left component of Eq. (5.9), it can be shown that

$$B = -c\text{Col}_k\left(\bar{X}_{i.}\right). \tag{5.10}$$

When Eq. (5.10) is substituted into the bottom right component of Eq. (5.9), it follows that

$$c = \frac{1}{S_{XX}} \quad \text{and} \quad B = -\frac{1}{S_{XX}}\text{Col}_k\left(\bar{X}_{i.}\right), \tag{5.11}$$

where $S_{XX} = \sum_{ij}\left(X_{ij} - \bar{X}_{i.}\right)^2$.

When the two expressions in Eq. (5.11) are substituted into the upper left component of Eq. (5.9) and the entire equation is right-multiplied by $\text{Diag}_k(1/n_i)$, it follows that

$$A = \text{Diag}_k\left(\frac{1}{n_i}\right) + \frac{1}{S_{XX}}\text{Col}_k\left(\bar{X}_{i.}\right)\text{Row}_k\left(\bar{X}_{i.}\right). \tag{5.12}$$

Next it is simple to verify that

$$Z'Y = \begin{bmatrix} \text{Col}_k\left(n_i\bar{Y}_{i.}\right) \\ X'Y \end{bmatrix}. \tag{5.13}$$

From Eqs. (5.13) and (5.9) with $\boldsymbol{A}$, $\boldsymbol{B}$, and c replaced by Eqs. (5.12) and (5.11), it follows that

$$\hat{\gamma} = \begin{bmatrix} \hat{\mu} \\ \hat{\beta} \end{bmatrix} = \begin{bmatrix} \text{Col}_k\left(\bar{Y}_{i.} - \hat{\beta}\bar{X}_{i.}\right) \\ \dfrac{S_{XY}}{S_{XX}} \end{bmatrix}, \tag{5.14}$$

where S_{XY} is the sum of cross products given by

$$S_{XY} = \sum_j \left(X_{ij} - \bar{X}_{i.}\right)\left(Y_{ij} - \bar{Y}_{i.}\right). \tag{5.15}$$

This shows that the treatment means have been adjusted for the covariate and the slope relative to the covariate is given by the standard regression calculation. Note that it follows from Eq. (5.5), using $\text{var}(\mathbf{M} * \mathbf{W}) = \mathbf{M} * \text{var}(\mathbf{W}) * \mathbf{M}'$ for constant matrix $\mathbf{M}$, that

$$\text{Var}(\hat{\gamma}) = \sigma^2 (Z'Z)^{-1}, \tag{5.16}$$

so that when the solutions for $\boldsymbol{A}$, $\boldsymbol{B}$, and c from Eqs. (5.11) and (5.12) are substituted into the inverse matrix in Eq. (5.16), the following facts can be verified:

$$\begin{aligned} \text{Var}\left(\hat{\mu}_i\right) &= \sigma^2\left[\frac{1}{n_i} + \frac{\bar{X}_{i.}}{S_{XX}}\right], \\ \text{Cov}\left(\hat{\mu}_i, \hat{\beta}\right) &= -\sigma^2 \frac{\bar{X}_{i.}}{S_{XX}}, \\ \text{Cov}\left(\hat{\mu}_i, \hat{\mu}_j\right) &= \sigma^2 \frac{\bar{X}_{i.}\bar{X}_{j.}}{S_{XX}}, \\ \text{Cov}\left(\bar{Y}_{i.}, \hat{\beta}\right) &= 0. \end{aligned} \tag{5.17}$$

To test for treatment effects and the significance of the regression term $\boldsymbol{\beta}$, the proper variance term is needed. From linear models theory given in Appendix 2 (A2.62), the residual sum of square is given by

$$\text{RSS} = Y'\left[\mathbf{I}_n - Z(Z'Z)^{-1}Z'\right]Y. \tag{5.18}$$

The matrix inside the square brackets of Eq. (5.18) is idempotent, meaning when that matrix is multiplied by itself, the result is the same matrix. Thus, Eq. (5.18) can be written as

$$\text{RSS} = (Y - Z\hat{\gamma})'(Y - Z\hat{\gamma}), \tag{5.19}$$

and $Y - Z\hat{\gamma}$ will be recognized as the vector of residuals from the model, so RSS is the sum of squared residuals, as expected. With some matrix manipulation, it follows that

$$\text{RSS} = S_{YY} - \hat{\beta}S_{XY} = S_{YY} - \frac{S_{XY}^2}{S_{XX}}. \tag{5.20}$$

From Eq. (5.20), it can be seen that a result of the covariate is to reduce the simple ANOVA error term S_{YY}. ANCOVA has achieved its purposes: to adjust the treatment means for the covariate and reduce the unexplained error, the combined influence of which makes the detection of treatment effects more sensitive and less influenced by variation explained by the covariate. The RSS has degrees of freedom (DF) equal to the rank of the idempotent matrix in Eq. (5.18), which is $n - k - 1$.

The total sum of squares (corrected) is given by

$$\text{SSTOTC} = \sum_{i,j}\left(Y_{ij} - \bar{Y}_{..}\right)^2. \tag{5.21}$$

The model sum of squares is obtained by subtracting the error sum of squares, RSS, from SSTOTC:

$$\begin{aligned} \text{SSModel} &= \text{SSTOTC} - \text{RSS} \\ &= Y'\left[Z(Z'Z)^{-1}Z' - \frac{1}{n}J_nJ_n'\right]Y. \end{aligned} \tag{5.22}$$

It is a simple matter to verify that the defining matrices for the model and residual sums of squares are orthogonal, so the sums of squares (or quadratic forms) are independent. It is customary to decompose SSModel into two parts, corresponding to the contributions of the covariate and treatments. To do that, it is easily verified that

$$\begin{aligned} \text{SSTOTC} &= T_{YY} \\ &= \tilde{\beta}T_{xy} + \sum_i n_i\left(\bar{Y}_{i.} - \bar{Y}_{..}\right)^2 + \hat{\beta}S_{xy} - \tilde{\beta}T_{xy} + S_{yy} - \hat{\beta}S_{xy}, \end{aligned} \tag{5.23}$$

where

$$\tilde{\beta} = \frac{\sum_{i,j}\left(Y_{ij} - \bar{Y}_{..}\right)\left(X_{ij} - \bar{X}_{..}\right)}{T_{xx}} = \frac{T_{xy}}{T_{xx}}, \tag{5.24}$$

$$T_{xx} = \sum_{i,j} \left(X_{ij} - \bar{X}_{..}\right)^2,$$
$$T_{yy} = \sum_{i,j} \left(Y_{ij} - \bar{Y}_{..}\right)^2, \quad (5.25)$$
$$T_{xy} = \sum_{i,j} \left(Y_{ij} - \bar{Y}_{..}\right)\left(X_{ij} - \bar{X}_{..}\right).$$

See Table 5.5 for the construction of the ANOVA table from Eqs. (5.22) to (5.25).

$\tilde{\beta}$ will be recognized as the estimate of the common slope relative to the covariate if treatments are ignored, the T_{uv} in Eq. (5.23) are the sums of squares or cross products relative to the total or grand means for X and Y ignoring treatments, whereas the S_{uv} given by Eqs. (5.11) and (5.15) are the sums of squares or cross products relative to treatment means. While not explicitly stated above, the formula for S_{yy} is the same as that for S_{XX} with Xs replaced by Ys. It is not obvious that the last three sums of squares in Table 5.5 are independent, which is a necessary condition to formulate tests of the hypotheses implied by the table. The reader is referred to Searle (1987, especially chapter 11) for the required mathematical development. A test of hypothesis of no treatment differences after adjusting for the covariate is obtained by dividing the adjusted treatment sum of squares by the corresponding DF, $k-1$, then comparing the ratio of that over the residual mean square to the $F(k-1, n-k-1)$ distribution at the desired significance level, usually 0.05. A test for the significance of the covariate is obtained similarly using the covariate sum of squares.

The expression in Table 5.5 to test for significance of the covariate is not the most intuitive one, though it is traditional (see Searle, 1987, for example). It would be more intuitive to base a test on the maximum likelihood estimate of the slope given in Eq. (5.14), where it was shown that

$$\hat{\beta} = \frac{1}{S_{xx}} \sum_{i,j} \left(X_{ij} - \bar{X}_{i.}\right)\left(Y_{ij} - \bar{Y}_{i.}\right)$$
$$= \frac{1}{S_{xx}} \mathbf{X}' \left[\mathbf{I}_n - \mathrm{Diag}_k \left(\frac{1}{n_i} J_{n_i} J_{n_i}' \right) \right] \mathbf{Y}, \quad (5.26)$$

which has the form $\hat{\beta} = \mathbf{M}\boldsymbol{Y}$ for the indicated matrix $\mathbf{M}$ in Eq. (5.26) that is constant, given $\boldsymbol{X}$. It follows that

$$\mathrm{Var}\left(\hat{\beta}\right) = \mathbf{M}\mathrm{var}\left(\boldsymbol{Y}\right)\mathbf{M}' = \sigma^2 \mathbf{M}\mathbf{M}' = \frac{\sigma^2}{S_{XX}}. \quad (5.27)$$

A simple t-test can be used to test for significance of slope once an estimate of σ^2 is verified independent of $\hat{\beta}$. It is interesting to note that the F-test for significance of the covariate produced by SAS Proc Mixed is based on this more intuitive calculation rather than the expression in Table 5.5.

Note how the independence of $\boldsymbol{X}$ from the treatment is implicit in these calculations. The observations $\boldsymbol{Y}$ are random variables dependent on the treatment levels. $\boldsymbol{X}$ is assumed independent of treatments. The common slope already implies this. These calculations will now be illustrated with the fetal body weight data from Table 5.1 using sex ratio as the covariate. See Table 5.6 for the unadjusted and adjusted mean fetal weight and sex ratio (expressed as proportion male) by concentration. It is evident from the table that there is a clear downward trend in the fetal weight data. The purpose of the ANCOVA is to determine whether the apparent decrease is significant and related to the treatment or attributable to differences in sex ratio. See Tables 5.7 and 5.8 for calculations to create the ANCOVA tables using sex ratio as covariate.

It is left as an exercise to verify that the ANOVA table from SAS for these data agrees with the calculations in Table 5.8. A program, *ANOVA_Calc.sas*, to accomplish that is in Appendix 1. The R program *FETWGT PRELIM.r* also does this. Note that the F-test for adjusted treatment means is significant. As discussed at some length in earlier chapters, that is not important or especially helpful. What is of interest is whether one or more adjusted treatment means is significantly different from the control. A Dunnett–Hsu test answers that question, finding a significant treatment-related effect, but only in the high concentration.

The mathematical development and concepts are easily extended to more complex experimental designs. Exercise 5.10 gives an example with three factors in a factorial design with a single covariate.

5.2.2 Nonparametric ANCOVA

If no normalizing transformation can be found and the nonnormality is severe, a nonparametric ANCOVA can be performed. There are several ways to implement a nonparametric ANCOVA. There are also dangers of inflated false positive rates and care is needed. In this section, these issues will be explored and illustrative examples given.

In the simplest situations in ecotoxicology studies, there is a single explanatory variable, dose or concentration, and one or more covariates. When the within-treatment distribution of the response deviates markedly from normality (usually assessed through residuals from a simple one-factor ANOVA as described in Chapter 4) and no suitable normalizing transformation can be found, a rank transform might be considered. A question to consider is whether a rank transform is applied only to the response of interest or to the covariate(s) as well. Our interest is on the error rates (type I and II) in determining whether treatment means differ from the control mean. However, the basic assumption in most ANCOVAs is that the slope relative to the covariate, $\boldsymbol{X}$, is the same in all treatment groups (indicated by the dose or concentration variable $\boldsymbol{U}$). This is implicit in the equation

$$Y_{ij} = \mu_i + \beta X_{ij} + e_{ij} \tag{5.28}$$

where Y_{ij} is the response on the jth subject in the ith treatment, X_{ij} is the value of the covariate associated with that subject, μ_i is the mean response in the ith treatment, and β is the common slope. The form given by Eq. (5.28) assumes a single covariate but can easily be modified in matrix terms to include two or more covariates.

The idea is simple: Merely replace the $\boldsymbol{Y}$ by their *normalized* ranks, where response values from all treatment groups (including control) are ranked together. Then analyze the data using one of the tests described in Chapter 3. As Mansouri and Chang (1995) state, "The rank transform (RT) technique is a popular but controversial technique which is recommended by Conover and Iman (1981) for a wide range of statistical testing problems." It may be helpful to define normalized ranks to avoid misunderstanding. Once the data have been rank-transformed to create ranks r_i, $i=1$ to n, n being the total number of observations, then define the normalized ranks by Eq. (5.29):

$$\rho_i = \Phi^{-1}\left(\frac{r_i - c}{n - 2c + 1}\right), \tag{5.29}$$

where Φ is the standard normal distribution and $0 \le c \le 1$ is selected. One common choice of c is 3/8, and this defines the Blom (1958) normalized ranks. Bliss (1967) used $c=1/2$, while Tukey used $c=1/3$, and van der Waerden (1952) used $c=0$. A development of these scores and how they relate to the normal or alternative distributions was developed by Filliben (1975). In this text, the Blom formulation is used. An interesting discussion and literature review of rank transform, including normalized ranks, is given by Beasley et al. (2009). Further information on Eq. (5.29) and its variations is given in Chambers et al. (1983).

Some variations on this simple idea are provided below. A relevant question is whether to replace the covariates as well as the response by rank-transformed values. Conover and Iman (1982) describe the use of standard tests, i.e. those presented in Chapter 3, based on both $\boldsymbol{U}$ and $\boldsymbol{X}$ being replaced by their rank transforms. Stephenson and Jacobson (1988) transformed only $\boldsymbol{U}$. According to a simulation study conducted by Klockars and Moses (2002), both approaches control the false positive rates (type 1 error) when the distribution of $\boldsymbol{Y}$ is heavy tailed, but not when it is light tailed. The use of the ANCOVA model in Eq. (5.28) is predicated on the slope of the response relative to the covariate being the same for all treatment groups. Another way of expressing this is that there is no treatment by covariate interaction. It is prudent to assess the validity of this model assumption. Headrick and Sawilowsky (2000) presented simulation evidence that indicated that the Conover and Iman approach to testing differences in slopes can have very elevated type I error rates. Their work dealt with studies with two explanatory variables with a factorial structure and is in general agreement with the work cited in Section 3.5 by Bathke and Lankowski (2005), Akritas et al. (1997), and Akritas (1990). These references do not deal with the simple situation of treatment as the sole explanatory variable. It is tempting to interpret Headrick and Sawilowsky (2000) as concluding that if a test for equal slopes is not significant, then the approach to ANCOVA of Conover and Iman is justified. The work of Headrick and Sawilowsky (2000) does not address this idea. Mansouri and Chang (1995) demonstrated that many of the problems associated with analysis of ranks can be avoided if normalized ranks are used instead, and that is what is recommended here. Note that responses are rank-ordered using all observations for the ranking and observations having the same value are replaced by their average rank. This approach will be illustrated with two examples. Neither dataset requires a nonparametric analysis but such are provided so the parametric and nonparametric results can be compared.

5.2.2.1 Nonparametric ANCOVA of Mouse Fetal Weight Data

Two analyses were done, one using untransformed sex ratio as covariate and the other using normalized ranks of sex ratio as covariate (see Table 5.9). In both cases, normalized ranks of fetal weight were the response values. The conclusions are unchanged from those reported in Table 5.4 and subsequent discussion. For easy comparison, the ANCOVA summary reported in Table 5.4 is also reported here, and the SAS program *NPAR ANCOVA FETALWT. SAS* is given in Appendix 2. A modest adaptation of the R program *ANCOVA_SHIRLEY DATA.R* carries out the same analysis in that software.

5.2.2.2 Example: Shirley Antidote Data

Shirley (1981) described a study in which three groups of rats were tested for the time to enter a chamber. Initially, these rats were not treated and the entry time, called Before, was recorded. Subsequently, all rats received a treatment that was intended to delay their entry time. Groups 2 and 3 were also then given an antidote that was intended to counter the effects of the treatment. The time these rats took to enter was then recorded as After. (See Table 5.10 for the data.) An ANCOVA was done using the Before value as covariate with the object of determining whether the antidote was effective. As pointed out by Shirley, the

Table 5.9 Nonparametric ANCOVA of Mouse Fetal Weight

Response	Source	DF	SS	MS	*F* value	ProbF
FetalWt	Group	3	177.5351	59.17837	3.14	0.0302
	Sexratio	1	6.115885	6.115885	0.32	0.5708
	Residual	77	1453.094	18.87136	.	
			Effect	Group	Adjustment	Adj*P*
			Group	2	Dunnett–Hsu	0.9976
			Group	3	Dunnett–Hsu	0.7291
			Group	4	Dunnett–Hsu	0.0192
Response	Source	DF	SS	MS	*F* value	ProbF
NRank	Group	3	9.358191	3.119397	3.45	0.0205
FetalWt	Sexratio	1	0.145381	0.145381	0.16	0.6894
	Residual	77	69.57287	0.903544		
			Effect	Group	Adjustment	Adj*P*
			Group	2	Dunnett–Hsu	0.9988
			Group	3	Dunnett–Hsu	0.8636
			Group	4	Dunnett–Hsu	0.0145
Response	Source	DF	SS	MS	*F* value	ProbF
FetalWt	Group	3	9.252602	3.084201	3.41	0.0217
NRank	SexRatioRnk	1	0.007854	0.007854	0.01	0.926
	Residual	77	69.7104	0.90533		
			Effect	Group	Adjustment	Adj*P*
			Group	2	Dunnett–Hsu	0.9964
			Group	3	Dunnett–Hsu	0.8678
			Group	4	Dunnett–Hsu	0.0145

Response, actual or normalized rank transform of fetal weight; Source, SexRatio or SexRatioRnk indicates the covariate, either actual or normalized rank-transform.

Table 5.10 Shirley's Antidote Data

Group	Before	After	Group	Before	After	Group	Before	After
1	1.8	79.1	2	1.6	10.2	3	1.3	14.8
1	1.3	47.6	2	0.9	3.4	3	2.3	30.7
1	1.8	64.4	2	1.5	9.9	3	0.9	7.7
1	1.1	68.7	2	1.6	3.7	3	1.9	63.9
1	2.5	180	2	2.6	39.3	3	1.2	3.5
1	1	27.3	2	1.4	34	3	1.3	10
1	1.1	56.4	2	2	40.7	3	1.2	6.9
1	2.3	163.3	2	0.9	10.5	3	2.4	22.5
1	2.4	180	2	1.6	0.8	3	1.4	11.4
1	2.8	132.4	2	1.2	4.9	3	0.8	3.3

Group, treatment group; Before, entry time (seconds) before all treatments; After, entry time after treatment and, in groups 2 and 3, after the antidote was given.

After data were normally distributed (after removing the treatment means), but the variances were heterogeneous. Normality was preserved and variance was stabilized following a log transform. Nonetheless, a nonparametric analysis was done to illustrate Shirley's approach, briefly described in Section 5.2.2.3. The results of three ANCOVAs of these data are presented (see Table 5.11). The first is an ANCOVA of Ln(AFTER) with covariate Before.

Table 5.11 MEORGT Data for Fish Exposed to 4-Tert-octylphenol – Generation F1

Dose	Conc	Rep	WK	Eggs	Dose	Conc	Rep	WK	Eggs	Dose	Conc	Rep	WK	Eggs	Dose	Conc	Rep	WK	Eggs
1	0	A	1	69	1	0	V	3	120	3	13	E	1	156	5	50	C	2	208
1	0	A	2	105	1	0	W	1	71	3	13	E	2	188	5	50	C	3	241
1	0	A	3	116	1	0	W	2	89	3	13	E	3	131	5	50	D	1	127
1	0	C	1	146	1	0	W	3	83	3	13	F	1	135	5	50	D	2	135
1	0	C	2	115	1	0	X	1	145	3	13	F	2	138	5	50	D	3	89
1	0	C	3	118	1	0	X	2	126	3	13	G	1	0	5	50	E	1	159
1	0	E	1	179	1	0	X	3	171	3	13	G	2	2	5	50	E	2	128
1	0	E	2	127	2	6	B	1	158	3	13	G	3	7	5	50	E	3	40
1	0	E	3	219	2	6	B	2	80	3	13	H	1	161	5	50	G	1	145
1	0	F	1	184	2	6	B	3	101	3	13	H	2	166	5	50	G	2	37
1	0	F	2	164	2	6	C	1	176	3	13	H	3	118	5	50	G	3	19
1	0	F	3	178	2	6	C	2	170	3	13	I	1	175	5	50	H	1	132
1	0	G	1	152	2	6	C	3	167	3	13	I	2	180	5	50	H	2	132
1	0	G	2	139	2	6	D	1	201	3	13	I	3	210	5	50	H	3	125
1	0	G	3	137	2	6	D	2	209	3	13	J	1	199	5	50	I	1	40
1	0	H	1	138	2	6	D	3	227	3	13	J	2	182	5	50	I	2	60
1	0	H	2	163	2	6	E	1	208	3	13	J	3	151	5	50	I	3	4
1	0	H	3	196	2	6	E	2	216	3	13	K	1	181	5	50	K	1	0
1	0	I	1	180	2	6	E	3	180	3	13	K	2	185	5	50	L	1	132
1	0	I	2	172	2	6	F	1	223	3	13	K	3	197	5	50	L	2	128
1	0	J	1	31	2	6	F	2	227	3	13	L	1	201	5	50	L	3	79
1	0	J	2	60	2	6	F	3	231	3	13	L	2	223	6	100	A	1	97
1	0	J	3	69	2	6	G	1	90	3	13	L	3	256	6	100	A	2	101
1	0	K	1	87	2	6	G	2	74	4	25	A	1	202	6	100	A	3	35
1	0	K	2	92	2	6	G	3	79	4	25	A	2	175	6	100	B	1	31
1	0	K	3	70	2	6	H	1	74	4	25	A	3	154	6	100	B	2	126
1	0	L	1	178	2	6	H	2	102	4	25	B	1	142	6	100	B	3	162
1	0	L	2	192	2	6	H	3	71	4	25	B	2	60	6	100	C	1	88
1	0	L	3	185	2	6	I	1	121	4	25	B	3	27	6	100	C	2	25
1	0	M	1	183	2	6	I	2	119	4	25	C	1	0	6	100	C	3	17
1	0	M	2	214	2	6	I	3	110	4	25	D	1	184	6	100	D	1	0
1	0	M	3	206	2	6	J	1	161	4	25	D	2	197	6	100	E	1	0
1	0	N	1	29	2	6	J	2	143	4	25	D	3	121	6	100	F	1	64
1	0	N	2	31	2	6	J	3	160	4	25	E	1	133	6	100	F	2	53
1	0	N	3	10	2	6	K	1	191	4	25	E	2	13	6	100	F	3	44
1	0	O	1	197	2	6	K	2	182	4	25	F	1	62	6	100	G	1	40
1	0	O	2	198	2	6	K	3	30	4	25	F	2	110	6	100	G	2	2
1	0	O	3	161	2	6	L	1	178	4	25	F	3	99	6	100	G	3	6
1	0	P	1	134	2	6	L	2	201	4	25	G	1	95	6	100	H	1	0
1	0	P	2	145	2	6	L	3	210	4	25	G	2	44	6	100	I	1	13
1	0	P	3	154	3	13	A	1	178	4	25	G	3	11	6	100	I	2	0
1	0	Q	1	166	3	13	A	2	78	4	25	H	1	0	6	100	I	3	1
1	0	Q	2	165	3	13	A	3	56	4	25	I	1	0	6	100	J	1	0
1	0	Q	3	202	3	13	B	1	184	4	25	J	1	153	6	100	J	2	0
1	0	S	1	162	3	13	B	2	216	4	25	J	2	154	6	100	J	3	0
1	0	S	2	143	3	13	B	3	231	4	25	J	3	173	6	100	K	1	102
1	0	S	3	59	3	13	C	1	157	4	25	L	1	109	6	100	K	2	94
1	0	U	1	155	3	13	C	2	180	4	25	L	2	121	6	100	K	3	82
1	0	U	2	174	3	13	C	3	133	4	25	L	3	127	6	100	L	1	73
1	0	U	3	156	3	13	D	1	172	5	50	A	1	0	6	100	L	2	49
1	0	V	1	121	3	13	D	2	183	5	50	B	1	0	6	100	L	3	14

WK, week of egg collection; Eggs, total number of eggs laid (collected) during indicated week.

The second is an ANCOVA of the normalized rank transform of After with covariate BEFORE. And the third is of the normalized rank transform of After with covariate the normalized rank transform of BEFORE. All three analyses give the same result as Shirley's method.

An interesting feature of Shirley's data is that the two After times of 180 in group 1 were actually right-censored. Shirley did not consider these two observations a barrier to the analysis, and the decision to ignore the censoring in this example is followed here as well.

It will be left as an exercise to show that the adjustment for the covariate is significant, there are significant differences among the treatment groups, groups 2 and 3 are significantly different from group 1 but not from each other, and the residuals from the ANCOVA are normally distributed in the ANCOVA of After, Ln(After), and normalized rank transform of After with covariate Before or the normalized rank transform of Before. Furthermore, with the exception of the raw After values, the residuals have homogeneous variances. The SAS program *ANCOVA_SHIRLEY Data.sas* to verify this is given in Appendix 1 and contains the data. These results are the same as those reached in Shirley (1981) by a different nonparametric method.

5.2.2.3 Alternative Nonparametric ANCOVA

The approach taken by Shirley (1981) transformed only the treatment variable, U, but used a chi-squared test to determine treatment effects rather than the methods of Chapters 3 and 5. Shirley builds on work by Bennett (1968), who investigated a simple linear regression model:

$$Y_{ij} = \mu + \beta_1 X_{1j} + \beta_2 X_{2j} + \beta_3 X_{3j} + \cdots + \beta_k X_{kj} + e_{ij}, \tag{5.30}$$

where the X_{kj} are independent variables.

Bennett showed that a modification of the usual least squares approach could be applied to the rank-order transform of the response only to estimate the slopes, β_i, and test hypotheses of the form that some set of these slopes, i.e. those for one or more covariates, are zero. Shirley takes some of the independent variables X_i to be dummy variables defining the treatment levels in a typical ecotoxicology experiment, with the remaining one or more X_j being the covariate(s). Then Bennett's procedure can be used to test hypotheses of treatment effects by taking μ to be the control mean, so a significant slope relative to the dummy X_i is equivalent to a significant effect of the ith treatment. Alternatively, the μ term could be omitted and one of the X_i taken to represent the control so the corresponding β_i is the control mean.

While the above description of the Shirley and Bennett approach is conceptually simple, the details are much more involved than the standard analysis on normalized ranks. The reader is referred to those papers for further investigation.

5.3 EXPERIMENTS WITH MULTIPLE FACTORS

Many experiments done to meet regulatory test guidelines are simple in structure, with a single explanatory variable, the concentration or dosage of the test substance. However, there are also studies where multiple explanatory variables are present and the resulting models require care. Examples are given in Sections 5.3.2.1 and 5.3.2.2. Such studies often involve interactions of factors, nested factors, and both fixed and random factors. Some of these main effects and interactions may not be statistically significant, and the question arises of whether such factors or interactions should be retained in the model when testing for treatment effects or estimating an effects concentration. In typical ecotoxicity studies for regulatory risk assessment, where there are only two or three factors, it will often be best to retain interactions even if not significant to be sure that the subsequent hypotheses to be tested are actually testing what they are intended to test. Hocking (1985) and its references contain good discussions of the actual hypotheses being tested under various models, and it is made clear that formulating correct tests of some hypotheses can be challenging. The example in Section 5.3.2.1 contains interactions and nested factors that are not statistically significant but are retained so that the formulation and interpretation of hypothesis tests are clear.

However, in environmental modeling studies, it is common that there are a large number of potential explanatory variables, and some of these are highly correlated, so no meaningful model can be created that uses all or even most of these variables. An example is given in Chapter 13. In such situations, some reduction in the number of variables or some other reduction in the dimensionality of the model is needed. There is a large body of literature on model selection criteria, and no attempt will be made here to cover it. The discussion in Section 5.3.1 will highlight a few relevant topics and provide references for further study.

When there are two or more explanatory variables, it is usually important to consider the possibility that the effect of one of these variables varies according to the value of the other variable. Including an interaction term allows for this possibility and provides for a formal test that can be further explored through relevant contrasts. The examples in the following sections illustrate this.

5.3.1 Model Selection: Evaluation of Terms to Include/Exclude

When there are a large number of possible explanatory variables and some reduction in their number is needed to develop a useful model, there are numerous techniques that have been developed. For linear models (i.e. linear in the parameters), these include adjusted R-square, Mallow's C_p, stepwise regression, and best subset using some criterion such as adjusted R-square or Mallow's C_p. An alternative is principal component analysis (PCA), where all variables are retained, but a reduction in dimensionality can be obtained by determining a few linear combinations of the explanatory variables that account for most of the variability in the data. Each of these techniques will be briefly described with references for further exploration.

R-square is perhaps the simplest idea to explain. A linear model has the form $\boldsymbol{Y}=\boldsymbol{X\beta}+\boldsymbol{e}$, where $\boldsymbol{Y}$ is the response to be modeled, $\boldsymbol{X}$ is a design matrix, $\boldsymbol{\beta}$ is a vector of model parameters to be estimated from the data, and $\boldsymbol{e}$ is the error vector that has some specified variance–covariance structure. It can be shown that the total sum of squares can be decomposed into two independent sums of squares, one expressing the reduction in the total sum of squares explained by the model and the other the error sum of squares indicating the variability in the response not accounted for by the model. That is,

$$\text{SSTOT} \underline{\underline{\text{def}}} \sum\left(Y_i-\bar{Y}\right)^2=\sum\left(\hat{Y}_i-\bar{Y}\right)^2+\sum\left(Y_i-\hat{Y}_i\right)^2, \tag{5.31}$$

where the two sums of squares on the right are the model, SSModel, and error sums of squares (SSERR), respectively, and the sums. Then R-square (or R^2) is defined as SSModel/SSTOT, making it the proportion of the total sum of squares "explained" by the model. This idea is simple but applies only to linear models in general, because for nonlinear models Eq. (5.31) generally does not hold.

The use of R^2 for variable selection is limited. The simple idea is to find the model (i.e. subset of potential variables) that gives the highest R^2. But it is trivial to verify that adding more variables cannot reduce R^2, so the highest R^2 always results from inclusion of all variables. It is necessary to make an adjustment to R^2 to make it useful for variable selection. That is done by penalizing or discounting R^2 for the number of variables included. The adjusted R^2 value is given by

$$R^2_{\text{adj}}=1-\left(1-R^2\right)\frac{n-1}{n-p}=1-\frac{\text{MSError}}{\text{MSTot}}, \tag{5.32}$$

where n is the number of observations and p is the rank of the design matrix. One variable selection criterion is to select the set of variables that results in the highest adjusted R^2. Alternatively, one can select the best 2-, 3-, …, k-variable model, each based on the highest R^2 or adjusted R^2. One then makes a judgment as to which, if any, of the models so obtained is useful. For example, if the best 4-variable model improves the adjusted R^2 value only very slightly compared with the best 3-variable model, it may be best to select the model with only three variables following the maxim of using the simplest model that is useful (a paraphrase of Occam's razor). Like any automatic variable selection process, expert judgment is required to select a useful model. By useful is meant a model that makes sense biologically or physically and gives predictions that are consistent with the data. For example, a model that predicts nonmonotone dose–response when all other factors are held constant may not make sense biologically unless the model is therefore capturing hormesis or some other understandable biological phenomenon. If two 3-variable models have similar adjusted R^2 values, but one of them is much more understandable scientifically than the other even if its adjusted R^2 is lower, it is often preferable to select the more understandable model.

Mallow's C_p criterion is an alternative model evaluation tool, defined by Eq. (5.33):

$$C_p=\frac{\text{SSErr}_p}{\text{SSErr}_k}-n+2p, \tag{5.33}$$

where n is the number of observations, p is the number of terms included in the model, and k is the total number of potential terms considered for inclusion. The expected value of C_p is p under modest assumptions, so one selects a model with C_p near that value. References include Mallows (1973), Gilmour (1996), and Giraud (2015).

Stepwise regression proceeds in a different manner. Two critical p-values, SLEntry and SLStay, are defined, for example, SLEntry=0.25 and SLStay=0.15. In step 1, for each possible variable, a model is fit with just that variable, and the significance level of that 1-variable model is computed. Of all variables for which the significance level (p-value) of the 1-variable model is less than SLEntry, the one that is most significant (lowest p-value) is selected. In case there are two variables with equal significance, one is selected arbitrarily or according to some other criteria. If no variable has significance level<SLEntry, the process stops, and no model is fit. At step 2, for each variable not selected in step 1, a model with that variable and the one chosen in step 1 is fit, and among all such models for which the significance level of the new variable is less than SLEntry, the one that makes the most significant improvement (lowest p-value) in the model is selected. Each variable now in the model is examined to determine whether the significance level (p-value) of that variable is ≥SLStay. If such a variable is found, then the one with the highest

significance level (highest *p*-value) is removed from the model. In step 3, every potential variable not already in the model is added to the model, and its significance level in the resulting model is calculated. If any variable added has *p*-value < SLEntry, then the variable resulting in the lowest *p*-value is added to the model. If no such variable is found, the process stops, and the model found at the end of step 2 is reported. Otherwise, the significance of each variable now in the model is calculated, and if any one of them has *p*-value ≥SLStay, then the one with the highest significance level (highest *p*-value) is removed from the model. The procedure continues until no variable meets the entry or the exit criterion.

Variations on stepwise selection include forward and backward elimination. References for stepwise regression include Draper and Smith (2012) and Morrison (1983). All of these automatic variable selection methods were developed for linear regression but can be applied to hypothesis testing models by the well-known device of using dummy variables to identify the levels of a factor. All of these methods have come in for serious criticism, and the reader is advised to consult the work of Giraud (2015), Harrell (2015), Chatfield (1995), and Anderson and Burnham (2010). Alternatives have been proposed by Efron et al. (2004), Tibshirani (1996), and Hooda et al. (2008). The information criteria AICc and BIC discussed in Chapter 2 can also be used for variable selection.

5.3.2 Repeated Measures and Other Examples of Time in Models

Perhaps the most common added complexity in ecotoxicology occurs when the same subject is measured for the same response at several different times over the course of a single study. Growth and growth rate, food consumption, egg production, and mortality are often measured or recorded daily or weekly or on some other schedule. Data of this sort are referred to as repeated measures, and there are models for some types of repeated measures. In some experiments, the response changes so much over the course of the experiment that separate analyses at each time point are thought the best way to proceed. For example, in an 18- or 24-month study with rats, the study begins when the animals are only a few days old and they grow to many times their initial size. Shorter studies with some other species (e.g. daphnid, algae, and other short-life species) can have the same issue. Consequently, the mean and variance of the response within a treatment are far from constant. While it is possible to use a complex variance structure that captures the time-related changes in variance, the resulting analyses are more difficult to interpret than many study directors feel comfortable handling. If a qualified statistician is actively involved in the study, such difficulties can be overcome.

There are other studies in which changes in the treatment effect over time are of considerable interest. The action of suspected endocrine-disrupting chemicals is thought to be more subtle than typical toxicity responses and may not show up until the second or third generation of constantly exposed species. A repeated measures analysis can be quite powerful in this type of study. This will be illustrated below with several types of regulatory guideline studies.

Another type of study where time is a factor is when subjects are sacrificed and examined at intermediate times in the study. Obviously, no animal is measured twice in this way, so while time may be important, this does not constitute a repeated measures experiment unless animals are housed together in replicate cages and the cage is the unit of analysis.

5.3.2.1 MEOGRT: Multigeneration Studies in Regulatory Guidelines

OECD TG 240 (Medaka Extended One-Generation Reproduction Test (MEOGRT)) calls for three generations of medaka to be conducted. A related study design is under development (as of 2016) for zebrafish, and a multigeneration avian study has been published by the US Environmental Protection Agency (USEPA 890_2100JQTT, the Japanese quail toxicity test). Similar study designs are contemplated for Daphnia, and other species may be added in the future. The design and analysis of TG 240 will be presented in detail.

TG 240 calls for three generations of medaka, F0, F1, and F2, with four or five test concentrations plus control. During the reproduction phase, there are to be 12 replicates per concentration in each treatment group and 24 in the control, though these numbers are cut in half after the reproductive phase. In addition, there is a single mating pair per replicate. (Initial studies conducted in the validation phase of this test guideline had three mating pairs per replicate and six replicates per treatment, including control. The change was made to increase power.) The principal responses are fecundity (eggs laid) and histopathology. Histopathology will be described in detail in Chapter 10. In this section, the focus is on analysis of egg production. Egg production is measured daily for 3–5 weeks per generation during peak egg laying time for each generation. Egg production can be analyzed as a daily, weekly, or total count. A complication is that in some testing facilities, eggs are not counted on the weekend, so analysis of daily counts can be problematic. Also, some scientists are not concerned with effects of chemicals that change the pattern of egg production over time and are only interested in the effect on total egg count. The analysis presented here allows for analysis of effects on the time of egg production as well as totals.

The study begins with the exposure of the F0 generation. The young produced by the F0 parents constitute the F1 generation, and the young produced by the F1 fish constitute the F2 generation. There is a fundamental difference between the exposure history of the F0 generation from that of F1 and F2. Namely, the F1 and F2 are exposed from conception, whereas the F0 are only exposed at the embryonic stage and beyond. For that reason, some prefer to analyze F1 and F2 together in one model and analyze F0 separately. It must also be acknowledged that in spite of the ability to compare generations for effects, the current thinking within the USEPA is to analyze each generation separately. In this section, models for including multiple generations will be illustrated, but they can be simplified easily to model a single generation.

A complication with medaka and this experimental design is that some replicates in F0 may not produce enough eggs to populate F1 and even fewer will allow sufficient numbers of fish in F2. This would lead to an unbalanced design, possibly with complete loss of some replicates. While analysis of unbalanced designs is certainly feasible, the possible loss of power to detect effects was used to justify the decision for all eggs from one treatment group in F0 to be pooled and randomly allocated to replicates in F1 for that same treatment and a similar reallocation occurs to create F2 from F1. Thus, it is not possible to trace a single replicate from F0 through F2. In theory, a larger design, i.e. one with more replicates or more fish per replicate, would eliminate the issue with possible loss of power, though it might become too sensitive to be useful. In reality, the test design is already very large and very expensive, and a larger design was not considered feasible or practical.

It should be pointed out that the test guideline for zebrafish being developed (ZEOGRT) has a different experimental design in which replicates are preserved. This is possible because of physical differences in the two species. In the ZEOGRT design, for example, rep 1 of F1 treatment 2 consists solely of fish from rep 1 of F0 treatment 2. This is not the case for medaka in TG 240.

If Y is the egg count, perhaps square-root transformed, an applicable ANOVA model is given by

$$E\left[Y_{\text{gwtr}}\right] = \text{Gen}_{\text{g}} + \text{Week}_{\text{w}} + \text{Gen}_{\text{g}} * \text{Week}_{\text{w}} + \text{Trt}_{\text{t}} + \text{Gen}_{\text{g}} * \text{Trt}_{\text{t}} + \text{Week}_{\text{w}} * \text{Trt}_{\text{t}} + \text{Gen}_{\text{g}} * \text{Week}_{\text{w}} * \text{Trt}_{\text{t}},$$

with random effects

$$\text{Rep}_{\text{r}}\left(\text{Gen}_{\text{g}} * \text{Trt}_{\text{t}}\right) \text{ and } \text{Week}_{\text{w}} * \text{Rep}_{\text{r}}\left(\text{Gen}_{\text{g}} * \text{Trt}_{\text{t}}\right) \quad (5.34)$$

allowing for unequal variance components of both types across generations. The second indicated random effect is just the error term and need not be entered into the model specification. This is a repeated measures model, with week within generation the repeated factor. It should be noted that Week is treated as a fixed effect, since the same 3-to-5 week period is measured in each generation and there is interest in whether Week w in F2 is different from Week w in F1 or Week w in F0 across treatments or within the same treatment. In the ZEORGT design, generation would also be a repeated factor. Hocking (1985) is a good reference for this way of handling repeated measures.

A normal error variance–covariance error structure can be modeled, possibly after a transform, such as the square root. Alternatively a Poisson errors model can be used on untransformed egg counts. Both error structures will be illustrated and compared. Checks for normality and for variance homogeneity within each generation are advisable. This would also include identification of possible outliers.

Example data are given for 4-tert-octylphenol, a common environmental pollutant that has been shown to have weak estrogenic effects and to cause harm to the male reproductive system of vertebrates. (See Tables 5.11 and 5.12 for the F1 and F2 generations, respectively.) The MEOGRT design is intended to give definitive indications of endocrine disruptive effects in chemicals.

Included in the development of the test guideline was a need to determine whether the experimental design and statistical methodology had adequate power to detect effects if they were present. Since reproductive effects were expected for this chemical, the example provides some confidence in the guideline. A computer simulation study was done to support the conclusion of adequate power, but it will not be presented here. It will be evident that some F1 replicates did not produce enough eggs to carry into F2 without the reallocation of pooled replicates indicated above. Also apparent is that some mating pairs were removed for some reason, such as mortality or failure to mate. F0 was not included in the analysis for the reasons indicated in the above discussion. A histogram and a QQ-plot of the residuals from the normal errors model provide strong visual evidence to support a normal errors model. (See Figure 5.1.) With this visual evidence, a formal test is probably irrelevant, but an Anderson–Darling test found no statistically significant evidence of nonnormality. Demonstrating that is left as an exercise.

The QQ-plot indicates two small residuals with values less than −3 (F1 concentration 6, week 3, rep A and F2, concentration 0, week 2, rep P) and two large residuals with values greater than 3 (F1, concentration 50, week 3, Rep C, and F2, concentration 3, week 1, rep K). Omission of these four observations does not alter the conclusions, and the data with and without these four values is consistent with a normal distribution by the Anderson–Darling test. The studentized residual is obtained by dividing the residual by its standard deviation. The utility of

Table 5.12 MEORGT Data for Fish Exposed to 4-Tert-octylphenol – Generation F2

Dose	Conc	Rep	WK	Eggs
1	0	B	1	183
1	0	B	2	212
1	0	B	3	264
1	0	C	1	0
1	0	C	2	0
1	0	C	3	0
1	0	D	1	210
1	0	D	2	247
1	0	D	3	246
1	0	E	1	251
1	0	E	2	251
1	0	E	3	248
1	0	F	1	157
1	0	F	2	221
1	0	F	3	198
1	0	G	1	190
1	0	G	2	165
1	0	G	3	248
1	0	H	1	12
1	0	H	2	41
1	0	H	3	10
1	0	I	1	170
1	0	I	2	100
1	0	I	3	136
1	0	J	1	150
1	0	J	2	98
1	0	J	3	107
1	0	K	1	114
1	0	K	2	72
1	0	K	3	152
1	0	L	1	172
1	0	L	2	186
1	0	L	3	163
1	0	M	1	191
1	0	M	2	192
1	0	M	3	180
1	0	N	1	69
1	0	N	2	9
1	0	N	3	
1	0	O	1	120
1	0	O	2	74
1	0	O	3	154
1	0	P	1	131
1	0	P	2	0
1	0	P	3	189
1	0	Q	1	175
1	0	Q	2	195
1	0	Q	3	159
1	0	R	1	111
1	0	R	2	175
1	0	R	3	171
1	0	S	1	150
1	0	S	2	149
1	0	S	3	175
3	13	B	2	188
1	0	T	1	102
1	0	T	2	132
1	0	T	3	139
1	0	U	1	183
1	0	U	2	189
1	0	U	3	204
1	0	V	1	103
1	0	V	2	41
1	0	V	3	66
1	0	W	1	174
1	0	W	2	60
1	0	X	1	180
1	0	X	2	201
1	0	X	3	175
2	6	A	1	173
2	6	A	2	169
2	6	A	3	111
2	6	B	1	215
2	6	B	2	150
2	6	B	3	215
2	6	C	1	46
2	6	C	2	33
2	6	C	3	47
2	6	D	1	180
2	6	D	2	189
2	6	D	3	196
2	6	E	1	132
2	6	E	2	167
2	6	E	3	181
2	6	F	1	168
2	6	F	2	177
2	6	F	3	164
2	6	G	1	133
2	6	G	2	173
2	6	G	3	228
2	6	H	1	121
2	6	H	2	5
2	6	H	3	75
2	6	I	1	120
2	6	I	2	22
2	6	I	3	137
2	6	J	1	201
2	6	J	2	157
2	6	J	3	176
2	6	K	1	125
2	6	K	2	75
2	6	K	3	209
2	6	L	1	106
2	6	L	2	91
2	6	L	3	171
3	13	A	1	151
3	13	A	2	50
3	13	A	3	216
3	13	B	1	156
4	25	I	2	77
3	13	B	3	229
3	13	C	1	202
3	13	C	2	143
3	13	C	3	216
3	13	D	1	212
3	13	D	2	204
3	13	D	3	218
3	13	E	1	132
3	13	E	2	39
3	13	E	3	46
3	13	F	1	31
3	13	F	2	26
3	13	F	3	8
3	13	G	1	121
3	13	G	2	0
3	13	G	3	3
3	13	H	1	15
3	13	H	2	0
3	13	H	3	0
3	13	I	1	176
3	13	I	2	122
3	13	I	3	239
3	13	J	1	147
3	13	J	2	150
3	13	J	3	195
3	13	K	1	167
3	13	K	2	0
3	13	K	3	0
3	13	L	1	150
3	13	L	2	0
3	13	L	3	175
4	25	A	1	144
4	25	A	2	130
4	25	A	3	118
4	25	B	1	101
4	25	B	2	145
4	25	B	3	154
4	25	C	1	194
4	25	C	2	187
4	25	C	3	166
4	25	E	1	16
4	25	E	2	9
4	25	E	3	44
4	25	F	1	191
4	25	F	2	232
4	25	F	3	227
4	25	G	1	93
4	25	G	2	0
4	25	G	3	1
4	25	H	1	88
4	25	H	2	6
4	25	H	3	2
4	25	I	1	80
4	25	I	3	123
4	25	J	1	97
4	25	J	2	0
4	25	J	3	131
4	25	K	1	126
4	25	K	2	184
4	25	K	3	198
4	25	L	1	111
4	25	L	2	141
4	25	L	3	186
5	50	A	1	0
5	50	A	2	0
5	50	B	1	16
5	50	C	1	83
5	50	C	2	126
5	50	C	3	150
5	50	D	1	160
5	50	D	2	204
5	50	D	3	216
5	50	E	1	0
5	50	E	2	0
5	50	E	3	0
5	50	F	1	83
5	50	F	2	127
5	50	F	3	125
5	50	G	1	0
5	50	G	2	0
5	50	H	1	148
5	50	H	2	150
5	50	H	3	166
5	50	I	1	132
5	50	I	2	38
5	50	I	3	161
5	50	J	1	20
5	50	K	1	69
5	50	K	2	0
5	50	K	3	49
6	100	A	1	93
6	100	A	2	129
6	100	A	3	147
6	100	B	1	52
6	100	B	2	158
6	100	B	3	152
6	100	C	1	102
6	100	C	2	129
6	100	C	3	157
6	100	D	1	0
6	100	D	2	0
6	100	E	1	81
6	100	E	2	120
6	100	E	3	119
6	100	F	1	0
6	100	F	2	0
6	100	F	3	78

WK, week of egg collection; Eggs, total number of eggs laid (collected) during indicated week.

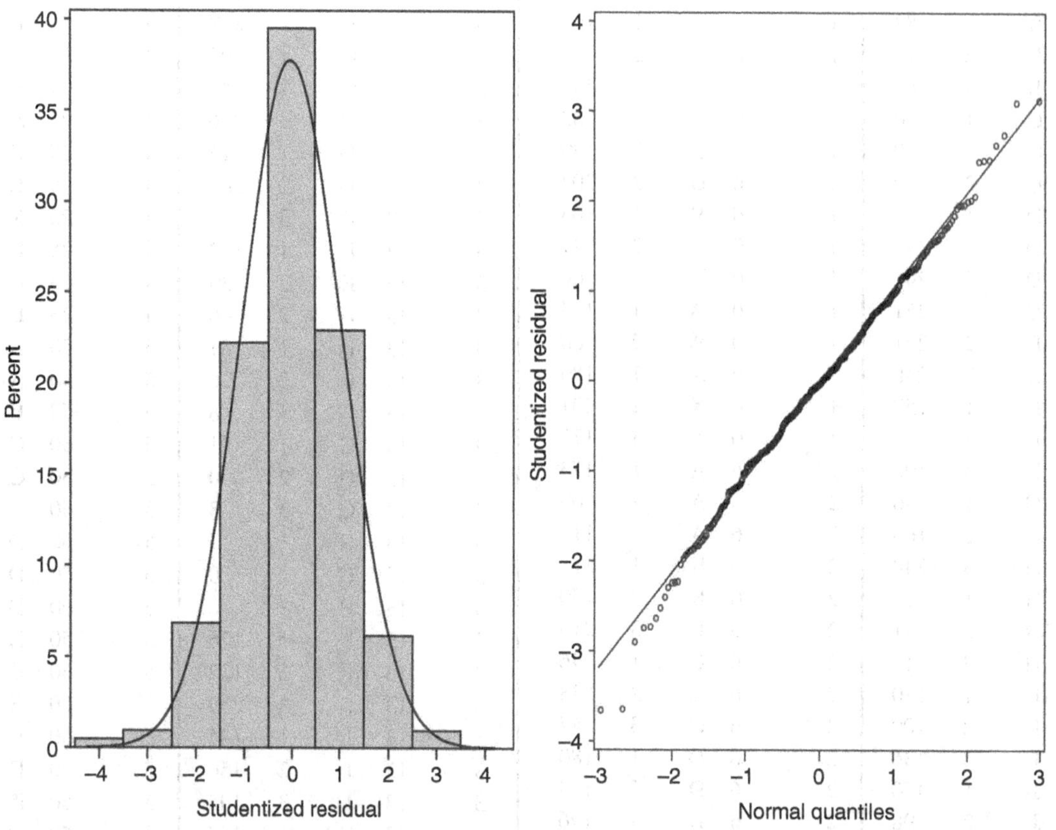

Figure 5.1 Plots of residuals from normal errors ANOVA of 4-tert-octylphenol data. Histogram and QQ-plots of normal errors ANOVA studentized residuals provide visual evidence of normality.

studentized over simple residuals is to provide an aide in identifying those that may signal outliers. Generally, a studentized residual from a normal errors model whose absolute value exceeds 3 is a value of interest. Having four such values out of 423 residuals is not unusual and does not, in itself, indicate a problem with normality.

When a Poisson errors model is fit, it is still appropriate to evaluate the studentized residuals in this way (Hilbe, 2007; Cameron and Trivedi, 2013; Agresti, 2015). See Figure 5.2 for the relevant figures.

The QQ-plot of studentized residuals from the Poisson model indicates numerous residuals with larger absolute values than is consistent with an adequate model. The histogram indicates some skewness, but more importantly, it shows a much wider distribution than expected. These problems could arise from overdispersion, in which case a negative binomial errors model might be better. See Figure 5.3 for the corresponding residual plots from a negative binomial errors ANOVA.

The QQ-plot from the negative binomial errors model still shows some large residuals, but only four studentized residuals exceeding 3. The histogram is now restricted to the expected range of −3 to 3, except as noted, but the percentage of values around zero is inflated relative to a normal distribution, and there is modest skewness apparent. There are 38 zero eggs recorded out of 423 observations across the two generations. Perhaps a zero-inflated negative binomial model would improve matters, but that will not be pursued here.

Three SAS programs are given in Appendix 1 to analyze these data. The first of these programs is *meogrt_anova.sas* for the normal errors case. See Tables 5.13–5.15 for the results from this program, which runs quickly, at least on a computer with ample memory and processing power. The second program, *GLMMix_anova.sas*, allows normal, Poisson, or negative binomial error structure. This program takes considerably longer to run even on a powerful computer. Calculations for the normal errors models differ somewhat under the two programs because of different algorithms used but agree on what is significant. An R version of the normal errors model is available (StatCharrms, 2017), though it only analyzes egg totals per group, not weekly totals. The program *MEORGT prelim.r* generates a basic ANOVA table but not the contrasts or slices. The two versions of StatCharrms are also used for analyses of histopathology ordinal data discussed in Chapter 9.

The normal errors and negative binomial errors models agreed that there were significant differences in egg

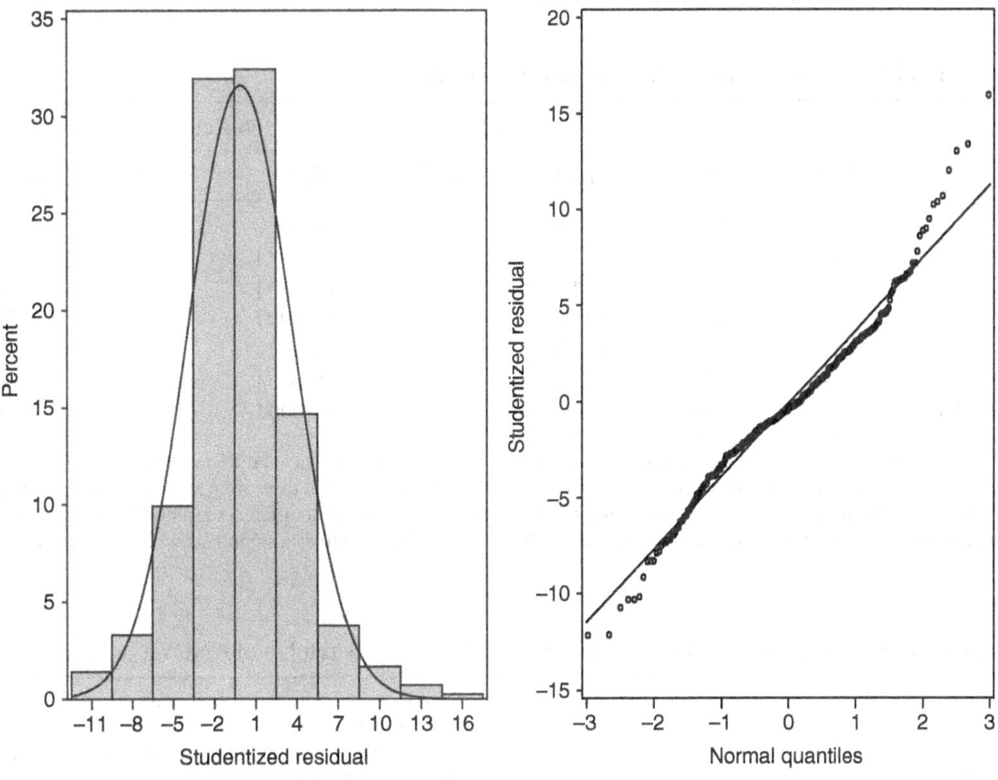

Figure 5.2 Plots of residuals from Poisson errors ANOVA of 4-tert-octylphenol data. Histogram and QQ-plots of Poisson errors ANOVA studentized residuals provide visual evidence of deviation from normality, indicating model inadequacy.

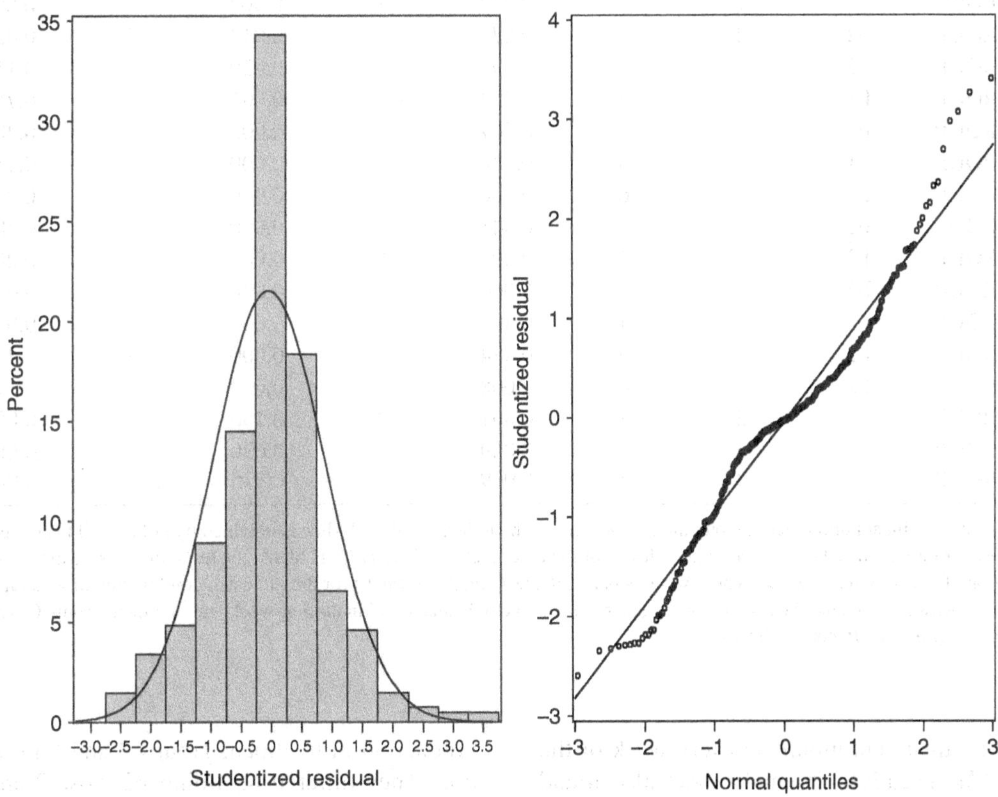

Figure 5.3 Plots of residuals from negative binomial errors ANOVA of 4-tert-octylphenol data. The QQ-plot still shows some large residuals, but only four studentized residuals exceeded 3. The histogram is now restricted to the expected range of −3 to 3, except as noted, but the percentage of values around zero was inflated relative to a normal distribution, and there was modest skewness apparent. There were 38 zero eggs recorded out of 423 observations across the two generations. Perhaps a zero-inflated negative binomial model would improve matters, but that will not be pursued here.

Table 5.13 Summary of F-tests of 4-Tert-octylphenol ANOVA Models

			Normal		Poisson		NegBin	
Effect	NDF	DDF	ProbF	Signif	ProbF	Signif	ProbF	Signif
GEN	1	139	0.674		0.263		0.434	
WEEK	2	248	0.034	*	<0.000 1	*	0.011	*
GEN*WEEK	2	248	0.000	*	<0.000 1	*	0.000	*
GROUP	5	139	0.000	*	<0.000 1	*	0.000	*
GEN*GROUP	5	139	0.169		0.212		0.205	
WEEK*GROUP	10	248	0.208		<0.000 1	*	0.441	
GEN*WEEK*GROUP	10	248	0.061		<0.000 1	*	0.006	*

NDF and DDF, numerator and denominator degrees of freedom, respectively; Signif, *means significant differences found among levels of Effect at the $p=0.05$ level. Generation abbreviated as GEN to conserve space. The three models agree that there were significant differences among weeks and among groups and the effects on weeks varied by generation. They agree there were significant differences ($p<0.062$) in the generation by week interaction among treatment groups. Only the least satisfactory of these three models (Poisson) finds a significant week by group interaction.

Table 5.14 Summary of Secondary Model Interactions (Slices) for 4-Tert-octylphenol ANOVA Models

				Normal		Poisson		NegBin	
Effect	Gen	Week	Dose	ProbF	Signif	ProbF	Signif	ProbF	Signif
GEN*WEEK*GROUP	F1	1		0.000	*	0.000	*	0.001	*
GEN*WEEK*GROUP	F1	2		0.000	*	0.000	*	0.000	*
GEN*WEEK*GROUP	F1	3		0.000	*	0.000	*	0.000	*
GEN*WEEK*GROUP	F2	1		0.003	*	0.002	*	0.003	*
GEN*WEEK*GROUP	F2	2		0.073		0.012	*	0.021	*
GEN*WEEK*GROUP	F2	3		0.192		0.020	*	0.112	
GEN*WEEK*GROUP	F1		2	0.354		0.001	*	0.783	
GEN*WEEK*GROUP	F1		4	0.267		0.000	*	0.488	
GEN*WEEK*GROUP	F1		5	0.128		0.000	*	0.158	
GEN*WEEK*GROUP	F1		6	0.700		0.000	*	0.050	*
GEN*WEEK*GROUP	F2		1	0.041	*	0.000	*	0.482	
GEN*WEEK*GROUP	F2		2	0.008	*	0.000	*	0.201	
GEN*WEEK*GROUP	F2		3	0.000	*	0.000	*	0.000	*
GEN*WEEK*GROUP	F2		4	0.300		0.000	*	0.076	
GEN*WEEK*GROUP	F2		5	0.084		0.000	*	0.257	
GEN*WEEK*GROUP	F2		6	0.008	*	0.000	*	0.012	*
GEN*WEEK*GROUP		2	3	0.001	*	0.156		0.054	
GEN*WEEK*GROUP		3	4	0.069		0.020	*	0.082	
GEN*WEEK*GROUP		3	6	0.009	*	0.015	*	0.003	*

Signif, *means there are significant differences among the components of the indicated slice. A slice is obtained by holding all factors except one constant and then testing for significant differences among levels of the other factor. Where Dose is blank, the test compares treatment groups for the indicated combination of generation and week. Where Week is blank, the test compares weeks for the indicated combination of generation and treatment group (dose). Where generation is blank, the test compares generations in the indicated combination of week and treatment group. Only slices where at least one model found significant effects are shown.

production among treatment groups in every week of the F1 generation. The negative binomial model also found significant differences among treatment groups in the F2 generation in weeks 1 and 2. These two models agreed that there were significant differences among weekly egg production in treatment groups 3 and 6 in the F2 generation. They differed regarding F2, dose 2 and F1, dose 6. These two models agreed that there were significant generational differences in week 3 and treatment group 6. They differed regarding week 2, treatment group 3.

Table 5.15 Summary of Contrast Tests for 4-Tert-octylphenol ANOVA Models

	Normal		Poisson		NegBin	
Label	Adj*P*	Signif	Adj*P*	Signif	Adj*P*	Signif
GEN=F1, WK=1, GRP 6-1	0.000	*	0.000	*	0.002	*
GEN=F1, WK=2, GRP 4-1	0.137		0.048	*	0.122	
GEN=F1, WK=2, GRP 5-1	0.273		0.038	*	0.144	
GEN=F1, WK=2, GRP 6-1	0.000	*	0.000	*	0.000	*
GEN=F1, WK=3, GRP 4-1	0.015	*	0.016	*	0.051	
GEN=F1, WK=3, GRP 5-1	0.006	*	0.009	*	0.017	*
GEN=F1, WK=3, GRP 6-1	0.000	*	0.000	*	0.000	*
GEN=F2, WK=1, GRP 5-1	0.007	*	0.003	*	0.010	*
GEN=F2, WK=1, GRP 6-1	0.016	*	0.074		0.070	
GEN=F2, WK=2, GRP 5-1	0.041	*	0.004	*	0.008	*
GEN=F2, WK=3, GRP 5-1	0.135		0.010	*	0.056	

Only those contrast tests that were significant at the $p=0.05$ level are shown. All contrast tests were two-sided. All models agree that treatment group 6 had significantly reduced egg production in Fl compared with the control every week. They also agreed that in F1, week 3, and in F2, week 2, there was significantly reduced egg production in treatment 5 compared with the control. The normal errors model found a significant reduction in F1, week 3, treatment 4. As with slices, with one exception (F2, week 1, dose 6), the Poisson model found significant all contrasts that were found significant by either of the other models. Table does not indicate whether the significant difference found was an increase or decrease. That requires a look at the observed or estimated mean egg production in the groups being compared.

Except for the week 2 by dose 3 slice, the Poisson model found significant all slices found significant by either of the other models. However, given the problems indicated with that model, these results are of dubious merit.

All models agree that treatment group 6 had significantly reduced egg production in F1 compared with the control every week. They also agreed that in F1, week 3, and in F2, week 2, there was significantly reduced egg production in treatment 5 compared with the control. The normal errors model found a significant reduction in F1, week 3, treatment 4.

As with slices, with one exception (F2 week 1, dose 6), the Poisson model found significant all contrasts that were found significant by either of the other models. The table does not indicate whether the significant difference found was an increase or decrease. That requires a look at the observed or estimated mean egg production in the groups being compared.

The normal errors and negative binomial models are in general agreement, though there were some differences, as noted. The Poisson model is suspect for the reasons discussed above. The purpose of exhibiting those results along with those from the normal and negative binomial errors models is to demonstrate what should be obvious from previous chapters: namely, the choice of model is important, and not all models give comparable results. Of the three models present, the normal errors model is arguably best due to the greatest consistency with model requirements as judged by residual analysis. However, the negative binomial errors model is conceptually sounder in the sense that it models counts as counts, rather than through a normal approximation.

5.3.2.2 ZEOGRT Experimental Design

The model for the ZEOGRT design is given by Eq. (5.35):

$$E\left[Y_{gwtr}\right] = \text{Gen}_g + \text{Week}_w + \text{Gen}_g * \text{Week}_w + \text{Trt}_t + \text{Gen}_g * \text{Trt}_t + \text{Week}_w * \text{Trt}_t + \text{Gen}_g * \text{Week}_w * \text{Trt}_t,$$

with random effects

$$\text{Rep}_r\left(\text{Trt}_t\right), \text{Week}_w * \text{Rep}_r\left(\text{Trt}_t\right), \text{ and } \text{Gen}_g * \text{Rep}_r\left(\text{Trt}_t\right) \tag{5.35}$$

if the same variances and covariances are expected across generations. If allowance is to be made for differences in the variance structure in different generations, then the third random effect term must be omitted. Exercise 5.4 provides an example ZEOGRT dataset and program to illustrate this topic.

5.3.2.3 Degradation Rate of Chemicals in the Soil

Another repeated measures experimental design arises in the study of degradation rates of chemicals in the soil that might arise from contamination or applications of pesticides or herbicides to crop land. The procedure described in this section for this purpose is to determine regression models to characterize the degradation of parent chemicals and their metabolites in soil. Model fitting criteria will also be discussed. Background information is given in Boesten et al. (2006), and a related regulatory guideline is OECD TG 307.

Some of the discussion in this section is given in Gao et al. (2011). That reference and Goerlitz et al. (2011) also present a Bayesian model that gives very similar results.

A simple four-parameter first-order model (SFO) will be presented for modeling such data. While there are theoretical reasons for believing a weighted analysis would be appropriate, there appears to be little or no advantage to doing so for the example presented below or the additional dataset presented in Exercise 5.11. Furthermore, there were compelling reason not to use weights for one of these datasets. Weighting seems theoretically the preferable procedure based on an expected reduction in variance when the amount of parent or metabolite in the soil approaches zero. However, abstract considerations should not overrule practical results, and the examples support the use of unweighted analysis.

A computer simulation study was conducted to further evaluate the advisability of using a weighted analysis and to explore alternative sampling designs. Based on this study, there is no advantage to increasing the number of reps beyond three per sampling time. For reasons stated in the following paragraph, three is also the minimum number of reps recommended. The spacing of sampling times appears to have little impact on the ability to fit a model or the quality of the fit, except as noted in this paragraph. The range of sampling times is very important, as are some key times. These two aspects of experimental design depend, however, on the unknown degradation rates, and rather different recommendations are made for the two examples. This is a result of the small degradation rate constant for the parent and the relationship between that and the higher degradation rate for the metabolite. Sampling up to 240 days is recommended and sampling to 360 days offers some advantage. See Table 5.5 for the first example dataset. For data with the general characteristics of the data in Exercise 5.11, a smaller design to 84 days with fewer intermediate sampling times gives essentially the same quality results as the design used. A notable feature of the Exercise 5.11 dataset is the higher parent degradation rate and the fact that the parent degrades faster than the metabolite.

To evaluate the need for a weighted analysis, it is necessary to have replicated observations at most sampling times. Three replicates are sufficient for this purpose, though certainly more would provide better evidence. For fitting a nonlinear model, it is preferable to have observations at numerous times. The optimum range and spacing of sampling times depend on the degradation rates of the parent and metabolite in the soil. Further discussion of experimental design is provided below. In general, there is more advantage to increased sampling times than to increased replication. A curve fit to replicated data will use directly only the mean result at each time point. The value of replication is that it (i) allows determination of the need for weighting, (ii) facilitates goodness-of-fit measures, and (iii) provides improved estimates of the mean result at each time point. The benefit of (iii) is reduced as more sampling times are included by virtue of the smoothing effect of the model. To some extent, the same can be said of (ii).

5.3.2.3.1 Description of Model The general model for metabolite degradation is given by the following equation:

$$M(t) = -C\int_0^t \frac{dP(u)}{du} F(t-u)du, \tag{5.36}$$

where $P(t)$ is the amount of the parent compound in the soil or sediment at time t, $M(t)$ is the amount of the metabolite formation at time t, $F(t-u)$ is the proportion of the metabolite remaining in the soil at time $t-u$ after formation at time u, and C is a degradation rate constant for the metabolite, so that $-C(dP(u)/du)$ is the formation rate of metabolite at time u due to degradation of the parent pesticide.

The models for the parent and metabolite have the forms

$$P(t) = M_0 G(t, k_p), \tag{5.37}$$

$$M(t) = CM_0 G(t, k_m) \tag{5.38}$$

where M_0 is the amount of parent initially in the soil at time zero, C is the proportion of the parent that metabolizes over time, $G(t,k)$ is an appropriate function of time, and parameter(s) k_p and k_m are estimated from the data. The simple first-order (SFO) model has the form

$$G(t, \beta) = \exp(-\beta * t), \tag{5.39}$$

so that

$$P(t) = M_0 e^{-k_p t} \text{ and} \tag{5.40}$$

$$M(t) = CM_0 k_p \frac{e^{-k_m t} - e^{-k_p t}}{k_p - k_m}. \tag{5.41}$$

Since parent degradation is assumed independent of the metabolite present, the modeling begins by fitting the parent data using Eq. (5.37) to estimate M_0 and k_p. Then Eq. (5.38) can be used to estimate C and k_m using the estimates of the other parameters found in fitting the parent. This model and the calculations by which to assess it are illustrated using data in Table 5.16. See Table 5.17 for a summary of the results.

Programs to fit these models are given in Appendix 1. In developing these programs, a difference between the capabilities of SAS and R came to light. The SAS procedure NLIN allows a weight function to be defined, that is, a function of the model being fit. In the example, the weight

Table 5.16 Degradation Dataset 1

Time	Parent	Met1	Time	Parent	Met1	Time	Parent	Met1
0	93.2	0	14	83.7	4.3	120	34.1	9.3
0	95.8	0	30	77.6	8.4	120	31.1	8.7
0	93	0	30	70.6	11.9	181	23.7	6.8
7	89.8	3.1	30	74	12.3	181	20	6
7	90.8	3	62	55.6	13.3	181	21.1	5.8
7	87.5	3.1	62	58	12.2	360	10.7	1.4
14	83.1	6.9	62	54.8	11.4	360	8.9	1.5
14	81.6	5.4	120	41.4	10.4	360	7.8	1.3

Time, days since chemical entered soil; Parent, amount of parent chemical ($\mu g\,l^{-1}$); Met1, amount of first metabolite ($\mu g\,l^{-1}$).

Table 5.17 Weighted and Unweighted Models Fit to Data from Table 5.16

Weighted				Unweighted			
M_0	94.0742	92.5572	95.5912	M_0	94.0757	91.8426	96.309
k_p	0.00819	0.00764	0.00873	k_p	0.0082	0.0075	0.0089
C	0.7734	0.6609	0.8859	C	0.7451	0.6215	0.8687
k_m	0.0296	0.0245	0.0358	k_m	0.026	0.0224	0.0359

Columns 1 and 5 give the parameter name. Columns 2 and 6 give the point estimates. Columns 3 and 7 give the lower 95% confidence bound for the estimate. Columns 4 and 8 give the upper 95% confidence bound for the estimate.

function used in the SAS program, called *Table516.sas*, is exp(−k * Time), where $k=k_p$ or k_m. We have been unable to find an R function that will accept such weights. This is unfortunate, since the variance of the predicted response is functionally related to the predicted response in this and some other applications. The references R-blogger (2017a, b) mentioned in Section 4.3.3 indicate progress in addressing this need. The R program *table 516.R* in Appendix 1 provides an unweighted analysis that gives similar results.

5.3.2.3.2 Experimental Design Considerations

The range of sampling times should optimize the parameter estimates in the sense of minimizing the variance of the parameter estimates. For the parent compound in an SFO model, there are two parameters, M_0 and k_p. Let θ represent either of these. Consider the partial derivative of $P(t)$ with respect to θ, $\partial P/\partial\theta$. If the absolute value of this partial derivative is small, then a change in t will have little impact on the value of the parent function. This in turn means the variability in the estimate of θ will be large. On the other hand, if the absolute value of this partial derivative is large, then a change in t will have high impact on the value of the parent function. This in turn means the variability in the estimate of θ will be small. So, the range of sampling times should be selected to provide high values of the partial derivative. For the parent, it is quite simple to find the maximum, as indicated below, but a plot of the absolute partial derivative is more important for finding the range of t values that provide high values of this derivative:

$$\frac{\partial P}{\partial M_0} = e^{-k_p t}, \tag{5.42}$$

which has its maximum when $t=0$. One should include values of t that make the partial derivative within, say, 20% of this maximum, which works out to be $0.223/k_p$. For the data in Table 5.5, where $k_p=0.008$, this is 28 days, while for the data in Exercise 5.11, where $k_p=0.09$, this is 2.5 days. Thus, the best estimates of M_0 are obtained using sampling times in the interval (0, 28) and (0, 2.5) for the two datasets, respectively. Also,

$$\frac{\partial P}{\partial k_p} = -M_0 t e^{-k_p t}, \tag{5.43}$$

To find the value of t that maximizes this function, it helps to find the partial derivative of it with respect to t, i.e.

$$\frac{\partial^2 P}{\partial k_p \partial t} = -M_0 e^{-k_p t}\left[tk_p - 1\right], \tag{5.44}$$

which is zero only at $t=1/k_p$, making the maximum of $\partial P/\partial k_p$ at that value of t. For example, if $k_p=0.008$, then the maximum occurs at $t=125$ days. Finding the values of t making the absolute value of this partial derivative within

20% of its maximum is more challenging. It is helpful to plot the absolute first partial derivative and obtain the approximate result from the plot. This is illustrated for the two example datasets below.

Of course, the metabolite must also be considered. Here M_0 and k_p are taken as given (the estimates from fitting the parent are used in fitting the metabolite), and there are again two parameters, C and k_m, to estimate. The required partial derivatives are given by

$$\frac{\partial M}{\partial C} = \frac{k_p M_0}{k_p - k_m}\left[e^{-k_p t} - e^{-k_m t}\right]$$
$$\frac{\partial M}{\partial k_m} = \frac{C k_p M_0}{\left(k_p - k_m\right)^2}\left\{\left[1 + t\left(k_m - k_p\right)\right]e^{-k_m t} - e^{-k_p t}\right\}. \tag{5.45}$$

While the maxima of these can be solved using the partial derivatives with respect to t, there are still equations that can only be solved numerically. It is simpler to plot these functions and visually find the times that maximize them. Of course, this requires preliminary estimates of the rate constants. An iterative process can be set up to refine the estimate, but this is often not necessary, as exact solutions are not required.

Once the maximum range is found in this way (and the process is in Figures 5.8 and 5.9 for the data in Table 5.5), then samples should also be taken at intermediate time points. A computer simulation study will explore the optimal spacing and number of sampling times and number of replicates per sampling time. One simple "dilution" design takes the maximum found from both parent and metabolite and then adds one-half that value, one-half of *that* value, and so on for five "dilutions." For example, if 120 days is the maximum range found, then samples are also taken at 60, 30, 15, and 7.5 days and, of course, at 0 days. Modifications of this would be to use a geometric spacing with a common ratio other than one-half. The appropriateness of this dilution design will be explored. Each sampling point should be replicated, so there are three samples at each sampling time. Additional sampling times can be included without increasing the variance of the estimated parameters. As a practical matter, one cannot take a sample at 120 days and then work backward to 60, 30, etc. Given this and uncertainty about the rate constant to be estimated, it may be prudent to include additional sampling times beyond those indicated.

There are other more sophisticated designs, but a simulation study is a reasonably simple way to evaluate various alternative designs. As a practical matter, if once data are collected and evaluated, these calculations indicate that a wider time interval should be used: it may still be possible to add a sample at a later time.

The method of fitting is through iteratively reweighted least squares with weights proportional to the mean predicted value. The weights are taken as inverse elements of the diagonal variance–covariance matrix of the dependent variable.

As noted by Huet et al. (2004, p. 94), "The most frequent mistake made in modeling the variance function is to assume that the variance of errors is constant, $\mathrm{Var}\,\varepsilon_{ij} = \sigma^2$, when, in actuality, heteroscedasticity exists, $\mathrm{Var}\,\varepsilon_{ij} = \sigma_i^2$." On the other hand, according to Motulsky and Christopoulos (2004), mis-specification of weights often has little impact on the fitted model. For the data in Table 5.6, a standard weighting scheme was employed, where the weights were inversely proportional to the predicted mean of the modeled response (parent or metabolite). This is equivalent to assuming the variability between reps is proportional to the predicted mean. It is not recommended to use the observed variances (or their inverses) as weights, as these are subject to high sampling variability, especially when there are few reps at each measurement time or varying numbers of reps at different times. Numerous authors have pointed this out (e.g. Motulsky and Christopoulos, 2004, pp. 86ff). For example, Huet et al. (2004, p. 61) state the following, where the reference to weighted least squares means using the inverse of the observed variance as weights.

> Although one can feasibly use the weighted least squares method based on within-treatment sample variances for a typically small number of replications, it is not the best option because the accuracy of the empirical variance as an estimator of the true variance is rather bad: With four replications, the relative error is roughly 80%. The weighting by such inaccurate estimators can be very misleading.

Huet et al. (2004, p. 63) specifically discuss modeling weights inversely proportional to the function being estimated, as in the following quote:

> Although the relative position of the data point corresponding to the fourth harvesting date is troublesome, we can represent the variance of the observations by an increasing function of the response function, $\sigma_i^2 = \sigma^2 f(x_i, \theta)$, for example.

An SFO model was fit to the parent and metabolite degradation data in Table 5.5. See Figure 5.4 for an overlay plot of the parent and metabolite data. See Figures 5.5 and 5.6 for displays of absolute and relative replicate variability.

There is some visual evidence in Figure 5.5 that the variance of the parent is greater than that of the metabolite and that each tends to decrease with time. Since the amount of metabolite also decreases with time after an initial increase (Figure 5.4), there is some graphical evidence for the weighting scheme for metabolite.

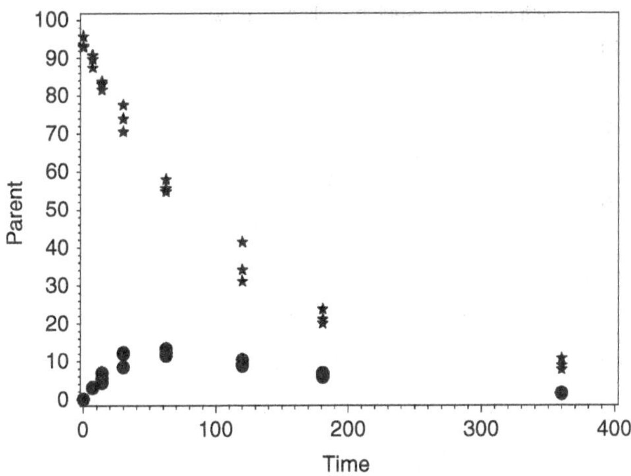

Figure 5.4 Parent and metabolite vs. time. Amount ($\mu g\,l^{-1}$) of parent (star) and metabolite (dot) vs. days post exposure.

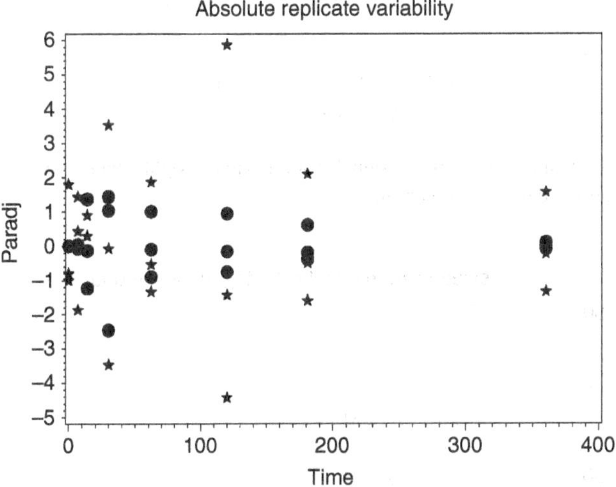

Figure 5.5 Absolute variability of parent and metabolite vs. time. Deviation from mean ($\mu g\,l^{-1}$) of parent (star) and metabolite (dot) vs. days post exposure.

If between-rep variance is proportional to the response mean, then there should be no relationship between the spread of these parent or metabolite normalized values and time of measurement. Except very close to time zero, this appears to be true. See Figure 5.7 for histograms of normalized pure error residuals for parent and metabolites. These are obtained by subtracting the mean at each time point from each observation at that time. That for metabolites is reasonably consistent with a normal distribution by the Shapiro–Wilk or Anderson–Darling test. That for the parent is not. Formal tests agree with these visual assessments. See Figure 5.8 for histograms of residuals about the fitted regression curves. It will be observed that these are both consistent with a normal distribution and formal tests support that perception.

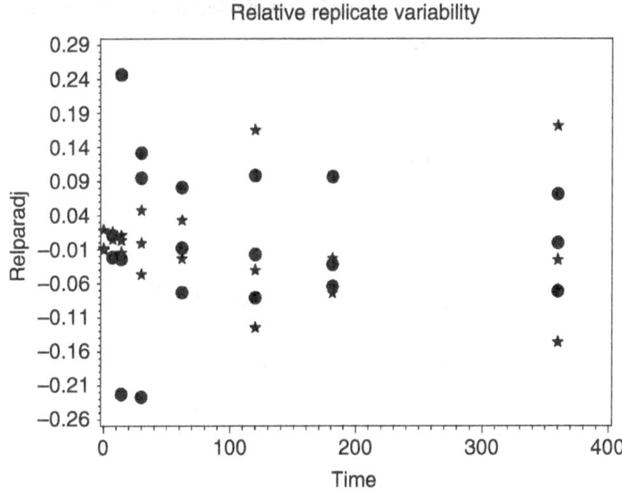

Figure 5.6 Relative variability of parent and metabolite vs. time. Adjusted deviation from mean ($\mu g\,l^{-1}$) of parent (star) and metabolite (dot) vs. days post exposure obtained at each time point by dividing the deviation from the mean by the mean.

See Figure 5.9 for a visual way to estimate the maxima for the metabolite. One plot shows the first partial derivatives of the model function with respect to the model parameters C and k_m. It will be seen that the two maxima are approximately 60 and 100 and values of t in the interval (0, 180) give values of the derivatives within 20% of the maximum. Also, the maximum for k_p is seen to be around 125 in the other panel, and values of t up to 240 give values of the derivative within 20% of the maximum. For M_0, it was shown in Section 5.3.2.3.1 using Eq. (5.38) that the interval (0, 28) was appropriate. The full sampling range used was (0, 360) with sampling times 0, 7, 14, 30, 62, 120, 181, and 360. Since 30 is close to 28, M_0 should be well estimated, and since 181 is close to 180, k_m should be well estimated. Since 62 was also a sampling time, the variance of C should likewise be near optimal. See Figure 5.10 for the resulting models for parent and metabolite.

5.3.2.4 Two Repeated Factors

It is possible to have two or more repeated factors. In this section, an example with two fixed repeated factors will be illustrated. The example is from a neurotoxicology study where rats are exposed to a chemical that is being evaluated for possible neurological damage. The response to be discussed is the duration of the reaction to a stimulus. Each animal is measured on each of several days following exposure. On each measurement day, each animal receives the stimulus (a mild shock or moderate sound) at six 10 s intervals, called bins, and the duration of reaction (in milliseconds) is recorded. The two repeated factors are

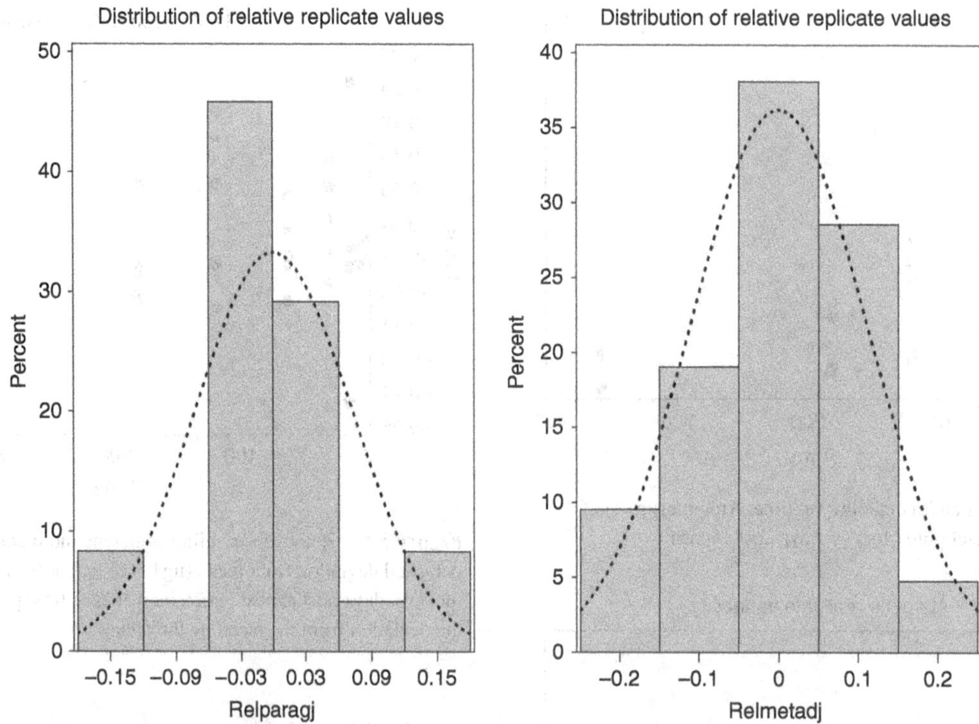

Figure 5.7 Histograms of normalized pure error residuals. Left panel shows normalized pure error residuals for the parent. Right panel is for metabolite. Dotted curve is overlaid normal density function with same mean and standard deviation as data.

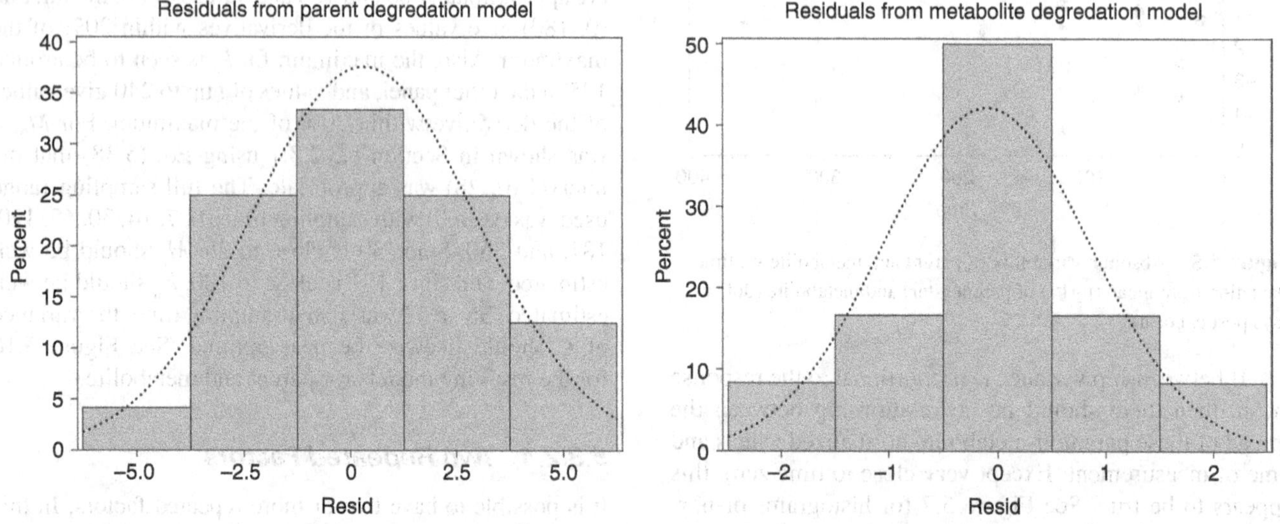

Figure 5.8 Histograms of regression residuals. Regression residuals from models for parent (left panel) and metabolite (right panel).

day and bin. These are standard, defined times of measurement, not random times, so these factors are treated as fixed. The other factor is dose. The model is given by Eq. (5.46):

$$\text{Response} = \text{Day} + \text{Bin}(\text{Day}) + \text{Dose} + \text{Day*Dose} + \text{Dose*Bin}(\text{Day}) + \text{Animal}(\text{Dose}) + \text{Day*Animal}(\text{Dose}) + \text{Error}$$

$$Y_{ijkr} = \text{Day}_i + \text{Bin}(\text{Day})_{ji} + \text{Dose}_k + \text{Day*Dose}_{ik} + \text{Dose*Bin}(\text{Day})_{kji} + \text{Rat}(\text{Dose})_{rk} + \text{Day*Rat}(\text{Dose})_{irk} + e_{ijkr}. \quad (5.46)$$

Random effects are error (e_{ijkr}) and the two terms in Eq. (5.46) that include Rat. A suggested variance–covariance structure uses restricted maximum likelihood (REML)

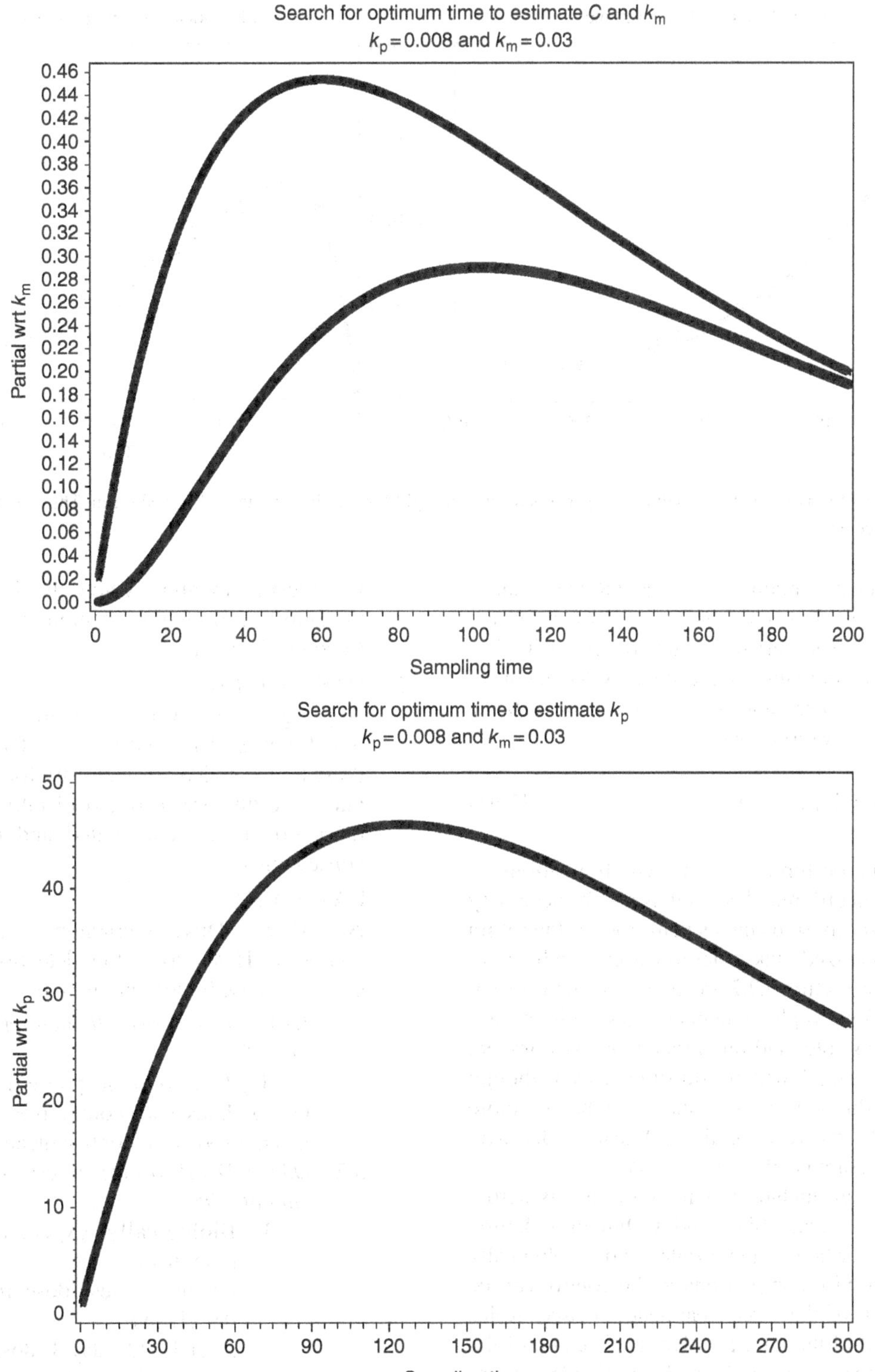

Figure 5.9 Optimizing variance estimates for model parameters. The top and bottom curves in the left panel are the partial derivatives of the metabolite degradation curve with respect to C and k_m, respectively, and the curve in the right panel is the partial derivatives of the parent degradation curve with respect to k_p.

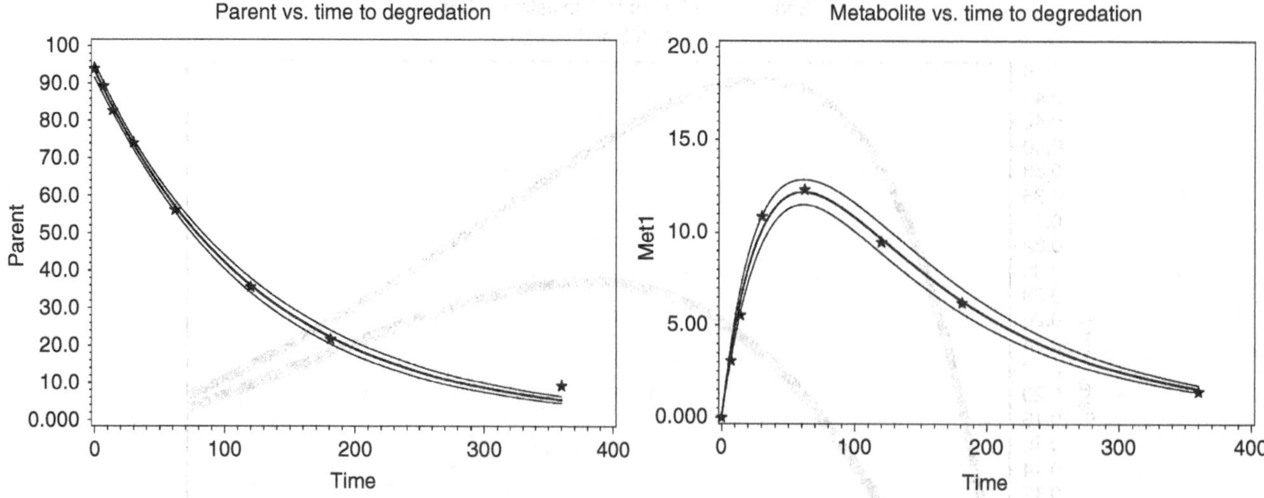

Figure 5.10 Models for parent and metabolite. Left panel is for parent, right for metabolite. Star = observed mean. Thick line = model. Thin lines = 95% confidence bounds.

estimates of variance components with an AR(1) variance–covariance structure, but alternative error structures are sometimes needed. The variance structure AR(1) implies that observations on the same subject are correlated, but the correlation between observations decreases as the distance (time) between observations increases:

$$\operatorname{cov}\left(Y_{ijkr}, Y_{ij^*kr}\right) = \sigma^2 \rho^{|j-j^*|}. \tag{5.47}$$

A similar relationship exists between observations on the same subject on different days. It may also be necessary to transform the response using a square-root or logarithm transform. A normalized rank-order transform can be used, but as discussed in Section 5.2.2, the literature on the use of rank transforms for complex models suggests a high false positive rate is possible, and this practice is discouraged. A normal errors model will be illustrated, even though conceptually a Poisson errors model might be more appealing. An alternative generalized linear model with Poisson error structure is left as an exercise.

The analysis can include test for dose effects across times, within days across bins, and within bins. Dose–response trends, where appropriate and biologically expected, and pairwise comparisons to the control can be tested within each of the three treatments of time. Rules are needed to control the false positive rate because of the large number of tests that are done. A set of rules implemented in one toxicology lab are stated in Sections 5.3.2.4.1.

5.3.2.4.1 Control of False Positive Rate

1. DOSE
   To assess the results of the ANOVA, the overall dose effect and trend in dose mean responses and Dunnett's test are all done at the 0.05 significance level. The trend test is important where the dose–response relationship is expected to be monotone. Otherwise, the trend test is ignored.
2. DAY and BIN(DAY)
   Any significant overall day effect or day trend is noted, but is of limited interest. The model includes these terms, which are necessary for proper modeling, but no results are recorded of either formal test, as these effects are anticipated and of no biological consequence.
3. DAY * DOSE
   Any Day * Dose interaction is explored through contrasts. This allows the detection of differential effects of dose in different time periods.
   3.1 *DAY * DOSE overall interaction is significant at 0.05.*
   On day I, test dose–response trend at 0.05.
   On day I, test each comparison (contrast) of dose mean response to control mean response at 0.05.
   3.2 *DAY * DOSE overall interaction is not significant at 0.05.*
   3.2.1 Biologically, expect monotone dose–response.
   On day I, test dose–response trend at 0.05/DAYS.
   3.2.1.1 On day I, dose–response trend is significant.
   On day I, test each comparison (contrast) of dose mean response to control mean response at 0.05/DAYS level.
   3.2.1.2 On day I, dose–response trend is not significant.
   On day I, do no further tests.

3.2.2 *Biologically, do not expect monotone dose–response.*
On day I, test dose–response trend at 0.05/DAYS.
On day I, test each comparison (contrast) of dose mean response to control mean response at 0.05/(DAYS * (DOSES – 1)) level.

4. DOSE * BIN(DAY)
   4.1 *DOSE * BIN(DAY) interaction significant at the 0.05 level.*
   On day I, bin J, test dose–response trend at 0.05.
   On day I, bin J, test each comparison (contrast) of dose mean response to control mean response at 0.05.
   4.2 *DOSE * BIN(DAY) interaction not significant at the 0.05 level.*
   4.2.1 *DAY * DOSE overall interaction significant at 0.05*
   On day I, bin J, test dose–response trend at 0.05.
   On day I, bin J, test each comparison (contrast) of dose mean response to control mean response at 0.05.
   4.2.2 *DAY * DOSE overall interaction not significant at 0.05.*
   4.2.2.1 *Biologically, expect monotone dose–response.*
   On day I, bin J, test dose–response trend at 0.05/(DAYS * BINS).
   4.2.2.1.1 On day I, bin J, dose–response trend is significant.
   On day I, bin J, test each comparison (contrast) of dose mean response to control mean response at 0.05/(DAYS * BINS) level.
   4.2.2.1.2 On day I, bin J, dose–response trend is not significant.
   On day I, bin J, do no further tests.
   4.2.2.2 *Biologically, do not expect monotone dose–response.*
   On day I, bin J, test dose–response trend at 0.05/(DAYS * BINS).
   On day I, bin J, test each comparison (contrast) of dose mean response to control mean response at 0.05/(DAYS * BINS * (DOSES – 1)) level.

This type of analysis will be illustrated with data given in Appendix 1 in the Excel file Motion.xlsx. See Figure 5.11 for a graphical check of normality and variance homogeneity. The residuals in the right panel of Figure 5.11 appear consistent with a normal distribution, and an Anderson–Darling formal test confirms this. A formal test of variance homogeneity that takes the AR(1) variance structure into account is more challenging. The variance component diagnostics in Hocking et al. (1989), Grynovicki and Green

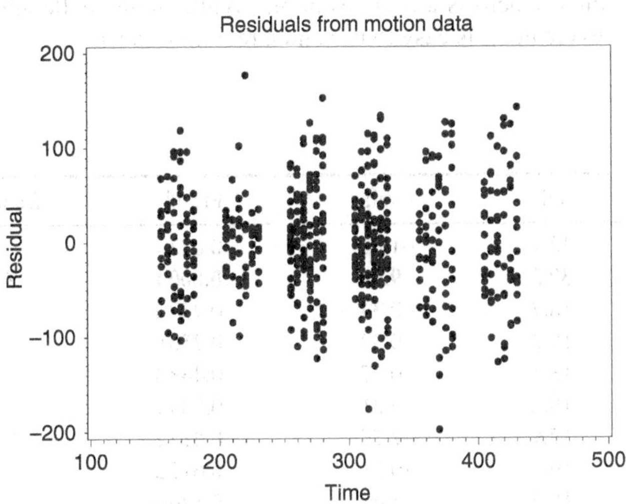

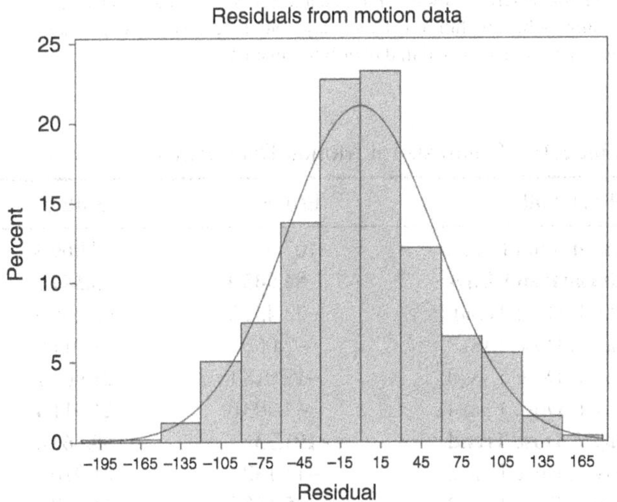

Figure 5.11 Check of normal errors assumption for motion data. Right panel = histogram of model residuals with overlaid normal density function. Left panel = residuals with horizontal axis given by time = 100 * day + 50 * dose + 5 * bin.

(1988), Green and Hocking (1988), Green et al. (1987), and Green (1985) could be used, but are not developed here. The visual representation in the left panel of Figure 5.11 should be sufficient and indicates that the variance in day 1 and control and low dose appear slightly smaller than at other times and doses, but the amount of reduction is unlikely to affect the model.

Consistent with the discussion in Section 5.3.2.4 for simple dose–response experiments, the omnibus tests (see Table 5.18) do not indicate whether there are significant differences between treatment groups and the control across or within times, so further testing is needed. The pathologist did not expect a monotone dose–response for these data, and all tests were done 2-sided with significance assessed at the 0.05 level except as otherwise noted. Dunnett's test for main effect of dose effect is done at the 0.05 significance level. Since none of the interactions involving dose were significant, the adjustments to the contrasts to determine dose effects within day or bin follow steps 3.2.2 and 4.2.2.2 in the protocol given above in Section 5.3.2.4.1. See Tables 5.19 and 5.20 for a summary of the analysis. A SAS program to carry out the analysis is discussed later in this section.

Six outliers were found using the Tukey outlier rule (See Table 5.21.) When these six observations are omitted, the DAY * DOSE interaction is significant at the 0.065 level, DOSE * BIN(DAY) is significant at the 0.01 level in addition to the terms found significant with the full data. Additional contrasts were also found significant. Several of these appear spurious, such as a dose 2 effect when no effect was found at higher doses in the same time, but the existence of an effect in the high dose is somewhat strengthened.

Table 5.18 ANOVA Table for Male Motion Data

Effect	NDF	DDF	*F* value	Pr>*F*
DAY	1	44	98.77	<0.0001
DOSE	3	44	1.2	0.3217
DAY*DOSE	3	44	1.84	0.153
BIN(DAY)	10	440	233.77	<0.0001
DOSE*BIN(DAY)	30	440	1.29	0.1416

Fixed effects terms indicate significant differences on the two test days and among bins within or between days. No differences among doses effect across time or within day or bin is evident.

Appendix 1 contains the SAS program *motime2.sas* that carries out the basic analysis without the details of applying Tukey's outlier rule or checking normality or variance homogeneity or making adjustments to *p*-values for number of comparisons. Those details can be obtained from the program *aovgenrl_hetvar.sas* using the SAS dataset Motion and making appropriate changes in the model and random statements and calling the Motion dataset. A much simplified program *motime2.sas* calls on the file *motime contrasts.sas* that contains all the contrast and estimate statements to generate Tables 5.19 and 5.20. It would be instructive to compare the tediously constructed contrast statements called by this simpler program to the effort required for calling the *Make_estimates_3var.sas* macro within the more complex program *MEOGRT_ANOVA.sas*. That macro is general and automatically constructs the contrasts needed for any dataset of the same general structure but with possibly different numbers of treatment groups and measurement times, and it could be employed in the *motime2.sas* program with some care. Manually constructing a set of contrasts for each study can be quite tedious. If there are empty cells, i.e. combinations of dose, day, and bin with no observations, defining contrasts correctly requires great care. General macros to automate the construction of all such contrast and estimate statements, including adjusting for empty cells, have been developed, and the cited macro is a basic example. While quite challenging to create, it is easy to use and a real time saver.

Table 5.19 Contrasts for Motion Data, Part 1

AR(1), Full	Estimate	StdErr	DF	*t* Value	Pr>|*t*|	Signif
Linear trend:Dose	−107.38	57.9694	15.1	−1.85	0.0837	
Linear trend:Days	**84.9453**	**8.5471**	**32.3**	**9.94**	**<0.0001**	*
Day 1, Dose Trend	−23.4322	69.3955	18.7	−0.34	0.7394	
Day 1, Dose 1 vs. 2	−7.0667	21.9119	19.2	−0.32	0.7506	
Day 1, Dose 1 vs. 3	−17.0181	21.9677	18.3	−0.77	0.4484	
Day 1, Dose 1 vs. 4	−4.4936	21.9114	19.2	−0.21	0.8397	
Day 2, Dose Trend	**−191.33**	**69.5634**	**17.9**	**−2.75**	**0.0132**	*
Day 2, Dose 1 vs. 2	−11.132	21.9107	19.2	−0.51	0.6172	
Day 2, Dose 1 vs. 3	−35.1552	21.967	18.3	−1.6	0.1266	
Day 2, Dose 1 vs 4	−55.7679	21.9704	18.3	−2.54	0.0205	

p-Value (Pr>|*t*|) does not show adjustment by Section 3.2.2, but significance flag does.

Table 5.20 Contrasts for Motion Data, Part 2

AR(1), Full	Estimate	StdErr	DF	t Value	Pr>|t|	Signif
DY 1, B 1, DS TREND	40.9983	101.57	80.6	0.4	0.6875	
DY 1, B 1, DS 1 v 2	23.1972	32.0971	82.9	0.72	0.4719	
DY 1, B 1, DS 1 v 3	18.6346	32.1353	79	0.58	0.5636	
DY 1, B 1, DS 1 v 4	15.187	32.0968	83	0.47	0.6373	
DY 1, B 2, DS TREND	−68.085	101.57	80.6	−0.67	0.5046	
DY 1, B 2, DS 1 v 2	25.1139	32.0971	82.9	0.78	0.4362	
DY 1, B 2, DS 1 v 3	−11.032	32.1353	79	−0.34	0.7323	
DY 1, B 2, DS 1 v 4	−10.6464	32.0968	83	−0.33	0.741	
DY 1, B 3, DS TREND	−21.7517	101.57	80.6	−0.21	0.831	
DY 1, B 3, DS 1 v 2	−14.5528	32.0971	82.9	−0.45	0.6514	
DY 1, B 3, DS 1 v 3	−66.1154	32.1353	79	−2.06	0.0429	
DY 1, B 3, DS 1 v 4	9.937	32.0968	83	0.31	0.7576	
DY 1, B 4, DS TREND	−73.5017	101.57	80.6	−0.72	0.4714	
DY 1, B 4, DS 1 v 2	−52.8861	32.0971	82.9	−1.65	0.1032	
DY 1, B 4, DS 1 v 3	−47.4487	32.1353	79	−1.48	0.1438	
DY 1, B 4, DS 1 v 4	−26.313	32.0968	83	−0.82	0.4147	
DY 1, B 5, DS TREND	−26.835	101.57	80.6	−0.26	0.7923	
DY 1, B 5, DS 1 v 2	−23.7194	32.0971	82.9	−0.74	0.462	
DY 1, B 5, DS 1 v 3	−4.1154	32.1353	79	−0.13	0.8984	
DY 1, B 5, DS 1 v 4	−15.4797	32.0968	83	−0.48	0.6309	
DY 1, B 6, DS TREND	8.5817	101.57	80.6	0.08	0.9329	
DY 1, B 6, DS 1 v 2	0.4472	32.0971	82.9	0.01	0.9889	
DY 1, B 6, DS 1 v 3	7.968	32.1353	79	0.25	0.8048	
DY 1, B 6, DS 1 v 4	0.3536	32.0968	83	0.01	0.9912	
DY 2, B 1, DS TREND	28.0063	101.69	76.8	0.28	0.7837	
DY 2, B 1, DS 1 v 2	8.8125	32.0963	83.1	0.27	0.7843	
DY 2, B 1, DS 1 v 3	6.4142	32.1348	79	0.2	0.8423	
DY 2, B 1, DS 1 v 4	10.1349	32.1371	78.6	0.32	0.7533	
DY 2, B 2, DS TREND	−79.2437	101.69	76.8	−0.78	0.4382	
DY 2, B 2, DS 1 v 2	6.8958	32.0963	83.1	0.21	0.8304	
DY 2, B 2, DS 1 v 3	6.2476	32.1348	79	0.19	0.8463	
DY 2, B 2, DS 1 v 4	−26.1985	32.1371	78.6	−0.82	0.4174	
DY 2, B 3, DS TREND	−219.66	101.69	76.8	−2.16	0.0339	
DY 2, B 3, DS 1 v 2	−7.6042	32.0963	83.1	−0.24	0.8133	
DY 2, B 3, DS 1 v 3	−64.1691	32.1348	79	−2	0.0493	
DY 2, B 3, DS 1 v 4	−54.3651	32.1371	78.6	−1.69	0.0947	
DY 2, B 4, DS TREND	−220.41	101.69	76.8	−2.17	0.0333	
DY 2, B 4, DS 1 v 2	−22.6875	32.0963	83.1	−0.71	0.4816	
DY 2, B 4, DS 1 v 3	−39.7524	32.1348	79	−1.24	0.2197	
DY 2, B 4, DS 1 v 4	−67.7818	32.1371	78.6	−2.11	0.0381	
DY 2, B 5, DS TREND	**−391.41**	**101.69**	**76.8**	**−3.85**	**0.0002**	*
DY 2, B 5, DS 1 v 2	−33.3542	32.0963	83.1	−1.04	0.3017	
DY 2, B 5, DS 1 v 3	−89.4191	32.1348	79	−2.78	0.0067	
DY 2, B 5, DS 1 v 4	**−111.78**	**32.1371**	**78.6**	**−3.48**	**0.0008**	*
DY 2, B 6, DS TREND	−265.24	101.69	76.8	−2.61	0.0109	
DY 2, B 6, DS 1 v 2	−18.8542	32.0963	83.1	−0.59	0.5585	
DY 2, B 6, DS 1 v 3	−30.2524	32.1348	79	−0.94	0.3494	

p-Value (Pr>|t|) does not show adjustment step 4.2.2.2 from Section 5.3.2.4.1, but significance flag does.

Table 5.21 Outliers Found by Tukey Rule in Motion Data

Dose	Day	Bin	Rat	Motime	Predicted	StdErr	Resid	UB	LB
4	1	3	1	2	178.3	27.3	−176.3	135.7	−137.9
4	2	6	1	207	64.5	27.3	142.5	135.7	−137.9
3	2	4	5	52	250.0	27.3	−198.0	135.7	−137.9
1	2	6	14	333	180.4	27.2	152.6	135.7	−137.9
2	1	4	28	267	89.8	27.2	177.2	135.7	−137.9
3	2	4	48	5	145.7	27.2	−140.7	135.7	−137.9

Motime and predicted are the observed and model predicted values, respectively. StdErr, standard error of predicted value. LB and UB are 95% confidence bounds on the predicted value. Resid, residual = Motime-predicted.

EXERCISES

In all datasets in the text and exercises, if a concentration is given with the value −1, that means a solvent used in the study and the water control is recorded as −1 and the solvent control is listed as 0. The reader should determine whether to combine the two controls for NOEC determination or ECx estimation. If the two controls are to be combined, then the water control should be changed to 0 and the replicates renamed so that replicates in the two controls can be distinguished. If only the solvent control group is to be used, then the water control should be discarded. If only the water control should be used, then the solvent control should be discarded and the concentration for the water control should be changed to 0. Under no circumstances should the concentration value −1 be used for NOEC or ECx purposes. Guidance on this is provided in Chapters 2 and 3.

5.1 Compute the probability of a random observation from a standard normal exceeding 3, and from that estimate the number of such values exceeding 3 out of 423 random values.

5.2 **(a)** Fit a normal errors model to the square root of egg counts from the 4-tert-octylphenol data and compare results with those reported here for untransformed counts assuming normal errors.

(b) Plot eggs vs. dose for each week and generation in a single figure and assess the formal conclusions for consistency with the plots.

(c) Plot studentized residuals vs. predicted value for the normal errors models for eggs and sqrt(Eggs) and for the Poisson and negative binomial errors models. Discuss any patterns that are apparent in each plot.

(d) Compute the mean egg production in every combination of generation, treatment group, and week and verify the direction of effect for the significant effects reported in Table 5.4.

(e) In the analysis of MEOGRT data in Section 5.3.2.1, contrasts to compare each treatment to the control were presented for each week and generation. These contrasts were produced by the SAS macro BCCNTRST(DATAIN=,RESPONSE=,FactorT=,FactorB=,FactorC=, BVAR=,BLabel=,IGNOREB=,CVAR=,CLabel=,TVAR=,TLabel=) given in Appendix 1 as part of the program file make_estimates_3var.sas. This was done by evaluating the macro as indicated below:

%BCCNTRST(DATAIN=DATAIN,RESPONSE=&RESPONSE,FactorT=GROUP,FactorB=GENERATION, FactorC=WEEK, BVAR=GENERATION, BLabel=GEN, IGNOREB=NO, CVAR=WEEK, CLabel=WK, TVAR=GROUP,TLabel=GRP).

This macro is for general contrasts for a three-factor ANOVA in a factorial model. Only pairwise comparisons on the third factor, FactorT, for fixed levels of the other two factors are created.

This can be used for contrasts on the other factors by reordering class variables in the ANOVA to make the variable of interest the third factor. Do this macro to compare the responses in the two generations for each treatment group and week. Note that this requires a separate call to Proc GLMMIX with class variables ordered so that generation is the third listed class variable after GROUP and WEEK. So FactorT=GENERATION, Factor B=Group, FactorC=Week. The labels should be adjusted to correspond to the altered factor names.

5.3 **(a)** Fit a normal errors model to the trenbolone egg production data (Appendix 1) after a square-root transform and assess it along the lines in the example in Section 5.3.

(b) Fit a Poisson errors model to the trenbolone egg production data after a square-root transform and assess it along the lines in the example in Section 5.3.

(c) Compare the models from Exercises 5.3a and 5.3b and, if necessary, from other models, and assess which provides the best analysis.

5.4 **(a)** Analyze the trenbolone egg data as though it followed a ZEOGRT experimental design with different variance structure in different generations. The SAS program GLMMix_anova_zeogrt_ALLWAYS.sas can be used.

(b) Analyze the trenbolone egg data as though it followed a ZEOGRT experimental design with the same variance structure in different generations. Assess which model is better for these data (assuming the experiment followed the ZEOGRT design).

Table 5.22 Soil Degradation Dataset 2

Time	Parent	Met1	Time	Parent	Met1	Time	Parent	Met1
0	97.2		14	28.5	7.5	56	1.3	3.8
0	96.8		14	27.3	7.1	56	1	3.1
0	97.5		14	27.5	7.5	56	1.1	3.1
1	82.9	0.7	21	14.8	8.4	70	0.9	2.7
1	86.7	0.7	21	13.4	6.8	70	0.7	2.3
1	87.4	0.2	21	14.4	8	70	0.7	2.1
3	72.8	2.2	28	7.7	7.2	84	0.6	1.6
3	69.9	1.8	28	7.3	7.2	84	0.4	1.1
3	71.9	1.6	28	8.1	6.9	84	0.5	1.3
7	51.4	4.1	42	2	4.9	120	0.4	0.4
7	52.9	4.2	42	1.5	4.3	120	0.3	0.4
7	48.6	4.2	42	1.9	4.5	120	0.3	0.3

Time, days since chemical entered soil; Parent, amount of parent chemical ($\mu g\, l^{-1}$); Met1, amount of first metabolite ($\mu g\, l^{-1}$).

5.5 **(a)** Do an ANOVA and a trend test on the sperm dataset (Appendix 1) to determine whether the test chemical had an effect on testis weight.

(b) Use ANOVA to determine whether the test chemical had an effect on sperm per gram testis (spgtestis). Then do an ANCOVA of spgtestis using testis_wt as covariate.

(c) Repeat Exercise 5.5b for sperm per testis (sptestis).

5.6 As part of a study on rats to determine possible reproductive effects, pregnant rats were exposed to varying levels of a test chemical. Each fetus from a rat litter was examined and the number of fetuses with bipartite ossification of the thoracic centrum was counted. Do an ANCOVA on the proportion fetuses with this finding after an arc-sine square-root transform using litter size as covariate. The data are in the Excel file Thoracic.xls.

5.7 Compute the mean, median, standard error, and quartiles for each treatment of the fetal weight and possible covariates for the mouse reproduction data from Table 5.1. Verify the results presented in Tables 5.2–5.4. Programs for this purpose are given in Appendix 1, including *Fetwgt prelim.sas* and *npar ancova fetalwt.sas*.

5.8 Verify the calculations in Table 5.8 using the SAS program *ancova_calc.sas* in Appendix 1.

5.9 An experiment was done exposing pregnant rats to a pesticide. The response of interest was the incidence of thoracic centrum, bipartite ossification. The data are in the Excel file *Thoracic*, where the litter size is labeled "Fetuses" and the number of fetuses with the indicated defect is labeled "Affected."

(a) Do a normal errors ANCOVA of the arc-sine square-root transform of the proportion affected fetuses using litter size as a covariate.

(b) Do a binomial errors ANCOVA of the number of affected fetuses using litter size as a covariate.

5.10 Two experiments were done exposing pregnant rabbits to a chemical to determine whether there was an effect on fetal weight. The data are in the Excel file *Rabbit* and SAS dataset *Rabbits*. Litter size is known to be inversely related to litter weight and was not expected to be affected by the chemical. Do an ANCOVA of pup weight using experiment, dose, and pup sex as factors in a factorial design and litter size as covariate to determine whether the chemical did have an apparent effect on male or female pups or both. Assess the data for normality and for homogeneity of variance across the two experiments. Adjust the variance structure to allow for different variances in the two experiments. A program, *AOVGENRL_hetvar.SAS*, is provided in Appendix 1.

5.11 Fit an SFO model to the second parent and metabolite soil degradation dataset (see Table 5.22). These data are also in the SAS dataset *soil2* in Appendix 1, and the programs *Table516.sas* and *table516.r* can be used with the obvious minor editing to call the indicated dataset.

5.12 Show that for the entry time data in Table 5.10, the adjustment for the covariate is significant, there are significant differences among the treatment groups, groups 2 and 3 are significantly different from group 1 but not from each other, and the residuals from the ANCOVA are normally distributed of in the ANCOVA of After, Ln(After), and normalized rank transform of After with covariate Before or the normalized rank transform of Before. Furthermore, with the exception of the raw After values, the residuals have homogeneous variances. Programs to verify this are given in Appendix 1.

Chapter 6

Analysis of Quantal Data: NOECs

6.1 INTRODUCTION

As with continuous responses addressed in Chapters 3 and 4, some regulatory test guidelines call for the determination of an NOEC for one or more quantal responses, and it is important to understand the statistical models that are available for this purpose. This chapter will address tests that are appropriate for analysis of quantal responses that arise from toxicity studies and other types of studies as well. It is important to keep in mind the distinction between quantal and count data, since different models are needed for these two types of responses. Recall from Section 1.1.2 that quantal data starts from a fixed number of subjects in each treatment group and the measure is binary (0–1 or yes/no). Each subject is classified as having or not having some characteristic. For each subject, the possible values of the response can be recorded as 0 (does not have the characteristic of interest) or 1 (has the characteristic of interest). For example, 10 seeds are planted in soil treated with a pesticide. The measure of interest is the number of plants that emerge. This number can only be integer in the range 0–10.

As with continuous responses, there are two broad categories of hypothesis tests used for NOEC determination, namely, pairwise and trend tests. All tests are based on models that have requirements, and the use of a test for data that do not meet the requirements of that test can lead to erroneous conclusions, sometimes dramatic. Informal tools were introduced in Chapter 2 to help determine whether the data meet model requirements such as independence of observations and overdispersion or extra-binomial variance. In this chapter, additional tools will be developed, including some to assess whether the data are consistent with a monotone dose–response, which is a requirement of the trend tests.

As always, the basic unit of analysis is the replicate, not the individual animal or plant. That said, it will soon be evident that some widely used tests for quantal responses ignore replicates. The same holds for regression models that are widely used for these responses. Alternative models that do take replicates properly into account exist and will be discussed in detail.

6.2 PAIRWISE TESTS

While trend tests are the preferred type of test for most NOEC determinations, there are situations and datasets for which only pairwise tests are appropriate. When the data do not follow a monotone dose–response, the use of a trend test based on an assumed monotone dose–response can lead to a nonsensical NOEC.

If $\pi_0, \pi_1, \pi_2, \pi_3, \ldots, \pi_k$ are the proportions of the control and treatment populations that respond (i.e. develop the characteristic of interest) to the various levels of the test substance, then the hypotheses to be tested can be stated as in Eq. (6.1):

$$H_0: \pi_0 = \pi_1 = \pi_2 = \pi_3 = \cdots = \pi_k \text{ vs. } H_1: \pi_0 \neq \pi_j \text{ for at least one value of } j. \tag{6.1}$$

To prepare for testing, incidence or quantal data from a standard toxicity test can be displayed in table form (see Table 6.1).

Medaka Sex Ratio Example

In developing OECD Test Guideline 234 for Fish Sexual Development Test (FSDT), medaka were exposed to the

Statistical Analysis of Ecotoxicity Studies, First Edition. John W. Green, Timothy A. Springer, and Henrik Holbech.

Companion website: www.wiley.com/go/Green/StatAnalysEcotoxicStudy

Table 6.1 Quantal Data from Toxicity Study

	Dose					
Feature	0	1	2	...	k	Row totals
Present	X_{10}	X_{11}	X_{12}		X_{1k}	R_1
Absent	X_{20}	X_{22}	X_{23}		X_{2k}	$R_2=N-R_1$
Col totals	$C_0=X_{.0}$	$C_1=X_{.1}$	$C_2=X_{.3}$		$C_k=X_{.k}$	N

Treatments are labeled as 0 (for the control) and 1, 2, 3, …, k for test concentrations or rates in increasing order. X_{1j}/C_j is the sample estimate of π_j. The overall proportion responding is R_1/N.

Table 6.2 Medaka Exposed to a Potentially Endocrine-Disrupting Chemical

	Conc						
Sex	0	6.2	12.3	23.5	50.4	100.6	Total
Female	18	17	23	15	28	25	126
Male	19	22	17	24	7	0	89
Total	37	39	40	39	35	25	215

Concentrations are measured in µg l^{-1}. Sex is the phenotypic sex at the end of the study. While the study was done with replicate tanks of fish in each treatment group, replicate information is suppressed in this table, and only treatment summary information is shown. We will analyze these data in several ways. Initially, a simple pairwise analysis will be done.

chemical 4-tert octylphenol. The interest in this study was whether this chemical affected the phenotypic sex of fish exposed at the embryonic stage. Every fish at the end of this study was observed to be phenotypically male or female. While that might seem obviously true, fish can be intersex or undifferentiated rather than male or female, so this is a relevant observation. The sex of the medaka in the embryonic stage cannot be determined. So while the intent was to have 50% genetic male and 50% genetic female in every treatment group, that cannot be guaranteed. Also, some fish died before the end of the study, so sample sizes are not equal. See Table 6.2 for data from one participating lab. That table does not show replicates. The full dataset with replicates is provided in Appendix 1 with the name *medaka sexratio data.xlsx*. The question to be addressed is whether there is a change in the proportion male resulting from exposure to the test chemical and, if so, at what concentration(s) does this occur.

6.2.1 Fisher's Exact Test

Pairwise analysis of quantal data compares each treatment with the control independent of other treatments except perhaps, to assess variability. For Fisher's exact test, it is convenient to create k 2×2 tables by extracting in turn one treatment column (group) and the control column to obtain a restricted form of Table 6.1 (see Table 6.3).

Table 6.3 Example 2×2 Incidence Table

	Group 1	Group 2	Total
Present	X_{11}	X_{12}	R_1
Absent	X_{21}	X_{22}	R_2
Total	C_1	C_2	N

Group j is a random sample of size C_j from population j. X_{ij} is the number of subjects in Group j that have the characteristic of interest. In the medaka example, "Present" could be taken to mean male. Alternatively, "Present" could be taken to mean female. It will make no difference in the analysis, though of course, the interpretation will depend on which definition of "Present" is used. Here $R_1=X_{11}+X_{12}$ is the total number of subjects from just these two groups that have the characteristic, while R_2 is the number of subjects that do not have the characteristic. Sample sizes are related to table entries by $C_j=X_{1j}+X_{2j}$.

Fisher's exact test is the test most commonly used historically to analyze a 2×2 table of quantal responses. This test was introduced by Fisher (1934) and apparently independently by Yates (1934) and Irwin (1935). Discussions of this test are found in numerous other sources, and extensive development of the properties of this test and the hypergeometric distribution are given in Johnson et al. (2005). The columns are treated as independent random samples from two populations.

The column totals C_1 and C_2 are the sample sizes and are fixed, as discussed in Section 6.1. The row totals R_1 and R_2 are generally not fixed, but that fact is widely ignored in analysis and specifically is ignored in applying Fisher's exact test to data of this sort. More accurately, the probabilities associated with this design are conditioned on the row and column totals. The hypotheses being tested are usually stated in terms of the proportion of each population that has the characteristic of interest. Let π_j be the proportion of population j having the property under investigation. Then the hypotheses can be stated as follows:

$$H_0: \pi_1 = \pi_2 \text{ vs. } H_1: \pi_1 \neq \pi_2 \text{ (two-sided)} \qquad (6.2)$$

Of course, one-sided alternative hypotheses exist. The population proportions π_1 and π_2 are estimated by $p_1=X_{11}/C_1$ and $p_2=X_{22}/C_2$, respectively. Exact (conditional) probabilities associated with all possible arrangements of the X_{ij} can be obtained from the hypergeometric distribution. To understand the hypergeometric distribution, it should first be recognized that all row and column totals are determined by N, C_1, and R_1 (or, equivalently by N, C_2, and R_2) since $C_2=N-C_1$ and $R_2=N-R_1$. Furthermore, given the row and column totals, all table entries are determined by any one of them. For example, given X_{11} it follows that $X_{12}=R_1-X_{11}$ and $X_{21}=C_1-X_{11}$ and $X_{22}=R_2-X_{21}$. The 2×2 table thus has only 1 degree of freedom, given the row and column totals. A test statistic to test H_0 versus H_1 can be based on the distribution of X_{11} (for example) whose pdf is given by

$$p(x \mid C_1, R_1, N) = \frac{\binom{C_1}{x}\binom{N-C_1}{R_1-x}}{\binom{N}{R_1}}. \tag{6.3}$$

The application is straightforward by computing the probability that a value of X_{11} as great as that observed could be obtained by chance alone:

$$P[X \geq x_{11} \mid C_1, R_1, N) = \sum_{x=x_{11}}^{R_1} p(x \mid C_1, R_1, N). \tag{6.4}$$

A similar expression can be written for the probability of getting as small a value as the observed X_{11}. For small values of the parameters C_1, R_1, and N, tables of these probabilities have long been published (e.g. Finney, 1948; Latscha, 1955). These tables often require relabeling groups and redefining "Present" and "Absent" to reduce the size of the table, but that need not concern us since software is now readily available to generate these distributions (e.g. SAS function PROBHYPR (SAS Institute Inc., 2011) and R function PHYPER) and even to apply Fisher's exact test without overt reference to the underlying hypergeometric distribution. Programs using these functions and test are provided in Appendix 1 as *FisherExact.sas* and *FisherExactTest.r*.

To apply Fisher's exact test to the data from a standard toxicity experiment with a control and k treatment groups, one would perform k instances of Fisher's exact test, consisting of the control data and the data from one of the treatments. To control the overall false positive rate, a Bonferroni–Holm adjustment could be made to the individual p-values. The data from Table 6.2 will be used to illustrate the test. See Table 6.4 for the results. Fisher's exact test makes no use of the replicate nature of the study.

Table 6.4 Two-Sided Fisher's Exact Tests Applied to Medaka Sex Ratio Data

		BHolmAdj	
Direction	DoseVal	p-Value	Significance
Two-sided	6.2	0.8182	
Two-sided	12.3	0.9943	
Two-sided	23.5	1	
Two-sided	50.4	0.029	*
Two-sided	100.6	0.0001	**

Significane = "*" or "**" indicates p-value <0.05 or <0.01 with Bonferroni–Holm adjustments.

Observations are treated as mutually independent regardless of whether they are in the same replicate.

6.2.2 Chi-square Test

A natural question to ask is whether there is a way to assess all the treatment groups in a single model, as is done for continuous responses. There are several ways in which this can be done. A starting point is the chi-square test, which for quantal responses is analogous to the F-test and Kruskal–Wallis test in that it gives an overall assessment of whether there are differences among the control and k treatment groups indicated in Table 6.1. Using the notation of Table 6.3, the test statistic is

$$\chi^2 = \sum_{i,j} \frac{\left(X_{ij} - \hat{X}_{ij}\right)^2}{\hat{X}_{ij}}, \tag{6.5}$$

where the sum is over both rows and all columns and

$$\hat{X}_{ij} = \frac{R_i C_j}{N}.$$

The distribution of the test statistic is χ^2 with k degrees of freedom. A good reference for the chi-square test is Agresti (2012).

The idea is simply that if there are no differences among treatments (i.e. columns), then the proportion responding in each column should be the proportion of all subjects that respond across the entire study. Like the F- and Kruskal–Wallis tests, if the test is significant, there is no indication of which groups differ, and all groups are compared with one another, not just treatments to the control. Unlike the F- and Kruskal–Wallis tests, however, a significant χ^2 test can result from heterogeneity (overdispersion) as well as differences in treatment proportions. Thus, this is of limited value for toxicity studies. As will be developed in Section 6.5.1, it is possible to obtain information more relevant to toxicity studies from the χ^2 test.

6.2.3 Analysis of Replicate Proportions

Rather than ignore replicates, as done in Fisher's exact test, it is possible to analyze replicate proportions using the methods developed in Chapter 3. The replicate proportions can be treated as a continuous response. For small samples, it has to be acknowledged that the possible values a replicate proportion can have are severely limited and treating them as a continuous response is only an approximation. It is important to understand the impact of this approximation on power to detect effects compared with Fisher's exact test and other alternatives to be explored later in this chapter, and that will be done in Section 6.6.

In this approach, applied to pairwise testing of the hypotheses in Eq. (6.1), one can use Dunnett's test on the treatment proportions, where replicate proportions are the observations. The DunnettTest.r and Dunnett Test.SAS programs developed for Chapter 3 can be used for these data to determine the same NOEC, 23.5 μg l^{-1}, as found by Fisher's exact test.

The novelty in applying the methods of Section 3.2.4 to these data is the likely need to apply an arcsine-square-root transform to the replicate proportions to stabilize the variances. A complication arises when all subjects are affected in some treatment group, as in Table 6.2 with the 100 μg l^{-1} concentration. For then the sample variance for that group is zero and no transformation will make the data conform to variance homogeneity. In that case, the application of Dunnett's test to these data is of dubious merit, and that is not improved by the use of the Tamhane–Dunnett robust version of Dunnett's test. That is a reason to prefer treating the data as quantal rather than continuous. Fisher's exact test does that, as does the Cochran–Armitage trend test to be discussed in Section 6.5.1. Before turning to trend tests, there is another issue to consider.

6.3 MODEL ASSESSMENT FOR QUANTAL DATA

The techniques discussed in previous chapters to assess models for continuous responses are often inappropriate for quantal data, and different techniques are needed. Each subject either has, or does not have, the phenomenon of interest. This means the observed incidence for a single subject is 0 or 1. It is not very helpful to plot zeros and ones. Given that most toxicity experiments are designed to have multiple replicate vessels of multiple subjects each, it is generally helpful to display in tabular or graphical form the proportion in each replicate having the characteristic of interest. The simplest models assume that the true probability of being affected by the treatment is the same for all replicates at the same nominal treatment level. It is well known, however, that sometimes there is more variation among replicate proportions than simple binomial probability would predict. Such a phenomenon is called extra-binomial variance or overdispersion. For example, Oris et al. (2012) found that in fish early life stage experiments, overdispersion in mortality exists in 14–40% of studies, depending on the species of fish tested. Thus, a display of replicate proportions might suggest the presence of overdispersion and the consequent need to modify the analysis or model to accommodate it.

To recognize overdispersion, it is first necessary to understand how much variation among replicate proportions is expected. From introductory statistics, it will be recalled that if every subject in a sample of size n has the same probability, p, of being affected, then the variance of the sample proportion is given by

$$\sigma_{\hat{p}}^2 = \mathrm{var}(\hat{p}) = \frac{p(1-p)}{n}, \tag{6.6}$$

and the distribution of $\hat{p}$ is approximately normal. If there are k replicates of the same size, n, and $\hat{p}_i$, $i = 1, 2, \ldots, k$ are the observed sample proportions and $\hat{p}$ is the pooled sample proportion, i.e.

$$\hat{p} = \frac{1}{k}\sum_i \hat{p}_i, \tag{6.7}$$

then the sample variance of the $\hat{p}_i$ should not be much larger than what is obtained from Eq. (6.6) by replacing p by $\hat{p}$.

See Table 6.5 for an example in summary form. The full data are in the spreadsheet *Tamxsex* in the Excel file *Chapter 2 Data.xlsx* in Appendix 1. The data are from an FSDT based on OECD Test Guideline 234 with zebrafish. In each treatment group, there were three tanks of 40 fish each. The treatments were concentrations of a suspected endocrine disrupting chemical that has the potential to change the phenotypic sex of fish exposed as embryos. It is not possible to determine the genetic sex of these fish at the embryonic stage and impractical, with this species, to determine genetic sex at the termination of the study. Thus, only phenotypic sex is available. Furthermore, while fish are placed into tanks randomly and we expect about 50% males and 50% females in the general population, random sampling with fish of unknown sex at study start may result in notable deviation from this expectation.

The Tamxsex spreadsheet provides the replicate proportions. Section 6.4.2 develops Bayesian criteria for how

Table 6.5 Proportion Female from FSDT Assay

Group	Mean	ObsVar	BinVar	Ratio
1	0.773	0.009	0.004	2.153
2	0.642	0.023	0.006	3.950
3	0.625	0.002	0.006	0.282
4	0.561	0.005	0.006	0.749
5	0.249	0.004	0.005	0.959

Mean is the pooled proportion female at study end. ObsVar is the sample variance of the observed replicate proportions. BinVar is the variance from Formula (2.16). Ratio is the ratio of the observed to binomial variance. Group 1 is the control, where one might expect a proportion of 0.5 females. Do groups 1 and 2 exhibit enough variability to suggest overdispersion? Is group 3 an example of underdispersion? A plot of the replicate proportions (Exercise 2.2) is not as helpful as one might hope in assessing overdispersion in this example.

large the sample variance can be before overdispersion is evident and how to address it when encountered and generalizes (Eq. 6.7) for unequal sample sizes. Further information is in Exercise 6.11. A formal test for overdispersion, Tarone's $C(\alpha)$ test (Tarone, 1979), is presented in Section 6.3.1. Overdispersion will be extended to multinomial random variables in Chapter 9.

6.3.1 Tarone $C(\alpha)$ Test for Overdispersion

Suppose there are m replicates of size $n_1, n_2, n_3, \ldots, n_m$ from a nominally binomial population with common incidence rate, p. To test whether there is overdispersion of the replicate proportions than is expected from a binomial distribution, Tarone's $C(\alpha)$ test statistic is given by

$$Z = \frac{\sum_{j=1}^{m} \frac{\left(x_j - n_j\hat{p}\right)^2}{\hat{p}\left(1-\hat{p}\right)} - \sum_{j=1}^{m} n_j}{\left[2\sum_{j=1}^{m} n_j\left(n_j - 1\right)\right]^{1/2}}, \tag{6.8}$$

where $\hat{p}$ is the mean proportion for a given concentration, m is the number of replicate tanks, n_j is the number of subjects in replicate j, and x_j is the number of subjects in that replicate responding, e.g. not hatched or dead. This test can be extended to multiple treatment groups with possibly different numbers of replicates in different group by first applying the test to each concentration separately. Since the square of each z-statistic is a chi-squared statistic with 1 degree of freedom, these independent chi-squared statistics can be added to define a chi-square statistics with k degrees of freedom, where k is the number of treatment groups including the control. This test can be seen as an adjusted chi-squared test, but limited power simulations done by Tarone have shown it to be more powerful than a chi-squared test.

If overdispersion is detected, the question then arises as to what to do about it. Sections 6.4 and 6.5.2 provide some answers.

Tarone's test will be illustrated with data from a snail reproduction test, OECD TG 243, in which adult mud snails (*Lymnaea stagnalis*) were exposed to cadmium (see Table 6.6). The experimental design called for 4 replicates of 10 adult snails each. Among other responses, the numbers of shelled and unshelled embryos were counted at the end

Table 6.6 Shelled and Unshelled Mud Snail Embryos

Group	Conc ($\mu g\,l^{-1}$)	Rep	Shell length (mm)	Shelled	Unshelled	Total
1	0	1	4.03	59	59	118
1	0	2	3.98	50	62	112
1	0	3	3.95	63	57	120
1	0	4	4.05	61	73	134
2	1.56	1	3.84	49	80	129
2	1.56	2	4.05	50	92	142
2	1.56	3	4.14	60	69	129
2	1.56	4	4.17	37	59	96
3	3.13	1	4.34	70	67	143
3	3.13	2	4.01	40	36	76
3	3.13	3	4.07	47	71	118
3	3.13	4	4.25	36	80	126
4	6.25	1	4.24	52	36	88
4	6.25	2	4.05	62	71	133
4	6.25	3	4.07	33	64	97
4	6.25	4	4.02	43	76	119
5	12.5	1	4.30	49	4	53
5	12.5	2	4.04	40	11	51
5	12.5	3	4.21	16	49	65
5	12.5	4	3.83	20	50	70
6	25	1	4.09	18	5	20
6	25	2	4.23	9	1	10
6	25	3	4.03	4	1	4
6	25	4	4.13	4	7	11

Group, treatment group with 1 = control; Shelled and unshelled, number of embryos that were or were not shelled.

of the study. For purposes of this example, these counts are summarized by replicate, ignoring the individual snails. A more complete investigation taking individual snails into account will be presented in Chapter 8. Tarone's $C(\alpha)$ test will be applied to determine whether the proportion of total embryos that were unshelled varied more across replicates within test concentration than binomial theory would predict. See Table 6.7 for summary calculations from this test. There is strong evidence of overdispersion.

6.3.2 Outliers in Quantal Data

Outliers in quantal data occur but can be difficult to detect and an analog of Tukey's outlier rule is not available. As an example, consider intersex in the species zebrafish, which is defined for an individual fish as exhibiting both male and female sex characteristics. The incidence of intersex is often very low unless the test substance is a powerful endocrine disruptor. See Table 6.8 for an experiment with a control and four test concentrations of the test chemical.

In this experiment, 20 zebrafish embryos were exposed to the test concentration in each replicate vessel. The fact that no intersex fish were observed in the control and low test concentration makes the tests for significant increase very sensitive to incidence in higher test concentrations. A step-down Cochran–Armitage test (Section 6.5.1) gives an NOEC = 10. Had there been a single intersex fish in the control, the same test would have found NOEC = 32. There is no way to know whether such a single observation from one experiment is an outlier. If there were a large control database of zebrafish embryos studied under the same conditions of temperature and water chemistry, it might be possible to address that question. A regression analysis of these data will be given in Chapter 7, where it will be shown that the EC10 estimates are 143.1 (77.0, 209.2) for the data in Table 6.8 and 131.8 (72.3, 191.2) in the alternative, so there is little impact of the potential outlier on these estimates.

Table 6.7 Summary Calculations from Tarone's Test

Group	z	TSqr	Probz	TotTsqr	ProbTot
1	−0.7153	0.512	0.23722	0.512	
2	0.0597	0.004	0.47618	0.515	
3	4.2257	17.857	0.00001	18.372	
4	3.3859	11.465	0.00035	29.837	
5	27.7501	770.067	0	799.903	
6	3.5705	12.748	0.00018	812.651	0

z is calculated from Eq. (6.8). Tsqr = z^2, a 1 − df χ^2-statistic. Probz is the probability Tsqr exceeds the critical value of the χ^2-distribution. TotTsqr is simply the sum of all Tsqr values at equal or lower concentrations, and probtot is the probability of a larger value than TotTsqr in a χ^2-distribution with df = 5. It is left as an exercise to verify these and some related calculations.

Table 6.8 Incidence of Intersex Zebrafish Assay 1

Rep	Conc	Intersex	Notintsex	Rep	Conc	Intersex	Notintsex
1	0	0	1	1	32	1	0.95
2	0	0	1	2	32	0	1
3	0	0	1	3	32	1	0.95
4	0	0	1	4	32	0	1
1	10	0	1	1	100	1	0.95
2	10	0	1	2	100	1	0.95
3	10	0	1	3	100	2	0.9
4	10	0	1	4	100	2	0.9

Intersex, Number of phenotypically intersex fish in replicate; Notintsex, Proportion of fish in replicate that are not intersex.

6.4 PAIRWISE MODELS THAT ACCOMMODATE OVERDISPERSION

Overdispersion can occur for several reasons. It could be that the experimental setup does not eliminate some sources of variability or that the model ignores such sources. In the two subsections of the current section, two models that take overdispersion into account will be described. Another possible source of overdispersion is that the distribution of the response is a mixture of two distributions. In particular, it can happen that there are far more zero responses than a binomial model would explain. Such zero-inflated models will be discussed in Section 6.7.

6.4.1 Generalized Linear Model

Generalized linear mixed models (GLMMs) offer a way to treat quantal responses as quantal while including the replicate structure in the analysis and adjusting for the number of comparisons with the control in a less conservative fashion, potentially increasing the power to detect effects. GLMMs were discussed in Section 3.7 with a Poisson error structure. For quantal responses, GLMMs are applied with a binomial error structure and can accommodate the cases of 0 or 100% effect. Also, there is a natural way to take sample sizes into account. As will be demonstrated in Chapter 7, the fixed effect term can be modeled to fit the nonlinear dose–response shapes typical of toxicity studies. The replicate is a random effect and treatments, of course, are fixed. So the general model can be stated as

$$X_{ij} = \pi_j + R_{i(j)}, \tag{6.9}$$

where the $R_{i(j)}$ are binomial random errors. The parameters π_j are estimated by minimizing the deviance, analogous to minimizing the residual sum of squares for a normal errors model. For a binomial errors model the deviance is defined by

$$\text{Deviance} = 2\sum_j X_j \text{Ln}\left(\frac{X_j}{n_j\hat{\pi}_j}\right) + \left(n_j - X_j\right)\text{Ln}\left(\frac{n_j - X_i}{n_j - n_j\hat{\pi}_j}\right), \tag{6.10}$$

where $\hat{\pi}_j$ is the estimate of π_j, n_j is the number of subjects sampled in the jth treatment, and X_j is the number of those subjects that are affected. When $Y=0$, the expression of the type $Y * \text{Ln}(Y)$ in Eq. (6.10) is replaced by 0. In the simple case at hand, with no regression model relating π_j to treatment level, the minimum is achieved when $\hat{\pi}_j = X_j/n_j$, in which case the deviance is zero. A good discussion of GLMMs for quantal responses is given by Kerr and Meador (1996). If Dunnett's test (called Dunnett–Hsu in the GLMM context) is applied to the medaka sex ratio data, the same NOEC, 23.5 µg l^{-1}, is obtained as found by Fisher's exact test and by the normal errors ANOVA of replicate proportions. Programs to implement this approach are given in Appendix 1. The SAS program is called *GLMM_Quantal.sas*. An R program, *GLiMM binomial.r*, provides the same conclusions for the medaka data modified to add one male to one replicate in the highest group. This program does not handle the existing data where there were no males in the highest group. Advantages of the GLMM model are that it handles replicates appropriately, unlike the Fisher' exact test and, at least in the SAS version, handles zero variances caused by 0 and 100% affected samples rather than ignoring them, unlike the normal errors ANOVA of proportions. Furthermore, as will be demonstrated in Chapter 8, GLMMs can be extended to count data.

6.4.2 Beta-binomial Model

The beta-binomial model is an example of the Bayesian approach to statistical analysis. It was introduced by Williams (1975) and refined by Williams et al. (1988). Programs are given in Appendix 1 for this model. Of these, the SAS program is based on Martinez et al. (2015) and Nelson et al. (2006). This model is useful in describing incidence rates where the probability of an event or incident is unknown or random. For example, the incidence of some fetal abnormality in a rodent study may vary according to the litter size even though all dams are exposed to the same environmental conditions, such as the level of some chemical in a toxicity experiment. A simple binomial density function with a fixed incidence rate may not describe the data well. Other situations where the beta-binomial might be useful are in a traditional aquatic toxicity experiment where multiple organisms (e.g. fish) are placed in each tank and there are multiple tanks at the same nominal test concentration. There often is more variation among incidence rates in tanks than a simple binomial density function would predict. Such "overdispersion" is common in many ecotoxicity experiments, as discussed by Oris et al. (2012) for fish early life stage studies. The FSDT example underlying Table 6.5 shows that overdispersion is not confined to ELS studies.

The binomial density function (or likelihood) with parameters n and θ is given by Eq. (A2.7), which is reproduced here with a change of notation to facilitate the current discussion:

$$p(x \mid n,\theta) = \binom{n}{x}\theta^x(1-\theta)^{n-x} \quad \text{if } 0 \le x \le n \text{ and } x \text{ is an integer}$$
$$p(x \mid n,\theta) = 0 \quad \text{otherwise.} \tag{6.11}$$

A conjugate prior density function is the beta density function with hyperparameters α and β, given by

$$p(\theta|\alpha,\beta) = \text{Beta}(\alpha,\beta) = \frac{\Gamma(\alpha+\beta)}{\Gamma(\alpha)\Gamma(\beta)}\theta^{\alpha-1}(1-\theta)^{\beta-1}, \tag{6.12}$$

where

$$\Gamma(\alpha) = \int_0^\infty t^{\alpha-1}e^{-t}dt. \tag{6.13}$$

It should be noted that for positive integer values of α, the gamma function simplifies to

$$\Gamma(\alpha) = (\alpha-1)!. \tag{6.14}$$

It is simple to verify that the mean and variance of a binomial are given by $n\theta$ and $n\theta(1-\theta)$ and the mean and variance of a beta distribution are given by

$$\mu = \frac{\alpha}{\alpha+\beta} \tag{6.15}$$

and

$$\sigma^2 = \frac{\alpha\beta}{(\alpha+\beta)^2(\alpha+\beta+1)}, \tag{6.16}$$

respectively. The mean and variance of the beta-binomial can be shown to be, respectively,

$$E[X \mid n,\alpha,\beta] = \frac{n\alpha}{\alpha+\beta}, \tag{6.17}$$

and

$$\mathrm{Var}\left[X \mid n,\alpha,\beta\right]=\frac{n\alpha\beta\left(\alpha+\beta+n\right)}{\left(\alpha+\beta\right)^{2}\left(\alpha+\beta+1\right)}. \tag{6.18}$$

Since the mean of the binomial is $n\theta$ and the mean of the beta-binomial is $n\alpha/(\alpha+\beta)$, if we let $\theta=\alpha/(\alpha+\beta)$ be the "realized" or observed binomial probability, then the variance of the beta-binomial random variable can be written as

$$\mathrm{Var}\left[X \mid n,\alpha,\beta\right]=n\theta\left(1-\theta\right)\left[1+\left(n-1\right)\rho\right], \tag{6.19}$$

where $\rho=1/(\alpha+\beta-1)$. The term ρ is called the overdispersion parameter. Values of ρ less than 0 or near 0 correspond to distributions that are essentially binomial, while large values of ρ correspond to distributions with notable overdispersion. In Chapter 9, these concepts will be extended to multinomial distributions. In Chapter 8, this concept is extended to a Poisson distribution.

It can be shown that the larger the dataset, the smaller the effect of the prior. That is, when we do not know much about the data (e.g. the sample size is small), then prior knowledge is very important. As we gain more extensive knowledge of the data (e.g. the dataset is large), the prior assumptions regarding the data become less important.

If y is the random variable representing the sample and θ is the model parameter (or vector of parameters), then

$$E\left[\theta\right]=E\left[E\left[\theta \mid y\right]\right] \tag{6.20}$$

and

$$\mathrm{Var}\left(\theta\right)=E\left[\mathrm{var}\left(\theta \mid y\right)\right]+\mathrm{var}\left(E\left[\theta \mid y\right]\right). \tag{6.21}$$

Equation (6.20) states that the variance of θ is the expected value of the conditional variance (conditioned on the data). Equation (6.21) implies that the conditional variance (i.e. posterior variance) is on average (first term of right-hand side) smaller than the unconditional variance (left-hand side). Another way to interpret (Eq. 6.21) is that the greater the uncertainty in the data (expressed by the second term on the right), the greater is the potential for reducing the posterior variance. Additional development of the Bayesian approach is given in Appendix 2.

One of the earliest applications of the GLMM approach is the beta-binomial model introduced in Section 6.4.2. This could be used as a basis for a pairwise analysis of another zebrafish sex ratio dataset (see Table 6.9), once it is verified by Tarone's $C(\alpha)$ test that the male data exhibit overdispersion. The SAS program ***betabinom.sas*** is provided in Appendix 1 to compare the treatment groups with the control after making allowance for the overdispersion.

Table 6.9 Zebrafish Sex Ratio Data from FSDT Assay 2

Rep	Conc	Group	Female	Male	Intersex	Undiff	Species	Chemical
1	−1	1	20	6	0	14	Zebrafish	4-Tert octylphenol
2	−1	1	22	16	0	1	Zebrafish	4-Tert octylphenol
3	−1	1	13	24	0	3	Zebrafish	4-Tert octylphenol
4	−1	1	20	19	0	1	Zebrafish	4-Tert octylphenol
1	0	2	17	20	0	3	Zebrafish	4-Tert octylphenol
2	0	2	20	15	0	5	Zebrafish	4-Tert octylphenol
3	0	2	21	10	0	9	Zebrafish	4-Tert octylphenol
4	0	2	20	8	0	12	Zebrafish	4-Tert octylphenol
1	32	3	24	7	0	9	Zebrafish	4-Tert octylphenol
2	32	3	15	17	0	8	Zebrafish	4-Tert octylphenol
3	32	3	19	14	0	7	Zebrafish	4-Tert octylphenol
4	32	3	21	5	0	5	Zebrafish	4-Tert octylphenol
1	100	4	19	12	0	9	Zebrafish	4-Tert octylphenol
2	100	4	19	7	0	14	Zebrafish	4-Tert octylphenol
3	100	4	20	6	0	14	Zebrafish	4-Tert octylphenol
4	100	4	19	5	0	11	Zebrafish	4-Tert octylphenol
1	200	5	19	4	0	17	Zebrafish	4-Tert octylphenol
2	200	5	21	2	0	17	Zebrafish	4-Tert octylphenol
3	200	5	18	5	0	17	Zebrafish	4-Tert octylphenol
4	200	5	26	1	0	13	Zebrafish	4-Tert octylphenol

Conc = −1 is water control, and Conc = 0 is solvent control, other concentrations in $\mu g\, l^{-1}$. Female, Male, Intersex, and Undiff, number of embryos with the indicated phenotypic sex at study end.

Table 6.10 Results from Beta-Binomial Model Fit to Zebrafish Assay 2 Sex Ratio Data

Label	Direction	Estimate	StdErr	DF	TStat	Probt	Adj*P*
Diff $\theta_2-\theta_1$	DECR	−0.102	0.096	20	−1.06	0.303	0.303
Diff $\theta_3-\theta_1$	DECR	−0.214	0.086	20	−2.49	0.022	0.044
Diff $\theta_4-\theta_1$	DECR	−0.333	0.086	20	−3.85	0.001	0.003

Label indicates which dose group is compared with the control. Estimate is a standardized difference in the estimated values of the parameter θ for the two indicated dose groups. The TStat and Probt values are for a decrease in θ, corresponding to a decrease in the proportion of phenotypic males. The table shows the Bonferroni–Holm adjustment to the p-values as ADJP. The NOEC = Dose 2, or 32 μg l^{-1}.

In R, the reader is directed to Hilbe (2013). See Table 6.10 for a summary of the results. It follows from Eq. (6.20) that

$$E\left[X_i \mid n_i,\alpha_i,\beta_i\right]=\frac{n_i\alpha_i}{\alpha_i+\beta_i}=n_i\theta_i,$$

where X_i is the number of ph_males, n_i is the sample size, and θ_i the location parameter for the ith dose group. A difference in the two groups is indicated by a difference in the estimates of θ, so the hypotheses to test regard those differences without reference to the sample sizes, i.e. H_0: $\theta_1=\theta_2=\theta_3=\theta_4$ versus H_1: $\theta_1\neq\theta_i$ for some i.

The estimates of the θ_i indicated in Table 6.10 have been obtained by three different SAS programs, differing at most in the third decimal place. These include the program given in Appendix 1 based on the nlmixed procedure, the logistic procedure using scale = Williams (following Agresti, 2002), and the finite mixture model (FMM) procedure. The proportions estimated by the R program betabinomai.r using the betareg function differ from those from the three SAS procedures. For example, $\theta_4-\theta_1$ was estimated as 0.247 using the betareg function or 0.293 using a numeric optimizer function. The p-values for the differences from control vary according to which SAS procedure is used, but all agree that the high dose group is significantly different from the control and the p-value for the second highest difference is less than 0.06, though not always less than 0.05. The betareg procedure also found the second highest dose group significant ($p=0.021$). This is a result of different model formulations giving rise to different approximate variance–covariance matrices of the parameter estimates. To get the estimated proportions from the output of the R program or the SAS logistic procedure, it is necessary to convert the logit values using the relationship

$$\text{Logit}(x)=y \text{ means } x=\frac{\exp(y)}{1+\exp(y)}.$$

As will be developed in Chapter 7, the beta-binomial model and the programs referenced above can be extended to incorporate nonlinear regression models for the proportions, so that ECx can be estimated from models for quantal responses that allow for overdispersion.

6.5 TREND TESTS FOR QUANTAL RESPONSE

As with continuous responses discussed in Chapter 3, the preferred way to analyze count data for NOEC determination is through the use of an appropriate trend model. This section will develop two such models.

6.5.1 Cochran–Armitage Test

The Cochran–Armitage (Cochran, 1954; Armitage, 1955) test decomposes the Pearson χ^2 test into a test for linear trend for the dose–response and a test for lack of monotonicity. Specifically,

$$\chi^2_{(k)}=\chi^2_{(1)}+\chi^2_{(k-1)}$$

where $\chi^2_{(1)}$ is the *1* df calculated Cochran–Armitage linear trend statistic and $\chi^2_{(k-1)}$ is a $k-1$ df χ^2 test statistic for lack of monotonicity. It remains to specify how this is done.

Suppose the jth treatment sample has n_j subjects, of which X_j are affected, and $d_0, d_1, d_2, \ldots, d_k$ are measures of the dose or concentration. That is, d_j is the dose metric: d_j could be the dose, log(dose), or dose rank (i.e. $d_j=j$). The proportion affected in the jth treatment is $p_j=X_j/n_j$. The model has the form $p_j=H(\alpha+\beta d_j)$, where H is some twice differentiable, monotone link function, such as the logistic. The Cochran–Armitage test is a test of $\beta=0$. The estimate of β from standard weighted regression is given by Eq. (6.22):

$$b=\frac{\sum_{j=1}^{k}n_j\left(p_j-\bar{p}\right)\left(d_j-\bar{d}\right)}{\sum_{j=1}^{k}n_j\left(d_j-\bar{d}\right)^2}, \tag{6.22}$$

where $\bar{d}$ is the mean of the d_j weighted by sample size and $\bar{p}$ is the similarly weighted mean of the p_j. A test statistic for $\beta=0$ has the form $Z=b/\text{Std}(b)$ that is approximately normally distributed, so Z^2 has a $\chi^2_{(1)}$ distribution. The test statistic can be written as Eq. (6.23):

$$Z = \frac{\sum_j X_j d_j - N\bar{p}\bar{d}}{\left[\bar{p}(1-\bar{p})\left(\sum_j n_j d_j^2 - N\bar{d}^2\right)\right]^{1/2}}, \quad (6.23)$$

where $N = n. = \Sigma n_j$ is the total sample size of the study.

The z-statistic has a standard normal distribution under the null hypothesis of no trend. The simplest and, in some ways, most attractive, dose metric d_j is simple rank order of doses (i.e. equally spaced dose scores or $d_j = j$). This reduces the dose dependence in the test. While the use of the actual doses is possible, doing so can emphasize the highest dose level rather than the trend and consequently distort the test.

A general linear trend model for quantal data is

$$P_j = H(a + bd_j),$$

where H is some monotone function (referred to as a link function), e.g.

$$H(u) = \frac{e^u}{(1+e^u)} \quad \text{logistic,}$$

$$H(u) = M(u) \quad \text{probit,}$$

$$H(u) = 1 - \exp(e^u) \quad \text{extreme value,}$$

$$H(u) = 1 - e^{-u} \quad \text{one-hit.}$$

Tarone and Gart (1980) showed that for any link function likely to be of practical use, the same test statistic for significance of trend always arises from likelihood-type considerations, namely, the Cochran–Armitage test. Thus, it is not necessary to postulate the particular form of the link function. They also showed that this test is in general the most efficient test of trend for any monotone model. For this reason, some regard the Cochran–Armitage test as inherently nonparametric. Hirji and Tang (1998) discuss the power properties of the Cochran–Armitage test compared with various alternatives. They argue this test can be used for small samples and sparse data despite it being developed as an asymptotic test. We have done an extensive power simulation study of this test for the sample sizes typical in guideline toxicity tests and found it adequately powered in many applications. Further discussion of the power properties of the Cochran–Armitage test in comparison with other tests discussed in this chapter is provided in Section 6.6.

The Cochran–Armitage test assumes subjects are independent within and among treatment groups and group proportions are monotonic with respect to the dose score. While this is formally a test for linear trend in the response in relation to the dose score used, it is generally the most powerful test against any monotone dose–response alternative, as shown by Tarone and Gart (1980).

Table 6.11 Results of Cochran–Armitage Test Applied to Medaka Sex Ratio Data

Conc	Doses	Z	Probz	Signif
100.6	6	−4.37965	0.00001	**
50.4	5	−2.10849	0.01749	*
23.5	4	0.45678	0.32391	

The column labeled Doses indicates the number of doses used in the step-down application of the Cochran–Armitage test. Thus, Doses = 6 when all doses are used, while Doses = 5 indicates that the high dose was set aside and the test was done on doses 1–5. Z is the Z-statistic from Eq. (6.23). Probz is the p-value associated with that Z-statistic. Signif = "**" indicates significance at the 0.01 level, while "*" indicates significance at the 0.05 level.

See Table 6.11 for a summary of the Cochran–Armitage test applied to the medaka data from Table 6.2. The resulting NOEC, 23.5 μg l^{-1}, is the same as obtained by Fisher's exact test and GLMM. An SAS program (*STDCOCHARM.sas*) to carry out this test is provided in Appendix 1. Also, an R program for this purpose is included in the file *Chapter 6A.r*.

There is no reason to expect the three approaches so far discussed always to determine the same NOEC, nor is it good statistical practice to perform multiple tests on the same data. We do so here only to illustrate the various tests. When multiple tests are done on the same dataset, it can be difficult to know how to interpret the results when the tests produce different NOECs. This is somewhat similar to fitting multiple regression models to the same data. In that case, we choose the model that best fits the data according to specified criteria and report only the results from that model. In the NOEC context, the analogous situation is to pick the test that best conforms to the experiment as performed and for which the data meet the requirements of that test. In general, we prefer using a test that models the expected monotone dose–response. Of the tests considered thus far, this would be the Cochran–Armitage. However, as noted, the Cochran–Armitage test ignores replicates. While the GLMM approach takes replicates into account, it ignores the monotone dose–response nature of the experiment, and we consider that the more serious error. A comparison of the power of the Cochran–Armitage test to detect effects compared with GLMM and analyses of replicate proportions by the methods of Chapter 3 will be demonstrated in Section 6.6. Regardless of power considerations, important as they are, there are situations where the failure to take replicates into account is troubling. Fortunately, there is a way to modify the Cochran–Armitage test so as to include the replicate structure. That is done in Section 6.5.2. Before turning to that development, another example will be considered where the three approaches so far considered do not lead to the same NOEC.

Example: Onions from a Nontarget Terrestrial Plant (NTTP) Study

In an NTTP vegetative vigor study done under OECD TG 227, four onion seeds were planted in each of six pots in the control and each treatment group. (See Table 6.12 for a summary of the data and of the analysis. Full data showing replicates is given in the SAS dataset *onion* in Appendix 1.) Treatment was exposure to some level of a pesticide, measured in g ha^{-1}. The response of interest is survival to the end of the 21-day study. Because of the large number of application rates at which there was no mortality, analysis of replicate proportions is not possible using a normal errors model. However, GLMM, Cochran–Armitage, and Fisher's exact tests can be done. One of the clear conclusions from this example is that Fisher's exact test with Bonferroni–Holm adjusted *p*-values has low power when there are so many treatment groups, while the Cochran–Armitage and GLMM tests can be extremely sensitive when there are many zero observations. There may be temptation to forego the Bonferroni–Holm adjustment for Fisher's exact test when there are so many treatments, but defining a scientifically valid rule for when to drop that adjustment is very challenging. We think Fisher's exact test is simply the wrong test for data of this sort.

Table 6.12 Summary of Onion Data and Analyses

Group	AtRisk	Dead	%Dead	Rate (g a.s. ha^{-1})	Signif
1	24	0	0	0	
2	24	0	0	0.08	
3	24	0	0	0.15	
4	24	0	0	0.3	
5	24	0	0	0.6	
6	24	0	0	1.2	
7	24	0	0	2.4	
8	24	0	0	4.8	
9	24	2	8.3	9.5	GLMM, CA
10	24	2	8.3	19	GLMM, CA
11	24	6	25	38	GLMM, CA
12	24	4	16.7	75	GLMM, CA
13	24	11	45.8	150	FE, GLMM, CA

Analysis can be done in terms of the number dead or the number alive at 21 days post exposure. The column marked Signif indicates the treatment groups at which one or more tests found a significant increase in mortality, with FE, Fisher's exact test; CA, Cochran–Armitage test; and GLMM, generalized linear model.

6.5.2 Rao–Scott Robust Cochran–Armitage Test

As noted in Sections 2.4 and 6.4.2, overdispersion is common in some studies, notably fish early life stage studies done under OECD TG 210. Methods for adjusting the Cochran–Armitage test to account for the replicate structure of quantal data, in particular overdispersion and differences in sample size, have been developed (Rao and Scott, 1992, 1999; Fung et al., 1994).

According to Rao and Scott, the natural estimator of the overall sample proportion for the *j*th concentration is $\hat{p}_j = x_j/n_j$, where n_j is the total sample size or number at risk in the *j*th concentration and x_j is the total number responding. If there are m_j subgroups in *j*th concentration, n_{ij} observations in subgroup *i* of treatment *j*, and x_{ij} responding in that subgroup, then the residual for that subgroup is $r_{ij} = x_{ij} - n_{ij}\hat{p}_j$ and the variance of $\hat{p}_j$ is

$$v_j = m_j\left(m_j - 1\right)^{-1} n_j^{-2} \sum_{i=1}^{m_j} r_{ij}^2. \quad (6.24)$$

The ratio of this variance to the sample binomial variance $n_j^{-1}\hat{p}_j(1-\hat{p}_j)$ is denoted d_j and is called the "design effect" or variance inflation due to clustering. The effective sample size is given by $\tilde{n}_j = n_j/d_j$. Finally, let $\tilde{x}_j = x_j/d_j$. Then the points $(\tilde{x}_j, \tilde{n}_j)$ are analyzed using the standard Cochran–Armitage test statistic modified only by using $(\tilde{x}_j, \tilde{n}_j)$ in place of (x_j, n_j). Thus,

$$Z = \frac{\sum \tilde{x}_j c_j - \hat{p}\sum \tilde{n}_j c_j}{\left[\hat{p}(1-\hat{p})\sum \tilde{n}_j\left(c_j - \bar{c}\right)^2\right]^{1/2}}, \quad (6.25)$$

where c_j is the dose score and $\bar{c} = \sum \tilde{n}_j c_j / \sum \tilde{n}_j$ and $\hat{p} = \sum \tilde{x}_{ii} / \sum \tilde{n}_j$. Generally, c_j is taken to be *j* to give equally spaced dose scores, as discussed elsewhere. The statistic *Z* has an asymptotically standard normal distribution under the null hypothesis. It is left as an exercise to verify that if there is no design effect, i.e. all the $d_j = 1$, then Eq. (6.25) reduces to Eq. (6.23).

Rao and Scott (1999) state that the false positive rate associated with this test statistic is close to the nominal level. However, extensive simulations have shown that the false positive rate can be inflated. There are several minor variations on this test discussed in Rao and Scott (1999), but all are subject to inflated false positive rates under some conditions. This is perhaps an acceptable price in order to take overdispersion into account. More discussion of this point will be given in Chapter 9 in conjunction with the RSCABS test. To illustrate the Rao–Scott Cochran–Armitage test, data from an FSDT with zebrafish will be used. The data appear in Table 6.9.

As indicated in Table 6.9, initially 40 fish embryos were placed in each replicate tank, but some fish died prior to the end of the study and were not sexed. Tarone's $C(\alpha)$ test found significant ($p<0.001$) evidence of overdispersion, with the water and solvent controls and lowest test concentration exhibiting the most overdispersion. The assessment of overdispersion was done with the two controls distinguished, although Tarone's test is still significant with the two controls combined.

See Table 6.13 for a summary of results of the standard and Rao–Scott versions of the Cochran–Armitage test. See Table 6.14 for some calculations used in computing the Rao–Scott Cochran–Armitage test statistic.

It will be seen that the standard Cochran–Armitage test, which ignores replicates, found the proportion phenotypic males in the lowest test concentration to be significantly reduced compared with the control, while the Rao–Scott Cochran–Armitage test, which takes replicates as well as overdispersion into account, did not find the lowest test concentration to have a significant effect.

Table 6.13 Summary of Trend Tests for Zebrafish Phenotypic Sex

Conc	TotRisk	PH_Male	Phati	Reps	Dose	Signif
0	319	118	0.37	8	1	
32	151	43	0.28	4	2	CA
100	155	30	0.19	4	3	CA, RSCA
200	160	12	0.08	4	4	CA, RSCA

TotRisk is the total number of fish at the end of the study. PH_Male is the number of phenotypic males. Phati is the proportion of males (PH_Male/TotRisk). Reps is the number of replicates after the two controls are combined. Signif indicates whether the decrease in proportion males is significant, and if so, by which test. For example, CA indicates the (standard) Cochran–Armitage test was significant.

Table 6.14 Intermediate Calculations for Rao–Scott Test of Zebrafish Data

Dose	BINvari	VARi	di	Ntilda	Xtilda
1	0.000 730 644	0.003 128 277	4.28	74.5	27.6
2	0.001 348 843	0.004 222 206	3.13	48.2	13.7
3	0.001 007 016	0.001 379 878	1.37	113.1	21.9
4	0.000 433 594	0.000 520 833	1.20	133.2	10.0

The notation follows that used in Eq. (6.25). SAS programs to carry out both versions of the Cochran–Armitage test are given in Appendix 1 as *RSCA_test.sas* and *STDCOCHARM.sas* and R programs are provided in *#Cochran–Armitage and Rao–Scott Cochran–Armitage tests* in *Chapter 6A.r.*

6.6 POWER COMPARISONS OF TESTS FOR QUANTAL RESPONSES

The previous three sections have discussed several pairwise and trend tests by which to analyze quantal data, as well as methods for taking overdispersion into account. The choice of which test to employ should be based on relevant factors. We have discussed in several places (e.g. Sections 2.2.3.2 and 3.3) the conceptual advantages of trend-based tests when the data are consistent with a monotone dose–response. There are also conceptual advantages to using a binomial error structure rather than a normal error structure for such data. In addition, there are clear conceptual advantages to including the replicate structure of studies in the tests used. Finally, if there is evidence of overdispersion, then it seems unwise to ignore that in tests. When overdispersion is evident, these conceptual issues would seem to favor the Rao–Scott modified Cochran–Armitage test for data consistent with a monotone dose–response and the beta-binomial test otherwise. If there is no compelling evidence of overdispersion, the choices are less clear. The Cochran–Armitage test addresses the quantal and dose–response trend nature of the data but ignores replicates. The step-down Jonckheere–Terpstra test applied to replicate means or medians addresses the replicate and dose–response trend nature of the data but ignores the quantal nature of the response. Where the data are not consistent with dose–response monotonicity, Fisher's exact test ignores replicates, and Dunnett's and Williams' tests ignore the quantal nature of the response, but the GLMM model with binomial error structure with replicates as a random term would appear to be best, at least conceptually.

One issue missing above is an understanding of which of these issues (quantal vs. normalized approximations, treating or ignoring replicates, treating or ignoring overdispersion, treating or ignoring monotonicity in the dose–response) matters in terms of finding treatment effects. The power to detect effects is an important tool in deciding which test to use. However, it must be understood that high power alone is not the deciding factor in test selection. The false positive rate is also important, and selecting a test that reflects the way the study was actually conducted is important. Still, understanding power is important, and this section will present some power properties of the tests so far considered.

Factors that affect power of course include the number of treatment groups, number of replicates per group, and number of subjects per replicate. Perhaps less obviously, the power of a test to detect an effect on a quantal response also depends on the background incidence rate and, at least for trend tests, the shape of the dose–response curve. Some power properties presented in Table 6.15 are for quantal responses from a fish early life stage experiment following

Table 6.15 Power Comparison: No Overdispersion, Mild Trend

ECPB	Base	HET	%Eff	Reps	WL	JT	DT	MW	Dunn	GQ	RS	CA	BB
1	80	0	5	4	8	11	4	0	3	17	19	40	20
1	80	0	10	4	16	23	8	0	5	33	30	54	36
1	80	0	15	4	33	38	18	0	11	46	49	68	47
1	80	0	20	4	47	56	31	0	21	63	65	80	61
1	80	0	30	4	73	84	61	1	41	87	86	94	79
1	80	0	35	4	85	92	73	1	51	93	91	97	84
1	80	0	40	4	90	97	85	1	61	97	95	98	87
1	80	0	45	4	93	98	92	1	71	99	97	99	89
1	80	0	50	4	96	100	95	1	78	100	98	99	91
1	80	0	55	4	97	100	98	2	85	100	99	100	90
1	80	0	60	4	97	100	100	2	92	100	99	100	91
1	80	0	65	4	98	100	100	3	96	100	100	100	90
1	80	0	70	4	98	100	100	4	98	100	100	100	89
1	80	0	75	4	99	100	100	5	99	100	100	100	90
1	80	0	80	4	97	100	100	7	100	100	100	100	89
1	80	0	85	4	97	100	100	9	100	100	100	100	85
1	80	0	90	4	96	100	100	8	100	100	100	100	81

ECPB defines the steepness of the curve; Base, %survival in control; HET, %overdispersion simulated; %EFF, %effect to be found significant; WL–BB, Power of Williams, Jonckheere–Terpstra, Dunnett, Mann–Whitney, Dunn, GLMM Dunnett, Rao–Scott C–A, Cochran–Armitage, and beta-binomial tests to detect %eff. The beta-binomial powers flatten out after 45% effect, which is rather puzzling. The behavior of this test is sensitive to the priors or initial conditions on the parameters, but extensive alternative initial conditions have not improved the power calculations. None of the other tests explored required initial conditions, which is a practical advantage for those tests.

the current 2013 version of OECD TG 210, where four replicates per treatment were required, compared with only two in the 1992 version of the guideline. To understand this table, it is necessary first to consider the range of dose–response shapes encountered in these studies for quantal responses. See Figure 6.1 for a range that includes most such studies, based on a database of well over a hundred studies with different fish species. A fifth shape labeled ECPB90 but not shown is more extreme than that labeled ECPB50, where the concentration–response is flat until there is a sharp drop-off in the high concentration.

All of these curves are generated from a Bruce–Versteeg model with a shape parameter labeled ECPB that controls the rapidity with which the effect begins. The maximum magnitude of the effect is modeled to range approximately from 5 to 90%, and the control percentage effect simulated was 0, 10, and 20%. Test guidelines generally do not permit more than 20% mortality in the control. However, mortality is not the only quantal response. Sex ratio, the percent males, females, intersex, or undifferentiated, can have a control percent "affected" around 50%, and the power properties and regression estimate quality are very different from what the simulations discussed in this section indicate. This is an issue of importance when the genetic sex of fish cannot be determined in a practical manner (which is true for some species tested) and only the phenotypic sex is known. If the genetic sex is known, then one can model the incidence of phenotypic sex differing from genetic sex and then the control incidence rate will be close to 0% (or 100%, depending on how the response is defined).

Power of various statistical tests to detect an effect was calculated from simulations under a variety of conditions. A sampling of the results is presented here. These power results are taken from a study reported on in Green (2017a). Three figures will be presented to display power properties. For datasets with only a mild trend and no overdispersion, see Figure 6.2 and Table 6.15. For datasets with a mild trend but a delayed response (i.e. initial dose–response is flat, then a decrease occurs), see Figure 6.3. For datasets with a mild trend but 50% overdispersion, see Figure 6.4.

For a mild concentration–response shape (ECPB = 1, 5, or 25) with no overdispersion, the standard Cochran–Armitage has a clear power advantage over all alternative tests, while the Rao–Scott and GLMM tests have very similar power properties, as do Williams and Jonckheere. The GLMM and Rao–Scott C–A tests make allowances for overdispersion even though it is not present in this instance. While no special consideration is given to overdispersion in the remaining tests, they do not assume binomial variances.

For a concentration–response shape (ECPB = 50 or 90) with delayed response and no overdispersion, the

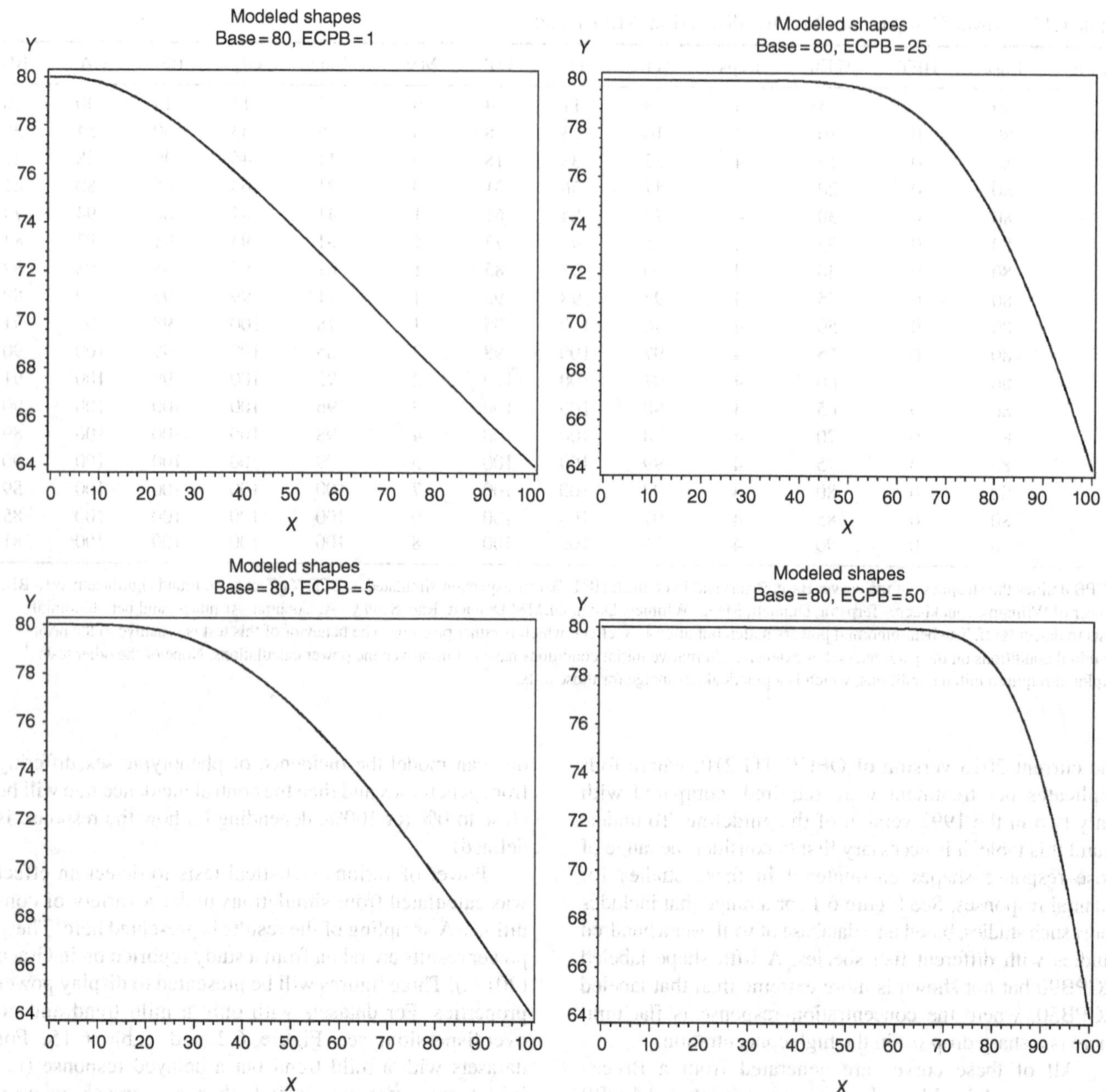

Figure 6.1 Simulated concentration–response shapes. Example shapes for simulated dose–response curves with 20% background incidence and a maximum of 36% incidence.

standard Cochran–Armitage has a clear power advantage over all alternative tests, while the GLMM test is next most powerful and the Rao–Scott loses some power and the Jonckheere test loses much power. The performance of the other tests is little changed from than discussed in relation to Figure 6.2.

When there is overdispersion, the power of the Cochran–Armitage test is still considerably greater than that of all other tests, but it is not appropriate because it ignores overdispersion. Of course, C–A also ignores the replicate structure. While that might be tolerable when there is no overdispersion, it seems especially inappropriate when it is present. The power characteristics of the Rao–Scott and GLMM tests are similar, with the former having slightly higher power under some conditions and the latter having a greater power advantage under other conditions. The tests compare in much the same way as when there is no overdispersion, though all of the tests have somewhat

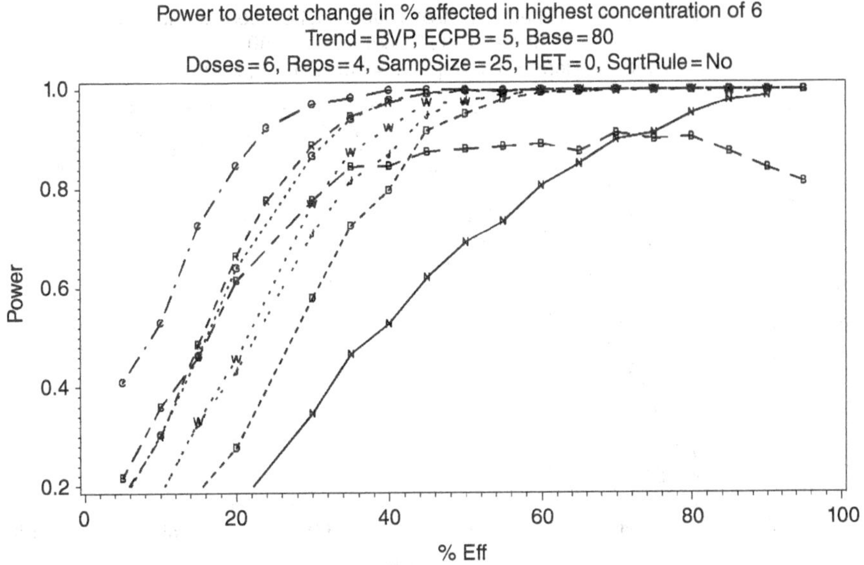

Figure 6.2 Power comparison for no overdispersion, mild trend. C, Cochran–Armitage; R, Rao–Scott Cochran–Armitage; G, GLMM with Dunnett; W, Williams; J, Jonckheere–Terpstra; D, Dunnett (standard); B, beta-binomial; N, Dunn test; 20% background incidence; four reps per treatment; and five test concentrations plus control.

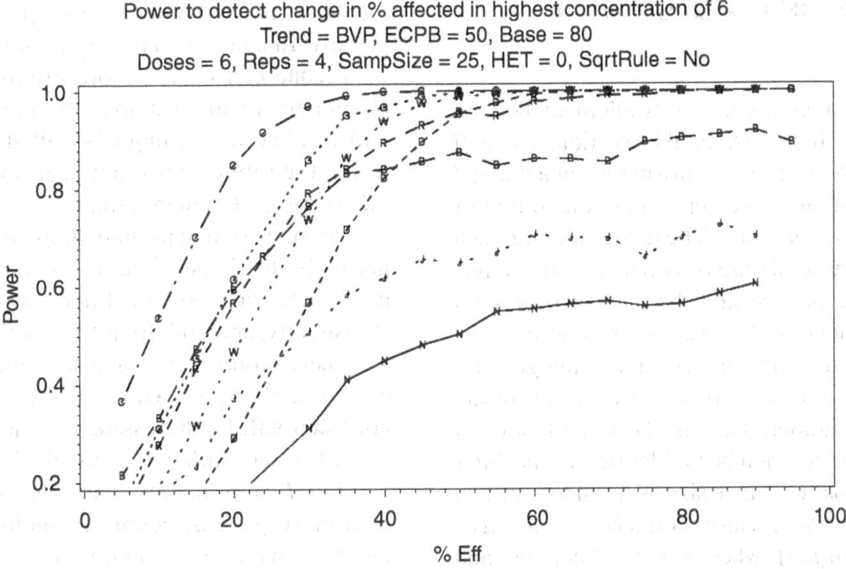

Figure 6.3 Power comparison for no overdispersion, but delayed response. C, Cochran–Armitage; R, Rao–Scott Cochran–Armitage; G, GLMM with Dunnett; W, Williams; J, Jonckheere–Terpstra; D, Dunnett (standard); B, beta-binomial; N, Dunn test; 20% background incidence; four reps per treatment; and five test concentrations plus control.

reduced power when overdispersion is present. This holds true for the very flat shapes (ECPB = 50, 90) as well. Overall, the standard Cochran–Armitage test is the best choice, in terms of power, when there is no indication of overdispersion, and GLMM is best when there is evidence of overdispersion. GLMM is a reasonable choice in all cases. The Rao–Scott modification of the Cochran–Armitage test is also a credible choice. These three are recommended over the tests based on treating replicate proportions as a continuous response, since all of those show considerably lower power in some, but not all, situations.

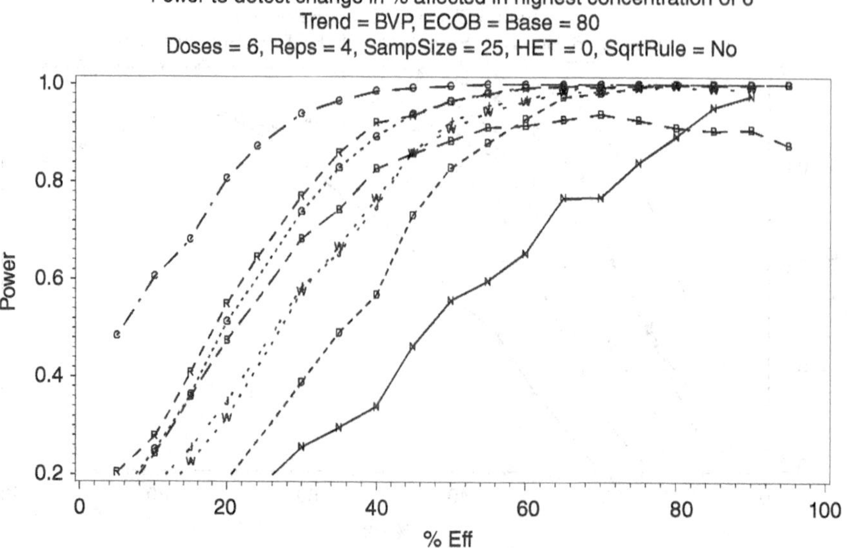

Figure 6.4 Power comparison for 50% overdispersion, mild trend. C, Cochran–Armitage; R, Rao–Scott Cochran–Armitage; G, GLMM with Dunnett; W, Williams; J, Jonckheere–Terpstra; D, Dunnett (standard); B, beta-binomial; N, Dunn test; 20% background incidence; 50% overdispersion; four reps per treatment; and five test concentrations plus control.

6.7 ZERO-INFLATED BINOMIAL RESPONSES

Chapter 13 will consider a general problem of how to analyze mixtures of chemicals. In this section, we will explore a different kind of mixture, arising when a quantal or count variable contains more zeros than a binomial or Poisson distribution can describe. The distribution of such data can be modeled as a mixture of two distributions, one describing a binomial process and the other describing a surplus of zero incidences. An example response is the presence of some fetal abnormality in a teratogenicity study in rats. Pregnant rats are exposed to a test substance. The litter size varies among dams in the same treatment group. The variable is the number of fetuses in the litter exhibiting the abnormality. Litter size is a random effect, and the number of abnormal fetuses could be modeled as a binomial response Bin(n,p), where n is the litter size and p is the probability of abnormality in that litter. Overdispersion is sometimes an issue in such studies, but the focus in this section is on another common problem in many such studies, namely, that the incidence of some abnormalities is low, perhaps rare. Rare events may be difficult to model well with a simple binomial. See Table 6.16 for an example, where there were four treatment groups of 21 or 22 pregnant rats each and litter sizes vary between 6 and 17. There are two responses of possible interest: the prevalence of litters with some fetal abnormality and the prevalence of fetuses with an abnormality. The first is a simple binomial response, which may also exhibit the low incidence phenomenon, though perhaps not as problematically. In addition to having a random component (litter), the second can have a more pronounced low incidence challenge. In the example, the incidence rate of fetuses with an abnormality might be called rare, only 6 out of 265 in the high treatment group and no more than a third of that in lower treatment groups.

No transform of the litter proportion could be considered normally distributed. The Dunn and Jonckheere–Terpstra nonparametric tests of litter proportions are justified statistically, and both found a significant effect in the high treatment group, 5, and nowhere else. It is left as an exercise to verify those two results. In this section, a different approach will be discussed, in which the litter counts are treated as zero-inflated binomials (ZIB).

Let T be a discrete variable, which can be quantal or a more general count. To include the possibility of more zeros than a binomial or Poisson distribution would support, the distribution of T can be described as in Eq. (6.26):

$$\Pr[T=t]=\begin{cases}(1-\omega)+\omega f(0;\beta), & \text{if } t=0\\ \omega f(t;\beta), & \text{if } t\neq 0\end{cases} \tag{6.26}$$

where β is a vector of parameters, f is a distribution function over the nonnegative integers (0, 1, 2, 3, …), and $1-\omega$ is the probability of a zero count over and above that described by f. If $\omega=1$, then there is no excess of zeros, while if $\omega=0$, there are only zeros. The present

Table 6.16 Incidence of Shortened Cervical Ribs in Rat Fetuses

Dam	Group	Affected	Litter	Dam	Group	Affected	Litter	Dam	Group	Affected	Litter
101	1	0	14	215	2	0	11	408	4	0	14
102	1	0	10	216	2	0	13	409	4	0	15
103	1	0	13	217	2	0	12	410	4	0	11
104	1	0	13	218	2	0	13	411	4	1	12
105	1	0	14	219	2	1	11	412	4	0	12
106	1	0	14	220	2	0	9	413	4	0	14
107	1	0	9	221	2	0	15	414	4	0	11
108	1	0	12	301	3	0	12	415	4	0	15
109	1	0	13	302	3	0	10	416	4	0	13
110	1	0	12	303	3	0	14	417	4	0	12
111	1	0	12	304	3	0	13	418	4	0	14
112	1	0	11	305	3	0	12	419	4	0	11
113	1	0	12	306	3	0	16	420	4	0	14
114	1	0	12	307	3	0	8	421	4	0	14
115	1	0	11	308	3	0	10	501	5	0	14
116	1	0	14	309	3	0	12	502	5	1	11
117	1	0	13	310	3	0	12	503	5	3	12
118	1	0	12	311	3	0	15	504	5	0	13
119	1	0	12	312	3	0	11	505	5	0	13
120	1	0	14	313	3	2	12	506	5	1	10
121	1	0	8	314	3	0	14	507	5	0	13
122	1	0	11	315	3	0	12	508	5	0	10
201	2	0	13	316	3	0	13	509	5	0	13
202	2	0	11	317	3	0	14	510	5	0	9
203	2	1	11	318	3	0	14	511	5	0	13
204	2	0	11	319	3	0	13	512	5	0	12
205	2	0	11	320	3	0	14	513	5	0	10
206	2	0	11	321	3	0	12	514	5	0	11
207	2	0	6	322	3	0	12	515	5	0	12
208	2	0	13	401	4	0	12	516	5	1	15
209	2	0	15	402	4	0	13	517	5	0	13
210	2	0	14	403	4	0	14	518	5	0	10
211	2	0	12	404	4	0	14	519	5	0	11
212	2	0	12	405	4	0	11	520	5	0	17
213	2	0	12	406	4	0	8	521	5	0	12
214	2	0	12	407	4	0	14	522	5	0	11

Dam, Code name for pregnant parent; Affected, Number of fetuses in litter with abnormality; Litter, Litter size; Group, Code for treatment group, with group 1 the control.

section is concerned with the case that f is a simple binomial defined by

$$f(t;m,p) = \binom{m}{t} p^t (1-p)^{m-t}, \tag{6.27}$$

and unless otherwise specifically indicated, that is the meaning of $f(t;m,p)$ in this section.

Cohen (1966) developed this general approach and gave special attention to the negative binomial distribution. It has also been applied to Poisson distributions, and that is covered in Chapter 8. Morel and Neerchal (2012) give nice developments of all three distributions just named and provide detailed instructions of how to apply this model in SAS. Yee (2015) gives some information on how to apply such models in R. Hothorn (2016) also gives R implementation for the zero-inflated Poisson distribution, but does not specifically address the binomial case of interest in this section.

Some basic properties of the ZIB will be given followed by an analysis of the data in Table 6.16. Following Cohen (1966), it will be helpful to reparameterize Eq. (6.23) using

$\Theta = \omega\,[1-f(0)]$, where for convenience $f(0;m,p)$ is abbreviated as $f(0)$. Here Θ must be in the range $0 \le \Theta \le 1-f(0)$ and is the probability that some fetus in the litter has an abnormality. Then (1) can be expressed as

$$\Pr[T=t]=\begin{cases}1-\Theta, & \text{if } t=0\\ \dfrac{\Theta}{1-f(0)}f(t;\beta), & \text{if } t\neq 0\end{cases}. \qquad (6.28)$$

Suppose $t_1, t_2, t_3, \ldots, t_N$ is a random sample from this distribution. For the data in Table 6.16, N is the number of litters in a treatment group and can vary across treatments. The likelihood function is given by

$$F=(1-\Theta)^{n_0}\left[\frac{\Theta}{1-(1-p)^m}\right]^n \prod_{t_i\neq 0}\binom{m}{t_i}p^{t_i}(1-p)^{m-t_i}, \qquad (6.29)$$

where n_0 is the number of sample values where $t_i=0$, $n=N-n_0$, and we are suppressing for the moment that the subscript on m that indicates it may vary by litter. From this, the log-likelihood function can be written as

$$\begin{aligned}L=\mathrm{Ln}(F)&=n_0\mathrm{Ln}(1-\Theta)+n\mathrm{Ln}(\Theta)-n\mathrm{Ln}\left[1-(1-p)^m\right]\\ &+\sum_{t_i\neq 0}\mathrm{Ln}\binom{m}{t_i}+\mathrm{Ln}(p)\sum_{t_i\neq 0}[t_i]\\ &+\mathrm{Ln}(1-p)\sum_{t_i\neq 0}(m-t_i).\end{aligned} \qquad (6.30)$$

The normal equations (Appendix A2.3) are obtained by setting to zero the partial derivatives of L with respect to the two parameters Θ and p:

$$\frac{\partial L}{\partial \Theta}=-\frac{n_0}{1-\Theta}+\frac{n}{\Theta}. \qquad (6.31)$$

From setting Eq. (6.31) to zero, one gets

$$\hat{\Theta}=\frac{n}{N}. \qquad (6.32)$$

Also, from Eq. (6.31), it follows that

$$\frac{\partial^2 L}{\partial^2\Theta}=-\frac{n_0}{(1-\Theta)^2}-\frac{n}{\Theta^2} \qquad (6.33)$$

and

$$\frac{\partial^2 L}{\partial\Theta\partial p}=0. \qquad (6.34)$$

The (asymptotic) variance of the estimated parameter vector (Θ, p) is the negative inverse of the matrix of expected values of the second partial derivatives (e.g. Serfling, 1980). From Eq. (6.34) it can be seen that this matrix is diagonal, so the elements of the inverse are just the reciprocals of the diagonal elements. From this, it follows that

$$\mathrm{Var}(\hat{\Theta})=\frac{\hat{\Theta}(1-\hat{\Theta})}{N}. \qquad (6.35)$$

We will also compute the partial derivative of L with respect to p, even though solving the resulting normal equation for p to estimate its mean and variance is more challenging and will not be pursued here.

$$\frac{\partial L}{\partial p}=-\frac{nm_i-(1-p)^{m_i-1}}{1-(1-p)^{m_i}}+\left(\frac{1}{p}-\frac{1}{1-p}\right)\sum_{t_i\neq 0}t_i-\frac{nm_i}{1-p}. \qquad (6.36)$$

Of more interest are the expected value and variance of T.

$$E[T]=\sum_{t_i\neq 0}\omega t\binom{m}{t}p^t(1-p)^{m-t}=\omega E[X]=\omega mp, \qquad (6.37)$$

where $t=0$ makes no contribution to the expected value and $X\sim\mathrm{Bin}(m,p)$. (Here, the subscript i on t has been suppressed but is still implied.) Similarly,

$$\begin{aligned}E[T^2]&=\sum_{t_i\neq 0}\omega t^2\binom{m}{t}p^t(1-p)^{m-t}=\omega E[X^2]\\ &=\omega\left[\mathrm{var}(X)+(E[X])^2\right]=\omega\left[mp(1-p)+(mp)^2\right].\end{aligned} \qquad (6.38)$$

From Eqs. (6.36) and (6.37), it follows that

$$\mathrm{Var}[T]=\omega mp\left[1-p+mp(1-\omega)\right]. \qquad (6.39)$$

Following Morel and Neerchal (2012), if the notation $\pi=\omega p$ and $p=\pi+\alpha(1-\pi)$ is introduced, then Eqs. (6.38) and (6.39) can be expressed as

$$E[T]=m\pi, \qquad (6.40)$$

$$\mathrm{Var}[T]=m\pi(1-\pi)\left[1+\alpha(m-1)\right], \qquad (6.41)$$

which makes clear that this reduces to the ordinary binomial case when $\alpha=0$ (or $p=\pi$). Having laid the theoretical groundwork, attention now turns to implementing the zero-inflated binomial or ZIB model.

To be consistent with the approaches used throughout this text, Equation (6.30) should be applied to the litters in each treatment group separately. Thus N is the number of litters in the treatment group, n_0 is the number of litters in

Table 6.17 Results from ZIB Model for Shortened Cervical Ribs in Rat Fetuses

Dose	Pred	StdEff	Z	Probz	LCCB	UCB	Rank	Adj*P*
5	0.021	0.009	2.394	0.008	0.004	0.039	1	0.042
3	0.009	0.006	1.549	0.061	−0.002	0.020	2	0.243
2	0.008	0.006	1.399	0.081	−0.003	0.019	3	0.243
4	0.004	0.004	1.001	0.158	−0.004	0.011	4	0.317
1	0.000	0.000	0.000	0.500	0.000	0.000	5	0.500

Pred, Predicted proportion affected fetuses in dose; StdErr, Standard error of pred; *Z*, Normal statistic to test hypothesis pred, 0; Probz, Significance of *z*-test; Adj*P*, Bonferroni–Holm adjusted *p*-value; LCB and UCB, Lower and upper 95% confidence bounds on pred; Rank, Rank order of probz values, used in Adj*P* calculations.

that group with no abnormal fetuses, n is the number of litters in that group with at least one affected fetus, t_i is the number of affected fetuses in the ith litter in that treatment group, m_i is the litter size of that litter, Θ is the probability of a litter with a nonzero observation, and p_i is the probability a given fetus in the ith treatment will have an abnormality. Implementing this model presents some challenges. A simple way to implement the ZIB model in SAS is provided by the FMM procedure. See Table 6.17 for the results from that procedure. The program *zib_fmm_macro.sas* using the FMM procedure for this purpose is given in Appendix 1. An alternative program, *zib_nlmix_macro.sas*, is based on the SAS procedure NLMIXED directly using the log-likelihood function in Eq. (6.30) gives similar results and can add a restriction on the probabilities that implies a monotone dose–response. An R program is given in Hilbe (2013).

The results of the ZIB model analysis agree with the results from the step-down Jonckheere–Terpstra and Dunn tests. The ZIB model approach is conceptually more appealing since it takes the litter sizes into account, which neither alternative approach does.

6.8 SURVIVAL- OR AGE-ADJUSTED INCIDENCE RATES

In chronic carcinogenicity studies, some animals die before the end of the study, and because of the long latency period associated with certain types of cancer, incidence is correlated with age, so that older animals are more likely to have one of these cancerous lesions than younger animals independent of any outside agent that might induce such tumors. When animals that die before the end of the experiment are examined for the presence of one of these tumors, some adjustment for the age at death is justified so as not to underestimate the incidence rate associated with exposure to a potential carcinogen. There have been several modifications to the Cochran–Armitage test to make such adjustments that will be discussed in this section.

The poly-3 test developed in Section 6.8.1 and the Peto test in Section 6.8.2 are examples of time to event tests and logically probably belong to Chapter 10. The poly-3 test is presented here because of its connection to the Cochran–Armitage test. The Peto test is presented in this chapter, perhaps even less justifiably, because of its relation to the poly-3 test. The reader may choose to defer these two sections until Chapter 10.

6.8.1 Poly-3 Trend Test

The poly-3 test was developed by Bailer and Portier (1988) and Portier and Bailer (1989). The presentation given here is based in part on Peddada and Kissling (2006), as well as the two cited papers of Bailer and Portier. The notation of Section 6.3.1 is used here. Let y_{ij} be the response on the ith animal in the jth-treatment group, so that $y_{ij}=1$ if that animal has the lesion, and $y_{ij}=0$ otherwise. A weight, w_{ij}, is associated with this observation, defined as 1 if the animal had the lesion at necropsy and as t_{ij}^3 otherwise, where t_{ij} is the proportion of the total duration of the study at the time of death. To justify these weights Bailer and Portier (1988) cite work of Portier et al. (1986) in which evidence is presented that many tumors tend to appear at the rate of a polynomial in time of order 3–5, and this also gave rise to the name of the test they proposed.

The adjusted sample size of the jth treatment is given by

$$\tilde{n}_j = \sum_{i=1}^{n_j} w_{ij}, \tag{6.42}$$

and the poly-3 survival-adjusted tumor incidence rate is given by

$$\hat{\pi}_j = \sum_{i=1}^{n_j} \frac{y_{ij}}{\tilde{n}_j}. \tag{6.43}$$

The poly-3 trend test statistic is then given by Eq. (6.44), which will be recognized as a modified Cochran–Armitage test statistic similar to the Rao–Scott modification but with different weights to address a different issue:

$$T_{\text{poly-3}} = \frac{\sum_{j=0}^{k} a_j d_j \hat{\pi}_j - \dfrac{\sum_{j=0}^{k} a_j d_j \sum_{j=0}^{k} a_j \hat{\pi}_j}{\sum_{j=0}^{k} a_j}}{S\sqrt{\sum_{j=0}^{k} a_j d_j^2 - \dfrac{\left(\sum_{j=0}^{k} a_j d_j\right)^2}{\sum_{j=0}^{k} a_j}}}, \tag{6.44}$$

where

$$a_j = \frac{\tilde{n}_j^2}{n_j},$$

$$S^2 = \frac{\sum_{j=0}^{k}\sum_{i=1}^{n_j}\left(r_{ij} - \bar{r}_j\right)^2}{\sum_{j=0}^{k}\left(n_j - 1\right)},$$

$$\hat{\pi} = \frac{\sum_{j=0}^{k}\sum_{i=0}^{n_j} y_{ij}}{\sum_{j=0}^{k}\tilde{n}_j},$$

$$r_{ij} = y_{ij} - \hat{\pi} w_{ij}$$

$$\bar{r}_j = \frac{\sum_{i=0}^{n_j} r_{ij}}{n_j}. \quad (6.45)$$

According to Bailer and Portier (1988), $T_{\text{poly-3}}$ has an asymptotically standard normal distribution. Bailer and Portier (1988) present some power properties of the poly-3 test and show that the false positive rate can be higher or lower than the nominal under certain conditions. Peddada and Kissling (2006) discuss some alternatives to the poly-3 trend test. The simplest adds a continuity correction factor to the numerator of Eq. (6.44). Another uses a PAVA-type estimator of the treatment proportions, and a third is a hybrid of this last alternative and the poly-3. Moon et al. (2006) provide a software download to implement the last mentioned modification that runs in Windows OS and is self-contained. However, the version associated with the paper does not seem to work. A related R version is available in the MCPAN package, which offers a Dunnett-type comparison of treatments with the control and what the MCPAN documentation calls Williams-type contrasts. Hothorn (2016) provides further discussion of these contrasts. These Williams-type contrasts compare the high dose with the control, then the mean of the two highest doses to the control, then the mean of the three highest doses to the control, and so on. It is not clear how to interpret these contrasts to provide an NOAEL. Consider the situation where the only increase is in highest dose group. All of these contrasts could be significant simply due to the high dose effect. Hothorn (2016) recommends doing both the Dunnett and Williams contrasts. The initial contrast comparing the high dose with the control is reasonably close to what the true Williams' test indicates. The other contrasts are harder to interpret. Young and Morfeld (2016) provide an example of the poly-k test and some power comparisons with other tests.

The poly-3 test will be illustrated with data from a 2-year toxicity test with rats, where initially 72 male rats were exposed to a control and three doses of a chemical and the organ of interest was the liver. A few animals were removed from the experiment for reasons unrelated to the exposure. See Table 6.18 for a summary of the data. The full dataset is given in Appendix 1 in the SAS dataset *mtumors* and provides days on test of each animal. Also in Appendix 1 are programs to carry out the poly-3 test. The SAS program for this purpose is *poly3test_macro.sas*, and the R program follows #Poly-3 test in the code file *Chapter 6A.r*.

See Table 6.19 for illustrations of the calculations in the poly-3 test. The value of the poly-3 test statistic given by Eq. (6.44) was 3.628 when all four dose groups were present, with a p-value of 0.000 143. The test is highly

Table 6.18 Summary of 2-Year Rat Toxicity Test for Liver Tumors

Dose	Dosage	FA	FAM	AC	ACM	N
1	0	14	2	9	1	72
2	25	8	4	8	1	63
3	250	9	3	11	2	62
4	1250	9	0	15	11	71

Dosage is ppm. FA = fibroadenoma, single. FAM = fibroadenoma, multiple. AC = adenocarcinoma, single, ACM = adenocarcinoma, multiple. N is the number of animals observed for tumors.

Table 6.19 Intermediate Calculations for the Poly-3 Test Applied to ACM

Dose	Y_j	$\tilde{n}_j$	n_j	d_j	a_j	π_j	a_jd_j	AD_j	AD_jSqr	$AD\pi_j$	$A\pi_j$	R_j
1	1	48.003	72	1	32.005	0.021	0.021	32.005	32.005	0.667	0.667	−2.675
2	1	46.022	63	2	33.620	0.022	0.022	67.239	134.478	1.461	0.731	−2.523
3	2	47.591	62	3	36.531	0.042	0.042	109.594	328.781	4.606	1.535	−1.643
4	11	54.316	71	4	41.553	0.203	0.203	166.212	664.848	33.661	8.415	6.842

The column headers correspond, for the most part, to the formulas in Eqs. (6.42)–(6.45), except AD_j is the sum of the values of the a_jd_j, AD_jSqr is the sum of the values of the $a_j(d_j)^2$, $AD\pi_j$ is the sum of the values of the $a_jd_j\pi_j$, $A\pi_j$ is the sum of the values of the $a_j\pi_j$, and R_j is the sum over all i of the values of r_{ij} for fixed j. The notation is also consistent with the poly-3 program, *poly3test_MACRO.sas*, provided in Appendix 1. The R function for this purpose, poly3test, is part of the MCPAN package.

significant with all doses present. It is left as an exercise to show that when the high dose is omitted, the resulting poly-3 test statistic is not significant, so the NOEC is dose 3 or 250 ppm. The same result is obtained from the standard Cochran–Armitage test that ignores survival time. That result should not be taken as justification for ignoring survival time. It merely indicates that in this example, survival time had no impact on the conclusion.

6.8.2 Peto Cause of Death Test

There are several ways to adjust tumor incidence to take into account survival rates. One such method, the poly-3 trend test, was presented in Section 6.8.1. An alternative older test, due to Peto et al. (1980), will be presented here. It continues to be used in some toxicology studies, in part because it allows adjustments to tumor incidence not only for survival but also for whether the tumor was incidental or the cause of death.

The following development of Peto's test is based on Gart et al. (1986) and Selwyn (1988). Since there are other tests in the statistical literature also called Peto's test, it is probably best to refer to this as Peto's survival-adjusted cause of death test. Suppose there are J treatment groups with doses $d_1, d_2, \ldots, d_J$ and a negative control (dose $= d_0 = 0$). Denote the times death occurs $t_1 < t_2 < t_3 < \cdots < t_K$. At death time t_K, the data can be summarized in a contingency table (see Table 6.20).

The distribution of $(x_{0k}, x_{1k}, x_{2k}, \ldots, x_{Jk})$ is multivariate hypergeometric under the assumptions that the column totals, n_{jk}, are fixed and the null hypothesis that there are no differences in death rates among the $J+1$ groups. It follows that the expected number of deaths in group j is $e_{jk} = x_{.k}^* n_{jk}/n_{.k}$, and the covariance of x_{jk} and $x_{j'k}$ is given by

$$V_{jj'}(k) = \begin{cases} \dfrac{n_{jk}\left(n_{.k} - n_{jk}\right) x_{.k}\left(n_{.k} - x_{.k}\right)}{n_{.k}^2\left(n_{.k} - 1\right)} & \text{if } j = j' \\ \dfrac{-n_{jk} n_{j'k} x_{.k}\left(n_{.k} - x_{.k}\right)}{n_{.k}^2\left(n_{.k} - 1\right)} & \text{if } j \neq j' \end{cases}. \tag{6.46}$$

Table 6.20 Contingency Table for Death Time t_k

Dose	d_0	d_1	d_2	d_3	…	d_J	Total
Deaths	x_{0k}	x_{1k}	x_{2k}	x_{3k}		x_{Jk}	$x_{.k}$
AtRisk	n_{0k}	n_{1k}	n_{2k}	n_{3k}		n_{Jk}	$n_{.k}$

x_{jk}, number of deaths in group j and time t_k; n_{jk}, number of subjects in group j just prior to time t_k; $x_{.k}$, number of deaths in all groups at time $t_k = x_{0k} + x_{1k} + x_{2k} + \cdots + x_{Jk}$; $n_{.k}$, number of subjects in all groups just prior to time $t_k = n_{0k} + n_{1k} + n_{2k} + \cdots + n_{Jk}$.

The observed and expected numbers of deaths over the entire study in group j are given by

$$O_j = \sum_k x_{jk} \text{ and } E_j = \sum_k e_{jk}. \tag{6.47}$$

The difference between the total observed and expected number of deaths in group j is thus $D_j = O_j - E_j$. If O and E are the column vectors of the O_j and E_j, respectively, then the variance matrix of the zero-mean column vector $O-E$ is given by the matrix V with (j, j') element $\sum_k V_{jj'}(k)$. From this, the log-rank statistic can be calculated as

$$\text{LR} = \left(O - E\right)' V^{-1} \left(O - E\right). \tag{6.48}$$

Under the null hypothesis of no difference in death times among the $J+1$ groups, LR has a chi-squared distribution with k degrees of freedom. To test for a monotone trend in the survival distributions, the statistic

$$T = \frac{D'\left(O - E\right)}{D'VD}, \tag{6.49}$$

where D is the column vector of doses (or dose scores if some other dose metric is used), has an approximate standard normal distribution under the null hypothesis of no difference in survival distributions against the alternative that the survival functions decrease faster with increasing dose. A test for departure from monotone trend in death times is given by

$$Q = \text{LR} - T^2, \tag{6.50}$$

which can be shown to have a chi-squared distribution with $J-1$ degrees of freedom.

In the present context, the test statistics Eqs. (6.48)–(6.50) can be used to test for differences among treatment groups in the incidence of rapidly lethal tumors, provided deaths are restricted to those caused by the tumors. Deaths from other causes would be excluded from the analysis. Obviously, an assessment of the cause of death by appropriately trained toxicologists or pathologists is required, as this is not a statistical judgment.

Equations (6.48)–(6.50) can be modified by giving different weights to different death times, for example, weighting early deaths more heavily than later deaths. Alternatively, one could define strata based on death times. The US National Toxicology Program has recommended intervals 0–52, 53–78, 79–92, and 93–104 plus the terminal kills for a 2-year rodent carcinogenic test (OECD, 2014b), while the USFDA (2001) suggested partitions of 0–50, 51–80, and 81–104 weeks, plus the terminal and any interim kills. Another way to define strata is by mode of death, for example, interim sacrifice, sacrificed in extremis, found dead, and terminal sacrifice. Some combination of these could also be used, as indeed the FDA and NTP methods do. OECD (2014b) contains very interesting discussion of these topics, as well as a comparison of the poly-3 and Peto tests. For the present, we will consider the statistics of including strata in the tests.

Suppose death times have been divided into K_P intervals (strata). Table 6.20 is modified to apply to these intervals and the numbers of subjects at risk is interpreted to mean the number that die in the interval, and the number of deaths is interpreted to be the number of subjects that died in the interval that were found at autopsy to have a tumor of the type of interest. The resulting statistics are LR_P, T_P^2, and Q_P. It may be helpful to use the subscript L for the lethal tumors first developed. So D_L is the vector of differences between observed and expected deaths from the tumor with variance matrix V_L, and D_P and V_P are the corresponding expressions for the sublethal tumors. Then the analysis taking both types of tumors into account is developed from $D_C = D_L + D_P$ with variance $V_C = V_L + V_P$ giving rise to test statistics LR_C, T_C, and Q_C.

This development should make clear some of the issues and, indeed, controversies associated with the Peto test. It is often difficult to determine whether a given tumor was the cause of death for a subject that died and is subsequently found to have a tumor. Comparisons of false positive rates have been made, and professional societies of pathologists have debated the relative merits of these two tests. Another issue is the somewhat arbitrary nature of the strata. The poly-3 test does not suffer from either issue. On the other hand, the poly-3 test makes no distinction between lethal and incidental tumors. The interested reader is referred to OECD (2014a, b, c) and its references for more discussion. Also, further discussion of survival functions, dose scores, and what a trend in survival functions means is deferred to Chapter 10. The present section will indicate how the Peto test can be carried out in SAS and R.

The Peto test will be illustrated with a 2-year rat study, where the survival time of each animal is recorded along with the incidences of two types of renal tumor, labeled bilateral and unilateral. In this study, the presence of these tumors was considered incidental to the death of the rat. It will be shown in Chapter 10 that there was no significant difference in survival times in any treatment group compared with the control. Nonetheless, the possibility that time on test might influence the tumor incidence was explored. In SAS, the MULTTEST procedure can be used to implement the Peto test. The program *Peto test.sas* indicates how this can be done.

There were 66–69 rats per treatment group and four treatment groups of 0 (negative control), 100, 500, and 2500 ppm. All animals that did not die earlier were sacrificed at the end of the study (104 weeks). There were no interim sacrifices, but a few rats were sacrificed in extremis before the end of the study. Rats were individually housed, and because of the rotation of cages over the course of the study, there were no cage effects and individual animal observations were considered independent. See Table 6.21 for the tumor incidence.

Table 6.21 Renal Tumor Incidence in 2-year Rat Study

		Group				
Bilateral	Tumor?	1	2	3	4	Total
	0	62	66	63	65	256
	1	4	2	3	4	13
	Total	66	68	66	69	269
		Group				
Unilateral	**Tumor?**	1	2	3	4	**Total**
	0	62	64	61	57	244
	1	4	4	5	12	25
	Total	66	68	66	69	269

Tumor? = 0 means no tumor; Tumor? = 1 means tumor present; Group = 1 is control.

There is an apparent increase in unilateral tumors in group 4 and no apparent difference in the incidence of either type of tumor otherwise. It will be left as an exercise to apply the Cochran–Armitage test to determine whether that test, which does not take survival times into account, finds a significant effect in either tumor type. For the purpose of illustration, the Peto test will be applied to the tumor data treating all deaths of tumor bearing rats as incidental, not the cause of death. This is done by assigning the value of 1 to each tumor, which is the same as recording the presence of a tumor. As an exercise, repeat the Peto test by assigning a value of 2 to each tumor, that is, replacing each instance of 1 in the two tumor variables by 2.

The SAS code to implement the Peto test has the following syntax. Here the strata variable is commented out, but it can be used to specify additional adjustments. For example, stratum by be defined as 1, 2, 3, 4, 5, or 6 according as DOT ≤ 365, 365 < DOT ≤ 455, 455 < DOT ≤ 546, 546 < DOT ≤ 637, 637 < DOT ≤ 720, or DOT = 728, respectively. Or strata could be defined according as the cause of death was found dead, sacrificed in extremis, terminal sacrifice, or other, respectively. It is interesting to do several versions of this test, according as whether adjustments are made for both types of tumors simultaneously, as in the code shown, or unilateral and bilateral tumors are tested separately, and whether with or without strata to see how sensitive this test can be to the sort of adjustments made. This means the statistician must work closely with the subject matter specialist to assure that the choices made can be supported scientifically.

```
proc multtest data=Chap12.rats notables
out=rats2 stepbon;
  test peto(unilateral bilateral /
  permutation=20 time=DOT uppertailed
  continuity=0.5);
  class Group;
  /*strata stratum;*/
  contrast 'Hi vs Ctrl'       -1  0  1;
  contrast 'Int vs Ctrl'   -1 1  0;
  contrast 'Lo vs Ctrl'   -1  1  0;
  ods output pvalues=rats_pvalues;
  title "Peta Analysis Renal Tumors
  from Rat Data";
run;
```

EXERCISES

In all datasets in the text and exercises, if a concentration is given with the value −1, that means a solvent used in the study and the water control is recorded as −1 and the solvent control is listed as 0. The reader should determine whether to combine the two controls for NOEC determination or EC*x* estimation. If the two controls are to be combined, then the water control should be changed to 0 and the replicates renamed so that replicates in the two controls can be distinguished. If only the solvent control group is to be used, then the water control should be discarded. If only the water control should be used, then the solvent control should be discarded and the concentration for the water control should be changed to 0. Under no circumstances should the concentration value −1 be used for NOEC or EC*x* purposes. Guidance on this is provided in Chapters 2 and 3.

6.1 **(a)** Apply Fisher's exact test to the two renal tumor types in Table 6.21.

(b) Apply the Cochran–Armitage test to these same data.

(c) Apply the poly-3 test to these data.

(d) Repeat (a)–(c) for the tumor data in the datasets mtumor3.sas7bdat and mtumor3c.sas7bdat, which differ only in whether interim sacrifice animals are included. Discuss whether these animals should be included in the analysis.

6.2 **(a)** Do the Peto test for the data in Exercise 6.1 where adjustments are made to both tumor types in Table 6.21 simultaneously, with no strata.

(b) Repeat (a) but with the first definition of strata defined in Section 6.8.2.

(c) Repeat (a) but using the second definition of strata.

(d–f) Repeat the above for unilateral tumors only.

(g–m) Repeat (a)–(g) replacing the Unilateral=1 codes with Unilateral=2, thereby treating the tumor as cause of death.

6.3 A study was done on the Collembola *Folsomia candida* in artificial soil with 5% peat following OECD TG 232. Adult Collembola were exposed to a chemical for 28 days. Survival and reproduction were the responses of interest. This example concerns only survival. There were four replicate test sites of 10 adults in each treatment group and eight replicates in the control. Test concentrations were measured in mg a.i./kg soil. The data are in the spreadsheet Collembola in the Excel file mortality in Appendix 1.

(a) Use Tarone's $C(\alpha)$ test to test the survival data for overdispersion.

(b) Use Fisher's exact test with a Bonferroni–Holm adjustment to determine the NOEC.

(c) Use the Cochran–Armitage or Rao–Scott Cochran–Armitage test (as appropriate) on the counts to determine the NOEC.

(d) Use a GLMM model with binomial error structure assuming overdispersion if appropriate and the GLMM version of Dunnett's test to determine the NOEC.

6.4 A prolonged sediment toxicity test with the midge (*Chironomus riparius*) using spiked natural sediment was done following OECD TG 218. Ten midge larvae were exposed in each replicate test chamber. The response of interest was the number of adults that emerged after 28 days of exposure. There were four replicate vessels per treatment and control and both a water and a solvent control. Test concentrations were measured in mg a.i./kg soil. See Table 6.22 for the data, which are also in the spreadsheet *midge* in the Excel file *mortality* in Appendix 1. A purpose of the study was to determine an NOEC for emergence. These data will be analyzed again in Chapter 7 to estimate EC*x* for several choices of *x*. As a reminder, if the controls are combined based on preliminary investigation, then the reps in one of the controls should be renamed 5, 6, 7, and 8 and the water control concentration should be changed to 0 before analysis to determine an NOEC.

Table 6.22 Midge Emergence

Conc	Group	Rep	Exposed	Emerged	Ratio1
−1	1	1	20	19	0.95
−1	1	2	20	17	0.85
−1	1	3	20	19	0.95
−1	1	4	20	20	1
0	1	1	20	20	1
0	1	2	20	20	1
0	1	3	20	20	1
0	1	4	20	20	1
7.1	2	1	20	19	0.95
7.1	2	2	20	17	0.85
7.1	2	3	20	18	0.9
7.1	2	4	20	20	1
15	3	1	20	20	1
15	3	2	20	20	1
15	3	3	20	20	1
15	3	4	20	19	0.95
25	4	1	20	19	0.95
25	4	2	20	19	0.95
25	4	3	20	20	1
25	4	4	20	14	0.7

Conc=−1 or 0 is the code for water or solvent control when a solvent is present.

(a) One lab analyzed these data as the emergence ratio, which merely means the proportion of the exposed midge that emerged. The test statistic used was called the binomial test using nonlinear interpolation. This very impressive sounding test uses the approximate normal distribution of the test statistic

$$Z = \frac{\hat{p}_1 - \hat{p}_2}{\sqrt{\frac{\hat{p}_1(1-\hat{p}_1)}{n_1} + \frac{\hat{p}_2(1-\hat{p}_2)}{n_2}}},$$

where $\hat{p}_i$ is the observed proportion of midges emerged in treatment i and n_i is the sample size for that treatment. Each treatment is compared with the control independently of the information in other treatment groups. Compare the two controls using the indicated test statistic and combine them or discard the water control for further analysis. Think carefully about the implications of zero mortality in the solvent control. Carry out this test to derive an NOEC.

(b) Do an arc-sine square-root or Tukey transform of the proportions before comparing controls or determining the NOEC. Then do a Dunnett, Tamhane–Dunnett, or Dunn test (depending on what you find regarding normality and variance homogeneity after the transform) to derive an NOEC by pairwise methods.

(c) Repeat but (b) use an appropriate trend test (Williams or Jonckheere–Terpstra) on the transformed proportions.

(d) Analyze the actual counts using a generalized linear mixed model (GLMM) accepting reps and binomial error structure, followed by the GLMM version of Dunnett's test. Be sure to compare the controls using a GLMM model.

(e) Use the Cochran–Armitage or Rao–Scott Cochran–Armitage test on the counts to determine the NOEC.

(f) Discuss the results of these various tests and which are preferable.

ELS Hatching Data

In Exercises 6.5–6.9, rainbow trout eggs were exposed to a test chemical in part to determine hatching success. The experimental design was four replicate tanks of 20 eggs in each test concentration and either four or eight reps in the control.

(a) Plot the proportion surviving joining the mean hatch rate by straight lines segments.

(b) Test for overdispersion using Tarone's $C(\alpha)$ test.

(c) Determine the NOEC using Fisher's exact test with a Bonferroni–Holm adjustment.

(d) Find the NOEC using whichever of the Cochran–Armitage or Rao–Scott Cochran–Armitage test is appropriate.

(e) Determine the NOEC using ANOVA on proportion hatching after a normalizing, variance stabilizing transform.

(f) Determine the NOEC using a GLMM model with binomial error structure and, if appropriate, modeling overdispersion.

The data are in spreadsheets in the mortality Excel file with names ELShatch*x*, where $x = 1, 2, 3, 10$, or 16. These data will be revisited in Chapter 7 using regression models to estimate EC10 and EC20.

6.5 Use ELSHATCH1.

6.6 Use ELSHATCH2.

6.7 Use ELSHATCH3. Note lost rep.

6.8 Use ELSHATCH10.

6.9 Use ELSHATCH16. Note 100% mortality in highest two concentrations and erratic behavior below that.

6.10 Fit zero-inflated binomial (ZIB) models to the data in the following datasets, where possible. The SAS datasets have the names indicated in (a)–(h). The Excel data are in spreadsheets of the same name in the Excel file ZIB Exercises.xls. In most cases, the response variable is labeled "affected," and a title column gives the name to what is counted. For ribs, the response is named fetuses. Critique the model fits. There are two SAS programs that can be used: zib_fmm_macro.sas and zib_nlmix_macro.sas. These do not produce equivalent results. The first is more reliable. The second incorporates a trend feature, but often does not work. The ZIB model does not provide a good fit for some of these data.

(a) Full,

(b) knobby,

(c) Mthoracic,

(d) ribss,

(e) short,

(f) thoracic2,

(g) thoracic3,

(h) wavy.

The Excel files are spreadsheets in the Excel file ZIB.xls.

6.11 Overdispersion

The data for this example are in the spreadsheet *Tamxsex* in the Excel file Chapter 2 data in Appendix 1.

(a) Verify the summary in Table 6.5.

(b) Plot the replicate proportions female versus concentration on an equally spaced *x*-axis scale, i.e. plot replicate proportions against what is called dose in that spreadsheet. Discuss any indications of overdispersion apparent from the plot.

(c) Apply Tarone's $C(\alpha)$ test for overdispersion to these data. The SAS program *TARONE_CALPHA.sas* and the R program *TaroneCAlpha.r* in the code file *Chapter 6a.r* are provided in Appendix 1 to implement Tarone's test.

6.12 Mud Snails

Verify the entries in Table 6.7.

Chapter 7

Analysis of Quantal Data: Regression Models

7.1 INTRODUCTION

This chapter will be concerned with the estimation of effects concentrations, EC/LCx, for quantal responses. The initial focus will be on regression models that can be used for this purpose. As will be demonstrated, not all quantal datasets can be fit by regression models in the traditional understanding of that term, so it will be necessary to develop alternative methods to estimate effects concentrations. Several such methods will be developed. Special attention will be given to the treatment of background incidence, since it is not unusual for control subjects to be affected. An extreme example of that is the analysis of the proportion of some fish population that is phenotypically male or phenotypically female when exposed to an endocrine-disrupting chemical at the embryonic stage. The control population proportions will often be in the range of 0.3–0.7, depending on species and strain, and this can have a profound effect on the kind of model that can be fit. Care will also be given to the frequently observed phenomenon of overdispersion or extra-binomial variance and its impact on analysis.

7.2 PROBIT MODEL

The probit model is based on the cumulative normal distribution function. In particular, the proportion, y, of subjects responding is related to the concentration, U, by means of a "link" function, Φ (the standard normal distribution), as follows:

$$y = \Phi(a + bU). \tag{7.1}$$

This can be rewritten as

$$\Phi^{-1}(y) = a + bU, \tag{7.2}$$

where Φ^{-1} is the function inverse of Φ. This inverse is called the probit of y and can be seen to be a linearizing transform of y. With a background incidence rate, c, included, the model is given by

$$y = c + (1 - c)\Phi(a + bU), \tag{7.3}$$

and

$$\Phi^{-1}\left(\frac{y - c}{1 - c}\right) = a + bU. \tag{7.4}$$

(7.3) and (7.4) are appropriate to model an incidence rate that increases with increased concentration. For example, if mortality is being modeled, then the background incidence rate, c, is the proportion of unexposed subjects expected to die. It is possible to model survival, which would typically be a decreasing function. In that case, the appropriate model is given by

$$y = c\left[1 - \Phi(a + bU)\right]. \tag{7.5}$$

Statistical Analysis of Ecotoxicity Studies, First Edition. John W. Green, Timothy A. Springer, and Henrik Holbech.

Companion website: www.wiley.com/go/Green/StatAnalysEcotoxicStudy

In the case of modeling survival, C is the proportion surviving in the unexposed population. Since our interest will usually be in estimating ECx for some x, it is helpful to determine the relationship between the slope term, b, and ECx. For survival or mortality, this is given by

$$\text{EC}x = \frac{\Phi^{-1}\left[1-(x/100)\right]-a}{b}. \tag{7.6}$$

In practice, U is usually replaced by Ln(U), in which case (7.5) is replaced by

$$\text{EC}x = \exp\left[\frac{\Phi^{-1}(x/100)-a}{b}\right], \tag{7.7}$$

where $\exp(z) = e^z$, or U is replaced by $\text{Log}_{10}(U)$ and

$$\text{EC}x = 10^{\left(\frac{\Phi^{-1}(x/100)-a}{b}\right)}, \tag{7.8}$$

if common logs are used.

It will be observed that the formulas (7.5) and (7.7) do not depend directly on c, but the estimates of the parameters a and b are impacted by c, so ECx does depend on c indirectly. While using Ln(U) usually gives a good fit to the data, one should be open to fitting a probit to untransformed concentrations as well. The discussion in Collett (2002) on this point is of interest. Individual subjects are not expected to have the same tolerance to a given level of the test chemical. Instead, these tolerances will have some distribution modeled to center on the fitted probit curve defined by (7.1) or (7.3). The probit model assumes these tolerances follow a binomial distribution for each concentration of the test substance. Since (7.1)–(7.4) are defined in terms of the proportion of subjects affected, it may not be evident how sample size or the binomial distribution is taken into account. The fitting of the probit model in some popular software, such as SAS Proc Probit, is based on a binomial distribution of the number affected at each test concentration and therefore captures the sample sizes and their influence on confidence interval width for model parameters and estimated probabilities. The log-likelihood function is given by

$$L = \sum_i \text{Ln}(y_i) = \sum_i \left[x_i \text{Ln}(y_i) + (n_i - x_i)\text{Ln}(1-y_i)\right], \tag{7.9}$$

where n_i is the number at risk in treatment i and x_i is the number affected, with $i = 0$ corresponding to the control. As usual, the parameter estimates are obtained by maximizing the log-likelihood function, given the observed values of n_i and x_i after substituting (7.3) for y_i evaluated at the concentration U_i. In the standard or "classical" probit model fit, the replicate structure is ignored, and n_i and x_i are treatment totals. Next, many software packages, such as SAS Proc Probit, compute the variance matrix of the parameter estimates as the "information matrix" of second partial derivatives of the likelihood function evaluated at the estimate parameter vector. This does not affect the point estimates, but does affect the standard error and confidence bounds. More information on this approach is given in Appendix 2. More realistic results can be obtained using the binomial error structure and replicate information, and this will be developed in Section 7.2.3.

Since our primary interest is in obtaining point and interval estimates of ECx, it will be convenient with some model fitting software to reparameterize the model so as to make ECx one of the parameters to be estimated. Other packages, e.g. SAS PROBIT and NLMIXED procedures, will allow estimation of ECx and its confidence bounds without that. It is left as an exercise to demonstrate that (7.1) can be reparameterized as

$$y = \Phi\left(a + \frac{\Phi^{-1}(x/100)-a}{\text{EC}x}U\right) \tag{7.10}$$

where x is given and a and ECx are to be estimated from the data. It is easy to show from (7.10) that $a = \Phi(y_0)$. Model (7.4) can be reparameterized as

$$y = c + (1-c)*\Phi\left(a + \frac{\Phi^{-1}(x/100)-a}{\log(\text{EC}x)}\log(U)\right) \quad \text{for } U > 0,$$

$$y = c \quad \text{for } U = 0. \tag{7.11}$$

It does not matter whether natural or common logs are used, except that certain regulatory requirements are written in terms of the slope term in (7.3) (with Log(U) in place of U) expecting a common (or base 10) logarithm to be used. (7.11) defines a continuous and differentiable function at $U \geq 0$. The value of $\Phi(y_0)$ is c and the meaning of the parameter a is no longer of much interest, but is given by (7.12). It is left as an exercise to verify that

$$a = \Phi^{-1}\left(\frac{y(1)-c}{1-c}\right). \tag{7.12}$$

To understand the quality of the estimated probabilities, it is necessary to compute confidence bounds for the predicted values of y from (7.4) and (7.2). The delta method, discussed in Section A2.6, can be used for this purpose. Some software does this without the need to do one's own calculations. The programs *Probit_mites.sas* and *probit.r* are such programs.

7.2.1 Probit Example

The data below were modified from an experiment on mites. Mites were exposed to varying concentrations (g a.s. ha^{-1}) of a herbicide as part of a risk assessment for product registration. Each housing unit (REP) consists of a frame with a glass plate at the top and bottom of the frame. The number (RISK) of mites in each rep was supposed to be 5, but a counting problem was discovered showing that two replicates actually contained 6 mites. Herbicide residue is sprayed on the inner side of each glass plate. For the control, water is sprayed on the inner plate surface. After the plates dry, mite protonymphs are placed between the plates. Fresh air is circulated within the frame by an air pump. The mites are examined seven days after exposure begins. See Table 7.1 for the data from this experiment. ALIVE is the number of mites alive at the end of the experimental period. The herbicide concentrations were in ppm. There were nominally 20 mites per concentration, including control. Due to initial counting problems, two groups actually included 21 mites. The only difference between this example and the first quantal example is the control mortality. The purpose of this example is to indicate the effect control mortality has on the power of tests and the models fit to the data.

For these data, there is a background mortality rate of 10%. It is advisable to assess these data also for overdispersion. Tarone's $C(\alpha)$ test developed in Chapter 6 cannot be used on the full dataset because all of the replicates in the high concentration had 100% mortality. It is left as an exercise to show that Tarone's test applied to the data without the high concentration is not significant. An alternative approach to assessing overdispersion will be discussed in Section 7.2.3 in the context of generalized linear models.

See Table 7.2 for a summary of the fit of the probit model with background mortality to the mite mortality data. SAS and R programs for fitting the model are given in Appendix 1. A SAS program for probit analysis of both mortality and survival is *probit_nlmixed_basic.sas*.

To assess the goodness of fit, the estimated mortalities are compared to the observed values at each test concentration. See Table 7.3, where it will be observed that the agreement between the two is excellent. Next, see Table 7.4 for the ECx estimates for various choices of x. It should be clear from Table 7.4 that the estimates of EC1, EC5, and EC10 are indistinguishable in the sense that the confidence intervals for all three have considerable overlap and in fact span the two lowest test concentrations. The confidence interval for the EC20 estimate also shows considerable overlap with those for

Table 7.1 Mites Exposed to Herbicide

Conc	Rep	Risk	Alive	Conc	Rep	Risk	Alive
0	1	5	3	75	1	5	4
0	2	5	5	75	2	6	2
0	3	5	5	75	3	5	3
0	4	5	5	75	4	5	2
18.75	1	5	5	150	1	5	1
18.75	2	5	5	150	2	5	1
18.75	3	5	5	150	3	5	0
18.75	4	5	3	150	4	5	0
37.5	1	5	5	300	1	5	0
37.5	2	6	6	300	2	5	0
37.5	3	5	5	300	3	5	0
37.5	4	5	2	300	4	5	0

Conc, concentration (g a.s. ha^{-1}); Risk, number of mites in indicated replicate vessel (Rep); Alive, number of mites alive at end of study.

Table 7.3 Comparison of Observed and Probit Predicted Incidence Rates for Mite Mortality

Conc	Risk	Alive	ObsP	Pred	LCB	UCB
0	20	18	0.10	0.10	0.01	0.19
18.75	20	18	0.10	0.10	0.10	0.22
37.5	21	18	0.14	0.14	0.10	0.40
75	21	11	0.48	0.47	0.28	0.69
150	20	2	0.90	0.91	0.74	0.98
300	20	0	1.00	1.00	0.92	1.00

LCB and UCB are the 95% lower and upper confidence bounds; ObsP and Pred are the observed and predicted mortalities.

Table 7.2 Parameter Estimates for Mite Probit Model for Mortality

Parameter	Est	LCB	UCB	ChiSq	Prob>ChiSq	DF	StdErr
Intercept	−9.46	−14.63	−4.30	12.89	0.0003	1	2.6358
Log_{10}(Conc)	4.93	2.36	7.50	14.09	0.0002	1	1.3134
_C_	0.10	0.01	0.19			1	0.0453

Est is the estimated coefficient of the indicated model parameter; StdErr, LCB, and UCB are the standard error and confidence bounds of the estimate; ChiSq is the test statistics used to test whether the coefficient is significantly different from zero; Prob>ChiSq is the probability of obtaining so large a value of ChiSq by chance, so values less than 0.05 indicate the estimate is significantly different from zero.

EC1, EC5, and EC10. There is clearly much uncertainty about ECx estimates for small values of x. To improve the estimates, one must either increase replication or find a better model. On this last point, it should be noted that probit analysis, like the Cochran–Armitage test, ignores replicates.

Table 7.4 Comparison of ECx Estimates from (7.3) Applied to Both Mortality and Survival

Mortality				Survival			
X	ECx	LCB	UCB	X	ECx	LCB	UCB
1	28	7	45	99	11	3	18
5	39	13	56	95	16	7	25
10	46	18	63	90	21	10	30
20	56	28	74	80	27	16	39
30	65	37	83	70	33	21	47
40	74	47	93	60	40	27	57
50	83	58	104	50	47	32	69

A 10% mortality rate is equivalent to a 90% survival rate; In general, an x% mortality rate is equivalent to a $(100-x)$% survival rate; Clearly ECx(mortality) is not the same as EC$(100-x)$ for survival; LCB and UCB are the lower and upper 95% confidence bounds on the indicated ECx estimate.

7.2.2 Model Mortality or Survival?

In the mite example of Section 7.2.1, mortality was modeled. A natural question is whether it makes any difference whether mortality or survival is modeled. Logically, the two approaches are equivalent, since if one knows the mortality rate, then one can calculate the survival rate by subtracting the mortality rate from 100 (assuming rates are expressed as percentages), and the reverse is also true. (7.3) cannot be used to model survival data, since the background survival rate is an upper bound (on survival), whereas in modeling mortality, the background incidence rate is a lower bound (on mortality). If (7.3) is used for survival, the algorithm used in fitting the model has a problem. See Figure 7.1 and Table 7.4 to illustrate this problem.

Figure 7.1 and Tables 7.3 and 7.4 are produced by the Probit procedure in SAS, and the variability in C is not fully reflected. To obtain plots and probability and parameter bounds that fully reflect the uncertainties in the data, it is necessary in SAS to use the NLMIXED procedure. This does not change the point estimates from a model of mortality, but it does have minor impact on the bounds. The SAS programs *probit_nlmixed_basic.sas* and *probit_mites.sas* are provided in Appendix 1. More importantly, Proc NLMIXED allows (7.5) to be implemented for survival. See Figure 7.2 and Table 7.5 for comparisons of NLMIXED

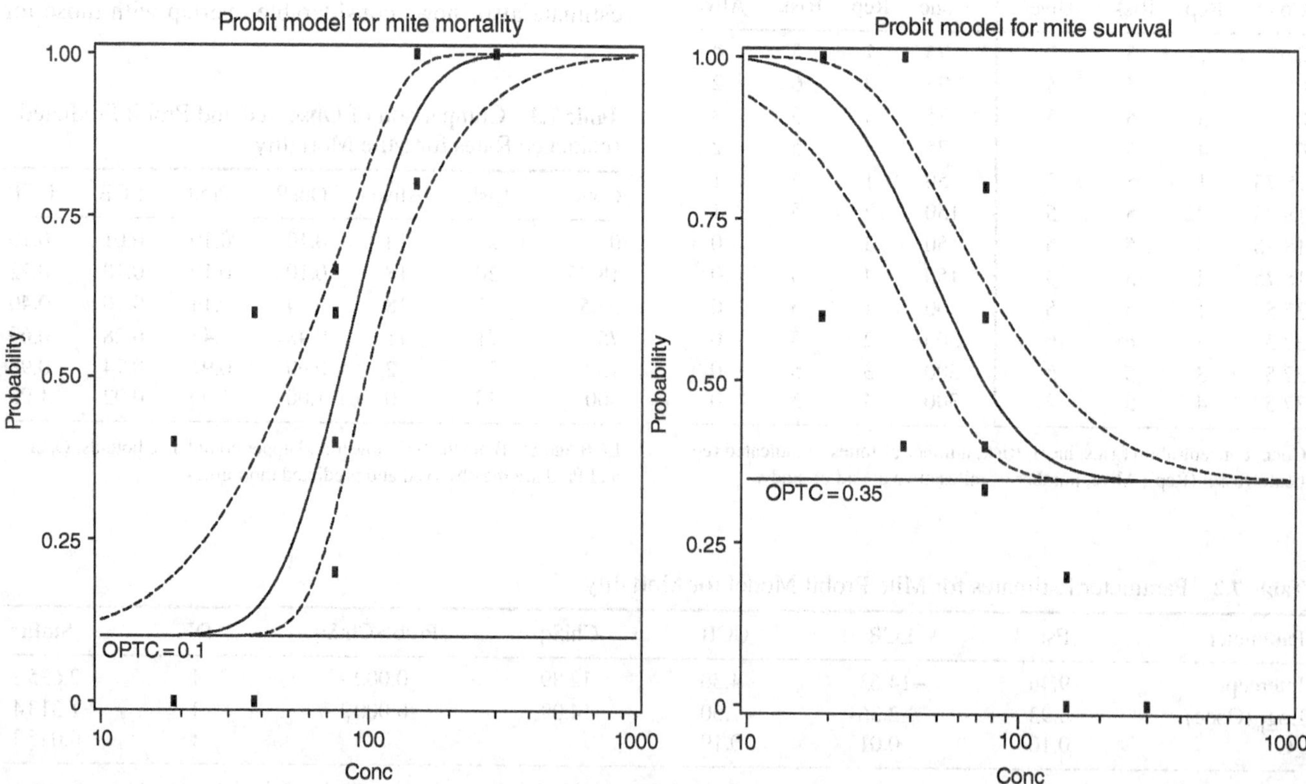

Figure 7.1 Model (7.3) used for both mortality and survival. Left panel models mortality and right panel models survival for the same mite data, both using model (7.3). This makes clear that (7.3) is not suitable for modeling survival.

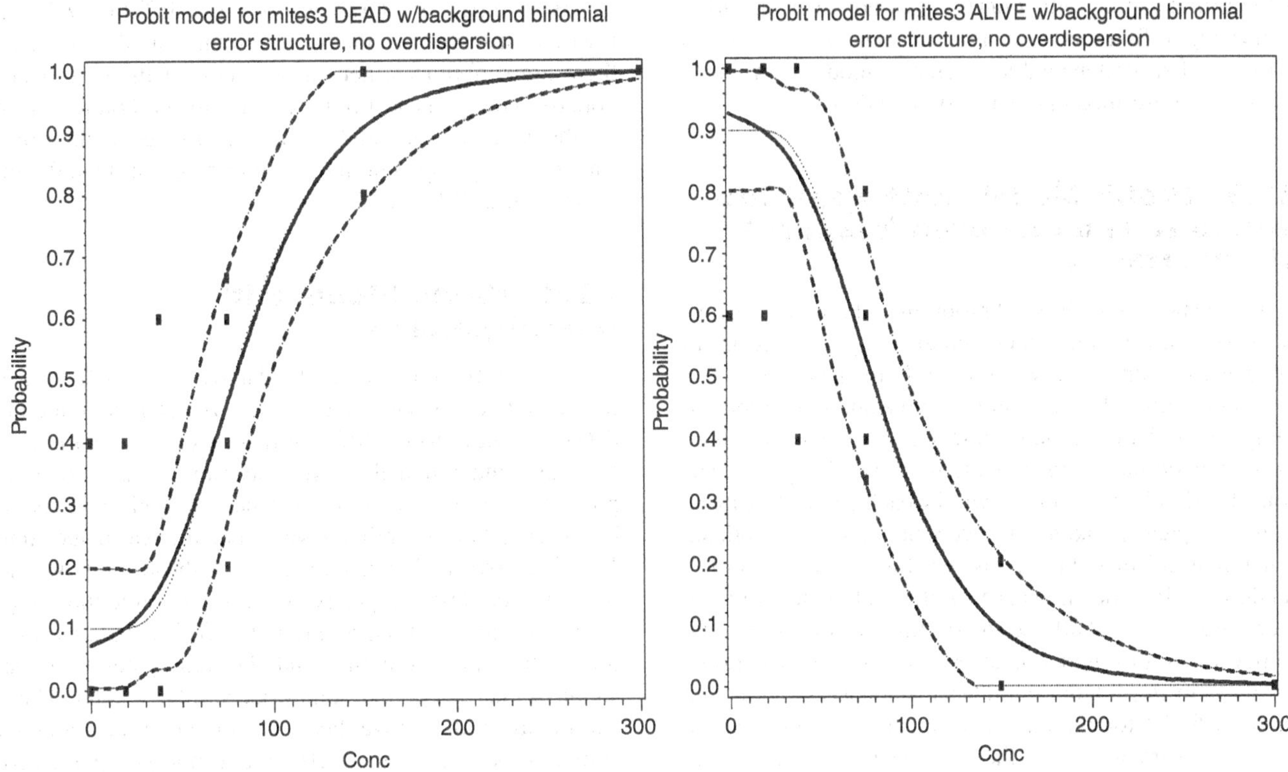

Figure 7.2 Comparison of probit model (7.3) for mite mortality and (7.5) for survival. Left and right panels for mite data mortality and survival using models (7.3) and (7.5), respectively. Model (7.5) captures background survival properly.

Table 7.5 Comparison of EC*x* Estimates from Model (7.3) for Mite Mortality and (7.5) for Survival

	Mortality				Survival		
X	EC*x*	LCB	UCB	*X*	EC*x*	LCB	UCB
1	28	7	49	1	28	7	49
5	39	16	61	5	39	16	61
10	46	22	69	10	46	22	69
20	56	33	79	20	56	33	79
30	65	42	87	30	65	42	87
40	74	52	96	40	74	52	96
50	83	61	105	50	83	61	105

A 10% mortality rate is equivalent to a 90% survival rate; In general, an *x*% mortality rate is equivalent to a (100−*x*)% survival rate; EC*x*(mortality) is the same as EC*x*(survival); LCB and UCB are the lower and upper 95% confidence bounds on the indicated EC*x* estimate.

output for models (7.3) and (7.5) applied to mite mortality and survival, respectively.

When there is background incidence, a term for it should be included. Note that these statements apply to probit models of quantal data. Similar statements apply to logit models for the same type of data. Figure 7.2 and Table 7.5 are derived from (7.3) for mortality and (7.5) for survival.

The results in Figure 7.2 and Table 7.5 are produced by the NLMIXED procedure in SAS, and the uncertainties in all parameters, including c, are captured. It should be noted that SAS Proc Probit can correctly model only mortality, and in doing so, the uncertainty of the parameter c is not fully captured. Proc Probit reports the same point estimates as NLMIXED for mortality (but not for survival!), but the confidence bounds, while similar, are not the same.

It is also possible to fit a probit model using U rather than Log(U). This is rarely done, and for good reason, though a few datasets have been encountered over a career spanning several decades where this proved advantageous.

If the probit model with Log(U) does not fit well, rather than trying to fit a model with U, it is generally wise to go with one of several available alternative models. Two such models are described in the next two sections.

7.2.3 Probit Models with Replicate Structure and Improved Treatment of Uncertainty

As noted above, the classical probit model ignores the replicate structure of the experimental design and treats all subjects as independently observed. In the past, this was a limitation caused by difficulty to carry out some types of computation. Developments in recent years make possible more realistic modeling. In particular, the binary distribution of individual responses within each replicate can be more completely taken into account, as is the replicate structure itself. In addition, it is possible to test for overdispersion and include allowance for that in the model where necessary. As we shall see, these have no impact on the point estimates of model parameters and ECx calculations, but they do impact the confidence bounds for both. See Tables 7.6–7.8 for the parameter estimates, probabilities, and ECx estimates when the improved treatment of the binomial error structure is taken into account. In Section 7.2.4, overdispersion will be examined. The previously cited programs using the NLMIXED procedure in SAS provide this functionality.

By comparison of Tables 7.6–7.8 with Tables 7.2–7.4, it will be observed that the point estimates are the same, but there are differences, generally minor, in the confidence bounds. The greatest differences are in the bounds given for the estimated probabilities at very low concentrations. That is readily apparent in the comparisons of the left panels in Figures 7.1 and 7.2.

7.2.4 Probit Model with Overdispersion

Oris et al. (2012) investigated a database of 102 fish early life stage (ELS) studies done under OECD Test Guideline (TG) 210 with several different fish species using Tarone's $C(\alpha)$ test and found that overdispersion "was frequently present, consistent with the presence of significant chamber-to-chamber variability. Overdispersion was detected in 14.3%, 28.6%, and 40.0% of Danio, fathead minnow, and rainbow trout tests, respectively." The present authors supplemented this with an additional 110 ELS studies done under the same guideline and 47 multigeneration fish studies done and found somewhat similar results. These additional studies have been presented at professional conferences (e.g. Green, 2013) but not yet published. Furthermore, data from other OECD guideline studies also show that overdispersion is not uncommon. It is therefore important to accommodate overdispersion in the analysis of quantal data, at least where the data indicate its presence.

Table 7.6 Parameter Estimates for Mite Probit Model for Mortality with Replicates

Param	Est	LCB	UCB	t Value	Prob > |t|	StdErr	DF
a	−9.46	−15.08	−3.84	−3.59	0.003	2.64	15
Log_{10}(Conc)	4.93	2.13	7.73	3.75	0.002	1.31	15
c	0.10	0.004	0.2	2.22	0.042	0.05	15

Est is the estimated coefficient of the indicated model parameter (Param); StdErr, LCB, and UCB are the standard error and confidence bounds of the estimate; t Value is the test statistics used to test whether the coefficient is significantly different from 0. Prob > |t| is the probability of obtaining so large a value of the t-statistic by chance, so values less than 0.05 indicate the estimate is significantly different from zero.

Table 7.7 Comparison of Observed and Probit Predicted Incidence Rates for Mite Mortality

Conc	Risk	Alive	ObsP	Pred	LCB	UCB
0	20	18	0.10	0.10	0.004	0.2
18.75	20	18	0.10	0.10	0.007	0.2
37.5	21	18	0.14	0.14	0.03	0.25
75	21	11	0.48	0.47	0.27	0.67
150	20	2	0.90	0.91	0.79	1.00
300	20	0	1.00	1.00	0.99	1.00

LCB and UCB are the 95% lower and upper confidence bounds; ObsP and Pred are the observed and predicted mortalities.

Table 7.8 ECx Estimates from Probit Model with Replicates Taken into Account

X	ECxEst	LCB	UCB
1	28.01	7.36	48.66
5	38.51	16.39	60.63
10	45.63	23.21	68.05
20	56.04	33.82	78.26
30	64.99	43.20	86.78
50	83.02	61.46	104.58

Binary error structure fully implemented with replicates included in model; X is % effect; ECxEst is the estimated effects concentration; LCB and UCB are the lower and upper bounds of the 95% confidence interval.

Table 7.9 Emergence Data from a Nontarget Terrestrial Plant Study (OECD TG 208) of Peas

Rate	Dose	Rep	Emerge	Survive	Wgt
0	1	1	10	10	2.29
0	1	2	10	10	1.43
0	1	3	10	8	1.98
0	1	4	10	9	1.46
0.051	2	1	10	10	2.1
0.051	2	2	8	8	1.88
0.051	2	3	8	7	1.72
0.051	2	4	7	5	0.99
0.103	3	1	8	8	1.65
0.103	3	2	8	8	1.86
0.103	3	3	9	8	1.23
0.103	3	4	9	5	1.22
0.205	4	1	10	10	0.77
0.205	4	2	10	10	1.76
0.205	4	3	9	8	1.95
0.205	4	4	10	10	1.83
0.41	5	1	8	7	1.4
0.41	5	2	6	6	1.57
0.41	5	3	9	8	1.54
0.41	5	4	6	3	0.51
0.82	6	1	9	8	1.37
0.82	6	2	6	5	0.53
0.82	6	3	3	3	0.96
0.82	6	4	7	5	0.37
1.64	7	1	5	5	0.91
1.64	7	2	4	4	0.52
1.64	7	3	6	3	0.66
1.64	7	4	7	3	0.44
3.28	8	1	4	4	0.58
3.28	8	2	7	7	0.34
3.28	8	3	3	3	1.27
3.28	8	4	4	4	0.65
6.56	9	1	2	2	0.06
6.56	9	2	1	0	0
6.56	9	3	0	.	.
6.56	9	4	0	.	.
13.13	10	1	1	1	0.1
13.13	10	2	5	2	0.15
13.13	10	3	4	2	0.04
13.13	10	4	1	1	0.09
26.25	11	1	0	.	.
26.25	11	2	3	1	0.57
26.25	11	3	0	.	.
26.25	11	4	1	0	0

Rate, application rate in g a.s. ha^{-1}; Emerge, number of seedlings that emerged out of 10 planted; Survive, number of merged seedlings that survived to the end of the study (=. if none emerged); Wgt, total dry weight of plants at end of study in grams.

Table 7.10 Parameter Estimates from NTTP Study of Peas, with Overdispersion

Parameter	Estimate	StdErr	Lower	Upper	DF	t Value	Prob>|t|
a	−0.20	0.13	−0.49	0.08	10	−1.57	0.15
B	0.99	0.15	0.65	1.33	10	6.48	<0.0001
c	0.00	0.00	0.00	0.00	10	0.03	0.98
s2	0.11	0.08	−0.08	0.30	10	1.3	0.22

Estimate is the estimated coefficient of the indicated model parameter (parameter); StdErr, Lower, and Upper are the standard error and 95% confidence bounds of the estimate; t Value is the test statistics used to test whether the coefficient is significantly different from 0; Prob>|t| is the probability of obtaining so large a value of the t-statistic by chance, so values less than 0.05 indicate the estimate is significantly different from zero.

Table 7.11 Comparison of Observed and Probit Predicted Emergence Rates When Overdispersion Is Modeled

Conc	ObsP	Pred	LCB	UCB
0	0.00	0.00	0.00	0.00
0.051	0.18	0.13	0.02	0.24
0.103	0.15	0.14	0.03	0.24
0.205	0.03	0.08	0.00	0.18
0.41	0.28	0.28	0.14	0.41
0.82	0.38	0.38	0.23	0.53
1.64	0.45	0.46	0.31	0.62
3.28	0.55	0.57	0.41	0.72
6.56	0.93	0.86	0.74	0.99
13.13	0.73	0.75	0.62	0.89
26.25	0.90	0.89	0.80	0.99

LCB and UCB are the 95% lower and upper confidence bounds; ObsP and Pred are the observed and predicted mortalities.

Such analysis will be illustrated with a nontarget terrestrial plant study with peas. See Table 7.9 for the data, Table 7.10 for model parameter estimates, Table 7.11 for a comparison of observed and predicted emergence rates, and Table 7.12 for ECx estimates for selected values of x.

Table 7.12 indicates that the estimates for ECx for $x \leq 10$ are unreliable, given their lower confidence bounds are negative. It is also interesting that in Table 7.11, the predicted emergence at the application rate 0.205 g a.s. ha^{-1} stands out as violating the otherwise monotone dose–response. The observed response at that rate was likewise very different from those on either side of it. This could result from overdispersion. On the other hand, the overdispersion parameter s2 is not significantly different from zero.

Table 7.12 ECx Estimates from Probit Model When Overdispersion Is Modeled

X	ECxEst	Lower	Upper
1	0.01	−0.01	0.02
5	0.03	−0.01	0.08
10	0.08	−0.01	0.17
20	0.23	0.03	0.42
30	0.47	0.12	0.82
50	1.60	0.52	2.69

ECxEST is the estimated effects concentration; Lower and upper are the lower and upper bounds of the 95% confidence interval.

7.3 WEIBULL MODEL

The Weibull distribution is not necessarily symmetrical and is usually applied to the concentrations themselves (not their logarithms). This model (without the background mortality parameter *a*) may be expressed as

$$y = 1 - \exp\left[-\left(\frac{U}{b}\right)^d\right], \tag{7.13}$$

where as with the probit model, *y* is the proportion affected in each replicate. The Weibull model has two parameters, a location parameter *b* and a shape parameter *d* (high values of *d* give steep slope). The EC*x* is related to *b* and *d* by

$$\text{EC}x = b\ln\left(\frac{1}{r}\right)^{1/d}, \tag{7.14}$$

where $r = x/100$. While it is possible to obtain an approximate variance of the EC*x* estimate from (7.14) using the delta method, it is simpler to reparameterize the model to make EC*x* a model parameter and obtain the variance and confidence interval during the model fitting process. The reparameterized model is given by (7.15):

$$y = 1 - \exp\left[-\left(\frac{U}{\text{EC}x}\right)^d \log\left(\frac{1}{r}\right)\right]. \tag{7.15}$$

When background incidence rate is to be estimated, the model is

$$y = 1 - (1 - c)\exp\left[-\left(\frac{U}{\text{EC}x}\right)^d \log\left(\frac{1}{r}\right)\right]. \tag{7.16}$$

(7.15) and (7.16) must be modified in a straightforward way if *U* is replaced by Log(*U*), but this is not commonly done for this model. See Tables 7.13 and 7.14 for results

Table 7.13 Weibull Model Summary for Mite Mortality

Parameter	Estimate	LCB	UCB
d	−3.22	−5.39	−1.06
EC20	56.28	36.61	75.95
C	0.11	0.03	0.20

LCB and UCB are the 95% lower and upper confidence bounds, respectively, of the model parameter estimates.

Table 7.14 Comparison of Observed and Weibull Predicted Incidence Rates for Mite Mortality

Conc	ObsP	Pred	LCB	UCB
0	0.10	0.11	0.03	0.20
18.75	0.10	0.11	0.03	0.20
37.5	0.15	0.12	0.03	0.20
75	0.47	0.47	0.24	0.70
150	0.90	0.92	0.81	1.00
300	1.00	0.99	0.97	1.00

ObsP and Pred are the observed and Weibull model predicted proportion dead; LCB and UCB are the 95% lower and upper confidence bounds, respectively, of the predictions.

from a fit of the Weibull model to the mite mortality data. It will be observed from Table 7.13 that all model parameters are significantly different from zero. The confidence bounds all are based on a binary error structure that includes proper treatment of replicates.

Table 7.14 gives the mortality proportions estimated from the Weibull model. These estimates are comparable to those from the probit model. See Figure 7.3 for a display of the data, model, and confidence bounds, taking the uncertainty of the estimated background incidence into account. Programs to generate the model fit, estimates, and plot are given in Appendix 1. In addition to the programs previously cited, the programs *Weibull_MITES.sas* and Weibull.r fit the Weibull model. Alternative ways to fit a Weibull model can be based on the R function *survreg* from the package *Survival* for fitting a Weibull model illustrated in https://stat.ethz.ch/R-manual/R-devel/library/survival/html/survreg.html (accessed 3 September 2017).

It is left as an exercise to compute EC*x* estimates and their confidence bounds for the Weibull model of the mite data.

7.4 LOGISTIC MODEL

In the logistic model, the proportion, *y*, of subjects responding is related to the concentration, *U*, by means of a "link" function as follows:

$$\text{Logit}(y) = \log\left(\frac{y}{1-y}\right)$$

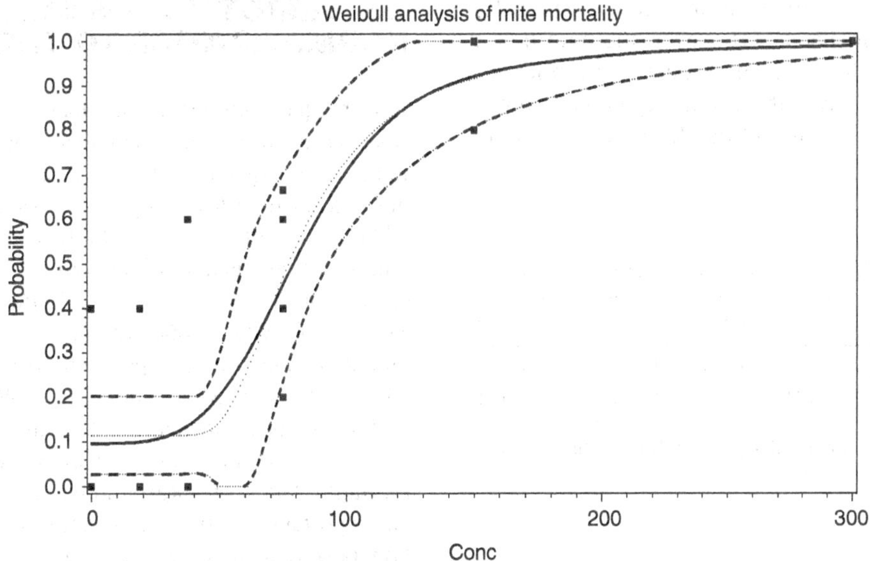

Figure 7.3 Weibull model for mite mortality. Solid squares are actual replicate proportions dead. Dashed lines are 95% lower and upper confidence bounds for model predictions. Solid line is model prediction.

and the model is taken to be

$$\text{Logit}(y) = a + bU, \tag{7.17}$$

where U is concentration.

The logit model is based on the cumulative logistic distribution function. The logistic distribution has slightly wider tails than the normal distribution underlying the probit model but is similar otherwise. Just as the probit model, the logit model is usually applied to the log concentrations. The "logit" is another linearizing transform of the proportion of the population responding, analogous to the Probit.

In theory, not the logit or the probit can linearize the proportion responding for the same dataset, as they are different functions. In application, it is often difficult to determine which model better describes a given dataset. We must choose between them (and among other models as well). The logistic model is much simpler mathematically and is favored by theorists. Nevertheless, the probit is by far the most commonly used model in ecotoxicology for quantal data. It is simple to rewrite (7.17) as

$$y = \frac{e^{a+bU}}{1+e^{a+bU}}. \tag{7.18}$$

It is left as an exercise to show that

$$y_0 = \frac{e^a}{1+e^a} \tag{7.19}$$

$$\text{EC}x = \frac{1}{b}\text{Ln}\left(\frac{1-p}{1+pe^a}\right),$$

where $p=x/100$. (7.18) can be solved for a and b to show that

$$a = \text{Ln}\left(\frac{y_0}{1-y_0}\right) \quad \text{and}$$
$$b = \frac{1}{\text{EC}x}\text{Ln}\left(\frac{1-p}{1+pe^a}\right). \tag{7.20}$$

(7.20) can be substituted into (7.18) to reparameterize the model in terms of y_0 and ECx. Background incidence is built in since

$$y_0 = \frac{e^a}{1+e^a}, \quad \text{which cannot } be \text{ zero.}$$

However, this does not assure an estimated background incidence rate that agrees with the observed rate, as will be seen with the mite data. As with the previous two models, this is useful to obtain both point and interval estimates of ECx from model fitting programs. Appendix 1 provides *probit_nlmixed_basic.sas* and *logit.r* for this purpose. It will be noted that y_0 cannot be 0 or 1.

As with the probit model, the concentration U in (7.18) and (7.19) can be replaced by Ln(U), but there is less reason for doing so than with the probit. The logistic model will be illustrated with the mite data using the SAS program *probit_nlmixed.sas*. See Tables 7.15 and 7.16 and Figure 7.4 for summaries of the results. It will be observed that the point estimates of mortality probabilities at tested concentrations are not as close to the observed proportions as with the probit model, especially for the background incidence. However, the confidence bounds, which all include proper treatment of replicates, tend to be tighter, and given the variability in the control, the lower predicted background incidence may be more reasonable.

Table 7.15 Logistic Model Summary for Mite Mortality

Parameter	Estimate	LCB	UCB	StdErr	DF	*t* Value	Prob>|*t*|
*y*0	0.07	0.01	0.12	0.03	24	2.29	0.0311
EC10	30.52	20.14	40.91	5.03	24	6.07	<0.0001

LCB and UCB are the 95% lower and upper confidence bounds, respectively, of the model parameter estimates

Table 7.16 Comparison of Observed and Logistic Predicted Incidence Rates for Mite Mortality

Conc	Risk	Alive	ObsP	Pred	LCB	UCB
0	20	18	0.10	0.07	0.01	0.12
18.75	20	18	0.10	0.11	0.04	0.19
37.5	21	18	0.14	0.19	0.09	0.29
75	21	11	0.48	0.45	0.30	0.59
150	20	2	0.90	0.90	0.79	1.00
300	20	0	1.00	1.00	1.00	1.00

LCB and UCB are the 95% lower and upper confidence bounds; ObsP and Pred are the observed and predicted mortalities.

7.5 ABBOTT'S FORMULA AND NORMALIZATION TO THE CONTROL

A technique common in the analysis of some ecotoxicity studies is to normalize data in treatment groups to the control. In the case of quantal data (e.g. mortality, emergence, hatching), the method for doing this is sometimes referred to as Abbott's formula. ISO 22030 (2005) advises expressing the number of emerged seedling counted for each pot as a percentage of the mean emergence in the control pots. OECD TG 221 (OECD, 2006a) offers normalized lemna growth inhibition data as an acceptable basis for regression. Similar advice is found in Moore and Caux (1997), Christensen et al. (2009), and Nyholm et al. (1992), among others.

The original purpose of Abbott's formula was to adjust mortality data for background incidence, and that is indeed an important practice. A question deserving attention is how best to accomplish this purpose. When Abbott (1925) first proposed his method, most statistical analysis was done manually, so there was a need for simple numerical calculation methods. It has subsequently been adapted for continuous responses, where the object is to estimate the concentration at which a specified percent effect relative to the control mean response occurs. In the latter situation, there is no need to adjust for background, and the primary purpose in normalizing to the control is to modify the data so that a probit analysis (which assumes quantal data) software package can be used. Others fit a continuous model to the normalized data, though the reasons for doing so are not clear. In any event, both practices violate statistical

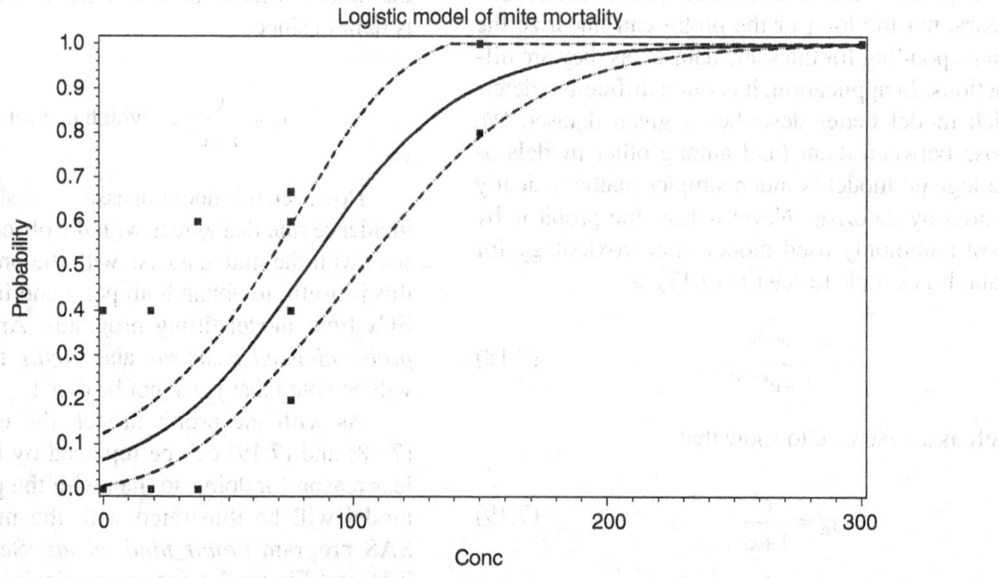

Figure 7.4 Logistic model for mite mortality. Solid squares are actual replicate proportions dead. Dashed lines are 95% lower and upper confidence bounds for model predictions. Solid line is model prediction.

theory and can lead to large errors in estimated EC*x* values, as will be demonstrated in this section.

Abbott proposed replacing the proportion, P_i, of subjects affected in each treatment group with $P_{ni} = (P_i - P_0)/P_0$, which is the proportion decrease from control in treatment *i*. Here P_0 is the proportion affected in the control. This could be applied on individual treatment replicates, but the entire control proportion is used in all cases. One concern that should be immediately apparent is that, as defined, the P_{ni} can be negative or greater than 1. To avoid that, the definition is modified to replace negative values by zero and values exceeding 1 by 1. Doing so reduces the observed variance, which in turn alters the width of confidence bounds for parameter estimates.

It should be noted that P_0 in Abbot's formula is not a constant but a random variable that will vary from study to study, just as the treatment proportions vary. Thus, the normalized proportions are all correlated by virtue of having the random variable P_0 in the denominator. This approach therefore violates a fundamental statistical assumption underlying all models likely to be used, namely, that observations be independent. Nor is there any useful way to take these correlations into account. Furthermore, there is no reason to think these correlations are trivial. Fleming and Retnakaran (1985) pointed out problems with Abbott's formula but did not investigate it in depth, and their paper has not deterred many scientists in ecotoxicology from using it. Research that has been presented by the first author of this book at several professional conferences and workshops will be summarized here in the hope of stopping this unfortunate, misleading, and unnecessary practice. ISO/TS 20281-2006 (2006) also noted the fallacy in this formula.

Better ways to take background mortality into account or model continuous responses have been known for several decades (e.g. Collett, 2002), and software is widely available to carry out these mathematically correct methods. For probit analysis, the model to take background mortality into account is given by (7.11).

It will be shown that not only is the analysis of normalized data theoretically unsound and unnecessary but also is misleading. In the context of several types of guideline studies, it will be shown that EC*x* estimates from analysis of normalized data (quantal or continuous) can over- or underestimate the true value by large factors, while other readily available approaches provide accurate estimates.

First, theoretical problems with normalized data are made clear. Second, data from fish ELS, nontarget plant, and fish sexual development studies are analyzed using probit models fit to normalized data and more appropriate models to indicate the difference in conclusions (EC*x* point and interval estimates, principally). Third, Monte Carlo simulation studies, summarized in Section 7.7, based on experimental designs, variability, and dose–response shapes typical of these studies have been used to develop distributions for EC*x* estimates from these models. In the simulation studies, the true EC*x* is known, so these distributions can reveal the quality of estimates to be expected from the different modeling approaches.

Actual studies yield EC10 and EC25 estimates that differ by a factor of as much as 6X using probit analysis of normalized data compared with a theoretically sound method of analysis. The simulation studies demonstrate that the estimates based on normalized data often err from the truth by large factors. Distributions of EC*x* estimates from the simulated data also allow a comparison of the relative variability of EC*x* estimates from the alternative approaches.

Readily available software allows the use of statistically valid regression models to analyze quantal and continuous data arising from ecotoxicity studies. Analysis of normalized responses is unreliable, with both under- and overestimation of EC*x* values by large factors occurring with little way to determine in a particular study that whether such mis-estimation has occurred.

See Table 7.17 for results from numerous nontarget plant studies. Studies labeled 1–4 are from vegetative vigor studies done under OECD TG 227, and response is the survival of plants to the end of their study. Studies labeled A–E are from seedling emergence studies done under OECD TG 208, and the response is emergence of seedlings after planting. The studies were done by various labs and are part of a large database of nontarget terrestrial plant studies compiled from numerous companies and testing laboratories. Slopes are given of the fitted probit curves where the test concentrations were common-log transformed. All slopes will be seen as low to moderate and are typical for such studies, where high mortality is uncommon. Two ratios are calculated to ease understanding. The variable named Ratio is the ratio of the probit EC*x* estimate divided by the EC*x* estimate from a probit model fit to the normalized data. RatioN is the reciprocal. It will be observed that estimates (EC*x*N) from normalized models can over- or underestimate EC*x* by factors as high as 26 in one extreme case. Data were excluded if either model failed to converge or produced an EC*x* estimate without confidence bounds or the upper confidence bound of either estimate was more than 10X the highest tested concentrations or the estimated slope was negative, as all of these are indications of such poor fitting models that no reasonable data analyst would accept them.

It will be observed that substantial differences (2X or greater) occur for EC10, EC25, and EC50, though certainly EC10 is the most common. It is not possible to state with any certainty whether EC*x* or EC*x*N is a better indicator of the true effects concentration. However, probit models that treat the background incidence in a mathematically correct manner are certainly preferred to models that violate

Table 7.17 Comparison of ECx Estimates from Normalized and Nonnormalized Quantal Data

Study	Species	Slope	Bkgr	X	LCB	UCB	ECx	ECxN	NLCB	NUCB	RatioN	Ratio
1	Onion	3.22	0.02	10	0.01	0.02	0.02	0.01	0.00	0.01	0.45	2.23
1	Onion	3.22	0.02	25	0.02	0.03	0.02	0.01	0.01	0.02	0.62	1.62
1	Onion	3.22	0.02	50	0.03	0.05	0.04	0.03	0.02	0.05	0.88	1.14
2	Cucumber	3.25	0.01	10	0.01	0.02	0.02	0.01	0.01	0.01	0.48	2.09
2	Cucumber	3.25	0.01	25	0.02	0.03	0.03	0.02	0.02	0.04	0.87	1.15
2	Cucumber	3.25	0.01	50	0.04	0.06	0.04	0.07	0.05	0.16	1.68	0.60
3	Bean	3.73	0.02	10	0.07	0.51	0.37	0.15	0.08	0.24	0.40	2.52
3	Bean	3.73	0.02	25	0.23	0.67	0.54	0.70	0.41	1.62	1.31	0.76
3	Bean	3.73	0.02	50	0.65	1.18	0.81	4.04	1.73	18.88	**4.97**	0.20
3	Corn	1.68	0.01	10	0.10	0.31	0.20	0.11	0.05	0.19	0.56	1.79
3	Corn	1.68	0.01	25	0.28	0.63	0.45	0.69	0.41	1.20	1.53	0.65
3	Corn	1.68	0.01	50	0.82	1.60	1.13	5.29	2.75	13.44	**4.68**	0.21
3	Oat	2.86	0.01	10	0.08	0.12	0.10	0.08	0.03	0.13	0.81	1.24
3	Oat	2.86	0.01	25	0.13	0.18	0.16	0.67	0.37	2.43	**4.22**	0.24
3	Oat	2.86	0.01	50	0.23	0.33	0.27	7.22	2.11	214.48	**26.64**	0.04
4	Soybean	1.63	0.01	10	0.01	0.05	0.03	0.01	0.01	0.02	0.47	2.13
4	Soybean	1.63	0.01	25	0.04	0.10	0.07	0.05	0.03	0.09	0.73	1.38
4	Soybean	1.63	0.01	50	0.12	0.32	0.18	0.22	0.12	0.50	1.18	0.85
A	Ryegrass	0.46	0.03	10	0.00	0.10	0.02	0.01	0.00	0.03	0.42	2.37
B	Oat	4.22	0.05	10	0.12	0.31	0.23	0.09	0.06	0.13	0.39	2.54
B	Oat	4.22	0.05	25	0.22	0.40	0.33	0.21	0.15	0.28	0.63	1.58
B	Oat	4.22	0.05	50	0.38	0.60	0.47	0.51	0.36	0.86	1.07	0.93
B	Sorghum	2.44	0.34	10	0.00	0.02	0.01	0.00	0.00	0.00	0.28	**3.61**
B	Sorghum	2.44	0.34	25	0.00	0.03	0.02	0.01	0.01	0.01	0.46	2.19
B	Sorghum	2.44	0.34	50	0.02	0.05	0.04	0.03	0.02	0.04	0.80	1.25
C	Buckwheat	3.48	0.16	10	0.20	1.09	0.72	0.30	0.18	0.42	0.41	2.42
C	Buckwheat	3.48	0.16	25	0.47	1.46	1.08	0.70	0.50	0.95	0.65	1.54
C	Buckwheat	3.48	0.16	50	1.12	2.16	1.68	1.80	1.29	2.80	1.07	0.93
C	Oilseed rape	3.00	0.08	10	0.36	1.57	1.10	0.84	0.48	1.22	0.76	1.31
C	Oilseed rape	3.00	0.08	25	1.00	2.31	1.75	1.93	1.33	3.15	1.10	0.91
C	Oilseed rape	3.00	0.08	50	2.22	4.95	2.94	4.85	3.00	12.45	1.65	0.61
C	Sugar Beet	0.59	0.10	10	0.00	1.10	0.21	0.02	0.00	0.08	0.08	**12.95**
D	Ryegrass	3.19	0.12	10	3.12	8.59	6.10	1.05	0.46	1.75	0.17	**5.78**
D	Ryegrass	3.19	0.12	25	6.04	12.28	9.45	3.18	1.96	4.58	0.34	2.97
D	Ryegrass	3.19	0.12	50	11.74	19.57	15.38	10.85	7.57	17.20	0.71	1.42
E	Onion	1.83	0.20	10	0.02	0.14	0.08	0.06	0.03	0.09	0.68	1.47
E	Onion	1.83	0.20	25	0.08	0.28	0.18	0.20	0.13	0.37	1.15	0.87
E	Onion	1.83	0.20	50	0.26	1.02	0.41	0.85	0.44	2.78	2.08	0.48

ECx and ECxN are the estimates of the indicated effects concentrations from nonnormalized and normalized responses, respectively; LCB and UCB are the 95% confidence bounds for ECx, and NLCB and NUCB are the 95% confidence bounds for the estimated from models fit to the normalized responses; Bkgr is the background incidence rate; Slope is for the untransformed response; RatioN, ECxN/ECx; Ratio, ECx/ECxN.

fundamental requirements for modeling. That favors results from nonnormalized data analysis.

It should be noted that in the vegetative vigor studies, the background incidence is very low, and it might be surprising that the two approaches can give such different results. Partly this is because of the way normalized incidence is forced to be between 0 and 1 in each replicate. This distorts the data and is another reason to question the use of normalized data.

Simulations were done to determine whether the examples in Table 7.9 are unusual or typical of analyses done on normalized quantal data. Data was simulated with binomial distribution with mean incidence following a probit curve with slopes relative to $\text{Log}_{10}(U)$ of 1, 5, or 10,

Table 7.18 Comparison of ECx Estimates from Probit and Normalized Probit Analysis

Model	NFIT	Slope	C	Ratio10	Mid10	Sprd10	Ratio20	Mid20	Sprd20
NProbBE	1000	5	0.05	2.9	1.4	2.2	3.6	1.4	1.8
NProbBE	1000	5	0.1	4.8	1.4	2.0	5.9	1.3	1.7
NProbBE	1000	5	0.2	10.5	1.6	2.4	12.8	1.4	1.9
NProbBE	1000	5	0.3	23.7	1.7	2.6	29.1	1.5	2.1
NProbBE	1000	5	0.4	52.4	1.7	2.9	64.2	1.5	2.3
ProbBE	1000	5	0.05	1.0	1.3	1.6	1.2	1.2	1.5
ProbBE	1000	5	0.1	1.0	1.3	2.3	1.2	1.3	2.0
ProbBE	1000	5	0.2	1.0	1.5	2.5	1.2	1.4	2.2
ProbBE	1000	5	0.3	1.0	2.1	2.7	1.2	1.9	2.3
ProbBE	1000	5	0.4	1.0	2.2	3.1	1.2	1.9	2.5

NProbBE, normalized probit with binomial errors; ProbBE, untransformed probit with binomial errors; NFIT, number of models out of 1000 simulated datasets that converged; Slope and C are the true slope and background simulated; Ratio$X$, median estimated ECx/TrueECx; Mid$X$, ratio of 75th percentile of ECx estimates to 25th percentile; Sprd$X$, ratio of 90th percentile of ECx estimates to 10th percentile.

with control and four, five, or six test concentrations with geometric space and common ratio of 1.6, 2.3, or 3.2, four replicates of 10–40 subjects each, and background incidence of 10, 20, 30, or 40%. These cover a wide range of test guideline recommendations. See Table 7.18 for results from studies with four test concentrations of four reps of size 10 and common ratio of 3.2. It is clear from Table 7.18 that ECx estimates from probit analysis of normalized quantal data have a strong tendency to seriously underestimate the true value and the degree of underestimation increases as the background incidence increases. More details are given in Green (2017c).

7.5.1 Normalized Continuous Responses

This same idea has been applied to continuous responses as well and is even less justified than with quantal data. A simulation study was done to address this issue in a way to understand whether the comparisons in Table 7.17 are aberrations from peculiar datasets or inherent in the normalization process. In the simulations, the true ECx is known, so the relative performance of probit models of normalized and nonnormalized data can be assessed.

See Figure 7.5 for a comparison of ECx estimates from probit models fit to normalized continuous response (dry weight of terrestrial plants) with those from two mathematically sound models, the Bruce–Versteeg model (BVP) and a three-parameter exponential model (OE3).

With regard to Figure 7.5, it is worth noting that the data were simulated to follow a Bruce–Versteeg curve, so the comparison of models is skewed in favor of models with that shape, and nonetheless, OE3 models to the same data performed as well as BVP and much better than the normalized probit. Data simulated to follow other shapes showed similar bias can result from a normalized probit analysis. Results vary depending on the shapes simulated, and under some conditions, the normalized probit analysis performs as well as other models. In a particular dataset, it is not possible to know the correct model or ECx, so the wisest course of action is to fit a model that is mathematically and statistically sound. Fortunately, it is no more difficult to fit such a model than it is to normalize the data and fit a probit model to the transformed response.

7.5.2 Fish Sexual Development Data

Another example of normalized versus untransformed quantal data may also be of interest. In developing the OECD fish sexual development test guideline 234, zebrafish were exposed to the endocrine-disrupting chemical tamoxifen citrate. The issue of concern was whether this chemical might alter the phenotypic sex of fish that were exposed to this chemical in the embryonic stage. The genetic sex cannot be changed, but the fish can take on almost all aspects of the genetically opposite sex, including reproductive ability, so they are for most purposes, of that sex. For some species, genetic markers are available, and for those species, analysis can be done using difference between genetic and phenotypic sex as the response. At the time of the study, there was no feasible way to determine the genetic sex of the zebrafish embryos or even of the juvenile fish at the end of the study, so only phenotypic sex was available to analyze. Such data can be problematic for probit modeling, since the expectation might be that the background incidence rate is around 50% corresponding to half males and half females in the undisturbed environment. Proportions have their greatest variance at 0.5 and this can result in high variability in the control. The particular strain of zebrafish used in the study of the example were predominantly female, around 70%, which does not do much to alleviate the concern about variability, and the

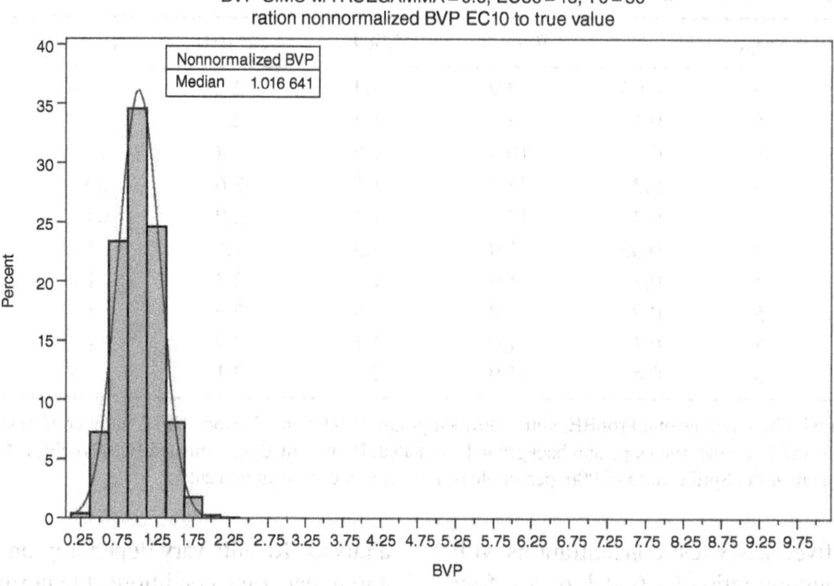

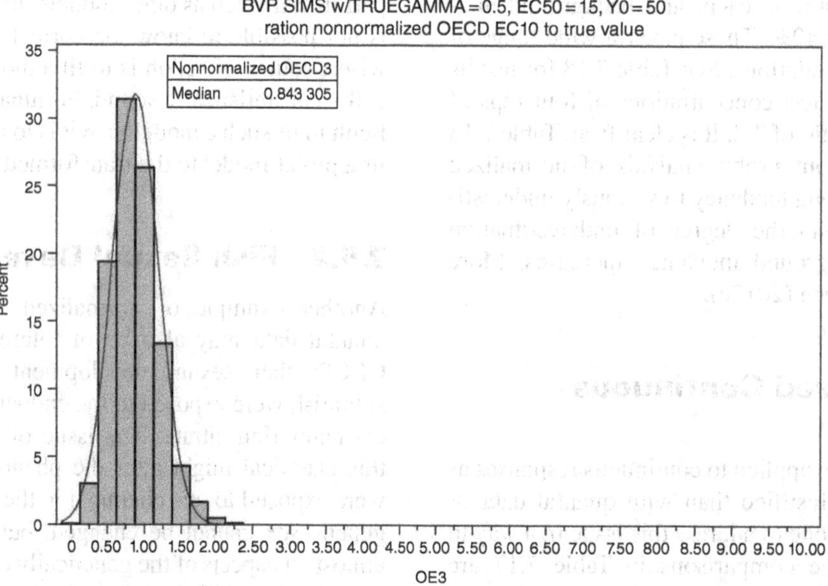

Figure 7.5 Comparison of EC*x* estimates from normalized probit models fit to continuous response to estimates from mathematically sound models. Bottom right is simulated dose–response curve from a BVP model with γ=0.5 (see Eq. 4.1*). Histograms of estimated EC10 values: upper left=BVP, upper right=OE3, lower left=normalized probit. Dose–response curve not unusual, but normalized probit estimates greatly overestimate EC10, whereas estimates from the other models are centered near the true value and are much less variable.

issue was whether the chemical would effectively change them into males. In this study, there were three replicates of 40 fish each and the concentrations were in μg l^{-1}. The data were summarized by replicate (see Table 7.19).

The response of interest is the incidence of phenotypic females. Following the strategy discussed in Section 7.2.2, a probit analysis was done on the incidence of males, as this would be an increasing response and the observed background incidence was 0.23. Tarone's $C(\alpha)$ test did not find significant overdispersion. However, it is important to realize that the halfwidth of the confidence interval (UCB-Mean) for the control proportion female is 14% of the control proportion female, while the corresponding figure for proportion males is 48%, so that on its face, estimation of EC10 is not sensible. Several models were fit to the incidence of males: standard probit, probit with a random effect (generalized nonlinear model [GNLM]), and standard probit on normalized incidence of males. In addition, several models of the proportion female treat those proportions as continuous. Of those, only an OE3 is reported, as it

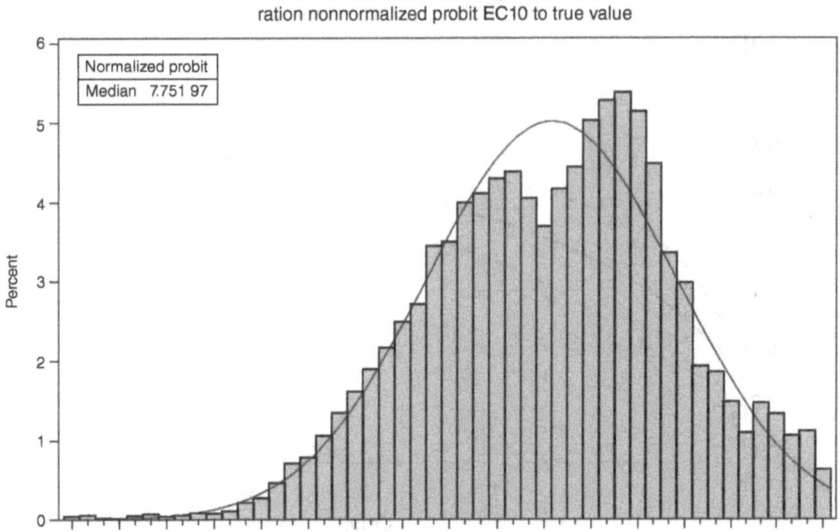

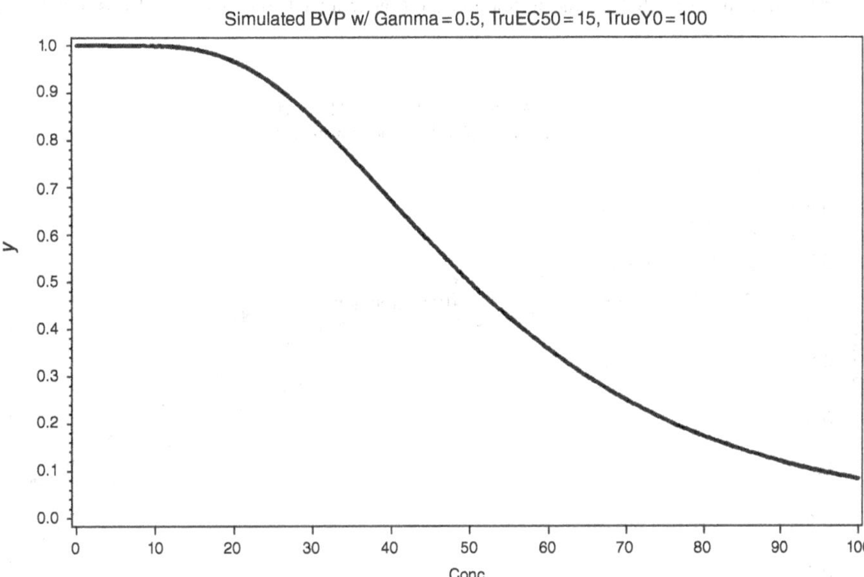

Figure 7.5 (Continued)

Table 7.19 Fish Sexual Development Test Data

Conc	Group	Rep	Females	Conc	Group	Rep	Females
0.00	1	A	30	0.22	3	B	25
0.00	1	B	28	0.22	3	C	27
0.00	1	C	35	0.77	4	A	24
0.09	2	A	20	0.77	4	B	24
0.09	2	B	32	0.77	4	C	19
0.09	2	C	25	2.70	5	A	7
0.22	3	A	23	2.70	5	B	12

Females, number of phenotypic females in random sample of 40 fish.

arguably gave the best fit among four continuous models considered. For all models, estimates of model parameters and ECx values for a range of x values are reported and compared. See Figure 7.6 for the GNLM probit model of proportion males, and see Figure 7.7 for the OE3 model of proportion female. See Table 7.20 for a comparison of the model fits, where all model predictions are expressed in term of proportion female regardless of which response was actually analyzed. See Table 7.21 for a comparison of ECx estimates from these models.

It is interesting that the probit model using replicate information and based on a binomial error structure, which

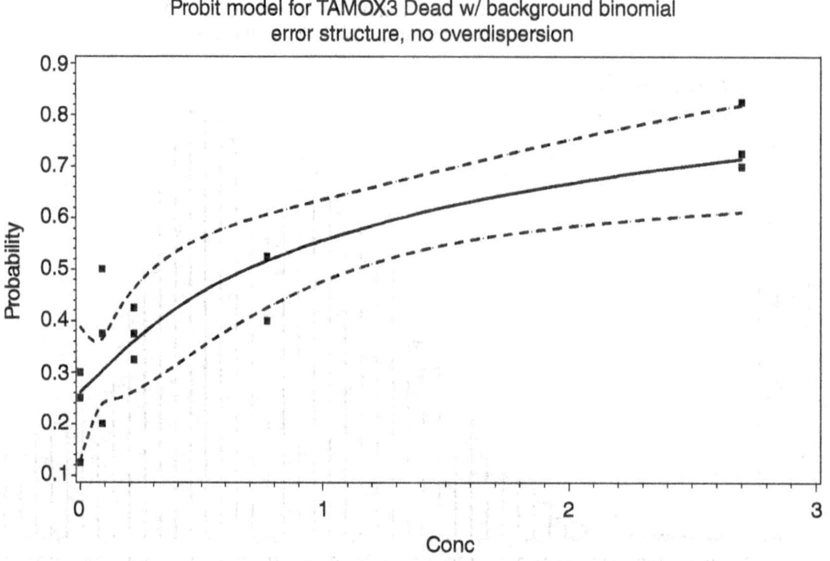

Figure 7.6 GNLM probit model of tamoxifen-citrate data with replicates considered. Solid squares are observed replicate proportions. Dashed lines are 95% confidence bounds. Solid line is predicted concentration mean. A generalized nonlinear model was fit with binomial error structure and replicates modeled.

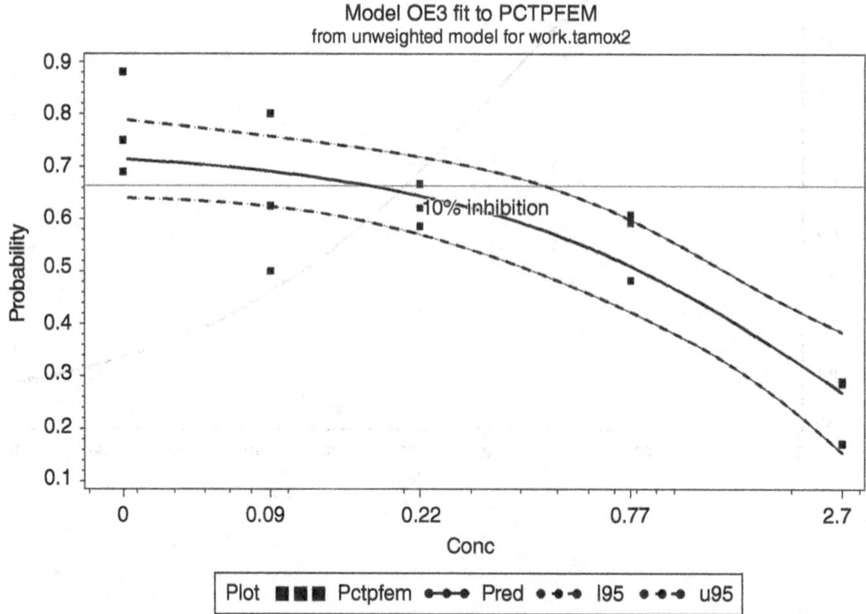

Figure 7.7 Three-parameter exponential model fit to female tamoxifen-citrate proportions. Solid squares are observed replicate female proportions. Dashed lines are 95% confidence bounds. Solid line is predicted concentration mean. A three-parameter exponential model was fit with normal error structure and replicates modeled.

is ostensibly the most theoretically appealing model, gives the widest confidence intervals for the ECx estimates. Only EC40 and EC50 are significantly different from zero. This suggests that the variability in these data render estimation of ECx for values of $x \leq 30$ unreliable. Moreover, all of these models except OE2 and the normalized probit show a lower confidence bound on EC20$\leq$0. This is consistent with the initial observation that the halfwidth of the confidence interval for the control proportion female is 14% of the control proportion female, while the corresponding figure for proportion males is 48%. So, both from the models themselves and the uncertainty of the control proportions, estimation of EC10 is absurd, and that of EC20 is at least questionable. It is noteworthy that the NOEC for proportion females by

Table 7.20 Comparison of Models Fit to Tamoxifen-Citrate Data

Model	Conc	0	0.09	0.22	0.77	2.7
Probit GNLM	**ObsP**	**0.23**	**0.36**	**0.38**	**0.44**	**0.75**
	Pred	**0.26**	**0.30**	**0.36**	**0.52**	**0.71**
	LCB	0.13	0.24	0.26	0.43	0.61
	UCB	0.39	0.36	0.46	0.61	0.82
BVP	**Pred**	**0.29**	**0.30**	**0.34**	**0.49**	**0.72**
	LCB	0.18	0.23	0.26	0.39	0.60
	UCB	0.39	0.37	0.42	0.59	0.85
OE3	**Pred**	**0.26**	**0.31**	**0.36**	**0.49**	**0.73**
	LCB	0.16	0.24	0.28	0.40	0.62
	UCB	0.37	0.38	0.43	0.58	0.85
OE2	**Pred**	**0.29**	**0.31**	**0.34**	**0.47**	**0.75**
	LCB	0.21	0.25	0.29	0.41	0.64
	UCB	0.36	0.37	0.40	0.53	0.85

Conc = test concentration in $\mu g\,l^{-1}$. ObsP is the observed proportion female in the indicated concentration; Pred is the proportion estimated by the indicated model; LCB and UCB are the 95% confidence bounds on that estimate; Model OE3 provides a marginally better fit to the data at lower test concentrations, especially to the control; However, model OE2 has the lowest AICc value of the continuous models; Since the probit model is based on a different response and error structure, the AICc value is not comparable to those of the continuous models.

the Jonckheere–Terpstra test was 0.09, the lowest test concentration, where a 17% decrease in proportion females was observed.

It is also useful to consider the estimates from the normalized data carefully. To that end, it is informative to express the data from Table 7.21 in terms of how the estimates from the mathematically appropriate models compare with those from the normalized data. (See Table 7.22.)

There is much consistency in the results from mathematically correct models to each other and their relationships to estimates from normalized data. While that does not in itself prove the results based on normalized data are wrong, it is very suggestive. The simulations discussed above provide strong evidence that models using normalized responses can be very wrong. The information in Tables 7.21 and 7.22 suggest there is no great difference in results from the probit with binary error structure and replicates modeled and OE3. So while the conceptual advantage of the former is compelling, the penalty in using the latter is small in this example.

7.6 PROPORTIONS TREATED AS CONTINUOUS RESPONSES

Replicate proportions can be modeled as continuous and then all the methods of Chapter 4 can be employed. In this approach, the between-replicate variability is taken directly into account as is overdispersion if present. Unlike some

Table 7.21 Comparison of EC*x* Estimates from Models for Tamoxifen-Citrate Data

X	5	10	20	30	40	50
LCL	0.00	0.00	0.00	0.02	0.11	0.50
UCL	0.33	0.49	0.79	1.15	1.65	2.57
EC*x*	0.08	0.15	0.34	0.60	0.99	1.58
LCLn	0.01	0.03	0.12	0.29	0.58	1.04
UCLn	0.05	0.10	0.26	0.55	1.11	2.31
EC*x*N	0.03	0.06	0.19	0.41	0.79	1.46
LCLG	−0.15	−0.21	−0.24	−0.14	0.13	0.59
UCLG	0.31	0.51	0.92	1.35	1.86	2.56
EC*x*G	0.08	0.15	0.34	0.60	0.99	1.58
LCBB	−0.14	−0.14	−0.05	0.13	0.40	0.71
UCBB	0.44	0.67	1.06	1.49	2.01	2.80
EC*x*B	0.15	0.26	0.51	0.81	1.21	1.76
LCB3	−0.10	−0.14	−0.10	0.07	0.37	0.77
UCB3	0.23	0.46	0.93	1.41	1.94	2.63
EC*x*O3	0.07	0.16	0.41	0.74	1.16	1.70
LCB2	0.08	0.16	0.34	0.34	0.78	1.05
UCB2	0.19	0.39	0.83	0.83	1.90	2.58
EC*x*O2	0.13	0.28	0.58	0.58	1.34	1.81

EC*x* is the estimate from a standard probit model fit to males; EC*x*N is the estimate from a standard probit model fit to normalized males; EC*x*G is the estimate from a probit with replicates included (GNLM); EC*x*B is the estimate from a BVP model fit to females replicate proportions; EC*x*O3 is the estimate from a three-parameter exponential fit to females replicate proportions; EC*x*O2 is the estimate from a two-parameter exponential fit to females replicate proportions. LC and UC prefixes refer to the lower and upper confidence limits of the indicated estimates.

Table 7.22 Further Comparison of Estimates Based on Normalized and Untransformed Data

X	RatioP	RatioG	RatioB	RatioO3	RatioO2
5	2.95	2.95	5.85	2.49	5.08
10	2.36	2.36	4.12	2.52	4.30
20	1.81	1.81	2.70	2.20	3.11
30	1.49	1.49	1.99	1.82	1.44
40	1.26	1.26	1.53	1.47	1.70
50	1.08	1.08	1.20	1.16	1.24

RatioP, EC*x*/EC*x*N; RatioG, EC*x*G/EC*x*N; RatioB, EC*x*B/EC*x*N; RatioO3, EC*x*O3/EC*x*N; RatioO2, EC*x*O2/EC*x*N;

implementations of quantal models, when modeling replicate proportions as continuous, survival should be modeled rather than mortality. To understand this, an *x*% effect is judged against the control. That is, the EC/LC*x* is the concentration that is associated with an *x*% reduction in survival relative to the control survival. If, for example, there is 80% survival in the control, one estimates a 50% reduction (i.e. a decrease from 80 to 40% survival), from 0.8, rather than a 50% increase from 0.2 (i.e. an increase from 20 to 30% mortality). Since these methods have been

Table 7.23 Bruce–Versteeg Model for Mite Data: Comparison of Observed and Predicted Survival

Conc	N	Mean	Pred	%Chng	Pred %Chng
0	20	0.90	0.90	0	0
18.75	20	0.90	0.90	0	0
37.5	21	0.86	0.86	−5	−4
75	21	0.52	0.53	−42	−40
150	20	0.10	0.09	−89	−90
300	20	0.00	0.00	−100	−100

Mean, mean of observed data; Pred, mean predicted value; %Chng, observed percent change from control; Pred %Chng, percent change in predicted mean from predicted control mean.

Table 7.24 Bruce–Versteeg Model ECx Estimates

X	ECx	LCB	UCB
1	27	2	52
5	38	19	74
10	45	20	70
20	56	37	85
30	65	46	91
40	74	56	98
50	84	65	108

discussed in detail in Chapter 4, only a simple example will be provided here. The mite data from Table 7.1 will be fit using the BVP. See Table 7.23 for a comparison of observed and predicted survival, Table 7.24 for ECx estimates, and Figure 7.8 for a plot of the fitted model.

There is little difference in the estimated EC50 values from the Bruce–Versteeg, probit, and Weibull models for the mite data, though the Bruce–Versteeg method gives slightly shorter confidence intervals. Also, models were fit to proportion females treating that response as a continuous response. The best fit by that response was obtained from a simple two-parameter exponential model.

7.7 COMPARISON OF MODELS

Probit data was simulated with binomial distribution at each treatment group, with slopes 1, 5, or 10; background incidence of 5, 10, 20, 30, or 40%; and samples sizes 10, 20, 30, or 40 per replicate. The maximum mortality simulated was background plus 20% (25% for background = 5%) up to 100% in increments of 10%. In all simulations presented there were four replicates in the control and in each of four, five, and six test concentrations. Concentrations were simulated with geometric spacing and common ratios of 1.6, 2.3, or 3.2, which covers the range of ratios in OECD test guidelines. Only the five test concentration results with spacing 1.6 and sample size 10 are discussed in this section, but the other simulations followed similar patterns, except the spread for larger sample sizes was somewhat smaller. Additional results are given in Green (2017a).

Eight models were fit to the simulated data. These were probit model fit to mortality assuming binomial errors (ProbBE) or normal errors (Prob) both with background; Bruce–Versteeg (BVP) and log-logistic (LLog) fit to survival assuming continuous response; normalized probit fit

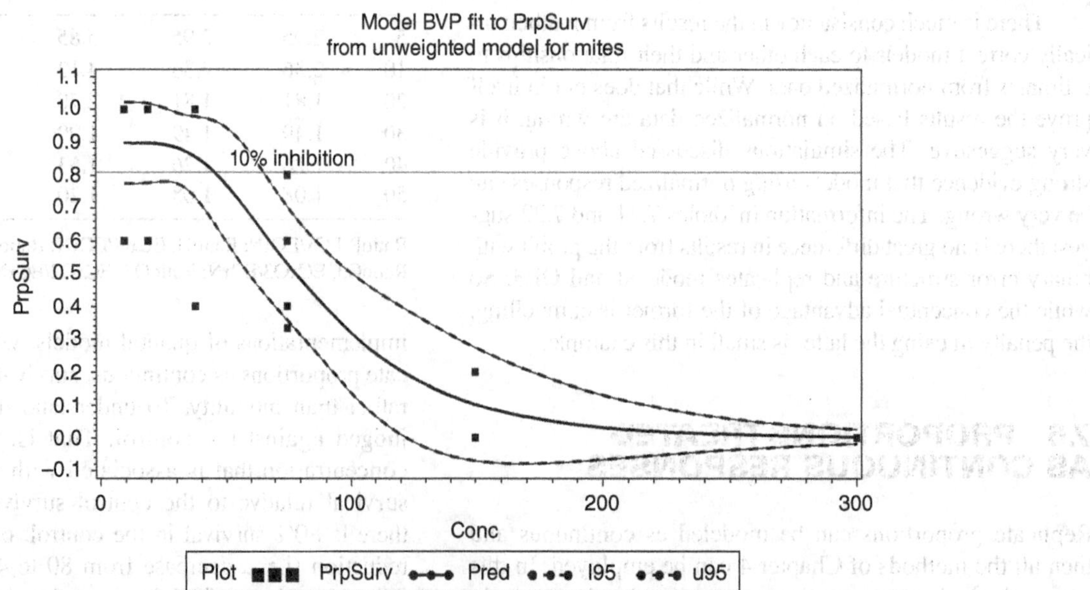

Figure 7.8 Bruce–Versteeg model fit to mite data. PRPSURV is the proportion surviving. Conc = concentration in g a.s. ha^{-1}.

to mortality assuming normal (NProb) and binomial errors (NProbBE), respectively; and OE2 and OE3 fit to survival assuming normal errors.

The normal errors models treat the observed replicate proportions as normally distributed with mean given by the formulas presented in (7.3) and (7.18) above and Eqs. (4.1), (4.4), and (4.5) from Chapter 4. The binomial errors model treats the observed number of deaths as a binomial variable with probability of incidence given by the same formulas.

The normalized probit model replaces the observed replicate proportions, y, by $z=(y-y_0)/y_0$, where y_0 is the observed control proportion mortality. Modifications are made to avoid negative values or values exceeding 1. These normalized proportions are no longer independent since the random variable y_0 is common to all observations. This contributes to the poor performance of this approach documented here.

Shallow slope dose–response (TSlope = 1) presents challenges for all models considered. This is expected and careful evaluation of model fit is necessary. No single approach is clearly superior for all datasets. In terms of agreement of median EC10 estimates to the true value, the normalized probit model with normal errors tends to underestimate EC10, with worse agreement as the background mortality increases. The simple exponential model OE2 with normal errors strongly tends to overestimate EC10 independent of the background mortality rate, while OE3 strongly tends to underestimate EC10. The median EC10 estimate is not the only consideration, however. In terms of spread of estimates, the normalized probit model with binomial errors and simple exponential both compare favorably with the alternatives, for both the middle range (Q75/Q25) and wider range (Q90/Q10), the consequence being that both models are consistent in their mis-estimation of EC10. The Bruce–Versteeg, probit with normal errors, and probit with binomial errors all provide good estimates of EC10 "on average" but have considerable spread by either measure.

For EC50 estimation from shallow slope models, all models strongly tended to underestimate, but that is mostly due to the shallow slope dose–response never achieving a 50% response. The normalized error structure is sharply degraded for background incidence of 20% or greater. The simple exponential model, OE2, actually performs best in terms of median performance and spread, while the standard probit with either error structure, Bruce–Versteeg, and log-logistic perform about the same but worse than OE2, and both error structures on normalized probit models perform less well.

For medium and high slope dose–response (TSlope = 5), the two normalized probit models and OE2 estimate EC10 and EC50 markedly worse than the alternatives. These alternatives perform approximately equally well.

Only under restrictive conditions do normalized probit models provide good estimates. The best results overall are from binomial errors probit models of untransformed data. The two-parameter exponential model is useful for modeling survival only in one narrow set of conditions, but the Bruce–Versteeg and log-logistic models are often acceptable alternatives. Also acceptable for survival is the three-parameter exponential but only for slope 5 or greater. See Tables 7.25–7.28 for more detail.

7.8 INCLUDING TIME-VARYING RESPONSES IN MODELS

It is common in guideline studies with mortality as an endpoint to measure mortality at different times. For example, in OECD TG 211 *Daphnia* reproduction studies, mortality is usually recorded daily even though the only required analysis is of total adult mortality by the end of the study. In OECD TG 203 fish acute test, mortality is recorded at four times during the study, but again only analyzed at the end of the 96 h study. A similar requirement exists in TG 204 (Fish, Prolonged Toxicity Test: 14-day Study), where LC50 for cumulative mortality at 14 days (or 21 or 28 days if the test is extended an extra 1 or 2 weeks) is required and LC50 at other 1-week interval is optional. In TG 202 (Daphnia sp. Acute Immobilization Test), cumulative mortality is recorded at 24 and 48 h, with the LC50 at 48 h required and LC50 at 24 h optional. In all of these cases, it has been traditional to do separate LC50 calculations at different times and ignore the obvious lack of independence of the results across time. To get a more complete understanding of the relationship between toxicity and duration of exposure, an analysis that considers the entire range of observation times would be helpful, and there is increasing interest in doing this. This section will describe one way to incorporate time into models for quantal response using a method introduced by Petkau and Sitter (1989). Ohara Hines and Lawless (1993) modified this approach to allow for a more precise accommodation of overdispersion. Chen (2007) developed a more general framework for these models, and Li et al. (2008) improved the computation of confidence bounds. Chapter 13 will describe a more holistic approach using TKTD models and DEBtox. However, the approach of Petkau and Lawless remains of interest and has the appeal of simplicity as well as utility.

The experimental design considered is that there are k test concentrations (excluding control) each in L replicate vessels and mortality is recorded for all vessels at the same m times, $t_0=0, t_1, \ldots, t_m$.

Let U_i, $i=1, \ldots, k$ denote the test concentrations, n_{il} the number of subjects in replicate l of test concentration U_i, y_{ijl} the number of subjects in replicate l of test concentration U_i

Table 7.25 Summary of Simulation Results for Slope 1, Sample Size 10

Model	C	Ratio10	Mid10	Sprd10	Ratio50	Mid50	Sprd50	NFIT
BVP	0.05	1	12.6	201.2	18.5	2.7	6.5	985
BVP	0.1	1.2	14.7	406.5	22.2	3.0	9.0	986
BVP	0.2	0.9	12.4	254.5	16.6	2.6	5.9	963
BVP	0.3	0.7	11.3	385.2	13.7	2.5	6.1	972
BVP	0.4	0.8	13.5	812.1	14.6	2.6	7.6	946
LLog	0.05	0.8	11.6	349.4	14.4	2.6	8.3	1000
LLog	0.1	0.8	16.0	756.7	14.4	2.8	10.1	1000
LLog	0.2	0.9	13.8	1871.5	18.1	2.7	14.1	1000
LLog	0.3	1.1	22.7	2126.0	20.4	3.0	14.5	1000
LLog	0.4	0.9	26.9	>10000	17.0	3.2	22.8	1000
NProb	0.05	1.1	12.3	81.9	21.5	2.8	4.1	1000
NProb	0.1	1.4	13.6	77.2	27.2	3.0	4.2	1000
NProb	0.2	2.8	18.4	86.4	53.3	3.2	4.5	1000
NProb	0.3	3.8	20.8	83.5	73.5	3.6	4.4	1000
NProb	0.4	7.9	17.8	85.6	150.4	3.5	4.6	1000
NProbBE	0.05	0.7	3.1	7.7	14.0	1.6	1.7	1000
NProbBE	0.1	0.8	2.9	7.3	16.0	1.6	1.6	1000
NProbBE	0.2	1.1	2.7	6.0	21.9	1.6	1.5	1000
NProbBE	0.3	1.6	2.5	6.2	29.7	1.6	1.5	1000
NProbBE	0.4	2.2	2.3	5.2	41.5	1.5	1.4	1000
OE2	0.05	0.1	1.3	1.6	2.8	1.3	1.2	1000
OE2	0.1	0.2	1.3	1.7	2.9	1.3	1.2	1000
OE2	0.2	0.2	1.4	1.8	3.1	1.4	1.3	1000
OE2	0.3	0.2	1.4	2.0	3.2	1.4	1.5	1000
OE2	0.4	0.2	1.5	2.3	3.4	1.5	1.7	1000
OE3	0.05	9.7	81.3	>10000	185.1	4.6	28.5	1000
OE3	0.1	11.1	109.7	>10000	212.2	5.1	48.7	1000
OE3	0.2	7.8	105.1	>10000	149.6	4.6	63.5	1000
OE3	0.3	4.3	64.6	>10000	83.1	3.8	62.8	1000
OE3	0.4	4.7	155.7	>10000	88.9	4.6	979.0	1000
Prob	0.05	0.5	7.9	63.4	9.4	2.4	4.4	931
Prob	0.1	0.7	8.6	106.4	14.1	2.4	5.3	996
Prob	0.2	1.1	19.0	1741.6	20.2	3.3	14.8	1000
Prob	0.3	0.9	30.4	9185.8	16.4	3.6	31.4	999
Prob	0.4	0.9	21.0	>10000	18.1	3.2	50.3	1000
ProbBE	0.05	0.8	9.7	52.2	15.9	2.6	3.7	1000
ProbBE	0.1	0.8	13.2	111.3	16.0	2.8	4.9	1000
ProbBE	0.2	1	20.6	619.7	18.8	3.2	12.1	1000
ProbBE	0.3	1	20.8	2471.1	18.7	3.2	20.7	999
ProbBE	0.4	0.8	22.8	4925.1	15.2	3.2	27.1	999

Ratio*X*, ratio of the true (i.e. simulated) EC*x* to the median estimate from 1000 simulated experiments; Mid*X*, ratio (75th percentile of estimated EC*x* values/25th percentile); Sprd*X*, ratio (90th percentile of estimated EC*x* values/10th percentile); NFIT is the number (out of 1000) of models that produced an EC*x* estimate; *C* is the simulated background incidence.

that die in the time interval $(t_{j-1}, t_j]$, and r_{ijl} the number of subjects in replicate l of test concentration U_i that die by time t_j, so

$$r_{ijl} = \sum_{v \le j} y_{ivl}. \tag{7.21}$$

A distribution for time to death is given by an extension of the Weibull distribution discussed in Section 7.3:

$$F(t,U,l) = 1 - \exp\left(-\lambda_l(U)t^{\gamma_l}\right), \tag{7.22}$$

Table 7.26 Summary of Simulation Results for Slope 5, Sample Size 10

Model	C	Ratio10	Mid10	Sprd10	Ratio50	Mid50	Sprd50	NFIT
BVP	0.05	1	1.2	1.4	1.8	1.1	1.1	1000
BVP	0.1	1	1.2	1.5	1.8	1.1	1.1	996
BVP	0.2	1	1.3	1.7	1.8	1.1	1.2	988
BVP	0.3	1	1.4	1.9	1.8	1.2	1.2	962
BVP	0.4	1	1.5	2.1	1.8	1.2	1.3	951
LLog	0.05	1	1.2	1.5	1.8	1.1	1.1	999
LLog	0.1	1	1.3	1.5	1.8	1.1	1.1	998
LLog	0.2	1	1.3	1.7	1.8	1.1	1.2	994
LLog	0.3	1	1.4	2.0	1.8	1.2	1.2	977
LLog	0.4	1	1.5	2.4	1.8	1.2	1.3	967
NProb	0.05	1.1	1.2	1.4	2.0	1.1	1.0	1000
NProb	0.1	1.3	1.2	1.5	2.4	1.1	1.0	1000
NProb	0.2	2.1	1.4	1.8	3.7	1.1	1.0	1000
NProb	0.3	3.6	1.5	2.3	6.5	1.2	1.0	1000
NProb	0.4	7.1	1.8	3.1	12.8	1.2	1.1	1000
NProbBE	0.05	1.3	1.2	1.4	2.3	1.1	1.0	1000
NProbBE	0.1	1.6	1.2	1.4	2.9	1.1	1.0	1000
NProbBE	0.2	2.4	1.3	1.6	4.3	1.1	1.0	1000
NProbBE	0.3	3.7	1.4	2.0	6.6	1.2	1.0	1000
NProbBE	0.4	6.2	1.7	2.7	11.3	1.2	1.1	1000
OE2	0.05	4.2	1.1	1.2	7.5	1.1	0.9	1000
OE2	0.1	4.2	1.1	1.3	7.5	1.1	0.9	1000
OE2	0.2	4.2	1.2	1.4	7.5	1.2	1.0	1000
OE2	0.3	4.1	1.2	1.4	7.4	1.2	1.1	1000
OE2	0.4	4	1.3	1.5	7.2	1.3	1.1	1000
OE3	0.05	1.1	1.3	1.6	2.0	1.1	1.1	1000
OE3	0.1	1.1	1.3	1.7	2.1	1.1	1.1	989
OE3	0.2	1.1	1.4	1.9	2.0	1.1	1.2	975
OE3	0.3	1.2	1.5	2.2	2.1	1.2	1.2	939
OE3	0.4	1.2	1.6	2.4	2.1	1.2	1.3	930
Prob	0.05	1	1.3	1.5	1.8	1.1	1.1	909
Prob	0.1	1	1.2	1.4	1.8	1.1	1.1	993
Prob	0.2	1	1.3	1.7	1.8	1.1	1.1	991
Prob	0.3	1	1.4	1.9	1.8	1.2	1.2	967
Prob	0.4	1	1.5	2.3	1.8	1.2	1.3	940
ProbBE	0.05	1	1.2	1.4	1.8	1.1	1.1	1000
ProbBE	0.1	1	1.2	1.5	1.8	1.1	1.1	1000
ProbBE	0.2	1	1.3	1.7	1.8	1.1	1.2	1000
ProbBE	0.3	1	1.4	1.8	1.8	1.2	1.2	1000
ProbBE	0.4	1	1.5	2.2	1.8	1.2	1.3	1000

Ratio*X*, ratio of the true (i.e. simulated) EC*x* to the median estimate from 1000 simulated experiments; Mid*X*, ratio (75th percentile of estimated EC*x* values/25th percentile); Sprd*X*, ratio (90th percentile of estimated EC*x* values/10th percentile); NFIT is the number (out of 1000) of models that produced an EC*x* estimate; *C* is the simulated background incidence.

where $\lambda_l(U)$ is a function defined by

$$\lambda_l(U) = \exp(\alpha_l + \beta_l Ln(U)). \quad (7.23)$$

It will be noted that $F(t|U,l)$ has the form of an extreme value distribution, though that is not relevant for our present purposes. A model for mortality is then given by (7.24):

$$r_{ijl} = n_{il}\left[1 - \exp\left(-\lambda_l(U)t^{\gamma_l}\right)\right], \quad (7.24)$$

Model (7.24) suggests a binomial error structure and this will be illustrated. Petkau and Sitter actually modeled r_{ijl}/n_{il} using a normal error structure. Both approaches will be illustrated for comparison. The distribution (7.22)

Table 7.27 Summary of Simulation Results for Slope 10, Sample Size 10

Model	*C*	Ratio10	Mid10	Sprd10	Ratio50	Mid50	Sprd50	NFIT
BVP	0.05	1	1.1	1.2	1.3	1.1	1.1	999
BVP	0.1	1	1.1	1.2	1.3	1.1	1.1	991
BVP	0.2	1	1.1	1.3	1.3	1.1	1.1	963
BVP	0.3	1	1.2	1.4	1.4	1.1	1.1	911
BVP	0.4	1	1.2	1.4	1.4	1.1	1.2	824
LLog	0.05	1	1.1	1.2	1.4	1.1	1.1	991
LLog	0.1	1	1.1	1.3	1.4	1.1	1.1	988
LLog	0.2	1	1.2	1.3	1.3	1.1	1.1	958
LLog	0.3	1	1.2	1.4	1.4	1.1	1.1	891
LLog	0.4	1.1	1.2	1.5	1.4	1.1	1.2	808
NProb	0.05	1	1.1	1.2	1.4	1.1	1.1	998
NProb	0.1	1.1	1.2	1.4	1.5	1.1	1.1	1000
NProb	0.2	2.1	2.3	3.4	2.8	1.2	1.2	997
NProb	0.3	5.8	1.6	2.6	7.7	1.2	1.0	1000
NProb	0.4	13.5	2.1	4.6	18.2	1.2	1.1	1000
NProbBE	0.05	1.8	1.2	1.4	2.4	1.1	1.1	1000
NProbBE	0.1	2.3	1.2	1.4	3.1	1.2	1.1	1000
NProbBE	0.2	3.7	1.3	1.6	5.0	1.2	1.0	1000
NProbBE	0.3	6.1	1.5	2.2	8.2	1.2	1.0	1000
NProbBE	0.4	11	1.9	3.2	14.8	1.2	1.0	1000
OE2	0.05	5.2	1.1	1.2	7.0	1.1	0.9	1000
OE2	0.1	5.2	1.1	1.3	7.0	1.1	0.9	1000
OE2	0.2	5.1	1.2	1.4	6.9	1.2	1.0	1000
OE2	0.3	5.1	1.2	1.5	6.8	1.2	1.1	1000
OE2	0.4	4.9	1.3	1.6	6.6	1.3	1.2	1000
OE3	0.05	1.1	1.2	1.3	1.4	1.1	1.1	976
OE3	0.1	1.1	1.2	1.4	1.4	1.1	1.1	979
OE3	0.2	1.1	1.2	1.5	1.4	1.1	1.1	951
OE3	0.3	1.1	1.3	1.6	1.4	1.1	1.2	897
OE3	0.4	1.1	1.3	1.7	1.4	1.1	1.2	809
Prob	0.05	1	1.1	1.2	1.4	1.1	1.1	888
Prob	0.1	1	1.1	1.2	1.3	1.1	1.1	974
Prob	0.2	1	1.1	1.3	1.4	1.1	1.1	948
Prob	0.3	1	1.2	1.4	1.4	1.1	1.1	887
Prob	0.4	1	1.2	1.5	1.4	1.1	1.2	815
ProbBE	0.05	1	1.1	1.2	1.3	1.1	1.1	1000
ProbBE	0.1	1	1.1	1.3	1.3	1.1	1.1	1000
ProbBE	0.2	1	1.1	1.3	1.3	1.1	1.1	999
ProbBE	0.3	1	1.2	1.5	1.3	1.1	1.2	1000
ProbBE	0.4	1	1.3	1.8	1.3	1.2	1.3	1000

Ratio*X*, ratio of the true (i.e. simulated) EC*x* to the median estimate from 1000 simulated experiments; Mid*X*, ratio (75th percentile of estimated EC*x* values/25th percentile); Sprd*X*, ratio (90th percentile of estimated EC*x* values/10th percentile); NFIT is the number (out of 1000) of models that produced an EC*x* estimate; *C* is the simulated background incidence.

allows for different or equal parameters λ_l and γ_l for different replicates and successive models to explore the possibilities, following Petkau and Sitter. However, replicate-wise analysis is problematic except for some exploratory purposes, since replicate l of concentration i does not correspond in any meaningful way to replicate l in concentration $i^* \neq i$. It is also possible to model a common within-replicate by time error structure or allow for overdispersion or variance heterogeneity in various forms. The interest is in estimating LC50 (or LCx for some other x) as a function

Table 7.28 Summary of EC20 and EC30 Simulation Results for Slope 10, Sample Size 10

Model	*C*	Ratio20	Mid20	Sprd20	Ratio30	Mid30	Sprd30	NFIT
BVP	0.05	1.1	1.1	1.2	1.2	1.1	1.1	999
BVP	0.1	1.1	1.1	1.2	1.2	1.1	1.2	991
BVP	0.2	1.1	1.1	1.2	1.2	1.1	1.2	963
BVP	0.3	1.1	1.1	1.3	1.2	1.1	1.2	911
BVP	0.4	1.1	1.2	1.4	1.2	1.1	1.3	824
LLog	0.05	1.1	1.1	1.2	1.2	1.1	1.1	991
LLog	0.1	1.1	1.1	1.2	1.2	1.1	1.2	988
LLog	0.2	1.1	1.1	1.3	1.2	1.1	1.2	958
LLog	0.3	1.1	1.1	1.3	1.2	1.1	1.2	891
LLog	0.4	1.2	1.2	1.4	1.3	1.1	1.3	808
NProb	0.05	1.1	1.1	1.2	1.2	1.1	1.1	998
NProb	0.1	1.2	1.1	1.3	1.3	1.1	1.2	1000
NProb	0.2	2.3	1.8	2.4	2.5	1.5	1.9	997
NProb	0.3	6.4	1.4	2.0	6.9	1.3	1.7	1000
NProb	0.4	15.0	1.7	3.0	16.1	1.5	2.2	1000
NProbBE	0.05	2.0	1.1	1.3	2.1	1.1	1.3	1000
NProbBE	0.1	2.6	1.1	1.3	2.8	1.1	1.3	1000
NProbBE	0.2	4.1	1.2	1.4	4.4	1.2	1.3	1000
NProbBE	0.3	6.7	1.3	1.7	7.2	1.2	1.5	1000
NProbBE	0.4	12.2	1.5	2.3	13.1	1.4	1.8	1000
OE2	0.05	5.8	1.1	1.2	6.2	1.1	1.2	1000
OE2	0.1	5.8	1.1	1.3	6.2	1.1	1.3	1000
OE2	0.2	5.7	1.2	1.4	6.1	1.2	1.4	1000
OE2	0.3	5.6	1.2	1.5	6.1	1.2	1.5	1000
OE2	0.4	5.5	1.3	1.6	5.9	1.3	1.6	1000
OE3	0.05	1.2	1.1	1.2	1.3	1.1	1.2	976
OE3	0.1	1.2	1.1	1.3	1.3	1.1	1.2	979
OE3	0.2	1.2	1.2	1.3	1.3	1.1	1.3	951
OE3	0.3	1.2	1.2	1.4	1.3	1.2	1.3	897
OE3	0.4	1.2	1.2	1.5	1.3	1.2	1.4	809
Prob	0.05	1.1	1.1	1.2	1.2	1.1	1.1	888
Prob	0.1	1.1	1.2	1.2	1.2	1.1	1.2	974
Prob	0.2	1.1	1.1	1.2	1.2	1.1	1.2	948
Prob	0.3	1.1	1.1	1.3	1.2	1.1	1.2	887
Prob	0.4	1.1	1.2	1.4	1.2	1.1	1.3	815
ProbBE	0.05	1.1	1.1	1.2	1.2	1.1	1.1	1000
ProbBE	0.1	1.1	1.1	1.2	1.2	1.1	1.2	1000
ProbBE	0.2	1.1	1.1	1.3	1.2	1.1	1.2	999
ProbBE	0.3	1.1	1.1	1.4	1.2	1.1	1.3	1000
ProbBE	0.4	1.1	1.2	1.6	1.2	1.2	1.5	1000

Ratio*X*, ratio of the true (i.e. simulated) EC*x* to the median estimate from 1000 simulated experiments; Mid*X*, ratio (75th percentile of estimated EC*x* values/25th percentile); Sprd*X*, ratio (90th percentile of estimated EC*x* values/10th percentile); NFIT is the number (out of 1000) of models that produced an EC*x* estimate; *C* is the simulated slope background incidence.

of time. This is done by setting the distribution function F to 0.5 (or, more generally, to $x/100$) and solving the resulting equation for U. It is left as an exercise to verify

$$\text{LC}x = \exp\left\{\frac{\text{Ln}\left[\text{Ln}\left(x/100\right)\right] - \gamma\text{Ln}\left(t\right) - \alpha}{\beta}\right\}, \quad (7.25)$$

where common values of α, β, and γ are modeled. Confidence bounds can be obtained by the delta method.

Model (7.22) will be illustrated using the data from a daphnia chronic study given in Appendix 1 Table A1.7.12, where the concentrations are in mg l^{-1}, Day$x$ = survival after x days exposure, and Risk$x$ = number of daphnia exposed in indicated replicate. In addition to standard

Table 7.29 Parameter Estimates for Models Fit to Daphnia Survival Data

Normal errors					Binomial errors			
Est	StdErr	LCB	UCB	Parameter	Est	StdErr	LCB	UCB
−20.06	1.66	−23.33	−16.80	Alpha	−20.26	1.81	−23.92	−16.60
3.29	0.28	2.75	3.84	Beta	3.29	0.33	2.62	3.96
2.48	0.23	2.03	2.93	Gamma	2.55	0.23	2.08	3.02
				s1	0.87	0.20	0.46	1.28

Est, parameter estimate; StdErr, standard error of estimate; LCB and UCB, lower and upper 95% confidence bounds for estimate; s1, random error for overdispersion in binomial errors model.

probit models at each time, two Petkau–Sitter models were fit, one assuming a normal error structure and the other a binomial error structure. In the latter, overdispersion among replicates was modeled, as that gave a better fit than the alternative. Model parameter estimates were similar. In addition, probit models were fit to the 14- and 21-day data for comparison. A program to fit both models is *petkau data3.sas*, given in Appendix 1. Little difference in parameter estimates and predicted proportions between normal and binomial errors models was found for these data (see Tables 7.29 and 7.30). See also Table 7.31 for the LCx estimates from probit and binomial errors Petkau–Sitter model for 21-day survival not in agreement for $x<50$.

It should be noted that the Petkau–Sitter model performs poorly for some data, and even for the data illustrated here, that model could be fit only when several low concentrations exhibiting little or no mortality were omitted. This is one illustration to support a claim that the broader the application of a model, the less likely it is to fit a particular dataset. There is a trade-off between attempts to generalize and to describe the particular. Regulatory risk assessment is an ongoing attempt to bridge the gap between the particular and the broader ecological implications.

Table 7.30 Estimated Proportions from Models Fit to Daphnia Survival Data

Time	Conc	Atrisk	Died	ObsProp	PredNrml	PredBnml
1	16	4	0	0.000	0.000	0.000
2	16	4	0	0.000	0.000	0.000
4	16	4	0.1	0.025	0.001	0.001
7	16	4	0.1	0.025	0.002	0.003
14	16	4	0.1	0.025	0.012	0.019
21	16	4	0.2	0.050	0.033	0.052
1	36	4	0	0.000	0.000	0.000
2	36	4	0	0.000	0.001	0.001
4	36	4	0	0.000	0.008	0.006
7	36	4	0.4	0.100	0.032	0.026
14	36	4	0.7	0.175	0.164	0.139
21	36	4	0.7	0.175	0.387	0.336
1	81	4	0	0.000	0.004	0.005
2	81	4	0	0.000	0.020	0.031
4	81	4	0.2	0.050	0.109	0.166
7	81	4	3.2	0.800	0.370	0.502
14	81	4	3.7	0.925	0.924	0.915
21	81	4	3.7	0.925	0.999	0.983
1	180	4	0.1	0.025	0.050	0.039
2	180	4	0.2	0.050	0.249	0.206
4	180	4	3	0.750	0.797	0.709
7	180	4	4	1.000	0.998	0.986
14	180	4	4	1	1	1
21	180	4	4	1	1	1

PredNrml and PredBnml, predicted proportion dead from normal and binomial errors models, respectively; ObsProp, mean proportion dead in indicated concentration and time; Atrisk, number of Daphnia in replicate.

7.9 UP-AND-DOWN METHODS TO ESTIMATE LC50

Another technique for estimating LC50 is a sequential strategy that had its origins in explosives research (Dixon and Mood, 1948). The basic concept is very simple to describe in one of its early applications, determining the breaking strength of auto glass or the ability of armor to withstand a projectile. In the former, one starts with a weight of some magnitude and dropped from a fixed height. If the glass does not break, then the weight is dropped from a greater height. This continues until the glass breaks. (Actually a new piece of glass is used each time to prevent glass fatigue from biasing the result.) Once the glass breaks, a tentative upper bound of the height required to break the glass is set. Then lower heights are used until a tentative lower bound is found. Smaller changes in the heights are used at this stage than in the initial increases. Then greater heights are used (again in smaller increments) until once more the glass breaks. The process is repeated with the expectation that the breaking strength of the glass, as measured by the height of the drop, is spiraled into. With suitable refinements of the increments in height used, this can provide a good estimate of LC50, the median breaking

Table 7.31 LCx Estimates from Models Fit to Daphnia Survival to Day 21

Probit model				Binomial errors model		
LCx	LCB	UCB	X	LCx	LCB	UCB
5.62	3.64	7.86	10	22.58	17.62	27.54
10.78	7.69	14.33	20	28.37	23.02	33.72
17.24	12.89	22.64	30	32.72	27.07	38.37
25.76	19.60	34.20	40	36.50	30.55	42.45
37.48	28.45	51.27	50	40.05	33.77	46.33

Probit model results are for the standard probit model that ignores replicates; Binomial errors model results also follow a probit curve for treatment means but account for between-replicate variability.

strength. For armor piercing experiments, the height is replaced by the explosive charge in the projectile. A downside is also rather obvious, namely, that the process can go on for a very long time. In the case of breaking strength of glass, this may not be a major concern. However, in the ecotoxicity context, if one is trying to determine the median lethal dose of a chemical to birds, fish, mice, etc., it can be completely infeasible to maintain a stock of the organisms at the same stage of development, age, and growth to complete the study. Also, the length of time required to complete the study if one has to wait 14 days after exposure to determine whether a bird dies, for example, makes such an experiment impractical. One of the authors of this textbook, T.A. Springer, was an influential member of an expert team working with OECD that developed a very efficient sequential process that captures the "flavor" of the up-and-down method without the potential of a long sequence of steps. The resulting avian acute toxicity test guideline, OECD TG 223, will be described below. To put the revised TG 223 in perspective, a description of the previous avian acute test guideline will be given.

7.9.1 Previous Avian Acute Test Guideline

EPA OCSPP 850.2100 defines the definitive test as requiring of at least five test concentrations geometrically spaced plus a control, with 10 birds per treatment, for a total of 60 birds. This number does not include any range finding experiment. The birds are all dosed simultaneously, and a probit curve is fit to the resulting mortality data, if possible, with the goal of providing point and interval estimates of LC50 and the slope of the probit curve. Allowance is made for the case that a probit curve cannot be fit and the binomial or binary method of Section 7.10.3 is specifically supported, and the following general guidance is provided as well: "Statistical procedures for modeling quantal data are available and should be used." The guideline reference OCSPP 850.2000 for statistical methods that might be used and the various methods discussed in this chapter are included, including Section 7.10. For the test to be valid, there must be no more than 10% mortality (i.e. no more than one death) in the control.

Section (f) Paragraph 2 of the guideline contains the sentence "Test substance treatment data should be adjusted to account for any control mortality." This seems to indicate the normalization to the control that was severely criticized in Section 7.5. It is distressing how pervasive this erroneous and unnecessary normalization method is in ecotoxicity.

7.9.2 Revised Avian Acute Test Guideline: TG 223

The guideline describes three types of studies that can be used for avian acute lethality studies. The simplest is the **limit dose test.** This is appropriate when toxicity is expected to be low and lethality is unlikely at the limit dose, usually $2000\,\mathrm{mg\,kg^{-1}}$ – body weight. The guideline calls for 5 or 10 birds at the limit dose and 5 or 10 in a control group. The limit dose test begins with five control birds and five birds at the limit dose.

The most comprehensive of the three is the **LD50-slope test**. This is appropriate when the slope of the dose–response curve and both a point estimate and confidence interval are required for LD50. This is a three- or four-stage test with 24 or 34 birds in addition to the control group.

Intermediate between these is the **LD50-only test**, which is appropriate when only the point estimate of LD50 is required but not the slope or confidence interval. This test has two stages, with 14 birds in addition to the control.

For both of LD50-slope and LD50-only procedures, the data from the full study is analyzed, preferably by probit analysis or by one of the alternatives developed in this chapter, as determined by the data and model fitting criteria. What remains is to describe the different stages of these two procedures, that is, the dose selection and number of birds per dose at each stage. This can be rather tedious, and some details will be deferred to the free Excel-based software, SEDEC (2012), and its documentation, developed by T.A. Springer and available on the OECD website for downloading.

In determination of whether and, if so, how to proceed in the sequential strategy of this guideline, it is necessary to understand the terms "reversal" and "partial kill" as used here. A reversal occurs when the percent mortality is lower at the next higher dose than at a given dose. A partial kill occurs when multiple birds are given a single dose, and the percent mortality at that dose is greater than 0% and less than 100%. The rules for determining proceeding among the stages are given in Section 7.9.2.1.

7.9.2.1 Description of Stages

Limit Test (Five Birds at Limit Dose + Five Birds in Control)

If there is no death in limit dose, then LD50 > limit dose.
If there is only one death, dose five more birds at limit dose.
If there is only one death and no sign of toxicity in survivors, then LD50 > limit dose.
If there is sign of toxicity in survivors, go to stage 2.
If there are two to fourt deaths at limit dose, proceed to stage 2.
If all five birds die at the limit dose, proceed to stage 1.
Stage 1. Estimate LD50 from prior knowledge. Dose one bird at each of four doses, recalculate LD50 from all available data, and go to stage 2.
Stage 2. Dose 10 birds, 1 bird per dose. (Include stage 1 data if that stage was done.)
If there are two or more reversals, calculate LD50 from stages 1+2. Proceed to stage 3a.
If there is no more than one reversal, calculate LD50 from stages 1+2 and proceed to stage 3b.
Stage 3a. Add two doses, five birds per dose. Estimate LD50, slope, and confidence bounds from all data. No further testing required.
Stage 3b. Add five doses with two birds per dose.
If there are two or more reversals or two or more partial kills, then estimate LD50, slope, and confidence bounds from all data. No further testing required.
If there is no more than one reversal and no more than one partial kill, then calculate LD50 from stages 1+2+3b and proceed to stage 4.
Stage 4. Add five doses, with two birds per dose. Estimate LD50, slope, and confidence bounds from all data. No further testing required.

What is missing in the above description is how doses are selected at each stage and under what conditions the number of control birds is increased beyond the initial five. These two decisions are tedious and their justification is demanding technically. The reader is referred to the guideline and SEDEC (2012) for further discussion.

TG 223 is superficially similar to the mammalian test guideline, OECD TG 425. A critical difference between the two is that TG 425 is used for creating what are called labels, which describe how a plant protection product is to be used safely. TG 425 is not used for risk assessment, as the estimates coming from it are not sufficiently definitive for that purpose. TG 223 is used for risk assessment.

7.10 METHODS FOR EC*x* ESTIMATION WHEN THERE IS LITTLE OR NO PARTIAL MORTALITY

The models discussed above require multiple test concentrations with partial mortalities. It is common in toxicity studies for only one or no concentration to have partial mortality. (By a partial mortality meant the mortality rate is neither 0 nor 100%.) For example, Rufli and Springer (2011) reported on two large databases of fish acute toxicity tests done under OECD guideline 203 or USEPA guideline OPPTS 850.1075. They found that 75% of all studies had fewer than two test concentrations with partial mortalities and many had no partial mortalities. When there are fewer than two partial mortalities, the models discussed in this chapter will often fail. It is thus clear that methods are needed to estimate EC*x* that do not rely on the models presented in earlier sections of this chapter. Fortunately, there are alternative methods, some of which will be described in this section.

7.10.1 Moving Average Angle Estimation of LC50

This section describes how moving average and moving average angle estimates of LC50 and its approximate confidence bounds are calculated. The method is based on numerous published articles (Armitage and Allen, 1950; Bennett, 1952, 1963; Thompson, 1947; Thompson and Weil, 1952; Harris, 1959; Stephan, 1972; Engleman et al., 1986; Hoekstra, 1989, 1991; Peltier and Weber, 1985). Moving average angle estimates are generally used only for LC50 but are, in theory, possible for any percentile bracketed by observed probabilities. There must be successive groups whose mortality rates have 0.5 between them. For simplicity, this description will refer to mortality. However, the methodology applies equally well to nonlethal effects. Also, for clarity, mortality *rates* are expressed as proportions. It will be convenient to refer on occasion to mortalities in percentage terms, and whenever this is done, it will be done so explicitly.

Let U_i, $i=1, 2, 3, \ldots$ be the doses or concentrations in increasing order and let $L_i=\log(U_i)$. Since the computations will be in terms of L_i, U_1 corresponds to the first nonzero dose. Let p_i be the observed proportion affected at dose U_i.

The general case of possibly unequal sample sizes and unequal dose spacing will be developed using an "angular" transformation of the sample proportions that improves the performance of the estimator. More detail is provided in this section than might be expected since there is a perception in some recent publications (e.g. Morgan, 1992; Gad, 2006) that the moving average method is restricted to equal sample sizes or equal geometric spacing of doses. It is also appropriate to mention that while the focus in this section is on LC50 estimation, there is nothing in the method that prohibits estimation of LCx for values of x other than 50. Only in (7.20) and comments on it is there any implicit use of $x=50$, and that equation can be modified easily for other values of x.

The angular transformation is the familiar arc-sine square-root transform

$$y_i = \sin^{-1}\left(\sqrt{p_i}\right) \equiv g(p_i). \tag{7.26}$$

Here and in subsequent development in this section, if $p_j=0$, then it is replaced by some small positive value, and if $p_j=1$, then it is replaced by 1 – the same small positive value to prevent numerical problems. In Table 7.32, this value is taken as $1/n_j$, where n_i is the sample size in dose i. In the calculations for the Fieller bounds and the program *MovAvgAngle_LC50.SAS*, the value used in $1/(4n_j)$. It might be good to review the discussion of plotting locations in Section 5.2.2. The choice primarily affects the confidence bounds, not the point estimate.

From Eq. A2.73 in Appendix 2, it follows that

$$y_i \approx g(\theta_i) + g'(\theta_i)(p_i - \theta_i), \tag{7.27}$$

where θ_i is the true population proportion affected, for which p_i is the estimate, and $g'(x)$ denotes the derivative of the function g at x. From this it follows that

$$Var(y_i) = \frac{\theta_i(1-\theta_i)/n_i}{4\theta_i(1-\theta_i)} = \frac{1}{4n_i} \equiv \sigma_i^2. \tag{7.28}$$

Note the variance-stabilizing effect of the transformation, so that only the effects of differing sample sizes remain. The basic idea to be described is to compute a moving average of the (transformed) observed proportions y_i. At dose i, consider a weighted average of the observed proportions at that dose and the k doses closest to it on either side. Thus the moving average has the form

$$\hat{y}_i = \hat{\theta}_i = \sum_{j=-k}^{k} \lambda_{i+j} y_{i+j}. \tag{7.29}$$

In ecotoxicity studies with a small number of test concentrations, k is usually taken to be 1, which generates a moving average of three consecutive proportions, but the technique is more general. The coefficients λ are chosen to sum to 1 and to provide an unbiased estimator of $g(\theta_i)$ and to do so with minimum variance among all such estimators. It is left as an exercise to show that this is achieved if the λ are inversely proportional to the binomial weights of the sample proportions, that is,

$$\lambda_{i+j} = \frac{n_{i+j}\theta_i/\theta_{i+j}(1-\theta_{i+j})}{\sum_j \left[n_{i+j}/\theta_{i+j}(1-\theta_{i+j})\right]}, \tag{7.30}$$

so

$$\lambda_{i+j} = \lambda_i \frac{\sigma_i^2}{\sigma_{i+j}^2}, \tag{7.31}$$

and

$$\lambda_i = \frac{4n_i}{\sum_j 4n_{i+j}}. \tag{7.32}$$

It follows that

$$\mathrm{Var}(\hat{y}_i) \equiv V_i = \sum_j \lambda_{i+j}\,\mathrm{var}(y_{i+j}) = \sum_j \left(\frac{4n_{i+j}}{S_i}\right)^2 \frac{1}{4n_{i+j}} = \frac{\sum_j 4n_{i+j}}{S_i^2} = \frac{1}{S_i}, \tag{7.33}$$

where all sums, except as explicitly indicated otherwise, are from $j=-k$ to $j=k$ and

$$S_i = \sum_j 4n_{i+j}. \tag{7.34}$$

It follows that

$$\lambda_i = \frac{4n_i}{\sum_j 4n_{i+j}} \tag{7.35}$$

and

$$\hat{y}_i = \sum_{j=-k}^{k} \lambda_{i+j} y_{i+j} = \frac{\sum_j 4n_{i+j}y_{i+j}}{\sum_j 4n_{i+j}} = \frac{\sum_j 4n_{i+j}y_{i+j}}{S_i}. \tag{7.36}$$

The need is to bracket $p=0.5$ by successive proportions. Since the arc-sine square-root transform has been used, this means successive y values should bracket $\pi/4=0.785398$. So if $\hat{y}_i \le \pi/4 \le \hat{y}_{i+1}$ and $\hat{y}_i < \hat{y}_{i+1}$, then given that the doses have been (natural) log-transformed, an estimator of LC50 is given by exp(T^*), where

$$T^* = L_i^* + \frac{(\pi/4) - \hat{y}_i}{\hat{y}_{i+1} - \hat{y}_i}\left(L_{i+1}^* - L_i^*\right), \tag{7.37}$$

where

$$L_i^* = \frac{\sum_j n_{i+j}y_{i+j}}{\sum_j n_{i+j}}. \tag{7.38}$$

The variance of T^* can be developed from Eq. A2.73 in Appendix 2 and is given by

$$\operatorname{Var}(T^*) = \left(L_{i+1}^* - L_i^*\right)^2 \left[(1-\tau)^2 V_i + \tau^2 V_{i+1} + 2\tau(1-\tau)C_i\right], \tag{7.39}$$

where

$$\tau = \frac{(\pi/4) - \theta_i}{\theta_{i+1} - \theta_i} \approx \frac{(\pi/4) - \hat{y}_i}{\hat{y}_{i+1} - \hat{y}_i}, \tag{7.40}$$

$$C_i = \frac{4\sum_{j=-k+1}^{k} n_{i+j}}{S_{i+1}S_i}. \tag{7.41}$$

So approximate confidence bounds on LC50 are given by

$$\exp\left(T^* \pm 1.96\sqrt{\operatorname{var}(T^*)}\right). \tag{7.42}$$

These bounds are from Bennett (1963). Under some conditions, better bounds can be obtained using Fieller's theorem (Collett, 2002). However, Fieller's theorem requires the ability to estimate the intercept and slope of a probit model fit to the data, and the main use of the moving average angle estimator is in situations with no or only one partial mortality, in which case those parameter estimates are sometimes unavailable. Furthermore, as Harris (1959) demonstrated, Fieller bounds are unstable under some conditions. The program *mav example.sas* obtains Fieller bounds where they are stable and Bennett bounds otherwise and is given in Appendix 1. It is derived in part as a modification of a FORTRAN program written by J. Dryer of US Environmental Monitoring and Support Laboratory and edited by W.H. Peltier and C.I. Weber. A current link to that FORTRAN program has not been found.

These calculations will be illustrated with an example (see Table 7.32) using a span of three doses or test concentrations, i.e. $k=1$. In the special case of $k=1$, the following formulas apply:

$$\begin{aligned}
L_i^* &= \frac{n_{i-1}L_{i-1} + n_iL_i + n_{i+1}L_{i+1}}{n_{i-1} + n_i + n_{i+1}}.\\
\hat{y}_i &= \frac{n_{i-1}y_{i-1} + n_iy_i + n_{i+1}y_{i+1}}{n_{i-1} + n_i + n_{i+1}}.\\
S_i &= 4\left(n_{i-1} + n_i + n_{i+1}\right).\\
C_i &= \frac{4\left(n_i + n_{i+1}\right)}{S_iS_{i+1}}.\\
\tau &= \frac{(\pi/4) - \hat{y}_i}{\hat{y}_{i+1} - \hat{y}_i}.
\end{aligned} \tag{7.43}$$

Table 7.32 Example Moving Average Angle Calculations

i	Dose	L_i^*	p_i	y_i	$\hat{\theta}_i$
1	2	–	0	0.32175	.
2	4	1.36919	0	0.32175	0.32175
3	7.6	2.04524	0	0.32175	0.63085
4	15.2	2.71014	1	1.24905	0.93995
5	29.4	3.38757	1	1.24905	1.24905
6	58	–	1	1.24905	.

Each sample size is 10 and all p_i are 0 or 1.

From Table 7.32, it can be observed that $\hat{\theta}_3 s = \hat{y}_3 = 0.63085 < \pi/4 = 0.785398 < \hat{\theta}_4 = 0.93995$, so Log(LC50) is between 7.6 and 15.2. More precisely, the reader should verify that $\tau=0.5$, $\hat{\theta}_4 - \hat{\theta}_3 = 0.3091$, $S_3 = S_4 = 120$, $C_3 = 0.00555$, $T^* = 2.37777$, the estimated LC50 = 10.7809, and the 95% confidence interval for LC50 (by Bennett's method) is (8.3122, 13.9805). Normally, results would only be reported to one or two decimal places but are given here in more detail to allow checking of user calculations.

7.10.2 Fieller's Theorem

Fieller's theorem will be introduced as a general method of estimating the variance of a ratio and then will be applied to probit analysis and moving average angle estimation. Consider the problem of estimating the ratio of two possibly correlated random variables

$$\rho = \frac{\beta_1}{\beta_2} \tag{7.44}$$

and providing confidence bounds on that estimate. In particular, we want a solution in terms of estimates of β_1 and β_2. Let

$$\psi = \hat{\beta}_1 - \rho\hat{\beta}_2.$$

Then

$$E[\psi] = \beta_1 - \rho\beta_2 = 0.$$

If $\hat{\beta}_1$ and $\hat{\beta}_2$ are normally distributed, $Z_{\alpha/2}$ is the upper $\alpha/2$ – critical value of the standard normal distribution – and $V = \operatorname{var}\left(\hat{\beta}_1 - \rho\hat{\beta}_2\right)$, then

$$P\left[\left|\frac{\hat{\beta}_1 - \rho\hat{\beta}_2}{\sqrt{V}}\right| \le z_{\alpha/2}\right] = \alpha. \tag{7.45}$$

So, with probability $1-\alpha$, it follows that

$$\frac{\hat{\beta}_1 + z_{\alpha/2}\sqrt{V}}{\hat{\beta}_2} \ge \rho \ge \frac{\hat{\beta}_1 - z_{\alpha/2}\sqrt{V}}{\hat{\beta}_2}, \tag{7.46}$$

provided $\hat{\beta}_2>0$. However, the provision that $\hat{\beta}_2$ be positive is troubling, since a confidence interval for β_2 could very well contain zero. More troubling is the realization that V is itself a function of ρ, so this is not a real solution to the problem. To avoid these problems, with probability $1-\alpha$, we have

$$\left|\hat{\beta}_1-\rho\hat{\beta}_2\right|\leq z_{\alpha/2}\sqrt{V}$$

and

$$\left(\hat{\beta}_1-\rho\hat{\beta}_2\right)^2\leq z_{\alpha/2}^2V.$$

If the inequality is replaced by an equality, this becomes a quadratic equation in ρ, giving the extremes of the inequality:

$$\hat{\beta}_1^2+\rho^2\hat{\beta}_2^2-2\hat{\beta}_1\hat{\beta}_2\rho-z_{\alpha/2}^2V=0. \tag{7.47}$$

Now V is itself a function of ρ, namely,

$$V=\operatorname{var}\left(\hat{\beta}_1\right)+\rho^2\operatorname{var}\left(\hat{\beta}_2\right)-2\rho\operatorname{cov}\left(\hat{\beta}_1,\hat{\beta}_2\right) \tag{7.48}$$

or

$$V=V_{11}+\rho^2V_{22}-2\rho V_{12}. \tag{7.49}$$

Thus, the equation to solve is

$$\rho^2\left(\hat{\beta}_2^2-z_{\alpha/2}^2V_{22}\right)+\rho\left(-2\hat{\beta}_1\hat{\beta}_2+z_{\alpha/2}^2V_{12}\right)+\left(\hat{\beta}_1^2-z_{\alpha/2}^2V_{11}\right)=0. \tag{7.50}$$

The solutions are

$$\rho=\frac{\hat{\beta}_1\hat{\beta}_2-z_{\alpha/2}^2V_{12}\sqrt{\left(\hat{\beta}_1\hat{\beta}_2-z_{\alpha/2}^2V_{12}\right)^2-\left(\hat{\beta}_2^2-z_{\alpha/2}^2V_{22}2\right)\left(\hat{\beta}_1^2-z_{\alpha/2}^2V_{11}\right)}}{\hat{\beta}_2^2-z_{\alpha/2}^2V_{22}}. \tag{7.51}$$

Alternatively, if (7.50) is divided by $\hat{\beta}_2^2$ and m is defined by

$$m=\hat{\rho}=\frac{\hat{\beta}_1}{\hat{\beta}_2},$$

then the equation to solve can be written as

$$\rho^2\left(1-z_{\alpha/2}^2\frac{V_{22}}{\hat{\beta}_2^2}\right)+2\rho\left(-m+\frac{z_{\alpha/2}^2V_{12}}{\hat{\beta}_2^2}\right)+\left(m^2-z_{\alpha/2}^2\frac{V_{11}}{\hat{\beta}_2^2}\right)=0. \tag{7.52}$$

Let

$$g=\frac{z_{\alpha/2}^2V_{22}}{\hat{\beta}_2^2}. \tag{7.53}$$

Then (7.52) can be written as

$$\rho^2\left(1-g\right)+2\rho\left(-m+\frac{gV_{12}}{V_2}\right)+\left(m^2-g\frac{V_1}{V_2}\right)=0, \tag{7.54}$$

yielding the solutions for ρ as

$$\rho=\frac{m-g\left(V_{12}/V_{22}\right)\pm\left(z_{\alpha/2}/\hat{\beta}_2\right)\sqrt{V_{11}-2mV_{12}+m^2V_{22}-g\left[V_{11}-\left(V_{12}^2/V_{22}\right)\right]}}{1-g}. \tag{7.55}$$

Some statistical software uses Fieller bounds in reporting estimates from probit models. This includes SAS Proc Probit and the R-module GLM with family=binomial(link="probit"). The development of Fieller bounds can be related to moving average angle estimation by through the following observations. From (7.40), the ratio to be estimated is

$$\rho^*=\frac{(\pi/4)-\hat{\theta}_i}{\hat{\theta}_{i+1}-\hat{\theta}_i},$$

and the relevant variance matrix is given by

$$\operatorname{Var}(\hat{\theta})=\operatorname{var}\begin{pmatrix}\hat{\theta}_i\\ \hat{\theta}_{i+1}\end{pmatrix}=\begin{bmatrix}V_1 & C_i\\ C_i & V_2\end{bmatrix}, \tag{7.56}$$

So if

$$\operatorname{Var}\begin{pmatrix}\frac{\pi}{4}-\hat{\theta}_i\\ \hat{\theta}_{i+1}-\hat{\theta}_i\end{pmatrix}=\begin{bmatrix}V_{11} & V_{12}\\ V_{12} & V_{22}\end{bmatrix}, \tag{7.57}$$

then

$$V_{11}=V_1=\frac{1}{S_1}, \tag{7.58}$$

$$V_{22}=\operatorname{var}\left(\hat{\theta}_i\right)+\operatorname{var}\left(\hat{\theta}_{i+1}\right)-2\operatorname{cov}\left(\hat{\theta}_i,\hat{\theta}_{i+1}\right)=V_i+V_{i+1}-2C_i, \tag{7.59}$$

$$V_{12}=V_1-C_i. \tag{7.60}$$

Thus, if LCB and UCB are the confidence bounds on ρ^*, then the bounds on T^* are

$$L_i^*+\text{LCB}\left(L_{i+1}^*-L_i^*\right)\quad\text{and}\quad L_i^*+\text{UCB}\left(L_{i+1}^*-L_i^*\right).$$

According to Harris (1959), these Fieller bounds are unstable for $g>0.85$. These calculations will be illustrated with the same data as in Table 7.11. It can be shown that

$$\hat{\beta}_2=\hat{\theta}_{i+1}-\hat{\theta}_i=0.93994-0.63084=0.309,\quad\text{so}$$

$$m=\hat{p}=\frac{(\pi/4)-\hat{\theta}_i}{\hat{\theta}_{i+1}-\hat{\theta}_i}=\frac{\hat{\beta}_1}{\hat{\beta}_2}=0.5.$$

$$\text{Var}\begin{pmatrix}\hat{\theta}_i\\ \hat{\theta}_{i+1}\end{pmatrix}=\begin{bmatrix}V_i & C_i\\ C_i & V_{i+1}\end{bmatrix}=\begin{bmatrix}0.008\overline{3} & 0.00\overline{5}\\ 0.00\overline{5} & 0.008\overline{3}\end{bmatrix}.$$

$$g=\frac{V_{22}}{\hat{\beta}_2^2}z_{0.025}^2=0.22338.$$

$$\text{Var}\begin{pmatrix}\hat{\beta}_1\\ \hat{\beta}_2\end{pmatrix}=\begin{bmatrix}V_{11} & V_{12}\\ V_{12} & V_{22}\end{bmatrix}=\begin{bmatrix}0.008\overline{3} & 0.00\overline{5}\\ 0.00\overline{5} & 0.008\overline{3}\end{bmatrix}.$$

$$\rho=\frac{m-g\left(V_{12}/V_{22}\right)\pm\left(z_{\alpha/2}/\hat{\beta}_2\right)\sqrt{V_{11}-2mV_{12}+m^2V_{22}-g\left[V_{11}-\left(V_{12}^2/V_{22}\right)\right]}}{1-g}$$
$$=-0.09961,\ 1.09961\equiv(\rho_L,\rho_U)$$

$$\text{LB}=L_i^*+\rho_L\left(L_{i+1}^*-L_i^*\right)=2.045-0.09961^*(0.6649)$$
$$=1.97901.$$

$$\text{UB}=L_i^*+\rho_U\left(L_{i+1}^*-L_i^*\right)=2.045-1.09961^*(0.6649)$$
$$=2.77637.$$

And finally, the 95% confidence interval for LC50 (Fieller method) is (7.24, 16.06). (Note: With the other choice of adjusting the p^* values, the bounds are [8.17, 14.23].) The Fieller interval is thus wider than Bennett's and presumably better reflects the uncertainty in the estimate.

7.10.3 Binary Method of Estimating LC50

The binary method for estimating LC50 is the simplest and is only available in the case there are no partial mortalities, that is, in every treatment group (including control), either 0 or 100% of the subjects die, and there is some observed positive concentration below which every group has 0% mortality and some observed concentration above which there is 100% mortality. References to mortality can be replaced by references to the incidence of some other characteristic, such as seedling emergence or egg hatching or a rat having a specific type of tumor. See Table 7.3 for an example.

Binary estimates of LC50 are based on simple binomial probabilities and are available in the case that at least one group exhibited 0% effect, at least one group exhibited 100% effects, and no group exhibited partial effects. Furthermore, each group with 100% effects had a higher exposure level than all groups with 0% effects. Under these conditions, there is exactly one concentration, U_i, with $p_i=0$ and $p_{i+1}=1.0$. Let n_i, n_{i+1} be the number of subjects exposed at doses U_i, U_{i+1}, respectively. The estimate of EC50 is the geometric mean of the two concentrations, U_i and U_{i+1}. The probability of observing 0 effects at concentration U_i if the true incidence rate is 0.5 or greater is at most $(1/2)^{n_i}$. With n_i replaced by n_{i+1}, this is also the probability of observing total mortality at concentration U_{i+1} if the true incidence rate is 0.5 or less. Thus, if n_i and n_{i+1} are both at least 6, then the interval $[U_i, U_{i+1}]$ contains the 95% fiducial bounds for LC50. If $n_i<6$, then the lower fiducial bound will drop to U_{i-1}, provided $n_i+n_{i-1}\geq 6$. If this sum is still less than 6, the lower fiducial bound drops to the next lower concentration and so on until at least six subjects total are obtained. Similar observations apply to the upper fiducial bound.

Table 7.33 Example Data with No Partial Mortality

Group	Dose	p_i
1	58	0
2	82	0
3	118	0
4	168	0
5	240	0
6	343	0
7	490	0
8	700	1.0

p_i is the observed proportion mortality in the ith treatment; The LC50 estimate from these data will be between groups 7 and 8, assuming $n_i=10$ subjects in each concentration is estimated to be $\sqrt{490*700}=585.6620$.

Assuming $n_i=10$ subjects in each concentration, the fiducial bounds for the data in Table 7.33 are [490, 700]. The actual probability of observing 0 effects at 490 if the true incidence rate is exceeds 0.5 is 0.000976556. This is also the probability of observing 10 dead at dose 700 if the true incidence rate is less than 0.5. Next suppose the sample size per concentration is 5.

Then the estimated LC50 is unchanged. However, the 95% fiducial bounds are then [240, ∞). The actual probability of observing 0 effects at 490 if the true incidence rate exceeds 0.5 is 0.03125. The probability of observing five dead at dose 700 if the true incidence rate is less than 0.5 is also 0.03125, so the interval [343, 700] is a 93.75% fiducial interval. A program to carry out the binomial LC50 calculations is given by *Binomial_LC50.sas*. It would also be very simple to set this up in Excel or R.

7.10.4 Trimmed Spearman–Karber Estimates

This is a nonparametric procedure that has been widely used. Hamilton et al. (1977) point out several issues with probit and logit models that they claim make their use to estimate LC50 problematic. They motivate their trimmed Spearman–Karber method in part by criticizing the two leading parametric models, logit and probit, in the following way: (i) The maximum likelihood (ML) method used in these two models may not converge. (ii) The ML method may converge but may depend on the starting value used in the fitting routine. (iii) Repeated experiments under the same conditions can produce data leading to notably different LC50 estimates under these two models. They state, for example, that "Even if the population tolerance distribution does correspond to the probit or logit models, it is certainly possible that an especially sensitive fish could be randomly assigned to a low concentration tank; that fish would die although no fish die in the tank of next highest concentration. The statistical methodology should not be so delicate that it is seriously affected by such a mortality."

The first two of these criticisms apply to all models fit to biological data. The third also applies if LC50 is replaced by a generic EC*x*, since a 50% effect is not always the focus of an analysis, especially for sublethal effects. One must be careful in evaluating model fits, and criteria for doing so were provided in Chapter 4 for continuous responses. The heart of their preference for their nonparametric approach seems, to us, to be the potential sensitivity of the models to expected variation in biological data. We think there is merit in the criticism and have presented computer modeling results that quantify the variation in EC*x* estimates that can result. However, it is not entirely clear that a method that is insensitive to such data variations is an improvement. Regarding the sensitive fish in the quote, is this a matter of one sensitive fish, or might it be that the EC*x* for some small x is in range between the two "offending" treatment groups, so that the chance of a lethality occurring is small, but not zero? If so, should that be ignored in model fitting? The width of the confidence interval for LC50 is some indication of the uncertainty in the estimate. A method that has tight confidence bounds may seem attractive, but does it really mask the uncertainty in the data? In our opinion, overinterpreting model results is a danger in risk assessment. It is one of the reasons regulatory requirements require several types of studies on many species (e.g. limit tests, acute and chronic studies, multigeneration studies) and tests of many different species in the risk assessment process. Furthermore, how serious the effect of one sensitive fish might have on the LC50 estimate would depend on the sample size as well as the model fit.

It is also important to realize that the trimmed Spearman–Karber method requires zero mortality in the control, which, as discussed above, is not true in a nontrivial proportion of studies. Hamilton et al. (1977) state that when there is control mortality, Abbott's formula should be used first to normalize the data to the control. In Section 7.5, it has been shown that very large errors are frequently introduced when this is done. Thus, there is a fundamental problem in the use of the Spearman–Karber or trimmed Spearman–Karber when there is control mortality. When there is no control mortality, the method can be used and is therefore described below.

The object is to describe the function $P(U)$ that gives the probability of death (or occurrence of whatever nonlethal response is of interest) as a function of the test concentration, U. Let $p_i = P(U_i)$, $i = 1$ to k, where U_i is the ith positive test concentration. Next, these proportions p_i must be adjusted so as to make $p_1 \le p_2 \le p_3 \le \cdots \le p_k$. This is done using the PAVA algorithm discussed in connection with Williams' test. (Some sources refer to this as the ABERS algorithm after the original authors of the techniques: Ayer et al., 1955.) In the remaining discussion, it will be assumed that p_i are the result of this amalgamation process. Next, the points (U_i, p_i) are joined by line segments to form a polygonal curve that, perhaps crudely, describes the cumulative relative proportions. Next, choose a percent trim, α, to use, where $0 \le \alpha < 50$, and trim (i.e. remove) the lowest and highest α% of the p_i. If $\alpha = 0$, then one obtains the (untrimmed or conventional) Spearman–Karber method. Next, replace p_i by

$$\hat{p}_i = \frac{p_i - \alpha/100}{1 - 2\alpha/100}. \quad (7.61)$$

Discard values less than 0 or greater than 1. The resulting polygon, $F(u)$, defined by connecting the points $(U_i, \hat{p}_i)$ is the empirical tolerance distribution function of the middle $(100 - 2\alpha)$% of the tolerances. The final step is to compute an appropriate mean of this distribution, that is, estimate $E[U]$. One estimate is given by the rightmost member of (7.62):

$$E[U] = \int_{-\infty}^{\infty} u\,dF(u) \approx \sum_j \frac{u_j + u_{j-1}}{2}\left[\hat{p}_j - \hat{p}_{j-1}\right]. \quad (7.62)$$

Since the tolerance distribution is best realized in terms of Ln(U), (7.62) is replaced by (7.63):

$$E[\mathrm{Ln}(U)] = \int_{-\infty}^{\infty} \mathrm{Ln}(u)\,dF(u) \approx \sum_j \frac{\mathrm{Ln}(u_j) + \mathrm{Ln}(u_{j-1})}{2}\left[\hat{p}_j - \hat{p}_{j-1}\right]. \quad (7.63)$$

Then LC50 is estimated as exp($E[U]$). These calculations will be illustrated using the mite data, ignoring the mortalities in the control. This will be done with $\alpha = 0$ and then with $\alpha = 10$. See Table 7.34 for the modified mite data with additional calculations. It should be observed that the LC50 estimate is very similar to that reported in Table 7.5 from the probit model.

Table 7.34 Spearman–Karber Calculations for Mite Data with $\alpha = 0$

U_i	$\text{Ln}(U_i)$	N_i	X_i	p_i	$(p_i - p_{i-1})$	$[\text{Ln}(U_i) + \text{Ln}(U_{i-1})]/2$	Product
18.75	2.931194	20	2	0.1	.	.	
37.5	3.624341	21	3	0.1428	0.0428	3.277767	0.140
75	4.317488	21	10	0.4762	0.3334	3.970915	1.324
150	5.010635	20	18	0.9	0.4238	4.664062	1.977
300	5.703782	20	20	1	0.1	5.357209	0.536
Sum							3.977
Antilog							53.332

U_i and $\text{Ln}(U_i)$ are the test concentrations and natural logs, respectively; N_i and X_i are the number of mites exposed and dead, respectively, and $p_i = X_i/N_i$; Product $= (p_i - p_{i-1}) * [\text{Ln}(U_i) + \text{Ln}(U_{i-1})]/2$; Sum, sum of products; Antilog, exp(sum) = LC50 estimate.

Table 7.35 Spearman–Karber Calculations for *Daphnia magna* Chronic Data with $\alpha = 0$

U_i	$\text{Ln}(U_i)$	N_i	X_i	p_i	$(p_i - p_{i-1})$	$[\text{Ln}(U_i) + \text{Ln}(U_{i-1})]/2$	Product
0	.	10	1	0.1	.		
0.65	−0.431	10	0	0	.		
1.3	0.262	10	2	0.15	0.15	−0.084	−0.013
2.5	0.916	10	1	0.15	0	0.589	0.000
5	1.609	10	9	0.9	0.75	1.263	0.947
10	2.303	10	10	1	0.1	1.956	0.196
Sum							1.130
Antilog							3.096

U_i and $\text{Ln}(U_i)$ are the test concentrations and natural logs, respectively; N_i and X_i are the number of daphnia exposed and dead, respectively; p_i, adjusted mortality proportions after PAVA amalgamation; Product $= (p_i - p_{i-1}) * [\text{Ln}(U_i) + \text{Ln}(U_{i-1})]/2$; Sum, sum of products; Antilog = exp(sum) = LC50 estimate.

The LC50 estimate of 53.3 is inconsistent with the observed probabilities and compares poorly with the results obtained by probit, logit, and Weibull models. It is left as an exercise to show that a 10% trimmed Spearman–Karber LC50 estimate is 73.8, which is more reasonable but still seemingly low compared with the observed mortalities. A 15 or 20% trim eliminates all but one observation and produces an estimate quite at odds with the overall data. So it would seem that this approach is not satisfactory for this one example. Of course, one should not conclude on the basis of a single example that trimmed Spearman–Karber is a poor LC50 estimator. For another example, see Table 7.35, which comes from a chronic *Daphnia magna* study (OECD TG 211).

The LC50 estimate appears consistent with the observed mortalities. As before, the control mortality has been ignored in the calculations. See Table 7.36 for the calculations for these data with a 10% trim.

The LC50 estimate appears consistent with the observed mortalities. As before, the control mortality has been ignored in the calculations. There is no basis for preferring the 0 or 10% trim over the other. The more relevant observation is that there is little difference between the two estimates.

Confidence bounds for the LC50 estimate are needed. To obtain them, it is helpful to rearrange (7.63) to get (7.64):

$$E[\text{Ln}(U)] \approx -\hat{p}_1 \frac{\text{Ln}(u_1) + \text{Ln}(u_2)}{2} + \sum_{j=2}^{k-1} \hat{p}_j \frac{\text{Ln}(u_{j-1}) - \text{Ln}(u_{j+1})}{2} + \hat{p}_k \frac{\text{Ln}(u_k) + \text{Ln}(u_{k-1})}{2}. \tag{7.64}$$

Since $\hat{p}_j$ are independent (after the PAVA amalgamation), then the variance of the estimator of $E[\text{Ln}(U)]$ is obtained easily from (7.64) to be

$$\text{Var}(\hat{\mu}) \approx \frac{\hat{p}_1(1-\hat{p}_1)}{n_1}\left[\frac{\text{Ln}(u_1) + \text{Ln}(u_2)}{2}\right]^2 + \sum_{j=2}^{k-1} \frac{\hat{p}_j(1-\hat{p}_j)}{n_j}\left[\frac{\text{Ln}(u_{j-1}) - \text{Ln}(u_{j+1})}{2}\right]^2 + \frac{\hat{p}_k(1-\hat{p}_k)}{n_k}\left[\frac{\text{Ln}(u_{k-1}) + \text{Ln}(u_k)}{2}\right]^2, \tag{7.65}$$

Table 7.36 Spearman–Karber Calculations for *Daphnia magna* Chronic Data with $\alpha = 10$

U_i	$\text{Ln}(U_i)$	N_i	X_i	p_i	$\text{Adj}p_i$	$(pi - p_{i-1})$	$[\text{Ln}(U_i) + \text{Ln}(U_{i-1})]/2$	Product
0	.	10	1	0.1	0			
0.65	−0.431	10	0	0	−0.125			
1.3	0.262	10	2	0.15	0.0625			
2.5	0.916	10	1	0.15	0.0625	0.000	0.589	0.000
5	1.609	10	9	0.9	1	0.938	1.263	1.184
10	2.303	10	10	1	1.125			
Sum								1.184
Antilog								3.267

U_i and $\text{Ln}(U_i)$ are the test concentrations and natural logs, respectively; N_i and X_i are the number of daphnia exposed and dead, respectively; p_i, adjusted mortality proportions after PAVA amalgamation; $\text{Adj}p_i = p_i$ adjusted for trim: $(p_i - 0.1)/0.8$; Product = $(\text{Adj}p_i - \text{Adj}p_{i-1})$ * $[\text{Ln}(U_i) + \text{Ln}(U_{i-1})]/2$; Sum, sum of products; Antilog = exp(sum) = LC50 estimate.

Table 7.37 Spearman–Karber Confidence Interval for Daphnia LC50 with 0% Trim

$\text{Ln}(U_i)$	p_i	$\text{Var}(p_i)$	Coef	Product	Halfwidth	LCB	UCB
.	0.1						
−0.43078	0						
0.262364	0.15	0.006	0.876	0.006			
0.916291	0.15						
1.609438	0.9	0.009	0.480	0.004			
2.302585	1	0.000	3.826	0.000			
Sum				0.010	0.195	2.547	3.763

$\text{Ln}(U_i)$, natural log of test concentration; p_i, adjusted mortality proportions after PAVA amalgamation; $\text{Var}(p_i)$ is the variance of the indicated proportion from (7.64); Coef is the coefficient of that variance term in (7.64); Product = Coef * $\text{var}(p_i)$; Sum = $\text{var}(\hat{\mu})$ from (7.65); Halfwidth = square root of $\text{var}(\hat{\mu})$* 1.96 obtained by assumed normal distribution of $\hat{\mu}$; LCB and UCB = $\exp(\hat{\mu} \pm \text{halfwidth})$.

Table 7.38 Spearman–Karber Confidence Interval for Daphnia LC50 with 10% Trim

$\text{Ln}(U_i)$	p_i	$\text{Adj}p_i$	$\text{Var}(\text{Adj}p_i)$	Coef	Product	Halfwidth	LCB	UCB
.	0.1	0						
−0.43078	0	−0.125						
0.262364	0.15	0.0625						
0.916291	0.15	0.0625	0.006	2.590	0.015			
1.609438	0.9	1	0.000	100.000	0.000			
2.302585	1	1.125						
Sum					0.015	0.241	2.566	4.160

$\text{Ln}(U_i)$, natural log of test concentration; p_i, adjusted mortality proportions after PAVA amalgamation; $\text{Var}(p_i)$ is the variance of the indicated proportion from (7.64); Coef is the coefficient of that variance term in (7.65); Product = Coef * $\text{var}(p_i)$; Sum = $\text{var}(\hat{\mu})$ from (7.64); Halfwidth = square root of $\text{var}(\hat{\mu})$ * 1.96 obtained by assumed normal distribution of $\hat{\mu}$; LCB and UCB = $\exp(\hat{\mu} \pm \text{halfwidth})$.

where $\hat{\mu}$ is the estimate of Ln(U) given by the right-hand side of (7.64).

It is important to apply (7.64) to the amalgamated proportions and the test concentrations and sample sizes restricted to those that remain after the amalgamation procedure, since otherwise the variance terms do not add. To avoid ambiguity of the coefficient of the variance of the $\text{Var}(\hat{p}_j)$ term, it is to be understood that u_{j-1} is the largest test concentration less than those amalgamated to obtain $\hat{p}_j$ and u_{j+1} is the smallest test concentration greater than those amalgamated to obtain $\hat{p}_j$. This is illustrated with the *D. magna* data (see Tables 7.37 and 7.38).

An R package that includes the Spearman–Karber trimmed LD50 estimator is found at https://r-forge.r-project.org/R/?group_id=1169 (accessed 9 October 2017). The command to install this package, which is not in the CRAN library, is

Table 7.39 Daphnia Chronic Mortality Data 2

Conc ($mg\,l^{-1}$)	Rep	Survive1	Survive2	Survive4	Survive7	Survive14	Survive21	Atrisk
0	1	4	4	4	4	4	4	4
0	2	4	4	4	4	4	4	4
0	3	4	4	4	4	4	4	4
0	4	4	4	4	4	4	4	4
0	5	4	4	4	4	4	4	4
0	6	4	4	4	4	4	4	4
0	7	4	4	4	4	4	3	4
0	8	4	4	3	3	3	3	4
0	9	4	4	4	4	4	4	4
0	10	4	4	4	4	3	3	4
0.0066	1	4	4	4	4	4	3	4
0.0066	2	4	4	4	4	4	4	4
0.0066	3	4	4	4	4	4	4	4
0.0066	4	4	4	4	4	4	4	4
0.0066	5	4	4	4	4	3	3	4
0.0066	6	4	4	3	3	3	3	4
0.0066	7	4	4	4	4	4	3	4
0.0066	8	4	4	4	4	4	4	4
0.0066	9	4	4	4	4	4	4	4
0.0066	10	4	4	4	3	3	3	4
0.0027	1	4	4	4	4	4	4	4
0.0027	2	4	4	4	4	4	4	4
0.0027	3	4	4	3	3	3	3	4
0.0027	4	4	4	4	4	4	4	4
0.0027	5	4	4	4	4	4	4	4
0.0027	6	4	4	4	4	4	4	4
0.0027	7	4	4	4	4	4	4	4
0.0027	8	4	4	4	4	4	4	4
0.0027	9	4	3	3	3	3	3	4
0.0027	10	4	4	4	4	4	4	4
0.021	1	4	4	4	4	4	3	4
0.021	2	4	4	4	4	4	4	4
0.021	3	4	4	4	4	4	4	4
0.021	4	4	4	4	4	4	4	4
0.021	5	4	4	4	3	3	3	4
0.021	6	4	4	4	4	4	4	4
0.021	7	4	4	4	4	4	4	4
0.021	8	4	4	4	4	4	4	4
0.021	9	4	4	4	4	4	4	4
0.021	10	4	4	4	4	4	4	4
0.065	1	4	4	4	4	4	4	4
0.065	2	4	4	4	4	4	4	4
0.065	3	4	4	4	4	3	2	4
0.065	4	4	4	4	4	4	4	4
0.065	5	4	4	3	3	3	3	4
0.065	6	4	4	4	3	3	3	4
0.065	7	4	4	4	4	4	4	4
0.065	8	4	4	4	4	4	3	4
0.065	9	4	3	3	3	3	3	4
0.065	10	4	4	4	4	4	4	4
0.18	1	4	4	4	4	4	3	4

Table 7.39 (Continued)

Conc (mg l^{-1})	Rep	Survive1	Survive2	Survive4	Survive7	Survive14	Survive21	Atrisk
0.18	2	4	4	4	4	4	4	4
0.18	3	4	4	4	4	4	3	4
0.18	4	4	4	4	4	4	4	4
0.18	5	4	4	4	4	4	4	4
0.18	6	4	4	4	4	4	4	4
0.18	7	4	4	4	4	4	4	4
0.18	8	4	4	4	4	4	4	4
0.18	9	4	4	4	4	4	4	4
0.18	10	4	4	4	4	4	4	4
0.54	1	4	2	1	1	1	1	4
0.54	2	4	4	4	4	4	4	4
0.54	3	4	4	4	4	4	4	4
0.54	4	4	3	2	2	2	2	4
0.54	5	4	2	1	1	1	1	4
0.54	6	3	3	2	2	2	2	4
0.54	7	4	3	3	2	2	2	4
0.54	8	4	2	2	2	2	1	4
0.54	9	4	4	3	3	3	3	4
0.54	10	4	3	2	2	2	2	4
1.5	1	4	3	0	0	0	0	4
1.5	2	3	2	0	0	0	0	4
1.5	3	4	1	0	0	0	0	4
1.5	4	3	2	0	0	0	0	4
1.5	5	3	1	0	0	0	0	4
1.5	6	4	3	0	0	0	0	4
1.5	7	4	2	0	0	0	0	4
1.5	8	4	3	1	0	0	0	4
1.5	9	3	0	0	0	0	0	4
1.5	10	4	2	0	0	0	0	4

Survive*x*, survival after *x* days exposure; Atrisk, number of daphnid exposed.

R install command: install.packages("tsk", repos= "http://R-Forge.R-project.org"). A simple program giving an alternative way to install this and several other necessary packages is given in Appendix 1 with the name *SpearmanKarber.r*. It is illustrated with a different dataset. No SAS program is provided for this.

EXERCISES

In all datasets in the text and exercises, if a concentration is given with the value −1, that means a solvent used in the study and the water control is recorded as −1 and the solvent control is listed as 0. The reader should determine whether to combine the two controls for NOEC determination or EC*x* estimation. If the two controls are to be combined, then the water control should be changed to 0 and the replicates renamed so that replicates in the two controls can be distinguished. If only the solvent control group is to be used, then the water control should be discarded. If only the water control should be used, then the solvent control should be discarded and the concentration for the water control should be changed to 0. Under no circumstances should the concentration value −1 be used for NOEC or EC*x* purposes. Guidance on this is provided in Chapters 2 and 3.

7.1 **(a)** Verify the formula (7.10).

(b) Verify formula (7.16).

(c) Verify formula (7.13).

7.2 Show that Tarone's test applied to the data in Table 7.1 without the high concentration is not significant.

7.3 Compute EC*x* estimates x=5, 10, 20, 30, 40, 50 and their confidence bounds for the Weibull model of the mite mortality data in Table 7.1.

7.4 **(a)** For the *Daphnia magna* survival data (see Table 7.39), also provided as Table A1.7.39 in Appendix 1, fit appropriate models to estimate LC*x* for 21-day survival.

Table 7.40 Daphnia Chronic Mortality Data 3 (μg l^{-1})

Conc	Rep	Day1	Risk1	Day2	Risk2	Day4	Risk4	Day7	Risk7	Day 14	Risk 14	Day 21	Risk 21
0	1	10	10	10	10	10	10	10	10	10	10	8	10
0	2	10	10	10	10	10	10	10	10	10	10	10	10
0.05	1	10	10	10	10	10	10	10	10	10	10	10	10
0.05	2	10	10	10	10	10	10	9	10	4	10	4	10
0.1	1	10	10	10	10	10	10	9	10	3	10	3	10
0.1	2	10	10	10	10	10	10	10	10	10	10	10	10
0.589	1	10	10	10	10	10	10	10	10	10	10	10	10
0.589	2	10	10	10	10	10	10	10	10	10	10	10	10
4.14	1	10	10	10	10	10	10	10	10	10	10	10	10
4.14	2	10	10	10	10	10	10	10	10	9	10	9	10
20	1	10	10	10	10	10	10	10	10	10	10	8	10
20	2	10	10	10	10	10	10	10	10	9	10	9	10
105	1	10	10	9	10	9	10	9	10	9	10	9	10
105	2	10	10	10	10	10	10	10	10	9	10	6	10
403	1	10	10	10	11	4	11	4	11	0	11	0	11
403	2	11	11	11	11	5	11	4	11	0	11	0	11

Day*x*, survival after *x* days exposure; Risk*x*, number of daphnid exposed.

(b) Do the same for 14-day survival.

(c) Do the same for 7-day survival.

(d) Fit the Petkau–Sitter model to these data and discuss its fit to the data.

7.5 **(a)** For the *Daphnia magna* survival data (see Table 7.40), fit appropriate models to estimate LC*x* for 21-day survival.

(b) Do the same for 14-day survival.

(c) Do the same for 7-day survival.

(d) Fit the Petkau–Sitter model to these data and discuss its fit to the data.

7.6 Verify formulas (7.19) and (7.20).

7.7 **(a)** Verify formula (7.24).

(b) Use a program in Appendix 1 to verify the Petkau–Sitter results in Tables 7.11–7.13.

(c) Modify the binomial errors Petkau–Setter program to obtain LC*x* estimates at 7, 10, and 14 days.

(d) Compare the point and interval estimates obtained in (c) at days 7 and 14 with those obtained from probit analysis.

7.8 **(a)** Use a program in Appendix 1 to calculate the variance–covariance matrix of the parameters in the normal errors Petkau–Sitter model for the data in Section 7.8.

(b) Use the delta method to calculate the LC*x* estimates and their 95% confidence bounds for the Daphnia chronic mortality data presented in Section 7.8 from the normal errors model.

7.9 See Table 7.41 for the data for this example. Do a moving average angle estimate of LC50 for the Mysid mortality data in Table 7.41. It is necessary to combine the water and solvent controls for analysis.

Table 7.41 Mysid Mortality After 96 Hours' Exposure

Group	AtRisk	Dead	%Dead	Conc (μg l^{-1})
1	20	0	0	−1
2	20	0	0	0
3	20	2	10	12
4	20	0	0	24
5	20	20	100	46
6	20	20	100	110
7	20	20	100	190

Conc = −1 is label for water control; Conc = 0 is label for solvent control; As always, the controls should be compared; If no significant difference is found, then the controls should be combined and labeled as conc = 0 for both and replicates relabeled to avoid duplicate names; Otherwise, the water control is dropped, and further analysis should use only the solvent control.

7.10 Do a moving average angle estimate of LC50 for the stickleback mortality data (see Table 7.42). It is necessary to combine the water and solvent controls for analysis.

7.11 Do a moving average angle estimate of LC50 for the earthworm mortality data (see Table 7.43). It is necessary to combine the water and solvent controls for analysis.

7.12 Estimate LC50 for the fish acute mortality data (see Table 7.44).

(a) Do this by probit and moving average angle estimation.

(b) Fit a continuous model to the observed proportions or the arc-sine square-root transform of the observed proportions to estimate LC50. Be sure to estimate the 50% lethal concentration with respect to observed concentrations and proportions. For this, the raw data (see Table 7.45) may be useful.

Table 7.42 Stickleback Mortality After 96 Hours' Exposure

Group	AtRisk	Dead	%Dead	Conc (μg l^{-1})
1	20	0	0	−1
2	20	0	0	0
3	20	2	10	12
4	20	0	0	24
5	20	20	100	46
6	20	20	100	110
7	20	20	100	190

Conc = −1 is label for water control; Conc = 0 is label for solvent control. As always, the controls should be compared; If no significant difference is found, then the controls should be combined and labeled as conc = 0 for both and replicates relabeled to avoid duplicate names; Otherwise, the water control is dropped, and further analysis should use only the solvent control.

Table 7.43 Earthworm Mortality After 96 Hours' Exposure

Group	AtRisk	Dead	%Dead	Conc (mg a.i. kg^{-1} soil)
1	80	0	0	0
2	40	0	0	6.5
3	40	0	0	11.8
4	40	0	0	21.2
5	40	0	0	38.1
6	40	0	0	68.6
7	40	0	0	123.5
8	40	12	30	222.2
9	40	40	100	400

Table 7.44 Fish Acute Mortality Data

Group	Conc	Ln(Conc)	Dead	AtRisk	ObsProb	ArSin(Sqrt(Y))
1	37	3.61092	2	20	0.1	0.32175
2	61	4.11087	14	20	0.7	0.99116
3	103	4.63473	9	20	0.45	0.73531
4	180	5.19296	7	20	0.35	0.63305
5	310	5.73657	18	20	0.9	1.24905
6	400	5.99146	20	20	1	1.5708

Control data not shown, but there were 40 controls (water and solvent combined) with no mortality.

7.13 Estimate LC50 for the rainbow trout mortality data in Table 7.46.

7.14 Estimate LC50 for the daphnid mortality data (see Table 7.47).

7.15–7.20 For the data in Exercises 6.3–6.9, fit a regression model, if possible, and estimate EC10 and EC20 or explain why no meaningful estimate can be obtained.

Table 7.45 Fish Acute Mortality Raw Data

Rep	Risk	Conc	Dead	Rep	Risk	Conc	Dead
AA	5	0	0	A	5	103	0
BB	5	0	0	B	5	103	0
CC	5	0	0	C	5	103	1
DD	5	0	0	D	5	103	0
A	5	0	0	A	5	180	0
B	5	0	0	B	5	180	0
C	5	0	0	C	5	180	0
D	5	0	0	D	5	180	0
A	5	37	0	A	5	310	1
B	5	37	0	B	5	310	1
C	5	37	0	C	5	310	0
D	5	37	0	D	5	310	1
A	5	61	2	A	5	400	3
B	5	61	3	B	5	400	5
C	5	61	0	C	5	400	3
D	5	61	1	D	5	400	5

Reps AA, BB, CC, and DD are water control; Reps A, B, C, and D for conc = 0 are solvent control.

Table 7.46 Rainbow Trout Mortality Data

Group	AtRisk	Dead	%Dead	Conc (mg l^{-1})
1	7	0	0	0
2	7	0	0	0.63
3	7	0	0	1.3
4	7	0	0	2.53
5	7	0	0	4.57
6	7	7	100	10.7

Table 7.47 Daphnid Mortality

Group	AtRisk	Dead	%Dead	Conc (μg l^{-1})
1	20	0	0	−1
2	20	0	0	0
3	20	0	0	3.2
4	20	0	0	5.7
5	20	0	0	10
6	20	0	0	19
7	20	20	100	32
8	20	20	100	58
9	20	20	100	99

Conc = −1 is label for water control; Conc = 0 is label for solvent control; As always, the controls should be compared; If no significant difference is found, then the controls should be combined and labeled as conc = 0 for both and replicates relabeled to avoid duplicate names; Otherwise, the water control is dropped, and further analysis should use only the solvent control.

Chapter 8

Analysis of Count Data: NOEC and Regression

8.1 REPRODUCTION AND OTHER NONQUANTAL COUNT DATA

Count data arises in many ecotoxicity studies and as indicated in Chapter 1 is distinguished from quantal in that for each subject, the count is not just zero or one but can be any nonnegative integer. Example count variables include the number of eggs produced by a frog or daphnia, the number of flowers or stems per plant, the number of micronucleated polychromatic erythrocytes in a cell culture, and the number of days until the first egg hatches in an avian or fish study or the first fish embryo swims up to the top of the tank. The methods developed in Chapters 6 and 7 for quantal data do not apply in any of these examples.

For count data, an integrated approach will be developed that includes both hypothesis testing to establish an NOEC and regression to estimate an effects concentration, ECx. There are several reasons for this. In some instances, there is only minor technical adjustment needed in the model to handle the two approaches to analysis, though the studies that call for an NOEC will generally have fewer test concentrations so that the regression approach is less practical. In other applications, the same statistical tests and models as discussed in Chapters 3 and 4 can be applied, perhaps with alternative error structures, so there are in some ways fewer new tests and models to develop and thus less justification for separate chapters.

The work will start with the simplest approach and one that has been used for many years in this field. Namely, the data are modeled after a transformation that approximates a continuous variable with a normal error structure, so all the methods of Chapters 3 and 4 can be applied. Of course, it may not be necessary to transform the data at all, as the untransformed data may be well approximated by a normal error structure. Following the discussion of transformations will be a discussion of how developments in theory and statistical software have greatly improved the way count data can be analyzed in ways more consistent with the actual data using an alternative error structure. Following that discussion will be a comparison of how the two approaches work in terms of sensitivity, specificity, and robustness. Finally, examples from current test guidelines will both illustrate and expand the types of models discussed.

8.2 TRANSFORMATIONS TO CONTINUOUS

The most common transformation to apply to count data is the square root, which tends to normalize the data. If the counts span several orders of magnitude, then a log-transformation may be more helpful. In either case, should the transformed data be reasonably consistent with normality, as judged from the residuals from an appropriate ANOVA, then the methods of Chapters 3 and 4 can be applied. In the case of hypothesis testing, there is no difficulty interpreting the results in terms of the original measurement scale. For regression analysis, there is the additional need to express the percent effect in terms of the untransformed response, since an $x\%$ change in the square root or logarithm of the response does not back-transform into an $x\%$ change in the untransformed response. This is not a straightforward process, unfortunately, but also is not impossible. A few examples will illustrate the way forward. See Table 8.1 for a *Daphnia magna* chronic reproduction study done under

Statistical Analysis of Ecotoxicity Studies, First Edition. John W. Green, Timothy A. Springer, and Henrik Holbech.

Companion website: www.wiley.com/go/Green/StatAnalysEcotoxicStudy

Table 8.1 *Daphnia* Chronic Reproduction

Treatment	Conc	Group	Rep	RepDay	AtRisk	Surv14	Surv21	TLY14	TLY21
Water	0	1	1	8	10	9	8	326	681
Water	0	1	2	7	10	10	9	302	718
Water	0	1	3	9	10	10	8	214	624
Water	0	1	4	7	10	10	10	306	827
DMF	0	1	5	8	10	8	7	352	784
DMF	0	1	6	8	10	10	6	336	582
DMF	0	1	7	8	10	9	9	246	654
DMF	0	1	8	8	10	7	7	217	585
Treatment 1	0.113	2	1	8	10	10	10	308	643
Treatment 1	0.113	2	2	8	10	10	10	288	611
Treatment 1	0.113	2	3	8	10	10	10	361	837
Treatment 1	0.113	2	4	8	10	10	9	351	726
Treatment 2	0.4	3	1	8	10	8	8	239	581
Treatment 2	0.4	3	2	8	10	9	9	259	681
Treatment 2	0.4	3	3	8	10	6	6	98	351
Treatment 2	0.4	3	4	8	10	9	8	266	488
Treatment 3	1.39	4	1	8	10	7	7	295	478
Treatment 3	1.39	4	2	8	10	6	5	224	464
Treatment 3	1.39	4	3	8	10	7	7	300	623
Treatment 3	1.39	4	4	8	10	9	9	343	582
Treatment 4	4.86	5	1	8	10	9	7	344	607
Treatment 4	4.86	5	2	8	10	10	8	380	684
Treatment 4	4.86	5	3	8	10	8	6	336	627
Treatment 4	4.86	5	4	8	10	8	6	281	575
Treatment 5	17	6	1		10	7	7	0	0
Treatment 5	17	6	2		10	3	3	0	0
Treatment 5	17	6	3		10	1	1	0	0
Treatment 5	17	6	4	15	10	5	5	0	1

RepDay is the first day of reproduction among *Daphnia* in the indicted rep, Rep. AtRisk is the number of adult *Daphnia* at the beginning of the study. Surv14 and Surv21 are the number of adults alive after 14 and 21 days following exposure. TLY14 and TLY21 are the total number of live young produced over the first 14 and 21 days, respectively. Where RepDay is blank, the *Daphnia* did not reproduce by the end of the study.

OECD TG 211 that will be used for illustration. Adult *D. magna* were grouped into four replicate beakers of 10 adults each in every test concentration and in both water and DMF solvent controls. After 14 and 21 days' exposure, the total number of live young produced per beaker, TLY14 and TLY21, was counted, as was the number of surviving adults. TG 211 calls for analysis of the number of live young per surviving adult at the end of the study, TYS21. Different regulatory authorities have requested different ways to determine TYS21. The guideline itself calls for a simple calculation of TLY21/Surv21, where Surv21 is the number of adults surviving to the end of the study. Others request TLY21/REPSIZE, where REPSIZE is the number of adults at the beginning of the study, as this is deemed a better indication of the population effect. In Section 8.4.2, TG243 for the mud snail *Lymnaea stagnalis* reproduction takes a more sophisticated approach to a statistically similar response. The risk assessor should determine the specific calculation required before proceeding. Also computed were the numbers of young per surviving adult, TYS14 and TYS21. Units were in $\mu g\,l^{-1}$ of the test substance.

It is left as an exercise to verify that there is no statistically significant difference in TLY21 between the two controls. They are thus combined for further analysis.

The survival endpoints are, of course, quantal responses, and it will be left as an exercise to analyze those using the methods of Chapters 6 and 7. An analysis of the first day of reproduction will be given later in this chapter that accommodates the censoring evident in three of the four replicates in the highest test concentration. Several analyses of TLY21 will be illustrated. (i) A hypothesis testing approach with untransformed data uses the Tamhane–Dunnett and Jonckheere–Terpstra tests. While the data are found to be consistent with a normal distribution, the variances by Levene's test are heterogeneous, so Williams' test is not done. (ii) Following a square-root transform, the responses are found to be consistent with normality and variance homogeneity, and the Williams test along with Dunnett is done. The Jonckheere–Terpstra test is not affected by the transformation, so the results obtained in Part (i) still apply. The contrast in results should be noted. (iii) A nonlinear regression model assuming normally distributed and homogeneous response is fit to the untransformed response, and EC*x* is estimated for a range of effect levels, in spite of the observed variance heterogeneity. Methods for modeling the unequal variances will also be discussed. (iv) A nonlinear regression model is fit to the square-root transformed response again assuming a normal homogeneous error structure and EC*x* is estimated. In Section 8.3, the same data will be analyzed without transform using Poisson errors with accommodation for overdispersion.

8.2.1 NOEC for Untransformed *Daphnia* TLY21 Data

It is left to the reader to verify that while the data are consistent with a normal distribution, according to the Shapiro–Wilk test, the variances by Levene's test are heterogeneous, so Williams' test is not done. See Table 8.2 for a summary of the analysis of untransformed TLY21.

8.2.2 NOEC for Square-root Transformed *Daphnia* TLY21 Data

Following a square-root transform, the responses were found to be consistent with normality and variance homogeneity, and Williams' test along with Dunnett is done. The Jonckheere–Terpstra test is not affected by the transformation, so the results obtained in Section 8.2.1 still apply and

Table 8.2 Summary of Unweighted Analysis of Means

Treatment (μg l^{-1})	Mean	StdErr	DF	Control	ObsDiff	Crit	Signif
0.113	704.25	50.4503	7.58275	681.875	22.375	193.956	
0.4	525.25	70.1883	5.55151	681.875	−156.625	260.848	
1.39	536.75	38.9773	9.25032	681.875	−145.125	163.906	
4.86	623.25	22.907	9.6045	681.875	−58.625	137.98	
17	0.25	0.25	7.00044	681.875	−681.625	131.14	*

_JT_ IS JONCKHEERE STATISTIC
Z_JT IS STANDARDIZED JONCKHEERE STATISTIC
PR_JT IS *P*-VALUE FOR UPWARD TREND
PL_JT IS *P*-VALUE FOR DOWNWARD TREND
P-VALUES ARE FOR TIE-CORRECTED TEST WITH CONTINUITY CORRECTION FACTOR
SIGNIF RESULTS ARE FOR Decreasing ALTERNATIVE HYPOTHESIS

Jonckheere trend test on dose 0 + lowest five doses thru 17 μg l^{-1}

__JT___	Z_JT	PL_JT	PR_JT	Signif	Dose
101.5	−3.43561	0.0003	.	**	6

Jonckheere trend test on dose 0 + lowest four doses thru 4.86 μg l^{-1}

__JT___	Z_JT	PL_JT	PR_JT	Signif	Dose
100	−1.89575	0.029	.	**	5

Jonckheere trend test on dose 0 + lowest three doses thru 1.39 μg l^{-1}

__JT___	Z_JT	PL_JT	PR_JT	Signif	Dose
59	−2.39094	0.0084	.	**	4

Jonckheere trend test on dose 0 + lowest two doses thru 0.4 μg l^{-1}

__JT___	Z_JT	PL_JT	PR_JT	Signif	Dose
45	−1.43221	0.076	.		3

The Tamhane–Dunnett test found only the high treatment group significantly different from the control at the 0.05 significance level. These results are weighted by Surv21. The step-down Jonckheere–Terpstra test found the three highest treatment groups significantly different from the control. A weighted Jonckheere analysis is not available. The NOEC is determined to be 0.4 μg l^{-1} based on the Jonckheere–Terpstra test. "*" and "**" indicate significance at the 0.05 and 0.01 levels, respectively.

Table 8.3 Summary of Transformed Analysis of Means

ANOVA results

Contrast	Signif	*p*-Value	Mean
DOSE TREND	**	0.0001	.
DOSE QUAD	**	0.0001	.
DOSE 3 – 1		0.9486	703.692
DOSE 4 – 1	*	0.027	541.516
DOSE 5 – 1	*	0.0378	545.179
DOSE 6 – 1		0.4374	627.148
DOSE 7 – 1	**	0.0001	0.313

Amalgamated means from PAVA algorithm for DECR alternative

Obs	Level	NSTR	YSTR
1	0	8	681.875
2	0.113	4	704.25
3	0.4	12	561.75
4	17	4	0.25

Williams' test results

Conc	YBAR	y_0	DiffD	SEDifff	DF	WillD	pD	Crit2	NOAELD
17	0.25	681.875	681.625	53.905	22	12.6449	0	1.84576	.
4.86	561.75	681.875	120.125	53.905	22	2.2285	0.02218	1.83817	.
1.39	561.75	681.875	120.125	53.905	22	2.2285	0.02172	1.82521	.
0.4	561.75	681.875	120.125	53.905	22	2.2285	0.0208	1.79849	.
0.113	704.25	681.875	−22.375	53.905	22	−0.4151	0.65895	1.71714	0.113

DOSE TREND is a test for linear trend in the means. DOSE QUAD is a test for a quadratic curve in the means. DOSE $k-1$ is Dunnett's test comparing Dose k with the control. Level is the minimum concentration of treatments combined by the PAVA algorithm used in Williams' test. NSTR is the amalgamated sample size. YSTR, YBAR is the amalgamated mean. y_0 is the control mean. DiffD = YBAR − y_0. SEDiff, standard error of DiffD. WillD is the value of the Williams test statistic. *p*D is the Williams *p*-value WillD. Crit2 is the critical value of Williams' test. NOAELD is the no adverse effects level (NOAEL) using Williams' test for a decreasing trend. "*" and "**" indicate significance at the 0.05 and 0.01 levels, respectively.

will not be repeated. See Table 8.3 where the results are summarized.

8.2.3 Discussion of the *Daphnia* Reproduction Analyses with High Treatment Excluded

A comparison of the results in Sections 8.2.1 and 8.2.2 indicates an advantage of analyzing the square-root transformed data, since the more sensitive Dunnett test can be used rather than the robust Tamhane–Dunnett and Williams' test can be used and is found more sensitive for these data than the Jonckheere–Terpstra test. The NOEC is set by Williams to be 0.113 μg l^{-1}, where a 3% decrease from the control was observed, whereas the Jonckheere–Terpstra-derived NOEC is 0.4 μg l^{-1}, where a 23% decrease was observed. Indeed, the difference a transformation can make in the results is striking.

However, an important observation to make about this example is that the data from the high treatment groups are fundamentally different from the rest of the study. Adult survival in that group is low (50%) and reproductive output is almost nil. The latter will lower the pooled within-treatment variance used in the analysis of the transformed data. Even though Levene's test did not find significant heterogeneity in the transformed data, it is unwise to ignore the difference of the data from the high treatment to the remainder of the data. It will be left to the reader to verify that the NOEC for survival to day 21 is 0.4 μg l^{-1}. Since TLY21 is intended to be a sublethal response, an analysis with the high treatment omitted may be more appropriate. One could omit all treatments with significant adult mortality, but this is problematic, since Surv21 captures only total mortality, and if adults live only to day 20 or even less, they could have had substantial reproductive output. Omitting all groups with significant mortality reduces the degrees of freedom, and together these factors could underestimate the effect on reproduction.

It is left as an exercise to show that if the high treatment group is omitted, the transformed and untransformed data are both consistent with normality and variance homogeneity, and Williams' test leads to the same NOEC, 0.113 μg l^{-1} in both analyses. So, in this example, analysis of the transformed data with or without the high treatment group leads to the same NOEC.

8.2.4 Nonlinear Normal Errors Regression Model for Transformed and Untransformed *Daphnia* TLY21 Data

Numerous models were attempted but none fit the data well. The problem is the limited number of treatment groups plus the nonmonotone concentration–response (see Table 8.4). The data with and without the 17 μg l^{-1} concentration could not be fit well by any attempted regression model. This was true regardless of whether a normal or Poisson errors model was used. It seems evident from the observations that there is an effect on TLY21 at 17 μg l^{-1} and not at the NOEC, 0.113 μg l^{-1}. In between those extremes, regression is not helpful.

It is natural to ask whether these data are unusual in terms of difficulty to fit a regression model and estimate EC10. A database of seventeen *Daphnia* chronic studies from laboratories around the world has been analyzed. Of those, regression models could be fit to fifteen, though the quality of fit was highly questionable for one dataset and somewhat questionable for two others. An NOEC could not be obtained for four of the seventeen: That is, for four studies there was a significant effect in the lowest tested concentration, but an EC10 bounded away from 0 could be estimated for three of those four. The NOEC exceeded the highest tested concentration for only one of the 17 studies, and no regression model could be fit to those data. Where both NOEC and EC10 estimate were obtained (11 studies), the NOEC fell within the 95% confidence interval of EC10 in all but one study. The ratio of NOEC/EC10 was greater than 1 in six studies and less than 1 in five studies, and the ratio exceeded 2 only once and was less than 0.5 only once. What these results suggest is that if a meaningful EC10 estimate cannot be obtained but an NOEC can, then the NOEC is a plausible substitute for EC10 for this response. These results also suggest that it is usually possible to fit a useful regression model to this response. The exercises provide several additional examples of *Daphnia* TLY21 datasets, and the reader is asked to fit a regression model and determine an NOEC wherever possible. Another example may be informative. See Table 8.5 for another *Daphnia* study with a TLY21 response. In this study, there was only one adult in each beaker, so replicate and individual were the same. See Tables 8.6 and 8.7 for summaries of the NOEC and regression results for these data. See Figure 8.1 for a graphical representation of the data and regression model.

Table 8.4 Summary of Observed Effects in *Daphnia* Reproduction Data

Conc	Mean	%Chng
0	26.0642	0
0.113	26.4878	2
0.4	22.7564	−13
1.39	23.1221	−11
4.86	24.9525	−4
17	0.25	−100

The concentration–response is nonmonotonic. The small increase at the lowest tested concentration is not uncommon and does not present a serious problem. It is too small for a hormetic model with so few test concentrations. The rise in TLY21 at 4.86 μg l^{-1} is more troubling.

8.3 GLMM AND NLME MODELS

There is conceptual appeal as well as theoretical support to treat TLY21 as a Poisson-distributed response. Development of such models should probably be traced back to Nelder and Wedderburn (1972) but gained widespread attention through the publication of McCullagh and Nelder (1983). Since then there have been numerous articles and books on various aspects of GLMM. Notable among the textbooks are Hilbe (2007), Cameron and Trivedi (2013), and Agresti (2015). Programs to implement these models are described in several of the books cited above and in Kerr and Meador (1996), Hothorn et al. (2016), Hothorn (2016), Lehmann et al. (2016), Nielsen et al. (2004), and McCulloch et al. (2008).

A GLMM model can be fit to the *Daphnia* data of Table 8.5 and a form of Dunnett's test applied. The Jonckheere–Terpstra test can still be applied, since it does not depend on a particular distribution. A program in SAS (*GLMM_DUNNETT.sas*) is provided in Appendix 1 to carry out a generalized linear models approach to the NOEC. The R program, *Table 8_6.r*, provides the Dunn, Kruskal–Wallis, and Jonckheere–Terpstra tests for such data. The result for these data is the same as given by Dunn's test above, namely, NOEC = 0.091 mg l^{-1}. The Jonckheere–Terpstra test is preferred since it takes the

Table 8.5 Another Daphnia Chronic TLY21 Dataset

Conc	Rep	TLY21	Conc	Rep	TLY21	Conc	Rep	TLY21
0	1	173	0.051	1	186	0.165	1	62
0	2	186	0.051	2	205	0.165	2	2
0	3	160	0.051	3	140	0.165	3	2
0	4	58	0.051	4	132	0.165	4	69
0	5	166	0.051	5	186	0.165	5	6
0	6	158	0.051	6	154	0.165	6	9
0	7	168	0.051	7	162	0.165	7	105
0	8	159	0.051	8	15	0.165	8	115
0	9	167	0.051	9	12	0.165	9	105
0	10	147	0.051	10	169	0.165	10	4
0.027	1	222	0.091	1	19	0.304	1	5
0.027	2	179	0.091	2	160	0.304	2	2
0.027	3	175	0.091	3	88	0.304	3	27
0.027	4	215	0.091	4	131	0.304	4	1
0.027	5	216	0.091	5	169	0.304	5	0
0.027	6	202	0.091	6	12	0.304	6	0
0.027	7	17	0.091	7	158	0.304	7	0
0.027	8	133	0.091	8	14	0.304	8	15
0.027	9	99	0.091	9	10	0.304	9	3
0.027	10	161	0.091	10	155	0.304	10	16

TLY21, number of live young produced by day 21.

presumed monotone concentration–response into account unlike either the Dunnett in the GLMM model or Dunn in the nonparametric approach. Details are left as an exercise. While some numerical details from the R program differ from the results in Table 8.6, the conclusions are identical. Concerted effort to duplicate the numerical results from SAS failed. The SAS programs have been validated by manual calculations based on the underlying mathematical concepts. The R program has some additional nice features that warrant its inclusion.

A normal errors regression model was fit to these data. See Table 8.7 for the results. SAS and R programs from Chapter 4 can be used to obtain the results in Table 8.7. In addition, the R program, *Table 8.7b.r*, is provided in Appendix 1 to generate all the details in that table, though with some numerical differences from what is in the SAS-generated table. The reason for a different program for this example was to illustrate another approach that includes several interesting features not captured in the earlier programs.

It is also possible to fit the same Bruce–Versteeg model to these data with a Poisson error structure to explore what impact that might have on the results. Programs for that (*BVP POISSON DECR.r* and *NLME_BVP_model.sas*) are provided in Appendix 1 and can easily be modified if overdispersion is present. See Table 8.8 where the results are summarized. All other models for the mean from Chapter 4 can be fit with Poisson error structure.

8.3.1 Comparison of Poisson and Normal Errors Models for Count Data

It will be useful to compare the traditional normal errors modeling approach that treats counts as though they are continuous variables with a normal distribution with the approach that treats count responses as counts. There is no question about the conceptual preference for treating counts as counts. The issue to be explored in this section is whether the results from these alternative approaches differ in a noteworthy way. Examples have already been given comparing the two approaches. The question to be pursued here is whether there is objective reason to prefer one over the other besides the conceptual appeal. This task was attacked through a large computer simulation study reported in Green (2017c).

A database of *D. magna* chronic studies done by numerous labs under OECD TG 211 was used as a basis for the simulation work. There were 88 control datasets that had 10 reps per treatment group and 19 additional datasets with fewer reps. The response of interest was the number of total live young produced in a 21-day study. The control mean response was around 120. Overdispersion was found in most of these studies, with mean/variance ratio of around 0.5. Given that a Poisson response is expected within each treatment group, this indicated 100% overdispersion. Seven or 10 replicates per treatment group were simulated with overdispersion ranging from 0 to 300%

Table 8.6 NOEC Results for Example Data from Table 8.5

Dose	DoseVal	Count	Mean	Median	StdDev	StdErr
1	0	10	154.2	163	35.3327	11.1732
2	0.027	10	161.9	177	64.3073	20.3358
3	0.051	10	136.1	158	68.2307	21.5765
4	0.091	10	91.6	109.5	70.6717	22.3484
5	0.165	10	47.9	35.5	48.393	15.3032
6	0.304	10	6.9	2.5	9.267	2.9305

Shapiro–Wilk test of normality of TLY21

Obs	Std	Skew	Kurt	SWStat	*P*-Value	Signif
60	51.6103	−0.87208	0.46307	0.92973	0.0019	**

Modified Dunn's multiple comparisons (decreasing) on TLY21

Conc	Count	MRank	AbsDiff	Crit05	Signif	*P*-Value
0	10	42	0	18.1648		.
0.027	10	45.9	3.9	18.1648		1
0.051	10	39.15	2.85	18.1648		1
0.091	10	28.7	13.3	18.1648		0.2213
0.165	10	18	24	18.1648	**	0.0053
0.304	10	9.25	32.75	18.1648	**	0.0001

Monotonicity check of TLY21

Parm	Probt	Signif
TREND	0.0001	**
QUAD	0.0072	**

Jonckheere trend test on dose 0 + lowest five doses thru 0.304 mg l^{-1}

___JT____	Z_JT	PL_JT	PR_JT	Signif	Dose
422	−5.90282	0.0001	.	**	6

Jonckheere trend test on dose 0 + lowest four doses thru 0.165 mg l^{-1}

___JT____	Z_JT	PL_JT	PR_JT	Signif	Dose
365	−4.15846	0.0001	.	**	5

Jonckheere trend test on dose 0 + lowest three doses thru 0.091 mg l^{-1}

___JT____	Z_JT	PL_JT	PR_JT	Signif	Dose
297.5	−2.14901	0.0158	.	**	4

Jonckheere trend test on dose 0 + lowest two doses thru 0.051 mg l^{-1}

___JT____	Z_JT	PL_JT	PR_JT	Signif	Dose
212	−0.15217	0.4395	.		3

Shapiro–Wilk test (p-value = 0.0019) indicated nonnormality, so nonparametric analysis was done. Dunn's test is based on the mean rank, MRank, absolute difference, AbsDiff, between treatment and control mean rank. Crit05 is the critical value for Dunn's test, obtained by multiplying the critical value of the standard normal by the standard error of the difference being tested. TREND and QUAD are contrasts of treatment mean ranks. A significant QUAD together with a non-significant TREND is an indication of possible non-monotonicity in the concentration-response. _JT_ is Jonckheere statistic. Z_JT is standardized Jonckheere statistic. PR_JT is p-value for upward trend. PL_JT is p-value for downward trend. *P*-Values are for tie-corrected test with continuity correction factor. Signif results are for decreasing alternative hypothesis. NOEC = 0.051 mg l^{-1} based on Jonckheere-Terpstra test. Dunn test results indicated only for comparison. "**" indicates significance at the 0.01 level.

Table 8.7 Normal Errors Regression Results for Example Data from Table 8.5

Source	DF	Sum of squares	Mean square	*F* value	Approx Pr>*F*
Model	3	791 284	263 761	94.96	<0.000 1
Error	57	158 320	2 777.5		
USS	60	949 604			

Parameter	Estimate	StdErr	LCB	UCB
EC10	0.0444	0.3567	0.0217	0.0906
Gamma	0.694	0.2128	0.2679	1.1202
y_0	159	13.2811	132.4	185.6

Observed and predicted % change from control mean

Conc	Observed mean	Predicted mean	Observed % change	Predicted % change
0	154.2	159.028	0	0
0.027	161.9	155.383	5	−2
0.051	136.1	136.771	−12	−14
0.091	91.6	94.979	−41	−40
0.165	47.9	43.023	−69	−73
0.304	6.9	10.797	−96	−93

Bruce–Versteeg model evaluation measures

SSReg	SSErr	SSPE	FLOF	PLOF	SSTotC	RSquare	AICc
791 284.19	158 319.8	157 153.6	0.9544	0.569 64	949 604	0.833 28	481.422

NOEC=0.051 and EC10=0.044; estimates are comparable.

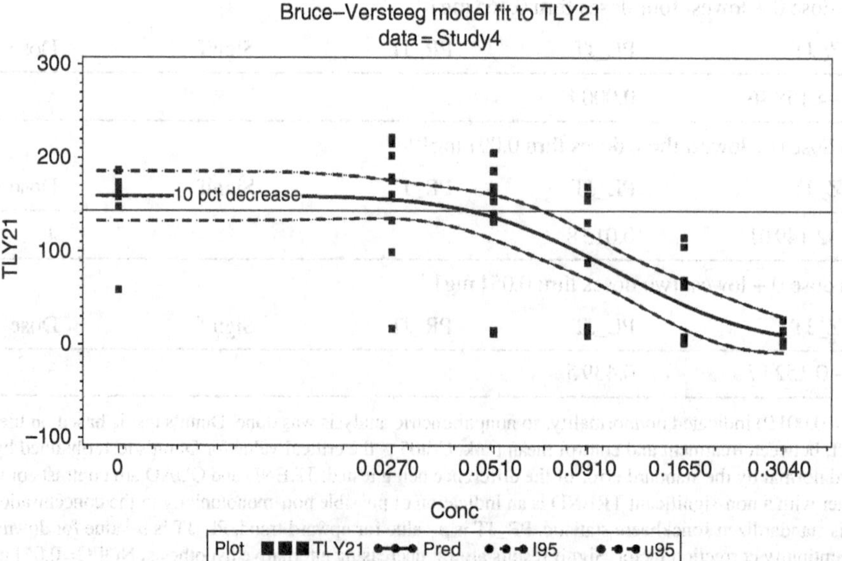

Figure 8.1 Bruce–Versteeg normal errors model fit to Table 2.4 *Daphnia magna* chronic TLY21 data. Adult *Daphnia* were housed one to a beaker. Solid curve is the fitted model. Dashed curves are the 95% confidence bonds for the predicted mean response. Solid squares are the observed values.

Table 8.8 Poisson Errors Bruce–Versteeg Model Fit to Example Data from Table 8.5

Parameter	Estimate	StdErr	DF	LCB	UCB
ECx	0.048	0.004	5	0.038	0.058
Gamma	0.636	0.039	5	0.535	0.737
y_0	158.140	3.005	5	150.420	165.860

Estimates, StdErr, and LCB and UCB 95% confidence bounds from Poisson errors model with means following the Bruce–Versteeg shape. The main differences are tighter confidence bounds on the parameters.

and a concentration–response relationship following a three-parameter log-logistic distribution. This shape was selected because it fit the limited number of study data with six test concentrations plus control. While many studies with four or five test concentrations were in the database and are allowed in TG 211, it was thought best to determine the shape of the concentration–response curve using the largest studies available. *Daphnia* are typically housed one per beaker, so rep and individual are the same. It is possible to house multiple daphnid per beaker, but this is not the most common procedure under the latest version of TG210 and was not simulated. Also, as few as seven replicates are permitted and frequently done, so that size was also simulated. Finally, studies with four, five, and six test concentrations plus control were simulated. The range of concentrations of test chemical in this database varied greatly because the chemistry varied. All data were rescaled to a maximum test concentration = 100 before the characteristics of the curves were analyzed, and that is the maximum concentration simulated. Next, these studies all had approximate geometric spacing of test concentration, as called for in TG 211. A maximum ratio of 3.2 of adjacent concentrations is allowed in TG210, and this spacing and common ratios of 2.3 and 1.6 were simulated to cover the range in numerous ecotoxicity test guidelines for many species. While these other values are not especially relevant to *Daphnia* studies, they were included to give the results greater generality.

The models fit to the simulated data were the Bruce–Versteeg, the three-parameter log-logistic, and OE2 and OE3 with both normal and Poisson errors. In addition, an increasing normal errors Bruce–Versteeg model was fit to the response normalized to the control following to the method in Moore and Caux (1997). (They refer to this as a probit model.) In addition, an increase normal errors model was fit to normalized response data. The normalization idea has been criticized in Chapter 7 for both quantal and continuous responses, and it was thought appropriate to explore it for count responses as well. 1000 datasets were simulated for each set of conditions and all models were fit to these data. See Table 8.9 for a summary of one set of simulations, that with 100% overdispersion, five test concentrations plus control, and 10 reps per test concentration.

Control of the shape comes from specifying the control mean, always 120 in the simulations, the exponent d, and the value of the response in the highest tested concentration, denoted by Ymax. To be clear, the three-parameter log-logistic model was simulated as

$$y = \frac{y_0}{1+b(U)^d}, \tag{8.1}$$

with $y_0 = 120$ and $d = 1$, 2, or 3. For fixed values of y_0 and d, b was varied to make the simulated curve (before Poisson random noise was added) have the value Ymax at $U = 100$.

One of the advantages of simulated data is that the true ECx values are known. It thus gives us an objective measure of how well the various models do in estimation. In all cases, Ymax was simulated to range from 30 to 100, and ECx was estimated for $x = 10$–50 in increments of 10. The distribution of each ECx was quantified to give the minimum, maximum, and qth quantiles, for $q = 10$, 25, 50, 75, and 90. The same statistics were calculated for the distribution of lower and upper confidence bounds for ECx and for all other model parameters (y_0, d, Ymax). In addition, the number of ECx estimates with zero or negative lower bound and bounds spanning the entire range of tested concentrations including zero was calculated as an aid in identify especially poor estimates. Finally, the number of simulated datasets for which each model could be fit was recorded, since a model that rarely fits but gives good estimated when it does would not be confused with a model that gives good estimates and always fits.

From Table 8.9, it can be seen that when there was a steep drop in the response (Ymax = 30), the two exponential models, OE2 and OE3, did poorly at estimating EC10 with both normal and Poisson errors models. OE3 improved at estimating EC20 and performed well for EC50. The error structure used with the Bruce–Versteeg model had little impact on the estimates, but the model based on data normalized to the control provided worse estimates. Unsurprisingly, PL3 gave the best estimates for ECx, $x = 10$, 20, and 30, regardless of error structure.

When Ymax = 70, the estimates from models based on data normalized to the control deteriorate. The Bruce–Versteeg model does well for either error structure, and there is little to choose between error structures. PL3 continues to do well, while the two exponential models improve compared with the case Ymax = 30, but are clearly inferior to PL3 and Bruce–Versteeg.

When Ymax = 100, so for a relatively shallow curve, one would expect poor estimation of EC50, and the table confirms that models PL3 and OE3 with normalized error structure are adequate, given the extrapolation. The results from models based on normalized errors perform very poorly for this shallow dose–response.

Table 8.9 Comparison of Models for Simulated *Daphnia* Chronic Reproduction Data

Error	Model	Ymax	LBLTO	Wide	NFit	Ratio10	Ratio20	Ratio50
Normal	3PL	30	0	0	1000	0.99	0.99	1.00
Normal	BVP	30	0	0	1000	1.10	1.01	1.01
Normal	BVPON	30	0	0	1000	0.90	0.87	0.98
Normal	OE2	30	0	0	1000	1.69	1.59	1.23
Normal	OE3	30	14	0	1000	0.69	0.87	1.06
Normal	PL3NO	30	0	0	1000	0.85	0.90	0.98
Poisson	BVPB	30	5	0	998	1.07	0.99	1.00
Poisson	OE2B	30	0	0	1000	1.97	1.85	1.44
Poisson	OE3B	30	0	0	1000	0.63	0.83	1.07
Poisson	PL3B	30	0	0	1000	1.01	1.00	1.00
Normal	3PL	70	284	0	1000	0.99	1.00	0.99
Normal	BVP	70	165	0	1000	1.04	0.98	1.04
Normal	BVPON	70	4	0	1000	0.68	0.80	1.24
Normal	OE2	70	0	0	1000	1.24	1.17	0.90
Normal	OE3	70	424	0	1000	0.94	1.01	0.94
Normal	PL3NO	70	7	0	1000	0.70	0.84	1.14
Poisson	BVPB	70	145	0	969	0.95	0.94	1.07
Poisson	OE2B	70	0	0	1000	1.22	1.15	0.89
Poisson	OE3B	70	338	11	1000	1.08	1.07	0.86
Poisson	PL3B	70	140	1	956	0.96	0.98	0.99
Normal	3PL	100	493	205	942	0.97	0.87	0.86
Normal	BVP	100	440	154	844	0.88	0.92	1.40
Normal	BVPON	100	480	153	994	0.44	1.75	12.92
Normal	OE2	100	40	40	1000	1.03	0.97	0.75
Normal	OE3	100	498	213	812	0.90	0.89	0.83
Normal	PL3NO	100	463	149	997	0.47	1.70	19.70
Poisson	BVPB	100	70	25	187	0.93	0.94	1.62
Poisson	OE2B	100	46	35	1000	0.92	0.87	0.67
Poisson	OE3B	100	694	15	1000	1.76	0.84	0.25
Poisson	PL3B	100	444	113	707	0.78	0.95	1.38

For all simulations in table, overdispersion=100%; 3PL, Log exponent simulated; Ymax, simulated response at *U*=100; Ratio*X*, Median EC*x* estimate/True EC*x* value simulated; NFit, number (of 100) for which model could be fit; LBLTO, number of EC10 estimate with zero or negative lower confidence bound; Wide, number of EC10 estimates where confidence interval spans entire range [0,100]; NO suffix means the model is fit to normalized data.

Of course, the ratios in Table 8.9 provide only a limited analysis of the simulations done, since they only indicate the median behavior of the distribution of EC*x* estimates. Green (2017c) explores the spread of the estimates and other measures of model performance.

8.4 ANALYSIS OF OTHER TYPES OF COUNT DATA

There are various complications that arise in some studies with count variables that warrant special attention. These include censored data, the subject of Section 8.4, where the actual value of some responses cannot be known, but also should not be ignored. Two examples giving rise to censored data are illustrated. Another complication is with reproduction studies, where some parents die before the end of the study. Clearly, after death they no longer reproduce. The question is how, or whether, to adjust mortality measures for parental mortality. This will be addressed in Section 8.4.2.

8.4.1 Censored Data

A different type of Poisson-distributed response arises in several guideline studies. This is the time to some event, such as time to hatch (avian, fish, daphnid), time to swimup (fish, daphnid), time to reach developmental stage 62 (LAGDA), or time to emerge (NTTP). These times are typically crudely measured in ecotoxicity studies as the number of days post exposure to the event. It is not uncommon for a study to end before some subjects have exhibited the characteristic of interest. For example, a fish embryo may not have swum up or an African clawed frog (*Xenopus laevis*) larva may not have reached developmental stage NF 62 by study end. Such observations are called censored: We do not know how long it would have taken for a fish embryo to swim up, if ever. All we know is that it was longer than the time at which the study was stopped. If such values are treated as equal to the end of the study, the effect of the test compound may be underestimated. Given the brief duration of many ecotoxicity studies and the crudeness of the measurement, the set of possible values tends to be small. As a result, some of the typical ways of dealing with censored data used in medical studies of survival or remission of illness perform poorly. Fortunately, nonlinear mixed effects (NLME) models can accommodate this situation. Two examples will illustrate the methods of analysis. The first is for the time for a fish egg from an early life stage experiment (OECD TG 210) to hatch and swim up to the surface, which is a measure of maturation and an indication of the population to survive. Regression models are found to be helpful in these studies. The second example is the time for African clawed frog (*X. laevis*) larvae in a study following OECD TG 241 (LAGDA) to reach developmental stage 62. These data are more problematic for regression modeling due to a small number of test concentrations.

8.4.1.1 Analysis of Time to Swimup: OECD 210

It is possible to take censoring into account when fitting a regression model. In SAS, the NLMIXED procedure allows censoring to be modeled as well as alternative distributions. This will be illustrated with first day of swimup in an early life stage study with rainbow trout following OECD TG 210. See Table 8.10 or Hatch2c.sas7bdat for this and several other endpoints that will be used in exercises.

An increasing normal errors Bruce–Versteeg model was fit to the swimup data after the water and solvent control data were combined. Censoring should be taken into account, and a method for doing that is described next.

8.4.1.2 Use of Censored Values in Determining Distribution

For a normal distribution, the likelihood function has the form

$$L(x;\mu,\sigma) = \prod_{i=1}^{L}\Phi\left(\frac{x_i-\mu}{\sigma}\right)\prod_{i=L+1}^{L+k}\varphi\left(\frac{x_i-\mu}{\sigma}\right)\prod_{i=L+k+1}^{nL+k+R}\left[1-\Phi\left(\frac{x_i-\mu}{\sigma}\right)\right] \tag{8.2}$$

where φ and Φ are the unit normal pdf and cdf, respectively, where there are L left-censored values, k uncensored values, and R right-censored values. This form of the likelihood function is given in numerous sources, for example, Helsel (2005, p. 14), Kendall and Stuart (1979, p. 552), and Johnson et al. (1994, p. 146). A good discussion is also provided by Frey and Zhao (2002). Left censoring is common in environmental monitoring in the form of nondetects, which is the focus of Helsel (2005). Right censoring is common in survival analysis, where a study is terminated before some subjects die, and is discussed in Collett (2014) and many other sources. Complete censoring, where both right and left censoring may occur, is also discussed in Wiki (2017). Equation (8.2) is easily modified to apply to a general distribution. Cook et al. (1999), Thiébaut and Jacqmin-Gadda (2004) and Vock et al. (2012) are good resources on censored data modeling, especially with regard to programming issues.

See Table 8.11 and Figure 8.2 for a summary of the model fit and estimates and a plot of the model and data.

Next for comparison, the same model was fit, but assuming Poisson errors in curve fitting, but still a normal error structure to capture overdispersion. For this, we need the density and distribution functions for the Poisson distribution. The density function is given by Eq. (8.3). The distribution function at y is obtained by summing the density function over all values of $x \le y$:

$$f(x;\lambda) = \frac{e^{-\lambda}\lambda^{x}}{x!}, \quad x = 0,1,2,\ldots. \tag{8.3}$$

For regression modeling, λ is evaluated at the value predicted at each concentration by the Bruce–Versteeg model fit to the response data. Interestingly, unlike the normal errors model, it was not possible to include an adjustment for overdispersion in the Poisson errors model except for the censored values in the highest concentration. See Table 8.12 and Figure 8.3 for summaries of the model fit.

By comparing Tables 8.11 and 8.12, it will be seen that the point estimates for EC10 and y_0 from the two models are similar, but the confidence intervals from the Poisson errors model for these two parameters are wider. This could be because the latter model more accurately reflects the nature of the data. The point estimates of gamma differ by a factor of approximately 6.7, and the widths of the confidence intervals differ by a factor of approximately 23. By comparing the agreement of the predictions to the observed values, the normal errors model appears to provide the better fit. With that in mind, the wider confidence intervals for two key parameters from the Poisson errors model may arise from the poorer fit of that model to the data rather than from more realistically capturing variability. A single

Table 8.10 Rainbow Trout First Day of Swimup

Conc	Rep	FHatch	LHatch	Hatch	Surv	Abnormal	SwimUp	Censor
−1	1	29	30	17	17	0	42	0
−1	2	29	31	15	15	0	42	0
−1	3	28	30	20	20	0	41	0
−1	4	28	30	19	19	0	44	0
0	1	28	29	16	16	0	41	0
0	2	28	29	17	16	0	42	0
0	3	28	30	17	17	0	42	0
0	4	28	30	16	16	0	42	0
0.031	1	28	29	18	18	0	41	0
0.031	2	28	30	18	18	0	42	0
0.031	3	28	29	19	19	0	42	0
0.031	4	27	29	17	16	0	42	0
0.044	1	28	30	19	18	0	42	0
0.044	2	28	29	14	14	0	41	0
0.044	3	28	29	17	17	3	40	0
0.044	4	28	30	19	19	0	42	0
0.11	1	28	30	16	16	0	40	0
0.11	2	28	30	19	19	0	42	0
0.11	3	28	30	18	18	0	42	0
0.11	4	29	30	16	16	0	42	0
0.25	1	29	30	18	18	0	46	0
0.25	2	29	30	16	16	6	46	0
0.25	3	28	30	17	17	0	44	0
0.25	4	28	30	19	18	0	44	0
0.59	1	29	30	14	14	0	46	0
0.59	2	29	30	16	16	0	46	1
0.59	3	29	31	17	16	0	46	1
0.59	4	28	31	16	16	2	46	0
1.5	1	29	31	17	11	0	46	1
1.5	2	29	31	15	13	2	46	1
1.5	3	27	31	14	10	0	46	1
1.5	4	28	30	15	9	0	46	1

Conc = −1 is water control; Conc = 0 is solvent control. FHatch and LHatch are the first and last days of hatching observed for eggs in the replicate. Surv, number of embryos (of 20) that survived. Swimup is the first day an embryo from the rep swam up to the surface. Censor = 0 means swimup was observed (i.e. not censored); Censor = 1 means observation was censored. Censoring applies only to swimup. Abnormal, number of embryos that were abnormal in some (unspecified) manner.

example cannot be used to make that determination. A computer simulation study is reported in Table 8.9 that does address that question. It should be noted that the AICc values cannot be used to compare model fits since these models are based on different error structures. Programs (*Table8_11b.r*, *normal test censor.sas*, and *BVPincr_censor_Poisson.sas*) for fitting both models are provided in Appendix 1.

8.4.1.3 NOEC for Swimup

The Jonckheere–Terpstra test is a possible choice for determining the NOEC for the swimup data, both because it is based on a monotone concentration–response and because it does not assume a particular error structure. However, it does not take censoring into account. In this case, the NOEC by the Jonckheere–Terpstra test was found to be 0.25 mg l^{-1}, which is below the two test concentrations containing all the censored data. Thus, this NOEC can be accepted, as higher values for the six censored observations could neither raise nor lower the test concentrations at which a significant effect is found.

It might also seem desirable to do a Poisson errors test that takes censoring into account. This is challenging since all the observations in the high test concentration are censored, and in the absence of an underlying model that

Table 8.11 Summary of Normal Errors BVP Fit to Swimup

Parameter	Estimate	StdErr	DF	LCB	UCB
EC10	0.4153	0.0643	31	0.2842	0.5464
Gamma	1.4433	0.4312	31	0.5639	2.3228
y_0	41.6388	0.2308	31	41.168	42.1097
S2	0.8266	0.2896	31	0.236	1.4172
S1	0.0391	0.0452	31	0	0.1312

Conc	Count	Observed mean	Predicted mean	Observed % change	Predicted % change
0	8	42	41.6388	0	0
0.031	4	41.75	41.6813	−1	0
0.044	4	41.25	41.7348	−2	0
0.11	4	41.5	42.0694	−1	1
0.25	4	45	45.1553	7	8
0.59	4	46	46.4251	10	11
1.5	4	46	56.1127	10	35
−2 log likelihood			76.9		
AICC (smaller is better)			89.2		
BIC (smaller is better)			94.2		

Parameter names follow the notation in Eq. (4.1*) with γ written as gamma. S1, regression standard error about the curve. S2, standard deviation of the error term capturing overdispersion. LCB and UCB, 95% confidence bounds for the indicated parameter.

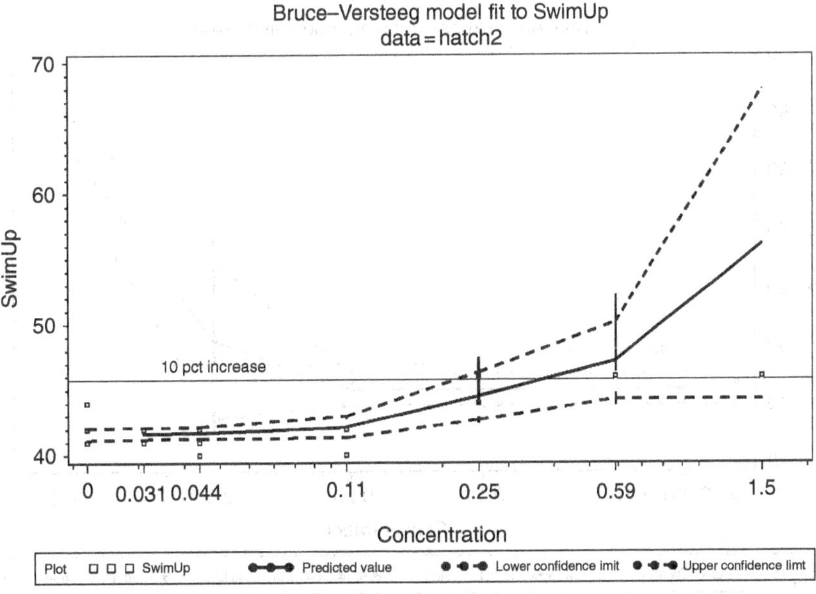

Figure 8.2 Normal errors BVP model fit to swimup data. Solid line is crude depiction of model predictions. Dashed lines are crude depictions of 95% confidence bounds for model predictions. Squares are observations. Horizontal scale is log scale. Observations do not show censoring.

constrains these censored values, there is no basis for estimating the mean response in that test concentration, and consequently, there is no way to test for a significant increase in response there. A solution would be to impose a constraint on the distribution of the true values for which only censored data are available. A Bayesian prior on the unknown time to swimup could provide such a constraint. Alternatively, in the example dataset, one could assume the true mean time to swimup in the high test concentration is the same as in the preceding test concentration. Either approach changes the issue to that of developing an appropriate prior or assumption. The authors are not aware of available information that supports a nonarbitrary procedure, so this will be left for a future project.

Table 8.12 Summary of Poisson Errors BVP Model Fit to Swimup

Parameter		Estimate	StdErr	DF	LCB	UCB
EC10		0.5072	0.08981	31	0.3241	0.6904
Gamma		0.2138	0.01826	31	0.1766	0.2511
y_0		42.7907	1.3244	31	40.0896	45.4918
S1		2.4521	4.8645	31	−7.4691	12.3733
Conc	**Count**	**Observed mean**	**Predicted mean**	**Observed % change**		**Predicted % change**
0	8	42	42.7907	0		0
0.031	4	41.75	42.7907	−1		0
0.044	4	41.25	42.7907	−2		0
0.11	4	41.5	42.7907	−1		0
0.25	4	45	42.7908	7		0
0.59	4	46	47.5455	10		11
1.5	4	46	85.5814	10		100
−2 log likelihood				147.5		
AICC (smaller is better)				157		
BIC (smaller is better)				161.3		

Parameter names follow the notation in Eq. (4.1*) with γ written as gamma. S1, regression standard error about the curve. LCB and UCB, 95% confidence bounds for the indicated parameter.

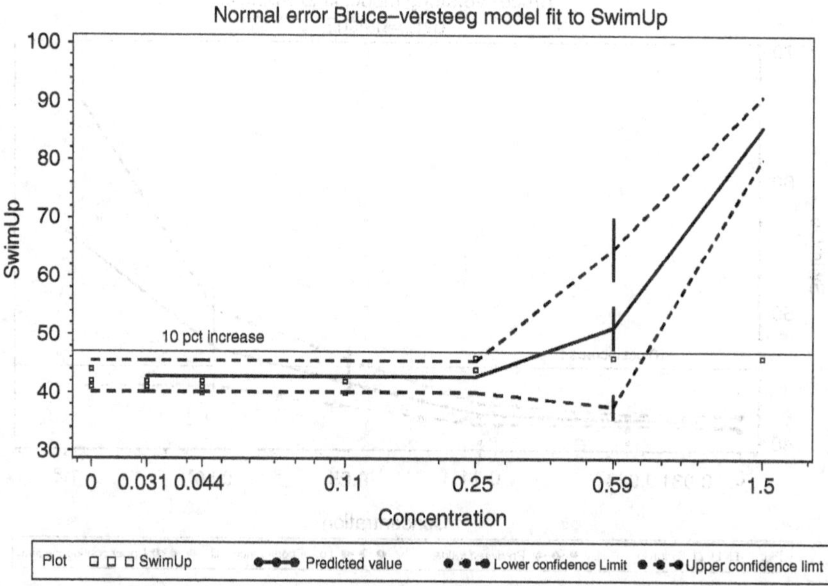

Figure 8.3 Poisson errors BVP model fit to swimup data. Solid line is crude depiction of model predictions. Dashed lines are crude depictions of 95% confidence bounds for model predictions. Squares are observations. Horizontal scale is log scale. Observations do not show censoring.

A normal errors model could be developed if one were willing to assume homogeneous variances across all treatment groups. It is not clear that this is appropriate for the censored data because it is not clear that swimup would occur within a few days if the experiment were not terminated, which is what the homogeneous error assumption would imply. However, if that assumption is made, it is interesting to find that the uncertainty in the high test concentration prevents it from being an effects concentration, but significant effects are found at both 0.25 and 0.59 mg l^{-1}. A program (*normal test censor.sas*) is given in Appendix 1 to explore this idea.

8.4.1.4 Analysis of Time to Reach Developmental Stage 62: OECD 241

OECD TG 241 is designed to explore the endocrine disruption potential of chemicals in African clawed frog (*X. laevis*) larvae. The focus of this guideline is on potential population relevant effects, such as survival, development, growth, and reproductive development. Analysis of survival is described in Chapters 6 and 7, while methods to analyze growth endpoints are covered in Chapters 3 and 4. The discussion here is on developmental stage. The Amphibian Metamorphosis Assay (OECD TG 231) requires analysis of the developmental stage reached by the end of a 21-day study. In the design of TG 241, the focus is on the time (in days) to reach a specific developmental stage, namely, stage 62 on the Nieuwkoop and Faber (1994) scale. Developmental stages are ordinal-scale measurements with integer labels (with a few exceptions) and should not be treated as numerical. Methods for analyzing ordinal-scale responses are given in Chapter 9. By changing the problem to time to reach stage 62, the response is numerical, though crudely measured, and this opens up the analysis to statistical techniques covered in this chapter. The experimental design calls for four test concentrations plus a negative control and, when a solvent is used, a solvent control, though three test concentrations are acceptable if there is also a positive control. In each test concentration, there are four replicate tanks of 15–20 animals each, while in each control, there are eight replicate tanks of 15–20 animals each. The concentrations are spaced so that the ratio of adjacent nominal concentrations should not exceed 3.2. It is recognized that four test concentrations limit the ability to do regression modeling and a NOEC is expected. However, if a suitable model can be found, it will provide good information and should be used. Both analyses will be illustrated. The response, days (time) to reach stage 62, can be modeled as Poisson variable, or, taking the possibility of overdispersion into account, a negative binomial distribution may be appropriate. Some animals have not reached developmental stage 62 at study termination, so right-censored responses are not uncommon.

TG 241 studies require large numbers of animals and are expensive. For that reason and because of animal welfare issues, a larger study with more test concentrations that would be better for regression modeling is not feasible. A dataset is given in Appendix 1, Table A1.8.10, which has a negative control, three concentrations (0.2, 1.6, and 6 $mg\,l^{-1}$) of the test chemical, and a positive control of a different chemical at a much higher concentration (1000 $mg\,l^{-1}$).

See Figure 8.4 for a plot of the data, in which it appears that there is delayed metamorphosis or time to reach developmental stage 62 in all three treatment groups compared with the control. It will also be evident that there is a reversal in effect between groups 2 (1.6 $mg\,l^{-1}$) and 3 (0.2 $mg\,l^{-1}$), with greater delay in group 2 than in group 3. A GLMM model with Poisson errors finds all treatment groups significantly delayed and finds a significant concentration–response trend in spite of the reversal mentioned above. The positive control was not included in the analysis or figure. It should be evaluated separately. A SAS program, *LAGDA.sas*, using the NLMIXED procedure is provided in Appendix 1 that can be used to carry out the analysis. This application of NLME modeling does not have the problems encountered in the swimup data of Section 8.4.1.1 because there are numerous uncensored observations in all treatment groups. An alternative analysis will be given in Chapter 12 using more traditional survival analysis methods.

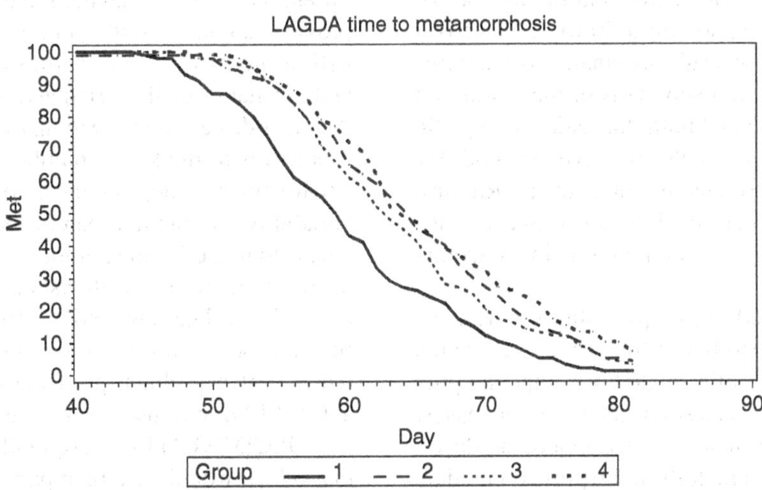

Figure 8.4 LAGDA time to metamorphosis. MET, percent of frogs that have reached metamorphosis (stage 62).

8.4.2 Snail Reproduction: OECD TG 243

A 1998 letter to Environmental Health Perspectives (James, 1998) stated, "There is overwhelming evidence that mammalian hormone levels around the time of conception are associated with the sex of resulting offspring." Underscoring this letter is the finding that men exposed to certain endocrine-disrupting chemicals (EDCs) are much more likely to have female children than male, so there are real consequences to humans as well as to other species. The US National Institutes of Health has been actively studying EDCs for over 20 years. OECD created the Validation Management Group for Ecotoxicity Testing (VMG-eco) in 2001 to develop test guidelines for the presence of EDCs, primarily in crop protection products. Of course, experiments on humans are not done, but survival of other species (bees, fish, frogs, etc.) does impact human survival as well as serving as surrogates for humans in risk assessment. The three authors of this text have served as members of VMG-eco since 2004 or earlier. Previous test guidelines have considered sex ratio in frogs and fish and other indicators of ED on reproduction.

Molluscs (or snails) are highly important organisms from both a global ecological and economic perspective. They are considered uniquely sensitive to the developmental and reproductive effects of certain groups of chemicals. An OECD team worked from 2013 to 2016 to develop a TG to explore the effects of possible EDCs on the mollusc species *L. stagnalis*, with a focus on reproduction. A different team developed a guideline for another species, *Potamopyrgus antipodarum*. The final TG calls for continuous exposure of adults in a reproductive state over 28 days, during which time egg clutches and eggs are counted daily. Both NOEC and ECx are expected statistical endpoints.

The average adult shell height varies in the range of 20–60 mm. The average life span of *L. stagnalis* is 1–2 years. This species is self-reproducing and the number of eggs per clutch generally varies from 50 to 120 and is a function of adult size. These snails are small, easy to cultivate in labs, and abundant in many parts of the world and serve as vectors for human and both domestic and aquatic animal parasitology. These molluscs serve as food for numerous other species (including humans in Russia). They are very adaptive to many environments, and any substance that alters their survival or reproduction warrants attention.

L. stagnalis are housed in groups within test vessels, with multiple replicate vessels per treatment. Treatments are different concentrations of a single test compound plus a control group. Generally, as test concentration increases, effects on organisms also increase, so a monotone dose–response is expected and trend tests and regression models are the expected statistical tools. The data are counts (number of egg clutches produced). Normal errors regression models can be developed for this response by a transformation to approximate a normal homogeneous response, but it is more natural to treat counts with a Poisson error structure, or possibly a negative binomial error structure to capture variance heterogeneity or overdispersion.

Adults are housed in groups of five per replicate with six replicates per treatment. Egg clutches in each replicate are counted frequently over the study and summed. Nreprocumul is the total reproduction per replicate. A challenge results from the fact that some adults die before the end of the study and, of course, reproduction stops at death of the parent. Nindtime is introduced as a measure of the cumulative reproduction days for the replicate. The number of days each mollusc is alive to reproduce can be calculated from its death date. From this, the sum of all such days in each replicate can be calculated. This sum, Nindtime, is the number of individual days for the rep.

To be clear if an adult is alive at measurement day t_1 and found dead at the next measurement day, t_2, it is treated as having died at time $t = (t_1 + t_2)/2$, which is thus its individual days on test. The number of individual days for the replicate is the sum of the individual days on test for each adult. To allow for the possibility that more than one adult might die during the time interval $(t_1, t_2]$, define $N(t_2) = (A(t_1) - A(t_2)) * (t_2 - t_1)/2$, where $A(t)$ = number of adults alive at time t. Then the Nindtime for replicate j of test concentration Ui, NID_{ij}, is the sum of the $N(t_2)$ over all measurement times for that replicate.

8.4.2.1 Bayesian Gamma-Poisson Model for Snail Reproduction

Bayesian models are increasingly used by regulators, though certainly at present, these are still much in the minority for ecotoxicology studies. Whether that will change in the future remains to be seen. It is certainly important to understand the Bayesian approach and TG 243 provides a nontrivial illustration. The discussion of the statistical model for TG 243 follows what is in the guideline and in Charles et al. (2016) and in particular owes much to the clear discussion in Delignette-Muller et al. (2013). The problem is to model the number of egg clutches produced per treatment group, taking into account among-replicate variability and random parent mortality. For the latter, the idea is to relate N_{ij}, the number of egg clutches in replicate j of treatment group i with test concentration U_i to the number of individual days, NID_{ij}. To that end, let $y(U;\beta)$ be a function describing the concentration–response shape for the rate of egg clutch production per individual day. See Table 8.13 or molluscs3.sas7bdat for an example dataset.

OECD TG 243 was originally conceived with the idea that $y(U;\beta)$ would be a three-parameter log-logistic function given by

$$y\left(U;p,d,\mathrm{EC}_{50}\right)=\frac{p}{1+\left(U/\mathrm{EC}_{50}\right)^{d}},\tag{8.4}$$

where p is the rate of egg clutch production per individual day in the control (i.e. where $U=0$).

So $\beta=(p,d,\mathrm{EC}_{50})$. This was later extended to include other shapes, such as those indicated in Eq. (8.12). The number, N_{ij}, of egg clutches produced in replicate j of treatment I is assumed to follow a Poisson distribution with mean $y(U_j;\beta)*\mathrm{NID}_{ij}$. This suggests

$$N_{ij}\sim\mathrm{Poisson}\left[y\left(U_i;\beta\right)*NID_{ij}\right].\tag{8.5}$$

To take into account the random survival of the parent snails and the variability among replicates in the same test concentration, $y(U_i;\beta)$ is replaced by a gamma variable y_{ij} with shape parameter $y(U_i;\beta)/\omega$ and scale parameter $1/\omega$, i.e.

$$N_{ij}\sim\mathrm{Poisson}\left(y_{ij}{}^{*}NID_{ij}\right),\tag{8.6}$$

where

$$y_{ij}\sim\mathrm{Gamma}\left(\frac{y\left(U_i;\beta\right)}{\omega},\frac{1}{\omega}\right).\tag{8.7}$$

The term ω is an overdispersion parameter. The gamma density function is defined by

$$g\left(x;a,b\right)=\frac{b^{a}x^{a-1}e^{-bx}}{\Gamma\left(a\right)}.\tag{8.8}$$

The mean and variance of the gamma are given by a/b and a/b^2. It follows that the mean and variance of the gamma distribution in Eq. (8.7) are $y(U_i;\beta)$ and $\omega y(U_i;\beta)$, respectively, making clear that ω indicates a measure of overdispersion, with larger value of ω indicating greater among replicate overdispersion. Further reading of the properties of the gamma distribution is given in Johnson et al. (1994). To complete the development of the gamma-Poisson distribution defined by Eqs. (8.6) and (8.7), it is helpful to define prior distributions for the parameters p, d, EC_{50}, and ω. Based on the data collected in development of the model and expert elicitation, the test guideline suggests for the TG 243 experimental design that

$$\begin{aligned}&\mathrm{Log}_{10}\left(d\right)\sim\mathrm{Uniform}\left(-2,\,2\right),\\&\mathrm{Log}_{10}\left(\omega\right)\sim\mathrm{Uniform}\left(-2,\,2\right)\\&p\sim\mathrm{Uniform}\left(0,\,1\right),\\&\mathrm{Log}_{10}\left(EC_{50}\right)\sim\mathrm{Normal}\left(\mu,\,\sigma^{2}\right),\end{aligned}\tag{8.9}$$

where μ is the average of the common logarithms of the minimum and maximum test concentrations and σ is one-fourth the difference of those two common logarithms. This would imply 95% confidence that EC50 lies between the minimum and maximum test concentrations. An alternative prior for p is indicated in Charles et al. (2016). Namely,

$$p\sim N\left(\mu_p,\sigma_p^2\right),\tag{8.10}$$

where

$$\mu_p=\frac{1}{r_0}\sum_j\frac{N_{0j}}{\mathrm{NID}_{0j}}\quad\text{and}\quad\sigma_p^2=\frac{\sum_j\left[\left(N_{0j}/\mathrm{NID}_{0j}\right)-\mu_p\right]^2}{r_0\left(r_0-1\right)}.\tag{8.11}$$

The mean in Eq. (8.11) will be seen as the overage rate of egg clutch production per individual day in the control. This alternative thus gives more structure to the prior but has a similar interpretation. In the alternative, the control is not used in fitting the model since it was used to define the prior. This Bayesian gamma-Poisson model was proposed in OECD TG 243 and is described in depth in Charles et al. (2016) as well as the previously cited Delignette-Muller et al. (2013). SAS programs using the MCMC procedure are provided in Appendix 1 to fit this model for several forms for $y(U_i;\beta)$. An R program is located on the free website (http://pbil.univ-lyon1.fr/software/mosaic/reproduction). TG 243 used the three-parameter log-logistic model for this purpose that proved satisfactory for most of the validation studies conducted in developing the guideline. However, there were some datasets where alternatives models gave a better fit to the data. The models that have been explored for the validation datasets are listed in Eq. (8.12), where the hormetic shape is model HORM1 or Eq. (4.2). SAS programs for gPLLog and gPOE3 are given in Appendix 1:

gPLLog = gamma-Poisson w/ Log-logistic shape
gPBVP = gamma-Poisson w/ Bruce-Versteeg shape
gPOE2 = gamma-Poisson w/ 2-parm. Exponential
gPOE3 = gamma-Poisson w/ 3-parm. Exponential
gPHorm = gamma-Poisson w/ hormetic shape
(8.12)

It is relevant to note the connection between the negative binomial distribution and the Poisson distribution with gamma mean (i.e. the gamma-Poisson distribution) discussed, for example, in Cook (2009) and Hilbe (2007).

8.4.2.2 Non-Bayesian Alternative Model for Snail Reproduction

An alternative way to model reproduction data for TG 243 is to fit one of the models from Chapter 4 to the ratio Nreprocumul/Nindtime. In all of the validation studies leading to the development of TG 243, these alternatives gave very similar point estimates of ECx to those from the gamma-Poisson. Two models were fit to the snail reproduction data in Table 8.13. gPLLog was fit to Nclutchcumul and OE3 was fit to Nclutchcumul/Nindtime. See Tables 8.14 and 8.15 for summaries of these two models. As Table 8.15 illustrates, the width of the confidence intervals for the non-Bayesian models tended to be somewhat shorter than those for the gamma-Poisson models. Whether this reflects an improved way to handle the true variability or is a result of inherently greater uncertainty in the Bayesian approach is not clear.

The two models summarized in Tables 8.14 and 8.15 produce similar ECx point estimates for $x \geq 20$ but differ for smaller values of x. It is important to note that the smallest observed percent effect in these data was 17% and a 20% decrease was observed in the lowest tested concentration. Estimation of ECx for $x<20$ is uncertain regardless of the model and reinforces the discussion in earlier chapters about the dangers of extrapolating beyond the range of tested positive concentrations. See Figure 8.5 for a visual comparison of the two models, in which there is no visual reason for preferring one of these models over the other.

Table 8.13 Snail Reproduction Data

Rep	Conc	Nsurv	Ncumul	Nindtime	Rep	Conc	Nsurv	Ncumul	Nindtime
A	0	5	93	280	D	0	5	66	280
A	53	4	67	257	D	53	5	57	280
A	78	5	71	280	D	78	4	53	274.5
A	124	0	21	182.5	D	124	0	32	221
A	232	0	6	95	D	232	0	5	119.5
A	284	0	2	42.5	D	284	0	0	32
B	0	5	89	280	E	0	4	72	236
B	53	4	57	229	E	53	5	59	280
B	78	4	65	274.5	E	78	5	65	280
B	124	0	36	203.5	E	124	2	30	221.5
B	232	0	9	116	E	232	0	11	144
B	284	0	0	25	E	284	0	4	77.5
C	0	4	72	253.5	F	0	5	75	280
C	53	5	68	280	F	53	5	64	280
C	78	5	58	280	F	78	4	80	243
C	124	3	38	269	F	124	4	51	250
C	232	0	6	95	F	232	0	16	140.5
C	284	0	0	53	F	284	0	4	74

Conc in mg l^{-1}. Cumulative clutch counts (Nclutchcumul or Ncumul) and Nindtime already calculated.

Table 8.14 Summary of Observations and Model Predictions for Snail Reproduction

Conc	Observed	gPLLog	OE3	Obs%Decr	gPLLog%Decr	OE3%Decr
0	0.290	0.286	0.288	0	0	0
53	0.233	0.261	0.249	−20	−9	−14
78	0.242	0.230	0.220	−17	−20	−24
124	0.153	0.163	0.164	−47	−43	−43
232	0.073	0.064	0.064	−75	−78	−78
284	0.025	0.043	0.036	−91	−85	−87

Observed, mean ratio Nclutchcumul/Nindtime at concentration; gPLLog, mean Nclutchcumul predicted by model gPLLog divided by Nindtime; OE3, mean ratio predicted by model OE3 fit to ratio directly; Obs%Decr, gPLLog%Decr, OE3%Decr, observed and predicted change from control mean.

Table 8.15 Comparison of Snail Model ECx Estimates

gPLLog			OE3 fit to ratio			
ECx	CLB	CUB	x	ECx	LCB	UCB
41.22	18.43	73.37	5	27.16	12.05	42.28
55.79	29.04	90.58	10	42.88	24.86	60.91
77.82	47.42	114.64	20	69.02	49.16	88.88
87.67	56.36	124.50	25	81.08	61.00	101.17
97.34	65.26	134.34	30	92.93	72.72	113.13
138.93	106.74	174.63	50	141.63	118.19	165.06

CLB and CUB, 95% lower and upper credible bounds; LCB and UCB, 95% lower and upper confidence bounds.

EXERCISES

In all datasets in the text and exercises, if a concentration is given with the value −1, that means a solvent used in the study and the water control is recorded as −1 and the solvent control is listed as 0. The reader should determine whether to combine the two controls for NOEC determination or ECx estimation. If the two controls are to be combined, then the water control should be changed to 0 and the replicates renamed so that replicates in the two controls can be distinguished. If only the solvent control group is to be used, then the water control should be discarded. If only the water control should be used, then the solvent control should be discarded and the concentration for the water control should be changed to 0. Under no

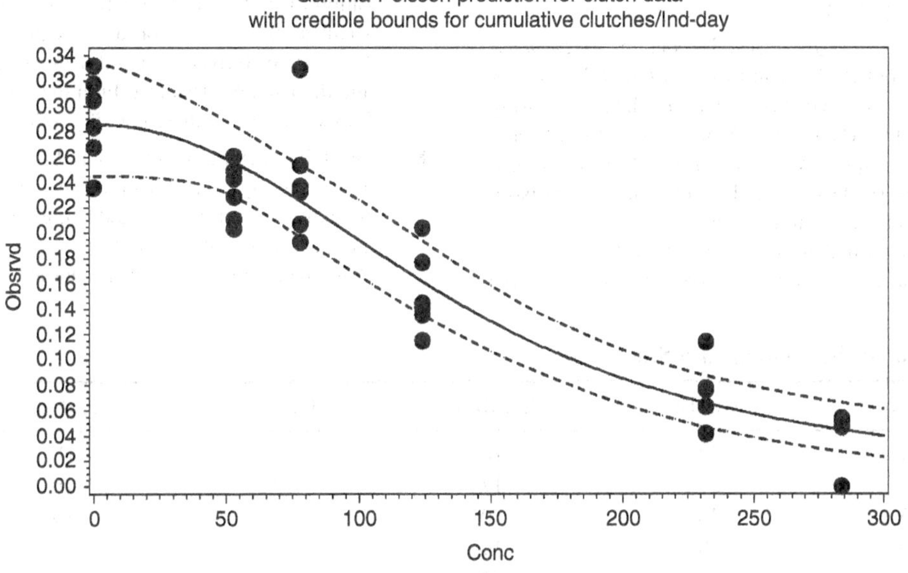

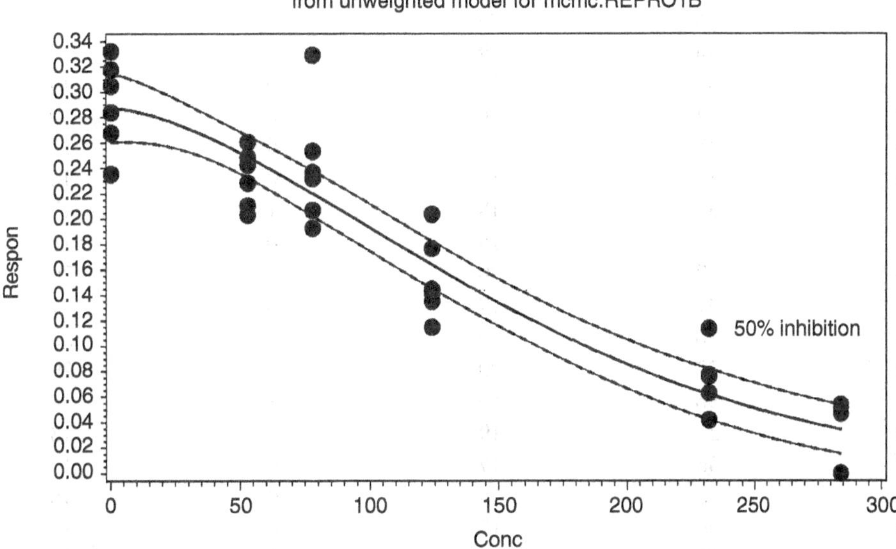

Figure 8.5 Comparison of models for snail reproduction. There is little or no visual reason to infer that does better at estimating the response below the lowest tested concentration.

circumstances should the concentration value −1 be used for NOEC or EC*x* purposes. Guidance on this is provided in Chapters 2 and 3.

8.1 Verify that there is no statistically significant difference in TLY21 between the two controls in Table 8.1.

8.2 Analyze the 14- and 21-day survival data in Table 8.1 to determine the NOEC and to estimate EC*x*. Determine for what values of *x* defensible estimates are possible.

8.3 Analyze the TLY21 data in Table 8.1 with the high treatment group omitted. Verify that both the untransformed and square-root transformed responses satisfy the normality and variance homogeneity requirements to use Williams' test. Compute the NOEC from that test and from the Jonckheere–Terpstra and Dunnett's tests. Determine whether suitable regression models can be fit to these data, and, if so, provide point and interval estimates of EC10 and EC*x* for other choices of *x* where defensible estimates are possible.

8.4–8.9 The Excel file dap_exercises.xls contains 6 *Daphnia magna* datasets from studies following OECD TG 211. For each dataset, do the following: (i) Determine the NOEC using a normal errors test (Williams + Dunnett) and Jonckheere–Terpstra test on the square-root transform of TLY21 and also without transform. (ii) Determine the NOEC using a Poisson errors model on TLY21 with no transform using a Dunnett–Hsu test. (iii) Fit a normal errors regression model to the data, if possible, and provide point and interval estimates of EC10 and EC20. Evaluate the model for goodness of fit and stability. (iv) Repeat (iii) with a Poisson errors model.

8.10 Fit a GLMM model to the data in Table 8.5 and verify the claims made in Section 8.3.

8.11 Analyze last day of hatching from Table 8.10 to determine the NOEC using the Jonckheere–Terpstra test. Repeat this using a Dunnett–Hsu test from a Poisson errors model.

8.12 Analyze survival from Table 8.10 to determine the NOEC using the Cochran–Armitage test.

8.13 Analyze first day of swimup from an ELS study. See Table 8.16 for the data.

8.14 The variable secondary sex, abbreviated as SecSex, is a count of male papillary processes. This variable is a measure of endocrine activity, with reduced counts indicative of feminization of male fish. See Table 8.17 for an example dataset. Analyze these data for a decrease in SecSex using the Jonckheere–Terpstra test and using a Poisson errors GLMM and the Dunnett–Hsu test. Fit a regression model (normal or Poisson errors) to these data if possible and estimate EC10.

8.15 See Table 8.18 for another dataset involving variable secondary sex. Analyze these data for a decrease in SecSex using the Jonckheere–Terpstra test and using a Poisson errors GLMM and the Dunnett–Hsu test. Fit a regression model (normal or Poisson errors) to these data if possible and estimate EC10.

Table 8.16 First Day of Swimup in ELS Study

Conc	Rep	SwimUp	TAtRisk	DLarvae	AtRisk	NumAbn
0	1	37	18	0	18	0
0	2	37	17	1	16	2
0	3	38	20	1	19	1
0	4	38	19	0	19	2
6	1	37	20	1	19	3
6	2	38	17	0	17	1
6	3	38	20	1	19	1
6	4	38	19	0	19	1
13	1	37	18	0	18	0
13	2	37	20	0	20	0
13	3	38	15	0	15	1
13	4	37	18	0	18	1
24	1	37	18	1	17	0
24	2	37	18	0	18	2
24	3	37	20	0	20	0
24	4	37	17	0	17	0
51	1	36	20	0	20	0
51	2	37	20	0	20	0
51	3	36	19	1	18	0
51	4	37	19	1	18	2
99	1	36	17	0	17	0
99	2	36	18	0	18	0
99	3	36	17	0	17	1
99	4	36	20	1	19	1

SwimUp, first day of swimup (last observation was made on day 46); TAtRisk, number at risk (cumulative dead plus live larvae); DLarvae, number of dead larvae; AtRisk, number at risk (live larvae only); NumAbn, number of abnormal larvae (from live larvae).

Table 8.17 Secondary Sex Characteristics for Medaka: Dataset 1

Group	SecSex	Rep	Conc	Group	SecSex	Rep	Conc
1	75	1	0	3	75	3	2.81
1	76	2	0	3	70	4	2.81
1	81	3	0	3	82	5	2.81
1	98	4	0	3	95	6	2.81
1	86	5	0	4	88	1	8.88
1	102	6	0	4	94	2	8.88
2	73	1	0.919	4	102	3	8.88
2	109	2	0.919	4	78	4	8.88
2	111	3	0.919	4	73	5	8.88
2	92	4	0.919	4	89	6	8.88
2	80	5	0.919	5	77	1	27.9
2	65	6	0.919	5	90	2	27.9
3	103	1	2.81	5	51	3	27.9
3	88	2	2.81	5	75	4	27.9
				5	60	5	27.9

Conc = mg l^{-1}. SecSex, number of papillary processes.

Table 8.18 Secondary Sex Characteristics for Medaka: Dataset 2

Group	SecSex	Rep	Conc	Group	SecSex	Rep	Conc
1	129	1	0	4	77	1	8.88
1	92	2	0	4	86	2	8.88
1	110	3	0	4	105	3	8.88
1	112	4	0	4	85	4	8.88
1	155	5	0	4	84	5	8.88
1	103	6	0	4	108	6	8.88
2	118	1	0.919	5	79	1	27.9
2	110	2	0.919	5	50	2	27.9
2	87	3	0.919	5	98	3	27.9
2	89	4	0.919	5	76	4	27.9
2	147	5	0.919	5		5	27.9
2	99	6	0.919	5	71	6	27.9
3	113	1	2.81	6	0	1	84.3
3	135	2	2.81	6		2	84.3
3	94	3	2.81	6	0	3	84.3
3	101	4	2.81	6	0	4	84.3
3	103	5	2.81	6	0	5	84.3
3	171	6	2.81	6	0	6	84.3

Conc = mg l^{-1}; SecSex, number of papillary processes.

8.16 See Table 8.19 for data from a rainbow trout ELS study. Evaluate the two controls for last day of hatching and decide whether to combine them for further analysis or use only one control group. Defend your choice. Then analyze last day of hatch both for NOEC using the Jonckheere–Terpstra test and from a Poisson errors GLMM model using the Dunnett–Hsu test. If possible, fit a Poisson errors regression model and estimate EC10.

8.17 See Table 8.20 for another rainbow trout ELS study. Evaluate the two controls for last day of hatching and decide whether to combine them for further analysis or use only one control group. Defend your choice. Then analyze last day of hatch, both for NOEC using the Jonckheere–Terpstra test and from a Poisson errors GLMM model. If possible, fit a Poisson errors regression model and estimate EC10. Repeat the above treating day 36 as a right-censored observation.

8.18 See Table 8.21 for a third rainbow trout ELS study. Analyze first and last day of hatch, both for NOEC using the Jonckheere–Terpstra test and from a Poisson errors GLMM model. If possible, fit a Poisson errors regression model and estimate EC10.

Table 8.19 Last Day of Hatch from Rainbow Trout ELS Study

Conc	Group	Rep	FHatch	LHatch	TotEgg	DeadEgg	LarHatch
−1	1	1	30	36	20	4	16
−1	1	2	30	35	20	4	16
−1	1	3	28	34	20	4	16
−1	1	4	28	35	20	5	15
0	2	1	28	36	20	4	16
0	2	2	30	34	20	5	15
0	2	3	29	38	20	5	15
0	2	4	30	34	20	4	16
1.5	3	1	29	34	20	4	16
1.5	3	2	30	32	20	5	15
1.5	3	3	29	34	20	5	15
1.5	3	4	30	35	20	4	16
3.8	4	1	29	34	20	4	16
3.8	4	2	30	31	20	5	15
3.8	4	3	28	33	20	4	16
3.8	4	4	29	33	20	5	15
9.6	5	1	30	34	20	6	14
9.6	5	2	30	36	20	5	15
9.6	5	3	30	32	20	5	15
9.6	5	4	30	31	20	4	16
24	6	1	30	32	20	5	15
24	6	2	30	32	20	4	16
24	6	3	30	32	20	3	17
24	6	4	29	32	20	3	17
60	7	1	30	32	20	3	17
60	7	2	30	32	20	3	17
60	7	3	29	32	20	2	18
60	7	4	29	31	20	0	20
150	8	1	30	33	20	5	15
150	8	2	29	34	20	3	17
150	8	3	29	34	20	5	15
150	8	4	29	33	20	2	18

Conc = −1 is water control; Conc = 0 is solvent control, other concentrations in μg l^{-1}; FHatch, first day of hatching; LHatch, last day of hatching; TotEggs, number of eggs exposed; DeadEggs, number of exposed eggs that did not hatch; LarHatch = TotEggs − DeadEggs.

Table 8.20 Last Day of Hatch from Rainbow Trout ELS Study: Dataset 2

Conc	Group	Rep	FHatch	LHatch	TotEgg	DeadEgg	NumHatch
−1	1	1	29	31	20	9	11
−1	1	2	29	31	20	10	10
−1	1	3	29	32	20	10	10
−1	1	4	29	34	20	7	13
0	2	1	29	31	20	8	12
0	2	2	29	33	20	9	11
0	2	3	29	32	20	11	9
0	2	4	29	34	20	10	10
1.5	3	1	28	32	20	9	11
1.5	3	2	28	34	20	9	11
1.5	3	3	29	31	20	9	11
1.5	3	4	29	32	20	8	12
3	4	1	29	34	20	10	10
3	4	2	29	33	20	7	13
3	4	3	28	32	20	9	11

Table 8.20 (Continued)

Conc	Group	Rep	FHatch	LHatch	TotEgg	DeadEgg	NumHatch
3	4	4	29	33	20	9	11
6	5	1	28	36	20	8	12
6	5	2	29	31	20	9	11
6	5	3	30	34	20	10	10
6	5	4	30	35	20	9	11
13	6	1	30	35	20	9	11
13	6	2	30	32	20	9	11
13	6	3	29	32	20	9	11
13	6	4	29	31	20	9	11
25	7	1	30	32	20	7	13
25	7	2	29	35	20	10	10
25	7	3	29	32	20	9	11
25	7	4	28	32	20	10	10
50	8	1	29	33	20	7	13
50	8	2	30	32	20	9	11
50	8	3	28	34	20	10	10
50	8	4	29	34	20	10	10
100	9	1	29	36	20	9	11
100	9	2	29	33	20	8	12
100	9	3	29	31	20	9	11
100	9	4	29	32	20	8	12

Conc = −1 is water control; Conc = 0 is solvent control; FHatch, first day of hatching; LHatch, last day of hatching; TotEgg, number of eggs exposed; DeadEgg, number of exposed eggs that did not hatch; NumHatch = TotEggs − DeadEggs.

Table 8.21 Last Day of Hatch from Rainbow Trout ELS Study: Dataset 3

Conc	Group	Rep	FHatch	LHatch	TotEgg	DeadEgg	NumHatch
0	1	1	25	27	20	1	19
0	1	2	25	28	20	3	17
0	1	3	24	27	20	3	17
0	1	4	25	27	20	4	16
7.6	2	1	24	27	20	5	15
7.6	2	2	24	27	20	4	16
7.6	2	3	24	26	20	2	18
7.6	2	4	25	26	20	4	16
15	3	1	25	27	20	4	16
15	3	2	24	25	20	1	19
15	3	3	24	26	20	2	18
15	3	4	24	26	20	3	17
29	4	1	23	25	20	4	16
29	4	2	24	25	20	6	14
29	4	3	24	26	20	3	17
29	4	4	24	25	20	3	17
52	5	1	23	25	20	4	16
52	5	2	24	25	20	3	17
52	5	3	24	24	20	4	16
52	5	4	24	25	20	3	17
95	6	1	23	24	20	4	16
95	6	2	23	24	20	4	16
95	6	3	23	24	20	4	16
95	6	4	23	24	20	3	17

Conc = −1 is water control; Conc = 0 is solvent control; FHatch, first day of hatching; LHatch, last day of hatching; TotEgg, number of eggs exposed; DeadEgg, number of exposed eggs that did not hatch; NumHatch = TotEggs − DeadEggs.

Table 8.22 First Day of Swimup from Rainbow Trout ELS Study: Dataset 2

Group	Conc (mg l^{-1})	Rep	SwimUp (days)	TAtRisk	DLarvae	AtRisk	NumAbn
1	−1	1	40	19	0	19	0
1	−1	2	40	17	0	17	0
1	−1	3	40	18	0	18	1
1	−1	4	41	17	0	17	1
2	0	1	40	17	2	15	0
2	0	2	41	15	0	15	0
2	0	3	40	17	1	16	0
2	0	4	41	18	1	17	1
3	0.11	1	40	18	0	18	0
3	0.11	2	41	20	0	20	0
3	0.11	3	40	18	1	17	0
3	0.11	4	40	17	0	17	0
4	0.329	1	40	18	0	18	1
4	0.329	2	40	17	2	15	0
4	0.329	3	40	18	1	17	0
4	0.329	4	40	15	0	15	0
5	0.956	1	42	16	0	16	0
5	0.956	2	41	16	3	13	1
5	0.956	3	40	18	2	16	0
5	0.956	4	41	19	3	16	3
6	2.81	1	40	16	2	14	0
6	2.81	2	42	14	3	11	4
6	2.81	3	41	14	0	14	0
6	2.81	4	41	16	0	16	6
7	9.46	1	40	13	1	12	2
7	9.46	2	41	16	1	15	3
7	9.46	3	42	17	2	15	4
7	9.46	4	41	14	2	12	3

SwimUp, first day of swimup (last observation was made on day 46); TAtRisk, number at risk (cumulative dead plus live larvae); DLarvae, number of dead larvae; AtRisk, number at risk (live larvae only); NumAbn, number of abnormal larvae (from live larvae).

8.19 See Table 8.22 for first day of swimup from a rainbow trout ELS study. Evaluate the two controls for first day of swimup and decide whether to combine them for further analysis or use only one control group. Defend your choice. Then analyze first day of swimup from the ELS study data in Table 8.22, both for NOEC using the Jonckheere–Terpstra test and from a Poisson errors GLMM model. If possible, fit a Poisson errors regression model and estimate EC10. Repeat the above treating day 42 as a right-censored observation.

8.20 Another snail reproduction dataset, Molluscs2, is given in Appendix 1. This is an SAS dataset. As indicated in several R programs in Appendix 1, the R package sas7bdat has a function, readsas7bdat, that reads an SAS dataset into R without the need to have SAS installed.

(a) Fit the model gPLLog to the Nclutchcumul data using the SAS program in Appendix 1 or the web application cited in this chapter.

(b) Fit a 3-parameter log-logistic model to Nclutchcumul/Nindtime from the same dataset.

(c) If SAS is available, analyze the Nclutchcumul data using gPOE3b.

(d) Analyze Nclutchcumul/Nindtime using model OE3.

(e) Modify the program fit_gPOE3.SAS to fit the model gPBVP and apply it to the molluscs2 data.

(f) Fit model BVP to Nclutchcumul/Nindtime from the molluscs2 data.

Chapter 9

Analysis of Ordinal Data

9.1 INTRODUCTION

Chapters 1–8 considered continuous, quantal, and count data. These three broad categories capture most responses that arise in ecotoxicity studies for regulatory risk assessment. There are nuances within these categories that will occupy later chapters, such as time-to-event and mesocosm studies. The present chapter considers a different category of responses that are increasingly important, especially in chronic and multigeneration studies. These are ordinal responses, which are not measured on a numerical scale, though numbers are often used as labels. The essential characteristic of ordinal-scale variables is that of order. For example, in histopathology assessment, damage to organ or other cell tissue is ranked in some manner, often with some subjectivity dependent on the scientist doing the ranking. The rankings none, minimal, mild, moderate, and severe are widely used for such data, and all that can be said statistically is that the rankings none < minimal < mild < moderate < severe indicate relative severity. There is no magnitude of damage that can be ascribed to these scores or rankings, and no distance between scores can be assigned. Thus a moderate score does not imply twice as much damage as a mild score or half as much damage as a severe score. There are biological indicators that guide the histopathologist in making a ranking, but these are more subjective than weighing a subject for weight or timing a subject for hatch or swimup. Different statistical techniques are needed to handle these types of response. Two important ordinal responses are considered in this chapter that appear in OECD and USEPA test guidelines.

9.2 PATHOLOGY SEVERITY SCORES

Several assays are being performed or proposed to screen for endocrine-disrupting chemicals. These include the following, which are difficult to run, are very expensive, and, at least under some conditions, require histopathologic assessments:

Fish Short Term Reproduction Assay (OECD TG 229 FSTRA)
Amphibian Metamorphosis Assays (OECD TG 231 AMA)
Japanese Quail Two-Generation Toxicity Test (USEPA TG 890.2100 JQTT)
Larval Amphibian Growth and Development Assay (OECD TG 241 LAGDA)
Medaka Extended One-generation Reproduction Test (OECD TG 240 MEORGT)

A severity score is assigned to a tissue sample by a trained pathologist. Scores depend on the type and extent of tissue damage found. The usual scale for these scores is:

0 = none (no tissue damage)
1 = minimal
2 = mild
3 = moderate
4 = severe

See Figure 9.1 for example tissue slides from which semiquantitative scores are determined for the response lamellar epithelial hyperplasia in the gills of fish. To understand the meaning of these slides and how they translate into severity scores requires expert knowledge.

Statistical Analysis of Ecotoxicity Studies, First Edition. John W. Green, Timothy A. Springer, and Henrik Holbech.

Companion website: www.wiley.com/go/Green/StatAnalysEcotoxicStudy

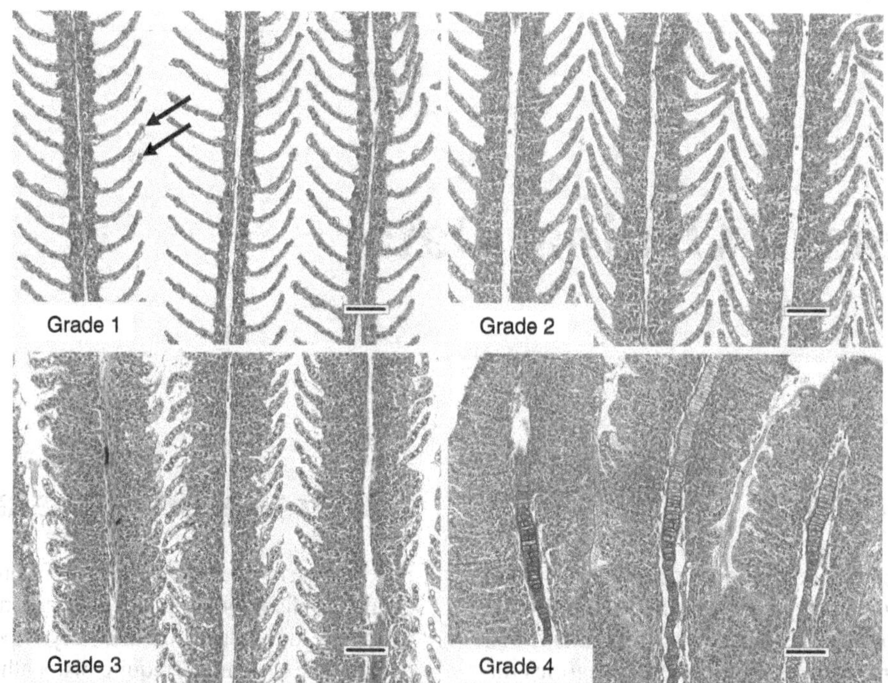

Figure 9.1 Semiquantitative severity grading of lamellar epithelial hyperplasia in the gills of fish. Grading is based on the approximate degree to which the spaces between adjacent gill lamellae (arrows) are filled in by proliferating cells. For example, grade 1 (minimal) = 10–25% filled, grade 2 (mild) = 26–50% filled, grade 3 (moderate) = 51–75% filled, and grade 4 (severe) = 76–100% filled. Cell proliferation less than 10% filled is recorded as "not remarkable." Hematoxylin and eosin stain. All bars = 50 μm. We are indebted to Jeff Wolf of Experimental Pathology Laboratories, Inc. who created this slide and caption expressly for this book.

None	Minimal	Mild	Moderate	Severe

Figure 9.2 Schematic for variation within and between severity scores. Width of interval indicates possible disparity in the range of injury to tissue that merits the indicated score.

The reader is cautioned about possible misinterpretation of the example in Figure 9.1. In that example, the scores are based on an approximate numerical measure, only visually assessed, of the space between adjacent gill lamellae that are filled. Severity scores need not have such a compelling quantitative interpretation. For example, severity scores in figures 5 and 6, pages 54–55 of USEPA (2015), are much more challenging to interpret quantitatively. Consequently, interpreting these scores statistically requires some care. A key characteristic of these scores, and ordinal data in general, is that the change in severity between scores need not be uniform. The difference between minimal and mild is not the same as the difference between mild and moderate or between moderate and severe. The range of severity within scores is not uniform. See Figure 9.2 for a schematic that might help visualize these points.

Few people would suggest that if half of the tissue samples have a *minimal* finding and half have a *moderate* finding, then on average, the finding is *mild*. Labeling these scores as 0, 1, 2, 3, 4 does not suddenly make it sensible to average these scores. A change from score 1 to 2 does not indicate a doubling of severity. A change from 3 to 4 may not indicate a change in severity equal to change from 1 to 2. A statistical method that is based on average scores thus is not justified. This rules out methods based on normality, for example. A *t*-test or Dunnett test would be entirely unjustified. It would be possible to use some nonparametric tests, such as the Jonckheere–Terpstra (JT) or Mann–Whitney, provided medians were used rather than means in summarizing replicates. However, as explained below, there are drawbacks to these tests on other grounds.

There are many histological endpoints of possible interest, and test guidelines vary as to which ones are of interest. A small sample of such endpoints is listed here. The reader may recognize maturation stage as another name for developmental stage, which was discussed in Chapter 8 in regard to the LAGDA test guideline.

Maturation stage
Some animals unable to mature, others mature too rapidly
Increased or decreased cells of

Spermatogonia, spermatocytes, spermatids, etc.
Oocytes (perinucleolar, early vitellogenic, etc.)
Proteinaceous fluid, interstitial, or intravascular
Hepatocyte basophilia, increased
Nephropathy
Interstitial fibrosis
Macrophage aggregates, increased
Egg debris, oviduct
Oocyte membrane folding

There are several complications to analyzing histopathology data arising from guideline ecotoxicology studies. In the development of OECD TG 240, there were multiple fish in each replicate tank. The goal was to include both sexes, but there is currently no feasible way to determine the sex of this species at the embryonic stage at which the study begins. Thus, there is variability in the number of males and number of females in each replicate, including the possibility of no fish of one sex. A second difficulty is that a test such as Mann–Whitney is based on a single summary measure on each rep, such as the median score. While this makes the response consistent with the requirements of that test, it ignores the variation of scores within the replicate and thus discards valuable information. Next, overdispersion, or extramultinomial variation, is common in these studies, so that needs to be accommodated in the test used. A final complication is the small sample sizes typical in TG 240 for histopathology data. It is not unusual to have only one or two subjects of each sex per replicate. Green et al. (2014) developed a method that is statistically sound, retains the replicate nature of the data, retains the within-replicate variation, and accommodates overdispersion and small samples. This method is also discussed in Green (2015a, b) and is recommended in TG 240. That test will be described in Section 9.2.1. That discussion leans heavily on Green et al. (2014).

There are several approaches available to analyze ordinal data, such as polychotomous or multicategory logit models (Agresti, 2007, 2012), that incorporate individual severity scores and are consistent with a monotone concentration–response, but in our view these methods tend to be more demanding to implement and interpret than many testing laboratories can handle, and it is not clear how to accommodate the replicate nature of the experiment.

The method to be described is a conceptually simple test with the following features: (i) It takes into account the fact that fish in the same test vessel may respond more similarly to each other than to fish in different vessels in the same treatment group. (ii) Individual fish severity scores are maintained. (iii) It gains power and biological appropriateness by incorporating the expectation that fish exposed to increasing test concentrations tend to have more severe effects. (iv) The statistical test displays the severity of any effect found statistically significant, so statistical significance can be judged simultaneously with biological importance. (v) The statistical test is easy to interpret.

Table 9.1 Contingency Table of Incidence of Findings

	Concentration group				
Finding	0	1	2	k	Row totals
Absent	X_{00}	X_{01}	X_{02}	X_{0k}	R_0
Present	X_{10}	X_{11}	X_{12}	X_{1k}	$N-R_0$
Column totals	C_0	C_1	C_2	C_k	N

k, number of concentration groups, not counting the control; X_0j, number of subjects in concentration j having a zero severity score; X_1j number of subjects in concentration j having a nonzero severity score; Cj, number of subjects in concentration group j; R_0, number of subjects in some treatment group with zero severity score. $N=C_1+C_2+C_3+\cdots+Ck$.

It will be helpful to start by considering a simple contingency table. Such a table could arise with severity scores if all scores greater than zero are lumped together, so that one is merely comparing the incidence of some adverse finding against the incidence of no finding. Clearly, this is not a suitable way to analyze severity scores, since it makes no distinction between minimal and severe effects. Nonetheless, it will be helpful to start with this simplification and then expand on it. See Table 9.1 for a basic contingency table, which in addition to ignoring how severe an effect is also does not use the replicate structure. The table merely shows the number of subjects with and without positive scores in each treatment group.

To analyze data from a table in the form of Table 9.1, the step-down Cochran–Armitage (CA) test from Section 6.5.1 could be used. This test takes the expected monotone concentration–response shape into account but ignores replicates. Preferable to that is the step-down Rao–Scott Cochran–Armitage (RSCA) test from Section 6.5.2, which is CA with the Rao–Scott adjustment for overdispersion. This test does use replicates and adjusts for differences in replicate sample sizes, as well as taking account of the expected monotone concentration–response shape. Tarone's $C(\alpha)$ test from Section 6.3.1 can be used to test for overdispersion. With this background, it is now possible to introduce a procedure that extends the RSCA test to include the range of severity scores.

9.2.1 Rao–Scott Cochran–Armitage by Slices (RSCABS)

The Rao–Scott Cochran–Armitage by Slices (RSCABS) test can be described in a stepwise manner as follows:

Step 1: In each replicate, count the number of fish in that replicate with severity 1 or greater and the number with score 0. This will define an incidence table. Apply the RSCA test to this incidence table to establish an LOEC and no-observed-effect concentration (NOEC) for increased severity, which will be referred to as NOEC1 and LOEC1. As usual, LOEC1 and NOEC1 will be reported as exceeding the highest test concentration if no effect is found.

Step 2: In each replicate, count the number of fish in that replicate with severity 2 or greater and the number with score less than 2 (i.e. slice the data between scores of 1 and 2). This will define a new incidence table. Apply the RSCA test to this incidence table to establish an LOEC and NOEC for test concentrations producing severity level 2 or higher, which will be referred to as LOEC2 and NOEC2. As before, the LOEC2 and NOEC2 will be reported as exceeding the highest test concentration if no effect is found.

Step J: In each replicate, count the number of fish in that replicate with severity *J* or greater and the number with score <*J* (i.e. slice the scores between severity scores <*J* and *J*). This will define a new incidence table. Apply the RSCA test to this incidence table to establish an LOEC*J* and NOEC*J* for test concentrations producing severity scores of *J* or higher. As before, the LOEC*J* and NOEC*J* will be reported as exceeding the highest test concentration if no effect is found.

Individual LOEC*J* values indicate where effects of severity *J* are observed. If desired, the lowest test concentration that is LOEC*J* for some *J*>0 is the LOEC for the response under investigation, and the value of *J* indicates the severity of the effect at that concentration. However, it is preferable not to take this additional step and instead report the individual LOEC*J* and NOEC*J*. In this way, there is a clear association between the presence of a statistically significant increase in incidence and the severity of that incidence. For example, it is helpful for the risk analysts to know that there is a significant increase in severity at concentration 2, but it corresponds only to an increase in minimal adverse effect, while at concentration 4, there is a significant increase in moderately severe effects, and at concentration 5, there is a significant increase in severe effects. In this example, the effect at concentration 2 probably does not call for any risk mitigation, while that at concentration 4 or 5 almost certainly would.

The theoretical considerations of RSCABS for each slice are precisely those of RSCA. The only novelties are the slicing procedure and interpretation. RSCABS is best understood by example and an example from Green et al. (2014) will be used. See Table 9.2. Other examples are given in the exercises.

There is notable variation in replicate sizes evident in Table 9.2, even one that is empty, namely, replicate A in the negative control. This happened because all fish in that replicate were discovered to be male at the end of the study period. What should also be noted is that there were no tissue samples that were scored 0. That is, all samples were judged to have at least minimal tissue damage. These issues are complications that any statistical method that might be employed must deal with. RSCABS was designed to handle such issues with no problem. Table 9.2 is summarized in Table 9.3 by summing across replicates in each treatment

Table 9.2 Liver Basophilia Severity Scores of Second-generation Female Medaka, Showing Replicate Information

		Severity						Severity			
Conc	Replicate	1	2	3	Total	Conc	Replicate	1	2	3	Total
1	A	0	0	0	0	3	D	0	1	0	1
1	B	1	3	0	4	3	E	2	0	0	2
1	C	1	0	0	1	3	F	4	0	0	4
1	D	2	1	0	3	4	A	1	1	0	2
1	E	2	0	1	3	4	B	1	0	0	1
1	F	3	0	0	3	4	C	0	1	0	1
2	A	0	4	0	4	4	D	0	3	0	3
2	B	0	2	0	2	4	E	1	3	0	4
2	C	0	0	0	0	4	F	0	2	1	3
2	D	2	1	0	3	5	A	1	0	1	2
2	E	1	1	0	2	5	B	2	0	0	2
2	F	1	0	0	1	5	C	1	1	0	2
3	A	0	2	0	2	5	D	1	1	2	4
3	B	1	2	0	3	5	E	1	2	0	3
3	C	0	1	0	1	5	F	0	0	1	1

Conc, concentration. Concentration 1 is the negative control, and concentrations 2–5 indicate increasing test substance concentrations. Total, number of individuals available for analysis in each replicate.

group to make concentration–response patterns more apparent, but the replicate structure is retained in the analysis.

The first step is to slice the data between scores 1 and 2. See Table 9.3 for a summary by treatment and the slice between scores 1 and 2.

It is left as an exercise to show that RSCA test applied to the right panel of Table 9.3 finds no significant increase in score 2+ tissue samples. Thus NOEC2 > 5. The next step is to slice the data between scores 2 and 3. See Table 9.4 where that is summarized. It is left as an exercise to show that Tarone's $C(\alpha)$ test found significant overdispersion in the data from the right panel of Table 9.2, demonstrating the importance of using the Rao–Scott adjustment to the CA test.

The RSCA test is applied to the right panel in Table 9.4. It is left as an exercise to verify that there was a significant increase in score 3 (moderate damage) tissue samples in the high treatment group but not below that, and consequently, NOEC3 = 4, LOEC3 = 5. The results for this example can be generated in two ways: (i) The RSCA test given in Appendix 1 for Chapter 6 can be applied to the right panel of each of Tables 9.3 and 9.4. (ii) The RSCABS SAS program given in Appendix 1 can be applied to the data in Table 9.2 as a whole to generate the entire step-down slicing process. The R package StatCharrms contains an RSCABS program and numerous other programs used in this book. Hothorn (2016) gives R programs for a limited version (CABS) of RSCABS under the heading of collapsing categories and also some alternative tests for this type of response. More information is given in Exercise 9.7.

9.2.2 Overdispersion in Histopathology Data

Overdispersion is common in histopathology data, as indicated for the data in liver basophilia data of Tables 9.2–9.4. This is one of the reasons for using RSCA rather than CA on each slice. Of course, another important reason is the variation in sample sizes. The problem of overdispersion in the current context can be related to the more general problem of incorporating overdispersion in a multinomial distribution by recognizing that the incidence rates of the various severity scores in a treatment group can be regarded as the probabilities in a multinomial distribution. If $x_0, x_1, x_2, \ldots, x_k$ are the counts of tissue samples with severity score 0, 1, 2, …, k, respectively, the pdf of a multinomial can be expressed as

$$f\left(x_0, x_1, x_2, \ldots, x_k; \theta\right) = \left(\frac{N!}{\prod_i x_i!}\right) \prod_i \theta_i^{x_i}, \qquad (9.1)$$

Table 9.3 Summary of Liver Basophilia Severity Scores and Slice 2

Frequency of severity scores in each treatment group							Frequency of severity scores <2 and ≥2 in each treatment group						
	Treatment							Treatment					
Score	1	2	3	4	5	Total	Score	1	2	3	4	5	Total
1	9	4	7	3	6	29	<2	9	4	7	3	6	29
2	4	8	6	10	4	32	≥2	5	8	6	11	8	38
3	1	0	0	1	4	6							
Total	14	12	13	14	14	67	Total	14	12	13	14	14	67

Cell entries on left are number of fish in indicated group having the indicated score. Cell entries on right are obtained by summing cell counts with score ≥2 (rows 2 and 3 in left panel summed to create row 2 in right panel) and score <2 (row 1 in left panel becomes row 1 in right panel). This is the first step of the RSCABS test since there is no score 0 fish.

Table 9.4 Summary of Liver Basophilia Severity Scores and Slice 3

Frequency of severity scores in each treatment group							Frequency of severity scores <3 and ≥3 in each treatment group						
	Treatment							Treatment					
Score	1	2	3	4	5	Total	Score	1	2	3	4	5	Total
1	9	4	7	3	6	29	<3	13	12	13	13	10	29
2	4	8	6	10	4	32	≥3	1	0	0	1	4	38
3	1	0	0	1	4	6							
Total	14	12	13	14	14	67	Total	14	12	13	14	14	67

Cell entries on left are number of fish in indicated group having the indicated score. Cell entries on right are obtained by summing cell counts with score <3 (rows 1 and 2 in left panel summed to create row 1 in right panel) and score ≥2 (row 3 in left panel becomes row 2 in right panel).

where

$$N = \sum_i x_i \quad \text{and} \quad \theta_k = 1 - \sum_{i=0}^{k-1} \theta_i. \tag{9.2}$$

It is not difficult to establish the following properties:

$$\mu_i = E[x_i] = N\theta_i,$$

$$\text{Var}(x_i) = N\theta_i(1-\theta_i) = \mu_i\left(1-\frac{\mu_i}{N}\right), \tag{9.3}$$

$$\text{cov}(x_i, x_j) = -N\theta_i\theta_j = -\frac{\mu_i\mu_j}{N}.$$

The multinomial is a natural generalization of the binomial. To incorporate overdispersion into the multinomial, the example of how a beta prior distribution was imposed on the binomial rate parameter to generate the beta-binomial distribution (Sections 6.4.2) will be extended to the multinomial setting. It would probably be helpful for the reader to review Section 6.4.2 before proceeding. The Dirichlet distribution with hyperparameters α is a generalization of the beta with density function given by

$$f(\theta|\alpha) = \frac{\Gamma\left(\sum_j \alpha_j\right)}{\prod_j \Gamma(\alpha_j)} \prod_j \theta_j^{\alpha_j - 1}. \tag{9.4}$$

The moments of the Dirichlet distribution are given by

$$E[\theta_i] = \frac{\alpha_i}{\sum_j \alpha_j},$$

$$\text{Var}(\theta_i) = \frac{\alpha_i\left(\sum_j \alpha_j - \alpha_i\right)}{\left(\sum_i \alpha_i\right)^2\left(1+\sum_i \alpha_i\right)}, \tag{9.5}$$

$$\text{cov}(\theta_i, \theta_j) = \frac{-\alpha_i\alpha_j}{\left(\sum_i \alpha_i\right)^2\left(1+\sum_i \alpha_i\right)}.$$

The first two formulas in Eq. (9.3) will be seen to be generalizations of Eqs. (6.12) and (6.13). The Dirichlet-multinomial distribution is the corresponding generalization of the beta-binomial, and its density function is given by

$$f(x|\alpha) = \frac{N!\,\Gamma\left(\sum_j \alpha_j\right)}{\left(\prod_j x_j!\right)\Gamma\left(\sum_j(\alpha_j + x_j)\right)} \prod_j \frac{\Gamma(\alpha_j + x_j)}{\Gamma(\alpha_j)}. \tag{9.6}$$

From this it can be shown that

$$E[x_j] = \mu_j = \frac{N\alpha_j}{\sum_i \alpha_i}, \tag{9.7}$$

and

$$\text{Var}(x_j) = \sigma^2\mu_j\left(1-\frac{\mu_j}{N}\right), \tag{9.8}$$

$$\text{Cov}(x_i, x_j) = -\sigma^2\frac{\mu_i\mu_j}{N},$$

where

$$\sigma^2 = \frac{N + \sum_j \alpha_j}{1 + \sum_j \alpha_j}. \tag{9.9}$$

A key observation is that the variance–covariance matrix of the Dirichlet-multinomial is a scalar multiple of the variance–covariance matrix of the multinomial. This scalar is a measure of the extent of overdispersion. The Dirichlet-multinomial can be traced to Mosimann (1962), which contains a detailed development. A particularly clear discussion of this distribution is given in Thorsén (2014).

9.2.3 Power of RSCABS to Detect an Increase in Severity

Use of the Dirichlet-multinomial distribution and the overdispersion factor σ^2 in Eqs. (9.8) and (9.9) was used in Green et al. (2014) to explore the power properties of RSCABS in the presence of varying amounts of overdispersion. As noted in Green et al. (2014), when there is no overdispersion, there is about 80% power to detect a 12% increase in the proportion of nonzero scores regardless of severity, when there is no positive severity score in the control. If 5% are affected in the control, then an increase of 20% affected subjects is required to achieve adequate power. With 10% of the controls having positive scores, it takes an increase of 25% affected subjects to achieve adequate power. As the amount of overdispersion increases, the power to detect an increase in the proportion of affected animals decreases. The impact of overdispersion observed here is consistent with what is seen when analyzing overdispersed binary response data, such as mortality or hatch rate.

9.2.4 Alternatives to RSCABS

If there is little variation in sample sizes among replicates, a standard CA test might be used in place of the RSCA in the slicing procedure to obtain what Green et al. (2014)

calls CABS. Alternatively, the JT test could be used to obtain the test called in the same source JTABS. Both of these alternatives ignore variation in replicate sizes and have weaker power properties than RSCABS. It should be acknowledged that Green et al. (2014) demonstrated that under some conditions, RSCABS can have inflated type I error. This is inherent in the RSCA test despite contrary indications in Rao and Scott (1992, 1999) and Fung et al. (1994).

A test that has been used in many mammalian histopathology analyses is the Mann–Whitney test. This ignores the expected monotone concentration–response and is applied to individual animals. The last is not generally a problem since, until recently, rodents were typically housed one per cage. Applying this test to fish studies means either ignoring replicates or using a summary measure, such as the median score, on the replicate. As pointed out above, this ignores or discards relevant information. There has been growing interest in joint housing for rodents used in laboratory testing (e.g. Baumans, 2005), and more recent regulatory guidance (OECD, 2012a, b) calls for paired housing for rodents in some studies, so the replicate-friendly methodology of RSCABS has received interest among some researchers in that field.

A SAS-based software package, StatCharrms, was developed by Green and Saulnier and is available in Appendix 1. Swintek (2016) adopted the SAS code for RSCABS into R, and an R version of the full StatCharrms software is also available (StatCharrms, 2017). A bug in the interface of StatCharrms (2017) has been reported, but the underlying code is correct. There are several other references such as Guimarães and Lindrooth (2005) and Farewell and Farewell (2013) that discuss regression models based on the Dirichlet-multinomial distribution. It might be possible to define a shift in the distribution of the multinomial to produce something akin to ECx estimates that is directly related to the severity scores, though just how that might be done and how useful it would be to the biologists and risk assessors are not clear.

9.3 DEVELOPMENTAL STAGE

One of the histopathology responses included in the discussion of RSCABS is developmental stage. This response also arises in OECD TG 231, the AMA, and in OECD TG 241, the LAGDA. In those assays, developmental growth is not assessed from histopathology tissue samples. Instead external examination of the larvae is done using well-defined characterizations of growth and development, such as the following from Nieuwkoop and Faber (1994):

Stage 56 typically occurs on day 38 post hatch. Forelimbs of stage 56 animals are visible beneath the skin of the tadpoles. The tadpoles are filter feeding.

Stage 57 typically occurs on day 41 post hatch. Stage 57 animals lack emerged forelimbs, and metamorphosis in the alimentary canal is just beginning.

Stage 58 typically occurs of day 44 post hatch. Stage 58 animals have emerged forelimbs, and there is significant histolysis of the duodenum (animals can no longer digest food).

Stage 59 typically occurs on day 45 post hatch. Stage 59 animal forelimbs now reach to the base of the hindlimb, and there is now histolysis of the nonpyloric part of the stomach (animals still cannot digest food).

Stage 60 typically occurs on day 46 post hatch.

In terms of development rates, a stage 57 animal is 3 days behind a stage 58 animal, whereas a stage 58 animal is only 1 day behind a stage 59 animal. Also, in terms of development rate, a stage 56 animal is 6 days behind a stage 58 animal, whereas a stage 58 animal is only 2 days behind a stage 60 animal.

The biological significance of moving between two stages might vary greatly depending on which stages are being considered. For example, a stage 56 animal can filter feed. None of the animals in the other stages listed above can. Clearly, the analysis of the stage data requires care, and it is important not to think of the stages as representing equal increments of development. And it should be clear that a shift in the stage of metamorphosis of a single stage *might* be, but need not be, meaningful.

While RSCABS could be employed, it would need to be modified (i) to accommodate more stages than the five used for histopathology severity scores and (ii) to allow a few intermediate stages, such as 9.5 or 12.5 reported, for example, in Xenbase (2017). These experiments have much larger replicate sizes and fewer replicates than do those arising in TG 240, and these facts suggest a different approach. In any case, the approach described here for TG 231 was developed before RSCABS and is still useful.

The JT test applied in step-down fashion to replicate medians could be used for amphibian developmental stage. However, the same objections as those raised about a replicate median score for histopathology responses apply here. What we (Green and Springer) eventually recommended to, and received approval from, the OECD Validation Management Group for Ecotoxicology was a refinement of the JT test, motivated by quantile regression, which we call the multiquantal Jonckheere–Terpstra (MQJT) test. This refinement has very similar power properties to the standard JT test, as will be demonstrated in Section 9.3.1. The advantage it offers is closer attention to the upper and lower tails of the distribution of developmental stage at each test concentration in order to reduce both the false positive and false negative rates.

It is expected that subjects in the same replicate vessel are not truly independent, but rather correlated by virtue of sharing exactly the same medium as other subjects in that vessel. The available data are ambiguous on the magnitude of these correlations. There are numerous references in the statistical literature to support the need for considering the rep as the experimental unit, as has been discussed elsewhere in this text (e.g. Section 1.3.2). Since developmental stage is a discrete measure, the nested variance structure available in standard ANOVA is not available, though it could be incorporated in a GLMM. However, even in the latter context, the incorporation of a nested variance structure has much the same effect as analyzing replicate means or medians. It seems important to more fully capture the variability of stages within replicate vessels than a single summary measure such as the median can do.

A chi-square test might be considered for developmental stage since that test is as much a test of change in spread (or variance) as a test of change in the mean or median developmental stage, but it is indirect and does not distinguish between the two. While an increase in variability might be of value to detect, a focused test for that as well as a focused test for change in median or mean stage would make clear what is being detected. Otherwise, the meaning of a statistically significant finding would be difficult to understand. Consequently, a chi-square test does not provide a satisfactory way to analyze severity score or developmental stage data.

9.3.1 The Multiquantal Jonckheere–Terpstra (MQJT) Test

This test was developed for OECD TG 231, the AMA. The species used for this test is *Xenopus laevis*. The guideline calls for a minimum of three test concentrations plus a negative control, with a constant ratio of each concentration to the next lower between 10 and 3.3. The guideline calls for four replicate tanks of 20 tadpoles in each test concentration and control. The primary responses of interest are developmental stage, snout to vent length, whole body length, hindlimb length, wet body weight, and thyroid gland histology. Thyroid gland histology is analyzed by RSCABS. The length and weight measurements are analyzed by the methods of Chapters 3 and 4. Noteworthy mortality is not expected in this sublethal test, but if it occurs, it can be analyzed by the methods of Chapters 6 and 7. This section is concerned with the analysis of developmental stage, the recommended test for which is the MQJT. Some of the example datasets included here had only two replicate tanks per treatment. These datasets were from the first phase of the validation of the test and illustrate the methodology. The increase in replication called for in the published guideline increases the power to detect effects of biological importance. Discussion of the power is given in Section 9.3.4.

The MQJT concept is quite simple. If a chemical affects an exposed population, that effect may first appear in the more sensitive or extreme parts of the population. Statistically, this would mean the effect might show up in a smaller quantile than the median or 50th quantile (or perhaps in a higher quantile, depending on the direction of change). On the other hand, if a chemical only affects the most sensitive members of the population, it may have little negative impact on the sustainability of the population. Thus merely analyzing some small quantile, such as Q20, might result in an unacceptable false positive rate, for the 20th quantile of a replicate of 20 subjects may concern as few as 4 subjects and an apparent effect on only 4 of 20 subjects may have limited meaning. A smaller quantile would involve even fewer subjects. Furthermore, essentially ignoring 16 of the 20 subjects in analysis is hardly justifiable. If there is truly an effect on the population, one would expect a shift in the entire distribution.

Assuming a common distribution in all test concentrations (a standard assumption in almost all statistical analyses), the difference between the nth quantile, Qn, and the median, Q50, is constant across concentrations. Thus, the shifts in Q20, Q30, Q40, Q50, …, Q80 all measure the same effect on test concentrations on a segment of the frequency distribution of stages. See Figure 9.3 for an illustration of this with a shift in a normal population. The same relationships hold for every distribution, not just the normal.

Under the stated assumption, the difference between Q20 and Q50 in the control is the same as that between Q20 and Q50 in each treatment group. Robustness is

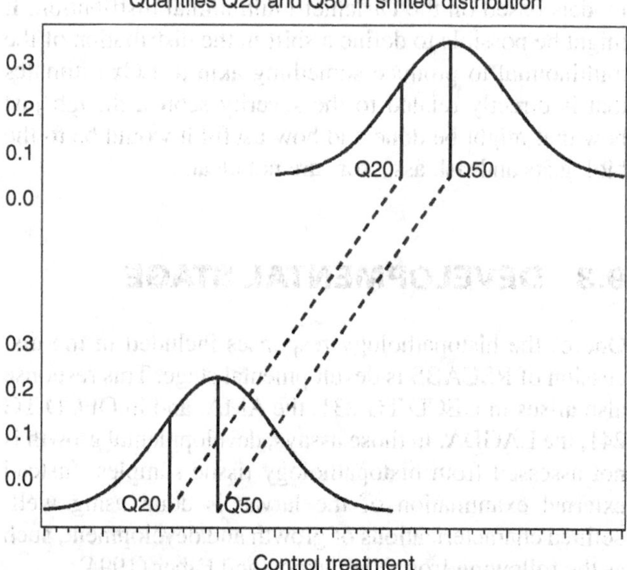

Figure 9.3 Shifted quantiles for MQJT. Control = lower left density curve. Treatment is upper right density curve. Vertical scales shown for the two density functions.

achieved because a slight perturbation in Q50 that is not evident in the other quantiles will not signal a significant effect, thereby avoiding a false positive. On the other hand, if by chance the shift in Q50 is not significant but all or most of the other quantiles do show a significant effect, the test will report a significant effect, thereby avoiding a false negative. Examples are presented below of each situation. In many datasets examined, this approach yields the same conclusion as the standard JT test.

9.3.1.1 Description of MQJT Test

If the entire distribution is shifted by an amount, each quantile is shifted by the same amount.

It follows that a robust way to estimate the significance of a shift in the distribution is to use the median *p*-value of the shifts in these quantiles. The proposed method proceeds as follows:

1. Determine the 20th percentile, Q20, of the stage distribution within each rep at each test concentration.
2. Perform a JT trend test on these Q20 values across all concentrations. Record the *p*-value for the trend test.
3. Repeat steps 1 and 2 for the 30th through 80th percentiles.
4. Determine the median of the *p*-values that were obtained in steps 1–3. This median *p*-value is the probability that a trend in distribution shifts across concentrations is due to chance. If this median *p*-value is 0.05 or greater, testing stops and the NOEC exceeds the highest tested concentration. Otherwise, proceed to step 5.
5. Drop the data from the highest test concentration, and repeat steps 1–4, so that a new median *p*-value is calculated for the reduced dataset. If this median *p*-value is 0.05 or greater, testing stops and the NOEC is the highest tested concentration *used in this* analysis. Otherwise, proceed to step 6.
6. Drop the data from the highest concentration used in the current analysis, and repeat steps 1–4, so that a new median *p*-value is calculated for the reduced dataset.
7. Continue dropping treatment groups until the new median *p*-value is no longer statistically significant. The highest concentration at which this occurs is the NOEC.

The above procedure is identical to the standard JT test except that it is applied to several quantiles in addition to the median and decisions are made using the median *p*-value. While the MQJT can be used for every developmental stage dataset arising from one of the named test guidelines, an alternative procedure could be to continue the use of the standard JT test on Q50 alone, employing MQJT as a supplement to increase understanding of challenging datasets.

9.3.1.2 Power of Modified Procedure

The power properties of the modified quantile regression procedure are very similar to the power properties of the Jonckheere test in many situations. Section 9.3.4 contains selected tables and figures showing the power of the standard and exact permutation JT test and the MQJT based on each of the table. The power curves for all four tests are quite similar for each design with 3 or 4 reps of 15 or 20 subjects each in an experiment with 4 test concentrations plus control. The similarity of the power curves does not mean that there is no value to the exact permutation test compared with the standard or to the nonparametric quantile regression procedure compared with the Jonckheere–Terpstra. Rather it means that the general or average ability to detect effects is approximately the same for all four tests. For a given individual dataset, one of these may hold distinct advantages over the others. In general, for small experiments, the exact permutation test has lower false positive and lower false negative rates than the standard JT test. With 4 test concentrations plus control of 4 reps each, there are only 20 observations for each JT test. The exact permutation test is expected to be more reliable for datasets of this size. Similarly, the MQJT approach provides deeper understanding of the data and will also lower the false positive and false negative rates. The best procedure for these types of data is likely to be MQJT based on exact permutation Jonckheere–Terpstra tests.

9.3.2 Relationship of MQJT and Quantile Regression

The MQJT is conceptually related to quantile regression and is motivated by it, although MQJT is much simpler and can be performed without any specialized software. Quantile regression is an interesting development in statistics with some application in ecotoxicity studies, but it will be only briefly addressed in this text. Quantile regression was introduced in Koenker and Bassett (1978) and expanded over the next 23 years in Gutenbrunner and Jureckova (1992), Koenker and d'Orey (1993), Koenker (1994), Koenker and Machado (1999), and Koenker and Hallock (2001), and software to do quantile regression is available (e.g. SAS Proc Quantreg (SAS Institute Inc., 2015)).

Standard quantile regression is more complex than MQJT. Standard and modified quantal regression share the following features: (i) They make no distributional assumptions. (ii) They can be applied to a single quantile, such as Q90, or a set of quantiles (such as Q20–Q80), or to all quantiles, simultaneously. (iii) They provide insight into the effect, if any, of a chemical on the entire exposed population rather than a summary measure such as the mean or median.

Differences between the MQJT and quantile regression include the following: (i) The justification for MQJT assumes homogeneous variances across test concentrations, while quantal regression can be applied when the variance is not uniform across the test concentrations. An advantage of basing the conclusion on the median result of several quantile regressions is additional robustness against the effects of outliers and also to strange patterns in the data. However, if the data are strongly heterogeneous, MQJT may be misleading. It is important that the data be plotted to detect such situations. This is good practice in any case. (ii) With quantile regression, there is no formal general way to assess the overall effect of the chemical when more than one quantile is analyzed. The primary novelty in the modified procedure is the use of the median *p*-value for the various quantiles analyzed to decide overall significance. Without some such summary measure of significance, the regulatory process would be problematic. There are also limitations on SAS PROC QUANTREG. For example, statistical inference is available only when a single quantile is specified. Again, this would complicate the regulatory process. (iii) In quantal regression, the quantiles are conditioned on the model. In the modified procedure, the unconditional quantile is computed for each rep. It is pointed out in the SAS documentation that "quantile regression cannot be carried out simply by segmenting the unconditional distribution of the response variable and then obtaining least-squares fits for the subsets. This approach leads to disastrous results when, for example, the data include outliers. In contrast, quantile regression uses all of the data for fitting quantiles, even the extreme quantiles." A problem with developmental stage data is that there are only a few replicates and it makes no sense to estimate quantiles from the replicate summary statistics. One is forced either to ignore replicates, which was criticized above, or to estimate the quantiles for each replicate separately. In the latter case, it is not clear, especially for discrete data, how to do this in a conditional fashion. While the danger mentioned in the quote is real in the general context of quantal regression, in the simple models relevant to developmental stage in environmental toxicology experiments with concentration as the only model term and quantiles restricted to the range 20–80, this is not a serious concern. Furthermore, using the JT test instead of ordinary least squares regression to model the quantiles makes the procedure relatively robust to the effects of outliers. (iv) Perhaps most important for the intended use, quantile regression is regression, that is, it is based on a concentration–response relationship assumed to be linear, whereas the modified procedure assumes only monotonicity. There is little reason to expect a specific mathematical concentration–response relationship and no reason to expect linearity.

The value of replacing a single analysis of the median, Q50, by some other single quantile is questionable in the case of developmental stage. It is also not at all clear that including all quantiles, Q1–Q99, is appropriate, as, in small samples typical in these experiments, there is little to distinguish Q23 from Q20, for example, and too little data for Q1 or Q99 to be meaningful. While the list Q20–Q80 is somewhat arbitrary, it should suffice to represent the general behavior of the distribution across the test concentrations. For larger datasets, Q10 or Q90 might be added, and for smaller datasets, Q20 and Q80 might be dropped.

9.3.3 Applications of MQJT

The Excel file ama_data.xls in Appendix 1 contains six datasets that can be used to explore the performance of the MQJT test. Two of these will be developed in this section. The other four are left as exercises. Most of these datasets contain continuous responses as well as stage data. Those responses can be analyzed by the methods of Chapters 3 and 4. These Excel files are too large to reproduce in this chapter, but three summary tables are produced by the *MQJTrev.sas* program in Appendix 1.

9.3.3.1 Frog4 Dataset

There were four test concentrations plus a water control, with four replicate tanks of 15 tadpoles each. See Table 9.5 for the calculated replicate quantiles. Those with noninteger values arise from the decision rules used in computation, not from noninteger stages. See Table 9.6 to see how the

Table 9.5 Quantiles for Frog4 Data

Conc	Rep	*N*	Q80	Q70	Q60	Q50	Q40	Q30	Q20
0	—	60	58	58	58	58	58	57	57
0.75	1	15	59	58	58	58	58	58	58
0.75	2	15	58.5	58	58	58	58	57	57
0.75	3	15	59	59	58	58	58	58	57.5
0.75	4	15	59	58	58	58	58	58	57.5
1.5	1	15	58.5	58	58	58	58	57	57
1.5	2	15	58	58	58	58	58	57	57
1.5	3	15	58	58	58	58	58	57	57
1.5	4	15	58	58	58	58	57	57	57
3	1	15	58.5	58	58	58	57	57	57
3	2	15	59	59	58	58	58	58	58
3	3	15	59	58	58	58	58	58	57
3	4	15	59	59	58.5	58	58	58	57
6	1	15	59.5	58	58	58	58	58	57
6	2	15	58	58	58	57	57	57	57
6	3	15	58	58	58	57	57	57	57
6	4	15	59	58	58	57	57	57	57

N, number of tadpoles in control (where conc = 0) or number in each tank (conc > 0).

treatment quantiles differ from the median quantile in the control. A positive value means the treatment quantile is larger than the control quantile. In Tables 9.5–9.6, the quantiles for the control are calculated across all replicate tanks, so there is just one control Qx value, x=20–80. See Figure 9.4, which shows all control replicate values. Finally, see Table 9.7 for the results of the MQJT test.

It should be noted that the JT test for Q50 is significant at the 0.01 level, but the quantiles as a whole do not support a significant change in stage in the high test concentration, as indicated by the median p-value. Figure 9.4 does indicate a flat trend in the median except for a sharp drop in the high concentration. The other quantiles do not suggest clear trends. Since the test is not significant with all treatment groups present, no further testing is required, and the NOEC>6mg l^{-1}.

9.3.3.2 Frog2 Dataset

See Tables 9.8 and 9.9 for quantiles and changes in quantiles from a second frog dataset and Table 9.10 for the results of the MQJT for these data.

Table 9.6 Change in Quantiles for Frog4 Data

Conc	Rep	N	Dif80	Dif70	Dif60	Dif50	Dif40	Dif30	Dif20
0.75	1	15	1	0	0	0	0	1	1
0.75	2	15	0.5	0	0	0	0	0	0
0.75	3	15	1	1	0	0	0	1	0.5
0.75	4	15	1	0	0	0	0	1	0.5
1.5	1	15	0.5	0	0	0	0	0	0
1.5	2	15	0	0	0	0	0	0	0
1.5	3	15	0	0	0	0	0	0	0
1.5	4	15	0	0	0	0	−1	0	0
3	1	15	0.5	0	0	0	−1	0	0
3	2	15	1	1	0	0	0	1	1
3	3	15	1	0	0	0	0	1	0
3	4	15	1	1	0.5	0	0	1	0
6	1	15	1.5	0	0	0	0	1	0
6	2	15	0	0	0	−1	−1	0	0
6	3	15	0	0	0	−1	−1	0	0
6	4	15	1	0	0	−1	−1	0	0

N, number of tadpoles in control (where conc=0) or number in each tank (conc>0); Dif$x$=Qx(treatment)−Qx(control).

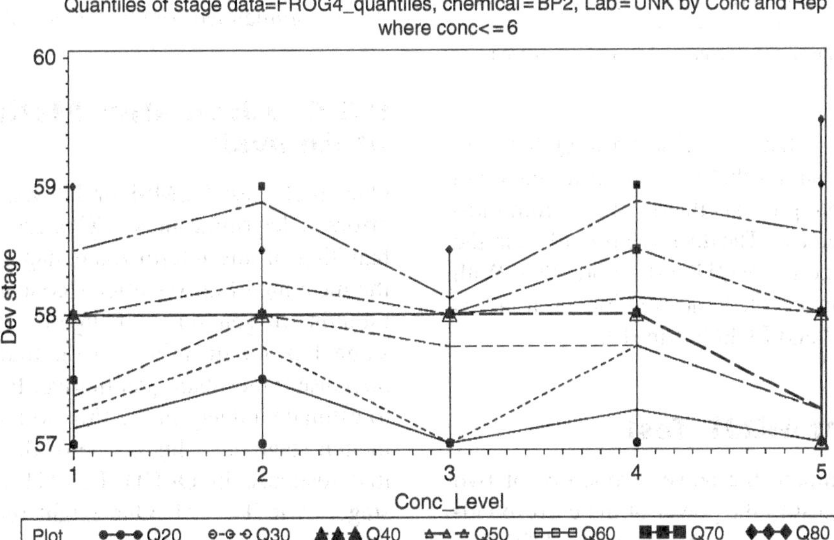

Figure 9.4 MQJT test for Frog4 dataset. Line segments join the median quantile Qx for each treatment group, x=20–80. A consistent drop from group 4 to 5 is evident. A drop is also evident from group 2 to 3, but it is followed by a rise or flat response from group 3 to 4.

Table 9.7 MQJT Results for Frog4 Data

Conc ≤ 6	JTTest	ZStat	Direction	P2	P2Med
Table Q20 * Conc	27	−0.670	Decrease	0.503	0.480
Table Q30 * Conc	53.5	0.646	Increase	0.518	0.480
Table Q40 * Conc	9.5	−2.491	Decrease	0.013	0.480
Table Q50 * Conc	1.5	−2.590	Decrease	0.010	0.480
Table Q60 * Conc	13.5	0.707	Increase	0.480	0.480
Table Q70 * Conc	29.5	0.432	Increase	0.666	0.480
Table Q80 * Conc	87	1.578	Increase	0.115	0.480

Table Qx * Conc, JT test for quantiles Qx; JTTest, value of JT test statistic; ZStat, normal approximation of JTTest; Direction, direction (increasing or decreasing) of trend; P2, p-value of two-sided JT trend test; P2Med, median of individual quantile p-values.

Table 9.8 Quantiles for Frog2 Data

Conc	Rep	*N*	Q80	Q70	Q60	Q50	Q40	Q30	Q20
0	–	60	62	61	61	60	60	60	60
0.75	1	15	61	61	61	61	60	60	60
0.75	2	15	62	62	60	60	60	60	60
0.75	3	15	62	61	61	61	61	60	60
0.75	4	15	61.5	61	61	61	60.5	60	60
1.5	1	15	62	62	62	61	61	60	60
1.5	2	15	62	62	61.5	61	60.5	60	60
1.5	3	15	62	62	62	61	60.5	60	60
1.5	4	14	62	61	61	60.5	60	60	60
3	1	15	60	60	60	60	60	60	59.5
3	2	15	62	62	61	61	60.5	60	60
3	3	15	61	60	60	60	60	60	59
3	4	15	62	62	61	60	60	60	60
6	1	15	60	59	58.5	58	58	58	56.5
6	2	14	58	58	58	57	57	57	56
6	3	14	59	59	59	57.5	57	57	56
6	4	14	60	60	60	59	59	58	58

N, number of tadpoles in control (where conc = 0) or number in each tank (conc > 0).

Table 9.9 Change in Quantiles for Frog4 Data

Conc	Rep	*N*	Dif80	Dif70	Dif60	Dif50	Dif40	Dif30	Dif20
0.75	1	15	−1	0	0	1	0	0	0
0.75	2	15	0	1	−1	0	0	0	0
0.75	3	15	0	0	0	1	1	0	0
0.75	4	15	−0.5	0	0	1	0.5	0	0
1.5	1	15	0	1	1	1	1	0	0
1.5	2	15	0	1	0.5	1	0.5	0	0
1.5	3	15	0	1	1	1	0.5	0	0
1.5	4	14	0	0	0	0.5	0	0	0
3	1	15	−2	−1	−1	0	0	0	−0.5
3	2	15	0	1	0	1	0.5	0	0
3	3	15	−1	−1	−1	0	0	0	−1
3	4	15	0	1	0	0	0	0	0
6	1	15	−2	−2	−2.5	−2	−2	−2	−3.5
6	2	14	−4	−3	−3	−3	−3	−3	−4
6	3	14	−3	−2	−2	−2.5	−3	−3	−4
6	4	14	−2	−1	−1	−1	−1	−2	−2

N, number of tadpoles in control (where conc = 0) or number in each tank (conc > 0); Dif$x$ = Qx(treatment) − Qx(control).

It should be noted that the JT test for Q50 is not significant. Figure 9.5 shows that there is an initial small rise in Q50. This is what prevents the trend test from indicating a significant decrease. The decrease is evident in the quantiles as a whole and the MQJT test reports that. With the high concentration omitted, no significant trend is found. Thus, NOEC = 3 and LOEC = 6 mg l^{-1}.

9.3.4 Power of MQJT Test

It is instructive to compare the power properties of two versions of the MQJT test to the power of the corresponding versions of the JT test on the median stage, Q50. See Tables 9.11 and 9.12. The versions refer to asymptotic and exact permutation version of the JT test. It will be observed that there is very little difference in power. The advantage claimed from MQJT over simple JT test on replicate medians is lower false positive and false negative rates by using more information from the dataset. See Coady et al. (2013) for a study reporting results using the MQJT.

Table 9.11 indicates high power to detect an increase or delay of one developmental stage by the end of the 21-day study with the experimental design of four reps of size 20. This should be compared with the design in Table 9.12 of three reps of size 15, where the power properties are noticeably lower. Power considerations of which these two tables are samples were among the reasons TG 231 recommended four replicates of 20.

9.3.5 Alternative Methods of Analysis

One might use GLMM or NLME models with Poisson errors to determine an NOEC or estimate ECx, with careful handling of any intermediate stages. For ECx estimation, the meaning of an x% effects concentration requires some thought. In general, a change of 10% in developmental stage has no meaning, given that stage is an ordinal response rather than quantitative. It certainly makes sense to estimate a delay in the time to reach a certain developmental stage, and this is one of the motivations for using that response in OECD TG 241 (LAGDA) rather than stage, as in TG 231. One might reason that EC10 refers to a 10% delay (or acceleration) in development, and that may sound intuitively reasonable, but it is challenging to interpret that in a scientifically sound manner. The use of

Table 9.10 MQJT Results for Frog2 Data

Conc≤6	JTTest	ZStat	Direction	P2	P2Med
Table Q20 * Conc	5	−3.6205	Decrease	0.00029	0.018242
Table Q30 * Conc	2	−3.06102	Decrease	0.00221	0.018242
Table Q40 * Conc	43	−1.82514	Decrease	0.06798	0.018242
Table Q50 * Conc	45.5	−1.76139	Decrease	0.07817	0.018242
Table Q60 * Conc	36	−2.36067	Decrease	0.01824	0.018242
Table Q70 * Conc	48.5	−1.60968	Decrease	0.10747	0.018242
Table Q80 * Conc	19.5	−3.05504	Decrease	0.00225	0.018242
Conc≤3	**JTTest**	**ZStat**	**Direction**	**P2**	**P2Med**
Table Q20 * Conc	−2.5	1.95872	Decrease	0.05015	0.43456
Table Q40 * Conc	40.5	0.70841	Increase	0.47869	0.43456
Table Q50 * Conc	42.5	0.73732	Increase	0.46093	0.43456
Table Q60 * Conc	−33	0.27072	Decrease	0.7866	0.43456
Table Q70 * Conc	46	0.82707	Increase	0.4082	0.43456
Table Q80 * Conc	−17	1.1725	Decrease	0.241	0.43456

Upper section of table is with all data. Lower section is with high concentration omitted. Table Qx * Conc, JT test for quantiles Qx; JTTest, value of JT test statistic; ZStat, normal approximation of JTtest; Direction, direction (increasing or decreasing) of trend; P2, p-value of two-sided JT trend test; P2Med, median of individual quantile p-values.

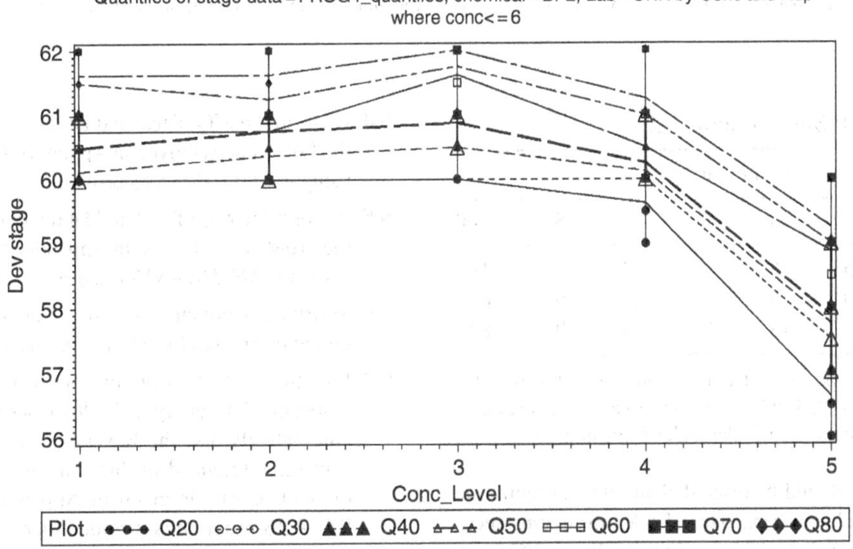

Figure 9.5 MQJT test for Frog2 dataset. A strong decrease from group 4 to 5 is evident in all quantiles. The trend from group 3 to 4 is less compelling.

GLMM with a nested variance structure to capture the replicate effect to determine an NOEC is justified.

It is left as an exercise to verify that for the Frog4 data, the NOEC using a GLMM Poisson model with Dunnett's test agrees with the MQJT result above. For the Frog2 data, Dunnett's test does not find a significant effect in any test concentration. The residual analysis suggests possible problems with the Poisson or negative binomial error structure. There may be too little variation in stages to make the Poisson model viable.

EXERCISES

In all datasets in the text and exercises, if a concentration is given with the value −1, that means a solvent used in the study and the water control is recorded as −1 and the solvent control is listed as 0. The reader should determine whether to combine the two controls for NOEC determination or ECx estimation. If the two controls are to be combined, then the water control should be changed to 0 and the replicates renamed so that replicates in the two controls can be distinguished. If only the solvent control group is to be used, then the water control should be discarded. If only the water control should be used,

Table 9.11 Power of JT and MQJT Tests for 3 Treatment Groups Plus Control w/ 4 Reps of 20

	Change in median Q50							
Test	0.3	0.5	0.8	1	2	3	4	5
JT	14	30	53	73	100	100	100	100
MQJT	9	25	53	78	100	100	100	100
JTX	12	26	49	70	100	100	100	100
MQJTX	9	23	50	76	100	100	100	100

JT and MQJT, asymptotic power of JT and MQJT tests; JTX and MQJTX, exact permutation power of JT and MQJT tests.

Table 9.12 Power of JT and MQJT Tests for 3 Treatment Groups Plus Control w/ 3 Reps of 15

	Change in median Q50							
Test	0.3	0.5	0.8	1	2	3	4	5
JT	14	30	53	73	100	100	100	100
MQJT	9	25	53	78	100	100	100	100
JTX	12	26	49	70	100	100	100	100
MQJTX	9	23	50	76	100	100	100	100

JT and MQJT, asymptotic power of JT and MQJT tests; JTX and MQJTX, exact permutation power of JT and MQJT tests.

Table 9.13 Heterogeneity Check of Between-REP Variances – Doses = 5 for Response = Liv_Basophilia with Severity Score ≥2

Statistic	Value	Conc	DoseNum	DF	Prob	Doses	Signif
Chi-square	7.247863	1	1	5	0.202849	5	
Chi-square	4.925926	2	2	5	0.424987	5	
Chi-square	10.27385	3	3	5	0.067838	5	
Chi-square	2.022222	4	4	5	0.846062	5	
Chi-square	6.461538	5	5	5	0.263862	5	

These are simple chi-square tests for variance heterogeneity. StatCharrms software implements Tarone's $C(\alpha)$ test. The R version of this software is given in the references as StatCharrms (2017). The SAS version is given in Appendix 1. Here no indication of overdispersion was found. (All p-values [Prob] exceed 0.05.)

Table 9.14 Rao–Scott Slice at Score = 2

	Treatment						
Adverse	1	2	3	4	5	6	Total
0	3	5	5	2	1	0	16
1	13	9	13	12	13	9	69
Total	16	14	18	14	14	9	85

Adverse = 0 is the count of scores <2. Adverse = 1 is the count of scores ≥2. Generation F1 with scores ≥2. The RSCABS test found a significant increase (p = 0.033214) in score 2+ in the high treatment group.

then the solvent control should be discarded and the concentration for the water control should be changed to 0. Under no circumstances should the concentration value −1 be used for NOEC or ECx purposes. Guidance on this is provided in Chapters 2 and 3.

9.1 Use the programs for the RSCA test from Chapter 6 to verify the NOEC2 and NOEC3 values for the data in Tables 9.2–9.4. Also, carry out Tarone's $C(\alpha)$ test for the data in the right panels of Tables 9.3 and 9.4 to determine whether overdispersion is present in these data.

9.2 Perform the MQJT test and Dunnett test in a GLMM model on the Frog1 dataset given in Appendix 1, Excel file ama_data.xls, using the *MQJTrev.SAS* program.

9.3 Perform the MQJT test and Dunnett test in a GLMM model on the Frog3 dataset given in Appendix 1, Excel file ama_data.xls, using the *MQJTrev.SAS* program.

9.4 Perform the MQJT test and Dunnett test in a GLMM model on the Frog5 dataset given in Appendix 1, Excel file ama_data.xls, using the *MQJTrev.SAS* program.

9.5 Perform the MQJT test and Dunnett test in a GLMM model on the Frog6 dataset given in Appendix 1, Excel file ama_data.xls, using the *MQJTrev.SAS* program.

9.6 Verify the Dunnett test results claimed from a GLMM model on the Frog2 and Frog4 in Sections 9.3.3.1 and 9.3.3.2.

9.7 Use the SAS version of the RSCABS program given in Appendix 1 to analyze the liver basophilia data in Table 9.2. Alternatively, use the R version of this program that is in R package referenced in StatCharrms (2017). The SAS version of StatCharrms is given in Appendix 1, Chapter 5, and the RSCABS program is included there also. Hothorn (2016, section 2.1.8.1) contains several ways severity scores can be analyzed and provides R programs. One of these is RSCABS, though it is referred to as Cochran–Armitage using collapsing categories and uses only the standard CA test, not the Rao–Scott modification. The same liver basophilia data discussed in this chapter and taken from Green et al. (2014) is presented.

The SAS output comes in two forms:

(i) There are files generated for each slice. For the file RSCA_LIVERBASOPH2RS2.LIS, see Tables 9.13–9.16 plus text, which has been edited to help interpretation. (ii) A SAS dataset, RSCAOUT, is created by the SAS version that summarizes the results in a simple form. This is a temporary dataset and needs to be moved to a permanent location if it is to be retained for future use.

Table 9.15 Heterogeneity Check of Between-REP Variances – Doses = 5 for Response = Liv_Basophilia with Severity Score ≥2

Statistic	Value	Conc	DoseNum	DF	Prob	Doses	Signif
Chi-square	7.247 863	1	1	5	0.202 849	5	
Chi-square	4.925 926	2	2	5	0.424 987	5	
Chi-square	10.273 85	3	3	5	0.067 838	5	
Chi-square	2.022 222	4	4	5	0.846 062	5	
Chi-square	6.461 538	5	5	5	0.263 862	5	

The test for overdispersion (variance heterogeneity) was repeated with the high treatment group omitted since a significant result was found when all doses were included.

Table 9.16 Rao–Scott Slice at Score = 2

	Treatment					
Adverse	1	2	3	4	5	Total
0	3	5	5	2	1	16
1	13	9	13	12	13	69
Total	16	14	18	14	14	85

Adverse = 0 is the count of scores <2. Adverse = 1 is count of scores ≥2. RSCABS with the high treatment group omitted was not significant, since the p-value exceeded 0.05. Treatment = 1 for control, =2 for lowest test concentration, etc.

Table 9.13 shows slimple chi-square tests for variance heterogeneity. StatCharrms software implements Tarone's $C(\alpha)$ test. The R version of this software is given in the references as StatCharrms (2017). The SAS version is given in Appendix 1. For these data, no indication of overdispersion was found. (All p-values [Prob] exceed 0.05.)

Table 9.14 is for the slice at score = 2. The RSCABS test found a significant increase in score 2+ in the high treatment group.

Table 9.15 gives a test for overdispersion (variance heterogeneity) with the high treatment group omitted.

Table 9.16 shows the slice at score = 2 with the high treatment group omitted. RSCABS with the high treatment group omitted was not significant, since the p-value exceeds 0.05. Thus NOEC2 = treatment 5 and LOEC2 = treatment 6.

(ii) A SAS dataset is generated with the name RSCA_Liv_basophilia that summarizes the RSCABS results from the four files named RSCA_LIVERBASOPH2RSx.LIS, where x = 1, 2, 3, 4. This file does not contain the tests for overdispersion or show the tables making the slices behind the test.

9.8 Run the RSCABS program (or RSCA test program) for the OCTOFEMF1W8_ECO dataset for the following responses: (a) Gon_Proteinaceous_fluid, (b) Kid_Nephropathy, (c) Kid_Proteinaceous_Fluid, (d) Kid_Tub_Dilation, and (e) Explore other responses of your choice from this dataset.

Chapter 10

Time-to-Event Data

10.1 INTRODUCTION

Survival analysis is one type of time-to-event analysis that has received considerable attention for many years. Chapter 8 introduced responses such as time to swimup and time to metamorphosis and developed some simple models for their analysis. In this chapter, other statistical methods for analyzing time-to-event data will be developed. These include the Kaplan–Meier product-limit test and Cox regression or proportional hazards models. Both parametric and nonparametric methods will be discussed, as well as Bayesian methods.

Some basic ideas will be used in all approaches to be discussed. For concentration group i, define the "survival" function $S_i(t)$ as the probability that a subject will not die by time t. In other words, $S_i(t)$ is the probability of survival past time t. While it will often be convenient to refer to death and survival in the development, the concepts and methods are more general. For example, $S_i(t)$ could be taken to mean that a daphnid in concentration U_i has not reproduced by time t. The empirical survival function is given by

$$\hat{S}_i(t) = \frac{\text{Number of subjects in group } i \text{ living beyond time } t}{\text{Number of subjects in group } i}. \quad (10.1)$$

The survival function is closely related to the distribution function of survival times, which is defined by $F(t) = \text{Probability}(T \le t)$, where T is survival. Thus $F(t)$ is the probability of death by time t (i.e. survival time at most t) and $S(t) = 1 - F(t)$. Mathematically, it is important to recognize that both functions $S(t)$ and $F(t)$ are continuous from the right, that is, at every time t,

$$S(t) = \lim_{x \to t-} S(x) \quad \text{and} \quad F(t) = \lim_{x \to t-} F(x). \quad (10.2)$$

Continuity from the right is a basic property of distribution functions and survival functions inherit that property.

The "hazard" function is the probability $h_i(t)$ that a subject will die exactly at time t given that it had not died before that time. The hazard function, $h(t)$, is the probability of death at time t, given that the subject survived up to that time. Thus if $S(t)$ is differentiable, then

$$h(t) = -\frac{dS(t)/dt}{S(t)}. \quad (10.3)$$

The cumulative hazard function, $H(t)$, is the probability of death up to (and including) time t. Simple calculus applied to Eq. (10.2) shows that

$$H(t) = -\text{Ln}\big(S(t)\big) \quad \text{or} \quad S(t) = e^{-H(t)}. \quad (10.4)$$

See Figures 10.1 and 10.2 for conceptual depictions of three relationships that might exist among survival curves for control and treatment groups. In examining these survival curves, one should compare the entire curves, not just the percent survival at a single point in time. Is there a tendency for one survival curve to fall more rapidly than another? If the treatments are ordered (e.g. by dose or concentration), is there a tendency for higher order treatments to fall more rapidly than lower order treatments?

Statistical Analysis of Ecotoxicity Studies, First Edition. John W. Green, Timothy A. Springer, and Henrik Holbech.

Companion website: www.wiley.com/go/Green/StatAnalysEcotoxicStudy

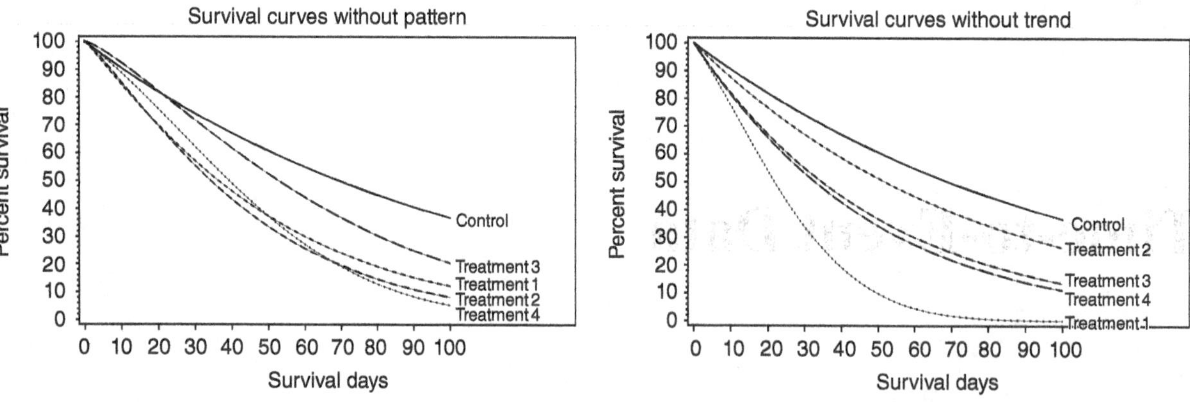

Figure 10.1 Two types of survival curves. Curves on the left cross and show no obvious treatment relationship, while the curves on the right do not cross one another, but treatment 1 curve is below that for the other treatments.

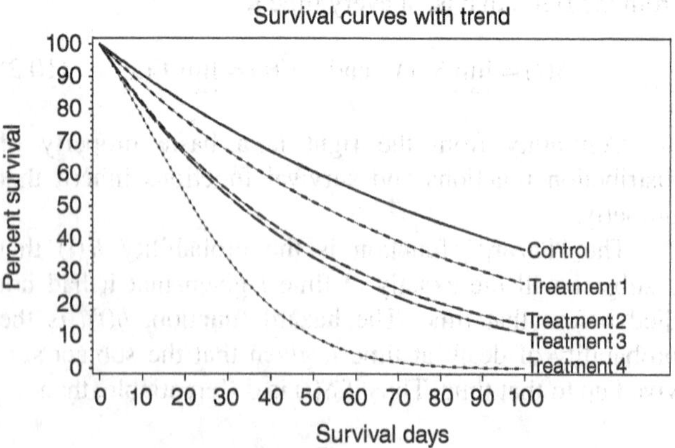

Figure 10.2 Survival curves showing concentration–response trends. The levels of the curves exhibit a trend, with lower curves corresponding to higher treatments.

To understand the survival function, it may help to first consider the situation that the times of death are known. In this situation, the times of death form a sequence t_1, t_2, t_3, …, t_k, where the first death occurs at time t_1 and the last at time t_k. Suppose there are d_i deaths at time t_i and n_i subjects alive just prior to that time. Then the (empirical) probability of survival beyond time t_1 is given by

$$S(t_1)=\frac{n_1-d_1}{n_1}. \quad (10.5)$$

At any subsequent death time, t_i, the probability of survival beyond time t_i is given by the recursive formula

$$S(t_i)=S(t_{i-1})\frac{n_i-d_i}{n_i}. \quad (10.6)$$

In terms of conditional probability, Eq. (10.6) may be rephrased in the following way, where the vertical bar | should be read as "given" or "conditional on."

$$\text{Probability(Survival beyond time}\, t_i) = \text{Probability (Survival beyond time}\, t_{i-1} | \text{Survival beyond time}\, t_{i-1}) \times \text{Probability(not dying in the interval}\, (t_{i-1}, t_i]). \quad (10.6^*)$$

A nonrecursive formula can be derived easily from Eq. (10.6):

$$S(t_i)=\prod_{j\le i}\frac{n_j-d_j}{n_j}=\prod_{j\le i}\left(1-\frac{d_j}{n_j}\right). \quad (10.7)$$

It is common in ecotoxicology experiments that the actual times of death are not known. Instead, observations are made at times $t_1<t_2<t_3<\cdots<t_k$, and the number, d_i, of subjects that died since the previous observation time, i.e. in the time interval $(t_i, t_{i+1}]$, is recorded, as is the number, n_i, alive just prior to time t_i, where the round parenthesis indicates that time t_i is not included and the square bracket indicates that time t_{i+1} is included. Then formula (10.6) still holds, provided d_i is taken to be the number of deaths that occurred in the ith time interval $(t_{i-1}, t_i]$ and n_i is taken to be the number that were alive at time t_{i-1}. However, it would be more realistic to somehow take into account that the d_i deaths now need not all occur at time t_i but rather are distributed in some manner throughout the ith time interval. Unless a death actually occurs at the time of observation, the actual times of death are not known, and, in the language of statistics, these observations are censored, in this case, interval censored. That is, the actual times are unknown but fall in the interval $(t_{i-1}, t_i]$. Various ways to handle such censoring and to estimate $S(t)$ for nonobservation times will be described in this chapter.

Appendix 1 contains several Excel files and SAS datasets that are used for examples and exercises in this chapter. The R package sas7bdat is available to read an SAS dataset into R without the need for SAS to be installed.

10.2 KAPLAN–MEIER PRODUCT-LIMIT ESTIMATOR

One of the standard models for survival is the Kaplan–Meier or product-limit model, which is nonparametric and easily handles censored data. Suppose the distinct survival times are $t_1 < t_2 < t_3 < \cdots < t_k$, some of which may be censored. The Kaplan–Meier estimate of the survival function is

$$\hat{S}(t) = \prod_{t_i \le t, t_i \text{ uncensored}} \left(\frac{n_i - d_i}{n_i} \right). \quad (10.8)$$

The variance of the Kaplan–Meier survival function can be derived by recognizing that the individual components of the product in Eq. (10.8) are independent proportions that can be turned into a sum by taking a log-transform of Eq. (10.8). Formulas for the variance of a binomial can be applied to a Taylor series expansion of that log-transform to derive Greenwood's formula:

$$\text{Var}\left(\hat{S}(t)\right) \approx \left[\hat{S}(t)\right]^2 \sum_{t_i \le t, t_i \text{ uncensored}} \left[\frac{d_j}{n_j \left(n_j - d_j\right)} \right]. \quad (10.9)$$

Details of the derivation can be found in Collett (2014) or Lee and Wang (2013).

The Kaplan–Meier calculations will be illustrated with data from a larval amphibian metamorphosis assay (LAGDA). The event in question is time to reach developmental stage 62. Observations were made daily and the number of larvae that had reached stage 62 was recorded. See Table 10.1 for the data from a single tank of tadpoles in the high concentration group along with the estimated survival function values and their standard errors. See Figure 10.3 for a typical display of the survival function.

In ecotoxicity studies, the interest in survival analysis is usually in comparing the survival functions in different treatment groups. Several ways to do this will be discussed in the sequel. The alert reader might question the example data in Table 10.1 for the reason that tadpoles in the same tank are not independent, while the Kaplan–Meier function implicitly assumes the observations are independent. That function was developed historically for the standard situation of independent times of death, and the purpose of the example was to illustrate the calculations in a very simple setting. In ecotoxicology, as we have illustrated in the previous chapters, the typical study is designed with several replicate vessels in each treatment groups and multiple subjects in each vessel. Section 10.4 will present developments in survival analysis that can handle the replicated nature of ecotoxicology experiments in an appropriate manner. Before introducing that added complexity, a simple example will be given in which the observations are independent.

10.2.1 Example: Daphnia First Day of Reproduction

A chronic *Daphnia magna* reproduction test was done with seven daphnia per treatment group in individual beakers, where the time for a daphnid to reproduce is measured in two ways: the first and last days of reproduction. The measurement is crude, since the daphnia are checked only once per day to determine whether reproduction has occurred. Also, these studies typically use only seven daphnia per test concentration. As a result, there is often very little variation in these measures. In typical studies (see Table 10.2), it will be observed that in the entire study, there were only six distinct values of first day of reproduction. Such data are poorly analyzed by the regression models discussed in earlier chapters. See Figures 10.4 and 10.5 for illustrations of the problem with such regression models.

Table 10.1 Time to Metamorphosis

Day	Count	Survival	StdErr	Censored	Day	Count	Survival	StdErr	Censored
51	1	0.96	0.04	0	68	1	0.43	0.10	0
53	1	0.91	0.06	0	69	1	0.39	0.10	0
54	1	0.87	0.07	0	70	2	0.30	0.10	0
56	1	0.83	0.08	0	74	1	0.26	0.09	0
57	1	0.78	0.09	0	75	1	0.22	0.09	0
59	2	0.70	0.10	0	76	1	0.17	0.08	0
62	1	0.65	0.10	0	77	1	0.13	0.07	0
63	1	0.61	0.10	0	80	1	0.09	0.06	0
64	2	0.52	0.10	0	81	1	0.04	0.04	0
65	1	0.48	0.10	0	81[a]	1	0.04	.	1

Count, number of larvae reaching stage 62; Day, day on which larvae reach developmental stage 62; Censored = 1 for censored value, Censored = 0 otherwise; Survival, Kaplan–Meier estimated survival function; StdErr, standard error of estimated survival from Greenwood formula.

[a] Censored value

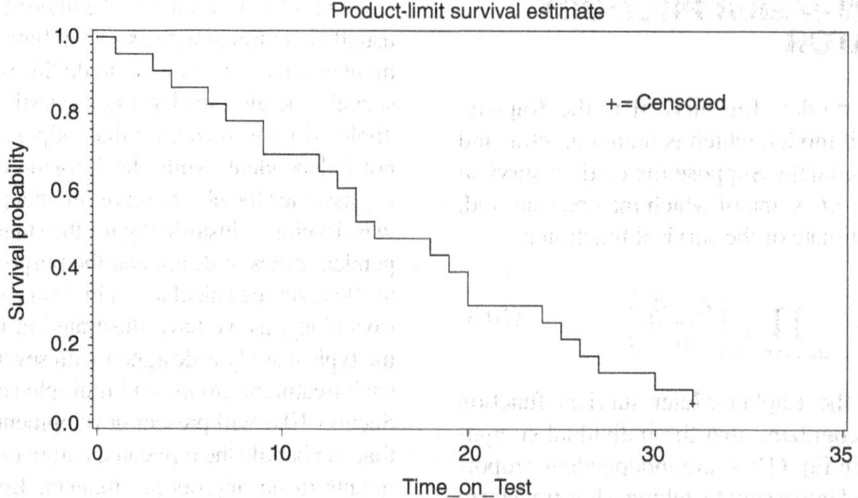

Figure 10.3 Survival function for metamorphosis. Time_on_Test=days till metamorphosis minus 50.

Table 10.2 Daphnia First Day of Reproduction

	Rep						
Conc	1	2	3	4	5	6	7
0	9	11	10	11	10	10	10
0.0016	10	12	10	12		10	12
0.0031	10	11	10	10	10	9	10
0.0066	10	10	11	11	10	10	10
0.012	10	11	11	10	11	10	10
0.027	10	10	10	10	11	10	10
0.05	14	13	13	13	14	14	14

The number in each cell in columns labeled reps 1–7 is the first day of reproduction of the daphnia in that replicate beaker. Concentrations were measured in mg l^{-1}.

Figure 10.4 shows data for the first day of reproduction for *Daphnia magna* for the data in Table 10.1. Figure 10.4 also highlights an additional challenge presented by these data, namely, the low within- and between-treatment variability.

In ecotoxicology, the question of most likely interest is in comparing the survival rates of the various treatment groups in order to determine whether a treatment effect exists. One simple way to do this is to analyze the survival times in the various treatment groups and control as such rather than in terms of the survival functions as a whole. Tests that can be used for this purpose have already been introduced. Since the data are unlikely to meet the requirements for an analyses requiring normally distributed, homogeneous responses, this can be done using Dunn's test or Mann–Whitney with an adjustment for the number of comparisons made to control. Alternatively, survival times arguably follow a Poisson distribution, so a generalized linear mixed model (GLMM) could be used, which allows a modified Dunnett's test to be performed. A conceptually preferred alternative, if the data are consistent with a monotone concentration–response, is the step-down Jonckheere–Terpstra test. All of these approaches will be illustrated for the example.

Yet another possibility is to compare the Kaplan–Meier survival functions at one or more times using the standard errors from Greenwood's formula. See Figure 10.6 for the survival curves for these data. The programs *daphnia.r* and *daphnia repro.sas* will draw Figure 10.6. Here "death" means reproduction occurred, so "time to death" in the foregoing development means time to first day of reproduction. There are concerns with that approach since there are multiple time points, so adjustments are needed to accommodate the multiplicity of comparisons, and the distribution of *t*-test-type statistics based on Greenwood's formula presents difficulties so another distribution is required. Furthermore, such comparisons at individual times do not really address the more important question of whether the overall survival experiences of the various groups differ and, if so, which have the greater chance of survival. Bland and Altman (2004) have a brief informative discussion of this point. Ways to assess survival functions overall will be explored in some detail in Sections 10.3 and 10.4.

Programs for the Jonckheere–Terpstra test, *Jonckheere macro.sas* and *JonckheereTest.r,* are given in Appendix 1, Chapter 3. An alternative R program appears in *Table8_6.r* under the heading *#Jonckheere Trend Test.* Both R programs are set up for a different dataset but can be modified easily to handle these data. From the Jonckheere–Terpstra test, the NOEC is the not-surprising 0.027 mg l^{-1}. Programs for the Mann–Whitney test, *MANN Whitney macro.SAS* and *MANN_WHIT_TEST.R*, and Dunn test, *Dunn Test*

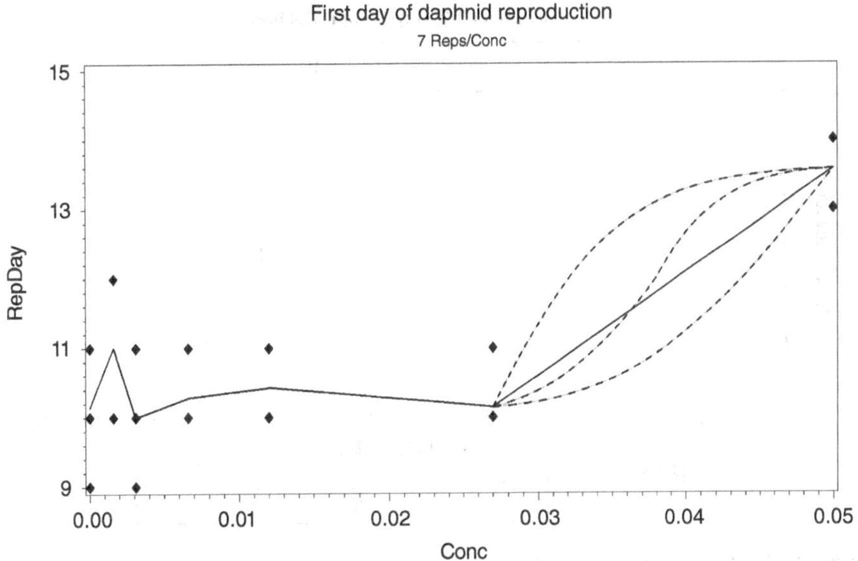

Figure 10.4 Daphnia first day of reproduction. The data consist of six integer values with many duplicate values. No standard model for continuous response or survival or time to event can be fit. The response is flat until the highest test concentration. There is no information from which to determine a "standard" regression curve to describe the data. Four possible dose–response shapes connecting the two highest test concentrations are shown to illustrate the range of possibilities. No basis for positing a regression model between highest two concentrations. Any EC*x* estimate from models introduced in previous chapters would be arbitrary.

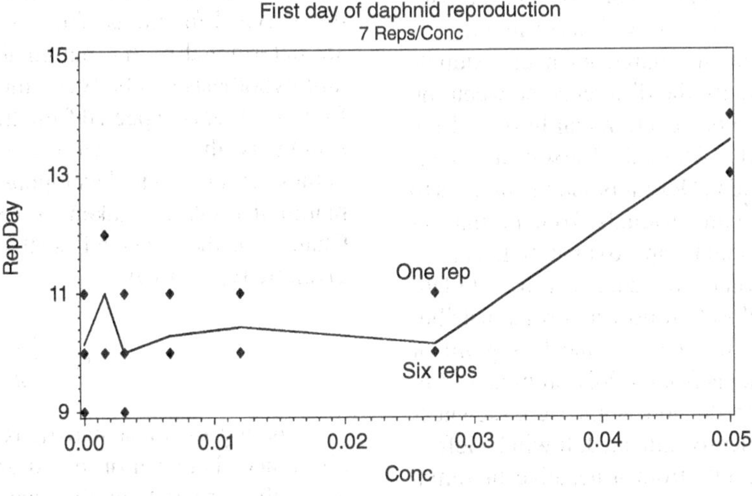

Figure 10.5 Another view of Daphnia first day of reproduction. There were six daphnid in separate beakers each treatment group. When only two or three points are displayed in some treatment group, then some daphnid had the same first day of reproduction.

Macro.sas and *DunnTest.r*, are in the same location and produce the same NOEC. These tests provide no insight beyond what has been described in Chapter 3, so it will be left as an exercise to verify these test results. Of possibly greater interest is the use of a generalized linear model with Poisson error structure for these data. A form of Dunnett's test is possible in such a framework. However conceptually appealing this approach may be, for the daphnia reproduction data in Table 10.2, no significant treatment- related effect is found, so the NOEC exceeds 0.05 mg l^{-1} by this method. It is helpful to assess the Poisson errors model for goodness of fit. To that end, an examination of residuals can be helpful.

There are several residuals of possible interest from count regression models. The simplest are Raw residuals: $r = y - \hat{y}$. These residuals are heteroscedastic and asymmetric for count regression models, which makes them difficult to use for goodness-of-fit assessment. Another type is the Pearson residual $p = (y - \hat{y})/\sqrt{\hat{w}}$, where $\hat{w}$ is an estimate of the variance of y. The sum of squared Pearson

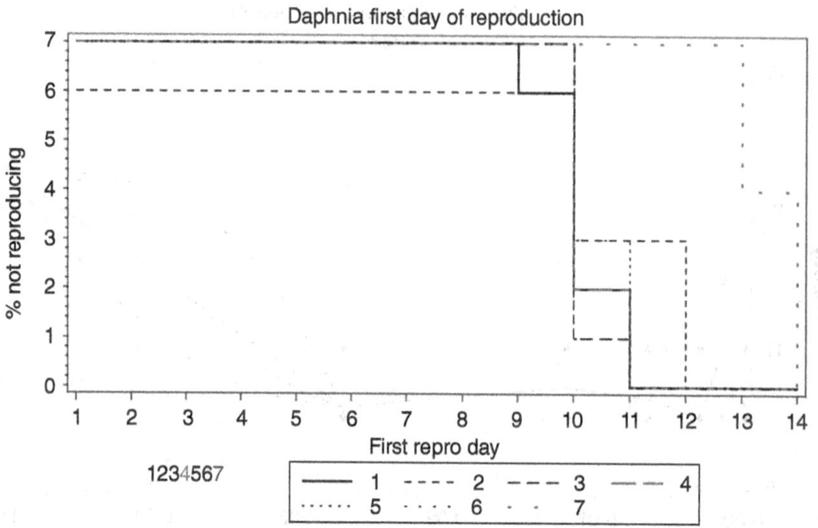

Figure 10.6 Survival curves for Daphnia first day of reproduction. Step function is typical presentation for time-to-event data. Each treatment has a unique pattern to display the survival function. Differences among survival functions are difficult to assess from this display.

residuals is the chi-square goodness-of-fit statistic. These residuals have zero mean and are homoscedastic but are also asymmetric. A third type is the deviance residual, $d = \text{sign}(y-\hat{y})\sqrt{2[L(y)-L(\hat{y})]}$, where $L(y)$ is the log-likelihood function of y and $\text{sign}(x)=\pm1$ according as x is positive or negative. The sum of squared deviance residuals is the deviance, which is twice the difference between the maximum value the likelihood function can have and the likelihood function actually obtained. Pearson and deviance residuals can be plotted in QQ-plots and histograms to give a visual check. As with residuals from normality-based models, QQ-plot should be approximately linear.

Both Pearson and deviance residuals can be "studentized" so that both more closely follow a normal distribution. See Figure 10.7 for the QQ-plot and histogram of studentized deviance residuals from a Poisson model fit to the daphnia first day of reproduction. Some right skewness is evident and the ties are clearly indicated. It will be left as an exercise to compare results from a negative binomial error structure to that with a Poisson. It is hardly surprising that these daphnia data present problems for GLMM. What may be surprising is that the standard nonparametric models do so well in identifying the intuitively clear delayed reproduction in the high test concentration.

This brings us back to the use of survival test methods, which are, after all, the subject of this chapter. The Kaplan–Meier product-limit survival functions allow the use of both trend tests and pairwise comparisons, making better use of the survival nature of these data. Probably the most widely used test for trend for survival data is the log-rank test. It may be helpful to start by comparing two survival functions and then generalize to an arbitrary number. So, suppose $t_1<t_2<t_3<\cdots<t_m$ are the times at which death occurs in either group, d_{ij} the number of deaths and n_{ij} the number at risk in group i and time t_j. The data at time t_j can be summarized in a contingency table (see Table 10.3).

Consider the null hypothesis of no difference in the two survival functions. Then deaths and survival numbers are determined by the marginal totals. Thus a test of the null hypothesis can be based on how the cell counts differ from the counts expected from the marginal totals. The reasoning is the same as discussed in connection with Tables 6.1 and 6.3 in developing the chi-squared test. If the marginal totals are taken as fixed, then as discussed in Chapter 6, the expected deaths in group i and time t_j is given by Eq. (10.10):

$$e_{ij} = \frac{n_{ij}d_j}{n_j}. \tag{10.10}$$

The log-rank test statistic is obtained by summing the differences between observed and expected deaths across all death times and dividing that sum by its variance. Thus,

$$\text{LR} = \frac{\sum_j (d_{1j} - e_{1j})}{\sqrt{V}}, \tag{10.11}$$

where V is the variance of the numerator. It will be observed that Eq. (10.11) differs from the chi-squared test statistic developed in Chapter 6. There, the interest was in whether the two groups differed. Here we are specifically interested in whether there is a trend evident in the survival functions. That is, does one survival function tend to decrease more rapidly than the other? If so, we would expect the terms in the numerator of Eq. (10.11) would tend to be positive or tend to be negative rather than be random. The variance can

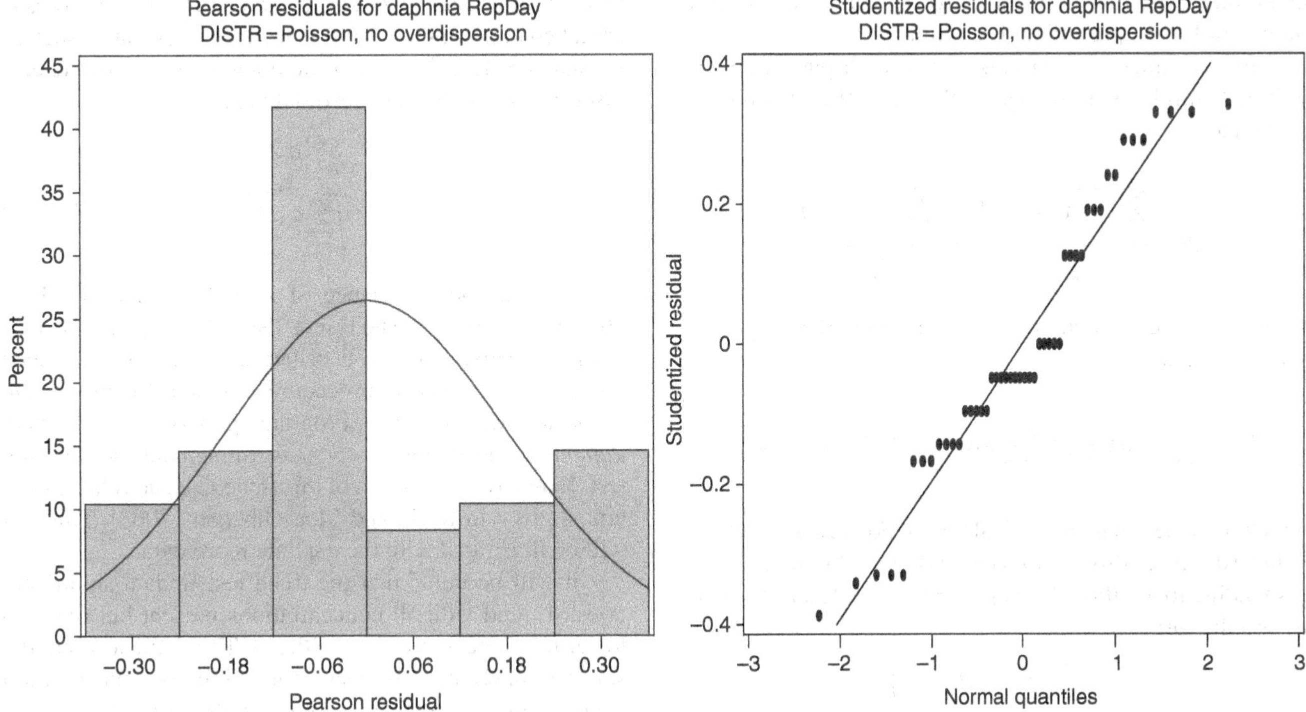

Figure 10.7 Studentized residuals from daphnia repday. Histogram and QQ-plot for studentized residuals from Poisson model of first day of reproduction. Some deviation from expected symmetry observed, suggesting the Poisson model may not fit well.

Table 10.3 Survival at Time t_j

	Group 1	Group 2	Totals
Deaths at t_j	d_{1j}	d_{2j}	d_j
Survival > t_j	$n_{1j}-d_{1j}$	$n_{2j}-d_{2j}$	n_j-d_j
At risk	n_{1j}	n_{2j}	n_1+n_2

d_{ij}, number of deaths; n_{ij}, number at risk in group i at time t_j.

be obtained from observing that d_{1j} follows a hypergeometric distribution. Thus,

$$\text{Var}\left(d_{1j}\right)=\frac{n_{1j}n_{2j}d_j\left(n_j-d_j\right)}{n_j^2\left(n_j-1\right)}. \tag{10.12}$$

It follows that

$$V=\sum_j \text{var}\left(d_{1j}\right). \tag{10.13}$$

The distribution of Eq. (10.11) is approximately standard normal under the null hypothesis provided the sample sizes are not too small. Equation (10.11) can be generalized in several ways. As already mentioned, we want to extend this to the comparison of several groups and that will be done below. It also can be generalized as in Eq. (10.14) by including a weight term. Thus,

$$T=\frac{\sum_j w_j\left(d_{1j}-e_{1j}\right)}{\sqrt{V}}, \tag{10.14}$$

where the w_j are nonnegative weights. The log-rank test takes $w_j=1$. The Wilcoxon test is obtained by setting $w_j=n_j$. Other weight functions lead to the Peto test (discussed below), and others that will not be explored here. Of course, the variance formula has to be changed to accommodate the weights. For the Wilcoxon test,

$$V_W=\sum_j n_j^2 \text{var}\left(d_{1j}\right), \tag{10.15}$$

and the resulting test statistic is

$$W=\frac{\sum_j n_j\left(d_{1j}-e_{1j}\right)}{\sqrt{V_W}}. \tag{10.16}$$

It is customary to use W^2 as the test statistic, which then has a chi-squared distribution with 1 degree of freedom. The log-rank test can similarly be expressed as a chi-squared test, but then the direction of the difference is lost. Further discussion of these test statistics can be found in numerous texts, such as Collett (2014), which served as a motivator for the preceding discussion.

To generalize the log-rank test to multiple groups, it will be helpful first to recognize that Eq. (10.11) could be written as

$$\mathrm{LR} = \frac{\sum_{i=1}^{2}\sum_{j}\left(d_{ij}-e_{ij}\right)}{\sqrt{V^*}} = \frac{2\sum_{j}\left(d_{1j}-e_{1j}\right)}{\sqrt{V^*}},$$

where it is left as an exercise to show that $d_{2j}-e_{2j}=-(d_{1j}-e_{1j})$ and

$$V^* = \sum_{i=1}^{2}\sum_{j}\mathrm{var}\left(d_{ij}\right)+2\sum_{j}\mathrm{cov}\left(d_{1j},d_{2j}\right)=4V, \quad (10.17)$$

which will be seen as equivalent to Eq. (10.11). When Table 10.3 is modified to include additional columns corresponding to additional groups, $i=1, 2, \ldots, k$, a test statistic has the form

$$Z = \frac{\sum_{i}\sum_{j} w_j\left(d_{ij}-e_{ij}\right)}{\sqrt{V^{\circ}}}, \quad (10.18)$$

where the outer sum is over the k treatment groups, the expected values are still given by Eq. (10.10), and the variance, V°, is given by the sum of variances and covariances of the terms in the numerator of Eq. (10.18), which in turn are given by

$$\mathrm{Var}\left(\sum_{j} w_j\left(d_{ij}-e_{ij}\right)\right) = \sum_{j} w_j^2 \frac{d_j\left(n_j-d_j\right)n_{ij}\left(n_j-d_{ij}\right)}{n_j^2\left(n_j-1\right)}, \quad (10.19)$$

$$\mathrm{Cov}\left(\sum_{j} w_j\left(d_{ij}-e_{ij}\right),\sum_{l} w_l\left(d_{il}-e_{il}\right)\right) = -\sum_{j} w_j^2 \frac{d_j\left(n_j-d_j\right)n_{ij}n_{il}}{n_j^2\left(n_j-1\right)}. \quad (10.20)$$

It should be noted that the weights, w_j, and expected values, e_{ij}, are considered fixed, so these are the variances and covariances of the sums of the weighted d_{ij}. The log-rank test statistic in this case is actually Z^2, which has a chi-squared distribution with $k-1$ degrees of freedom. The test statistic Eq. (10.18) can be modified to test for trend in the following way. Let $a_1, a_2, a_3, \ldots, a_k$ be scores for the different treatment groups (including control), and let Z_i be a test statistic for the ith group, for example,

$$Z_i = \sum_{j} w_j\left(d_{ij}-e_{ij}\right) \text{ is a test statistic for group } i,$$

which will be recognized from Eq. (10.18). If the scores a_i are taken to be an increasing sequence of constants, such as the integers 1, 2, 3, …, k or the linear contrast coefficients discussed in, then a test for trend is given by

$$Z = \frac{\sum_{i} a_i Z_i}{\sqrt{\sum_{i,l} a_i a_l V_{il}}}, \quad (10.21)$$

where V_{il} is the covariance of Z_i and Z_l, given by Eqs. (10.19) and (10.20). The test statistic Z in Eq. (10.21) has, asymptotically, a normal distribution and can be produced by the SAS LIFETEST procedure and the R function *comp* in the survMisc package. Programs previously mentioned, *daphnia.r* and *daphnia repro.sas,* implement the log-rank test. Extensive discussion of this trend test and related matters is given in Klein and Moeschberger (2003). This test will be illustrated with the daphnia repro data.

It will be noted that the trend test finds a significant upward trend with all concentrations present but finds no lower test concentration significant. This indicates that the survival function for subjects in the high test concentration declines more slowly than that for the control. In other words, subject in that group tend to reproduce later than the control daphnid. The ecological implication of delayed reproduction is a possible population decline resulting from increased risk to the parent from predators before reproduction can occur. One could also speculate that delayed reproduction might increase the risk of developmental problems in the young.

To summarize the results for the daphnia first day of reproduction (see Table 10.4), the simple Jonckheere–Terpstra trend test, the log-rank and Wilcoxon-based "survival" trend test, Mann–Whitney, and Dunn tests all find an increase in the first day of reproduction and, consequently, the same NOEC, 0.27 mg l^{-1}, whereas the GLMM Poisson errors model using pairwise comparisons finds the NOEC > 0.05. (The SAS program *Figure 10.10.sas* can be used for the GLMM analysis, and *Daphnia_repro.sas* contains code for the other tests named. The R program *daphnia.r* reproduces some of these tests.) For simple datasets such as in daphnia first day of reproduction, the use of survival functions for analysis is probably overly complex and the simple Jonckheere–Terpstra test, or if no trend is evident, the Dunn or Mann–Whitney test, should be sufficient. In Section 10.3, we will consider an example where analysis by survival functions is more appropriate.

10.3 COX REGRESSION PROPORTIONAL HAZARDS ESTIMATOR

Recall from Section 10.1 that the hazard function, $h(t)$, is the probability of death at time t. The control and treatment

Table 10.4 Trend Tests for Daphnia Reproduction Data

Test	Test statistic	Standard error	z-Score	Pr>|z|	Pr<z	Pr>z
Log-rank	−0.4001	0.0858	−4.6628	<0.0001	<0.0001	1
Wilcoxon	−11.3868	3.0471	−3.7369	0.0002	<0.0001	0.9999

			p-values	
Strata Conc	Comparison Conc	Chi-square	Raw	Dunnett–Hsu
0.0016	0	1.027	0.3109	0.8277
0.0031	0	0.0519	0.8197	0.9999
0.0066	0	0.0712	0.7896	0.9998
0.012	0	0.221	0.6382	0.9949
0.027	0	0.00281	0.9577	1
0.05	0	23.0707	<0.0001	<0.0001

Log-rank- and Wilcoxon-based trend tests find significant increasing trend in first day of reproduction. Dunnett-Hsu adjusted pairwise comparisons based on trend test find only the high concentration survival rate significantly different from the control survival rate.

groups have hazard functions. The proportional hazards model assumes

$$h_i(t) = \alpha_i h_0(t), \quad i = 1,2,3,\ldots,k, \tag{10.22}$$

where the α_i are constants and $h_0(t)$ is the hazard function for the control. Thus, the model posits that the hazard functions for the treatment groups are proportional to that for the control. It is possible to posit a mathematical form for h_0 but it is not necessary. The parameters α_i must be positive, so they can be written as

$$\alpha_i = e^{\beta_i}, \tag{10.23}$$

for some parameter β_i, and the proportional hazards model can be written as

$$h_i(t) = \exp(\beta_i) h_0(t). \tag{10.24}$$

The log-hazard ratio can then be written as

$$\text{Log}\left(\frac{h_i(t)}{h_0(t)}\right) = \beta_i. \tag{10.25}$$

We can test the hypothesis that $\beta_i = 0$, i.e. $h_i(t) = h_0(t)$, for all *i or* for a select value of *i or* a hypothesis that the β_i form a trend, $\beta_1 \le \beta_2 \le \beta_3 \le \cdots \le \beta_k$. To do such tests, it is necessary to have estimates of the parameters β_i, including their variances and covariances.

One implication of Eq. (10.22) is that the hazard functions for the control and treatments never cross. That is, if $h_i(t) < h_j(t)$ for some time *t* and groups *i* and *j*, the inequality cannot be the reverse for some other time *t* for the same two groups. While modest departures from this expectation need not be a problem for these models, it must be recognized that if the *observed* hazard functions do cross, then the proportional hazards model may not be appropriate. Of course, if the hazard functions never cross, then neither do the survival functions so a visual assessment of the proportional hazards model can be made from the survival functions. For short-duration studies, such as the 21-day daphnia reproduction, there are very few times from which to make an assessment. Figure 10.6 does not provide much visual evidence from which to decide whether a proportional hazards model is appropriate. A 2-year rat study is given below from which such an assessment is more reasonable. Also, a Larval Amphibian Growth and Development Assay (LAGDA) study is given in Section 10.4.1, which likewise more amenable to such an assessment.

One of the attractive features of proportional hazards models over the product-limit model is that they allow the easy incorporation of additional explanatory variables, such as age, sex, and prevalence of other medical conditions, similar to the covariate analysis discussed in Chapter 5. This can be done by replacing the single parameter, β_i, in Eq. (10.24) by a function of higher order. The most widely used models are of the form in Eq. (10.26):

$$h_i(t) = \exp\left(\sum_{m=1}^{k} \beta_m x_{mi}\right) h_0(t) = \exp\left(\beta' \mathbf{x}(i)\right) h_0(t), \tag{10.26}$$

where the X_m are explanatory variables, x_{mi} is the value of *m*th explanatory variable in group *i*, and $\mathbf{x}(i)$ is the column vector with *m*th component x_{mi}. In the special case that each X_m is an indicator variable for group *m*, i.e. $X_m = 1$ for observations in group *m* and $X_m = 0$ otherwise, this reduces to Eq. (10.24). The estimators of the parameters β_m in Eq. (10.26) are obtained by solving the corresponding likelihood equations. In general, there is no closed form solution and an iterative technique, such as Newton–Raphson (Appendix A2.5),

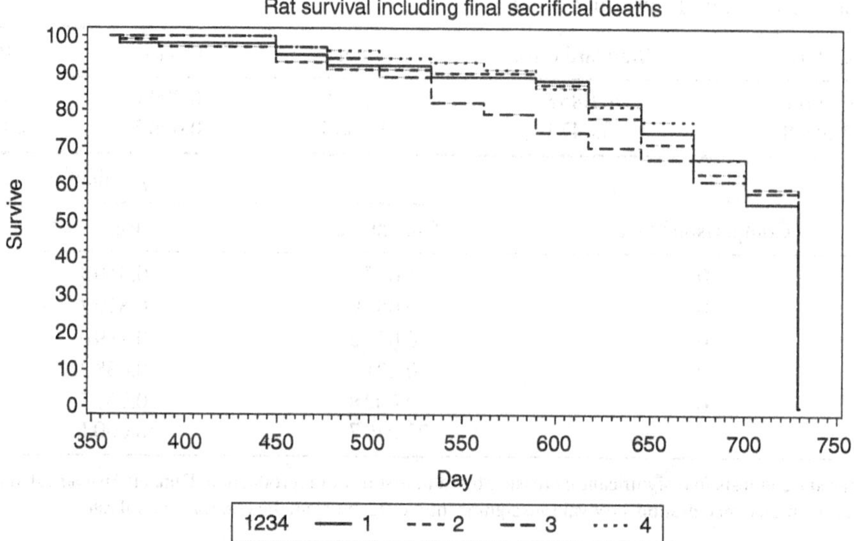

Figure 10.8 Survival curves for 2-year rat study. Only days >350 are shown for visual clarity. No deaths occurred on earlier days. Survival functions do cross, though there is little difference in the survival functions for groups 1, 2, and 4, so the proportional hazards model may still be appropriate. Group 3, 500 ppm, appears to have decreased survival between days 525 and 640, but after that, the differences with other groups go away.

must be used. The programs *rat phreg_2.sas* and *coxmodel.r*, to obtain these estimators are given in Appendix 1. More detail can be found in Collett (2014) and other standard texts on survival analysis, such as Lee and Wang (2013) and Klein and Moeschberger (2003). Log-rank tests can be used with the proportional hazards model.

Proportional hazard models will be illustrated with a 2-year rat study. The data are given in Appendix 1. There were 66–69 rats per treatment group, four treatment groups of 0 (negative control), 100, 500, and 2500 ppm. All animals that did not die earlier were sacrificed at the end of the study (104 weeks). There were no interim sacrifices, but a few rats were sacrificed in extremis before the end of the study. Rats were individually housed and because of the rotation of cages over the course of the study, there were no cage effects and individual animal observations were considered independent. See Figure 10.8 for the survival curves.

All sacrificial deaths were included but treated as censored. No treatment group had significantly higher mortality than the control. If the terminal sacrifices are excluded, treatment 3 was found to have affected mortality. However, such an analysis is considered a distortion, since it ignores almost 60% of the data. As Figure 10.8 suggests, no statistically significant difference was found in survival in any treatment group compared to the control. See Table 10.5 for a summary of the results from a proportional hazards model. These results can be obtained from the R program *coxmodel.r* and SAS program *rat phreg_2.sas*. Both programs include model diagnostics and formal tests for treatment effects. StatCHARRMS (2017) contains other R programs for Kaplan–Meier and Cox proportional hazards models.

Table 10.5 Summary of Proportion Hazards Model Fit to Rat Survival Data

Hazard ratios for GROUP			
		95% Wald confidence bounds	
Description	Point estimate	LCB	UCB
GROUP 2 vs 1	0.971	0.692	1.362
GROUP 3 vs 1	1.009	0.717	1.419
GROUP 4 vs 1	0.963	0.687	1.349

Point estimate, estimated ratio of treatment hazard function to control hazard function, based on Eq. (10.25). A ratio not significantly different from 1 indicates no significant difference in hazard and, hence, survival.

10.4 SURVIVAL ANALYSIS OF GROUPED DATA

As indicated in the introduction of this chapter, in many ecotoxicology studies, subjects are grouped, with multiple subjects housed together in vessels, e.g. tanks and multiple vessels per treatment group. The standard survival methods discussed in the previous sections are not applicable for such studies since they treat the individual subjects as independent. Subjects in the same replicate are correlated and the analysis should account for the correlations. PH models do this using "sandwich" estimators that were introduced by Lee et al. (1992). The term "sandwich" comes from the fact that the variance of the estimator can be written in the form of ABA, where A, the "bread," is a standard variance estimator, and B, the "meat," is a correction term. A thorough discussion of sandwich estimators is given in Therneau and Grambsch (2000). Such estimators are available in the SAS procedure PHreg and the R package sandwich.

An alternative approach is through what are called frailty models. The term frailty comes from failure-prone observations and can be thought of random effects survival models that capture variance heterogeneity that might arise from unexplained factors. In clustered designs, such as arise from multiple subjects housed together, the weakest individuals die first. The type of frailty models most appropriate for the grouped housing designs of interest in ecotoxicology experiments are called *shared* frailty models. To motivate shared frailty models and following the development in Jones (2017), we begin with the basic idea of individual frailty models, which is to add a term to Eq. (10.26) to get

$$h_i(t) = \exp\left(\beta' \boldsymbol{x}(i) + \gamma w_i\right) h_0(t) = h_0(t)\theta_i \exp\left(\beta' \boldsymbol{x}(i)\right), \tag{10.27}$$

where w_i is the frailty associated with deaths at time t_i and so $\theta_i = \exp(\gamma w_i)$. For simplicity, these frailties are assumed to be independent and have mean zero and variance 1, and θ_i is assumed to have mean 1 (by absorbing differences from 1 from the expression given for θ_i to be absorbed in $h_0(t)$). In the rightmost form of Eq. (10.27), it will be convenient (if confusing) to refer to the θ_i as the frailties.

In shared frailty models, subjects in the same "group" share the same frailty. In the context of ecotoxicology, this arises from treating subjects in the same test vessel as sharing the same frailty, but frailties may vary among vessels in the same treatment group. The relationship between this concept and the discussions in previous chapters of overdispersion or heterogeneity among replicates in the same treatment group should be clear. The model for shared frailty is a straightforward modification of Eq. (10.27) given by

$$h_{ij}(t) = h_0(t)\theta_i \exp\left(\beta' \boldsymbol{x}(ij)\right), \tag{10.28}$$

where θ_i is now interpreted to be the frailty shared by all subjects in replicate j of treatment i. Additional discussion of these models is given in Hougaard (2000). Frailty models are available in the SAS procedures PHREG and NLMIXED and in the R package frailtypack.

Further details of these two models are rather complex and will not be presented here. A third method is through GLMM. Two examples will be presented to illustrate the type of data and all three approaches to analysis.

10.4.1 LAGDA Time to Metamorphosis

In an instance of the LAGDA of OECD TG 241, there were three treatment groups and a water control, with eight tanks per treatment group and, initially, 25 tadpoles per tank in each treatment and control, though some were lost over the course of the study for various reasons (e.g. death or interim sacrifice). These are very large, expensive studies with numerous responses, including histopathology severity scores that were discussed in Chapter 9, length, weight, and other continuous responses that can be analyzed by the methods of Chapter 3 or possibly Chapter 4. More relevant to the current chapter, survival analysis is a required response. Rather than consider survival, this example will focus on the time to reach developmental stage 62. The reader may recall the example of developmental stage at the end of an experiment following the AMA (or OECD TG 231) given in Section 9.3. Instead of analyzing the developmental stage at the end of an AMA study, for a LAGDA study the measure of interest is the time to reach metamorphosis or developmental stage 62. The multiquantal Jonckheere–Terpstra test (MQJT) developed in Chapter 9 is not appropriate for this measure of development. Instead, the proportional hazards model with the sandwich estimator of variance or a frailty model can be used. The negative control data (see Table 10.6) is sufficient to demonstrate that there is unequal distribution of males and females and unequal total sample sizes within replicate vessels and right censoring when the study was terminated at 81 days postexposure. The full dataset is given in Appendix 1 with the name lagda.

A frailty model might be appropriate to capture a difference between the two sexes. A frailty model with a sandwich estimator might be suitable for the obvious potential for correlations among observations in the same vessel.

See Figure 10.9 for survival curves ignoring the correlation issue. The sandwich model gives the same survival curves. The impact of the sandwich estimator is in the estimated variance. The "survival" curves visually suggest that all three treatment groups had delayed metamorphosis. The formal analysis (see Tables 10.7 and 10.8) found all of these delays were statistically significant. SAS programs for this type of analysis include *lagda_sandwich and frailty by sex.sas* and *lagda_sandwich.sas* as well as *lagda_sandwich.r*. A product-limit test is given in *lagda survival example.sas* and *lagda_survival example.r*.

A frailty model failed to converge for these data. To explore further these data and the difference in results from the two models attempted, the possibility of differences between the two sexes was considered. The first attempt was to fit models with terms for sex and group-by-sex interaction included as well as group. In this event, neither model converged. Next, separate models were fit to each sex. Here the differences were very marked. For males, both models converged, but conclusions were different. See Table 10.9 for a summary of the results.

For females, the sandwich estimators gave the same conclusions as for males and for sexes combined, namely, that all three treatment groups had delayed time to

Table 10.6 LAGDA Negative Control Data

	Tank															
	1		2		3		4		5		6		7		8	
ID	Day	Sex	Day	Sex	Day	Sex	Day	Sex	Day	Sex	Day	Sex	Day	Sex	Day	Sex
1	47	F	44	F	44	M	45	M	47	M	49	M	47	F	47	F
2	49	F	47	M	51	F	47	M	47	M	49	F	47	F	48	M
3	49	F	47	M	51	F	51	F	47	M	49	M	52	M	52	M
4	49	F	48	M	51	M	53	F	49	M	51	F	53	M	52	M
5	53	F	53	F	51	F	53	M	52	M	54	F	55	F	53	M
6	53	F	53	M	52	F	53	M	53	M	54	F	55	F	53	M
7	55	M	54	F	52	F	54	M	55	F	55	F	55	F	54	F
8	57	M	54	F	52	F	54	F	61	M	55	M	57	F	54	M
9	57	F	54	F	54	F	55	M	62	F	56	M	58	F	55	M
10	60	M	58	F	54	M	56	M	65	F	57	F	59	F	56	M
11	61	F	60	F	54	M	56	M	67	F	58	F	59	F	57	M
12	63	F	60	M	55	F	56	F	70	M	59	F	59	F	58	M
13	65	M	61	M	57	F	59	M	73	F	59	F	60	F	58	F
14	65	F	61	M	58	M	59	F	73	F	.	.	61	F	59	F
15	68	F	61	F	58	M	59	M	81[a]	F	61	F	62	M	59	F
16	71	M	61	F	58	F	59	M	.		62	F	62	F	62	M
17	.		61	M	59	M	60	M	.		63	M	66	M	62	F
18	.		63	F	61	F	61	F	.		67	M	66	M	64	M
19	.		65	F	61	F	61	F	.		69	F	67	M	66	F
20	.		66	F	64	F	62	F	.		71	F	67	M	67	F
21	.		67	M	68	F	67	F	.		72	F	69	F	67	M
22	.		69	F	68	M	67	M	.		74	F	72	F	70	F
23	.		70	F	69	F	69	F	.		75	M	75	M	78	M
24	.		73	M	72	M	70	F	.		76	M	77	F	81	F

Tank, replicate test vessel; ID, animal number; Day, study day on which stage 62 was reached.
. = data missing (e.g. through death).
[a] Censored value.

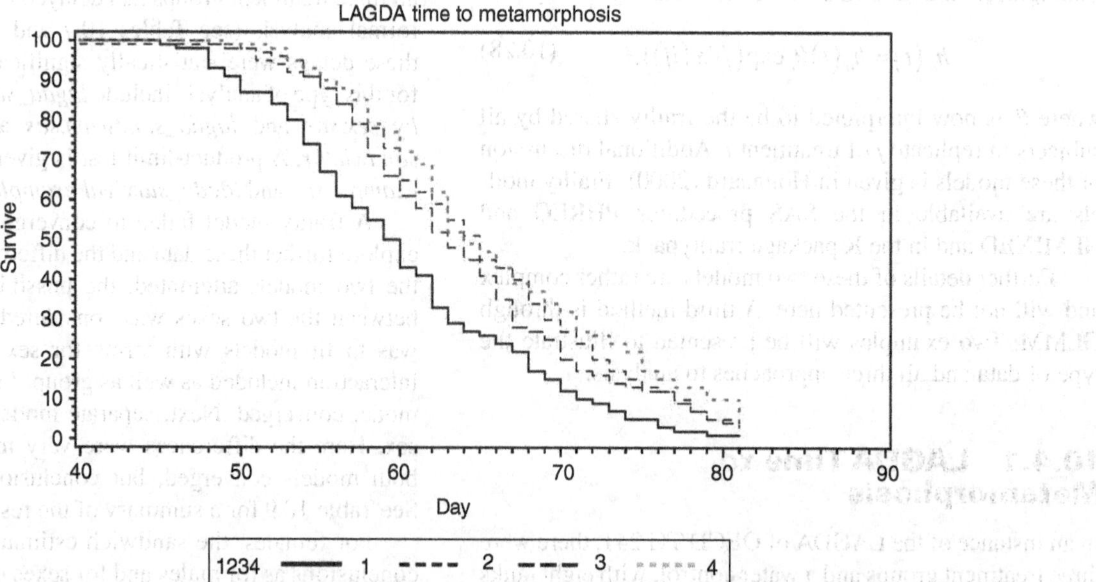

Figure 10.9 LAGDA negative control data. Survival here means not yet reached metamorphosis. It should be noted that the survival curves are not monotonic, since the delay in group 3 is less pronounced than that in groups 2 and 4.

Table 10.7 Sandwich Estimators for LAGDA Time to Metamorphosis

Hazard ratios for GROUP			
		95% Wald confidence bounds	
Description	Point estimate	LCB	UCB
GROUP 2 vs 1	0.59	0.528	0.659
GROUP 3 vs 1	0.652	0.571	0.745
GROUP 4 vs 1	0.512	0.408	0.643

Upper and lower confidence bounds are both less than 1 for each comparison to control, meaning significantly delayed time to metamorphosis in all treatment groups.

Table 10.8 GLMM Estimators for LAGDA Time to Metamorphosis

Differences of conc least squares means							
Adjustment for multiple comparisons: Dunnett							
Conc	Conc	Estimate	StdErr	DF	*t* Value	Pr>|*t*|	Adj*P*
0.2	0	0.07318	0.01478	28	4.95	<0.0001	<0.0001
1.5	0	0.06501	0.01481	28	4.39	0.0001	0.0004
6	0	0.09673	0.015	28	6.45	<0.0001	<0.0001

Conc, treatment compared to control; Adj *P*, *p*-values adjusted by the Dunnett-Hsu test; Pr>|*t*| is the unadjusted *p*-value for the comparison.

metamorphosis compared to the controls. But the frailty model failed to converge. Further exploration of the data is in order. See Figure 10.10, which reveals a major issue with the data, subsequently verified by examining the data: there was only one male in the high dose group. The frailty model for males indicated a problem with the high dose group by way of the very wide confidence limits on the group 4/control hazard ratio. The sandwich estimator was unable to reveal the uncertainty in that ratio. The lower panel of Figure 10.10 shows the survival curves for females. By comparing these two plots, it should be clear that in addition to the issues just mentioned, the treated males have much greater delays in time to metamorphosis than treated females do. Interestingly, the model diagnostic tools, such as deviance residuals and DFBETA, applied to the sandwich model do not reveal any serious problem with the data or model.

These observations highlight what should be obvious by now, namely, one should always explore the data before building models, and one should carefully evaluate models after they are fit for any problems that might cast doubt on the adequacy of the model. Statistical analysis requires critical thought.

Table 10.9 Comparison of Sandwich and Frailty Models for Males in LAGDA Study

	Hazard ratios for GROUP			
Model	Description	Point estimate	95% Wald robust confidence limits	
Sandwich	GROUP 2 vs 1	0.304	0.168	0.553
	GROUP 3 vs 1	0.206	0.116	0.367
	GROUP 4 vs 1	0.434	0.345	0.547
Frailty	GROUP 2 vs 1	0.304	0.145	0.636
	GROUP 3 vs 1	0.206	0.084	0.509
	GROUP 4 vs 1	0.434	0.06	3.147

The point estimates of the hazard ratios are the same, as expected, but the confidence limits are very different.

EXERCISES

In all datasets in the text and exercises, if a concentration is given with the value −1, that means a solvent used in the study and the water control is recorded as −1, and the solvent control is listed as 0. The reader should determine whether to combine the two controls for NOEC determination or EC*x* estimation. If the two controls are to be combined, then the water control should be changed to 0 and the replicates renamed so that replicates in the two controls can be distinguished. If only the solvent control group is to be used, then the water control should be discarded. If only the water control should be used, then the solvent control should be discarded and the concentration for the water control should be changed to 0. Under no circumstances should the concentration value −1 be used for NOEC or EC*x* purposes. Guidance on this is provided in Chapters 2 and 3.

10.1 Data from the F1 generation of a medaka extended one-generation reproduction test (MEOGRT) study are used for this exercise (see Table 10.10). Medaka embryos were exposed to various concentrations of a test chemical. The experimental design was six replicate tanks of 20 embryos each. The response of interest in this exercise is first day of hatch for the embryos. The study was terminated at the end of 14 days of exposure. Concentrations are in $\mu g\,l^{-1}$.

- **(a)** Analyze these data using the step-down Jonckheere–Terpstra test.
- **(b)** Analyze these data using a GLMM with the Dunnett-Hsu adjusted significance levels for pairwise comparisons to the control.
- **(c)** Create survival plots using a proportional hazards model.
- **(d)** Analyze these data with a sandwich estimator that incorporates the replicate nature of the data.
- **(e)** Repeat (d) using a frailty model.

10.2 A study was done to determine the effect of a chemical on the likelihood of midges to emerge from the egg state. There were 20 midges per replicate vessel composed of 10 males and 10 females, four replicates per treatment in six positive concentrations of the chemical, and a water control. Note that

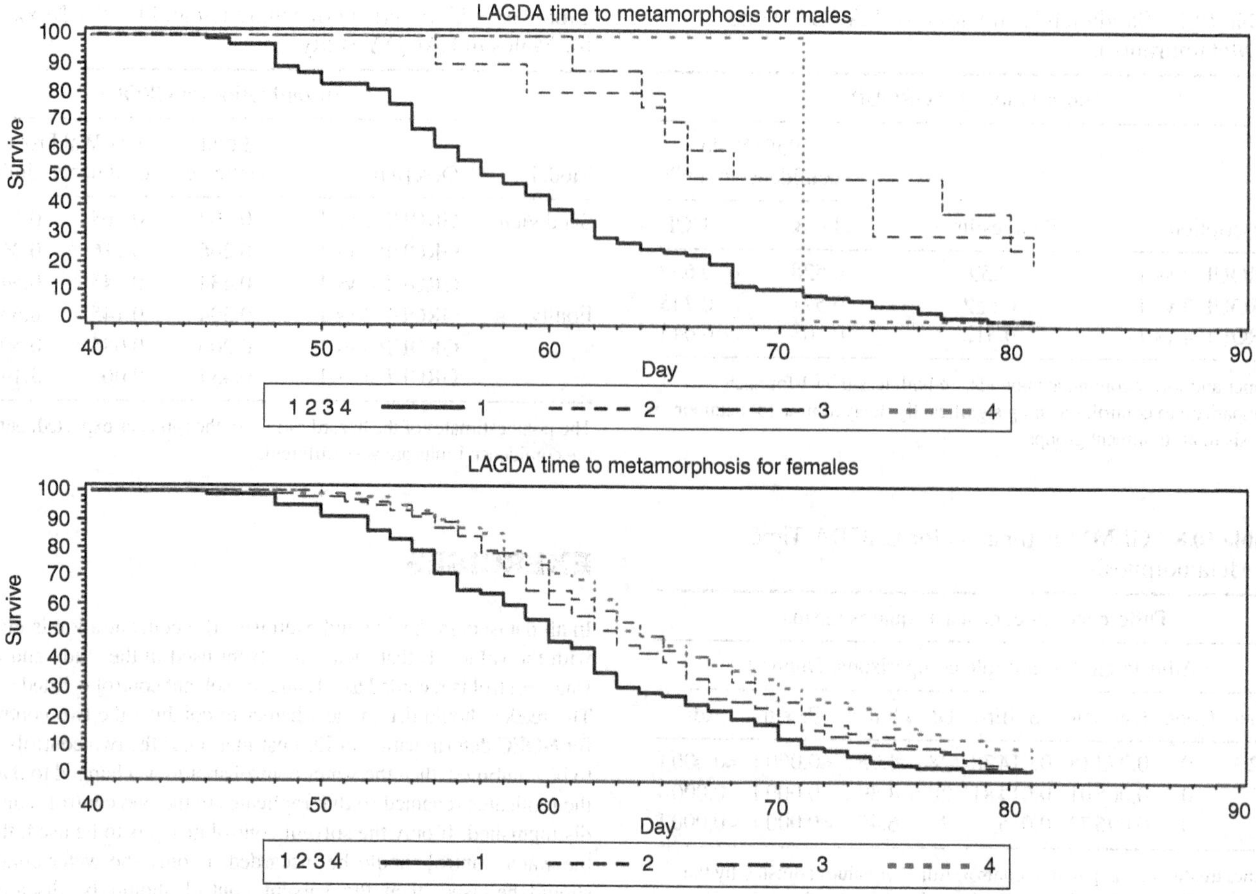

Figure 10.10 Survival curves by sex LAGDA study. Upper and lower panels are time to metamorphosis for males and females, respectively.

for each replicate vessel, where the number of a sex that emerged the value of Rep j, $j = 1, 2, 3$, or 4, does not sum to 10, the remaining number are censored, i.e. did not emerge by the end of the study. The data are in Appendix 1 in the Excel file Time to Emergence – midge.xls. Repeat the tasks in Exercise 10.1 for the midge data, paying attention to possible differences between the sexes.

10.3 An LAGDA study was done. The time to metamorphosis is given in the SAS dataset Alagda2 and the Excel file of the same name. The variable days is the number of days until developmental stage 62 was reached. All tadpoles in this study reach that stage before the end of the study, so there were no censored observations. Analyze the data to determine whether the test compound had an apparent effect on time to reach stage 62.

10.4 A study was done in the development of a potential test guideline for a multigenerational mysid assay. Data are provided in the Excel file mysid2gen.xlsx and the SAS dataset of the same name. There are two time-to-event responses in this dataset, TimeBrood1 and TimeBrood2. The first is the number of days for the mysid to brood during the first brooding period. The second is the time to brood during the second brooding period:

(a) Analyze TimeBrood1

(b) Analyze TimeBrood2

10.5 The Excel file firsthatch.xlsx contains reproduction data from three fish early life-stage studies (OECD TG 210) and six daphnia chronic studies (OECD TG 211). The names of datasets in these exercises refer to the names of spreadsheets in this file. The primary purpose for this exercise is to illustrate issues with time-to-event responses. Each dataset also contains other responses, such as survival, total live young, etc. that could be analyzed by methods from other chapters. As usual, if a solvent was used, the water control concentration is given the value −1, and the solvent control is given the value 0. Assess whether the controls should be combined and analyze the data accordingly.

(a) The dataset ELS1 contains two time-to-event responses, labeled FHatch and LHatch, which refer to the first and last days of hatching for trout embryos in the indicated replicate test vessels. Analyze these two responses to determine a NOEC using the Jonckheere–Terpstra test, the Dunn's test, a GLMM with Poisson error distribution and Dunnett-Hsu test, a proportional hazards model, and a Kaplan–Meier model and discuss the choices and compare results.

Table 10.10 Medaka First Day of Hatch

Conc	Rep	Exposed	Dead	First day of hatch							
				8	9	10	11	12	13	14	>14
Negative control	1	40	1	0	7	26	5	1	0	0	0
	2	40	0	6	16	16	1	1	0	0	0
	3	40	1	1	7	14	6	3	4	3	1
	4	40	1	1	8	11	4	4	6	3	2
	5	40	1	3	5	10	10	3	3	5	0
	6	40	3	1	8	13	9	1	2	2	1
12.5	1	40	0	3	5	7	0	1	9	15	0
	2	40	1	0	5	10	4	3	7	6	4
	3	40	1	3	4	8	1	0	6	12	5
	4	40	1	4	22	3	1	0	3	4	2
	5	40	3	0	9	10	3	2	0	11	2
	6	40	2	1	9	8	2	2	7	6	3
25	1	40	3	2	13	5	1	1	7	8	0
	2	40	0	2	8	7	3	1	9	8	2
	3	40	0	0	9	16	3	0	8	3	1
	4	40	0	4	8	11	5	1	6	4	1
	5	40	1	2	8	12	6	1	6	4	0
	6	40	2	2	12	14	4	1	4	1	0
50	1	40	1	2	7	21	7	0	2	0	0
	2	40	1	10	12	6	4	0	4	3	0
	3	40	1	5	6	7	4	3	8	5	1
	4	40	1	7	6	6	4	2	6	6	2
	5	40	0	1	2	10	4	0	13	10	0
	6	40	3	0	2	10	3	0	13	8	1

Dead = number of embryos that died before the end of the study. These were not considered censored. Individual cell values, such as 0, 7, 26, etc, =number of embryos that hatched on the indicated day.

(b) The dataset ELS2 contains the same two responses as in (a), but with labels FirstHat and LastHat, and also the response swimup, which is the first day of swimup, that is, the first day one of the hatchlings swam to the surface. Follow the instructions of (a) for these responses, especially swimup.

(c) The dataset ELS3 has one time-to-event response, Repday, which is the first day of reproduction for the indicated replicate test vessel. Follow the instructions of (a) for this response.

(d) The dataset Daphnia1 has one time-to-event response, Repday. Follow the instructions of (a) for this response.

(e) The dataset Daphnia2 has one time-to-event response, Repday. Follow instructions of (a) for this response.

(f) The dataset Daphnia2 has one time-to-event response, Repday1. Follow instructions of (a) for this response.

(g) The dataset Daphnia3 has one time-to-event response, Repday. Follow instructions of (a) for this response.

(h) The dataset Daphnia4 has one time-to-event response, Repday. Follow instructions of (a) for this response.

(i) The dataset Daphnia5 has one time-to-event response, Repday. Follow instructions of (a) for this response.

(j) The dataset Daphnia6 has one time-to-event response, Repday. Follow instructions of (a) where possible for this response. Note that there are censored values for Repday in this dataset.

Chapter 11

Regulatory Issues

11.1 INTRODUCTION

This chapter describes different structures of and approaches to regulatory testing of chemicals. Focus will be on standardized test guidelines (TGs) and testing strategies with the use of the Organisation for Economic Co-operation and Development (OECD) and US Environmental Protection Agency (USEPA) TGs as examples. If a more detailed and comprehensive overview of regulatory testing is required, information should be collected from regulatory bodies like the European Chemical Agency (ECHA) in Europe and USEPA in the United States, or information should be gathered from some of the excellent books that have been dedicated to hazard and risk assessment of chemicals, e.g. Van Leeuwen and Vermeire (2007) and Fisk (2014).

Testing of chemicals is required by regulatory bodies to identify hazards and assess derived risks through exposure for both human health and the environment. Tests on biotic systems can be used to investigate the inherent hazard of the chemical and establish the connection between dose and response. Abiotic tests can be used to investigate the risk of exposure in relation to the fate of the chemical, i.e. environmental distribution and persistence. The testing data and derived exposure data are used to set up rules to support safe use or to restrict or ban the production, import, and use of harmful chemicals. Chemical regulative legislations differ to some extent among types of chemicals, i.e. pesticides and industrial chemicals as well as among regions and countries. Fortunately, an increasing harmonization of testing, classification, labeling, and regulation of chemicals occurs worldwide. The United Nations (UN) Globally Harmonized System of Classification and Labeling of Chemicals (GHS) (UN, 2015) is an example of a harmonized system that will be presented in this chapter. Also, TGs are standardized and harmonized, and especially OECD plays an important role in this respect.

11.2 REGULATORY TESTS

Regulatory testing of chemicals can be divided into three different main categories: *In silico, in vitro*, and *in vivo*.

In silico testing is a term defined in the early 1990s that covers computer-based experiments like whole cell modeling and quantitative structure–activity relationship (QSAR). It was implemented in aquatic toxicology in the 1970s where Veith and Konasewich (1975) correlated the chemical structure to toxicity and bioconcentration in aquatic organisms. Today, *in silico* tests are widely used as screens to select among chemicals for further *in vitro* and *in vivo* testing and as alternatives to animal testing. Large QSAR databases including hundreds of thousands of chemicals are public available and allow to select chemicals for further testing based on physicochemical properties (e.g. Log_{Kow}), expected ecotoxicity, environmental fate, and organism's absorption, distribution, metabolism, and excretion (ADME) as well. Examples of such databases are the Danish QSAR database (http://qsar.food.dtu.dk/) and the OECD QSAR Toolbox (http://www.oecd.org/chemicalsafety/risk-assessment/theoecdqsartoolbox.htm).

In vitro (Latin for *within the glass*) testing defines testing on cells and tissue cultures outside an intact organism. *In vitro* testing allows very stable exposure conditions and higher throughput than *in vivo* testing and is used to screen chemicals for a variety of mechanisms and modes of action. Replacement of animal *in vivo* tests with *in vitro* tests has a

Statistical Analysis of Ecotoxicity Studies, First Edition. John W. Green, Timothy A. Springer, and Henrik Holbech.

Companion website: www.wiley.com/go/Green/StatAnalysEcotoxicStudy

high priority due to animal welfare issues, and several organizations are established to refine, reduce, and replace animal *in vivo* testing. Examples are the European Union Reference Laboratory for Alternatives to Animal Testing (EURL–ECVAM), the Japanese Center for the Validation of Alternative Methods (JaCVAM), and the US-based Interagency Coordinating Committee on the Validation of Alternative Methods (ICCVAM). Several new alternative methods have been developed recently. One example is the OECD Test No. 431: *In vitro* skin corrosion: reconstructed human epidermis (RHE) test method (OECD, 2016a), where no *in vivo* animal testing is needed to investigate the corrosive effect of a test chemical on human skin. Because *in vitro* testing occurs outside an intact organism, aspects like metabolism, bioavailability, and feedback mechanisms are often not taken into account in such tests, and therefore *in vitro* tests are in most cases supplemented with *in vivo* testing before regulatory decisions are taken. Efforts are taken to include metabolic activation in validated *in vitro* assays. For a recent review see Jacobs et al. (2013). OECD TG 476 (OECD, 2016b) and TG 487 (OECD, 2016c) are examples of *in vitro* assays already including metabolic activation. More advanced *in vitro* assay taking metabolism, bioavailability, and biotransformation into account will increase the usefulness of *in vitro* assays in regulatory decisions in the years to come.

In vivo (Latin for within the living) testing is defined as testing of chemicals on intact organisms. In ecotoxicology, *in vivo* tests have been developed to investigate the toxicity of chemicals in plant and animal phyla as well as bacteria and other microorganisms (although *in vivo* testing often refers to vertebrate and invertebrate animal testing). *In vivo* testing is in general the only way to study chronic effects like reproductive effects and multigeneration effects, but the use of historical *in vivo* data in combination with *in vitro* and *in silico* data as done in the ToxCast™ program (described below) can help reducing *in vivo* testing in the future.

Recently, data from all three categories of testing have been combined in databases for the prediction of chemicals toxicity. One example of a predictive database is the USEPA Toxicity Forecaster named ToxCast™. ToxCast uses more than 700 high-throughput *in vitro* screening assays that cover hundreds of signaling pathways and historical *in vivo* data to predict *in vivo* toxicity. Use of ToxCast is free, and information can be found at https://www.epa.gov/chemical-research/toxicity-forecasting. One of the strengths with ToxCast is that all data is available and can be downloaded free of charge (USEPA, 2016d). Examples of this free service (by September 2016) are presented (see Table 11.1) and include the R computer programming package used to process and model all EPA ToxCast and Tox21 chemical screening data. A current weakness of ToxCast is that most of the assays have not yet gone through substantial validation as the OECD and US OCSPP assays have.

11.3 DEVELOPMENT OF INTERNATIONAL STANDARDIZED TEST GUIDELINES

Standardized TGs are the key tool to obtain data on the potential hazards of chemicals and are used by manufacturers to fulfill data requirements and by regulating authorities to request data. OECD organizes and coordinates the development and update of the majority of the international accepted chemical TGs in the OECD TG program. OECD's 35-member countries and additional key partner countries like China, India, and Brazil account for about 80% of the world trade, and therefore the OECD guidelines will be used as examples of standardized tests.

11.3.1 OECD Test Guidelines

The OECD TGs are standardized and internationally validated tests that are developed to assess the effects (hazards) of chemicals on human health and the environment. The TGs comprise more than 150 individual guidelines and are divided into 5 main sections according to their use: Section 1: Physical–Chemical Properties, Section 2: Effects on Biotic Systems, Section 3: Degradation and Accumulation, Section 4: Health Effects, and Section 5: Other TGs. All OECD TGs and guidance documents are freely available at http://www.oecd.org/chemicalsafety/testing/oecdguidelinesforthetestingofchemicals.htm

To illustrate the different types of OECD TGs, the OECD conceptual framework for testing and assessment of endocrine disrupters is presented (see Table 11.2). The framework is organized in five levels ranging from *in silico* data and *in vitro* tests to three levels of *in vivo* testing. Most of the *in vivo* tests were developed recently, and many of the statistical cases used elsewhere in this book are taken from the international validation of these tests.

11.3.2 OECD Mutual Acceptance of Data (MAD)

One of the strengths of the OECD TGs program is the concept of Mutual Acceptance of Data (MAD) that was adopted in 1981 by the OECD Council. MAD states that "Data generated in accordance with OECD Test Guidelines and OECD principles of Good Laboratory Practice (GLP) shall be accepted in other Member countries (or non-member

Table 11.1 Examples of Toxcast Free Services

	Description	Release date and database version	Download
ToxCast and Tox21 chemicals distributed structure-searchable toxicity database (DSSTox files)	Chemical details for 8599 unique substances (GSIDs) and DSSTox standard chemical fields (chemical name, CASNR, structure, etc.) for EPA ToxCast chemicals and the larger Tox21 chemical list. Also includes chemical mapping files and quality control grades for chemicals	October 2015 DSSTox_ 20151019	ToxCast Chemicals: Data Management and Quality Considerations Overview Download ToxCast Chemical Information
ToxCast and Tox21 high-throughput assay information	ToxCast high-throughput assay information including assay annotation user guide, assay target information, study design information, and quality statistics on the assays	October 2015 invitrodb_v2	Assay Annotation User Guide Download Assay Information
R Package	The R computer programming package used to process and model all EPA ToxCast and Tox21 chemical screening data. The files include the R programming package as well as documents that provide overviews of the data analysis pipeline used and the R package. Users will need experience with R to use these files	May 2016 tcpl_1.2.2	Download R Package
ToxCast and Tox21 data spreadsheet	A spreadsheet of EPA's analysis of the chemicals screened through ToxCast and the Tox21 collaboration that includes EPA's activity calls from the screening of over 8000 chemicals	October 2015 invitrodb_v2	Download Data
ToxCast and Tox21 concentration response plots	Concentration response plots for all of the ToxCast and Tox21 assays	October 2015 invitrodb_v2	Download Concentration Response Plots
Collaborative estrogen receptor activity prediction project data	Data and supplemental files from CERAPP (a large-scale modeling project) that demonstrated the efficacy of using predictive computational models trained on high-throughput screening data to evaluate thousands of chemicals for estrogen-related activity. CERAPP combined multiple models developed in collaboration with 17 groups in the United States and Europe to predict ER activity of a common set of 32 464 chemical structures. *Quantitative structure–activity relationship* models and docking approaches were employed to build a total of 40 categorical and 8 continuous models for binding, agonist, and antagonist ER activity	January 2016 invitrodb_v1	Download Supplemental Files and Data
High-throughput screening data for estrogen receptor model	Estrogen receptor model data from the manuscript titled *Integrated Model of Chemical Perturbations of Biological Pathways Using 18 In vitro High-throughput Screening Assays for the Estrogen Receptor* (Judson et al., 2015)	August 2015 invitrodb_v1	Download Data
Animal toxicity studies: effects and endpoints (Toxicity Reference Database – ToxRefDB files)	Provides results from across the thousands of animal toxicity studies No effect level (NEL)/lowest effect level (LEL) and/or no observed adverse effect level (NOAEL)/lowest observed adverse effect level (LOAEL)	October 2014 toxrefdb_v1	Download Animal Toxicity Data

Modified from https://www.epa.gov/chemical-research/toxicity-forecasting.

Table 11.2 OECD Conceptual Framework for Testing and Assessment of Endocrine Disrupters

Level 1: Existing data and nontest information	Physical and chemical properties, e.g. MW reactivity, volatility, biodegradability All available (eco)toxicological data from standardized or nonstandardized tests Read across, chemical categories, QSARs and other *in silico* predictions, and ADME model predictions Existing data and Non-Test Information. Physical and chemical properties, e.g., MW reactivity, volatility, biodegradability. All available (eco)toxicological data from standardized or non-standardized tests. Read across, chemical categories, QSARs and other in silico and ADME model predictions.	
Level 2: *In vitro* assays providing data about selected endocrine mechanism(s)/pathways(s)	Estrogen or androgen receptor binding affinity Estrogen Receptor Transactivation (TG 455–TG 457) Stably Transfected Human Androgen Receptor Transcriptional Activation Assay for Detection of Androgenic Agonist and Antagonist Activity of Chemicals (TG 458) Steroidogenesis *In Vitro* (OECD TG 456) MCF-7 cell proliferation assays (ER ant/agonist)	
Level 3: *In vivo* assays providing data about selected endocrine mechanism(s)/pathway(s)	Uterotrophic Assay (TG 440) Hershberger Assay (TG 441)	Amphibian Metamorphosis Assay (TG 231) Fish Short Term Reproductive Assay (TG 229) Fish Screening Assay (TG 230) Androgenized female stickleback screen (GD 140)
Level 4: *In vivo* assays providing data on adverse effects on endocrine relevant endpoints	Repeated dose 28-day study (TG 407) Repeated dose 90-day study (TG 408) 1-generation reproduction toxicity study (TG 415) Male Pubertal Assay (see GD 150) Female Pubertal Assay (see GD 150) Intact Adult Male Endocrine Screening Assay (see GD 150) Prenatal Developmental Toxicity Study (TG 414) Chronic Toxicity and Carcinogenicity Studies (TG 451–53) Reproduction/Developmental Toxicity Screening Test (TG 421) Combined Repeated Dose Toxicity Study with the Reproduction/ Developmental Toxicity Screening Test (TG 422) Developmental Neurotoxicity (TG 426)	Fish Sexual Development Test (TG 234) Larval Amphibian Growth & Development Assay (LAGDA, TG 241) Avian Reproduction Assay (TG 206) Potamopyrgus Antipodarum Reproduction Test (TG 242) Lymnaea Stagnalis Reproduction Test (TG 243) Chironomid Toxicity Test (TG 218219) Daphnia Reproduction Test (with male induction) (TG 211) Earthworm Reproduction Test (TG 222) Enchytraeid Reproduction Test (TG 220) Sediment Water Lumbriculus Toxicity Test Using Spiked Sediment (TG 225) Predatory Mite Reproduction Test in Soil (TG 226) Collembolan Reproduction Test in Soil (TG 232)
Level 5: *In vivo* assays providing more comprehensive data on adverse effects on ED relevant endpoints over more extensive parts of the life cycle of the organism	Extended One-generation Reproductive Toxicity Study (TG 443) 2-Generation Reproduction Toxicity Study (TG 416 most recent update)	Medaka Extended One-generation Reproduction Test (MEOGRT, TG 240) Sediment Water Chironomid Life Cycle Toxicity Test (TG 233)

Modified from OECD (2012a).

Draft TGs Are Not Included in This Modification. The recommended statistical methods to analyze TG data are presented in Tables 11.4 and 11.7.

economies) for purposes of assessment and other uses relating to the protection of man and the environment." The decision was taken to prevent multiple testing of the same chemicals in different countries, and OECD estimated in 2010 that the harmonized work saves around 150 million € and thousands of experimental animals every year (OECD, 2010b). A monitoring program for GLP compliance was added in a council Act in 1989 to ensure that data are in compliance with the principles of GLP, and in 1997 the OECD Council decided that some non-OECD members could become full adherent to MAD with the same rights and obligations as OECD countries after on-site evaluation of their GLP compliance monitoring program.

11.4 STRATEGIC APPROACH TO INTERNATIONAL CHEMICALS MANAGEMENT (SAICM)

SAICM is a global policy framework adopted in 2006 and supported by UN Environment Programme (UNEP), the Intergovernmental Forum on Chemical Safety (IFCS), and the Interorganization Program for the Sound Management of Chemicals (IOMC) where the latter represent a wide range of international organizations including WHO, the International Labour Organization (ILO), OECD, and the World Bank. The SAICM secretariat is administered by UNEP and WHO. SAICM promotes chemical safety around the world, and the overall goal is the one agreed at the 2002 Johannesburg World Summit on Sustainable Development: "to achieve sound management of chemicals throughout their life cycle so that, by 2020, chemicals are produced and used in ways that minimize significant adverse impacts on human health and the environment" (UNEP, 2007). SAICM objectives are risk reduction, knowledge and information, governance, capacity-building and technical cooperation, and illegal international traffic (SAICM, 2017). For more information on the specific actions under SAICM, see the SAICM texts and resolutions of the International Conference on Chemicals Management (UNEP, 2007).

11.5 THE UNITED NATIONS GLOBALLY HARMONIZED SYSTEM OF CLASSIFICATION AND LABELLING OF CHEMICALS (GHS)

The work on GHS was initiated in 1992 where the development of the system was mandated by the UN. In 2002 the first edition of GHS was approved by UN, and currently the latest edition is No. 6 from 2015 (UN, 2015). The backbone for the elaboration of GHS was already existing requirements and directives from Canada, the US, Europe, and the UN recommendations on the transport of dangerous goods. Today GHS is adopted by a number of countries and unions representing the majority of the world trade including Australia, Brazil, Canada, China, the European Union (EU), Turkey, Japan, Korea, Malaysia, the Philippines, Taiwan, Vietnam, and the United States.

The mission for the GHS is to internationally standardize the classification and labeling of chemicals with the use of pictograms and hazard information and thereby increasing the safety of chemical handling as well as saving time, reducing costs, and facilitate international trade. The four main reasons for developing the GHS as described in (UN, 2015) are:

Enhance the protection of human health and the environment by providing an internationally comprehensible system for hazard communication.

Provide a recognized framework for those countries without an existing system.

Reduce the need for testing and evaluation of chemicals.

Provide clear guidelines for chemical labeling that can be used across borders.

In the context of regulatory testing, the GHS provide a standardized platform for translation of data from standardized TGs into safety data sheets (SDS). The GHS has developed harmonized criteria for classification of chemicals according to their hazards (physical, environmental, and health), and it has harmonized the hazard communication elements by classes, categories, and types. One example is standardized pictograms for specific hazards. See Table 11.3 for a pictogram covering some of the hazards to the aquatic environment.

11.6 STATISTICAL METHODS IN OECD ECOTOXICITY TEST GUIDELINES

For each OECD TG, more or less comprehensive guidance on how to perform the statistical evaluation of data is included. See Table 11.4 for ecotoxicity relevant OECD TGs, and which statistical tests and models are required or recommended. There are often more than one statistical option for a specific endpoint and more test design options, and it is recommended to seek advice with a statistician before OECD TG studies are started.

11.7 REGULATORY TESTING: STRUCTURES AND APPROACHES

Structures of and approaches to regulatory testing differ in some aspects among different countries and regions. The burden of proof of the safety of using a chemical is placed at the manufacturer in the EU Registration, Evaluation, Authorization, and Restriction of Chemicals (REACH), whereas in the United States, according to the Toxic Substances Control Act (TSCA), the USEPA has to identify the chemicals that could possess a risk and further prove the toxicity of the chemicals. In Japan, the government selects industrial chemicals for safety assessment as

Table 11.3 Hazard Class and Categories, Pictogram, Signal Word, Hazard Statement Code, and Description of Selected Hazard Classes for GHS Environmental Hazards

Hazard class	Hazard categories	Pictogram	Signal word	Hazard statement code	Description of hazard statement
Hazardous to aquatic environment short term (acute)	Acute 1		Warning	H400	Very toxic to aquatic life
Hazardous to aquatic environment long term (chronic)	Chronic 1		Warning	H410	Very toxic to aquatic life with long lasting effects
Hazardous to aquatic environment long term (chronic)	Chronic 2		None	H411	Toxic to aquatic life with long lasting effects

Modified from UN (2013).

described in the Chemical Substances Control Law (CSCL) and requests the manufacturer to submit information on hazards of the chemical. A number of countries including China, Malaysia, and Korea have implemented REACH-like chemical registrations and regulations.

United States: Chemical regulation and testing in the United States is governed by different authorities and different laws and acts dependent on the use of the chemical. Regulatory testing requirements of chemicals in the United States are described in the TSCA, which turned into force in 1976 and was updated in 2016 (USEPA, 2016a). TSCA provides USEPA with the authority to require testing of chemicals and mixtures. The TSCA mainly covers industrial chemicals and does not include food, drugs, pesticides, and cosmetics. The Federal Insecticide, Fungicide, and Rodenticide Act (FIFRA) implemented in 1996 (USEPA, 1996) regulates pesticides and provides the USEPA with authority to decide for the individual pesticide registration, which data and information are relevant in relation to the requirements found in the Code of Federal Regulations at 40 CFR Part 158 (USGPO, 2017). Food, drugs, and cosmetics are regulated by the Federal Food, Drug and Cosmetic Act (FFDCA).

The US testing structure is built on a number of standardized TGs, including the Office of Chemical Safety and Pollution Prevention (OCSPP) Harmonized TGs. The guidelines are divided into 11 groups (see Table 11.5). As an example, the group 890 – Endocrine Disruptor Screening Program (EDSP) TGs cover 14 guidelines (11 Tier 1 and three Tier 2 tests). Group 850 represents nearly 40 guidelines divided in 5 subgroups: Group A – Aquatic and Sediment-dwelling Fauna and Aquatic Microcosms; Group B – Terrestrial Wildlife; Group C – Terrestrial Beneficial Insects, Invertebrates, and Soil and Wastewater Microorganisms; Group D – Terrestrial and Aquatic Plants, Cyanobacteria, and Terrestrial Soil Core Microcosm; and Group F – Field Test Data Reporting Guidelines. Group 850 was updated in 2016 after a public commenting round (USEPA, 2016c).

Some of the OCSPP guidelines are identical to OECD TGs. For example, OCSPP 890.1350 fish short-term reproduction assay is identical to OECD TG 229 fish short-term reproduction assay (FSTRA) (OECD, 2012b), and OCSPP 890.2200 medaka extended one generation reproduction test (MEOGRT) is identical to OECD TG 240 medaka extended one generation reproduction test (MEOGRT) (OECD, 2015). Both tests were developed in collaboration between USEPA and OECD. Other OCSPP guidelines are developed by USEPA using material from other OCSPP guidelines as well as material from OECD TGs and, for example, American Society for Testing and Materials (ASTM) guides. An example of such guideline is the new OCSPP 850.1010: Aquatic Invertebrate Acute Toxicity Test, Freshwater Daphnids. A number of OCSPP guidelines are unique to the US regulatory system – OCSPP 850.1045: Penaeid Acute Toxicity Test, OCSPP 850.1025: Oyster Acute Toxicity Test (Shell Deposition) and OCSPP 850.1055 Bivalve Acute Toxicity Test (Embryo-Larval) are such examples. All final OCSPP 850 group A guidelines that are specifically relevant for ecotoxicity testing are presented (see Table 11.6) as updated by the USEPA in 2016. OCSPP 850.1000 that is a background document for the 850 group guidelines includes a general guidance on statistical treatment of

Table 11.4 Summarized Main Mechanistic and Apical Endpoints in All Nonmammalian Biotic OECD Test Guidelines (TG) and the Recommended Statistical Method(s) to Use to Evaluate the Data (Tests for Normality and Homogeneity of Variance as well as Data Transformation Methods Are Not Described in This Table)

OECD test guideline (TG)	Main endpoints	Statistical output	Statistical methods
TG 201 Freshwater Alga and Cyanobacteria, Growth Inhibition Test	Inhibition of growth	E_rCx (x% inhibition of growth rate) E_yCx (x% inhibition of yield is required by some regulations) NOEC and LOEC may be included	E_rCx and E_yCx: Nonlinear regression analysis is preferred. Details are described in Annex 5 of the TG: Guidance on Data analysis by nonlinear regression can be found in the literature (Bruce and Versteeg, 1992; Nyholm et al., 1992; OECD, 2006a) NOEC/LOEC: Treatment comparison by variance analysis (ANOVA). Different tests for multiple comparisons or trend tests are available, e.g. Dunnett's, Williams', Levene's test, or step-down Jonckheere test. The choice depends on data normality, homogeneity of variance, and monotonicity
TG 203 Fish acute toxicity test	Mortality	LC50 (95% confidence limits)	Several acceptable methods are available, e.g. probit, Litchfield–Wilcoxon, Trimmed Spearman–Karber, log-probability graph paper plot
TG 205 Avian dietary toxicity test	Mortality	LC50 (95% confidence limits)	Several acceptable methods are available, e.g. probit, Litchfield–Wilcoxon, trimmed Spearman–Karber, log-probability graph paper plot
TG 206 Avian reproduction test	Egg production Eggs set Viability Hatchability Survival of chickens Egg shell thickness	NOEC and LOEC	Not specified in the guideline but ANOVA is appropriate. Different tests for multiple comparisons or trend tests are available, e.g. Dunnett's, Williams', Levene's test, or step-down Jonckheere test. The choice depends on data normality, homogeneity of variance, and monotonicity
TG 207 Earthworm, Acute Toxicity Tests	Mortality	LC50 (95% confidence limits)	Log-probability graph paper plot or other probit methods
TG 208 Terrestrial Plant Test: Seedling Emergence and Seedling Growth	Visual assessment of seedling emergence Dry shoot weight (alternatively fresh shoot weight) In certain cases shoot height	EC*x*, NOEC, and LOEC	EC*x*: Regression analysis should be performed according to OECD (2006a). For weight and height as continuous endpoints, EC*x* can be estimated by, e.g. Bruce–Versteeg nonlinear regression analysis. For emergence as quantal data, logit, probit, Weibull, Spearman–Karber, trimmed Spearman–Karber methods, etc. could be appropriate NOEC/LOEC: Appropriate methods are described in OECD (2006a) and depend on data. Examples for nonmonotone dose–responses are Dunnett's and Mann–Whitney tests and for monotone dose–responses, Jonckheere–Terpstra test or Williams test
TG 210 Fish, Early Life Stage Toxicity Test	Growth (length) Hatching parameters Mortality	EC*x*, NOEC, and LOEC	Regression: Detailed recommendations are given in Annex 6 of the TG: Statistical guidance for regression estimates
TG 211 Daphnia magna Reproduction Test	Living offspring produced per surviving parental animal Total number of living offspring excluding parental accidental and/or inadvertent mortality Offspring sex ratio (optional)	LC50 (95% confidence limits) or NOEC and LOEC	NOEC/LOEC: For example, Student's *t*-test, Dunnett's test, Williams' test, or step-down Jonckheere–Terpstra test. Nonparametric Dunn's or Mann–Whitney's tests can also be used LC50: Regression methods, e.g. logistic or Weibull function, trimmed Spearman–Karber, or simple interpolation

(*Continued*)

Table 11.4 (Continued)

OECD test guideline (TG)	Main endpoints	Statistical output	Statistical methods
TG 212 Fish, Short-term Toxicity Test on Embryo and Sac-Fry Stages	Cumulative mortality Hatching parameters Length and weight Healthy larvae	LC*x*/EC*x* (95% confidence limits) or NOEC and LOEC	NOEC/LOEC: ANOVA. Different tests for multiple comparisons or trend tests are available, e.g. Dunnett's, Williams', Levene's test, or step-down Jonckheere test. Cumulative mortality and number of healthy larvae are not suitable for ANOVA and could be analyzed with probit methods LC*x*/EC*x*: Regression analysis is appropriate, and methods like least squares or nonlinear least squares could be used
TG 213 Honeybees, Acute Oral Toxicity Test	Mortality	LD50 (95% confidence limits)	Not specified in the TG but several acceptable methods are available, e.g. probit, Litchfield–Wilcoxon, trimmed Spearman–Karber. Log-probability graph paper plot
TG 214 Honeybees, Acute Contact Toxicity Test	Mortality	LD50 (95% confidence limits)	Not specified in the TG but several acceptable methods are available, e.g. probit, Litchfield–Wilcoxon, trimmed Spearman–Karber. Log-probability graph paper plot
TG 215: Fish, Juvenile Growth Test	Growth (tank-average specific growth rates)	EC*x* (*x*% variation in growth rate) or NOEC and LOEC	EC*x*: Not specified in the TG but several acceptable regression methods are available, e.g. logistic or Weibull function, trimmed Spearman–Karber, or simple interpolation NOEC/LOEC: Not specified in the TG but several acceptable methods are available, e.g. Dunnett's, Williams', Shapiro–Wilk's, Bartlett's, or step-down Jonckheere test
TG 216 Soil Microorganisms: Nitrogen Transformation Test	Nitrogen transformation (nitrate formation)	EC*x* (95% confidence limits)	Not specified in the TG but several acceptable methods are available dependent on the type of chemical tested For agrochemicals, e.g. *F*-test with 5% significance level For nonagrochemicals, regression methods like logistic or Weibull function, trimmed Spearman–Karber, or simple interpolation.
TG 217 Soil Microorganisms: Carbon Transformation Test	Respiration rates	EC*x* (95% confidence limits)	Not specified in the TG but several acceptable methods are available dependent on the type of chemical tested For agrochemicals, e.g. *F*-test with 5% significance level For nonagrochemicals, regression methods like logistic or Weibull function, trimmed Spearman–Karber, or simple interpolation.
TG 218: Sediment-Water Chironomid Toxicity Using Spiked Sediment	Emerging adults Time to emergence (development time)	EC*x* or NOEC and LOEC	A detailed statistical description is included in the TG. For example, emergence ratio is quantal data, and statistics are described in paragraph 46 of the TG: Use a Cochran–Armitage test applied in step-down manner where a monotonic dose–response is expected, and these data are consistent with this expectation. If not, a Fisher's exact or Mantel–Haentzal test with Bonferroni–Holm adjusted *p*-values can be used. If there is evidence of extrabinomial variation, a robust Cochran–Armitage or Fisher's exact test should be used
TG 219 Sediment-Water Chironomid Toxicity Using Spiked Water	Emerging adults Time to emergence (development time)	EC*x* or NOEC and LOEC	A detailed statistical description is included in the TG. For example, emergence ratio is quantal data and statistics are described in paragraph 46 of the TG: Use a Cochran–Armitage test applied in step-down manner where a monotonic dose–response is expected and these data are consistent with this expectation. If not, a Fisher's exact or Mantel–Haentzal test with Bonferroni–Holm adjusted *p*-values can be used. If there is evidence of extrabinomial variation, a robust Cochran–Armitage or Fisher's exact test should be used

TG 220 Enchytraeid Reproduction Test	Reproductive output Morphological changes	EC*x* and/or NOEC and LOEC	Some examples of statistical methods are given in the guideline: For EC*x*, e.g. probit, trimmed Spearman–Karber, or simple interpolation dependent on the data For NOEC/LOEC: If data are homogeneous, Williams' or Dunnett's test should be used or a Mann-Whitney-*U* test with Bonferroni-Holm adjusted *p*-values. When data are not homoscedastic but reveal monotone dose-response, a nonparametric test like Jonckheere-Terpstra or Shirley would be preferred
TG 221 *Lemna* sp. Growth Inhibition Test	Frond number (vegetative growth expressed as average specific growth rate) Yield	E_rCx (x% inhibition of growth rate) E_yCx (x% inhibition of yield is required by some regulations) NOEC and LOEC may be included	E_rCx and E_yCx: Nonlinear regression analysis is preferred. Guidance can be found in the literature (Bruce and Versteeg, 1992; Nyholm et al., 1992; OECD, 2006a) NOEC/LOEC: ANOVA. Different tests for multiple comparisons or trend tests are available, e.g. Dunnett's, Williams', Levene's test or step-down Jonckheere test
TG 222 Earthworm Reproduction Test (*Eisenia fetida/Eisenia andrei*)	Mortality Fecundity	LC50 (95% confidence limits) and/or NOEC and LOEC	A statistical overview is included as Annex 6 of the TG LC50: Probit analysis or logistic Regression is appropriate. If less than three concentrations with mortality occur, then the trimmed Spearman–Karber method or simple interpolation could be used For NOEC/LOEC: If data are homogeneous, Williams' or Dunnett's test should be used or a multiple *U*-test such as the Bonferroni-*U*-test*. When data are not homoscedastic but reveal monotone dose–response, nonparametric test like Jonckheere–Terpstra or Shirley would be preferred
223 Avian Acute Oral Toxicity Test	Mortality	LD50 (95% confidence limits)	Use a probit regression model with the logarithm of dose as the independent variable.
TG 224 Determination of the Inhibition of the Activity of Anaerobic Bacteria	Gas production	EC50 (percentage inhibition of gas production)	Assess the EC50 ($mg\,l^{-1}$) value visually or by regression analysis
TG 225 Sediment-Water Lumbriculus Toxicity Test Using Spiked Sediment	Reproduction (total number of surviving worms) Biomass (dry weight) at the end of the exposure	LC50 (95% confidence limits) and/or NOEC and LOEC	LC50: Methods like probit analysis, logistic or Weibull function, and trimmed Spearman–Karber are appropriate as well as simple interpolation NOEC/LOEC: ANOVA followed by pairwise comparison like Dunnett's or trend test like Williams' test. For nonparametric tests, Bonferroni-*U*-test according to Holm* or Jonckheere–Terpstra trend test are appropriate
TG 226 Predatory mite (*Hypoaspis (Geolaelaps) aculeifer*) reproduction test in soil	Reproductive output (number of juveniles)	EC*x* (95% confidence limits) and/or NOEC and LOEC	EC*x*: Methods like probit analysis, logistic or Weibull function, and trimmed Spearman–Karber are appropriate as well as simple interpolation NOEC/LOEC: ANOVA followed by pairwise comparison like Dunnett's or trend test like Williams' test. For nonparametric tests, Bonferroni-*U*-test according to Holm* or Jonckheere–Terpstra trend test are appropriate
TG 227: Terrestrial Plant Test: Vegetative Vigour Test	Dry shoot weight (alternatively wet shoot weight) Shoot height	EC*x* (95% confidence limits) and/or NOEC and LOEC	Statistics are not specified but examples of tests are: EC*x*: Logit, probit, Weibull, Spearman–Karber, and trimmed Spearman–Karber methods are all appropriate NOEC/LOEC: Appropriate methods according to OECD (2006a). For example, Dunnett's or trend test like Williams' test. For nonparametric tests, Bonferroni-*U*-test according to Holm* or Jonckheere–Terpstra trend test

(*Continued*)

Table 11.4 (Continued)

OECD test guideline (TG)	Main endpoints	Statistical output	Statistical methods
TG 228 Determination of Developmental Toxicity of a Test Chemical to Dipteran Dung flies (*Scathophaga stercoraria L. (Scathophagidae), Musca autumnalis De Geer (Muscidae)*)	Emergence (the sex and total number of emerged adult flies) Retardation of emergence, i.e. the number of emerged flies per day Morphological change	EC*x* (95% confidence limits) and/or NOEC and LOEC	Statistics are not specified but examples of tests are: EC*x*: Logit, probit, Weibull, Spearman–Karber, and trimmed Spearman–Karber methods are all appropriate NOEC/LOEC: Appropriate methods according to OECD (2006a). For example, Dunnett's or trend test like Williams' test. For nonparametric tests, Bonferroni-*U*-test according to Holm* or Jonckheere–Terpstra trend test
TG 229 Fish Short-Term Reproduction Assay	Fecundity Vitellogenin Secondary sex characteristics. Histopathology (optional)	NOEC and LOEC	ANOVA. Guidance is given in OECD (2006a). Appropriate methods depend on data. Examples for nonmonotone dose–responses are Dunnett's and Mann–Whitney tests and for monotone dose–responses, Jonckheere–Terpstra test or Williams test
TG 230 21-day Fish Assay A Short-Term Screening for Oestrogenic and Androgenic Activity, and Aromatase Inhibition	Vitellogenin Secondary sex characteristics	NOEC and LOEC	ANOVA. Guidance is given in OECD (2006a). Appropriate methods depend on data. Examples for nonmonotone dose–responses are Dunnett's and Mann–Whitney tests and for monotone dose–responses, Jonckheere–Terpstra test or Williams test
TG 231 231 Amphibian Metamorphosis Assay	Hind limb length Snout to vent length Developmental Stage Wet weight Thyroid histology Mortality	NOEC and LOEC	ANOVA. Guidance is given in OECD (2006a), and a statistical flow chart is included in the TG For continuous data with monotone dose–response, the step-down Jonckheere–Terpstra test should be used. For nonmonotone data there are different approaches dependent on the data, e.g. Dunnett's test when normality and homogeneous variance are achieved. For nonhomogeneous variance Tamhane–Dunnett, T3 test the Mann–Whitney–Wilcoxon *U*-test can be used
TG 232 Collembolan Reproduction Test in Soil	Mortality (parental) Reproductive output (number of juveniles)	LC*x* and EC*x* (95% confidence limits) Or NOEC and LOEC	LC*x* and EC*x*: Regression, e.g. logistic or Weibull function, trimmed Spearman–Karber method, or simple interpolation NOEC/LOEC (reproductive output): ANOVA and, e.g. Student's *t*-test, Dunnett's test, or Williams' test. Alternatively Bonferroni-*U*-test according to Holm* or Jonckheere–Terpstra trend test
TG 233 Sediment-Water Chironomid Life-Cycle Toxicity Test Using Spiked Water or Spiked Sediment	Emergence ratio Development rate	EC*x* (95% confidence limits) and/or NOEC and LOEC	EC*x*: For emergence ratios and development rate, EC*x* values can be calculated using regression analysis (e.g. probit, logit or Weibull models, or simple interpolation) NOEC/LOEC: ANOVA and, e.g. Student's *t*-test, Dunnett's test, or Williams' test. Alternatively Bonferroni-*U*-test according to Holm* or Jonckheere–Terpstra trend test
TG 234 Fish Sexual Development Test	Vitellogenin Sex ratio (proportions of sex) Genetic sex (three-spined stickleback and Japanese medaka) Histopathology (optional) Hatching Mortality Weight, length	NOEC and LOEC EC*x* (e.g. for hatching and mortality)	A detailed statistical flow chart is included as Annex 8 in the TG and describes the statistical methods to be used for vitellogenin and sex ratio data. For these two endpoints hypothesis testing (NOEC/LOEC) is required. Dependent on data monotonicity, normality, and homogeneity of variance, different methods are valid, e.g. Dunn's, Dunnett's, step-down Jonckheere and Williams', Mann–Whitney, and Tamhane–Dunnett's tests

TG 235 Chironomus sp., Acute Immobilization Test	Immobilization	EC50 (95% confidence limits) NOEC and LOEC	EC50 should be analyzed by, e.g. probit analysis. NOEC could be determined by, for example, Fisher's exact test. More details are given in OECD (2006a)
TG 236 Fish Embryo Acute Toxicity (FET) Test	Mortality based on 4 parameters: 1: coagulation of fertilized eggs; 2,3,4: lack of somite formation, detachment of the tail-bud from the yolk sac, and heartbeat, respectively	LC50 (95% confidence limits)	No specific statistical guidance is given in the TG. Relevant methods should be obtained from (OECD, 2006a) and (ISO/TS 20281-2006, 2006). Examples of such relevant methods could be probit analysis or logistic regression
TG 237 Honey Bee (*Apis Mellifera*) Larval Toxicity Test, Single Exposure	Mortality (% of the initial population)	LD50 (95% confidence limits)	Appropriate methods include probit analysis, moving average, and binomial probability as well as those recommended in TG 213 and 214, e.g. Litchfield–Wilcoxon, trimmed Spearman–Karber and log-probability graph paper plot
TG 238 Sediment-Free *Myriophyllum Spicatum* Toxicity Test	Inhibition of growth (shoot length, fresh weight, wet weight, and yield)	E_rCx (x% inhibition of growth rate) E_yCx (x% inhibition of yield is required by some regulations) NOEC and LOEC may be included	E_rCx and E_yCx: Nonlinear regression analysis is preferred. Guidance can be found in the literature (Bruce and Versteeg, 1992; Nyholm et al., 1992; OECD, 2006a) NOEC/LOEC: ANOVA. Different tests for multiple comparisons or trend tests are available, e.g. Dunnett's, Williams', Levene's test, or step-down Jonckheere test. Additional guidance can be found in OECD (2006a)
TG 239 Water-Sediment *Myriophyllum Spicatum* Toxicity Test	Inhibition of growth (shoot length and yield)	E_rCx (x% inhibition of growth rate) E_yCx (x% inhibition of yield is required by some regulations) NOEC and LOEC may be included	E_rCx and E_yCx: Nonlinear regression analysis is preferred. Guidance can be found in the literature (Bruce and Versteeg, 1992; Nyholm et al., 1992; OECD, 2006a) NOEC/LOEC: ANOVA. Different tests for multiple comparisons or trend tests are available, e.g. Dunnett's, Williams', Levene's test, or step-down Jonckheere test. Additional guidance can be found in OECD (2006a)
TG 240 Medaka Extended One Generation Reproduction Test (MEOGRT)	Mortality Growth Reproduction (fecundity) Vitellogenin Phenotypic secondary sex characteristics Genetic sex Sex ratio (proportions of sex) Histopathology	NOEC and LOEC EC*x* (rarely suitable because it require more test concentrations)	Guidance to statistical methods can be found in Annex 10 of the TG. Further advice can be obtained in OECD (2006a). Dependent on data monotonicity, normality, and homogeneity of variance, different methods are valid, e.g. Dunn's, Dunnett's, step-down Jonckheere and Williams' tests For histopathology a specific method has been developed where a severity score is included Rao–Scott Cochran–Armitage by Slices (RSCABS). More details can be found in (Green et al., 2014)

(*Continued*)

Table 11.4 (Continued)

OECD test guideline (TG)	Main endpoints	Statistical output	Statistical methods
TG 241 The Larval Amphibian Growth and Development Assay (LAGDA)	Mortality and abnormalities Time to NF stage 62 Thyroid gland Histo(patho)logy Weight and length Liver-somatic index (LSI) Genetic/phenotypic sex ratios Histopathology (gonads, reproductive ducts, kidney and liver) Vitellogenin (optional)	NOEC and LOEC EC*x* (95% confidence limits)	Guidance to statistical methods can be found in Annex 7 of the TG. Generally, advices in OECD (2006a) should be followed. The TG generates three forms of data: Continuous data like weight, vitellogenin and LSI, time-to-event data like days to NF stage 62 and ordinal data for histopathological derived severity score of developmental stages where the Rao–Scott Cochran–Armitage by Slices method (RSCABS) has been developed (Green et al., 2014).
TG 242 *Potamopyrgus antipodarum* Reproduction Test	Reproduction (embryo numbers in brood pouch)	EC*x* (95% confidence limits) or NOEC and LOEC	EC*x*: Appropriate regression models for count data, e.g. the logistic model with Poisson error structure NOEC/LOEC: Advices given in OECD (2006a) should be followed. Dependent on data monotonicity, normality, and homogeneity of variance, different methods are valid, e.g. Dunn's, Dunnett's, step-down Jonckheere and Williams' tests
TG 243 *Lymnaea stagnalis* Reproduction Test	Reproduction (cumulated number of egg clutches produced per snail)	EC*x* (95% confidence limits) or NOEC and LOEC	EC*x*: Appropriate regression models as described in OECD (2006a). The internet platform MOSAIC_repro was developed during the TG validation and is recommended for regression analysis. It is available at: http://pbil.univ-lyon1.fr/software/mosaic/reproduction/ NOEC/LOEC: Advices given in OECD (2006a) should be followed. Dependent on data monotonicity, normality, and homogeneity of variance, different methods are valid, e.g. Dunn's, Dunnett's, step-down Jonckheere and Williams' tests

* Several TGs call for a Bonferroni *U*-test according to Holm. What is intended is the Mann-Whitney *U*-test with Bonferroni-Holm adjusted *p*-values.

data in Section D: Experimental design and data analysis (Table 11.7).

All the OCSPP guidelines including more than 20 draft guidelines are freely available at https://www.epa.gov/sites/production/files/2016-12/documents/ocspp-testguidelines_masterlist-2016-12-28.pdf (OCSPP, 2016).

EU: In the EU, industrial chemicals are regulated under the REACH, which entered into force in 2007 (EU, 2006). REACH places the burden of proof on the registrant(s) of the chemicals (manufacturers and importers). The registrants must identify and manage the risks of the chemical for safe use. The standard information requirements in REACH are driven by the tonnage of the chemical. See Table 11.8 for the test requirement driven by tonnage and will be discussed further under testing strategies. As seen in Table 11.8, the guidelines required are mainly OECD TGs. Pesticides are regulated in the Plant Protection Product Regulation (PPPR) and biocides in the Biocidal Product Regulation (BPR).

Table 11.5 Grouping of the OCSPP Guidelines

810	Product Performance Test Guidelines
830	Product Properties Test Guidelines
835	Fate, Transport and Transformation Test Guidelines
840	Spray Drift Test Guidelines
850	Ecological Effects Test Guideline
860	Residue Chemistry Test Guidelines
870	Health Effects Test Guidelines
875	Occupational and Residential Exposure Test Guidelines
880	Biochemicals Test Guidelines
885	Microbial Pesticide Test Guidelines
890	Endocrine Disruptor Screening Program Test Guidelines

Japan: Chemical regulation and testing in Japan are governed by multiple ministries: the Ministry of the Environment; the Ministry of Health, Labor, and Welfare; the Ministry of Agriculture, Forestry and Fisheries; and the Ministry of Economy, Trade, and Industry (MITI). Also in Japan, different laws regulate different types of chemicals. The CSCL turned into force in 1973 and regulates industrial chemicals. CSCL was last updated in 2011. The Agricultural Chemicals Regulation Law turned into force in 1948 and regulated pesticides and biocides and was updated in 2007. Other laws in Japan regulate food additives, cosmetics, pharmaceuticals, etc. Japan also uses the OECD TGs to retrieve information of the hazards of chemicals. As an example, the Japanese Environment Agency, the Ministry of Health, Labor and Welfare, and MITI use OECD TGs as the minimum data requirement for initial screening of chemicals if these are persistent according to a biodegradation test (MITI, 2012). The ecotoxicity tests are algal growth inhibition test (OECD TG 201(OECD, 2011)), daphnia acute toxicity test (OECD TG 202 Part I (OECD, 2004)), daphnia reproduction test (OECD TG 202 Part II(OECD, 2004)), fish acute toxicity test (OECD TG 203 (OECD, 1992)), and earthworm acute toxicity test (OECD TG 207(OECD, 1984a)).

11.8 TESTING STRATEGIES

Strategies for the regulatory testing of chemicals can be either tiered – driven by toxic effect concentrations and predicted environmental concentrations (PEC), tiered – driven by tonnage or including specific requirements for certain types of chemicals, e.g. active substances in pesticides and biocides. The OECD has published guidance for

Table 11.6 OCSPP Group 850 Group A Guidelines: Aquatic and Sediment-dwelling Fauna and Aquatic Microcosms

850.1000	Background and Special Considerations-Tests with Aquatic and Sediment-Dwelling Fauna and Aquatic Microcosms (December 2016)
850.1010	Aquatic Invertebrate Acute Toxicity Test, Freshwater Daphnids (December 2016)
850.1020	Gammarid Amphipod Acute Toxicity Test (December 2016)
850.1025	Oyster Acute Toxicity Test (Shell Deposition) (December 2016)
850.1035	Mysid Acute Toxicity Test (December 2016)
850.1045	Penaeid Acute Toxicity Test (December 2016)
850.1055	Bivalve Acute Toxicity Test (Embryo-Larval) (December 2016)
850.1075	Freshwater and Saltwater Fish Acute Toxicity Test (December 2016)
850.1300	Daphnid Chronic Toxicity Test (December 2016)
850.1400	Fish Early Life Stage Toxicity Test (December 2016)
850.1710	Oyster Bioconcentration Factor (BCF) (December 2016)
850.1730	Fish Bioconcentration Factor (BCF) (December 2016)
850.1735	Spiked Whole Sediment 10-Day Toxicity Test, Freshwater Invertebrates (December 2016)
850.1740	Spiked Whole Sediment 10-Day Toxicity Test, Saltwater Invertebrates (December 2016)

Modified from USEPA (2016c).

Table 11.7 Statistical Recommendations and Requirements in the OCSPP Group 850 Group A Guidelines

OCSPP	Main endpoints	Statistical output	Statistical methods
850.1000 Background and Special Considerations –Tests with Aquatic and Sediment-Dwelling Fauna and Aquatic Microcosms	—	—	Describing experimental design and data analysis in general for all OCSPP 850 Group A guidelines
850.1010 Aquatic Invertebrate Acute Toxicity Test, Freshwater Daphnids	Immobilization as a surrogate for death	48 h EC50 (standard error and 95% confidence interval)	Quantal data: Regression-based methods (e.g. probit) that model the full concentration–response relationship and provide error estimates of the model parameters and point estimate(s) are desired. If data do not fit a regression-based model, other point estimator methods (e.g. binomial, moving average, trimmed Spearman–Karber, linear interpolation (e.g. Bootstrap ICp)) are available (OCSPP 850.1000 (USEPA, 2016b))
850.1020 Gammarid Amphipod Acute Toxicity Test	Mortality	96-h LC50 (standard error and 95% confidence interval)	Quantal data: Regression-based methods (e.g. probit) that model the full concentration–response relationship and provide error estimates of the model parameters and point estimate(s) are desired. If data do not fit a regression-based model, other point estimator methods (e.g. binomial, moving average, trimmed Spearman–Karber, linear interpolation (e.g. Bootstrap ICp)) are available (OCSPP 850.1000 (USEPA, 2016b))
850.1025 Oyster Acute Toxicity Test (Shell Deposition)	Shell deposition	96-h IC50 (Inhibitory concentration, median) (standard error and 95% confidence interval)	Continuous data: Regression-based methods (e.g. probit) that model the full concentration–response relationship and provide error estimates of the model parameters and point estimate(s) are desired. If data do not fit a regression-based model, other point estimator methods (e.g. binomial, moving average, trimmed Spearman–Karber, linear interpolation (e.g. Bootstrap ICp)) are available (OCSPP 850.1000 (USEPA, 2016b))
850.1035 Mysid Acute Toxicity Test	Mortality	96-h LC50 (standard error and 95% confidence interval)	Quantal data: Regression-based methods (e.g. probit) that model the full concentration–response relationship and provide error estimates of the model parameters and point estimate(s) are desired. If data do not fit a regression-based model, other point estimator methods (e.g. binomial, moving average, trimmed Spearman–Karber, linear interpolation (e.g. Bootstrap ICp)) are available (OCSPP 850.1000 (USEPA, 2016b))
850.1045 Penaeid Acute Toxicity Test	Mortality	96-h LC50 (standard error and 95% confidence interval)	Quantal data: Regression-based methods (e.g. probit) that model the full concentration–response relationship and provide error estimates of the model parameters and point estimate(s) are desired. If data do not fit a regression-based model, other point estimator methods (e.g. binomial, moving average, trimmed Spearman–Karber, linear interpolation (e.g. Bootstrap ICp)) are available (OCSPP 850.1000 (USEPA, 2016b))
850.1055 Bivalve Acute Toxicity Test (Embryo-Larval)	Failure to develop into normal larvae	48-h EC50 (standard error and 95% confidence interval)	Quantal data: Regression-based methods (e.g. probit) that model the full concentration–response relationship and provide error estimates of the model parameters and point estimate(s) are desired. If data do not fit a regression-based model, other point estimator methods (e.g. binomial, moving average, trimmed Spearman–Karber, linear interpolation (e.g. Bootstrap ICp)) are available (OCSPP 850.1000 (USEPA, 2016b))
850.1075 Freshwater and Saltwater Fish Acute Toxicity Test (comparable to OECD TG 203 Fish Acute Toxicity Test)	Mortality	96-h LC50 (standard error and 95% confidence interval)	Quantal data: Regression-based methods (e.g. probit) that model the full concentration–response relationship and provide error estimates of the model parameters and point estimate(s) are desired. If data do not fit a regression-based model, other point estimator methods (e.g. binomial, moving average, trimmed Spearman–Karber, linear interpolation (e.g. Bootstrap ICp)) are available (OCSPP 850.1000 (USEPA, 2016b))

850.1300 Daphnid Chronic Toxicity Test (comparable to OECD TG 211 Daphnia magna Reproduction Test)	Survival (immobilization as a surrogate for death), growth, and reproduction	NOEC, LOEC (Concentration–response regression-based models may also be fit, and point estimates (e.g., IC*x* and effect concentration (EC*x*)) may be calculated)	Quantal, discrete, and continuous data: Parametric and nonparametric analysis of variance (ANOVA) tests and multiple-comparison tests are often appropriate for continuous data and for count data Involvement of a statistician in both the design and analysis of the test results is recommended since OCSPP 850.1300 allows for variation in experimental design (e.g. number of test animals per replicate, number of replicates) (OCSPP 850.1000 (USEPA, 2016b))
850.1400 Fish Early Life Stage Toxicity Test (comparable to OECD TG 210 Fish Early Life Stage Toxicity Test)	Time to hatch, time to swim-up, hatching success, post-hatch success, overall survival, and early juvenile growth	NOEC, LOEC (Concentration–response regression-based models may also be fit, and point estimates (e.g. IC*x* and effect concentration (EC*x*)) may be calculated)	Quantal, discrete, and continuous data: Parametric and nonparametric analysis of variance (ANOVA) tests and multiple-comparison tests are often appropriate for continuous data and for count data All methods used for statistical analysis should be described completely Experimental units (replicates) are the individual test Vessels within each treatment level Additional discussion about endpoints and statistical procedures can be found in OCSPP 850.1000 (USEPA, 2016b)
850.1710 Oyster Bioconcentration Factor (BCF)	Steady-state bioconcentration factor (BCF_{SS}) and kinetic bioconcentration factor (BCF_K) for the test substance (and its major metabolites)	BCF_{SS} and BCF_k (95% confidence interval) The BCF is expressed in units of test solution volume per mass of the mollusk (or tissue thereof), defined here as liters of test solution per kilogram of tissue ($1 kg^{-1}$)	BCF_{SS}: The uptake curve of the test substance should be obtained by plotting its concentration in oyster in the uptake phase against time on arithmetic scales. If the curve has reached a plateau, that is, becomes approximately asymptotic to the time axis, calculate the BCFSS using Equation 13 of the guideline If steady state is not reached during 28 days of uptake, the BCF_{SS} should be calculated using nonlinear parameter estimation methods The BCF_K should be calculated as the ratio of the uptake rate constant (*k*1) to the depuration rate constant (*k*2) assuming first-order kinetics
850.1730 Fish Bioconcentration Factor (BCF) (comparable to OECD TG 305 Bioaccumulation in Fish)	Steady-state bioconcentration factor (BCF_{SS}) and kinetic bioconcentration factor (BCF_K) for the test substance (and its major metabolites)	BCF_{SS} and BCF_k (95% confidence interval) The BCF is expressed in units of test solution volume per mass of fish (or tissue thereof), defined here as liters of test solution per kilogram of fish ($1 kg^{-1}$)	BCF_{SS}: The uptake curve of the test substance is calculated by plotting its concentration in fish (or specified tissues) in the uptake phase against time on arithmetic scales. If the curve has reached a plateau, that is, becomes approximately asymptotic to the time axis, the BCF_{SS} should be calculated using Equation 26 of the guideline If steady state was not reached during the maximum 60-day uptake period, the maximum BCF_{SS} should be calculated using the mean tissue concentration from that and all the previous sampling days The BCF_K is calculated as the ratio of the uptake rate constant (*k*1) to the depuration rate constant (*k*2) assuming first-order kinetics (Equation 27 of the guideline)
850.1735 Spiked Whole Sediment 10-Day Toxicity Test, Freshwater Invertebrates	Survival and growth	10-d LC50, 10-day median growth EC (10-d EC50), or the 10-day NOEC) and 10-day LOEC) for survival and growth. For pesticides, the NOEC and LOEC are the preferred and for industrial chemicals, the LC50 and/or EC50 are the preferred endpoints	Statistical procedures are employed to calculate the 10-d LC50 (SE and 95% CI) based upon mortality and 10-d EC50 (SE and 95% CI) for growth (weight and/or length) when appropriate to the test design. The slope of the concentration–response curve, its standard error, and 95% CI should also be reported NOEC and LOEC values for survival and growth (weight, length) when appropriate to the test design. Hypothesis testing procedures described in OCSPP 850.1000 (USEPA, 2016b) can be used to determine NOEC and LOEC values
850.1740 Spiked Whole Sediment 10-Day Toxicity Test, Saltwater Invertebrates	Survival, reburial (optional for *E. estuarius, L. plumulosus*, and *R. abronius*), and growth (optional for *L. plumulosus*)	10-d LC50, 10-day median growth EC (10-d EC50), or the 10-day NOEC) and 10-day LOEC) for survival and growth. For pesticides, the NOEC and LOEC are the preferred and for industrial chemicals, the LC50 and/or EC50 are the preferred endpoints	Statistical procedures for modeling quantal data should be used for calculating the LC50. Statistical procedures for modeling continuous toxicity data should be used for calculating the EC50 Analysis of variance testing procedures are used to determine NOEC and LOEC values

Table 11.8 REACH Requirements for Ecotoxicological Information on Vertebrate and Invertebrate Animals (Abiotic Tests and Tests on Microorganisms and Plants Not Shown)

Examples of tests that cover the standard information demand	Reach				PPPR	BPR
	≥1 t	≥10 t	≥100 t	≥1000 t		
Acute Toxicity to Daphnia (TG 202)	X				X	X
Acute Fish Toxicity (TG203)		X			X	X
Acute Toxicity for Aquatic Gastropod, Insect, and Nondaphnian Crustacean (e.g. TG 235)					X	
Long-term or chronic toxicity in invertebrates (daphnia) (TG211)			X		X	X
Chronic Toxicity for Aquatic Gastropods and Insects					X	
Long-term or Chronic Toxicity in Fish			X		X	X
Fish Early Life Stage Toxicity Test (TG210)			X		X	X
Fish, Embryo, and Sac fry (TG212)[a]			X			X
Fish Short-time Reproduction Assay (TG 229/OPPTS 890.1350)			X		X	
21-day Fish Assay (TG 230)			X		X	
Amphibian Metamorphosis Assay (TG 231)			X		X	
Fish Sexual Development Test (TG234)			X		X	
Fish Juvenile Growth Test (TG215)[a]			X			X
Chronic Toxicity in Juvenile Fish					X	
Life Cycle Test with Fish (US EPA OPPTS 850.1500 (Benoit, 1982)					X	
Bioaccumulation in Fish (TG305)			X		X	X
Acute Earthworm Toxicity (TG207)			X		X	X
Long-Term or Sublethal Toxicity to Invertebrates (TG211 or TG222)				X	X	
Acute Toxicity to Bees (TG213/214)					X	X
Feeding Test for Bee Larvae					X	
Effects on Two Non-target Arthropods (besides the bees) (TG226/232)					X	
Long-term Toxicity to Sediment Organisms (TG218/219)				X	X	X
Acute Oral Bird Toxicity, 2 species (TG223)					X	X
5 or 8 days Oral Bird Toxicity (feeding) (TG205)					X	X
Long-term/subchronic or reproductive toxicity to birds (TG206)				X	X	X

Modified from DK-EPA (2013).

The OECD/US EPA validated test methods that may provide the information are indicated in parenthesis.

[a] Currently there appears to be limited demand for OECD TG 212, and TG 215 and they might be phased out or implemented in other guidelines (OECD, 2012b).

the strategic use of the OECD TGs in different guidance documents – OECD guidance document No. 150 on standardized TGs for evaluating chemicals for endocrine disruption (OECD, 2012a) and the OECD Fish Toxicity Framework (OECD, 2014c) as well as the OECD conceptual Framework for Testing and Assessment of Endocrine Disrupters (OECD, 2012a, Annex 1.4).

Although different testing strategies are developed among different regulatory bodies, a four-step approach as outlined in Figure 11.1 is the strategic backbone for most regulators.

Step 1: All existing information is collected

Step 2: Information needs are considered

Step 3: Gaps in the existing information are identified

Step 4: A testing strategy is proposed to generate missing data

Figure 11.1 A generalized stepwise approach to assess the safety of a chemical.

In the United States, a tiered testing strategy is the norm and described in the updated TSCA (USEPA, 2016a). Here it is stated that when new information is required results from screening tests and/or available information shall be used to decide if further testing is necessary (unless information available justifies more advanced testing of without first doing screening tests). Another example from the United States is the Endocrine Disruptor Screening Program (EDSP) that was outlined in two Federal Register Notices published in 1998 (USEPA, 1998a, b). The EDSP testing is divided into two tiers after a stage of priority setting: Tier 1 includes screening tests and Tier 2 includes confirmatory testing. Tier 2 is only initiated for a chemical that has responded positive in one or more Tier 1 screening tests. Testing requirements of pesticides under the FIFRA (USEPA, 1996) are described in the Code of Federal Regulations at 40 CFR Part 158 (USGPO, 2017), but there is flexibility in the requirements and dependent on the individual pesticide, additional data can be required, alternative approaches can be accepted, and data can be waived because data should only be required if they affect the decision making. This is consistent with the pesticide program's strategy of using integrated approaches to testing and assessment (IATA), which promotes "a hypothesis-based, systematic, integrative use of exposure and hazard information for assessing pesticide risk" (Strategic Vision for Adopting Twenty-first Century Science Methodologies, USEPA).

In the EU, different testing strategies are applied to different regulations. As previously mentioned, in REACH, the testing strategy is driven by the tonnage of the chemical, i.e. the early production and import. Other regulations have specific requirement to testing dependent on the nature and use of the chemical. This is the case for active substances in the EU PPPR, No 283/2013(EU, 2013) and active substances in the EU BPR No 528/2012(EU, 2012). Table 11.4 summarizes the standard information requirement according to REACH, PPPR, and BPR in the EU.

11.9 NONGUIDELINE STUDIES

It is generally agreed that all relevant available information on the effects of a chemical should be taken into account during a regulatory decision process. Very often relevant information is produced during scientific research via nonguideline studies and published in the peer-reviewed scientific literature. For some effects like endocrine disrupting effects getting attention only within the recent years, nonguideline studies could generate the majority of available information. Because nonguideline studies are not standardized and often not conducted according to GLP, the quality of these studies is validated on an individual basis by use of different – more or less recognized evaluation methods. The most widely accepted method for quality evaluation is the Klimisch score – "A Systematic Approach for Evaluating the Quality of Experimental Toxicological and Ecotoxicological Data" developed by Klimisch et al. (1997). Definitions are proposed for reliability, relevance, and adequacy of data. The outcome is data reliability differentiated into four categories/scores: (i) reliable without restriction, (ii) reliable with restriction, (iii) not reliable, and (iv) not assignable. Score 1 is generally assigned to data from validated guidelines performed according to the principles of GLP, and therefore nonguideline studies are rarely assigned Klimisch score 1. Most relevant data from the peer-reviewed scientific literature fall into Klimisch score 2 because they are not conducted according to GLP or a standardized guideline, but the test parameters are sufficiently described and documented. Score 3 and 4 studies are seldom used in regulatory decisions due to lack of documentation or nonrelevant exposure scenarios.

Recently other quality evaluation tools have been proposed because the Klimisch score has been claimed to lack space for details and focus too much on GLP and standardized methods. The software-based ToxRTool (Schneider et al., 2009) was developed in an EU project and added extra criteria and guidance to assigning Klimisch scores 1, 2, and 3. The tool consists of an *in vivo* and an *in vitro* part. Developed for toxicity testing, some modifications are needed when used for ecotoxicity study evaluation. Another tool called criteria for reporting and evaluating ecotoxicity data (CRED) specifically addresses aquatic ecotoxicity studies (Moermond et al., 2016). It defines 20 reliability and 13 relevance criteria as well as guidance, and it recommends 50 specific criteria divided into 6 categories to be reported. The relevance and reliability criteria are developed into an online tool that can be accessed and used for free at http://www.scirap.org/.

Chapter 12

Species Sensitivity Distributions

12.1 INTRODUCTION

Most of this book is devoted to statistical tests and models by which to explore the effects of a chemical on a single biological response on one species at one life stage. The primary task of risk assessors is to assess the potential for harm to the environment as a whole. One approach to this overall assessment comes from assembling the species data in a way to reflect the broader ecosystem. There are several parts to this assembly.

First, a single value is required for each species and study. This could the lowest NOEC or EC50 from all the responses taken on that species from a single study. Alternatively, there could be judgment of which response is most important, such as survival or reproduction. EFSA (2013) states: "The endpoints measured in the toxicity tests on which the SSD is based must be the most sensitive endpoints that are toxicologically and ecologically relevant. Acute toxicity data mostly address mortality and immobility as the most frequently studied endpoints for animals, while biomass and growth rate are mostly used for primary producers. Chronic toxicity data mostly address reproduction, feeding rate and growth as the most frequently studied endpoints in animals, and again biomass and growth rate are mostly addressed by chronic toxicity data for primary producers." ECx for some value of x other than 50 could also be used. However, there is generally no mixing: Use all NOECs, all EC50s, or all ECxs for the same x. However, guidance on this is in flux and EFSA (2013) indicates NOEC and EC10 values can be used equivalently. How much it matters which endpoint is used (NOEC or EC10) is of course of interest, and that question has been explored by Iwasaki et al. (2015) who found that "the choice of ECx (e.g., EC5, EC10, or EC20) or NOEC does not largely affect the resulting HC5. Therefore, use of NOECs would be acceptable particularly in regulatory contexts, although the NOEC has important shortcomings and should be used with caution."

Secondly, in situations where more than one study has been done on the same species under comparable conditions, the issue is how these study results are summarized by a single number. Where EC50 values on the same species are available from multiple studies, the generally accepted practice is to use the geometric mean of the EC50 values for each species in fitting the species sensitivity distribution (SSD). This practice is recommended in Duboudin et al. (2004), ECOFRAM (1999), EFSA (2005), Merrington and Schoeters (2010), Stephan et al. (1985), Wheeler et al. (2002), and Zajdlik et al. (2005), among others.

Other issues include what ecosystem to model, what distribution to fit to these species values, what to do with censored values (which are quite common). These are topics explored in this chapter.

The SSD was introduced by Kooijman (1987), although similar concepts had appeared earlier (e.g. USEPA, 1985). Kooijman proposed a way to quantify the risk to all organisms in an ecosystem and investigated the log-logistic distribution for this purpose from which he developed the hazard concentration for sensitive species. Subsequent authors have refined this concept, leading to what is now labeled HC5, the concentration hazardous to no more than 5% of species. Refinements include restrictions on the number and type of organisms to model and other distributions. It is now widely accepted that it is not sensible to model the entire ecosystem with one distribution. Instead, separate SSDs are developed for freshwater macrophytes, algae, fish, and invertebrates. The primary

Statistical Analysis of Ecotoxicity Studies, First Edition. John W. Green, Timothy A. Springer, and Henrik Holbech.

Companion website: www.wiley.com/go/Green/StatAnalysEcotoxicStudy

distribution used is the lognormal (e.g. Posthuma et al., 2002; Wheeler et al., 2002), but several other distributions have been used, and software to fit SSD and estimate HC5 from various distributions are available (e.g. SSD Master from Intrinsik used by Environment Canada, ETX 2.2 (Aldenberg et al. 2017) developed by RIVM, BurrliOZ (CSIRO, 2010) developed for the Australian Environmental Protection Agency and, in its first version, based on methodology developed by Shao, 2000), and MOSAIC_SSD (Kon Kam King et al., 2014), which gives a partial Bayesian implementation.

Distributions that have been used for SSDs include lognormal, log-logistic, exponential, Fisher–Tippett, Weibull, reciprocal Pareto, reciprocal Weibull, and Burr III. EUFRAM (Chapman et al., 2007) was a major effort by the regulatory, industry, and academic interests to provide guidance on SSD construction and interpretation, and the cited report was drawn on in numerous places in this chapter. Among the recommendations from EUFRAM was the need to take uncertainty into account in addition to variability, so that the SSD derived may not meet all of the requirements of a probability distribution, but still provides an appropriate HC5. Also discussed were the relative merits of HC5 point estimates compared with confidence bounds and the potential effect of mis-specifying the distribution to be fit. These two topics are explored in Section 12.5.

12.2 NUMBER, CHOICE, AND TYPE OF SPECIES ENDPOINTS TO INCLUDE

EFSA (2013) provides guidance on what species and endpoints to use for SSDs for plant protection products (PPPs). "The toxic mode of action of a PPP should be taken into account when constructing SSDs to derive a RAC. SSDs used in RA should always be constructed using toxicity data for the most sensitive taxonomic group (i.e. fish, invertebrates or primary producers). In the case of herbicides, vascular plants and/or algae usually constitute the most sensitive groups. For insecticides, arthropods (crustaceans and insects) usually are most sensitive. For fungicides, often a range of taxonomic groups are among the sensitive organisms." The general guidance is to use a minimum of eight species from different taxonomic groups. This general statement and the quoted statement capture only an indication, and the cited document should be consulted for extensive additional guidance. The remainder of this chapter will focus on statistical issues after the species have been selected.

Several authors have commented on biases inherent in the way SSDs are traditionally obtained from laboratory-cultured species (e.g. Newman et al., 2000; Dowse et al., 2013 and several of its references). It is not possible to defend the species used in regulatory filings as representative of all species, since, for example, large terrestrial species are never included and insects are vastly underrepresented. What is claimed is that the species used are representative of the more sensitive ones exposed in typical agricultural applications of crop protection chemicals. We will not explore this claim here, but think it is appropriate to draw attention to the issue.

12.3 CHOICE AND EVALUATION OF DISTRIBUTION TO FIT

Theoretical distributions allow for simple estimation of statistical parameters. Kooijman (1987) and Aldenberg and Slob (1993) suggested the lognormal and log-logistic distributions, respectively, as default distributions for SSDs. Aldenberg et al. (2002) argue for the lognormal distribution based, in part, on a high percentage of nonsignificant lack-of-fit tests for normality of numerous log-transformed datasets. From personal conversations, it is clear that Aldenberg recognizes that this is not a strong basis in itself for using the lognormal as the default approach. An appendix to Chapman et al. (2007), written by the first named author of this book, demonstrates the low power of tests of normality on small samples and how significant the impact on HC5 estimates can be from estimating HC5 from an assumed lognormal distribution when that is not the right model. Furthermore, this appendix shows data from product registration packages that do not follow a lognormal distribution. Examples are given of lognormal and nonlognormal data from nontarget plant studies. (See Figures 12.1 and 12.2.) Indeed, no single distribution fits all the data from these registration packages. Distribution-free methods have been

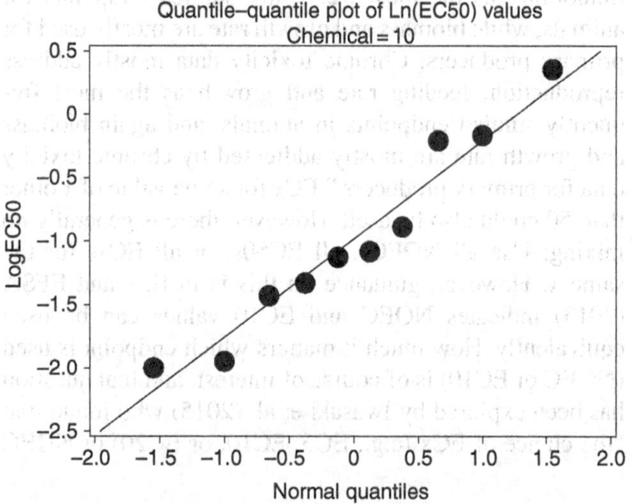

Figure 12.1 QQ-plot consistent with lognormality. Observations (dots) lie close to the line, indicating agreement with a normal distribution.

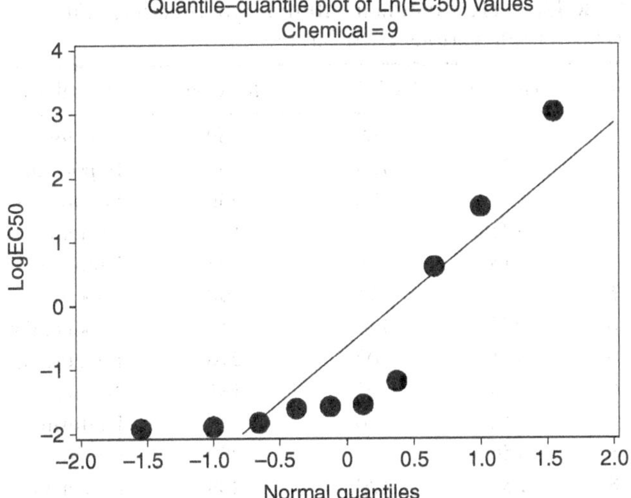

Figure 12.2 QQ-plot not consistent with lognormality. Observations (dots) deviate substantially from the line, indicating inconsistency with a normal distribution. Data is presented in Exercise 12.1.

described by Jagoe and Newman (1997), Newman et al. (2000), and Van der Hoeven (2001). However, these methods require larger numbers of species than are usually available, so parametric models are generally the only viable approach. This makes the need for selecting good models on a case-by-case basis all the more important.

12.3.1 Goodness of Fit

Goodness of fit can be assessed in various ways. The most straightforward formal method is to transform the data to normality and then apply a standard test of normality (e.g. Shapiro–Wilk or Anderson–Darling). This can be done by noting that if $F(x)$ is the assumed continuous distribution, then the distribution of the values $F(x_i)$, $i = 1$ to n (the number of species), is uniform(0,1) and $\Phi^{-1}(F(x_i))$ follows a normal distribution to which a test for normality can be applied (D'Agostino and Stephens, 1986, chapter 6). Here Φ is the standard normal distribution function.

It should be noted (Chapman et al., 2007) that the transformation approach requires estimating the parameters of the assumed distribution, usually by maximum likelihood methods, to obtain the presumed uniform $U(0,1)$ sample. Since a standard normal inverse function is applied to the $U(0,1)$ sample, there is no need to estimate the mean and standard deviation of the sample from the normal transform. As noted in D'Agostino and Stephens (1986), chapter 9, the mean and standard deviation of this normal have little bearing on the power of these tests, so the Shapiro–Wilk or Anderson–Darling test can be applied.

Table 12.1 Power of Tests to Detect Nonnormality in Samples from a Cauchy Distribution

Test				
SW	AD	KS	CM	Size
22	21	23	19	4
29	29	29	28	5
37	38	37	37	6
49	50	48	50	8
59	61	58	61	10
76	78	74	78	15
87	87	84	87	20
96	96	94	96	30

SW, power of Shapiro–Wilk test; KS, power of Kolmogorov test; AD, power of Anderson–Darling test; CM, power of Cramér–von Mises test.

Table 12.2 Power of Tests to Detect Nonnormality in Samples from an Exponential Distribution

Test				
SW	KS	AD	CM	Size
12	11	12	9	4
17	16	14	14	5
22	21	18	20	6
34	32	24	29	8
44	41	30	38	10
68	63	45	58	15
83	78	59	73	20
97	94	78	90	30

SW, power of Shapiro–Wilk test; KS, power of Kolmogorov test; AD, power of Anderson–Darling test; CM, power of Cramér–von Mises test.

As mentioned, for small datasets, tests for normality are not powerful, so it is quite possible to start with two distinct candidate distributions, F_1 and F_2, go through the above process, and find that the two transformed datasets are both consistent with normality (i.e. do not show statistically significant lack of normality). It is not possible for multiple distributions to be correct, so what this illustrates is the low power to detect departures from a presumed distribution. No claim is made that we will find "the correct" distribution for a given ecosystem. Instead, what should be attempted is to find a distribution that fits the data well enough to be useful. This will usually come down to visual inspection of the fit. Given the regulatory intent, particular attention should be given to the fit based on small samples. The power to detect nonnormality in small samples is low (see Tables 12.1 and 12.2), which underscores the caution in accepting lognormality for SSDs without careful consideration. The impact on HC5 estimation of fitting the wrong distribution can be pronounced (see Tables 12.3 and 12.4), further giving reason for care in deciding which distribution to fit. Additional related results are given in Chapman et al. (2007), based on smaller simulations.

Each power in Tables 12.1 and 12.2 is based on 5000 randomly generated samples from the indicated distribution. To be clear, when we are exploring the ability to detect nonnormality, we are investigating the data after any transformation was made in an effort to make the data normal. Thus, if we adopt the usual regulatory approach of assuming lognormality, then the data being assessed is that after a log-transform has been applied. Furthermore, in exploring the effect on HC5 or other model property, the alternative distribution simulated likewise refers to the data after a transform. For example, if the true log-transformed data is Cauchy and we fit a normal SSD to these data, we are assessing how likely we are to find significant nonnormality from a formal test of the log-transformed and the likely impact on HC5.

There is no expectation that a SSD would follow a Cauchy distribution. The purpose of Table 12.1 is to indicate the sensitivity (or lack thereof) of tests for normality in the presence of data from a heavy-tailed distribution. For SSDs based on eight species values, it will be observed that there is no more than 50% power to detect nonnormality in a dataset following the heavy-tailed Cauchy distribution. From Table 12.2, it can be seen that the power to detect nonnormality in a sample from an exponential distribution is notably worse. The more relevant question is whether it matters. That is, is there a notable impact on HC5 estimates that comes from mis-specifying the distribution? Tables 12.3 and 12.4 address this question.

The software package ETX from RIVM includes a formal test of lognormality, which is the only distribution it offers. SSD Master from Intrinsik, used by Environment Canada, offers formal goodness of fit for lognormal, log-logistic, Gompertz, Weibull, and Fisher–Tippett distributions. The BurrliOZ software does not offer goodness-of-fit tests but does do an undocumented assessment among the Burr III, reciprocal Weibull, reciprocal Pareto, lognormal, and log-logistic distributions and reports the HC5 estimate only for the distribution it deems best. The bias in version 1 of the BurrliOZ software is for the Burr III, and it reports one of the others only if its internal assessment is that the Burr III does not fit well. The software documentation and reference by Shao (2000) cited in the version 1 BurrliOZ software recognize technical conditions under which the Burr distribution cannot be fit. In particular, as Shao (2000) point out, when the parameter k is large and the parameter b is small, the Burr III model is unstable and is better replaced by the reciprocal Weibull. For very large values of the parameter c, again the Burr III model is unstable and is better replaced by the reciprocal Pareto. Version 2 of BurrliOZ offers only lognormal and log-logistic distributions.

The only novelty associated with Figure 12.1 is that it is applied to species EC50 values used to fit an SSD to contrast it with Figure 12.2, also of species EC50 values. These two figures demonstrate that the lognormal is not always the right distribution for SSD purposes. However, it should be noted that these 2 figures came from a small database of 12 nontarget plant studies and 11 of that datasets were consistent with a lognormal distribution. Before arguing that one should not be concerned with occasional deviation from the expected lognormal, it is important to understand the impact of HC5 estimation from a mis-specified distribution. That is discussed in Section 12.3.2.

Table 12.3 HC5 Estimates from Lognormal SSD Fit to Data from Various Distributions

Size	HC5est	HC5Q1est	HC5Q3est	Simulated
5	138	69	259	Cauchy
5	320	159	552	Exponential
5	166	98	256	Lognormal
5	121	89	177	Normal
5	13	6	24	Uniform
6	154	71	298	Cauchy
6	272	148	514	Exponential
6	151	102	216	Lognormal
6	123	93	181	Normal
6	12	6	21	Uniform
8	156	71	299	Cauchy
8	235	129	409	Exponential
8	134	94	193	Lognormal
8	124	96	167	Normal
8	11	6	18	Uniform
10	153	70	282	Cauchy
10	210	116	347	Exponential
10	133	94	184	Lognormal
10	121	96	158	Normal
10	10	6	16	Uniform
15	123	58	220	Cauchy
15	169	106	268	Exponential
15	120	92	158	Lognormal
15	116	97	141	Normal
15	8	5	13	Uniform
20	104	50	188	Cauchy
20	156	103	232	Exponential
20	116	91	148	Lognormal
20	113	97	134	Normal
20	8	5	12	Uniform
30	84	42	138	Cauchy
30	152	107	207	Exponential
30	109	90	134	Lognormal
30	110	97	126	Normal
30	7	5	10	Uniform

Size, number of species values used; HC5est, median HC5 estimate from lognormal SSD as percent of true HC5; HC5Q1est, lower quartile HC5 estimate from lognormal SSD as percent of true; HC5HC5Q3est, upper quartile HC5 estimate from lognormal SSD as percent of true HC5.

Table 12.4 HC5 Estimates from Exponential SSD Fit to Data from Various Distributions

Size	HC5est	HC5Q1est	HC5Q3est	Simulated
5	214	90	436	Cauchy
5	0	−173	266	Exponential
5	63	−10	157	Lognormal
5	122	89	185	Normal
5	6	−1	18	Uniform
6	188	81	373	Cauchy
6	22	−120	245	Exponential
6	66	4	146	Lognormal
6	118	88	174	Normal
6	5	−1	15	Uniform
8	148	63	296	Cauchy
8	34	−66	203	Exponential
8	73	25	132	Lognormal
8	109	83	149	Normal
8	5	0	12	Uniform
10	126	53	248	Cauchy
10	46	−35	179	Exponential
10	80	38	134	Lognormal
10	104	82	136	Normal
10	4	0	11	Uniform
15	86	36	174	Exponential
15	63	8	151	Lognormal
15	83	53	123	Normal
15	95	78	119	Uniform
15	3	1	8	Cauchy
20	66	27	138	Exponential
20	68	30	133	Lognormal
20	86	60	119	Normal
20	90	75	110	Uniform
20	3	1	6	Cauchy
30	47	19	92	Exponential
30	80	54	126	Lognormal
30	84	64	109	Normal
30	83	71	99	Uniform
30	3	2	5	Cauchy

Size, number of species values used; HC5est, median HC5 estimate from lognormal SSD as percent of true HC5; HC5Q1est, lower quartile HC5 estimate from lognormal SSD as percent of true; HC5HC-5Q3est, upper quartile HC5 estimate from lognormal SSD as percent of true HC5.

The log-transformed data in Figure 12.2 are consistent with an exponential distribution, but not with a normal distribution. It is left as an exercise to plot the data underlying Figure 12.2 after the transform to normality described above assuming an exponential distribution and to verify the subsequent normality by the Anderson–Darling goodness-of-fit test. The fifth percentile from a fitted exponential distribution is 0.146, but is 0.03 from a fitted normal and 0.024 from a fitted logistic. Clearly, selection of the distribution to fit matters. Other examples reveal greater disparities are possible.

Table 12.3 demonstrates the impact from estimating HC5 assuming a lognormal distribution when the true distribution is possibly different, and Table 12.4 does this when HC5 is estimated assuming an exponential distribution. Simulations allow one to know what the true HC5 is, so it is simple to compare the estimates against the true. 10 000 random samples of each sample size (5–30) were simulated from standard normal, lognormal, exponential, and Cauchy distributions. The medians, 25th and 75th percentiles ($Q1$ and $Q3$), of HC5 estimates from lognormal and exponential distributions fit to the simulated datasets were calculated and expressed as ratios to the true HC5. Table 12.3 makes clear that for samples of eight species values, HC5 estimates from an assumed lognormal SSD can underestimate the true HC5 by a factor of 9 *on average* and can overestimate HC5 by a factor of 2.3 *on average*. When the spread of the middle 50% of the distributions is considered, the potential for under- or overestimating HC5 is seen to be severe.

In Table 12.4, it can be seen that the HC5 estimates based on an exponential distribution exhibit a strong tendency to underestimate the true value. It is thus evident that choosing the right distribution is important and there is a potential for substantially under- or overestimating HC5 if the wrong distribution is assumed. The formal and visual tools for evaluating goodness of fit that have been discussed are thus important and should be used in selecting the distribution to fit to species data.

Including data from a simulated Cauchy distribution is not because a Cauchy distribution seems plausible for species data. Rather, it is meant to indicate the potential impact on HC5 estimation of a heavy-tailed observed distribution. In practical terms, this is likely to happen when there are one or two very small EC50 values that stand apart from the bulk of the data. Such occurrences are not rare. An example is given in Figure 12.3.

12.3.2 Example with Extreme Low EC50s

Species values were obtained for an herbicide. (See Table 12.5 for the data and Figure 12.3 for the associated SSD.) For our purposes, two characteristics of these data are noteworthy: (i) The seven species labeled aquatic plants are vascular plants, while the two algae species are not vascular. There could be a scientific issue as to whether vascular and nonvascular plants should be included in the same SSD due to possibly different modes of action or other concerns. (ii) The EC50 values for the two *Lemna* species were at least two orders of magnitude smaller than the other EC50s. This is an example of a heavy-tailed

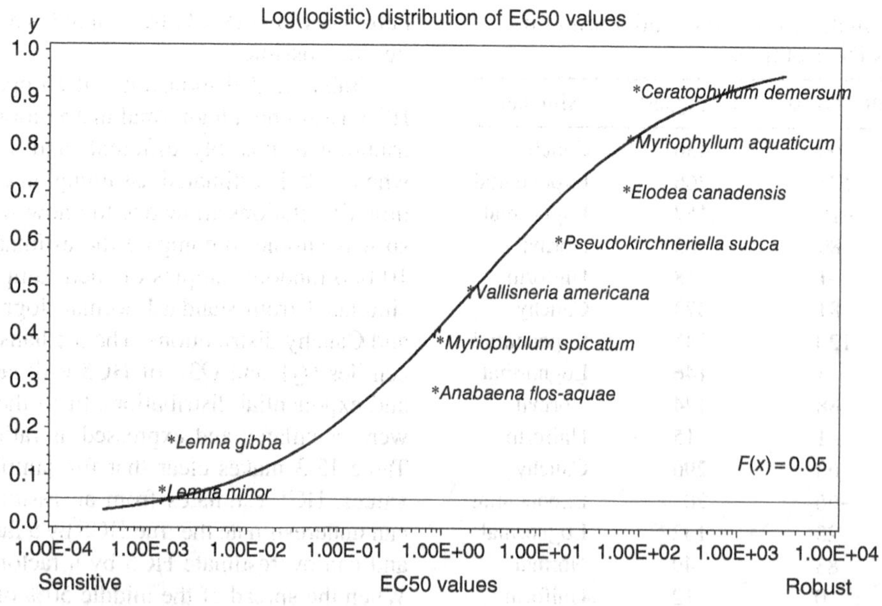

Figure 12.3 SSD with extreme low species values. Log-logistic SSD fit to nine species EC50 values. *Lemna* values 2+ orders of magnitude lower than all others. Log scale on horizontal axis.

Table 12.5 Aquatic Plant Data

Receptor	Species	EC50
Algae	*Pseudokirchneriella subcapitata*	14.5
Algae	*Anabaena flos-aquae*	0.84
Aquatic plants	*Lemna gibba*	0.0016
Aquatic plants	*Lemna minor*	0.0013
Aquatic plants	*Vallisneria americana*	1.9
Aquatic plants	*Myriophyllum spicatum*	0.94
Aquatic plants	*Ceratophyllum demersum*	88
Aquatic plants	*Myriophyllum aquaticum*	77
Aquatic plants	*Elodea canadensis*	75

EC50 in mg l^{-1}.

distribution, and that is true regardless of whether the two algae species are included. Several different SSDs and software packages were used to analyze these data. See Table 12.6 for a summary.

Clearly, SSD Master had a problem fitting the Fisher–Tippett and Weibull models to these data. Without intending to criticize that software, this example serves at least two objectives: (i) Demonstrate that one cannot take computer output at face value. It is important to evaluate model fits with a critical eye. (ii) Indicate the variability in HC5 estimates that sometimes occurs. It is unusual, but not rare, to find HC5 estimates from different plausible models that differ by orders of magnitude, and this calls for careful

Table 12.6 HC5 Estimates for Aquatic Plant Data

Program	Model	HC5	HC5LB	HC5UB
SSD Master	Lognormal	0.00599	0.000671	0.053471
SSD Master	Log-logistic	0.00341	0.000111	0.104131
SSD Master	Gompertz	0.00043	1.8E–05	0.010052
SSD Master	Weibull	112.4	—	—
SSD Master	Fisher–Tippett	210.9	—	—
ETX	Lognormal	0.000873	2.87E–06	0.019582
SAS	Lognormal	0.000465	2.56E–09	1.13783
SAS	Log-logistic	0.001763	6.95E–08	2.66354
BurrliOZ	Burr III	0.000335	—	—

Software and models used to fit species data, with 95% confidence bounds where possible.

evaluation. Aside from the two obvious problem estimates, there is a range of HC5 estimates. Excluding these two estimates and the one from BurrliOZ (version 1), all others came with goodness-of-fit assessments and were judged to fit. We assessed the log-logistic fits to be best visually and in terms of residual analysis. Of the two log-logistic fits, only the SAS program was based on EUFRAM recommendations to take uncertainty as well as variability into account (see Section 12.4). It is left as an exercise to verify the entries of Table 12.6.

12.3.3 Weighted Model Fitting

Since both SSD Master and BurrliOZ use weighted distribution fits, it may be helpful to review weighting schemes for nonlinear models typically used for species sensitivity modeling (and for linear models as well). The usual reason for a weighted fit is variance heterogeneity, where there tends to be less variability in the more sensitive species. In a weighted analysis, more emphasis is given to more precisely determined (i.e. less variable) observations than to less precisely determined observations. A proper weighting scheme along these lines can assure that small species values are given ample consideration in the model fit and subsequent HC5 estimation.

As noted by Huet et al. (2004, p. 94), "The most frequent mistake made in modeling the variance function is to assume that the variance of errors is constant, $\text{Var}\varepsilon_{ij} = \sigma^2$, when, in actuality, heteroscedasticity exists, $\text{Var}\varepsilon_{ij} = \sigma_i^2$." On the other hand, according to Motulsky and Christopoulos, 2004), mis-specification of weights often has little impact on the fitted model. A standard weighting scheme is to model the weights as inversely proportional to the predicted mean of the modeled response. This is equivalent to assuming the variability between reps is proportional to the predicted mean. Note that it is not recommended to use the observed variances (or their inverses) as weights even if available, as these are subject to high sampling variability, especially when there are few replicate values for each species or varying numbers of replicates for different species. Numerous authors have pointed this out (e.g. Motulsky and Christopoulos, 2004, pp. 86ff). For example, Huet et al. (2004, p. 61) state the following, where the reference to weighted least squares refers to using the inverse of the observed variance as weights.

"Although one can feasibly use the weighted least squares method for few replications, it is not the best option because the accuracy of the empirical variance as an estimator of the true variance is rather bad: With four replications, the relative error is roughly 80%. The weighting by such inaccurate estimators can be very misleading."

Huet et al. (2004, p. 63) specifically discuss modeling weights inversely proportional to the function being estimated, as in the following quote. It refers to a specific dataset regarding harvesting, but the concern is general.

"Although the relative position of the data point corresponding to the fourth harvesting date is troublesome, we can represent the variance of the observations by an increasing function of the response function, $\sigma_i^2 = \sigma^2 f(x_i, \theta)$, for example."

Residuals from a properly weighted model fit will more closely resemble a normal distribution than will residuals from an unweighted analysis, so that parametric methods, such as plotting histograms with overlaid normal density curves, stem-and-leaf or box-and-whisker plots, or applying formal checks of normality (e.g. Shapiro–Wilk or Anderson–Darling) may be useful than residuals based on unweighted model fits.

In most species sensitivity work, most of the species values are not replicated, so one must assume a relationship between variance and species value (LC_{50} or NOEC). The most common method of fitting is through iteratively reweighted least squares with weights inversely proportional to the mean predicted value, as previously indicated. These weights are assumed to be inverse elements of the diagonal variance–covariance matrix of the dependent variable. Other weighting schemes are also possible, such as using the square root of the aforementioned weights. Either has the effect of giving more weight to the more sensitive species, which may be acceptable to the regulatory agencies. The BurrliOZ software appears to use the first indicated weighting scheme. There is nothing in the BurrliOZ documentation regarding weights, but SAS programs have been written that duplicate the fits of that software using these weights and get very noticeably different fits using unweighted or different weights. SSD Master uses a different weighting scheme that can have the effect of ignoring the most sensitive species if such values are far smaller than most of the species data. This could make SSD Master problematic for regulatory work when there are such isolated small observations. When that is not the case, SSD Master is reliable. ETX appears not to use weighted model fitting. The SAS programs mentioned before allow the user to specify unweighted or weighted model fits, including the weights discussed above.

12.3.4 Impact on HC5 Estimates of Mis-specified Distribution

A simulation study was done by the first author of this book for the EUFRAM project (Chapman et al., 2007) and since updated (unpublished) to explore the effects on HC5 estimates of fitting an SSD of one distribution, such as lognormal, when the data were actually from a different distribution, such as log-logistic. At least 1000 datasets were simulated of each combination of simulated and fitted distributions and sample sizes. For each combination, a single distribution was simulated and the indicated SSD was fit to the simulated data. See Table 12.7 for fitting a lognormal SSD to data from several other distributions. Other simulations not reported here do the same when fitting an exponential SSD when the actual data come from a different distribution. The most accurate estimate of the true HC5, in terms of median and spread of estimates, would be from fitting a distribution of the same type as that simulated, for example, fitting a lognormal SSD to lognormal simulated data. No claim is made that Cauchy or uniform distributions are common for ecotoxicology species values. The distributions selected for simulation are merely intended to illustrate that the distribution fit does influence the HC5 estimate and that a poor choice of SSD to fit can have adverse implications for the quality of the HC5 estimate.

It will be evident from these two tables that even when an unusually large number of species values are available, the HC5 estimate is sensitive to the distribution fit to the data.

Table 12.7 HC5 Estimates from Lognormal SSD Fit to Data from Various Distributions

Size	Fit	HC5est	HC5Q1est	HC5Q3est	Simulated
5	Lognormal	138	69	259	Cauchy
5	Lognormal	320	159	552	Exponential
5	Lognormal	166	98	256	Lognormal
5	Lognormal	121	89	177	Normal
5	Lognormal	13	6	24	Uniform
6	Lognormal	154	71	298	Cauchy
6	Lognormal	272	148	514	Exponential
6	Lognormal	151	102	216	Lognormal
6	Lognormal	123	93	181	Normal
6	Lognormal	12	6	21	Uniform
8	Lognormal	156	71	299	Cauchy
8	Lognormal	235	129	409	Exponential
8	Lognormal	134	94	193	Lognormal
8	Lognormal	124	96	167	Normal
8	Lognormal	11	6	18	Uniform
10	Lognormal	153	70	282	Cauchy
10	Lognormal	210	116	347	Exponential
10	Lognormal	133	94	184	Lognormal
10	Lognormal	121	96	158	Normal
10	Lognormal	10	6	16	Uniform
15	Lognormal	123	58	220	Cauchy
15	Lognormal	169	106	268	Exponential
15	Lognormal	120	92	158	Lognormal
15	Lognormal	116	97	141	Normal
15	Lognormal	8	5	13	Uniform
20	Lognormal	104	50	188	Cauchy
20	Lognormal	156	103	232	Exponential
20	Lognormal	116	91	148	Lognormal
20	Lognormal	113	97	134	Normal
20	Lognormal	8	5	12	Uniform
30	Lognormal	84	42	138	Cauchy
30	Lognormal	152	107	207	Exponential
30	Lognormal	109	90	134	Lognormal
30	Lognormal	110	97	126	Normal
30	Lognormal	7	5	10	Uniform

Size, number of species values simulated; Fit, species sensitivity distribution fit to data; Simulated, distribution of species values simulated; HC5est, median estimated HC5; HC5Q1est, first quartile of HC5 estimates; HC5Q3est, third quartile of HC5 estimates.

12.4 VARIABILITY AND UNCERTAINTY

The primary use of SSDs in regulatory work is based on the estimated concentration, HC5, of the test substance deemed safe for all but the 5% most sensitive species. There is some inconsistency in how HC5 is estimated. Some software packages (e.g. Rodney et al., 2008; Shao, 2000) simply fit a distribution to the species data and estimate the fifth percentile of the fitted distribution. For example, if a normal distribution is fit to n distinct species values, then the fifth percentile is

$$\hat{p}_5 = \bar{x} - t(0.05, \text{df}) * S, \qquad (12.1)$$

where $\bar{x}$ and S are the sample mean and standard deviation, respectively, of the presumed normal distribution and $t(0.05, \text{df})$ is the fifth percentile of the t-distribution with degrees of freedom df=$n-1$. More appropriately, since S is a biased estimate of the population standard deviation, S should be replaced by S/c_4 (Bolch, 1968), where

$$c_4 = \sqrt{\frac{2}{n-1}} * \frac{\Gamma(n/2)}{\Gamma((n-1)/2)} \qquad (12.2)$$

An approximate 95% lower confidence bound on this HC5 estimate can be obtained using the approximately normal distribution of $\hat{p}_5$ (Serfling, 1980, p. 104) and the variance (Bolch, 1968):

$$\text{Var}(\hat{p}_5) = \frac{S^2}{n} + \left(t(0.05, \text{df}) * S\right)^2 * \left(\frac{1}{c_4^2} - 1\right) \qquad (12.3)$$

Other authors (e.g. Aldenberg and Jaworska, 2000; Posthuma et al., 2002; Chapman et al., 2007) take a Bayesian approach to incorporate the uncertainty of the data. The approach used by Chapman et al. (2007) was a two-dimensional (2-D) Monte Carlo process. Aldenberg and Jaworska (2000) argued that the same estimates could be obtained using different factors to replace $t(0.05,\text{df})$ in

Eq. (12.1), and they supplied a table of the factors for the point estimate and lower confidence bound for samples sizes ranging from 8 to 30. See Table 12.8 for a partial list of these factors.

The method developed in EUFRAM (Chapman et al., 2007) to incorporate uncertainty into the estimation of HC5 begins by specifying a family of distributions (e.g. lognormal) and the fit of that distribution to the existing data. This requires estimation of all model parameters. In the case of the lognormal, these are the mean, μ, and standard deviation, σ. For the log-logistic, the model parameters are the mean, μ, and a scale parameter, β, analogous to the standard deviation. For the Burr III distribution, there are three parameters to be estimated, called b, c, and k, in the BurrliOZ software documentation. Of these b is a scale parameter (analogous to standard deviation), and c and k are "shape" parameters not having an easy interpretation.

Table 12.8 Aldenberg–Jaworska Factors for HC5 Estimates from Lognormal SSD

N	AJ5LB	AJ5	AJ50LB	AJ50
8	3.1873	1.7187	0.6698	0
9	3.0312	1.7091	0.6198	0
10	2.9110	1.7016	0.5797	0
11	2.8150	1.6957	0.5465	0
12	2.7363	1.6910	0.5184	0
13	2.6705	1.6870	0.4943	0
14	2.6144	1.6837	0.4733	0
15	2.5660	1.6808	0.4548	0
20	2.3960	1.6712	0.3866	0
30	2.2198	1.6620	0.3102	0

AJ50LB, factor to use in Eq. (12.1) to estimate $HC_5(5)$; AJ5, factor to use in Eq. (12.1) to estimate HC_5; AJ5LB, factor to use in Eq. (12.1) to estimate $HC_{50}(5)$; AJ50, factor to use in Eq. (12.1) to estimate HC_{50}.

The HC5 values developed within EUFRAM are obtained by incorporating the uncertainty of the parameter estimates into the estimation process. EUFRAM describes in detail probabilistic methods starting with maximum likelihood estimates (MLEs) of the parameters themselves and proceeding based on 2-D Monte Carlo simulations of the distributions of the parameter estimates from a fully developed theory. In essence, this means for each pair of sampled parameter values from the appropriate distributions, a new lognormal or log-logistic distribution is fit, and the fifth percentile or quantile, $Q5$, is computed. The resulting set of $Q5$ values has a distribution, and the median of this distribution is the HC_5 value reported, sometimes labeled $HC_5(50)$. More generally, $HC_5(x)$ is the xth percentile of the distribution of HC5 estimates, so that a 95% lower bound of the distribution is variously labeled HC5LB or $HC_5(5)$. A plot of the collection of all these distributions functions is called a "spaghetti plot" (see Figure 12.4). The name was introduced by Aldenberg and Slob (1993).

To be clear, once the MLEs of the mean and variance of the basic distribution are obtained, the following procedure is used. Randomly sample from a chi-squared distribution with the appropriate degrees of freedom (obtained by subtracting 1 from the number, n, of species). Use the estimate $\sigma = \sqrt{(n-1)s^2/\chi^2(n-1)}$ for the standard deviation, where s^2 is the MLE estimated variance. Then randomly sample from a normal population with mean

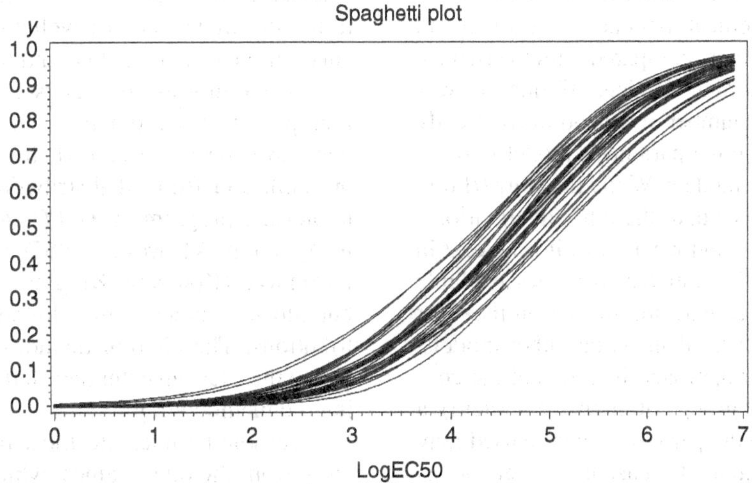

Figure 12.4 Spaghetti plot. The many lognormal distribution functions generated by the 2-D Monte Carlo process are illustrated by thin curves. The median curve is the heavy black curve. Where the horizontal line at $y=0.05$ intersects the median curve defines the median HC5 estimate. Where that same line intersects the collection of distribution curves defines the distribution of HC5 estimates. It will be apparent that the width of the middle 90 or 95% of this distribution can be very wide, especially when one realizes the horizontal axis is on a log scale. Consequently, the HC5LB can be a very small value.

equal to the MLE estimated mean above and the standard deviation just obtained.

Each such pair defines an outer loop in a 2-D Monte Carlo simulation. Some 100 000 or more such outer loops are simulated. The proportion of all species from the modeled ecosystem that are affected is then estimated by standard statistical methods for each such distribution. The median of this distribution is then estimated for each possible EC50 value. Where the "distribution" curve of such medians intersects the horizontal line at $p=0.05$ defines the median HC5 estimate.

Version 1 of the BurrliOZ software follows a different process. Parameter estimates are obtained in an alternative manner based on purely computational methods. No theoretical distributions for the model parameters are used. Given this, Monte Carlo methods do not appear possible. Instead, bootstrap methods are employed. What this means is that repeated samples are taken from the available EC_{50} values. Sampling is done with replacement (i.e. the same values may be included more than once), and a new distribution is fit to each such sample, and the fifth percentile, *Q*5, is computed. The resulting set of *Q*5 values is itself distributed in some fashion, and the median of this distribution is the HC_5 value reported. In general outline, the process used in the Burr III software is in the same spirit as that used by EUFRAM. However, there are two important differences. First, there are only a few EC_{50} or NOEC values, and there are well-known problems with bootstrap regression for such small samples that do not exist for the probabilistic methods used in the EUFRAM approach. Chernick (1999) recommends a minimum of 30 observations for bootstrap methods and prefers at least 50. This problem arises from the severely limited number of distinct samples, and it is possible to construct from a small set of numbers, even with replacement. Second, the Burr model has three parameters to estimate compared with the two for the log-logistic and lognormal models. Fitting a two-parameter model to a small number of data points is already questionable, and fitting a three-parameter model is more so and can lead to unstable models. While the Burr III distribution fits some data better than the other distributions, this is not true in every case, and where it is, it is at least in part a matter of overfitting. As more data becomes available (i.e. additional species are tested), the impact on the Burr model is likely be more severe than on the other models. Since the intent of the SSD approach is to model the ecosystem (or some taxonomic group within the ecosystem), a model that is highly tied to the particular sample used may have less value than one that fits reasonably well but is more robust to additional data. It should be observed that the BurrliOZ software issues a warning message when fewer than eight points are available for the fit stating that the estimates are unstable and caution should be used in interpretation. This is a more serious issue for the three-parameter Burr III model than for the two-parameter lognormal or log-logistic models.

12.5 INCORPORATING CENSORED DATA IN AN SSD

A challenge for SSDs is what to do with censored species values. Such values arise in three ways: extrapolation of LC/EC50 values far above the range of tested concentrations, extrapolations of NOECs below the lowest tested concentration, and data for which no regression model can be fit but for which it can be established with 95% confidence that EC50 exceeds the highest tested concentration. The first and third of these situations can be described as right censoring, while the middle example is left censoring. Censoring is ignored in some of the discussion in Kooijman (1987), and that was common practice until recently. Some regulatory guidance recommends excluding censored values from the SSD fit altogether (EFSA, 2013) or only with some qualifications. The EFSA report is based in part on Brock et al. (2011), and the reader is advised to consult those two sources for further information. However, as Aldenberg (2015) and Kon Kam King et al. (2014) indicate, that is changing. Dowse et al. (2013) have explored the use of censored data in characterizing the effects of salinity on ecosystems in Australia and found that including this information improves the quality of the information obtained. In Section 12.5.1, the effect of various ways of handling censored data is explored in a systematic way.

The only widely available software package that treats censored data as such is MOSAIC_SSD. This is somewhat surprising since statistical and mathematical ways of handling censored values are well known and routinely used in survival analysis (e.g. Lee and Wang, 2013) and environmental monitoring (e.g. Helsel, 2005). SAS programs to incorporate censored data in a mathematically sound manner have been developed for lognormal, log-logistic, Weibull, and Burr III distributions, among others, and a lognormal program, *SSD_LOGNORMAL.sas*, is included in Appendix A1. Mosaic_SSD software, based on R, was introduced (Kon Kam King et al., 2014) for this purpose, but allows a choice of only lognormal and log-logistic distributions. The citation includes a link to a free website where the user can enter data and obtain results implementing the methodology.

Censored values are most realistically treated differently from the other values, which are true estimates. The standard way to incorporate such censored values in fitting a distribution is to use MLEs of the mean and standard deviation of the presumed distribution taking the censored values directly into account, so as to reflect the uncertainty in these censored values. In the case of fitting a lognormal distribution, the likelihood function has the form

$$L(x;\mu;\sigma) = \prod_{i=1}^{L} \Phi\left(\frac{x_i - \mu}{\sigma}\right)$$
$$\prod_{i=L+1}^{L+k} \phi\left(\frac{x_i - \mu}{\sigma}\right) \prod_{i=L+k+1}^{nL+k+R}\left[1 - \Phi\left(\frac{x_i - \mu}{\sigma}\right)\right] \quad (12.4)$$

where (i) φ (used in the middle product in the above formula) and Φ are the unit normal pdf and cdf, respectively, and (ii) there are L left-censored values, (iii) k uncensored values, and (iv) R right-censored values. The MLEs of the mean and standard deviations can be obtained from SAS version 9 Proc Lifereg. They can also be incorporated in NLME models as demonstrated in Chapter 8. Censored data can also be modeled by MOSAIC_SSD. When there is a species with multiple studies reporting NOEC (or EC50) values, some of which are censored, then the same procedure is used to obtain a representative NOEC (or EC50) value for that species to be included in the overall SSD.

When there are only one or two censored values, there is generally little difference between estimated HC5 values ignoring censoring and those using MLEs based on censored values. A complication of censoring is that it makes little sense to test the data for normality using censoring since a presumed normal distribution (on log-transformed values) is used to estimate the mean and standard deviation of the distribution used. Furthermore, the method of obtaining MLEs of the mean and standard deviation does not estimate individual censored values, only the two named model parameters.

To understand the impact of censoring on a SSD, let μ_1 and σ_1 be the mean and standard deviation, respectively, of the data taking right censoring into account and μ_2 and σ_2 be the mean and standard deviation, respectively, of the data treating the right-censored data as uncensored. If uncertainty is ignored, the HC5 satisfies the following relationship:

$$\begin{aligned} &\text{Prob}[X < \text{HC5}] = 0.05 \\ &\text{Prob}\left[\frac{X-\mu}{\sigma} < \frac{\text{HC5}-\mu}{\sigma}\right] = 0.05 \\ &\text{Prob}\left[Z < \frac{\text{HC5}-\mu}{\sigma}\right] = 0.05 \\ &\frac{\text{HC5}-\mu}{\sigma} = \Phi^{-1}(0.05) \\ &\text{HC5} = \mu + \Phi^{-1}(0.05)\sigma \cong \mu - 1.645\sigma \end{aligned} \quad (12.5)$$

Thus in order that HC5(μ_1, σ_1) be less than HC5(μ_2, σ_2), it is necessary that

$$\begin{aligned} &\text{HC}_{5,1} < \text{HC}_{5,2} \\ &\text{or} \\ &\mu_1 - 1.645\sigma_1 < \mu_2 - 1.645\sigma_2 \\ &\text{or} \\ &\mu_1 - \mu_2 < 1.645(\sigma_1 - \sigma_2). \end{aligned} \quad (12.6)$$

Thus, if the change in the mean greatly exceeds the change in the standard deviation, then $\text{HC}_{5,1} > \text{HC}_{5,2}$. So the question is really when will the change in the mean greatly exceed the change in the standard deviation. The most general situation where this will occur is where the lower bounds of the right-censored values are widely dispersed. For in this case, there is likely to be a relatively small increase in variance and yet a large change in the mean when censoring is taken into account. In the case all or most of the right-censored values are the same (e.g. the limit of solubility), it is very likely that the change in variance will be large compared with the change in the mean and the HC5 will most likely decrease when right censoring is taken into account.

The greatest impact of taking censoring properly into account is on the confidence bounds for HC5. Censoring increases uncertainty and this in turn widens the confidence bounds. The point estimate of HC5 can increase or decrease, but the HC5 lower bound will almost certainly decrease, when censoring is included in the model fit. See Table 12.9 for an illustration of the impact of censoring in an algae study. The data from which Table 12.9 was developed is in Exercise 12.2.

Where only a point estimate of HC5 is needed, mathematical treatment of censoring provides useful results, but if a lower bound HC5 is required, the impact of such modeling can be extremely small bounds.

Typically, the estimated HC5 is lower when censoring is taken into account, but that is not what happens in this example. (See Figure 12.5.) The lower confidence bound, HC5LB, is lower when censoring is taken into account, reflecting the added uncertainty implied by censoring.

Table 12.9 Comparison of HC5 Point Estimates with Censored Data

Data	HC5	HC5 LB	Units	Comment
Algae EC50	14.48	0.008	mg l^{-1}	With censoring
Algae EC50	8.98	0.142	mg l^{-1}	Censoring ignored

With censoring, maximum likelihood fit treating censored data as censored; Censoring ignored, maximum likelihood fit treating censored data as not censored; HC5 LB, lower 95% confidence bound on HC5.

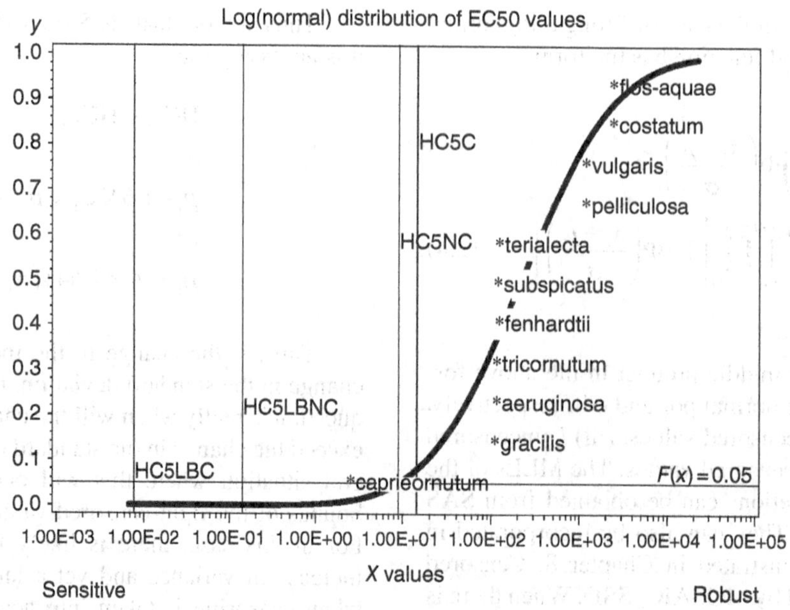

Figure 12.5 Censoring considered vs. ignored. HC5C = HC5 and HC5LBC = HC5LB when censoring is taken into account. HC5NC = HC5 and HC5LBNC = HC5LB when censoring is ignored.

12.5.1 A Simulation Study to Investigate Censoring

This section following is taken primarily from a study reported in Green (2017b). As noted above, the mathematics of how censored data is used in fitting a model is well known. What is probably of greater importance to the risk assessment process is a comparison of the three main approaches that have been used specifically for SSDs and the resulting HC5 estimates. To that end, an extensive computer simulation study has been undertaken and is reported in Green (2017b). As noted above, the lognormal distribution is the primary distribution used for SSDs for ecotoxicological risk assessment though other distributions are needed. Data follow a lognormal distribution if and only if the logarithms of the data values follow a normal distribution. Furthermore, all normal distributions have the same shape, differing only in the mean and standard deviation. Consequently, it suffices to explore the effect of censoring on fitting a standard normal distribution. Three approaches were considered based on recent practice or regulatory guidance. These were

1. Ignore censoring and treating "> X" values as equal to X.
2. Discard censored values and conduct additional studies on the same or new species until the number of uncensored values reaches the agreed minimum, usually 8.
3. Use the censored data in the mathematically sound way described in Section 12.5. A commonly accepted minimum number of EC50 or NOEC values for SSD purposes is eight distinct species. While these are animal welfare and cost considerations regarding method (2), Green (2017b) addresses only the scientific merits of these alternatives.

The primary use of SSDs in regulatory work is based on the estimated concentration, HC5, of the test substance deemed safe for all but the 5% most sensitive species. To investigate the relative merits of the three approaches to HC5 estimation in the presence of censored data, computer simulation studies were conducted, where eight values were randomly selected from a standard normal distribution subject to censoring. Three types of simulations were done. The simplest approach addressed the situation that the top 10% of the distribution was censored or the top 20, 30, 40, or 50% were censored. All three methods were applied to the resulting data, and fifth percentile of the fitted normal distribution was estimated following standard methods (Johnson et al., 1994). Methods (2) and (3) treat the species values as coming from a common normal distribution. Mathematically, this amounts to sampling from a truncated distribution. This would be appropriate if the highest tested concentrations for all species were at least approximately the same. This would be the case, for example, if the experiment for all species were done at concentrations up to the limit of solubility. If only survival ECxs or NOECs were used in constructing the SSD, this might be appropriate. See Tables 12.10 and 12.11 for results for this situation. See Figure 12.6 for the distribution of HC5 estimates underlying those tables.

Table 12.10 provides a brief summary of results related to Figure 12.6. Table 12.11 provides more information and includes other amounts of censoring. It is also informative to compare Tables 12.10 with Table 12.12, where a more complex sampling is tabulated.

For sublethal endpoints, an experiment on a given species is run only with concentrations that are not expected to have significant mortality for that species. The highest tested concentration, and consequently, the lower bounds on possible censored values, varies widely across species.

For a more realistic simulation program for sublethal responses, it is necessary to capture a range of censoring values, rather than merely sampling from a single truncated distribution. This was done by first generating a random sample of N (e.g. 8) observations from a standard normal. These were taken as the mean LC50 values for the N species. Then a random sample of N observations from a chi-squared distribution with N df was used to generate species variances. Following that, a random observation, X_i, from N(mu_i,v_i) was taken, where mu_i and v_i are the mean and variance of the ith species distribution as just described. If X_i exceeded the pth upper percentile of N(mu_i,v_i), it was considered censored. For method (2), each censored value was discarded, and another species was added in the same manner until N uncensored species values were obtained. The X_i, $i = 1, 2, \ldots, N$, constitute the N species ECxs (or NOECS) to which a single SSD is fit. The corresponding HC5 was estimated by the Aldenberg and Jaworska method. This process was repeated 100 000 times for each indicated censoring percentage indicated. As a reference, 1 000 000 HC5 values were estimated where no censoring was done. This constitutes the "true" HC5 being estimated by the three approaches.

The distribution of HC5 values under all three methods will now be described and summarized. Additional details are reported in Green (2017b). In summary, the mathematically sound way of using censored data tends to underestimate HC5, with greater underestimation associated with greater percentage of censoring, whereas continuing to sample until N uncensored species values are obtained tends to overestimate HC5, provided the censoring is less than 50%. Treating censored values as though they are uncensored consistently overestimates HC5 in all scenarios simulated. Thus, from a safety or regulatory perspective, there is little reason to require additional uncensored data except perhaps in extreme cases where most species values are censored and even in those cases the improvement is small.

Table 12.10 Comparison of HC5(P5) Estimates with 30% Censoring

Trim (%)	Type	Mean	Std	Median
30	IGNORE CENSORING	−1.533	0.472	−1.513
30	SAMPLE_MORE	−1.535	0.475	−1.515
30	TREAT CENSORED DATA AS SUCH	−1.797	0.599	−1.763

Little difference was observed in these summary measures between ignoring censoring and replacing censored values through additional sampling. The mean and median are lower when censoring is treated in a mathematically correct manner.

Table 12.11 Summary of Estimates for Normal SSDs with 8 Species Using fifth Percentile

DISTRIB	Type	%	Mean	Std	Max	Median	Min	THC5	Compare
NORMAL	UNCENSORED	10	−1.417	0.533	0.287	−1.411	−3.510	−1.645	OVER
NORMAL	UNCENSORED	20	−1.351	0.533	0.429	−1.343	−3.812	−1.645	OVER
NORMAL	UNCENSORED	30	−1.300	0.541	0.485	−1.286	−3.471	−1.645	OVER
NORMAL	UNCENSORED	40	−1.250	0.531	0.253	−1.239	−3.306	−1.645	OVER
NORMAL	UNCENSORED	50	−1.232	0.521	0.000	−1.213	−3.341	−1.645	OVER
NORMAL	KEEP SAMPLING	10	−1.450	0.513	0.258	−1.442	−3.423	−1.645	OVER
NORMAL	KEEP SAMPLING	20	−1.474	0.495	0.053	−1.450	−3.453	−1.645	OVER
NORMAL	KEEP SAMPLING	30	−1.533	0.472	−0.125	−1.513	−3.947	−1.645	OVER
NORMAL	KEEP SAMPLING	40	−1.608	0.457	−0.224	−1.588	−3.384	−1.645	OVER
NORMAL	KEEP SAMPLING	50	−1.684	0.441	−0.410	−1.655	−4.273	−1.645	UNDER
NORMAL	CENSOR	10	−1.597	0.568	0.332	−1.579	−4.201	−1.645	OVER
NORMAL	CENSOR	20	−1.705	0.586	0.267	−1.680	−4.055	−1.645	UNDER
NORMAL	CENSOR	30	−1.797	0.599	2.305	−1.763	−4.177	−1.645	UNDER
NORMAL	CENSOR	40	−1.886	0.619	3.088	−1.865	−5.116	−1.645	UNDER
NORMAL	CENSOR	50	−1.930	0.687	3.237	−1.935	−4.175	−1.645	UNDER

Last column indicates whether HC5 is over- or underestimated. Type = Uncensored means censored values were treated as uncensored. Type = Keep Sampling means censored values were discarded and additional species were studied until eight uncensored values were obtained. Type = Censor means the censored values were used in a mathematically sound manner. Percent is the upper percentage of the distribution that was simulated as censored. Number of simulated studies = 10 000 in each case. Mean, median, std, min, and max are the descriptive statistics for the HC5 estimates. THC5 = true HC5 value simulated.

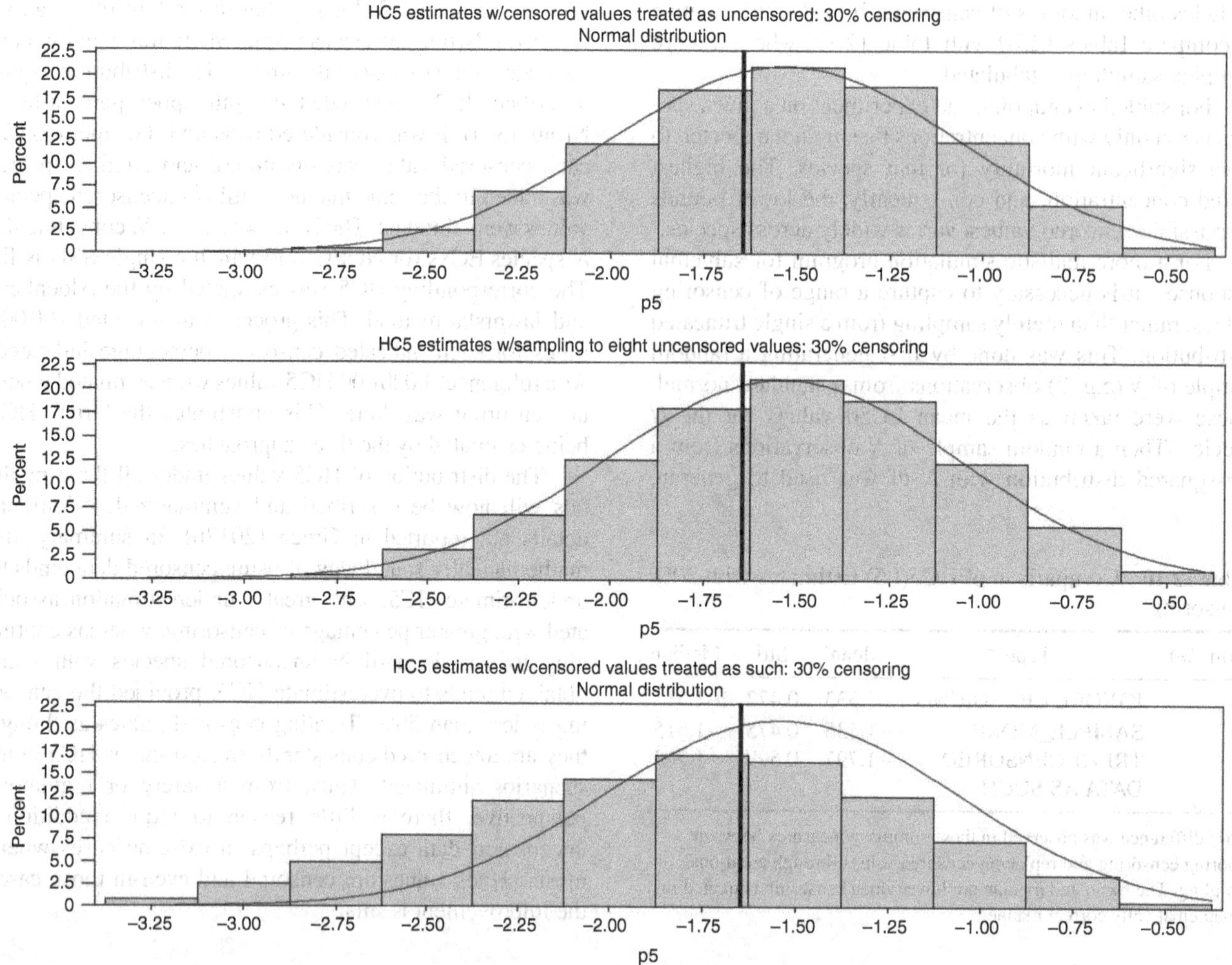

Figure 12.6 Comparison of three methods of estimating HC5 from a normal distribution with 30% censoring. Vertical line shows true HC5. Bottom figure with mathematically correct treatment of censored values shows a strong tendency to underestimate HC5, and the spread is slightly wider than with the other methods. Also, a tiny percentage (0.87%) of extreme underestimates occurs. One observation out of 10 000 grossly overestimated HC5. Middle figure shows results when censored values are discarded and sampling continues until eight uncensored values are obtained. There is a tendency to overestimate HC5. Top figure shows results when censored values are treated as uncensored. There is a tendency to overestimate HC5.

Simulations based on a total of eight species are reported first when a normal distribution is appropriate, as was the case in the study under discussion. Following that, the same methods were used for logistic distributions, the second most commonly used distribution for SSDs. Only the results for a normal distribution are reported here and only for 30% censoring. The overall distribution of HC5 estimates favors method 3 if there is no more than 30% censoring, while method 2 is favored when there is 40% or greater censoring. Since the bias in method 3 with 40% or more censoring is a tendency to underestimate HC5 and additional sampling would increase the use of animals (depending on which class of species is under investigation), there would appear to be little regulatory incentive to require continued sampling. Given the cost of additional species studies and the uncertainty of whether new studies would produce more censored values, registrants may consider the modest risk of underestimating HC5 by method 3 to be preferable to generating additional studies. However, additional sampling remains an option if the HC5 estimate is too low for product efficacy. These conclusions hold regardless of whether NOECs or ECx values are used and

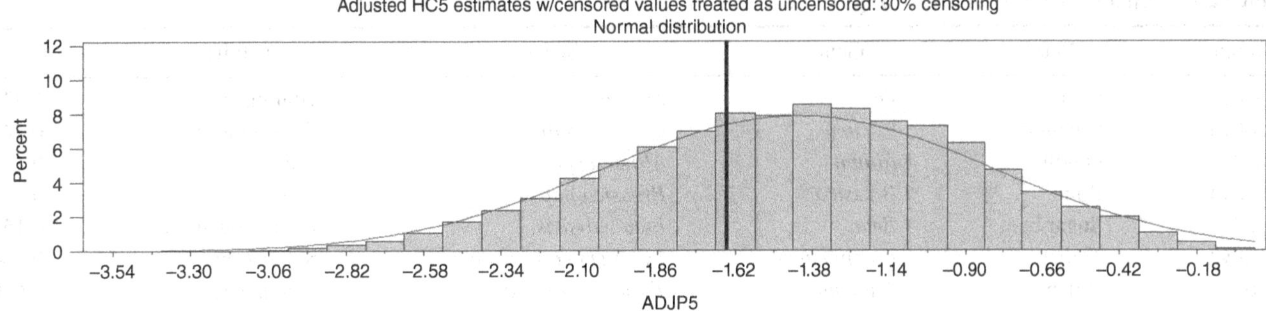

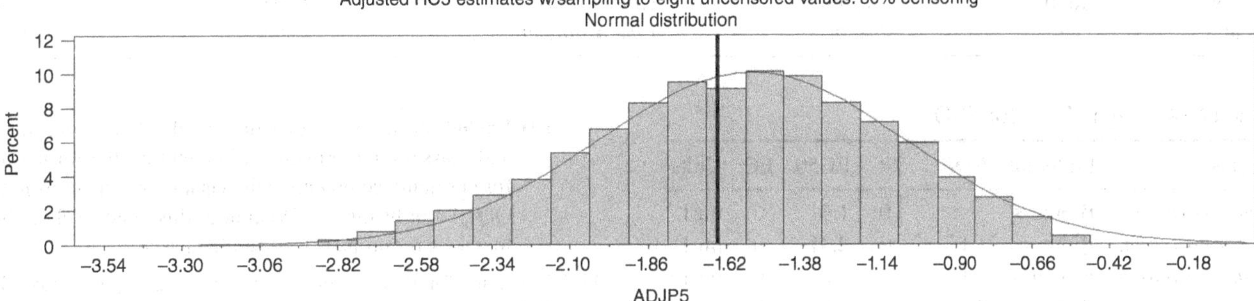

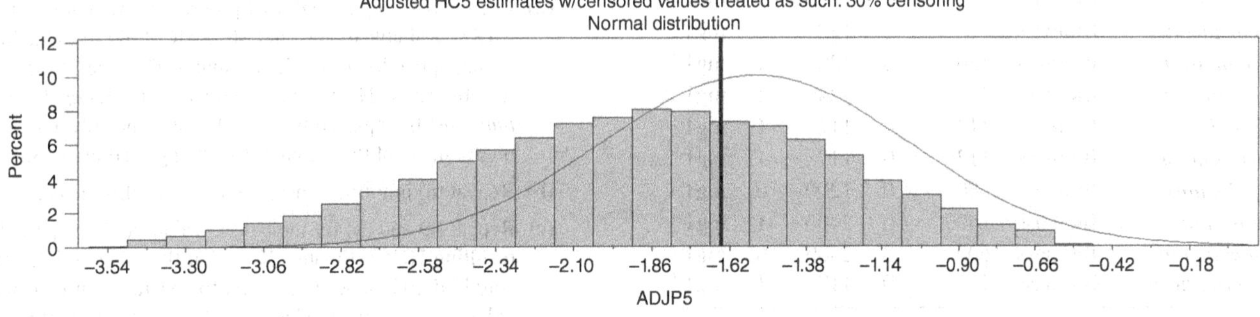

Figure 12.7 Sublethal HC5 simulation results (using the Aldenberg–Jaworska adjusted HC5 values). Normal distribution *w*/8 species, 30% censoring. Top panel shows distribution of HC5 estimates when censoring is ignored. Middle panel shows distribution of HC5 estimates when censored values are dropped and sampling continues until eight uncensored values are available. Bottom panel shows distribution of HC5 estimates when censored values are treated as censored using Eq. (12.4). Vertical line in each plot shows true HC5 value. The HC5 estimates from bottom panel are shifted toward lower values, and those in top panel are shifted toward higher values. A smaller shift toward higher values is also apparent in the middle panel.

Table 12.12 Comparison of Aldenberg–Jaworska HC5 Estimates with Eight Species and 30% Censoring

Trim (%)	TYPE	Mean	Std	Median
30	IGNORE CENSORING	−1.421	0.609	−1.382
30	SAMPLE_MORE	−1.537	0.477	−1.515
30	TREAT CENSORED DATA AS SUCH	−1.809	0.595	−1.781

regardless of the percent inhibition (x value) ECx describes, so long as one is consistent with NOEC or choice of x to use. See Figure 12.7 and Table 12.12 for some of the results from this simulation study.

EXERCISES

12.1 The data for this exercise are the full data underlying Figure 12.2. See Table 12.13.

(a) Fit a lognormal SSD to these data and verify the HC5 estimate claimed in Section 12.5.1.

(b) Next, fit a log-exponential SSD to these data, which merely requires fitting an exponential SSD to the logarithms to obtain the HC5 estimate reported in Section 12.5.1.

(c) Perform the transform to normality described in Section 12.5.1 and do an assessment of normality of the transformed data.

Table 12.13 Species EC50 Values for Figure 12.2

Endpoint	Plant	Genus	Species	Family	EC50
Shoot wt	Corn	*Zea*	*Zea mays*	Gramineae	1.834
Shoot wt	Cucumber	*Cucumis*	*Cucumis sativus*	Cucurbitaceae	0.146
Visual	Onion	*Allium*	*Allium cepa*	Liliaceae	0.160
Shoot wt	Rape	*Brassica*	*Brassica napus*	Cruciferae	0.305
Visual	Sugar beet	*Beta*	*Beta vulgaris*	Chenopodiaceae	0.151
Shoot wt	Tomato	*Lycopersicon*	*Lycopersicon esculentum*	Solanaceae	0.200
Shoot wt	Wheat	*Triticum*	*Triticum aestivum*	Gramineae	20.650
Shoot wt	Sorghum	*Sorghum*	*Sorghum bicolor*	Gramineae	4.704
Shoot wt	Soybean	*Glycine*	*Glycine max*	Fabaceae	0.212
Visual	Pea	*Pisum*	*Pisum sativum*	Fabaceae	0.206

Table 12.14 Algae Data for SSD

Species	Endpoint	NOEC	NC	EC50	EC	Units
S. capricornutum	Biomass		0	1.41	0	$mg\,l^{-1}$
S. capricornutum	Biomass	1.12	0	2.31	0	$mg\,l^{-1}$
S. capricornutum	Biomass	0.12	0	4.8	0	$mg\,l^{-1}$
S. capricornutum	Biomass	0.15	0	1.5	0	$mg\,l^{-1}$
C. vulgaris	Biomass	1200	0	1200	1	$mg\,l^{-1}$
S. subspicatus	Biomass		0	120	1	$mg\,l^{-1}$
C. reinhardtii	Biomass	120	0	120	1	$mg\,l^{-1}$
D. tertiolecta	Biomass	124	2	124	1	$mg\,l^{-1}$
E. gracilis	Biomass	112	0	112	1	$mg\,l^{-1}$
P. tricornutum	Biomass	117	0	117	1	$mg\,l^{-1}$
N. pelliculosa	Biomass	600	0	1200	0	$mg\,l^{-1}$
S. costatum	Biomass	2400	0	2400	1	$mg\,l^{-1}$
A. flos-aquae	Biomass	600	0	2400	0	$mg\,l^{-1}$
M. aeruginosa	Biomass	117	0	117	1	$mg\,l^{-1}$

NOEC and EC50 values for indicated species. NC=0 means NOEC was not censored. NC=2 means NOEC is left-censored. EC=0 means EC50 value was not censored. EC=1 means EC50 was right-censored.

Table 12.15 Aquatic Species EC50 Values

Common name	Species	Censored	EC50
Caddisfly	*Chimarra aterrima*		3.92
	Chironomus riparius	>	1.9
	Daphnia pulex		0.31
	Gammarus pseudolimnaeus		1.4
	Hyalella azteca		0.34
Mayfly	*Centroptilum triangulifer*	>	1.4
Mysid	*Americamysis bahia*	>	1.7
Mysid	*Americamysis bahia*	>	0.49
Crayfish	*Procambarus clarkii*	>	1.9
Stonefly	*Soyedina carolinensis*	>	1.8
	Daphnia magna		0.0596
	Daphnia magna		0.157
Oyster	*Crassostrea virginica*	>	0.223

Censored=> means value is right-censored. EC50 values in mg a.s. l^{-1}.

(d) Exclude all species with endpoint other than shoot weight and reassess lognormality. Comment on the formal test for normality compared to the visual assessment from the QQ-plot or histogram. What does this indicate about formal tests for normality on small datasets?

12.2 The data for this exercise come from an algae study. See Table 12.14.

(a) Use the SAS lognormal SSD program in Appendix A1, if SAS is available to you, and MOSAIC_SSD from the link in that appendix to fit a lognormal SSD to the EC50 data in Table 12.14. Note that the four values for *S. capricornutum* must be replaced by a single value, namely, the geometric mean of these four before being used in the SSD.

(b) Repeat (a) ignoring censoring to see the difference.

(c) Repeat (a) and (b) for the NOEC values and compare the resulting HC5 estimates. Note that there is one left-censored NOEC value in addition to several right-censored values. Assess the goodness of fit of the lognormal SSD.

12.3 The Excel file NTP_ssddata.xls given in Appendix A1 contains nontarget plant data for nine chemicals (identified only as chemical 1, 2, 3, etc.). The data for chemicals 9 and 10 were displayed in Figure 12.1 and 12.2 and that for Figure 12.2 is also given in Table 12.13. Select one or more other chemicals from this file and fit SSDs using MOSAIC_SSD (R based), SAS (if available), ETX (Excel based), BurrliOZ (version 2 is R based, and version 1 is self-contained), and SSD Master (Excel based) and make an assessment as to the preferable distribution and fit. For chemical 10, treat cucumber and pea as censored.

12.4 The data for this exercise are from an aquatic study with eleven aquatic species EC50 values. See Table 12.15. Note that there are two species with repeated EC50 estimates from different studies. Fit SSDs to these data after combining the repeated studies on the same species by a single value. Question: How is this done for *Americamysis bahia*, where both values are censored? Do this

(a) by ignoring censoring and

(b) taking censoring properly into account.

The SAS program and MOSAIC_SSD are the only known programs that do (b).

Chapter 13

Studies with Greater Complexity

13.1 INTRODUCTION

In this final chapter, some important topics will be covered that do not fit easily in previous chapters because of complexities in the experimental design or method of analysis. Section 13.2 deals with mesocosm and microcosm studies, a class of large, expensive studies that are sometimes considered the gold standard of ecotoxicity studies. As will be described in detail in that section, there are a great many responses measured in such studies, and it can be challenging making good use of the data because of low replication, sparse data, and occasional high variability. Just to give an idea of the scope of such studies, in a recently analyzed microcosm study, there were 174 phytoplankton, 35 periphyton, 22 lemna, and 168 zooplankton responses, each measured at multiple times.

At perhaps the opposite extreme are microplate studies, where minute quantities of chemicals and single organisms are placed in individual wells in a plate and observed, often for short time periods. These studies tend to be relatively inexpensive but present statistical challenges nonetheless. There is recent regulatory interest in these studies and new guidelines are in development. Section 13.4 considers what are called errors-in-variables regression models in which uncertainty in the explanatory variables as well as in the dependent variable is taken into account. Such models have been applied to some microplate experiments. Section 13.5 will give some attention to analysis of mixtures of chemicals. This is an important topic for environmental risk assessment, since few chemicals appear by themselves in the environment. It is much more common for mixtures of several chemicals to be present. The question of interest is whether toxicity studies on single chemicals adequately predict the toxicity of mixtures. Section 13.6 will give a brief discussion of benchmark dose (BMD) studies, which are strongly related to regression models to estimate an EC*x* value.

Section 13.7 concerns limit tests, which consist of a control and one test concentration only. Such tests might be done at a very early stage of evaluating a test substance. The idea is to expose the test organism to a very high level of the test chemical. If no adverse effect is found at this high level, no further testing may be needed to establish safety. If a significant effect is observed, then further testing with a dose–response study may be justified. An interesting statistical issue is how one proves that the effect at this high level does not exceed some regulatory limit. Section 13.8 extends the idea discussed in Section 13.7 to multiconcentration studies by flipping the usual idea of toxicity studies around. Rather than assume safety of a chemical as the default and look for significant evidence of an adverse effect, the starting point is to assume a chemical has an adverse effect and look for significant evidence that no adverse effect exists. While there are currently no regulatory guidelines calling for such evaluations, the methodology can be useful in some situations where no regression model is found but a sound statistical conclusion regarding toxicity is needed.

Finally, in Section 13.9, an introduction will be given to toxicokinetic–toxicodynamic (TKTD) models, which are increasingly important in ecotoxicology. An example of a continuous response measured over time will be given, followed by an indication of the recent attention being given to survival data from A TKTD perspective, including a reexamination of the biological principles underlying traditional modeling, such as the probit models discussed at length in Chapter 7.

Statistical Analysis of Ecotoxicity Studies, First Edition. John W. Green, Timothy A. Springer, and Henrik Holbech.

Companion website: www.wiley.com/go/Green/StatAnalysEcotoxicStudy

13.2 MESOCOSM AND MICROCOSM EXPERIMENTS

Statistically, there is no difference between mesocosm and microcosm studies, and after this paragraph, the term mesocosm will be used with the understanding that what is said applies to both types of study. The principal physical difference is that mesocosm studies are done outside and subject to influence from uncontrolled environmental factors, such as invasion by unplanned species. Microcosm studies are done in a laboratory setting, but attempt to mimic mesocosm studies by including numerous species in the same test vessel. In both cases, the intent is to determine both the direct and indirect effects of the toxic potential of the test substance. Indirect effects occur, for example, if one of the primary food sources of test organism A is reduced, forcing species A to adapt its diet to a possibly inferior food source or face population reduction or extinction unrelated to any direct toxic effect on species A. This is in sharp contrast to the single species studies that have been the subject of all previous chapters in this book. A mesocosm study employs a dose–response design, with multiple ponds at each of the several exposure concentrations and a control. The concentrations span the range of predicted environmental concentrations resulting from surface run-off and spray drift contamination. Under normal use, the application of the test material and exposure conditions result in direct contact between the introduced species and the test compound during application. The impact of a test compound typically is assessed at various times over weeks or months following treatment.

In older studies, a power analysis was often not reported prior to the start of a mesocosm study. When these studies are used in renewed applications for continued sale of a product, the sensitivity or ability to detect effects on abundance can be assessed by computing the minimum detectable difference percent, MDDEFF, that can be detected statistically from the data collected. The concept of the MDDEFF is quite simple and is described below. Indeed, even in a new study, an MDDEFF analysis might be substituted for formal power analysis.

A standard formula (e.g. Hogg and Tanis, 1988) to compare two sample means from normal populations with the same variance and sample sizes is

$$T = \frac{\text{Diff}}{s\sqrt{2/n}}, \tag{13.1}$$

where n is the common sample size, Diff is the magnitude difference ($\bar{x}_1 - \bar{x}_2$) between two sample means, s is the pooled estimate of the common standard deviation, and T is Student's t-statistic with $df = 2(n-1)$ degrees of freedom. If Eq. (13.1) is solved for Diff, the result is

$$\text{Diff} = Ts\sqrt{\frac{2}{n}}. \tag{13.2}$$

The minimum magnitude difference, MDD, that can be detected with 95% confidence is obtained by replacing T in Eq. (13.2) by $T(0.025, \text{df})$, which is approximately equal to 2. Thus,

$$\text{MDD} = Ts\sqrt{\frac{2}{n}} = s\sqrt{\frac{2T^2}{n}} \approx s\sqrt{\frac{8}{n}}. \tag{13.3}$$

The percent change from control ($\bar{x}_1$) that can be detected with 95% confidence is $100\text{Diff}/\bar{x}_1$.

Now $\text{CV} = 100s/\bar{x}_1$, so $\bar{x}_1 = 100s/\text{CV}$, and it follows that the minimum percent change from control that can be detected with 95% confidence is given by

$$\text{MDEFF} = \frac{100\text{Diff}}{\bar{x}_1} = \frac{100s\sqrt{8/n}}{(100s/\text{CV})} = \text{CV}\sqrt{\frac{8}{n}}. \tag{13.4}$$

For the small sample sizes in this microcosm study ($n=3$), T will be somewhat larger than 2, and in the formal analysis, the two-sided t-test will be replaced by the one-sided Williams (recommended by Brock et al., 2015) or Jonckheere–Terpstra or Dunnett test. The one-sided test will tend to reduce the minimum size effect that can be detected, but the lower degrees of freedom and the need to adjust for multiple comparisons of treatments to control rather than just the one used in the t-test will tend to increase it. Equation (13.4) is a balance between these offsetting trends and gives a reasonable indication of the actual minimum size effect that can be detected. Minor variations on Eq. (13.4) are available when the sample sizes or variances in the control and treatment group are different.

According to Brock et al. (2015), abundance in microcosm studies should be done after a log-transform of the data because abundance data is expected to be lognormally distributed. In particular, they follow the recommendation of Van den Brink et al. (2000) in using the transform

$$y = \text{Log}(ax+1), \tag{13.5}$$

where x is the abundance measure. The purpose of the +1 within the logarithm function is to handle incidents of zero abundance. Clearly the transformation given by Eq. (13.5) depends on the choice of the parameter a. Brock et al. advise $a = 1/(\text{minimum nonzero abundance value})$. Brock references van den Brink et al. (2000) for explanation, where the justification is "to avoid false discrepancy between zero abundance values and low abundance values." Van den Brink et al. (2000) refer to Cuppen et al. (2000) for further justification, where the reason given is to "downweight high abundance values and approximate a normal distribution of the data."

Cuppen et al. (2000) further references Van den Brink et al. (1995) for justification, and that reference justifies the factor a in the same way as given in Van den Brink et al. (2000). Sorting through the justification is somewhat unsatisfying. Whatever the merits of the justification, what is important is how much influence the choice of the term a has on the measures MDD and MDD%.

It is helpful to review the various MDD calculations in Brock et al. Reporting the MDDEF in terms of the transformed values can be confusing, so it is important to back-transform the results to the original units. Let MDD_{Ln} and MDD_{raw} be the minimum detectable difference on the log and untransformed scales, respectively, and let $Mean_{co,Ln}$ be the mean of the transformed controls. Then Brock et al. (2015) state that

$$MDD_{raw} = \frac{\exp\left(Mean_{co,Ln}\right) - \exp\left(Mean_{co,Ln} - MDD_{Ln}\right) - 1}{a} \tag{13.6}$$

and

$$MDDEFF_{raw} = 100 \frac{MDD_{raw}}{\left[\exp\left(Mean_{co,Ln}\right) - 1\right] / a}. \tag{13.7}$$

The factor a in the denominator of Eq. (13.6) cancels the a in the denominator of Eq. (13.7), but the dependence of $MDDEFF_{raw}$ on the term a persists through the dependence of $Mean_{co,Ln}$ on a. The calculations in Eqs. (13.6) and (13.7) are based on $\exp(Mean_{co,Ln})$ being an estimate of the control mean of the untransformed data. Unfortunately, this is not true. The relationship between the mean and variance of the untransformed data and the mean and variance of the transformed data is given by the following well-known formulas (Johnson et al., 1994):

$$\mu_{abu} = e^{\left(\mu_{Ln} + \frac{\sigma_{Ln}^2}{2}\right)} \tag{13.8}$$

$$\sigma_{abu}^2 = \left[e^{\sigma_{Ln}^2} - 1\right] e^{\left(2\mu_{Ln} + \sigma_{Ln}^2\right)}. \tag{13.9}$$

Unless σ_{Ln}^2 is zero, the back-transformed mean is never the mean of the untransformed data. The formulas in Brock et al. (2015) can be taken only as an approximation. If σ_{Ln}^2 is small relative to μ_{Ln}, then the approximation will be good. An alternative way to interpret the Brock et al. (2015) calculation is based on the symmetry of the normal density function about the mean. A consequence of that symmetry is that the mean and median of a normal distribution are the same. Furthermore, the log-transform is order-preserving so that the back-transform of the median is the median of the raw data. Thus, the Brock et al. MDD_{Raw} calculation can be regarded as the minimum detectable size effect on the median rather than the mean.

It should be noted that $a = 1/$(the minimum abundance value) and thus is not constant across treatments or families of organisms: It is a random variable with an extreme value distribution and highly data dependent and variable. The behavior of MDD_{Ln} and hence MDD_{abu} is highly sensitive to the value of a, but $MDD\%_{abu}$ is relatively insensitive. This can be illustrated with data from a microcosm study done for a regulatory submission. Macrophytes and micro-algae species were among those investigated. Minimum abundance values for these species ranged from 6 to 1401, so that $\mathrm{Log}_{10}(\mathrm{MMIN}) \le \mathrm{Log}_{10}(1401) < 3.15$. See Figure 13.1

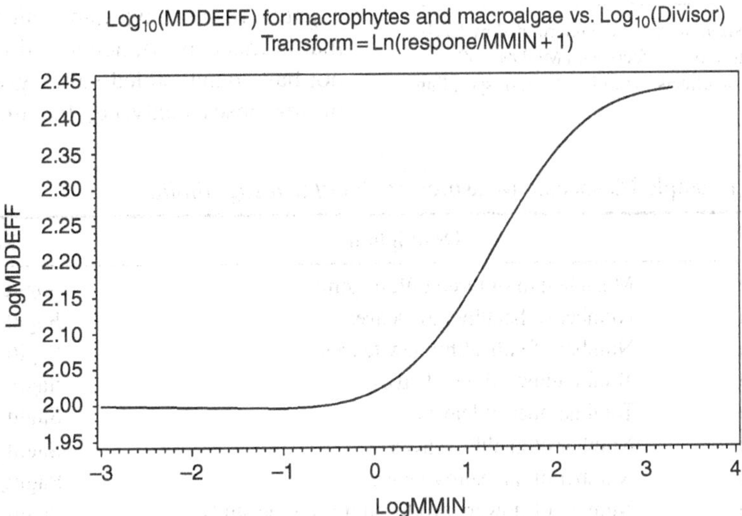

Figure 13.1 MDD% *vs*. minimum value for macrophytes and macro-algae. The increase in $\mathrm{Log}_{10}(\mathrm{MDD\%})$ over the range $0 < a < 2000$ is less than 22.5%.

for the relationship between Log_{10}(MDD%) and Log_{10}(MMIN) for minimum abundance values ranging from 0.001 to 2000. It is clear from the figure that over this range of possible minima, the impact on MDD% is small. It is only for very large values of the minimum that there is potential for a significant impact on MDD%. A subset of these data for the species *Sagittaria sagittifolia* is given in Table A1.13.1. The responses measured for this species and shown in spreadsheet Table 13.1 in the Excel file Sagittaria and a SAS dataset of the same name were (i) number of leaves, (ii) number of flowers produced during the study, (iii) number of new leaves produced by new plants during the study, and (iv) number of new plants produced during the study. Additional variables measured but not shown were the height of the three tallest stems (cm) and mean height. In the table, responses are listed as Sagitt_flowers, Sagitt_g, Sagitt_new_g, Sagitt_plants, Sagitt_stems, Sagitt_t, Sagitt_new_t, Sagitt_new_yb, and Sagitt_yb. Here flowers is the number of flowers per plant, the extension yb refers to unhealthy leaves, g refers to healthy leaves, t refers to total leaves, and new refers to new plants produced during the study. Thus new_t is the number of new leaves produced by new plants during the study. In the exercises, these responses are used to illustrate the calculations discussed in this section.

Table 13.1 Classes of Minimum Detectable Differences (MDD) in EFSA Aquatic Guidance Document

MDD class	%MDD comment
0 > 100%	No effects can be determined statistically
I 90–100%	Only large effects can be determined statistically
II 70–90%	Large to medium effects can be determined statistically
III 50–70%	Medium effects can be determined statistically
IV <50%	Small effects can be determined statistically

%MDD comment indicates how MDD% values are interpreted by EFSA. Thus, high value will be placed on tests of effects with MDD% < 50, whereas no value is given to tests on effects for which the corresponding MDD% exceeds 100.

Once the MDD% has been calculated, the question is what to make of it. See Table 13.1 for the interpretation used by EFSA in evaluating a mesocosm study.

The spreadsheet in the Excel file Sagittaria is a very small subset of a large mesocosm study and contains nine responses on a single species, *S. sagittifolia*. (See Table 13.2.) One reason for this example, besides illustrating the MDD calculations, is to indicate the variety of responses that might be recorded in one of these large mesocosm studies and the need for some way to summarize the huge number of analyses.

These responses were measured over 9 different days. (See Table 13.3 for days of measurement and some calculations.) It will be left as an exercise to verify the CV and MDD% calculations shown in Table 13.3. A SAS to do these calculations is *MDD_Calc.SAS*.

If these MDD% values are compared to the guideline in Table 13.1, it will be seen that 2 of the 63 calculations fall in the worst category (No effects can be determined statistically), 1 is in category I (Only large effects can be determined statistically), 3 are in category II (Large to medium effects can be determined statistically), 5 fall in category III (Medium effects can be determined statistically), and 52 fall in category IV (Small effects can be determined statistically). These quick summaries help the risk assessor evaluate how much credibility to give to the univariate statistical analyses of these data. To make the point clear, little credibility can be given to negative results from a study in which a large number of responses have so much variability as to make it likely that few or only very large effects can be found statistically significant. Conversely, if variability is low so that for most responses, small effects are likely to be statistically significant, then negative results have some value to the risk assessor. Note that in this context, actual estimates of effects (EC*x*) may not have much added value, given the very large number of responses analyzed, the uncertainty associated with

Table 13.2 Example Mesocosm Measures on *Sagittaria sagittifolia*

Species	Description	Response
S. sagittifolia	Mean height of three tallest stems	Sagitt_stems
S. sagittifolia	Number of healthy new leaves	Sagitt_new_g
S. sagittifolia	Number of unhealthy new leaves	Sagitt_new_yb
S. sagittifolia	Total number of new leaves	Sagitt_new_t
S. sagittifolia	Total number of leaves	Sagitt_t
S. sagittifolia	Number of healthy leaves	Sagitt_g
S. sagittifolia	Number of unhealthy leaves	Sagitt_yb
S. sagittifolia	Number of flowers produced during the study	Sagitt_flowers
S. sagittifolia	Number of new plants produced during the study	Sagitt_new_plants

Response is the variable name corresponding to the measure in the description.

Table 13.3 MDD% for *Sagittaria sagittifolia* Measures

Day	Name	MDD%	Std	CV	Normal
−7	SAGITT_FLOWERS	149	0.29	125.94	0.00
−1	SAGITT_FLOWERS	70	0.59	51.80	0.28
7	SAGITT_FLOWERS	67	0.40	41.63	0.44
14	SAGITT_FLOWERS	65	0.62	43.17	0.27
21	SAGITT_FLOWERS	56	0.90	46.60	0.15
28	SAGITT_FLOWERS	66	0.72	53.58	0.74
42	SAGITT_FLOWERS	34	0.17	14.87	0.92
56	SAGITT_FLOWERS	38	0.19	17.03	0.54
70	SAGITT_FLOWERS	43	0.30	16.37	0.75
−20	SAGITT_G	23	0.10	10.01	0.51
−14	SAGITT_G	25	0.11	10.71	0.51
−7	SAGITT_G	26	0.11	11.52	0.85
−1	SAGITT_G	23	0.09	10.36	0.62
7	SAGITT_G	24	0.09	10.76	0.47
14	SAGITT_G	21	0.08	9.18	0.93
21	SAGITT_G	24	0.10	10.24	0.00
28	SAGITT_G	30	0.13	13.75	0.95
42	SAGITT_G	24	0.10	10.74	0.79
56	SAGITT_G	36	0.15	16.89	0.23
70	SAGITT_G	27	0.14	10.56	0.56
84	SAGITT_G	118	0.19	84.02	0.00
−7	SAGITT_PLANTS	88	0.55	78.67	0.13
28	SAGITT_PLANTS	41	0.23	17.88	0.99
56	SAGITT_PLANTS	33	0.17	14.12	0.82
70	SAGITT_PLANTS	40	0.20	18.39	0.11
84	SAGITT_PLANTS	33	0.13	15.64	0.69
−27	SAGITT_STEMS	12	0.04	5.22	0.78
−20	SAGITT_STEMS	9	0.03	4.06	0.55
−14	SAGITT_STEMS	9	0.03	3.87	0.32
−7	SAGITT_STEMS	10	0.04	4.56	0.02
−1	SAGITT_STEMS	9	0.03	4.26	0.26
7	SAGITT_STEMS	7	0.02	3.32	0.25
14	SAGITT_STEMS	6	0.02	2.79	0.09
21	SAGITT_STEMS	7	0.02	2.96	0.46
28	SAGITT_STEMS	6	0.02	2.63	0.61
42	SAGITT_STEMS	6	0.02	2.82	0.24
56	SAGITT_STEMS	7	0.02	3.04	0.17
70	SAGITT_STEMS	11	0.04	4.90	0.61
84	SAGITT_STEMS	9	0.03	4.07	0.52
−20	SAGITT_T	18	0.07	7.98	0.57
−14	SAGITT_T	32	0.15	14.20	0.48
−7	SAGITT_T	26	0.11	11.47	0.95
−1	SAGITT_T	18	0.07	7.99	0.27
7	SAGITT_T	17	0.06	7.22	0.90
14	SAGITT_T	19	0.08	8.39	0.39
21	SAGITT_T	23	0.10	9.58	0.17
28	SAGITT_T	22	0.09	9.54	0.72
42	SAGITT_T	21	0.08	9.32	0.63
56	SAGITT_T	27	0.12	12.02	0.29
70	SAGITT_T	24	0.11	10.58	0.45
84	SAGITT_T	89	0.48	74.28	0.01
−20	SAGITT_YB	107	0.44	94.18	0.85
−14	SAGITT_YB	60	0.36	33.11	0.50
−7	SAGITT_YB	38	0.22	15.86	0.74
−1	SAGITT_YB	30	0.14	13.23	0.82
7	SAGITT_YB	33	0.20	12.84	0.28
14	SAGITT_YB	27	0.14	10.18	0.76
21	SAGITT_YB	39	0.24	15.78	0.51
28	SAGITT_YB	33	0.18	13.99	0.36
42	SAGITT_YB	28	0.13	12.10	0.27
56	SAGITT_YB	32	0.17	13.43	0.66
70	SAGITT_YB	35	0.15	16.49	0.89
84	SAGITT_YB	98	0.45	84.20	0.11

Normal is the p-value for the Shapiro–Wilk test of normality of the response indicated.

each such estimate, and the broad categories of interest defined by Table 13.1. That said, these univariate analyses ignore the interactions of the numerous species in each mesocosm, and a multivariate approach has more conceptual appeal.

13.2.1 Multivariate Analysis of Mesocosm Studies

Section 13.2 deals with univariate analysis of one response on a single study day. Such an analysis addresses at best indirectly the influence of changes in other species on the particular response on the particular species being analyzed. The intent of a mesocosm study is to investigate the effect on species in relation to their environment. A univariate analysis is certainly useful, expected, and readily understood, but does not really capture the full value of this type of study. The standard multivariate approach is through principal response curves (PRC), a concept introduced van den Brink and ter Braak (1999). PRC analysis is a type of principal component analysis (PCA), and that topic will be addressed only very sketchily here. There are numerous useful texts on PCA, and the reader is referred to Jolliffe (2002), Rencher and Christensen (2012), Morrison (2004), and Furuhashi (2015) for a development of PCA, and in the last reference, to relevant R programs. In this section, an introduction to PRC will be given that draws heavily on van den Brink and ter Braak (1999).

The idea of PCA is to reduce dimensionality in the presence of a large number of factors or variables, such as exists in a mesocosm study. To do this, linear combinations, called principal components, of the many variables are generated. The idea is that there are some unknown underlying mechanisms or causes that explain the various responses that have been measured directly. These principal components thus represent latent or hidden variables that are not measured directly and are most likely unknown at the time measurements were taken and can only be inferred from the data collected. Ideally, each of these components can be interpreted as corresponding to some "natural" grouping of the variables that capture in some way one of these latent variables. The actual statistical mechanism to generate these principal components is conceptually simple. Each principal component is a linear combination of the form given in Eq. (13.10):

$$P_i = c_{i1}X_1 + c_{i2}X_2 + c_{i3}X_3 + \cdots + c_{ik}X_k, \quad (13.10)$$

where $X_1, X_2, X_3, \ldots, X_k$ are the measured variables and the coefficients c_{ij} are constrained by requiring the sum of their squares is 1. This constraint can be stated in vector form as

$$C_i'C_i = 1, \quad (13.11)$$

where $C_i' = (c_{i1}, c_{i2}, c_{i3}, \ldots, c_{ik})$.

The first principal component, P_1, is determined by maximizing the variance of the linear combination. This combination accounts for the greatest proportion of the observed variation in the data among all possible linear combinations. The observed variation in the data can be quantified as the total variance, which is defined in terms of the variance–covariance matrix, **S**, of the response variables $X_1, X_2, X_3, \ldots, X_k$. If the matrix $[X_1 X_2 X_3 \ldots X_k]$ has rank k, then the diagonal elements of **S** are the variances of the individual response variables, and the total variance is simply the sum of these diagonal elements. A more general development can be given in terms of eigenvectors and eigenvalues, but that would take us too far afield from our present purpose.

The second principal component is determined in the same way but with the added constraint that it is orthogonal to the first component. That is, the constraints are

$$C_2'C_1 = 0 \text{ and } C_2'C_2 = 1. \quad (13.12)$$

The remaining principal components are obtained sequentially by requiring that each new component maximize it variance among all linear combinations that are orthogonal to all previous components. Statistically then, the principal components are independent linear functions of the response variables. The magnitude of c_{ij} indicates the relative importance of variable X_j to that component, and the sign of c_{ij} indicates whether the variable X_j is positively or negatively correlated with the component. Each component will have a numeric value or score for each replicate.

If Y_{sj} is a measure of abundance of species s in sample j and there is only one principal component, then the basic model for Y_{sj} is given by

$$Y_{sj} = \overline{Y}_s + b_s x_j + e_{sj}, \quad (13.13)$$

where x_j is the score (value) of the principal component for sample j and e_{sj} is a zero-mean error term, and sample here means a replicate observed at a particular time. Typical of mesocosm studies, Y_{sj} is taken to be $\text{Log}_{10}(u+1)$, where u is the measured abundance value, usually a count or perhaps a normalized count (e.g. number of leaves per unit length of stem). If there are more principal components to be included to explain a sufficient proportion of the total variance, then Eq. (13.13) is modified in an obvious way. The PRC model can be described next.

Let T_{dt} be the treatment effect of treatment d at time t. Then to PRC model is given by

$$Y_{djts} = \overline{Y}_{0ts} + b_s T_{dt} + e_{djts}, \quad (13.14)$$

where Y_{djts} is the abundance measure on species s at time t in replicate j and treatment d and $d=0$ denotes the control.

Equation (13.14) requires a value for T_{dt}. To obtain that value, we fit a regression model

$$Y_{sj} = \sum_t a_{ts} I_{ts} + \sum_t \sum_d \tau_{dts} I_{jdt} + e_{sj} \quad (13.15)$$

constrained so that

$$\tau_{dts} = b_s T_{dt}, \quad (13.16)$$

where I_{ts} and I_{jdt} are indicator variables. In the language of van den Brink and ter Braak (1999), this constrains species-specific treatment effects to have a "basic response pattern." A specialized computer program, CANOCO (ter Braak, 1988; ter Braak and Smilauer, 2002; Šmilauer and Lepš, 2015), can be used to carry out the analysis. Alternatively, the R function prc from the vegan package is also now available. The chlorpyrifos invertebrate data for the example in van den Brink and ter Braak (1999) are conveniently included with the vegan package and are discussed in http://cc.oulu.fi/~jarioksa/softhelp/vegan/html/prc.html, Van Wijngaarden et al. (1996), and Van den Brink et al. (1996). To obtain the data, install and load the package prc from the CRAN website. Then issue the commands

```
require(vegan)
data(pyrifos)
write.table(pyrifos,"c:\\temp\\pyrifos.
txt",col.names=F,row.names=F)
write.table(pyrifos,"c:\\temp\\pyrifos.
txt",col.names=T,row.names=T)
data <-read.table("c:\\temp\\pyrifos.
txt",header=T) write.
table(data,"clipboard",sep="\t",col.
names=NA)
```

Then open Excel and paste the data into an open spreadsheet. To do the analysis using the vegan package, the syntax is given in the documentation for the vegan package as follows. The structure of the data is given by the gl and factor functions. The function gl provides labels for weeks. The syntax is gl(n, k, labels), where n is the number of levels of the indicated variable and k is the number of replicates. Thus week <-gl(11, 12, labels) indicates 11 weeks in each of 12 ponds (replicates), with weeks labeled from −4 to 24 as indicated. So each week of data is replicated (observed) in ponds 1–12. Doses (concentrations) are in the order shown in the factor function. Thus pond 1 is the low dose, 0.1 μg l^{-1}; ponds 2 and 3 are control; pond 4 is 0.9 μg l^{-1}; pond 5 is another control; pond 6 is the high dose, 44 μg l^{-1}; etc.

```
require(vegan)
data(pyrifos)
week <-gl(11, 12, labels=c(-4, -1, 0.1,
1, 2, 4, 8, 12, 15, 19, 24))
dose <-factor(rep(c(0.1, 0, 0, 0.9, 0,
44, 6, 0.1, 44, 0.9, 0, 6), 11))
# PRC
mod <-prc(pyrifos, dose, week)
mod              # RDA
summary(mod)     # PRC
logabu <- colSums(pyrifos)
plot(mod, col=1, type = "b", scaling =
"symmetric", axis=1, select = logabu >
100)
# Permutations should be done only
within one week, and we only
# are interested on the first axis
anova(mod, strata = week, first=TRUE)
```

See Figure 13.2 for the plot produced by the plot function in this code. This is figure 3 of van den Brink and ter Braak (1999), except inverted. It is at least visually evident that the two highest treatment groups are different from the control from week 1 through week 12 or 13. The F-test produced by the indicated code indicates significant differences among the treatment groups across time but does not indicate what treatments are different from the control in what weeks. There are several ways to address that issue. One method proposed in van den Brink and ter Braak (1999) is to compute the first principal component of the data separately by week and then analyze the scores from that component in the different treatment groups and replicates by Williams' test. This is quite simple to do, and SAS and R programs (*pca4.sas* together with *pca_Will4.sas* and *pyrifos_will_updatedd.R*) to generate these scores and apply Williams' test are provided in Appendix 1. It is left as an exercise to carry out the analysis. One of the flaws in this method, duly noted in the cited paper, is that it ignores much of the variability in the data by using only the first principal component. In the chlorpyrifos data, the second or even the third principal component often accounts for a substantial proportion of the total variance. A completely univariate approach would be to analyze each species or selected groupings or abundance measures of species separately, but this approach ignores the interactions of species, which is a primary reason for doing a mesocosm study, and many species values are zero, making such an analysis problematic.

Van Wijngaarden et al. (1995) reported an ordination method for the multivariate analysis of mesocosm studies,

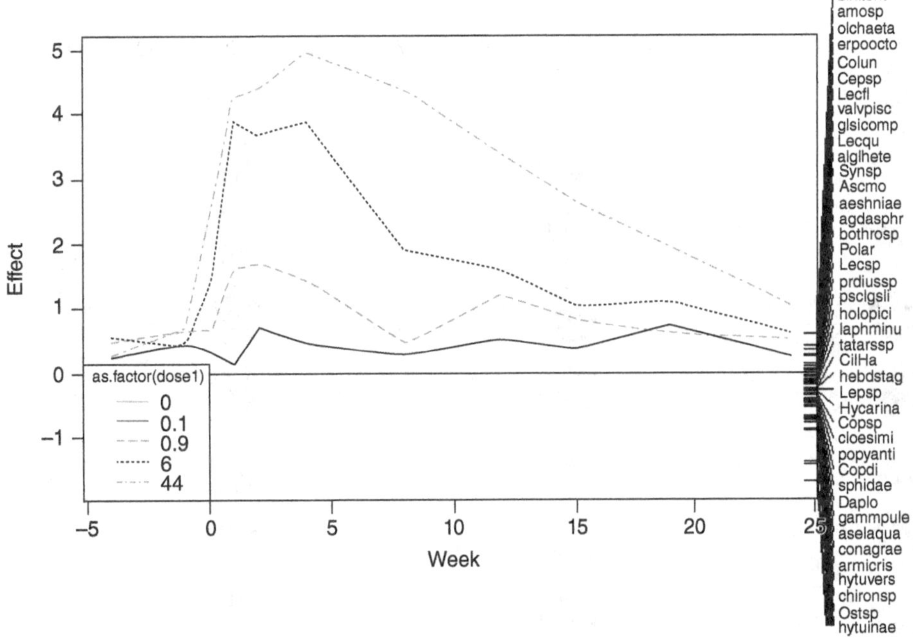

Figure 13.2 Principal response curve analysis of chlorpyrifos mesocosm data. Species codes on the right indicate species positively and negatively correlated with the first principle response curve.

but they state it has poor power properties when the number of replicates is small, which is almost always true for mesocosm studies. The reader is referred to that and Van den Brink and ter Braak (1999) among a series of articles and texts dealing with multivariate methods. In particular, Šmilauer and Lepš (2015) gives a thorough description of such methods.

13.3 MICROPLATE EXPERIMENTS

A type of experiment not previously discussed in this text is one in which very small quantities of biological material are placed in tiny chambers, called wells, in a plate that holds 6, 24, 96, 348, or even 1536 wells. Such well plates are examples of multiwell microtiter plates that are used in various types of studies. These are very commonly used in *in vitro* bioassays such as gene profile studies and are discussed in Schlain et al. (2001), Straetemans et al. (2005), and Murie et al. (2014), for example. In high-throughput greenhouse studies in the discovery phase of crop protection products, 96-well plate designs are also very common (e.g. Zhang et al., 2017). In general toxicology, the enzyme-linked immunosorbent assay (ELISA) (e.g. Pratheeshkumar et al., 2012) uses antibodies to detect detecting levels of cytokines, hormones, steroids, and other substances and is typically done with 96-well microtiter plates consisting of 12 columns of eight rows. For convenience, columns are labeled 1–12 and rows are labeled A–H. See Figure 13.3 for a schematic of such a 96-well plate.

There are known issues with studies done using these titer plates. As Zimmermann et al. (2009) states it, these include "Spatial heterogeneities caused by different positions within and between the plates that interfere with the measurements of the responses to different samples and dosages, e.g. differences between rows, columns, plates, days of measurements and the positions of the standards and samples." Proper experimental design has the objective of assuring an unbiased and stable estimate of the effect of the chemical under investigation. Lazic (2017) has a very clear discussion of potential problems and practical experimental designs to reduce their effect, including the one shown in Figure 13.3. There is also a website (http://labstats.net/articles/randomise_spatial.html) related to that book that, as of 7 March 2017, is readily accessible and informative. The basic issues are (i) a possible edge effect, where the outermost wells may dry out before the study is done, making the contents unusable; (ii) possible heat or light differences across the plate that affect the results; and (iii) mislabeling of wells can occur because of manual error or malfunction of automated equipment. Problem (iii) is most acute when there is an attempt to randomly locate treatments in each plate, especially if the location is manual. Zimmermann et al. (2009) found good results when the experimental design called for each plate dedicated to a single treatment with multiple plates for each treatment (including controls). Their design allowed for five dilutions of each treatment and standard in the same titer plate. These dilutions were systematically located in increasing order down the columns shown in Figure 13.3. To reduce

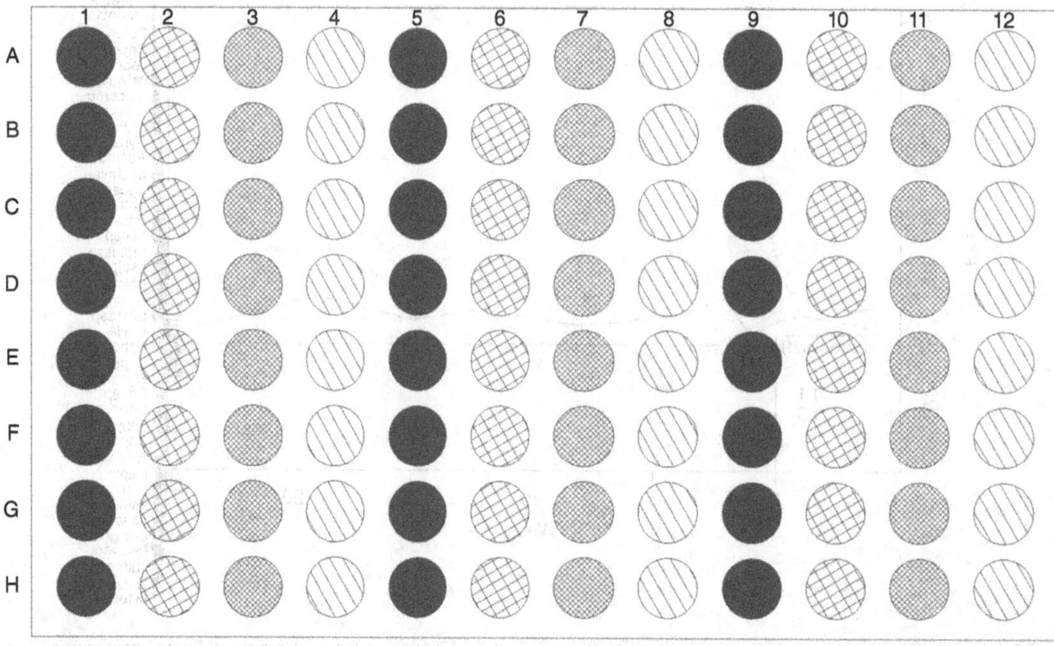

Figure 13.3 96-Well titerplate. Treatments are arranged in columns. Solids are controls. Other patterns indicate the three treatment groups.

the possible edge effects, rows A and H and columns 1 and 12 were blanks, and in fact they made row G and column 11 blanks as well. They found in their investigations that there was little position bias evident in other rows and columns.

Such a design will not always be feasible, however, and Lazic (2017) permits several treatments on each plate. An OECD test guideline (TG) under development in 2017 called Xenopus Embryonic Thyroid Signaling Assay (XETA) (discussed below) follows this approach.

The OECD TG 236, the Fish Embryo Acute Toxicity (FET) Test, has a design in which only one treatment is applied within a single titer plate. This design should remove any substantial concern about position bias. In the embryo stage, instead of fish swimming freely in a tank, the fish are immobile. They are very tiny at this stage of life, and the test chamber is a 24-well titer plate with a single embryo in each well. The test design calls for a separate titer plate for each treatment group and control. In each plate, there are five rows of four columns devoted to embryos in the same concentration of the test chemical or control but with row five containing only dilution water, making it an internal plate control. One titer plate is devoted to the negative control. If a solvent or a positive control is used, then another titer plate is devoted to each control. The embryos in different wells on the same plate are treated as independent observations. If any mortality is observed in the internal dilution water controls in a titer plate, the entire plate is rejected and not used in the evaluation. If this occurs in the plate devoted to the solvent control or positive control (if present), then the experiment must be repeated. If such mortality occurs in a plate devoted to a treatment concentration, there could be a negative impact on the quality of the concentration–response curve fit to the data.

A guideline for frog embryos under development is the XETA, which uses 96-well plates, which can be visualized as 8 rows of 12 wells for fluorescence quantification and 6-well titer plates to expose tadpoles. There are three controls (described below) and five test concentrations. The transgenic *Xenopus* tadpoles used carry a green fluorescent protein (GFP). The level of fluorescence detected is proportional to the level of transcription of the GFP gene, which in turn is an indicator of developmental delay or acceleration. An internal positive control is provided by the most biologically active form of thyroid hormone, T3 (triiodothyronine), and all test chemicals are tested with and without T3. A T3+T4 control is also included in each assay to confirm that the fluorescence obtained with T3 is not saturated, so the fluorescence measured includes the effect, if any, of the test chemical. The assay uses tadpoles at the same developmental stage and the same medium as the Frog Embryo Teratogenesis Assay Xenopus (FETAX) test that has been used in numerous laboratories, so there is also a FETAX control in which tadpoles are exposed only to the FETAX medium. Except for blanks, each well contains 10 *Xenopus* tadpoles initially.

As stated above, the assay measures GFP protein fluorescence in the transgenic THbZIP tadpoles. It does this by way of a robotic imaging system that transforms the fluorescence signal to a numerical format. Given the manual loading and the automatic measuring of the wells, some wells labeled as blank contain an embryo, and some labeled as containing an embryo are actually empty except for the medium, and a well label blank may actually be empty, that is, have no medium. Detection and handling of the resulting outliers in the data is thus a serious concern.

The 96-well titer plate design calls for four 96-well plates. Plate 1 contains the three controls and the highest level of the test chemical, each occupying 20 wells in adjacent parallel rows. Plate 2 contains the other four levels of the test compound, each again occupying 20 wells in adjacent parallel rows. Plate 3 contains the same four concentrations of the test chemical mixed with T3, again in 20 wells in parallel rows, while plate 4 contains 20 wells in parallel rows of the high test concentration mixed with T3. All other wells in each plate are either empty of blank. Except for empty and blank wells, each well also contains a single *Xenopus* embryo. Finally, there are three distinct runs (on different days) of the entire experimental setup.

In the validation studies done in developing the XETA TG, the between-plate (within-replicate) variance was smaller than the within-plate variance by a factor of 2–6, though from the design, the plate-to-plate variance could only be assessed for the blanks. Between-plate variance was usually about the same as between-rep variance, though between-plate variance was sometimes smaller by a factor of about 3. These results are somewhat different from those reported by Zimmermann et al. (2009), where between-replicate variance dominated other sources of variability. However, the experimental design investigated by Zimmermann et al. was different, and the results may not be comparable. Their modeling had a nested variance component structure, whereas the XETA experimental design leads to variance components for replicate, replicate-by-treatment, and well-nested within-replicate-by-treatment.

See Figure 13.4 for the setup on plate 1. The setup on plates 2 and 3 is similar with the three controls replaced by three test concentrations. Treatments in this experimental setup are confounded with plates, so it is not possible to separate plate differences from treatment differences. Since all wells with a given treatment are in the same plate, wells are treated as independent observations with no location-related variance components.

The analysis of XETA 96-well plate design deals with possible mislabeling of wells by trimming a small percent of the lowest and highest responses in each treatment group or control. The percent trim can be expressed as 1, 5, or 10%. A well whose response is less than or equal to $p\%$ of

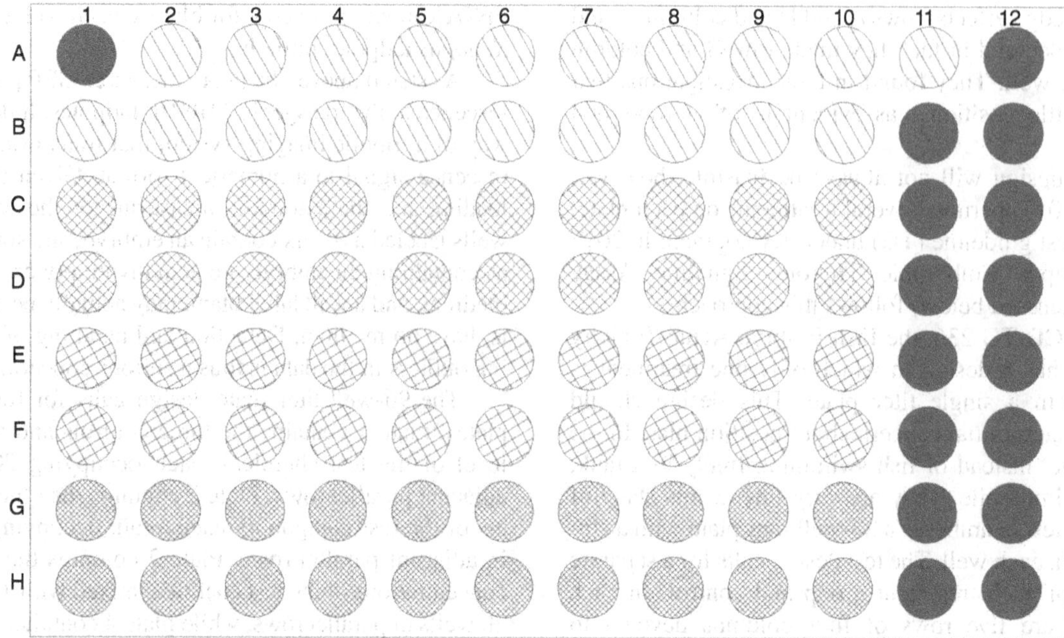

Figure 13.4 XETA 96-well design, plate 1. Dark gray wells are blanks, except well A1, which is empty due to a mechanical configuration. Otherwise, rows A and B = FETAX, C and D = T3, E and F = T3 + T4, G and H = test chemical low concentration.

the observed responses for the nominal treatment group is omitted. Given that there are 20 wells per group (excluding blanks), there is no difference between a 1% and a 5% trim: Either excludes the smallest and largest observations. A 10% trim would exclude the two smallest and two largest values. [In the unlikely event that several observations have exactly the same smallest or largest observed response, the number of excluded observations could be higher.] Such trimming would partially mitigate any edge effect, though clearly the top and bottom rows could still be affected. The data collected during validation of this guideline did not support the existence of an edge effect. If there were a column effect, it would apply equally to all groups. Once the trimming is done, the replicate means for each treatment were analyzed by two applications of Williams' test, one comparing unspiked treatments against the FETAX control and the other comparing treatments spiked by T3 against the FETAX + T3 control. In each case, care is given to how many treatment groups are amalgamated by the PAVA algorithm and whether a trend is evident in the concentration–response. If three or four of the five treatment groups are combined into one smoothed mean by PAVA, serious doubt exists whether there is actually a trend and a pairwise analysis by Dunnett's may be advised. As a check, the Jonckheere–Terpstra and Dunnett tests were done for all datasets. For the most part, these led to the same NOECs, though there were a few noteworthy exceptions. Examples from XETA experiments are provided to illustrate various concerns. See Figure 13.4 for a summary of one example.

Table 13.4 Williams' Test with Four Groups Amalgamated

DoseVar	N	YRAW	NSTR	YBAR
0	51	3.60432	51	3.60432
1	49	3.7079	193	3.62727
3	48	3.59048	.	3.62727
10	48	3.59799	.	3.62727
30	48	3.61104	.	3.62727
100	49	3.67015	49	3.67015

N = sample size YRAW, mean observed response (after log-transform); NSTR = amalgamated sample size; YBAR, amalgamated mean response.

Test concentrations 1, 3, 10, and 30 were combined in forcing the means, YRAW, to increase. While a significant high concentration effect may be plausible, Williams' test did not find the higher mean response in the low test concentration significant, making the NOEC = 100 mg l^{-1}. The monotonicity of the concentration–response is not clear. The Dunnett and Jonckheere–Terpstra tests likewise set the NOEC as 100 mg l^{-1}. The compound, CEF, was not expected to have an effect.

At least in principle, regression analysis can be done on XETA data. Data such as in Table 13.4 will not support a useful model, but in cases where there is a clear concentration–response trend, the quality of models fit to the data can be assessed using the techniques explored in Chapter 4. For example, confidence intervals for model parameters, especially EC10, that are extremely wide would indicate a model with little utility. Two examples are presented to illustrate the regression approach.

For the first example, summary data are presented in Table 13.5 and Figure 13.6, with the full dataset showing individual well responses given in Appendix A1 as aa spreadsheet in the Excel file TRIAC_Lab1. Table 13.5 is also provided as a SAS dataset and an Excel spreadsheet, both named TRIAC_LAB1_Repmeans, with the latter in the TRIAC_Lab1 Excel file. Another example is given in Exercise 13.4. SAS programs to carry out the required trimming and NOEC analyses are *normalize_CEF.sas* and *Williams macro.sas*, to be run in that order. For regression, the programs from Chapter 4 are applied after first running the *normalize_CEF.sas* program to trim the data as required. The tests and regressions are based on the replicate means, which are created from the data after the normalized program is run. A simple program, *make_mean_cef.sas*, is provided for this purpose. See Table 13.5 for the replicate data and Figure 13.6 for the treatment means. The Williams and Dunnett tests are based on the full data provided in the appendix in order to obtain the proper variance structure.

For model fitting, note that that if SQRTFLUOR is modeled, a 10% change in FLUOR corresponds approximately to a 5% change in SQRT for the data in this example. To justify this, a 10% from the control mean gives a mean response of and finding the percent change in SQRTFLUOR

Table 13.5 XETA Data with TRIAC

Conc	Rep	Fluor	SqrtFluor	LogFluor	*N*	StdFluor	StdSqrt	StdLog
0	1	4953.31	70.178	8.4964	16	786.187	5.5023	0.15491
0	2	5341.69	73.036	8.58053	16	412.665	2.81012	0.07666
0	3	3450.88	58.684	8.14228	16	326.66	2.75353	0.09301
0.01	1	4661	68.165	8.44062	16	526.084	3.94026	0.11838
0.01	2	6113.06	78.06	8.71167	16	714.357	4.59242	0.11843
0.01	3	3989.44	63.117	8.28857	16	309.469	2.45202	0.07784
0.1	1	4795.06	69.149	8.46961	16	515.524	3.7973	0.11234
0.1	2	6441.75	80.16	8.76559	16	671.195	4.13748	0.10235
0.1	3	4211.69	64.852	8.34281	16	327.487	2.51217	0.07719
1	1	5977.13	77.229	8.6914	16	567.776	3.69097	0.09615
1	2	8064.56	89.706	8.990086	16	765.189	4.31334	0.09748
1	3	5706.63	75.487	8.64648	16	458.472	2.98518	0.07785
2	1	6907.13	83.017	8.83585	16	662.398	4.0321	0.09841
2	2	8498.25	92.117	9.04464	16	685.201	3.68321	0.07932
2	3	6572.25	81.014	8.78786	16	502.816	3.10181	0.07664
5	1	8060.44	89.721	8.99208	16	594.921	3.35235	0.07565
5	2	10558.44	102.667	9.26134	16	914.258	4.36273	0.08342

Fluor, replicate mean of fluorescence values; SqrtFlour, mean of square-root transformed fluorescence values; LogFluor, mean of log-transformed fluorescence values; *N*, number of well values used for mean (after 10% trim); StdFluor, standard deviation of well values yielding indicated mean; StdSqrt, standard deviation of square-root transformed well values yielding mean = SqrtFluor; StdLog, standard deviation of log-transformed well values yielding mean = LogFluor.

Table 13.6 Treatment Means of TRIAC Data

Conc	Fluor	SqrtFluor	LogFluor	NSubJS	StdFluor	StdSqrt	StdLog
0	4581.96	67.30	8.406	48	982.150	7.341	0.222
0.01	4921.17	69.78	8.480	48	1041.130	7.276	0.205
0.1	5149.50	71.39	8.526	48	1083.270	7.384	0.203
1	6582.77	80.81	8.776	48	1220.860	7.354	0.178
2	7325.88	85.38	8.889	48	1044.750	6.036	0.140
5	8832.65	93.70	9.074	48	1414.200	7.336	0.153

Fluor, mean untransformed fluorescence value for concentration; SqrtFluor, mean of square-root transformed fluorescence values for concentration; LogFluor, mean of log-transformed fluorescence values for concentration; StdFluor, standard deviation of well values yielding indicated mean across wells, ignoring reps; NSubJS, number of well values used for mean (after 10% trim); StdSqrt, standard deviation of square-root transformed well values yielding mean = SqrtFluor across wells, ignoring reps; StdLog, standard deviation of log-transformed well values yielding mean = LogFluor across wells, ignoring reps.

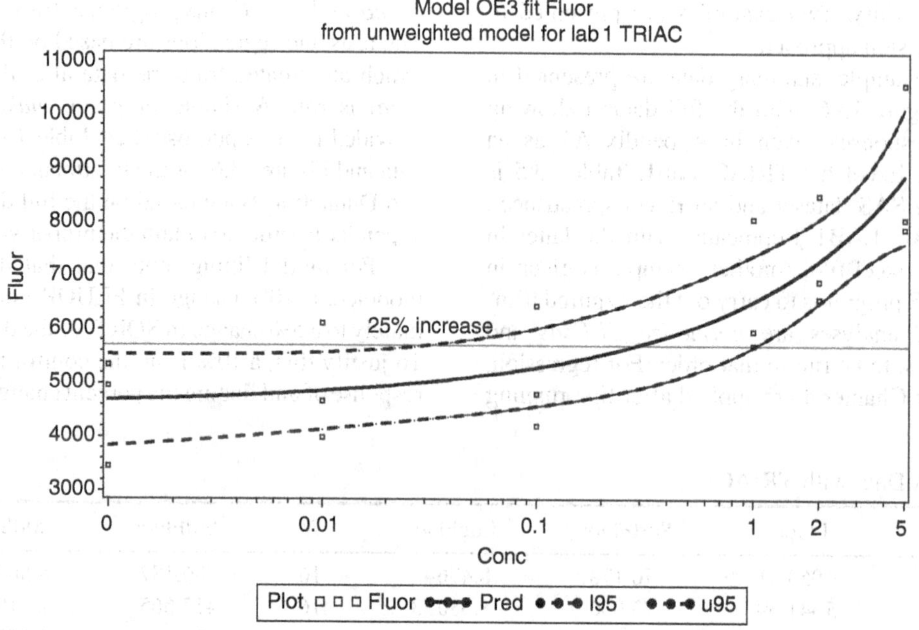

Figure 13.5 Regression model fit to data from Table 13.6. One of three models fit to data in Table 13.6. Wide confidence bounds for EC25 apparent.

that yields a mean response of 5040. Sqrt(5040)=71, which is about 5.5% of the control mean of SQRTFLUOR. Similar calculations can be done for the log-transform.

Three models were fit to the data in Table 13.6, namely, OE3, Bruce–Versteeg, and the three-parameter log-logistic (3PL), and it is rather interesting to compare the results of these models. Figure 13.5 gives the plot for model OE2. See Tables 13.7–13.9 for summaries of the model details. OE3 gives the best fit in terms of agreement between observed and predicted means and percent change from control (see Figure 13.5). The agreements of the observed and predicted control means are especially good, and this is generally critical to good estimation. However, the confidence interval for EC10 is very wide, encompassing all positive test concentrations.

The confidence interval for EC10 from the model BVP is not as wide, enclosing only two test concentrations, but the agreement of the predicted and observed control means is poor and the point estimate of EC10 is not plausible. Table 13.8 summarizes the fit of this model.

The 3PL model fit is rather interesting. The agreement of the predicted control mean to the observed is almost as poor as that for BVP, while the point estimate of EC10 is no more plausible than that for BVP. However, it is striking how short is the confidence interval for the EC10 estimate, being completely enclosed within the span between two adjacent test concentrations. The shortness of the confidence interval is striking given the apparent disconnect between the point estimate and the observed percent changes.

Table 13.7 Summary of OE3 Fit to TRIAC Data

Conc	Obs mean	Pred mean	Obs %Chng	Pred %Chng
0	4 581.96	4 567.46	0	0
0.01	4 921.17	4 855.82	7	6
0.1	5 149.5	5 295.9	12	16
1	6 582.77	6 531.43	44	43
2	7 325.88	7 282.46	60	59
5	8 832.65	8 861.95	93	94

Source	DF	SS	MS	*F* value	Prob *F*
Model	3	7.40E+08	2.47E+08	217.03	<0.000 1
Error	15	17 051 665	1 136 778		
Uncor total	18	7.57E+08			

Parameter	Estimate	LCB	UCB
Y0	4 567.5	3 460.2	5 674.7
EC10	0.031 736	0.000 136	7.432 75
D	0.383 3	0.010 6	0.756

Obs mean, mean fluorescence value for indicated test concentration; Pred mean, estimated mean fluorescence value for indicated test concentration; Obs %Chng, % change from observed control mean to observed treatment mean; Pred %Chng, % change from estimated control mean to estimate treatment mean; DF, degrees of freedom for indicated model terms; SS and MS, sum of squares, mean square, for indicated model terms; Uncor total, uncorrected total sum of squares; F value and Prob *F*, *F*-test for indicated model term and associated probability; LCB and UCB, lower and upper 95% confidence bounds for indicated parameter.

Table 13.8 Summary of BVP Fit to TRIAC Data

CONC	Obs mean	Pred mean	Obs %Chng	Pred %Chng
0	4581.96	4876.85	0	0
0.01	4921.17	4877.14	7	0
0.1	5149.5	4957.65	12	2
1	6582.77	6525.03	44	34
2	7325.88	7504.54	60	54
5	8832.65	8690.38	93	78

Source	DF	SS	MS	*F* value	Prob *F*
Model	3	7.40E+08	2.47E+08	211.33	<0.0001
Error	15	17501379	1166759		
Uncor total	18	7.57E+08			

Parameter	Estimate	StdErr	LCB	UCB
EC10	0.3131	0.05	0.0455	2.157
Gamma	1.3447	0.6	0.0657	2.6236
y0	4876.8	360.8	4107.9	5645.8

Obs mean, mean fluorescence value for indicated test concentration; Pred mean, estimated mean fluorescence value for indicated test concentration; Obs %Chng, % change from observed control mean to observed treatment mean; Pred %Chng, % change from estimated control mean to estimate treatment mean; DF, degrees of freedom for indicated model terms; SS and MS, sum of squares, mean square for indicated model terms; *F* value and Prob *F*, *F*-test for indicated model term and associated probability; Uncor total, uncorrected total sum of squares; LCB and UCB, lower and upper 95% confidence bounds for indicated parameter.

Table 13.9 Summary of 3PL Fit to TRIAC Data

Conc	Obs mean	Pred mean	Obs %Chng	Pred %Chng
0	4581.96	4847.93	0	0
0.01	4921.17	4864.36	7	0
0.1	5149.5	5011.68	12	3
1	6582.77	6377.13	44	32
2	7325.88	7510.8	60	55
5	8832.65	8787.13	93	81

Source	DF	SS	MS	*F* value	Prob *F*
Model	3	7.40E+08	2.47E+08	211.68	<0.000,1
Error	15	17472350	1164823		
Uncor total	18	7.57E+08			

Parameter	Estimate	LCB	UCB
D	0.7371	0.0349	1.4392
EC10	0.29885	0.12122	0.73677
Y0	4847.9	4068.9	5627

Obs mean, mean fluorescence value for indicated test concentration; Pred mean, estimated mean fluorescence value for indicated test concentration; Obs %Chng, % change from observed control mean to observed treatment mean; Pred %Chng, % change from estimated control mean to estimate treatment mean; DF, degrees of freedom for indicated model terms; SS, MS, sum of squares, mean square for indicated model terms; *F* value, Prob *F*, *F*-test for indicated model term and associated probability; Uncor total, uncorrected total sum of squares. LCB and UCB, lower and upper 95% confidence bounds for indicated parameter.

These three regression models have problems. Wide confidence bounds for EC10 from two of the models have already been mentioned, and the estimates for the other two parameters in all three model fits are wide. The range of point estimates of EC10 spans an order of magnitude from 0.03 to 0.3. AICc provides no help, as no two of the model AOC values differ by more than 1. Formal lack of fit tests likewise offers no meaningful distinction. The NOEC approach is likewise not especially satisfying, as the NOEC set by both the Williams and Dunnett tests is 0.1 mg l^{-1}, where the observed change from the control was 12%, while the Jonckheere–Terpstra test sets the NOEC as 1 mg l^{-1}, where the observed change from the control was 44%. Given that an analysis has to be provided in a risk assessment, the Williams NOEC of 0.1 mg l^{-1} is probably best, though the 3PL EC10 estimate of 0.3 mg l^{-1} is probably best among regression models. In the phase 1 validation of the XETA protocol, plausible regression models were obtained for all active compounds except one, which was not expected to exhibit an effect. A third of the datasets (8 of 24) generated gave EC10 estimates with acceptably tight confidence bounds, with the others (16 of 24) giving very wide confidence intervals, or no model could be fit.

An alternative regression approach has been proposed (Lammer et al., 2009) for TG 236, namely, an errors-in-variables approach, which has not been discussed previously. That is the subject of the next section. The XETA example will be revisited after that method is developed.

13.4 ERRORS-IN-VARIABLES REGRESSION

A standard regression models treats the independent variables such as concentration, as observed without error. In fact, there is often some uncertainty in these observations. Often the uncertainty is very small relative to the nominal differences in test concentrations, but this is not always true. In the present context, the automated insertion of small amounts of media to fill each well could lead to a larger relative error in the test concentration than is typically present in a more conventional experiment with a few tanks at each nominal test concentration. The effect of measurement error in the independent variable will be developed largely from Fuller (1987), and the reader is referred to that source for a much more complete discussion, though it does not address the specifics of titer plate experiments. The model for the response, y_i, as a function of the test concentration, x_i, is given by (13.17):

$$y_i = \alpha + \beta x_i + e_i, \quad (13.17)$$

where e_i is the usual random error with mean zero and variance σ_{ee} associated with repeated sampling. The notation for variances and covariances that has been used previously is modified in this section because of the proliferation of random variables involved in measurement error models. We now consider the idea that x_i is not observed directly, and instead we observe

$$U_i = x_i + w_i \quad (13.18)$$

where w_i is the error of the measurement with mean zero and variance σ_w^2. Suppose as well that the unknown true concentration x_i, is not fixed, as in ordinary regression, but is itself a random variable with mean μ_x and variance σ_{xx}. This can be understood easily for titer plate experiments such as described in Section 13.3 as capturing the anticipated variation of the actual concentration of the test chemical imparted to the wells at the same nominal concentration. So (x_i, w_i, e_i) is a random vector with variances $(\sigma_{xx}, \sigma_{ww}, \sigma_{ee})$. It is convenient to assume components of this random vector are independent, so the variance matrix of the indicated random vector is diagonal. The mean vector has the form $(\mu_x, 0, 0)$. If Y is the vector of observations defined by Eq. (13.17), then the vector (Y,U) can be described as follows:

$$E\left[\begin{pmatrix} Y \\ U \end{pmatrix}\right] = \begin{bmatrix} \mu_Y \\ \mu_U \end{bmatrix} = \begin{bmatrix} \alpha + \beta\mu_x \\ \mu_x \end{bmatrix}, \quad (13.19)$$

$$\text{var}\begin{pmatrix} Y \\ U \end{pmatrix} = \begin{matrix} \sigma_{YY} & \sigma_{XY} \\ \sigma_{YX} & \sigma_{XX} \end{matrix} = \begin{matrix} \beta^2\sigma_{xx} + \sigma_{ee} & \beta\sigma_{xx} \\ \beta\sigma_{xx} & \sigma_{xx} + \sigma_{ww} \end{matrix}. \quad (13.20)$$

If the x_i are fixed, rather than random, then Eq. (13.20) reduces to a diagonal matrix with entries σ_{ee} and σ_{ww}. Note that Y in Eq. (13.19) is defined in terms of the true, but unobserved, x_i, so it cannot be used to estimate the model parameters. If a standard regression is done on the measured or observed concentrations U_i (i.e. with the x_i replaced by U_i), then the resulting estimated slope is given by

$$\hat{\gamma} = \frac{\sum_i (y_i - \bar{y})(U_i - \bar{U})}{\sum_i (U_i - U)^2}. \quad (13.21)$$

The expected value of the standard slope estimator in Eq. (13.21) is given by

$$E[\hat{\gamma}] = \frac{\beta}{1 + (\sigma_{ww}/\sigma_{xx})} \stackrel{\text{def}}{=} \kappa\beta. \quad (13.22)$$

It follows that the slope, β, is underestimated when measurement error is present but ignored. An unbiased estimator can be obtained from Eq. (13.22) by dropping the expectation symbol and solving for β, replacing β by $\hat{\beta}$ in the process. If the x_i are fixed rather than random, so only the measurement error remains, then σ_{xx} should be interpreted as the mean sum of squared deviations of the x_i about their mean. The underestimation is negligible if σ_{ww} is small relative to σ_{xx}, but not otherwise. An unbiased estimator of the true slope is given by

$$\hat{\beta} = \kappa^{-1}\hat{\gamma}. \quad (13.23)$$

The expressions (13.22) and (13.23) are not very useful in themselves, since neither x_i nor w_i is observed, so that κ cannot be estimated unless additional assumptions are made to permit some useful approximation. For example, if the x_i are fixed, but still subject to measurement error, then one approach to obtaining estimators of the model parameters is to assume the ratio $\delta \equiv \sigma_{ww}/\sigma_{ee} = 1$, that is, the measurement error on the independent variables is the same as the error on the dependent variable. More general developments are based on other presumed known values of the ratio δ.

The above assumes the errors in x and w are homogeneous. Models vary according to the assumptions made for the distribution of w_i. Implicit in what was indicated just above, a common assumption is that w_i is independent of x_i, but this can be modified. An important issue is how to estimate σ_{xx}.

Errors-in-variables models appear to have been introduced in the ecology literature by McArdle (1988, 2003), though only for the linear model being discussed here. The approach followed by Lammer et al. (2009) was to assume that $\sigma_{xx} = \sigma_{ee}$, which is sometimes referred to as orthogonal regression. In that approach, Eq. (13.20) is solved for β, replacing the left-hand side by $\hat{\beta}$ and the ratio of variances by 1. It should be noted that Lammer et al. (2009) did not apply a nonlinear model to estimate EC10 from a toxicity study, but used a linear model to compare results on different fish species.

An alternative to the above development, which avoids making assumptions about the unknown σ_{xx}, is the use the generalized method of moments estimator of Pal (1980), given under reasonable conditions by

$$\hat{\beta} = \frac{\sum_i (y_i - \bar{y})^2 (U_i - \bar{U})}{\sum_i (y_i - \bar{y})(U_i - \bar{U})^2}. \quad (13.24)$$

The variance of this estimator is given in Pal (1980).

Errors-in-variables methods can be extended to more complex linear models (Dagenais and Dagenais, 1997; Erickson and Whited, 2002) and nonlinear models (Fuller, 1987; Carroll et al., 2006). The latter is needed for the XETA data and, indeed, for most applications in ecotoxicology. A simple application in toxicology was described in Carakostos and Green (1992).

13.5 ANALYSIS OF MIXTURES OF CHEMICALS

All of the discussion in previous chapters dealt with the effects of a single chemical. In fact, in many environmental settings, chemicals are present in mixtures, often with many other chemicals. For example Gilliom et al. (1999) concluded that in aquatic environments, chemicals almost always are present in mixtures. Consequently, the toxicity of a mixture cannot be assessed simply from the properties of a single constituent. Various methods have been developed to estimate the toxicity of a mixture from the properties of its constituents. As ably discussed for herbicidal *s*-triazines in Faust et al. (2001), it was concluded that "Predictions based on *independent action* tend to underestimate the overall toxicity of *s*-triazine mixtures. In contrast, the concept of *concentration addition* provides highly accurate predictions of *s*-triazine mixture toxicity, irrespective of the effect level under consideration and the concentration ratio of the mixture components. This also holds true when the mixture components are present in concentrations below their individual NOEC values." In this section, the terms "independent action" and "concentration addition" will be defined, and methods will be introduced to help assess the toxicity of mixtures. Statistical analysis of mixture is an important topic, and the scientific literature on it is large. The intention here is to provide an introduction and cite some of the relevant literature so the reader can pursue this topic in greater depth.

Relevant references on mixtures in the regulatory context include the Swedish biocide registry KEMI, which can be found at http://www.kemi.se/en/directly-to/databases, which includes mixtures of 2–17 chemicals; Backhaus and Karlsson (2014), which deals with pharmaceutical effluents and reports that the top 10 mixture components explain more than 95% of the mixture; Backhaus and Faust (2012), which presents data to conclude that additive models provide adequate risk assessment for the vast majority of environmental mixtures; and Posthuma et al. (2016), which deals with mixtures of 5–10 chemicals. Also of interest is a meta-analysis done by Cedergreen (2014), who reported that between 3 and 26% of two-component mixtures seen in ecotoxicology exhibit synergism with at most 10X additional effects compared with additivity. There is some apparent inconsistency between the two Backhaus papers and that of Cedergreen that underscores some uncertainty in the analysis of mixtures due in large part to limited detailed studies of specific mixtures. What is provided in this section is an introduction to this increasingly important topic for risk assessment.

Concentration addition is based on the somewhat ill-defined concept of what Faust et al. (2001) call similar action of the mixture components. The toxicity of a mixture described by concentration addition is simply stated by Eq. (13.25):

$$\mathrm{EC}x(\mathrm{mix}) = \frac{1}{\sum_i \left(p_i/\mathrm{EC}x(i)\right)}, \tag{13.25}$$

where p_i is the proportion of chemical i in a mixture causing an x% effect, EC$x(i)$ is the x% effects concentration of chemical i acting alone, and ECx(mix) is the x% effects concentration of the mixture. Where this formula is applicable, it is a simple matter to assess the toxicity of a mixture from that of its constituents. Eq. (13.25) can be rearranged by replacing p_i by c_i/ECx(mix), where c_i is the concentration of chemical i in the mixture, to show that

$$\sum_i \left(\frac{c_i}{\mathrm{EC}x(i)}\right) = 1. \tag{13.26}$$

Equation (13.26) appears more general than Eq. (13.25) in that it applies to any mixture without reference to a specific effect level, but Eq. (13.25) follows from (13.26) by reversing the substitution just defined.

It might be helpful for understanding Eq. (13.26) to consider a mixture of just two chemicals. In this simple mixture,

$$\frac{c_1}{\mathrm{EC}x(1)} + \frac{c_2}{\mathrm{EC}x(2)} = 1. \tag{13.27}$$

A plot of the relationship between c_1 and c_2 is called an isobologram. (See Figure 13.6.) In this simple case of a mixture of two chemicals, the isobologram is a straight line joining the ECx values of the two single components. In mixtures of k chemicals, an isobologram is a hyperplane in dimension $k-1$. All mixtures of these two chemicals whose components fall on the line have the same toxicity. A mixture whose toxicity (i.e. ECx) falls above the line is said to be synergistic. Christiansen et al. (2009) gave an interesting example of synergism. It should be noted that Cedergreen (2014) cited evidence that fewer than 5% of typical pesticide mixtures are synergistic. Older but still useful discussion of synergistic mixtures is given by Greco et al. (1995) and Berenbaum (1989). A mixture that falls below the line is said to be antagonistic. A mixture falling on the line is often said to be additive. Further discussion of isobolograms is given in Gessner (1995).

One of the challenges for mixture analysis is identifying which compounds are additive, which are synergistic, and which are antagonistic. It is very expensive to design

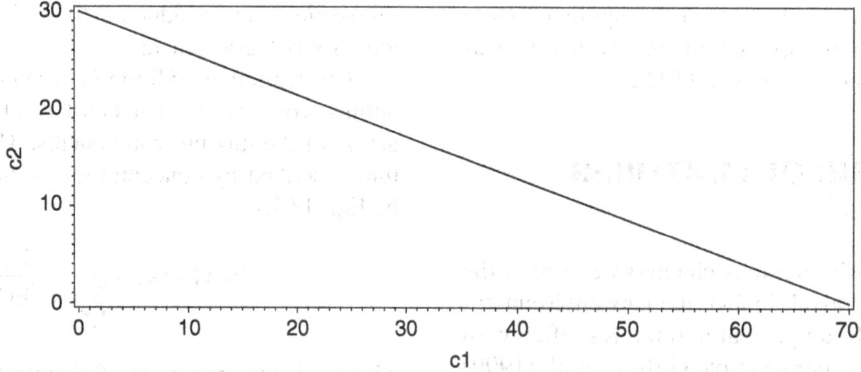

Figure 13.6 Isobologram for mixture of two chemicals with ECx(2) illustrated as 30 and ECx(1) illustrated as 70. The value of *x* is irrelevant.

and run studies that allow a thorough investigation of this. Examples show it is possible for a mixture of chemicals to be antagonistic in some regions, synergistic in others, and additive in yet others. In the absence of data to the contrary, risk assessment has usually been based on concentration additivity, and there is scientific literature to support this. For example, Deneer (2000), Belden et al. (2007), and Kortenkamp et al. (2009) all report evidence that typical chemical mixtures of pesticides act additively or slightly less than additive.

The independent action approach is designed for chemical mixtures whose components have different mechanisms of action. The effect of the mixture in this approach is described by Eq. (13.28):

$$E(\text{mix}) = 1 - \prod_i (1 - E_i), \tag{13.28}$$

where E(mix) and E_i are interpreted as probabilities. Thus, E_i is the probability of an effect in the *i*th component. If the effect is mortality, the meaning is obvious. For sublethal effects, Faust et al. (2001) describes E_i as the proportion of the maximum possible effect. What this mean in practice for an effect measured as an inhibition is just the proportion inhibition. Faust et al. found much better results using the concentration addition model than the independent action model, and it will not be pursued further here.

According to Kienzler et al. (2016), the two models thus far introduced are the main mixture models used in risk assessment. However, Howard and Webster (2009) introduced an alternative model called generalized concentration addition, and Hadrup et al. (2013) compared this to the above two models and found it gave the best results for the mixtures they considered. The model is given by Eq. (13.29):

$$E(\text{mix}) = \frac{\sum_i c_i \left(E_{\max}(i)/\text{EC}_{50}(i)\right)}{1 + \sum_i \left(c_i/\text{EC}_{50}(i)\right)}, \tag{13.29}$$

where $E_{\max}(i)$ is the concentration at which the *i*th chemical acting alone has its maximum effect. For regulatory risk assessment, Kienzler et al. (2016) is a valuable resource.

There have been several papers proposing ways to accommodate synergism, antagonism, and additivity in a single model. One such approach by Kong and Lee (2006) that applies only to a mixture of two chemicals will be presented here, and a program to implement it is given in Appendix 1. They begin with log-linear functions for the effect of each chemical acting alone. So if Y_j is the response of chemical *j* at concentration C_j, then

$$Y_j = \beta_{0j} + \beta_{1j}\text{Ln}(C_j). \tag{13.30}$$

This model is fit to the data for chemical C_j acting alone, which allows estimation of the four parameters β_{0j}, β_{1j}, $j=1, 2$.

Next, assuming additivity, solve Eq. (13.24) for C_1 to get

$$C_1 = y_1 + \rho y_2, \tag{13.31}$$

where $\rho = C_1/C_2$. Given C_1 and C_2 are the concentrations at which the two chemicals have *x*% effect, i.e. $C_j = \text{EC}x(j)$, ρ is the relative potency of chemical 1 relative to chemical 2.

Similarly,

$$C_2 = y_2 + \rho^{-1} y_1. \tag{13.32}$$

So C_1 and C_2 are the concentrations of chemicals 1 and 2 that, acting alone, produce a specific effect (i.e. *x*%) and y_1 and y_2 are the concentrations of those chemicals in the mixture. For these concentrations, the effects of the two chemicals acting alone are equal, so $Y_1 = Y_2$, and from Eq. (13.30) with some algebraic manipulation, it follows that

$$\rho = \exp\left[\gamma_1 + \gamma_2 \text{Ln}(C_2)\right], \tag{13.33}$$

where $\gamma_1=(\beta_{02}-\beta_{01})/\beta_{11}$ and $\gamma_2=(\beta_{12}-\beta_{11})/\beta_{11}$. So ρ is estimated from Eq. (13.33) subject to the constraint Eq. (13.32). If C_1 in Eq. (13.30) is replaced by Eq. (13.31), the effect Y at the mixture (y_1,y_2) is given by

$$Y = \beta_{01} + \beta_{11}\text{Ln}\left(y_1 + \rho y_2\right), \tag{13.34}$$

assuming additivity. To allow for the possibility of interaction of the two chemicals, Kong and Lee (2006) modify Eq. (13.31) to have the form

$$Y = \beta_0 + \beta_1\text{Ln}\left[y_1 + \rho y_2 + f\left(y_1,y_2;\gamma,\kappa\right)\left(y_1\rho y_2\right)^{\frac{1}{2}}\right], \tag{13.35}$$

where the as yet unspecified function f is intended to capture local synergy, antagonism, or additivity and $y_1\rho y_2$ captures the interaction of the two chemicals. The specific function f Kong and Lee proposed is

$$f\left(y_1,y_2;\gamma,\kappa\right)= \kappa_0 + \kappa_1 y_1^{\frac{1}{2}} + \kappa_2\left(\rho y_2\right)^{\frac{1}{2}} + \kappa_3 y_1 + \kappa_4 \rho y_2 + \kappa_5\left(y_1\rho y_2\right)^{\frac{1}{2}}. \tag{13.36}$$

It should be noted that a much simpler model was proposed by Plummer and Short (1990), which is obtained from Eq. (13.35) by setting $f(y_1,y_2;\gamma,\kappa)=\kappa_5$. It should be recognized that the function f is a quadratic in $(\rho y_2)^{1/2}$ and $y_1^{1/2}$, which relates this to response surface models. One problem with Eq. (13.35) is the large number, 10, of model parameters to be estimated in addition to the function ρ, which is a function of y_1 and y_2. In the full model fit to the example below from their paper, 8 of the 10 parameter estimates are indistinguishable from zero, making the model highly unstable. Kong and Lee used an ad hoc approach to parameter reduction, which they label backward elimination. This is done starting with the full model, for which AICc (or, in their case, AIC) is computed. The parameter with the smallest nonsignificant absolute t-statistic is omitted and AICc is recomputed. If the new AICc value is no larger than the previous AICc, then the same process is followed to eliminate one more parameter. This process is continued until either there are no nonsignificant parameter estimates or AICc increases. They also constrain the procedure by never eliminating a parameter for a first-order term, $(\rho y_2)^{1/2}$ or $y_1^{1/2}$, unless all higher-order terms involving that term have already been eliminated.

We will demonstrate how this works using the dataset Kong and Lee (2006) explored. (See Table 13.10.) It should be noted that parameter estimates that come from the program we developed (*NLINHONG.sas*, *NLINHONG_alt.sas*, and *NLINHONG_reduced.sas*) and present in Appendix 1 may not match theirs. In their final model, there were still five of seven parameters indistinguishable from zero. It is interesting that if AICc is used (Eq. 4.24), then a rather different model is obtained in which all retained parameters are significantly different from zero, still with predictions in good agreement with observations. Also noteworthy is that the AICc values for the full model and the final were −176.519 and −195.227, so they differed by more than 10, making the final model better than the initial by the criterion recommended in Chapter 4. Finally, it should be noted that in our final model, ρ is estimated to be 1, so synergism, antagonism, and additivity are captured in other model terms. While the modeling approach is interesting, caution is required in applying it for the small datasets likely to be encountered in ecotoxicity studies.

Cell lines of human squamous cell carcinoma were treated with various combinations of two chemicals (drugs) labeled SCH66336 and 4-HPR. Initially, the same number of cancer cells was in each combination of treatments. After 6 h, cells were counted again and the cancer cells surviving were counted. See Table 13.11 for the parameter estimates from our model and Table 13.12 for the model predictions.

Table 13.10 The Hong and Lee Human Squamous Cell Carcinoma Data

	4-HPR				
SCH66336	0	0.1	0.5	1	2
0	1	0.7666	0.5833	0.5706	0.4934
0.1	0.6701	0.6539	0.4767	0.5171	0.3923
0.5	0.6289	0.6005	0.4919	0.4625	0.3402
1	0.5577	0.5102	0.4541	0.3551	0.2851
2	0.455	0.4203	0.3441	0.3082	0.2341
4	0.3755	0.3196	0.2978	0.2502	0.1578

SCH66336 and 4-HPR are in μM. Entries in cells of the table are the proportion of cancer cells remaining after 6 h exposure to the indicated combination of drugs.

Table 13.11 Parameter Estimates From Our Fit of the Kong and Lee Model

Parameter	Estimate	StdErr	LCB	UCB
$\beta 0$	0.535	0.011	0.513	0.557
$\beta 1$	−0.087	0.008	−0.103	−0.071
$\kappa 1$	−7.369	3.312	−14.191	−0.548
$\kappa 2$	5.832	2.165	1.372	10.291
$\kappa 3$	5.871	2.718	0.274	11.468

Parameters shown are all retained in final model. LCB, UCB are lower and upper 95% confidence bounds on the parameter estimate.

Table 13.12 Predicted Results From Our Fit of the Kong and Lee Model

	4-HPR				
SCH66336	0	0.1	0.5	1	2
0	1.0000	0.7746	0.6052	0.5534	0.4810
0.1	0.6842	0.6544	0.5197	0.4833	0.3865
0.5	0.5734	0.5846	0.5015	0.4394	0.3352
1	0.5257	0.5164	0.4498	0.3720	0.2908
2	0.4780	0.4286	0.3578	0.3094	0.2375
4	0.4303	0.3323	0.2725	0.2305	0.1652

These results are in general agreement with the raw data in Table 13.10 and compare favorably to the predicted values in Kong and Lee (2006).

13.6 BENCHMARK DOSE MODELS

At its most basic, the BMD approach to risk assessment is just the regression approach to estimate ECx for a specified effects level x, though it is called the BMD (or sometimes BMR). A novelty is that it is recommended to fit a specific set of models, estimate BMD and its 95% lower confidence bound, BMDL, from each and either take the model that produces the lowest lower confidence bound or use some sort of model averaging to provide the "best" BMD, called the benchmark dose or reference point, or, more commonly, report the BMDL, the smallest lower confidence bound. The idea, as reported in EFSA (2017), is given "As the purpose of a BMD analysis is not to find the best estimate of the (true) BMD but rather to find all plausible values of the (true) BMD, given the data available, not only the best-fitting model but also the models resulting in a slightly poorer fit need to be taken into account. After all, it could well be that the second (or third, ...) best-fitting model is closer to the true dose–response than the best-fitting model. This type of uncertainty is called 'model uncertainty', and implies that the BMD confidence interval needs to be based on the results from various models, instead of just a single ('best') model."

EFSA (2017) provides extensive guidance on how to carry out this approach. For continuous responses, EFSA (2017) recommends four models. These are OE3, OE5, and 3- and 4-parameter Hill models, which are available in the BMD software from USEPA and PROAST (Slob, 2003), which was developed by one the authors of the EFSA document. As has been illustrated in earlier chapters of this book, many datasets arising from guideline studies do not permit fitting the four-parameter model OE5, and the same can be demonstrated for the four-parameter Hill model. Indeed, OE3 and the three-parameter Hill model are sometimes difficult to fit or produce unstable models with one or more model parameters indistinguishable from zero or very wide confidence intervals. This is one of the challenges of the BMD approach and is recognized by EFSA. The result can be wide disparities in the BMD and BMDL values.

For quantal data, the EFSA opinion recommends the probit, log-probit, logistic, log-logistic, and Weibull models that have been discussed here in previous chapters, plus a gamma model and what is called a two-stage or linearized multistage (LMS) model. The EFSA document hints at normalizing response data to percent change from control, though no details are provided. It would be unfortunate if this guidance document encouraged further use of this unnecessary and flawed methodology.

EFSA (2017) uses the AIC to compare models but makes claim that if two models have AIC values differing by more than 2, then the one with the lower value is the better fit. In probabilistic terms, this is rather liberal, as the probability that the model with the lower AIC value is better is approximately 0.73, much lower than the usual 95% confidence typically used in evaluating other probabilistic results. The cutoff of 10 discussed in Chapter 2 is associated with a 95% probability. To more fully appreciate the BMD approach, it is necessary to outline the model averaging process that is used. The following discussion is based on Wheeler and Bailer (2009), which is what is cited by the EFSA opinion.

The idea is simple: Compute a weighted average risk of all models used. The estimate of risk for a given model is given by Eq. (13.37):

$$\text{BMR} = \frac{\pi(\text{BMD}) - \pi(0)}{1 - \pi(0)}, \tag{13.37}$$

where $\pi(d)$ is the probability of an effect (or the estimated proportion effect) at dose d from the given model. Here BMR is the level of effect to be estimated, usually 10%. BMD is estimated using maximum likelihood methods. This is simply a matter of obtaining the ECx estimate and its confidence bounds, where $x = \text{BMR}$. This is done for all models in the list of models used, presumably those in the EFSA opinion. The model average estimate of BMD is given by Eq. (13.38):

$$\text{MBD} = \sum_i \text{BMD}_i^* w_i, \tag{13.38}$$

where the weights w_i are given by

$$w_i = \frac{\exp(-\text{AIC}_i/2)}{\sum_j \exp(-\text{AIC}_j/2)}, \tag{13.39}$$

where AIC_i is the AIC value for the ith model. The Bayesian information criterion (BIC) could be used in place of AIC if desired. While neither Wheeler and Bailer (2009) nor EFSA (2017) indicate it, AICc could be used instead of AIC.

Model averaging has been criticized by several authors (e.g. Gelman and Rubin, 1995). Sroka et al. (2006) contains some useful discussion and formal ways to do model averaging. They suggest eliminating models that fit the data very poorly or are overly complex when simpler models exist that perform reasonably well. These suggestions and others have been presented more recently by Wang et al. (2016) who criticized a 2015 EFSA opinion on this approach in part because they found that less reliable fits with larger confidence intervals are usually preferred over more reliable fits with smaller confidence intervals. In addition sometimes curves are selected that do not describe the data well or that are biologically not reasonable, for example, because parameter estimates or their bounds are inconsistent with biology, such as indicating body weight can reach zero. They recommend the following criteria: (i) From all mathematically reasonable models, exclude those that are biologically not reasonable (e.g. upper and lower limits). (ii) Evaluate the fits for all models based on goodness of fit (χ^2 test, R^2) and exclude all with insufficient fit (e.g. 15% criterion for χ^2 test as in FOCUS kinetics). (iii) Rank the remaining fits based on the reliability of parameter estimation (small confidence intervals). The approach is presented here since this is guidance from an important regulatory body.

An interesting alternative to BMD modeling and NOEC-based risk assessments using uncertainty factors is given by Chiu and Slob (2015).

13.7 LIMIT TESTS

A limit test is an example of a Tier 1 test in the EPA regulatory environment. The idea is simple. A group of subjects is exposed to a very high concentration of the test chemical. This concentration is set by regulations pertaining to the specific test organism, environment, and response. A negative control is also included in the study. It is customary to compare the treatment group with the control, and if there is no statistically significant difference, then under suitable conditions, it may not be necessary to do further testing on that organism in that environment. Such a test must be designed to provide adequate power to find significant an effect of biological importance. For a continuous response, a simple t-test will often suffice. Where there is concern that the test may be underpowered, it is possible to modify the t-test to test the null hypothesis that the treatment group mean is more than $x\%$ different from the control vs. the alternative that the difference is less than $x\%$. Here $x\%$ is the size effect deemed to be biologically relevant. If the null hypothesis is reject in favor of the alternative, then there is strong evidence that the test chemical does not have a significant effect. It should be noted that the usual formulations of the null and alternative hypotheses have been reversed, so that the null is that the chemical is unsafe and the alternative is that it is safe by indicated criterion.

A simple example will illustrate the topic. See Table 13.13 for data from an algal study, and the response used in the illustration is cell count.

The hypotheses to be tested can be stated as

$$H_0: \mu_1 \le \lambda\mu_0 \text{ vs. } H_1: \mu_1 > \lambda\mu_0,$$

where λ is the proportion decrease from control to be tested. That is, the question to be addressed is whether there has been a decrease of at least $100(1-\lambda)\%$. The null hypothesis is that there has been a decrease by at least that amount and the alternative is that whatever decrease, if any, may have occurred, it is less than $100(1-\lambda)\%$. The test statistic for a normally distributed homogeneous response is given by

$$T = \frac{\bar{x}_1 - \lambda\bar{x}_0}{s_p\sqrt{(1/n_1)+(\lambda^2/n_0)}}, \tag{13.40}$$

where s_p is the pooled within-group variance and $\bar{x}_i$ and n_i are the sample mean and number of replicates, respectively, of the ith treatment group. Positive values of T indicate a smaller than $100(1-\lambda)\%$ decrease, while negative values indicate a greater decrease. T has a t-distribution with n_1+n_2-2 degrees of freedom.

Table 13.13 Algae Cell Count

Conc (mg l^{-1})	Rep	Count (cells ml^{-1})	AUC	G_Rate (cells day^{-1} on a log scale)	Group	Species	Dose
0	1	3 420 000	57 240 000	0.081	1	Algae	1
0	2	3 220 000	48 360 000	0.0802	1	Algae	1
0	3	3 350 000	52 800 000	0.0808	1	Algae	1
100	1	2 710 000	40 320 000	0.0778	2	Algae	2
100	2	2 790 000	45 840 000	0.0782	2	Algae	2
100	3	3 060 000	46 200 000	0.0795	2	Algae	2

Count, number of cells at study end; AUC, area under growth curve; G_Rate, growth rate.

In the case of a normally distributed heterogeneous response, the test statistic is given by

$$T = \frac{\bar{x}_1 - \lambda \bar{x}_0}{\sqrt{\left(s_1^2/n_1\right) + \left(\lambda^2 s_0^2/n_0\right)}}, \tag{13.41}$$

where s_i^2 is the sample variance of the ith treatment group. In this case, an approximate degrees of freedom is given by Satterthwaite's formula:

$$df = \frac{\left[\left(\lambda s_0^2/n_0\right) + \left(s_1^2/n_1\right)\right]^2}{\left\{\left[\lambda^2 s_0^4/n_0^2\left(n_0 - 1\right)\right] + \left[s_1^4/n_1^2\left(n_1 - 1\right)\right]\right\}}. \tag{13.42}$$

The data in Table 13.13 will be tested for 10, 25, and 50% decrease in cell count. See Table 13.14 for the results. The program *Limit test.sas* is given in Appendix 1. It will be left as an exercise to carry out similar calculations for the other two responses given in Table 13.13.

Thus, at the 0.05 significance level, one cannot be confident that the true inhibition is less than 10%, but one can be confident that the true inhibition is less than 25% and, of course, that the inhibition is less than 50%.

Another application of this idea is given by experimental design to investigate the likelihood of a significant test for an increase in mortality in the treatment group in a limit test under various scenarios. Proposed experimental designs were one tank of 5–20 male and one tank of 5–20 female fish in the control and in the limit treatment concentration. The concern was to reduce the likelihood of a false positive, as that would trigger a more expensive test, while maintaining adequate power to detect a real increase in mortality.

A computer simulation study was done to investigate these designs with treatment mortality ranging from 0.01 to 0.5. The control mortality was assumed to be 0.01. The likelihood of a significant test would decrease if there is greater control mortality. The simulation study demonstrated 80% power to detect an 18% increase in mortality in the design with 10 fish of each sex in each group. The more relevant question is, if a significant test is found, what is the probability the increase in mortality in the limit concentration exceeded x%? This is a more challenging computation. See Table 13.15 for results from this simulation study that indicate what can occur in the control and limit treatment to obtain a significant test statistic.

Table 13.15 Probability of Significant Limit Mortality

n	$x0$	$x1$	Probz	n	$x0$	$x1$	Probz
5	0	1	0.146	15	0	1	0.155
5	0	2	0.057	15	0	2	0.072
5	0	3	0.019	15	0	3	0.034
5	0	4	0.005	15	0	4	0.016
5	0	5	0.001	15	0	5	0.007
5	1	1	0.500	15	1	1	0.500
5	1	2	0.245	15	1	2	0.271
5	1	3	0.098	15	1	3	0.141
5	1	4	0.029	15	1	4	0.071
5	1	5	0.005	15	1	5	0.034
10	0	1	0.152	20	0	1	0.156
10	0	2	0.068	20	0	2	0.073
10	0	3	0.030	20	0	3	0.036
10	0	4	0.013	20	0	4	0.018
10	0	5	0.005	20	0	5	0.008
10	1	1	0.500	20	1	1	0.500
10	1	2	0.266	20	1	2	0.274
10	1	3	0.132	20	1	3	0.146
10	1	4	0.061	20	1	4	0.076
10	1	5	0.025	20	1	5	0.038

n, fish in each treatment (control and limit). $x0$, $x1$ are the number of fish deaths in control and limit treatment, respectively. Probz is significance level of the corresponding test statistic. Probz<0.05 indicates a significant increase in mortality.

Table 13.14 Limit Test Results

Shapiro–Wilk test for Normality		Levene test for homogeneity	
Statistic	p-Value	Statistic	p-Value
0.9475	0.71998	0.3196	0.60205

Mean 0	Mean1	%Inhib observed	%Inhib tested	Test statistic	df	Pr(Inhib)>10%
3330000	2853333	14.3143	10	−1.24794	4	0.85994
Mean 0	**Mean1**	**%Inhib observed**	**%Inhib tested**	**Test statistic**	**df**	**Pr(Inhib)>25%**
3330000	2853333	14.3143	25	3.32671	4	0.014598
Mean 0	**Mean1**	**%Inhib observed**	**%Inhib tested**	**Test statistic**	**df**	**Pr(Inhib)>50%**
3330000	2853333	14.3143	50	12.4212	4	0.000120764

Mean0, control mean; Mean1, treatment mean; %Inhib observed, percent inhibition observed in Mean1 vs. Mean0; %Inhib tested, test for H_0: $\mu_1 \le (1-\lambda)\mu_0$ for 100λ=%Inhib tested; Pr(Inhib)>x, p-value associated with the null hypothesis.

Table 13.16 Probability of Control and Limit Mortalities

n	$x0$	$x1$	Prob0	Prob1	n	$x0$	$x1$	Prob0	Prob1
5	0	1	0.951	0.001	15	0	1	0.860	0.010
5	0	2	0.951	0.000	15	0	2	0.860	0.000
5	0	3	0.951	0.000	15	0	3	0.860	0.000
5	0	4	0.951	0.000	15	0	4	0.860	0.000
5	0	5	0.951	0.000	15	0	5	0.860	0.000
5	1	1	0.999	0.001	15	1	1	0.990	0.010
5	1	2	0.999	0.000	15	1	2	0.990	0.000
5	1	3	0.999	0.000	15	1	3	0.990	0.000
5	1	4	0.999	0.000	15	1	4	0.990	0.000
5	1	5	0.999	0.000	15	1	5	0.990	0.000
10	0	1	0.904	0.004	20	0	1	0.818	0.017
10	0	2	0.904	0.000	20	0	2	0.818	0.001
10	0	3	0.904	0.000	20	0	3	0.818	0.000
10	0	4	0.904	0.000	20	0	4	0.818	0.000
10	0	5	0.904	0.000	20	0	5	0.818	0.000
10	1	1	0.996	0.004	20	1	1	0.983	0.017
10	1	2	0.996	0.000	20	1	2	0.983	0.001
10	1	3	0.996	0.000	20	1	3	0.983	0.000
10	1	4	0.996	0.000	20	1	4	0.983	0.000
10	1	5	0.996	0.000	20	1	5	0.983	0.000

Prob0, probability of no more than $x0$ deaths in control; Prob1, probability of at least $x1$ deaths in limit. It is assumed that the true control and limit mortality rates are equal to 0.01 or less.

From Table 13.15, it can be seen that with 10 males per treatment, if there is no mortality in the control, it would take three deaths in the treatment group to be significant. But if there was one control death, it would take five deaths in the limit treatment to be significant. This indicates a low probability of a statistically significant test for increase in mortality when in fact there has not been a real increase.

See Table 13.16 for the probabilities of observing mortalities in the control and limit concentrations if there is no treatment effect and the underlying mortality rate is 0.01. This table makes clear that if a single mortality occurs in the limit concentration, then it is very likely the mortality rate at that concentration exceeds the assumed control rate of 0.01.

13.8 MINIMUM SAFE DOSE AND MAXIMUM UNSAFE DOSE

The idea presented in Section 13.6 for limit tests can be extended to multiconcentration studies. If a useful regression model can be fit to the concentration–response data, then as demonstrated in earlier chapters, it is possible to estimate the size effect at a range of test concentrations. Examples have been presented that demonstrate that sometimes, the data depart so much from monotonicity that a useful regression model cannot be fit, but the data are still suggestive of effects. The NOEC approach, of course, can be used to determine whether there is a significant effect at each test concentration, but that does not indicate the size effect that can be detected. Tamhane et al. (2001) developed a test that determines what is called the maximum safe dose. Tamhane and Logan (2004) extended that to include the case of heterogeneous variances.

Assuming a monotone concentration–response $\mu_0 \ge \mu_1 \ge \mu_2 \ge \mu_3 \ge \cdots \ge \mu_k$, the hypotheses to be tested are

$$H_{0i}: \mu_i \le \lambda\mu_0 \quad vs. \quad H_{1i}: \mu_i > \lambda\mu_0, \quad i = 1,2,3,\ldots,k.$$

Thus, a concentration is assumed unsafe, i.e. has an adverse effect, unless proven safe. The sample MAXSDλ is defined as the highest concentration for which $\mu_i > \lambda\mu_0$ can be demonstrated. It controls the FWE rate at a specified level, α. Some power properties are given in the cited papers.

There are two nonequivalent ways to implement the MAXSD idea. The simplest will be described first and is what is called SD2CP in the cited papers. Define test statistics for all treatment group using the formulation in Eq. (13.40) or Eq. (13.43):

$$t_i = \frac{\bar{x}_i - \lambda\bar{x}_0}{s_p\sqrt{(1/n_i) + (\lambda^2/n_0)}}. \tag{13.43}$$

MAXSDλ is the highest concentration U_i for which

$$\min_{1 \le j \le i}\{t_j\} > t_\nu^\alpha, \tag{13.44}$$

where t_ν^α is the critical value of Student's t-distribution at level α (usually 0.05) and degrees of freedom determined by standard ANOVA methods. The reader is referred to Tamhane et al. (2001) for the reason no adjustment to the p-values is made for the number of comparisons. This provides reasonable results provided there is no low concentration after which the response is flat. In such studies, the MAXSDλ may be too low. This will be illustrated with length measurements from a chronic *Daphnia magna* study with $\lambda = 0.9$ corresponding to a 10% effect. A summary of the data is provided in Table 13.17. The full dataset is given in Appendix 1 as maxsdxmpl.sas7bdat. There were 10 replicate beakers in each test concentration and control with four daphnid per beaker at study start. There was very little mortality during the study. A solvent was used, and preliminary analysis found no significant difference between the water and solvent controls, so they were combined for analysis of treatment effects. An ANOVA was done of replicate means using replicate size at study end as weights. See Table 13.17 for the indicated treatment standard deviations from that ANOVA. It is left as an exercise to verify that these almost the same standard deviations are obtained

Table 13.17 *Daphnia magna* Length Data Summary for MAXSDλ Test (SD2PC Version)

Conc	Dose	Length	%Decrease	Std	N2	t_i
0	1	4	0	0.18	20	.
0.3125	2	3.99	0.25	0.19	10	6.02
6.25	3	3.81	4.75	0.13	10	3.24
12.5	4	3.63	9.25	0.22	10	0.47
25	5	3.46	13.5	0.14	10	−2.16
50	6	3.21	19.75	0.14	10	−6.01

N2, number of replicates, initially with four daphnid per replicate tank; Std is computed from an ANOVA on replicate means weighted by number of living daphid at study end; %Decrease,% decrease of treatment mean from control; t_i=t-statistic from Eq. (13.43).

from an ANOVA on the full dataset using a nested variance structure with rep nested in treatment and error given by daphnid nested in rep. Tamhane et al. (2001) followed a different tact of ignoring replicates based on the nested ANOVA finding no significant replicate effect. A different procedure is followed here to be consistent with our overall approach of using the replicate as the unit of analysis. There is no difference in conclusions and only modest differences in calculations from what were published. The pooled within-treatment standard deviation from the full ANOVA was 0.1731 vs. 0.1735 from the weighted ANOVA of replicate means.

The critical value of the t-distribution with 64 degrees of freedom is approximately 1.96, so MAXSD0.9=6.25. While the observed effect at 12.5 ppm is less than 10%, it does not provide significant evidence that the true effect is less than 10%. It is left as an exercise to show that the Dunnett, Williams, and step-down Jonckheere–Terpstra tests find the 4.75% effect at 6.25 to be significant, so the NOEC is 0.3125. These are typical results for daphnia length because the variability in these measurements is very low. It is instructive to fit a regression model to these data and estimate EC10. It is left as an exercise to show that a Bruce–Versteeg model fit the replicate mean data well and gave an estimate of EC10=15.1 (12.5, 18.3).

The second way to implement the MAXSD test is also a step-down test but based on Helmert contrasts of the treatment means that take the full dose–response shape into account. This approach is called SD1HP in Tamhane et al. (2001). It is more complicated than SD2PC and is useful when the concentration–response shape is less regular in appearance (leaving that term undefined) and it has better power properties than SD2CP. The contrasts are defined by

$$C_{ij} = \left(\bar{y}_i + \cdots + \bar{y}_j\right) - \left(j - i + 1\right)\lambda\bar{y}_0 \quad (13.45)$$

for $j>i$.

A t-statistics defined by

$$t_{ij} = \frac{C_{ij}}{\text{stderr}\left(C_{ij}\right)} = \frac{C_{ij}}{s/n_0\sqrt{\left(j-i+1\right)^2\lambda^2 + \sum_{h=i}^{j}\left(n_0/n_h\right)}}, \quad (13.46)$$

where s is the usual pooled within-treatment standard deviation. H_{0i} is rejected in favor of H_{1i} if

$$t_{i,\max} = \max_{i\le j\le k}\left\{t_{ij}\right\} > t^{\alpha}_{m,\nu,R_m}, \quad (13.47)$$

where k is the number of test concentration compared with control, $m=k-i+1$, and t^{α}_{m,ν,R_k} is the upper α equicoordinate critical constant of an m variate central t-distribution with ν degrees of freedom and correlation matrix:

$$R_m = \left\{\rho^{(m)}_{jj'}\right\}, \quad (13.48)$$

where

$$\rho^{(m)}_{jj'} = \text{corr}\left(C_{ij}, C_{ij'}\right) = \frac{\left(j-i+1\right)\left(j'-i+1\right)\lambda^2 + \sum_{h=i}^{j}\frac{n_0}{n_h}}{\sqrt{\left[\left(j-i+1\right)^2 + \sum_{h=i}^{j}\left(n_0/n_h\right)\right]\left[\left(j'-i+1\right)^2 + \sum_{h=i}^{j}\left(n_0/n_h\right)\right]}},$$

for $i \le j < j' \le k$.

(13.49)

The critical values given in Tamhane et al. (2001), i.e. the rightmost term in Eq. (13.47), were derived from Schervish (1984) or approximated by Dunnett (1989). The procedure begins by testing H_{01} vs. H_{11} by computing

$$t_{1,\max} = \max_{1\le j\le k}\left\{t_{1j}\right\} \text{ and comparing } it \text{ to } t^{\alpha}_{m,\nu,R_m}.$$

If there is insufficient evidence to reject H_{01} in favor of H_{11}, then testing stops and MAXSD is less than the lowest tested concentration. Otherwise, one tests H_{0i} vs. H_{1i}, $i>1$ until the first integer i is encountered at which there is insufficient evidence to reject H_{0i} in favor of H_{1i}, at which point testing stops and MAXSD=$(i-1)$th concentration. For these data, the extra work made no difference, as $t_{i,\max}=t_i$ in every case. All calculations for both versions of the MAXSDλ test are given in Appendix 1.

The SD1HC test was developed using FORTRAN programs from the two cited articles (Schervish, 1984; Dunnett, 1989). Since FORTRAN is no longer widely used, it is fortunate that the critical values can be obtained from the more accessible R package mvtnorm available at http://mvtnorm.r-forge.r-project.org.

The reader is cautioned that Hothorn and Hauschke (2000) developed a different test with the same name and gave an R program for it. We have not studied that test.

13.9 TOXICOKINETICS AND TOXICODYNAMICS

Broadly speaking, toxicokinetics deals with uptake of a toxicant from the medium or exposure route to its distribution in the body, especially in the target organ or other tissue, and ultimately to elimination from the body. Toxicodynamics concerns mechanistic models that describe how the chemical interacts with the target organ or tissue, especially at the cellular level, and how the organ and whole animal respond. A focus of models in ecotoxicology is the effect of chemical exposure on a population rather than an individual. TKTD models are increasingly important in ecotoxicology, and numerous excellent papers have appeared since 2011 (e.g. Jager et al., 2011; Nyman et al., 2012; Ashauer et al., 2013, 2016; Ducrot et al., 2015; Gergs et al., 2016); workshops are being offered (summer schools held in Denmark in 2012, 2014, 2016, and announced for 2018; 2nd International workshop on TKTD modelling 2015 in Germany), as well as special sessions at SETAC conferences (e.g. Orlando 2016, Brussels 2017); and one or more textbooks are reported to be in preparation. Section 13.9.1 will describe a toxicokinetic (TK) model applied to subjects exposed to a single dose of chemical and the concentrations of that chemical in the blood over time. In Section 13.9.2, an example of degradation of chemicals in the soil with both parent compound and a metabolite in related models is discussed. Section 13.9.3 will describe models for bioconcentration studies. Finally, Section 13.9.4 will describe some of the aforementioned work on TKTD models.

13.9.1 Compartment Model for Absorption and Elimination of Toxicants in the Blood

There are several ways to model TK data. One with a long history in the closely related area of pharmacokinetics that has also been used for ecoxtocity data is compartment modeling. Typically, a compartment is some volume of the body or set of similar tissues or fluids. In the first example to follow, the compartment is the blood or blood circulation system, and the statistical issue is to develop a model to describe the concentration of the chemical as a function of time and initial exposure level in the blood following ingestion or some other means of exposure such as dermal or inhalation. Welling (1997) contains some good discussion of compartment models.

In the example to be described here, subjects were exposed to a single dose of 0, 0.005, 0.015, 0.030, 0.060, 0.090, or 0.15 $\mathrm{mg\,kg^{-1}}$ of the test chemical. Blood samples were then drawn repeatedly from each subject over the next 24 h, and RBC ChE levels were measured.

The model for this response was based on first-order kinetics for an open one-compartment model. The specific statistical approach followed is that described in Clayton et al. (2003), with minor adaptations. According to Welling (1997) and Clayton et al. (2003), a model in which absorption and elimination rates are unequal is the most plausible, and such models are seen to be consistent with the data observed. The form for this model is as follows:

$$Y_{rt} = A_r * \left[1 - K_r * d_r * \left(e^{-k_{er}*t} - e^{-k_{ar}*t}\right)\frac{k_{ar}}{k_{ar} - k_{er}}\right] \quad (13.50)$$

where k_{er} and k_{ar} are the subject-specific elimination and absorption rates, respectively, t is time in minutes, K_r is a proportionality constant, A_r is a subject-specific baseline ChE activity level, d_r is dose or concentration, and Y_{rt} is the measured response on subject r at time t. Now

$$\begin{aligned} K_r &= e^{\beta_1 + b_{1r}} \\ k_{ar} &= e^{\beta_2 + b_{2r}} \\ k_{er} &= e^{\beta_3 + b_{3r}} \end{aligned} \quad (13.51)$$

where β_1, β_2, β_3 are fixed parameters to be estimated and b_{1r}, b_{2r}, b_{3r} are subject-specific random deviates assumed to have a multivariate normal distribution with mean 0 and unstructured variance–covariance matrix to be estimated. The k_{ar} are inhibition rate constants reflecting gastrointestinal absorption, binding, and carbamylation (a reaction that creates ureas, otherwise known as carbamides). The k_{er} are recovery rate constants reflecting decarbamylation. Various simplifications can be made. Pinheiro and Bates (1995) indicate that the random effect term in the definition of one of the two rates, k_e and k_a, can be dropped. In Exercise 13.7, dropping the random effect in the elimination rate leads to a good fit. In the example to be discussed here, this was achieved when the random term was dropped from the absorption rate. In the latter case, the elimination rate was almost four times the magnitude of the absorption rate, and no adequate fit could be obtained using the alternative rate. This proved to be very helpful in obtaining good model fits that still account for the within-subject correlations.

This model can be further simplified if, instead of modeling the measured ChE level in absolute terms, we instead model, *for each subject*, the proportion of baseline each measurement represents. This does not violate the concerns raised in Chapter 7 about normalizing responses by expressing every observation as a percent change from the control mean. In that type of normalization, all observations become correlated because the same baseline estimate was used for all observations. In this modeling approach, each subject has its own baseline, and no intersubject correlations are introduced. Then the term A_{rt} is not needed. For the exercise, baseline was taken as the mean of the measured responses within 24 h prior to dosing. For the current example, baseline was taken as the mean of all measured responses prior to dosing, consisting of measurements at 72, 48, 16, and 0.5 h prior to dosing.

The model can reparameterize this by using

$$k_{ar} = k_e e^{e^{\beta_2 + b_{2r}}} \quad \text{or} \quad k_{er} = k_a e^{e^{\beta_3 + b_{3r}}}$$

to make sure the modeled absorption rate is greater than or less than, respectively, the modeled elimination rate. Another reason for using such a reparameterization is that it may improve the stability of the model fitting process. However, if the wrong assumption is made, say, that the absorption rate exceeds the elimination rate, then it may be very difficult to obtain a good fit and the model would be misleading in any event. Without a reparameterization, the final model will indicate which is greater, and, so, the indicated reparameterization may not be needed or even advisable. Other parameterizations can also be used, one of which is indicated in the exercise.

The multivariate normal variance–covariance structure specifically models the repeated measures nature of the measurements. This is necessary since repeated measurements on the same subject are correlated.

The unequal rates model was expected to describe these data. It is possible from this model to estimate the time for ChE levels of a typical person or the most sensitive person to recover to 80%, 90%, or *P*% for arbitrary $P<100$ of predosing value. In addition, the maximum concentration of the chemical in the blood, Cmax, and the time, Tmax, at which that occurs can be derived. Furthermore, the maximum decrease in ChE activity and the time that occurs can be estimated. The values for the "typical" person are provided. The following formulas are useful for these purposes. The formula for Cmax expresses the maximum proportion decrease in ChE. It is convenient below to transform this to a percent change from baseline:

$$\text{Tmax} = \frac{\text{Ln}(k_e/k_a)}{k_e - k_a} \tag{13.52}$$

$$\text{Cmax} = K * \text{Doseval} * \left(e^{-k_e * \text{Tmax}} - e^{-k_a * \text{Tmax}}\right) * \frac{k_a}{k_a - k_e}. \tag{13.53}$$

To obtain starting values for β_1, β_2, and β_3, a fixed effects model is fit, obtained by removing the random terms b_{1r}, b_{2r}, and b_{3r} from model Eq. (13.50). The initial estimates for that fit are obtained according to Welling (1997), as described next.

For the unequal rates model, estimate k_{er} for each subject as the slope of a regression of log(1 − ChE_Pct) on time (in minutes) for that subject. Estimate AUC from Simpson's rule (available from almost any calculus text) applied to that curve. Estimate K ($=K_r$) from equation $K_r = \text{AUC} * k_{er}/\text{Dose}$. Estimate k_{ar} from the same regression using the equation $\text{Int}_r = K_r * \text{Dose} * k_{ar}/(k_{ar} - k_{er})$, where Int_r is the intercept of the regression line. We now have K_r, k_{er}, and k_{ar} for each subject. From that, the initial variance–covariance matrix of the b_{1r}, b_{2r}, b_{3r} is obtained by first computing, for each subject, the values $b_{1r} = \log(K_r) - \beta_1$, $b_{2r} = \log(k_{ar}) - \beta_2$, and $b_{3r} = \log(k_{er}) - \beta_3$. Then the sample variance–covariance matrix of these values is used as starting values for the fit of the full model. Programs *uptake model.sas* and *depurate model2.sas* for this purpose using the NLMIXED procedure are provided in Appendix 1, as is an R program. Companion programs, *Helsel.sas* and *HelselW.sas* (for whole fish), are also given to handle values below the detection limit. The SAS datasets used by the two modeling programs were created by applying the Helsel programs. The original datasets, uptake and depurate, and those obtained from the Helsel programs, uptake_adj and whole_adj_depur, are both included in Appendix 1. Initial values of s_0 are obtained from the nonlinear regression just described or from a repeated measures ANOVA model.

Following Clayton et al. (2003), the response from a "typical" subject, together with 95% confidence bounds, is predicted from the above model by removing the random components from A_r, K_r, k_{ar}, and k_{er}. The data are given in Appendix 1 Table A1.13.2. See Table 13.18 for the parameter estimates, where it should be noted that all parameters were significantly different from zero. The variances for the random terms b1, b2, and b3 may seem very large, but this reflects the high intersubject variability evident in the data. This variability is clear (see Figure 13.7) for the highest treatment level. Plots for the other treatment groups are produced by the programs cited above and show the same level of intersubject variability. See Figure 13.8 for a QQ-plot for the residuals from the model showing reasonable consistency with normality.

13.9.2 Kinetic Model for Degradation of Chemicals in the Soil

The objective of this section is to investigate models and model fitting to characterize the degradation of parent chemicals and their metabolites in soil. See Table 13.19 for an example dataset. Related discussion was published in Gao et al. (2011). One issue of possible concern is that the variability of degradation rates may not be constant over the range of degradation times. While it was found not to be a concern for the example to be presented, it is possible to handle that using weights proportional to the predicted mean response. A simple four-parameter first-order model (SFO) was found quite adequate for modeling the data in the example.

Table 13.18 Parameter Estimates for Unequal Rates Model for RBC ChE

Parameter	Estimate	StdErr	*t*Value	Prob*t*	LCB	UCB
beta1	3.136	0.265	11.829	0.000	2.599	3.673
beta2	−2.881	0.195	−14.774	0.000	−3.276	−2.486
beta3	−3.595	0.160	−22.451	0.000	−3.920	−3.271
s0	−4.943	0.062	−80.196	0.000	−5.068	−4.818
vb1	43.896	0.830	52.892	0.000	42.215	45.578
cb13	18.264	4.162	4.388	0.000	9.830	26.698
vb3	11.375	4.404	2.583	0.014	2.452	20.299
cb12	−44.881	0.370	−121.386	0.000	−45.630	−44.132
vb2	45.950	0.855	53.741	0.000	44.218	47.683
cb23	−18.887	4.319	−4.373	0.000	−27.637	−10.136

LCB and UCB, 95% confidence bounds for parameter estimate.

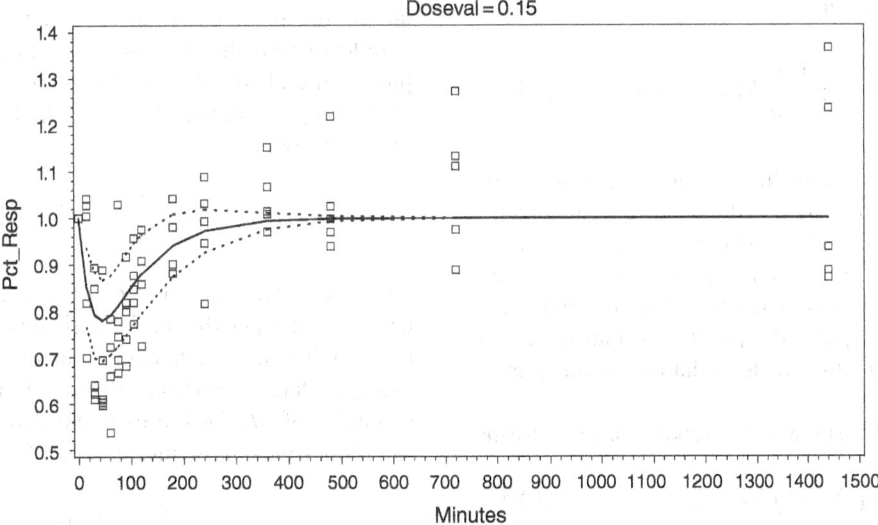

Figure 13.7 Plot of observed and predicted RBC ChE in high treatment group. Squares = observations. Solid line is fitted curve from model (13.50). Dashed lines are 95% confidence bounds for each predicted mean response.

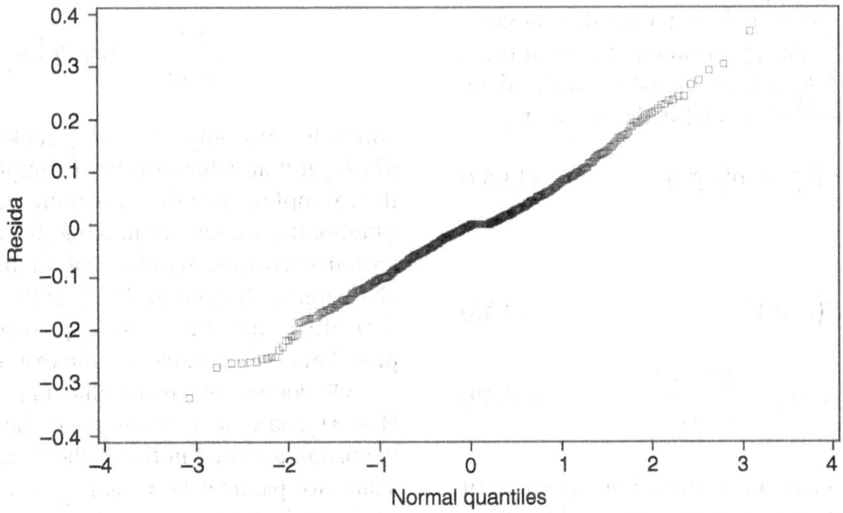

Figure 13.8 QQ-plot of residuals from model fit to RBC ChE. Observations appear close to linear, supporting a normal distribution. Label comes from programs fit_tkxmpl1.sas, fit_tkxmpl1_b1.sas, and QQplot RBC ChE.sas in Appendix 1.

Table 13.19 Soil Degradation Data

Time	Parent	Met1	Time	Parent	Met1	Time	Parent	Met1
0	93.2	0	14	83.7	4.3	120	34.1	9.3
0	95.8	0	30	77.6	8.4	120	31.1	8.7
0	93	0	30	70.6	11.9	181	23.7	6.8
7	89.8	3.1	30	74	12.3	181	20	6
7	90.8	3	62	55.6	13.3	181	21.1	5.8
7	87.5	3.1	62	58	12.2	360	10.7	1.4
14	83.1	6.9	62	54.8	11.4	360	8.9	1.5
14	81.6	5.4	120	41.4	10.4	360	7.8	1.3

Time, number of days' postexposure; Parent and Met1, amount.

The general model for metabolite degradation is given by the following equation:

$$M(t) = -C\int_0^t \frac{dP(u)}{du} F(t-u)du, \tag{13.54}$$

where $P(t)$ is the amount of the parent compound in the soil or sediment at time t, $M(t)$ is the amount of the metabolite formation at time t, $F(t-u)$ is the proportion of the metabolite remaining in the soil at time $t-u$ after formation at time u, and C is a degradation rate constant for the metabolite, so that $-C[dP(u)/du]$ is the formation rate of metabolite at time u due to degradation of the parent pesticide.

The models for the parent and metabolite have the forms

$$P(t) = M_0 G(t, k_p), \tag{13.55}$$

$$M(t) = CM_0 G(t, k_m), \tag{13.56}$$

where M_0 is the amount of parent initially in the soil at time zero, C is the proportion of the parent that metabolizes over time, $G(t,k)$ is an appropriate function of time, and parameters k_p and k_m are estimated from the data. The simple first-order (SFO) model is the following:

$$G(t, \beta) = \exp(-\beta t) \tag{13.57}$$

so that

$$P(t) = M_0 e^{-k_p t} \tag{13.58}$$

$$M(t) = CM_0 k_p \frac{e^{-k_m t} - e^{-k_p t}}{k_p - k_m}. \tag{13.59}$$

Since parent degradation is assumed independent of the metabolite present, the modeling begins by fitting the parent data using Eq. (13.58) to estimate M_0 and k_p. Then Eq. (13.59) can be used to estimate C and k_m using the estimates of the other parameters found in fitting the parent.

For the parent compound in an SFO model, there are two parameters, M_0 and k_p. Let θ represent either of these. Consider the partial derivative of $P(t)$ with respect to θ, $\partial P/\partial \theta$. If the absolute value of this partial derivative is small, then a change in t will have little impact on the value of the parent function. This in turn means that the variability in the estimate of θ will be large. On the other hand, if the absolute value of this partial derivative is large, then a change in t will have high impact on the value of the parent function. This in turn means that the variability in the estimate of θ will be small. So, the range of sampling times should be selected to provide high values of the partial derivative. For the parent, it is quite simple to find the maximum, as indicated below, but a plot of the absolute partial derivative is more important for finding the range of t values that provide high values of this derivative:

$$\frac{\partial P}{\partial M_0} = e^{-k_p t}, \tag{13.60}$$

which has its maximum when $t=0$. One should include values of t that make the partial derivative within, say, 20% of this maximum, which works out to be $0.223/k_p$. For the example data ($k_p=0.008$), this is 28 days. Thus, the best estimates of M_0 for the example data are obtained using sampling times in the interval (0, 28). Also,

$$\frac{\partial P}{\partial k_p} = -M_0 t e^{-k_p t}. \tag{13.61}$$

To find the value of t that maximizes this function, it helps to find the partial derivative of it with respect to t, i.e.

$$\frac{\partial^2 P}{\partial k_p \partial t} = -M_0 e^{-k_p t}\left[tk_p - 1\right], \tag{13.62}$$

which is zero only at $t=1/k_p$, making the maximum of $\partial P/\partial k_p$ at that value of t. For example, if $k_p=0.008$ (as in the example), then the maximum occurs at $t=125$ days. Finding the values of t making the absolute value of this partial derivative within 20% of its maximum is more challenging. It is helpful to plot the absolute first partial derivative and obtain the approximate result from the plot. This is illustrated for the example dataset.

Of course, the metabolite must also be considered. Here M_0 and k_p are taken as given (the estimates from fitting the parent are used in fitting the metabolite), and there are again two parameters, C and k_m, to estimate. The required partial derivatives are given by

$$\frac{\partial M}{\partial C} = \frac{k_p M_0}{k_p - k_m}\left[e^{-k_p t} - e^{-k_m t}\right]$$
$$\frac{\partial M}{\partial k_m} = \frac{C k_p M_0}{\left(k_p - k_m\right)^2}\left\{\left[1 + t\left(k_m - k_p\right)\right]e^{-k_m t} - e^{-k_p t}\right\}. \tag{13.63}$$

While the maxima of these can be solved using the partial derivatives with respect to t, there are still equations that can only be solved numerically. It is simpler to plot these functions and visually find the times that maximize them. Of course, this requires preliminary estimates of the rate constants. An iterative process can be set up to refine the estimate, but this is often not necessary, as exact solutions are not required.

See Table 13.20 for the results of fitting models (13.58) and (13.59) to the parent and metabolite degradation data in Table 13.19. All model parameters are significantly different from zero and the fit of the models to the data is good. (See Figure 13.9.)

A SAS program given in Appendix 1 to carry out the analyses in this section is *Soil degradation model.sas*, which optionally calls on *predicts.sas* and *predmins.sas*. The R program *Soil degradation model.r* is also provided.

13.9.3 Bioconcentration Models

Bioconcentration studies are covered by several regulatory TGs. OECD TG 305 and OPPTS 850-1730 concern flow-through designs for bioaccumulation in fish through aqueous and dietary exposure. OECD TG 315 concerns bioaccumulation in sediment-dwelling benthic oligochaetes. The purpose of the fish bioconcentration test (OPPTS 850-1730, OECD 305) is to determine whether a chemical will accumulate in the edible and nonedible fractions of fish and, if so, at what concentration. A TG proposed to OECD in 2017 to which the first author of this text has contributed replaces the whole fish testing with an *in vitro* bioconcentration estimate from bioassays using hepatocytes from the rainbow trout liver. The second author of this text was heavily involved in the revision of OECD TG 305 in 2012. This section will deal only with the whole fish assay and will not cover all the developments in the revised TG 305. It may be helpful to define two key concepts before exploring models. These definitions draw on clear descriptions given by van der Hoop (2013).

Biomagnification, also called bioamplification, refers to an increase in the concentration of a substance as it moves move up the food chain. The idea is that higher-order organisms take on the chemical burden from their

Table 13.20 Results of SFO Models Fit to Soil Degradation Data

			Parent degradation data				
Source	DF	SS	MS	*F*Value	Prob*F*		
Model	2	102 245	51 122.3	7984.8	<0.000 1		
Error	21	134.5	6.402 5	—	—		
USS	23	102 379	—	—	—		
Parameter	Estimate	StdErr	Alpha	LCB	UCB	*t*Value	Prob*t*
K	0.008 05	0.000 236	0.05	0.007 56	0.008 54	34.11	<0.000 1
M0	93.745 6	0.891 7	0.05	91.891 3	95.6	105.13	<0.000 1
			Metabolite degradation data				
Source	DF	SS	MS	*F*Value	Prob*F*		
Model	2	1 312	656	664.21	<0.000 1		
Error	21	20.741 3	0.987 7	—	—		
USS	23	1 332.8	—	—	—		
Parameter	Estimate	StdErr	Alpha	LCB	UCB	*t*Value	Prob*t*
U	1.281 8	0.080 3	0.05	1.114 9	1.448 7	15.97	<0.000 1
C	0.766 7	0.048	0.05	0.666 8	0.866 5	15.97	<0.000 1

All model parameters have tight confidence bounds and the model fits the data well, with the exception of the 360 day values for the parent. LCB and UCB, lower and upper 95% confidence bounds for indicated parameter.

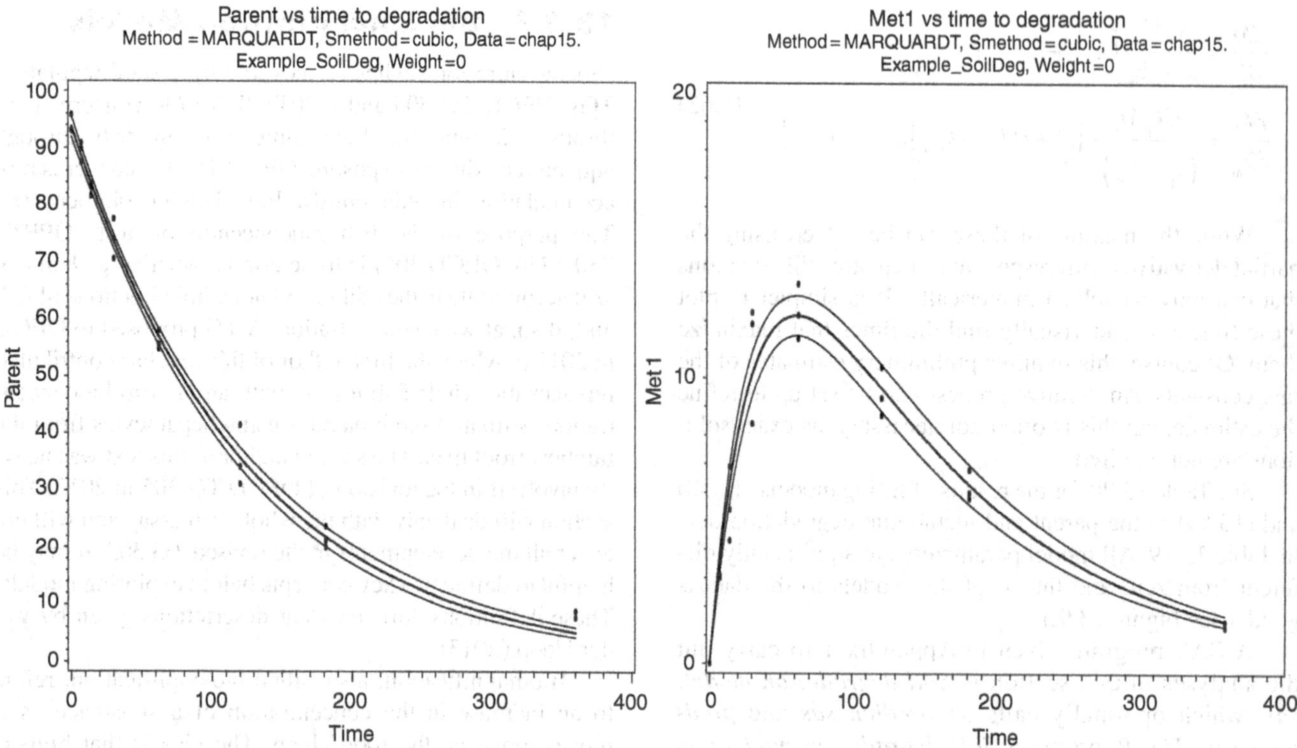

Figure 13.9 SFO models fit to soil degradation data. Left panel is for model (13.47) fit to the parent compound. Right panel is for model (13.48) fit to a metabolite.

prey and, as a result, tend to have higher levels of the chemical than their individual prey. The higher level organisms are thus exposed both directly to the chemical in the environment and indirectly from the diet. Biomagnification thus refers to entire food chains rather than individuals. Biomagnification occurs for chemicals that either do not breakdown (decompose) from natural environmental processes or do so only very slowly. As a consequence, such persistent pollutants are transferred up the food chain faster than they are broken down or eliminated from the body.

Bioaccumulation occurs within an organism, where a concentration of a substance builds up in the tissues and is absorbed faster than it is removed. Bioaccumulation in fish tissue occurs through the diet and by absorption from water. The older TG 305 dealt only with accumulation from water, and sometimes, especially in older literature, that is what is meant by bioconcentration. The 2012 revised version of TG 305 includes dietary sources as well as absorption from the water. This section will deal only with bioaccumulation and illustrates some of the challenges in guideline studies.

The object of these studies is to determine the time required for bioconcentration and, where appropriate, dietary biomagnification to reach steady state under constant exposure conditions and then the time for 50 and 95% depuration after exposure ends.

Some measurements in the depuration phase, especially in the low test concentration fillet data, can be below the detection limit. The true values are estimated, wherever possible, by Helsel's method (Helsel and Gilliom, 1986; See also Helsel (2005) and especially Helsel (2012), which provides R programs), and these estimated values are used in subsequent ANOVA and regression analyses. Further discussion of the treatment of values below detection limits is given in Newman (1995).

Steady state is defined as three consecutive measurement days over which there is no statistically significant increase in mean tissue residues in whole fish. A 2-factor factorial ANOVA model with day and concentration as factors and fish as random effect is used together with simple unadjusted t-tests to compare the mean tissue residues sampled on each fish sampling day.

Once the steady-state time period is determined, the mean tissue residue of fish sacrificed and tested on those three consecutive sampling dates is computed and used as the basis for calculation of bioconcentration factor (BCF) or biomagnification factor (BMF) based on maximum steady-state tissue residues.

Steady-state tissue residues for aqueous test substance concentrations are used to fit an open one-compartment model based on first-order kinetics to the depuration data for each aqueous test concentration. This model fitting can be done in two ways: fitting separate models for each test treatment or fitting a single model to all test treatments. The model fit to the depuration data is given by Eq. (13.64):

$$C_f = C_w \frac{k_1}{k_2} e^{-k_2(t-t_c)}, \quad t > t_c, \tag{13.64}$$

where t is time in days since exposure began, t_c is the time at the end of the uptake period, C_w is the mean measured concentration of the test substance in the aqueous test solution, C_f is the concentration in the fish tissue in μg kg^{-1}, and k_2 is the elimination rate constant. The parameters to be estimated were k_2 and k_1/k_2.

Either approach enables estimation of the elimination rate constant, k_2, and its confidence bounds. Using k_2, a simple exponential model can then be fit to the uptake data to estimate the absorption rate constant, k_1. The uptake model allows one to determine a conditional confidence interval for k_1, conditioned on k_2. In theory, the two constants can be determined simultaneously to provide unconditional confidence intervals for k_1 and k_2. The computational demands of that procedure can be rather onerous, but the effect on the final result ranges from minor to dramatic. An example is provided in Section 13.9.3.1.

The model for the uptake period was

$$C_f = C_w \frac{k_1}{k_2}\left[1 - e^{-k_2 t}\right]. \tag{13.65}$$

These models are described in Newman (1995) for the type of experiment in OECD TG 305 and more generally in Welling (1997).

See Table 13.21 for an example dataset, where the mean and standard deviation of bioconcentration values from four fish sampled on the indicated days are given for two portions of fish tissue, a fillet and the carcass. The third portion, whole fish, is defined as fillet plus carcass. Each of these three portions was analyzed separately. The full dataset is given in Appendix 1.

A simple factorial ANOVA model with concentration and day as fixed factors was fit to the uptake data summarized in Table 13.21, and the mean bioconcentration in each type of fish tissue on each day was compared with that in each other day in the same test concentration using unadjusted t-tests. The earliest day on which the mean concentration does not differ significantly from that in either of the subsequent two sampling dates is set as the day steady state was achieved. Programs are provided in Appendix 1 for this purpose. It is left as an exercise to demonstrate that steady state was achieved on day 3 for the high dose fillet and on day 10 for the low dose fillet. Some fillet measurements at the high concentration were below detection limit. Such values are called left-censored. However, there were enough data above detection limit to allow a statistically sound estimate of each desired mean. Three methods for doing this will be discussed.

One approach for treating censored values has already been developed in Chapter 8 using Eq. (8.2). This assumes a normal distribution, but can be modified for other distributions. If there are enough noncensored data, this is a viable and statistically justified approach. An alternative is to use a nonparametric approach discussed in Chapter 10 for the Kaplan–Meier survival analysis model. It has been applied to the nondetect problem by Helsel (2009) and will be described next.

The Kaplan–Meier method of handling nondetects begins by ranking all n observations from smallest to

Table 13.21 Uptake Period of Bioconcentration Study

Portion	Day	Conc	Bioconc	Std	Portion	Day	Conc	Bioconc	Std
C	1	4.2	1.1603	0.2097	F	1	4.2	0.138	0.0293
C	3	4.2	2.249	0.35272	F	3	4.2	0.2098	0.02998
C	5	4.2	2.3985	0.60603	F	5	4.2	0.2275	0.04312
C	7	4.2	2.0508	0.27365	F	7	4.2	0.1513	0.02595
C	10	4.2	2.6613	0.32444	F	10	4.2	0.2083	0.05108
C	12	4.2	2.6005	0.15916	F	12	4.2	0.242	0.08163
C	14	4.2	2.2933	0.45517	F	14	4.2	0.2003	0.03732
C	1	42	11.2478	1.88951	F	1	42	1.275	0.21785
C	3	42	17.166	2.53799	F	3	42	1.9068	0.27694
C	5	42	17.8958	1.4439	F	5	42	2.222	0.2071
C	7	42	19.0163	7.1078	F	7	42	1.9745	0.54053
C	10	42	21.7033	2.30485	F	10	42	2.0783	0.16799
C	12	42	19.7735	2.61984	F	12	42	1.8055	0.24018
C	14	42	17.5963	1.0325	F	14	42	1.8878	0.06913

C, carcass; F, fillet; Conc, concentration of chemical in water; Bioconc, concentration of chemical is fish tissue; Std, standard deviation of fish tissue bioconcentration in sample; Day, days of constant exposure.

largest. The largest observation has rank n. There are $n-1$ observations ranked less than it. In the case of ties, arbitrarily assign unique ranks to the tied observations so as to maintain the overall order. Assign the percentile $p_n = (n-1)/n$ to this largest observation. For the second highest ranked observations, there are $(n-2)$ observations ranked below it. Assign it the percentile $p_{n-1} = p_n * (n-2)/(n-1)$. Continue in this way to assign the kth largest ranking observation the percentile $p_{n-k} = p_{n-k+1} * (n-k-1)/(n-k)$. The lowest ranking observation is thus assigned the percentile 0. If there are no nondetects, this reduces to the ith smallest observation having percentile $(i-1)/n$. In this way, the distribution function, $F(x)$, of the bioconcentration, x, is defined. To make this concrete, consider the numbers arranged in increasing order (0.0072, <0.014, 0.033, <0.06, 0.077, 0.21) from Helsel (2009). For these numbers, ignoring for the moment the censoring indicated, the empirical distribution function is given by

$$F(x) = \begin{cases} 0 & \text{if } x < 0.0072 \\ \frac{1}{6} & \text{if } 0.0072 \le x < 0.014 \\ \frac{2}{6} & \text{if } 0.014 \le x < 0.033 \\ \frac{3}{6} & \text{if } 0.033 \le x < 0.06 \\ \frac{4}{6} & \text{if } 0.06 \le x < 0.077 \\ \frac{5}{6} & \text{if } 0.077 \le x < 0.21 \\ 1 & \text{if } 0.21 \le x \end{cases}. \quad (13.66)$$

The mean value of x, $E[X]$, is given by

$$E[X] = \int_0^{0.21} x dF(x). \quad (13.67)$$

The integral can be approximated by sums of the form

$$E[X] = \lim_{n\to\infty} \sum_{i=1}^{n} x_i^* \left[F(x_i) - F(x_{i-1})\right]. \quad (13.68)$$

where $x_i - x_{i-1} = 1/n$ and x_i^* is some value satisfying $x_{i-1} \le x_i^* \le x_i$. Since F is a step function, for n sufficiently large, every expression in square brackets in Eq. (13.68) is zero except when x_i is one of the breakpoints 0.0072, 0.014, etc. When x_i is one of the breakpoints, the expression in square brackets is 1/6. Furthermore, as $n \to \infty$, $x_i^* \to x_i$, so Eq. (13.68) approaches the sum of the x_i times 1/6, i.e. the mean of the x_i.

The above may appear to be a complex way to express the arithmetic mean of the observations. The purpose of expressing it this way is to set up the Kaplan–Meier way of handling censored values. For that, drop from the list of percentiles just calculated, those corresponding to nondetects. For each remaining observation, define $q_{n-k} = p_s * k/m$, where m is the number of observations (detected or not) that are at or below the observation in question, k is the number of observations (detected or not) that are definitely below, and p_s is the previous nonomitted percentile. The q's define a distribution function that accommodates censoring. In the case of the six numbers illustrated above and now using the indicated censoring, the initial percentiles are ($p_0 = 0$, $p_2 = 1/6$, $p_3 = 2/6$, $p_4 = 3/6$, $p_5 = 4/6$, $p_6 = 5/6$), so $q_6 = p_6$, $q_5 = p_5$, q_4 is omitted, $q_3 = 2/3 * p_5 = 4/9$, q_2 is omitted, $q_1 = 0$ and the corresponding empirical distribution function is given by

$$F(x) = \begin{cases} 0 & \text{if } x < 0.0072 \\ 0.4444 & \text{if } 0.0072 \le x < 0.033 \\ 0.6667 & \text{if } 0.033 \le x < 0.077 \\ 0.8333 & \text{if } 0.077 \le x < 0.21 \\ 1 & \text{if } 0.21 \le x \end{cases}. \quad (13.69)$$

The mean or expected value of x taking censoring into account is still given by Eqs. (13.67) and (13.68), but with this modified distribution function F. It is left to the reader to verify that $E[X] = 0.58$. An Excel spreadsheet, KMStats16.xls, is given in Appendix 1 to do these calculations. It is freely available from Helsel (2017), a website associated with Helsel (2012) that contains a wealth of information and software downloads for the handling of nondetects and other statistical issues.

A third method for handling nondetects is based on regression and assumes the response follows a normal distribution. What is done is to fit a linear regression of log(response) vs. normal scores and extrapolate below the detection limit to obtain estimates of the censored values. These estimates can then be back-transformed to the original scale, and summary statistics, such as the mean and variance, can then be calculated. This procedure works reasonably well if there is less than 50% censoring. A SAS program for this approach, based Helsel and Gilliom (1986) and also discussed in Helsel (2005) and Helsel and Hirsch (2002), is given in Appendix 1 with the name Helsel. SAS and was applied to the example partially summarized in Table 13.15.

Once the means for each treatment and day have been obtained taking censoring into account, the abovementioned ANOVA can be done to determine the day steady state is achieved and the bioconcentration at steady state. For fillet in the example, steady state occurs on day 10 for the low test concentration and on day 3 for the high test concentration. The steady-state concentrations were 0.21683 and 2.03442, respectively. The program *Bioconc_Background.SAS* does the calculation. The depuration models are fit next separately for the two test concentrations. The program *Depurate Model.SAS* estimates the elimination rate constant, k_2, to be 0.815 (0.676, 0.982) for the high test concentration. With that value, the uptake model can be fit using the program *Uptake Model.SAS*. The absorption rate constant derived from that model was 0.039 (0.037, 0.042).

See Figures 13.10 and 13.11 to see the fit of the depuration and uptake models for the high test concentration. It is left as an exercise to fit these models to the low dose data.

13.9.3.1 Simultaneous Estimation of k_1 and k_2

We are deeply indebted to T. Aldenberg for providing this section. The work on the revised TG 305 arrived at several important understandings about bioconcentration studies that were not captured in the previous version of the guideline. One of these was the importance of jointly estimating the parameters k_1 and k_2 from nonlinear regression of both the uptake and depuration phases simultaneously. One needs numerical software to accomplish that, and obtain statistical measures, as well as plots, to assess the quality of

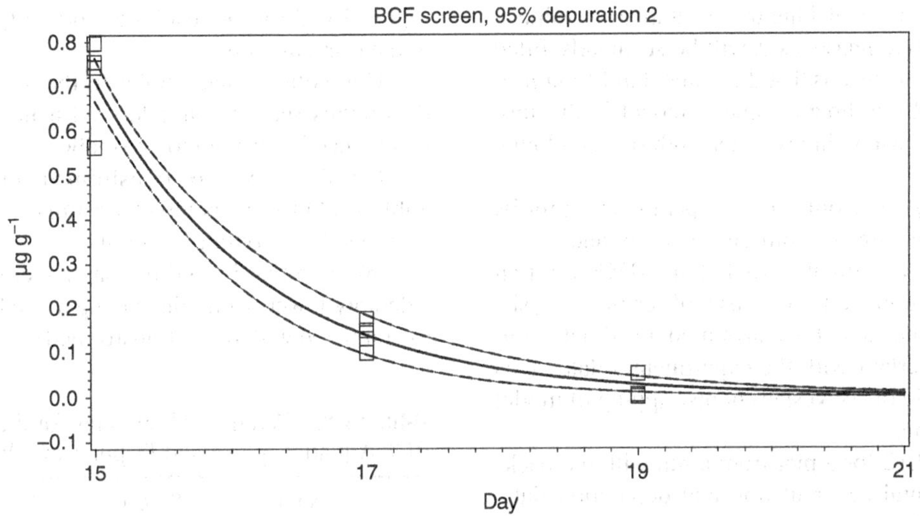

Figure 13.10 Depuration model for fish bioconcentration at high exposure concentration. Observations, adjusted for censoring, as squares, prediction as solid curve, 95% confidence bounds as dashed curves.

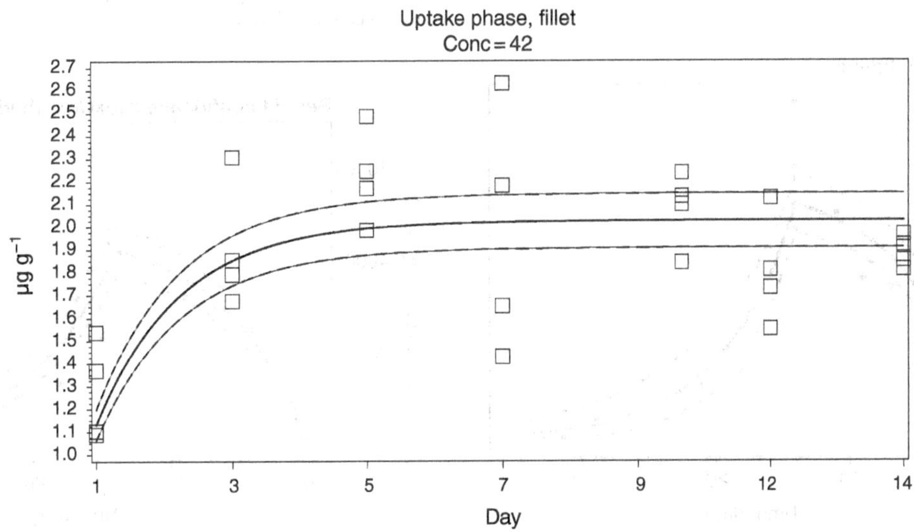

Figure 13.11 Uptake model for fish bioconcentration at high exposure concentration. Observations, adjusted for censoring, as squares, prediction as solid curve, 95% confidence bounds as dashed curves.

the results. These are goodness-of-fit measures, graphical overlays of data and model fits, and confidence limits of the parameter estimates and model predictions.

Aldenberg (2016) provides R software, developed specifically for OECD TG 305, to fit uptake and elimination phases simultaneously for aquatic exposure and also for dietary experiments, usually only comprising depuration data.

One advantage of jointly estimating k_1 and k_2 is that this model is more sparse than that described before: estimating k_2 from the elimination phase data alone and then using this point estimate to estimate the absorption rate constant k_1 from the uptake phase. Fitting the exponential depuration data involves a free constant for the intercept that will not guarantee matching the accumulation value at the end of the uptake phase, as it will be separately fitted with k_1 given k_2. This means that the sequential fitting procedure is essentially a three-parameter model in disguise that may involve a discontinuity, when both fits are plotted together.

The R package not only fits the parameters jointly, but it calculates parameter confidence limits and confidence limits of the estimated BCF (and BMF for that matter). This is done on the basis of error analysis. Moreover, it calculates and displays model estimate confidence limits, overlaid with the experimental data. This is done with the aid of (Bayesian) bootstrapping of model estimates over time.

See Figure 13.12 for a plot from a run with the package for experimental accumulation and depuration data, involving the nonlinear fitting process, and model confidence limit results. The picture is modified from the guidance document (Section 3.4.1) to include the model confidence limits. See Table 13.22 for the parameter estimates and their standard error and 95% confidence limits.

When fitting uptake and elimination curves jointly, the issue arises as to what data (and model) transformation to apply. For the sequential method, elimination curves are usually fitted with a logarithmic transformation. The same transformation may, or may not, be applied to the uptake data.

In the R package, it was decided to employ a Box–Cox analysis in order to suggest a common transformation exponent applicable to both phases simultaneously. Obviously, the transformation exponent encompasses both the untransformed ($\lambda = 1.0$) and the logarithmic ($\lambda = 0.0$) cases.

The analysis of the optimal Box–Cox transformation yields $\lambda = 0.3$, which is somewhere between the square root ($\lambda = 0.5$) and the logarithmic transformation. See Figure 13.13 for the resulting model fit, back-transformed to the original units.

One notices that, similar to the log-transform, higher data points seem to have less influence, while the model confidence limits tend to be smaller.

For the Box–Cox transformed joint model fit, see Table 13.23 for the parameter estimates and their standard error and 95% confidence limits.

We note that indeed the parameter estimates are considerably lower, while displaying a smaller standard error, as compared with the untransformed case. Remarkably, the

Table 13.22 Parameter Estimates for the Jointly Fitted BCF Aquatic Uptake and Depuration Model

	Estimate	StdErr	2.5%	97.5%
k_1	451.3	79.0	317.2	653.9
k_2	0.1784	0.0379	0.1123	0.2731
BCF	2529.0	180.0	2189.0	2982.0

k_1 and k_2 are the absorption and elimination rate constants. BCF is the bioconcentration factor.

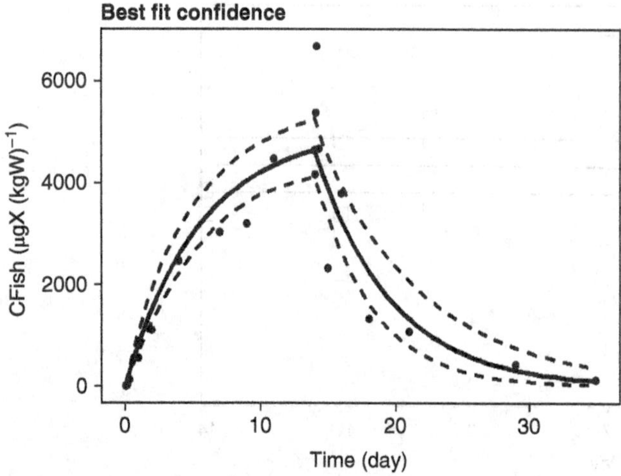

Figure 13.12 Joint nonlinear model fit of the uptake and depuration phases for the untransformed data.

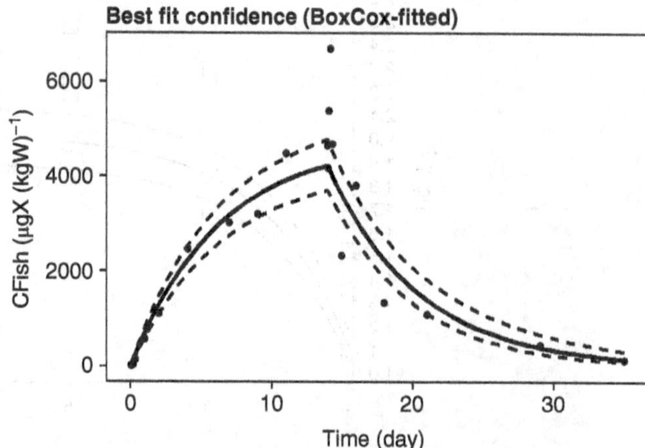

Figure 13.13 Joint nonlinear model fit of the uptake and depuration phases for the optimal Box–Cox transformation: $\lambda = 0.3$.

Table 13.23 Parameter Estimates for the Jointly Fitted BCF Aquatic Uptake and Depuration Model with the Optimal Box–Cox Transformation

	Estimate	StdErr	2.5%	97.5%
k_1	367.8	45.3	273.0	462.6
k_2	0.1565	0.0194	0.1158	0.1971
BCF	2351.0	183.0	1993.0	2709.0

k_1 and k_2 are the absorption and elimination rate constants. BCF is the bioconcentration factor.

BCF estimate is only somewhat lower, with approximately the same standard error. This is due to the fact that the BCF equals the ratio of the coefficients, while the estimates are correlated.

The reader may check the guidance document (OECD, 2017, section 3.4.2) that applying the logarithmic transformation to both phases leads to a joint curve estimate that fits the data very poorly, as shown by the fit diagnostics and graphically. The BCF becomes 1882 with a standard error of 264. The bad fit is due to the very low initial values of the uptake phase, as well as the very low final values of the depuration phase that pull the curve downward through the logarithmic transformation. This is not due to the joint fitting; it may happen to log-transformed fitting of the depuration data in the sequential method also. In the dietary model, the log-transformation is applied almost exclusively.

13.9.3.2 Dietary Exposure

A simple study that includes a dietary exposure could include a control group and a group with exposure only through the diet. An elimination rate constant is derived as a function of the total wet weight of the fish. In addition, the assimilation efficiency (α), the BMF, and its lipid-normalized value (BMF_L) can be estimated.

Individual fish weights for all time periods are tabulated separately for dietary test substance concentration and the control group for all days during the uptake and depuration phases. Weight data are converted to natural logs and plotted vs. day, separately for the dietary and control treatments. A linear least squares correlation is calculated for the ln(fish weight) vs. day for both dietary and control treatments using data for individual fish and standard statistical procedures. If necessary, the variances in the slopes of the test and control lines are evaluated to determine the statistical significance of the difference in the two slopes (growth rates) using Student's t-test. If there is no significant difference, the dietary and control treatment data are pooled and an overall fish growth rate for the study (k_{growth}) is calculated as the overall slope of the linear correlation. If statistical differences are observed, growth rates for control and treated fish are reported separately.

Individual fish test substance residue measurements expressed in terms of concentration (wet weight) as well as the lipid-normalized concentrations (if appropriate) are tabulated for test and control fish for individual sample times. The individual fish tissue residue data for the depuration period are converted to their natural logarithms and plotted versus time (day). If a visual inspection of the plot shows obvious outliers, a statistically valid outlier test may be applied to remove spurious data points. A linear least squares correlation is calculated for the ln(fish tissue residue) vs. depuration day data. The slope and intercept (day 0 of depuration) of the line are reported as the overall elimination rate ($k_{overall}$) and time zero concentration ($C_{0,depuration}$). If necessary, the variances in the slope and intercept of the line are calculated using standard statistical procedures and the 90% confidence intervals around these results evaluated. The mean fish concentration for the final day of uptake is also calculated and used as the value of $C_{0,depuration}$.

The growth-corrected depuration rate is calculated by subtraction of the growth rate from the overall elimination rate ($k_{depuration} = k_{overall} - k_{growth}$). The chemical assimilation efficiency (α) is calculated using the equation

$$\alpha = \frac{C_{0,depuration} \cdot k_{overall}}{I \cdot C_{food}} \cdot \left[1 - \exp\left(-k_{depuration} \cdot t\right)\right], \quad (13.70)$$

where I is the food ingestion rate ($g_{food} \cdot g_{fish}^{-1} \cdot d^{-1}$) and C_{food} is the chemical concentration in the food. The dietary BMF is calculated as $BMF = I \cdot \alpha \cdot k_{depuration}^{-1}$. A growth-corrected half-life can also be calculated. Aldenberg (2016) provides R software that was developed specifically for OECD TG 305 and a detailed description of how dietary and water exposure can be modeled simultaneously. Further exploration of the dietary model is deferred.

13.9.4 TKTD Models

TK models have been used for many years in bioconcentration studies, but the addition of TD to these models is a relatively new interest in ecotoxicity studies. As indicated in the introduction to this chapter, recent attention on TKTD models has focused on models for survival, though Ashauer et al. (2011) is an exception. This section is by no means a comprehensive treatment of TKTD models. It is intended merely to serve as a brief introduction to such models and indicate some interesting papers in the growing literature on this subject. Jager et al. (2011) describe traditional mortality data as based on the idea that an animal dies immediately upon reaching a critical internal concentration, called the critical body residue (CBR). Different animals have different sensitivities, which is why they do not all die at the same concentration of the toxicant. This gives rise to a species sensitivity distribution. Others refer to the CBR as the individual tolerance (IT). Under this concept, apart from natural (background) mortality, half of a population will survive prolonged exposure to a toxicant at the median tolerance concentration. This is a simplified view of IT that ignores variation of IT within the same organism over time and conditions, though it has been argued (e.g. Newman and McCloskey, 2000; OECD, 2006a, chapter 7; Zhao and Newman, 2007) that this is implicit in the models in common use. An alternative concept of stochastic death (SD) has been proposed (Bedaux and Kooijman, 1994; Widianarko and Van Straalen, 1996) as an entrée to hazard modeling. An examination of the form of the survival function proposed by Bedaux and Kooijman (1994) will reveal a strong connection to the one-compartment model given in Section 13.9.1, and the reader is referred to that paper for further insight. For our purposes, it is preferable to describe the more current development using the construct in Jager et al. (2011). The SD approach assumes that all individuals in a population are equally sensitive but death occurs as a stochastic process (i.e. random), or to paraphrase their words, death comes not to the weak, but to the unlucky. One of the consequences is that under prolonged exposure at any concentration of the toxicant that increases mortality, eventually the entire population will die.

The first step in describing a TKTD model is to describe the "TK" part, that is, the TK component that relates the external and internal concentrations of the test chemical as a function of time. What follows is strongly related to the discussion in Sections 13.9.1–13.9.3. A one-compartment first-order kinetic model has the form

$$\frac{\mathrm{d}C_{\mathrm{i}}(t)}{\mathrm{d}t}=k_{\mathrm{i}}C_{\mathrm{w}}(t)-k_{\mathrm{e}}C_{\mathrm{i}}(t), \tag{13.71}$$

where C_i is the internal concentration, C_w is the concentration in water (or other medium), k_i is the accumulation rate constant (analogous to the absorption rate described in Section 13.9.1), and k_e is the elimination rate constant. The internal concentration is generally not measured, and the elimination rate constant is usually poorly estimated. It is preferable to work with the ratio k_i/k_e, known as the bioconcentration factor. The internal concentration divided by the BCF is a scaled internal concentration that is assumed to be directly proportional to the actual, but unknown, internal concentration. Equation (13.71) then leads to

$$\frac{\mathrm{d}C_{\mathrm{i}}^{*}(t)}{\mathrm{d}t}=k_{\mathrm{e}}\left[C_{\mathrm{w}}(t)-C_{\mathrm{i}}^{*}(t)\right]. \tag{13.72}$$

As noted by Ashauer et al. (2011), the same TK approach can be used for all responses on a single species and chemical that are affected by the same mode of action, whereas the TD component will be specific to the type of damage done. The TD part of a TKTD model relates the internal concentration at the target site to the effect on the individual, most easily exemplified by mortality. Ashauer et al. (2011) describe the TK component as characterizing what the test subjects do with the chemical, whereas the TD component describes what the chemical does to the test subject. The description of the TK component fits the interpretation of the example discussed in Section 13.9.1. The aim of TKTD models is, in the words of Ducrot et al. (2015), to refine "exposure and effect patterns through the consideration of more realistic exposure scenarios and effect mechanisms." For quantal endpoints, e.g. mortality, this is done by including an assessment of "damage," an ill-defined concept that cannot be measured directly and that currently is not a working concept for sublethal effects. Whatever damage means, it is assumed to be proportional to the internal concentration, which leads to the equation for a scaled damage function, $D^*(t)$, of the form

$$\frac{\mathrm{d}D^{*}(t)}{\mathrm{d}t}=k_{\mathrm{r}}\left[C_{i}(t)-D^{*}(t)\right], \tag{13.73}$$

where k_r is a recovery rate constant.

The following discussion is based on Gergs et al. (2016). In the SD framework, the probability of survival decreases once a threshold concentration, z, is reached. They assume that the hazard, H, for an individual is a linear function of time and is determined from

$$\frac{\mathrm{d}H(t)}{\mathrm{d}t}=k_{\mathrm{k}}\max\left(D^{*}(t)-z,0\right)+h_{\mathrm{b}}, \tag{13.74}$$

where h_b is the background hazard rate and k_k is the killing rate. As in Chapter 12, the survival function can be derived from the hazard function by the equation

$$S(t)=\mathrm{e}^{-H(t)} \tag{13.75}$$

In the IT framework, the individual dies as soon as its IT concentration is exceeded. This implies that half of the exposed population dies once the median tolerance concentration, α, is reached. Gergs et al. (2016) assume the individual threshold has a log-logistic distribution:

$$F(t)=\frac{1}{1+\left[\left(\max\limits_{0<\tau<t}D^{*}(\tau)\right)\Big/\alpha\right]^{-\beta}}. \tag{13.76}$$

In Eq. (13.76), the parameter β is a shape parameter that is often described as the slope of the threshold distribution. [An alternative formulation is obtained by replacing $D^*(t)$ in Eqs. (13.74) and (13.76) by $C_i^*(t)$.] It follows that the survival function is given by

$$S(t)=(1-F(t))e^{-h_b t}. \tag{13.77}$$

The interested reader is referred to Gergs et al. (2016) for further development and examples related to Eqs. (13.75) and (13.77). Software to carry out this type of modeling is given in https://www.ecotoxmodels.org/guts/, site with many additional valuable resources.

One of the challenges for the adoption of TKTD models for regulatory risk assessment is the difficulty in validating such models. Two papers that discuss good modeling practice are Schmolke et al. (2010) and Grimm et al. (2014). Also relevant is the EFSA guidance document on good modeling practice (EFSA, 2014). Further examples of TKTD models applied to ecotoxicity data include Gillies et al. (2016) and Martin et al. (2012). A program and supplementary data related to the latter are available at http://cream-itn.eu/projects/wp-1/daphnia-2/deb-ibm. This is a readable introduction to DEBtox and mechanistic modeling.

EXERCISES

In all datasets in the text and exercises, if a concentration is given with the value −1, that means a solvent used in the study and the water control is recorded as −1 and the solvent control is listed as 0. The reader should determine whether to combine the two controls for NOEC determination or EC*x* estimation. If the two controls are to be combined, then the water control should be changed to 0 and the replicates renamed so that replicates in the two controls can be distinguished. If only the solvent control group is to be used, then the water control should be discarded. If only the water control should be used, then the solvent control should be discarded and the concentration for the water control should be changed to 0. Under no circumstances should the concentration value −1 be used for NOEC or EC*x* purposes. Guidance on this is provided in Chapters 2 and 3.

13.1 For each response in Table A1.13.1 and each day, compute (i) the minimum value of the response, (ii) the mean control value, (iii) MDD, and (iv) MDD% following the methods in Section 13.2. That is, verify the data in Table 13.3. Programs are given in Appendix 1.

13.2 Run the program pca2.sas to create weekly pyrifos scores datasets. Compute the replicate mean scores for each week and apply Williams' test to these eleven sets of weekly scores to find the NOEC for the chlorpyrifos data discussed in Section 13.2.

13.3 The following refer to the chlorpyrifos dataset, pyrifo1. Note that column Simve identifies week and pond using the following notation. Week are −4 and −1, indicated as W_4 and W_1, respectively, and 0.1, 1, 2, 4, 8, 12, 15, 19, and 24, indicated by W01, W1, W2, etc. Ponds are 1–12, indicated by c1–c12. Thus, W01C5 is week 0.1, pond 5. Next, ponds 2, 3, 5, and 11 are controls (conc = 0), 1 and 8 are low dose (conc = 0.1 μg l^{-1}), 4 and 10 are low intermediate (0.9 μg l^{-1}), 7 and 12 are high intermediate (6 μg l^{-1}), and 6 and 9 are high dose (44 μg l^{-1}). Use the Ln(1 + *x*) transform on each column (species abundance) in the dataset analyzed.

(a) For each week, calculate the MDD% for the species Daplo and do Williams' test to determine the NOEC for this species.

(b) Repeat the instructions in part (a) for the species Synsp.

13.4 Determine the NOEC for the data in Table A1.3.2. Be sure to explore normality and variance homogeneity of the residuals from ANOVA. For Dunnett and Williams' tests, the variance structure comes from treating replicate and replicate *X* treatment as random effects. SAS and R programs are provided in Appendix 1 that can be adapted easily to these data. The Jonckheere–Terpstra test should be done on the replicate means. Also, try to fit a useful regression model to the replicate means data and estimate EC10.

13.5 For AUC and growth rate in Table 13.8, do a limit test to determine whether the effect, if any, in the limit concentration exceeds (a) 10%, (b) 25%, and (c) 50%.

13.6 For the daphnia length data in Table A1.13.1, (a) do the Dunnett, Jonckheere, and Williams tests to determine the NOEC. (b) Fit the Bruce–Versteeg and other models to these data to estimate EC10.

13.7 Bioconcentration

The SAS datasets whole, LDWHOLE, and HDWHOLE contain the bioaccumulation measurements, BIOACC, on whole fish from a bioconcentration study. Spreadsheets of the same names are in the Excel file Wholefish. Some values may be below detection limit, indicated by a value of < for BDETECT, and need to be estimated.

(a) Use the Helsel regression approach to estimate BIOACC for all results below detection limit assuming a normal distribution. The SAS program HELSELW.SAS can be used for this purpose. Do this for both the uptake and depuration periods.

(b) Determine the 3-day period in which steady state was reached. The same SAS program does this. This program creates a dataset, DATA_ADJ_PERIOD where DATA = WHOLE, LDWHOLE, or HDWHOLE, and PERIOD = DEPUR or UPTAKE, containing any estimated values. Save that dataset for further work.

(c) Calculate the mean bioconcentration level, y_0, during the 3-day steady-state period, for example, by running the program *BIOCONC_BACKGROUND.SAS.*

(d) Using the value of y_0 obtained in (c), fit the model (13.64) to the depuration data created in (b), defined as day ≥ 15, and estimate k_2. Do this separately by concentration. The SAS program *depurate_model.sas* will do this.

(e) Using the value of k_2 obtained in (d), fit model (13.65) to the uptake data. The SAS program *uptake_model.sas* will do this.

13.8 Bioconcentration

Modify the programs used in Exercise 13.7 to model depuration and uptake so as to have one model that simultaneously fits the data from both concentrations and compare results with those obtained before.

13.9 Bioconcentration

Follow the instructions of Exercise 13.7 for fillet rather than whole fish using datasets hdfill and ldfill.

13.10 Bioconcentration

Follow the instructions of Exercise 13.7 for carcass rather than whole fish using datasets hdcarc and ldcarc.

Appendix 1

Dataset

Available in website: www.wiley.com/go/Green/StatAnalys
EcotoxicStudy

Statistical Analysis of Ecotoxicity Studies, First Edition. John W. Green, Timothy A. Springer, and Henrik Holbech.

Companion website: www.wiley.com/go/Green/StatAnalysEcotoxicStudy

Appendix 2

Mathematical Framework

A2.1 BASIC PROBABILITY CONCEPTS

The reader is assumed to be familiar with basic probability. However, to prepare for some concepts of primary interest to toxicity analysis, it may be helpful to review a few ideas. One begins with some phenomenon of interest. In toxicity, a simple example is the observation of a set, M, of subjects at an instant in time exposed to some environmental condition (such as a chemical), each of which is either alive or dead. For each subset of M, we record the subjects in that subset that are alive and those that are dead. Thus, for each subset, $R=\{R_1, R_2, R_3, \ldots\}$, of M, we define a set $X_R=\{X_1, X_2, X_3, \ldots\}$, where $X_i = 0$ if subject R_i is alive and $X_i = 1$ if subject R_i is dead. Each such set X_R is an *outcome* for this phenomenon and the set S of all possible outcomes defines the sample description space for this phenomenon.

More generally, a *sample description space*, S, is the set of all possible outcomes of some random phenomenon. An *event* is a set of outcomes. An event, E, is said to occur if the observed outcome of the underlying random phenomenon belongs to E. We define an associated family, F, of events. In general, we do not require F to be the set of all possible outcomes. We only require F to satisfy the following requirements:

S belongs to F.

If E belongs to F, then its complement, E^C, belongs to F.

If $E_1, E_2, E_3, \ldots$ belong to F, then their union, $\bigcup_i E_i$, belongs to F.

To this is added the notion of a probability function, P, which is defined for every outcome in F and has the following properties.

$$P \geq 0 \text{ for every event } E \text{ in } F. \quad \text{(A2.1)}$$

$$P[S] = 1. \quad \text{(A2.2)}$$

For every sequence $E_1, E_2, E_3, \ldots$ of mutually exclusive events in F,

$$P\left[\bigcup_i E_i\right] = \sum_i P[E_i]. \quad \text{(A2.3)}$$

A2.1.1 Conditional Probability

The probability that two events, A and B, both occur can be written as $P[A \cap B]$ or, simply, $P[AB]$. The conditional probability of event A, given event B, is defined by

$$P[A|B] = \frac{P[AB]}{P[B]}, \quad \text{(A2.4)}$$

provided $P[B] > 0$. If $P[B] = 0$, then $P[A|B]$ is not defined. It follows that if $P[A] \neq 0$ and $P[B] > 0$, then

$$P[B|A] = \frac{P[AB]}{P[A]}$$

and hence

$$P[A|B] = \frac{P[B|A]P[A]}{P[B]} \quad \text{(A2.5)}$$

Equation (A2.5) is known as Bayes' theorem or Bayes' rule. Intuitively, $P[A|B]$ is the probability that A occurs, given that we know already that B has occurred. If $B_1, B_2, B_3, \ldots, B_n$

Statistical Analysis of Ecotoxicity Studies, First Edition. John W. Green, Timothy A. Springer, and Henrik Holbech.

Companion website: www.wiley.com/go/Green/StatAnalysEcotoxicStudy

are mutually exclusive events and A is a subset of $\bigcup_i B_i$, that is, $A=\bigcup_i(A \cap B_i)$, then it follows from Bayes' rule that

$$P[A]=\sum_i P[A|B_i]P[B_i]. \qquad (A2.6)$$

A and B are said to be independent events if the probability of either one of those events is not affected by whether the other event occurs. That is, $P[A|B]=P[A]$ and $P[B|A]=P[B]$. Note that if $P[A|B]=P[A]$, then it follows that $P[B|A]=P[B]$. More generally, it is easily demonstrated that $E_1, E_2, E_3, \ldots$ are independent if and only if the probability of them all occurring is the product of their separate probabilities, i.e.

$$P\left[\bigcap_i E_i\right]=\prod_i P[E_i]. \qquad (A2.7)$$

A random variable is a real, finite-valued function, X, defined on a sample description space, S, and for which a probability function P is defined on the associated family, F, of events. There is an additional technical requirement that need not concern us. For these and other details, the reader is referred to one of the many texts on the subject, e.g. Parzen and Karlin (2013) or Parzen (1992).

It will be convenient to define notation for real number intervals. An interval has one of the forms $(a,b)=\{x: a<x<b\}$, $[a,b)=\{x: a\le x<b\}$, $(a,b]=\{x: a<x\le b\}$, $[a,b]=\{x: a\le x\le b\}$, $(a,\infty)=\{x: a<x\}$, $[a,\infty)=\{x: a\le x\}$, $(-\infty,b)=\{x: x<b\}$, or $(-\infty,b]=\{x: x\le b\}$. So a square bracket, [or], indicates the endpoint is included and a parenthesis, (or), indicates that the endpoint is not included. For our purposes, perhaps the most salient property of a random variable is its associated distribution function. The distribution function for a random variable X is a nonnegative, real-valued function F defined by

$$F(x)=P[X\le x] \text{ for all real numbers } x.$$

A2.2 DISTRIBUTION FUNCTIONS

A *distribution function* F has the following characteristic properties:

$$0\le F(x)\le 1 \text{ for all real numbers } x.$$
$$\text{If } x_1<x_2, \text{ then } F(x_1)\le F(x_2).$$
$$\lim_{x\to\infty} F(x)=1.$$
$$\lim_{x\to-\infty} F(x)=0.$$
$$\log_{x\to a+} F(x)=F(a) \text{ for each real number a.}$$

In Chapter 1, four types of responses were defined: continuous, quantal, count, and ordinal. These correspond to random variables of the same names. The simplest is perhaps the quantal random variable. The distribution for a quantal random variable is a step function. That is, it has the form

$$\begin{aligned} F(x)&=0 \text{ if } x<0\\ &=p_0 \text{ if } 0\le x<1\\ &=p_0+p_1 \text{ if } 1\le x<2\\ &=p_0+p_1+p_2 \text{ if } 2\le x<3 \end{aligned}$$

See Figure A2.1.

For quantal random variables, it is usually more convenient to define the *probability density function* (pdf). The pdf for a quantal random variable is defined as follows:

$$\begin{aligned} f(x)&=0 \text{ if } x<0\\ &=p_0 \text{ if } 0\le x<1\\ &=p_1 \text{ if } 1\le x<2\\ &=p_2 \text{ if } 2\le x<3 \end{aligned}$$

The relationship between the distribution function and the density function for a quantal random variable is given by

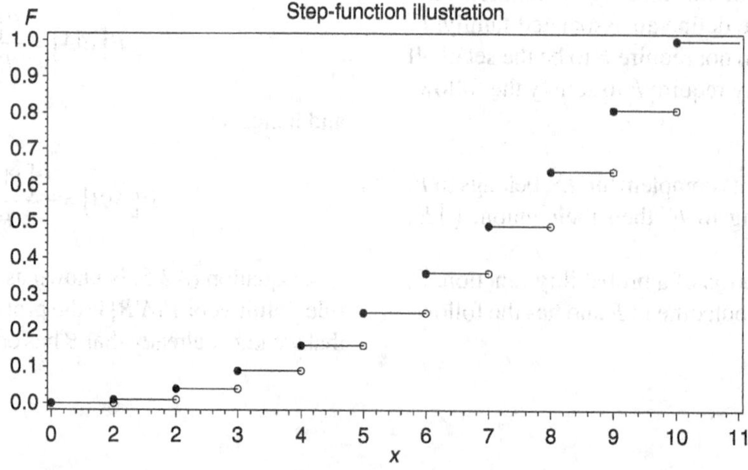

Figure A2.1 Step function. The end points are exaggerated for effect, with solid dots indicating the point is on the plot and the open circles indicate the point is not on the plot.

$$F(x)=\sum_{t\le x} f(x). \tag{A2.8}$$

Quantal distributions of some importance in toxicity include the binomial and Poisson, defined by their density functions below.

Binomial probability density function with parameters n and p:

$$\begin{aligned} f(x) &= \frac{n!}{x!(n-x)!}p^x(1-p)^{n-x} && \text{if } 0\le x\le n \text{ and } x \text{ is an integer} \\ f(x) &= 0 && \text{otherwise.} \end{aligned} \tag{A2.9}$$

To put this in a toxicity context, if n subjects are exposed to some substance and the probability of death for each individual subject is p, then $f(x)$ is the probability that exactly x subjects from the group of n die.

Poisson probability density function with parameters n and p:

$$\begin{aligned} f(x) &= e^{-np}\frac{(np)^x}{x!} && \text{if } 0\le x\le n \text{ and } x \text{ is an integer} \\ f(x) &= 0 && \text{otherwise.} \end{aligned} \tag{A2.10}$$

Usually, the term np in the Poisson density function is replaced by a single symbol, such as λ. The Poisson is often used when n is very large and p is very small, so that np is modest in size. Both density functions apply to situations where individual subjects all have the same probability, p, of exhibiting the random phenomenon of interest, e.g. death. These density functions would not be appropriate if different subjects have different probabilities. We will examine other quantal density functions later in context of specific types of toxicity experiments.

In the above two examples, the density function depends on the values of *parameters*, n and p, and it will often be useful to include those parameters in the functional form by writing $f(x;n,p)$ for the density function and $F(x;n,p)$ for the corresponding distribution function.

The discussion for quantal distributions and density functions can also be extended to include the discrete case. It is also possible to define probability distribution and density functions for continuous random variables, but the relationship between them is different from that for quantal or discrete variables. One critical difference is that for a continuous random variable, the probability that it has a specific value is generally zero. For example, if we are describing body weight of an adult male Sprague-Dawley rat, it makes little sense to try to calculate the probability that it weighs exactly 475 g. It is much more reasonable to calculate the probability that its weight falls in the interval 470–480. We might construct a histogram with weight intervals $340\le x<360$, $360\le x<380$, $380\le x<400$, …, where the heights of the histogram bars represent the probability of the weight falling in the indicated interval (see Figure A2.2.).

If there were sufficient data, we could draw histograms with shorter intervals that might do a better job of capturing the shape. This would lead to approximate slopes to the curve on each interval, i.e.

$$\frac{F(x_{i+1})-F(x_i)}{x_{i+1}-x_i}.$$

In the limit (and in the abstract, not for a simple example of real data), as x_{i+1} approaches x_i, we get the instantaneous

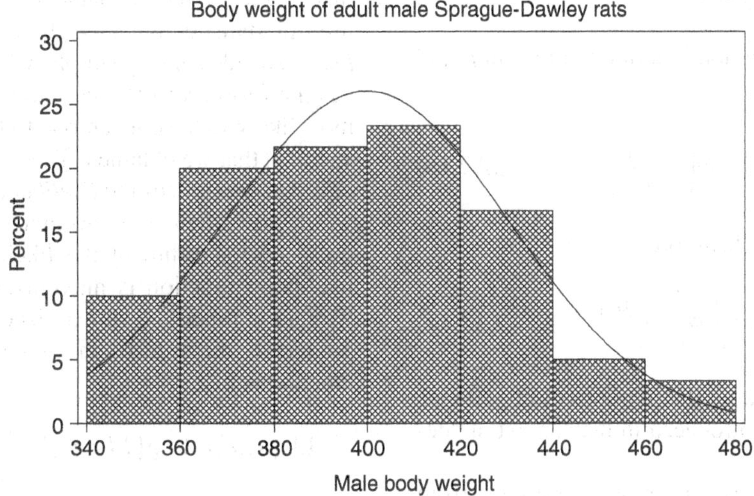

Figure A2.2 Histogram of Sprague-Dawley adult male body weights. Probabilities (vertical axis values) have been multiplied by 100. Overlaid on the histogram is a theoretical curve that attempts to capture the general shape of the histogram.

slope of the distribution function at $x = x_i$. So, provided the distribution function enjoys nice mathematical properties, the pdf for a continuous distribution is given by

$$f(x) = F'(x) = \frac{dF}{dx}. \qquad (A2.11)$$

That is, the pdf is the derivative of the distribution function. We can also describe the relationship between density function and distribution function in general by writing

$$F(x) = \int_{-\infty}^{x} f(t)dt. \qquad (A2.12)$$

Familiar examples of continuous density functions used in toxicity include the normal or bell-shaped curve, defined (for mean=μ, variance=σ^2) by

$$f(x;\mu,\sigma) = \frac{1}{\sqrt{2\pi}\sigma} e^{-(x-\mu)^2/2\sigma^2}. \qquad (A2.13)$$

The overlaid density curve in Figure A2.2 is a normal density function with mean and variance estimated from the data. It will be convenient to have a standard notation for this widely used density function. The standard normal density function is the normal with mean 0 and variance 1 and is denoted by

$$\varphi(x) = \frac{1}{\sqrt{2\pi}} e^{-x^2/2}, \qquad (A2.14)$$

and the corresponding standard normal distribution function is given by

$$\Phi(x) = \frac{1}{\sqrt{2\pi}} \int_{-\infty}^{x} e^{-t^2/2}dt. \qquad (A2.15)$$

Then, a normal distribution function with mean μ and variance σ^2 is given by

$$F(x;\mu,\sigma) = \Phi\left(\frac{x-\mu}{\sigma}\right), \qquad (A2.16)$$

and the density function is, therefore,

$$f(x;\mu,\sigma) = \frac{1}{\sigma}\varphi\left(\frac{x-\mu}{\sigma}\right). \qquad (A2.17)$$

There are other continuous distributions of importance in toxicology that will be introduced in the context of specific types of studies.

Statistics deals primarily with random samples from populations. By a random sample from a population, we mean a subset of the subjects from that population with the property that any two subjects from that population have the same chance, i.e. probability, of being in the sample. We are not generally interested in the subjects per se, but rather in some measurement made on those subjects, such as body weight, whether alive or dead at the end of the study, or when it reproduced. The measurement of interest will have some distribution, F, over the population. We record the value of that measurement on the jth subject as X_j. Each such measurement is the value of a random variable with distribution F, so that our random sample (of measurements, not subjects) is a sequence $X_1, X_2, X_3, \ldots, X_n$ of random variables with distribution F. Initially, we will be concerned with independent observations, so the X_j are *iid* random variables with distribution F. We will consider correlated observations later but will keep the discussion simpler for now.

Suppose the density function f for the population of measurements depends on some parameter $\boldsymbol{\alpha}$ (which may have several components, i.e. $\boldsymbol{\alpha}=(\alpha_1, \alpha_2, \alpha_3, ..)$, so f may be written as $f(x,\boldsymbol{\alpha})$. Given that the random variables X_j are independent, the density function for the random sample is

$$L(\boldsymbol{x};\boldsymbol{\alpha}) = \prod_i f(x_i;\boldsymbol{\alpha}). \qquad (A2.18)$$

Equation (A2.18) is called the likelihood function.

A2.3 METHOD OF MAXIMUM LIKELIHOOD

Generally in statistical applications, we do not know the values of the population parameters $\boldsymbol{\alpha}$ and will need to estimate them from a random sample from that population. There are several approaches to how best to estimate these parameters and, indeed, how to characterize the "best" estimate. One of the fundamental approaches to estimation used in statistics is called the *method of maximum likelihood*. The idea is very simple. Given that we believe the pdf has the form (A2.13), we search for the value of $\boldsymbol{\alpha}$ that is most likely to have given rise to the random sample $\boldsymbol{x}=(x_1, x_2, x_3, \ldots)$ that we obtained. That is done by finding the value of $\boldsymbol{\alpha}$ that maximizes the *likelihood function* (A2.18).

It will often be convenient mathematically to maximize the logarithm of the likelihood function, since the likelihood function is maximized if and only if the log-likelihood function is maximized. Since the logarithm of a product is the sum of the logarithms, the *log-likelihood function* is just

$$\mathrm{LL}(\boldsymbol{x};\boldsymbol{\alpha}) = \mathrm{Log}\left(L(\boldsymbol{x};\boldsymbol{\alpha})\right) = \sum_i \log\left(f(x_i;\boldsymbol{\alpha})\right). \qquad (A2.19)$$

Finding the value of $\boldsymbol{\alpha}$ that maximizes (A2.19) is a calculus problem that depends on the mathematical form of the density function f. This will be illustrated in the case the sample is from a normal distribution with unknown mean μ and variance σ^2, so $\boldsymbol{\alpha}=(\mu,\sigma^2)$ and f are given by Eq. (A2.17). So we must maximize the function

$$\begin{aligned}\mathrm{LL}(\boldsymbol{x};\mu,\sigma) &= \sum_i \mathrm{Log}\left(\frac{1}{\sigma}\mathrm{e}^{-(x_i-\mu)^2/2\sigma^2}\right)\\ &= -\frac{n}{2}\mathrm{Log}(\sigma^2) - \frac{1}{2\sigma^2}\sum_i (x_i-\mu)^2,\end{aligned} \tag{A2.20}$$

where n is the number of observations (i.e. $x=(x_1, x_2, x_3, \ldots, x_n)$) and Log means the natural logarithm. From many standard calculus or real analysis texts (e.g. Apostol, 1974; Rudin, 1974; Beyer, 2010), it is known that, provided the density function satisfy certain simple requirements, the maximum value will occur where all partial derivatives equal zero. (It should be noted that the minimum value also occurs where all partial derivatives equal zero, and, indeed, it is possible for all partial derivatives equal zero at some point and the function have neither a maximum nor a minimum value there. Nevertheless, for many distributions we will consider, the maximum value is often found by setting the partial derivatives to zero and solving the resulting equations, referred to as the normal equations regardless of whether the underlying distribution is normal.)

If we treat σ^2 as a parameter rather σ, then we obtain the following partial derivatives:

$$\frac{\partial \mathrm{LL}}{\partial \sigma^2} = \frac{1}{\sigma^4}\left[\sum_i (x_i-\mu)^2 - n\sigma^2\right] = 0 \text{ iff } \sigma^2 = \frac{1}{n}\sum_i (x_i-\mu)^2. \tag{A2.21}$$

$$\frac{\partial \mathrm{LL}}{\partial \mu} = \frac{1}{\sigma^2}\left[\sum_i x_i - n\mu\right] = 0 \text{ iff } \mu = \frac{1}{n}\sum_i x_i. \tag{A2.22}$$

Here the abbreviation iff means if and only if. If the estimate for μ from Eq. (A2.22) is inserted in Eq. (A2.21), then one obtains the maximum likelihood estimate for the sample variance. The traditional notation for the estimated sample mean is given by

$$\bar{x} = \hat{\mu} = \frac{1}{n}\sum_i x_i. \tag{A2.23}$$

While maximum likelihood estimation of model parameters is very important in statistics, it is not the only important idea and is, in fact, incomplete by itself. A critical property of a good estimator is that it be unbiased. Intuitively, that means that if we repeat the sampling process over and over and estimate the parameters each time in the same way, then "on average" the estimates of the parameters so obtained will equal their true value. We need to formalize this notion of unbiasedness before we can hope to determine whether the estimators above are unbiased. To do this, several concepts have to be introduced. The first to be addressed is the *expected value* of a random variable.

A2.3.1 Expected Value

The expected value of a random variable is a measure of the center of the distribution. If X is a quantal random variable with density function f, the expected value of X, or $E[X]$, is defined by

$$E[X] = \sum_x x f(x), \tag{A2.24}$$

where the sum is over all values of x for which $f(x) \neq 0$. The expected value of X is thus a *weighted average* of X, with weights equal to the probability or likelihood that X has the given value. If all possible values of X have the same probability of occurrence, then the expected value of X is simply the arithmetic mean of the possible values of X.

If X is a continuous random variable, Eq. (A2.24) must be modified:

$$E[X] = \int_{-\infty}^{\infty} x f(x)\mathrm{d}x. \tag{A2.25}$$

It is possible to give a single unified definition of expected value that encompasses both the quantal and continuous cases (and any other type of random variable), but that need not concern us. A very good reference for that and many other mathematical ideas of statistics and probability is Serfling (2008). The definition of expected value needs to be extended in several ways. The expected value of a $U(X)$ of a random variable X with density function f is given by

$$E[U(X)] = \sum_x U(x) f(x), \tag{A2.26}$$

or

$$E[U(X)] = \int_{-\infty}^{\infty} U(x) f(x)\mathrm{d}x. \tag{A2.27}$$

according as f is discrete or continuous.

One of the important simple functions of a random variable is its variance.

$$\mathrm{Var}(X) = E\left[(X-\mu)^2\right] \tag{A2.28}$$

where $\mu = E[X]$.

Other simple examples of functions of a random variable include X, X^2, e^X, and $X/(1-X)$. It is straight forward to verify the following basic properties of expectation. Additional properties of expectation can be found in many standard references, such as Parzen (1992).

$E[c]=c$ for any constant c.
$E[cU(X)]=cE[U(X)]$ for any constant c and function U whose expectation exists.
$E[U(X)+V(X)]=E[U(X)]+E[V(X)]$ for any two functions U and V whose separate expectations exist.
If $U(x)\le V(x)$ for all x, then $E[U(X)]\le E[V(X)]$ any two functions U and V whose separate expectations exist.
$|E[U(X)]|\le E[|U(X)|]$ where both expectations exists.

Next, we need to extend the notion of expected value to include sequences of random variables. If $X_1, X_2, X_3, \ldots, X_n$ are random variables, then the joint distribution function of $X_1, X_2, X_3, \ldots, X_n$ is a real-valued function $F(x_1, x_2, x_3, \ldots, x_n)$ defined by

$$F(x_1,x_2,x_3,\ldots,x_n)=P(X_1\le x_1,X_2\le x_2,\ldots,X_n\le x_n)$$
$$=P\left(\bigcap_i (X_i\le x_i)\right).$$

The joint pdf of $X_1, X_2, X_3, \ldots, X_n$ is a function f satisfying

$$F(x_1,x_1\ldots,x_n)=\int_{-\infty}^{x_n}\int_{-\infty}^{x_{n-1}}\int_{-\infty}^{x_{n-2}}\cdots\int_{-\infty}^{x_1}f(t_1,t_2,\ldots,t_n)\,\mathrm{d}t_1\mathrm{d}t_2\ldots\mathrm{d}t_n$$

for continuous random variables, with integrals replaced by sums for discrete variables.

If the joint pdf is $f(x_1, x_2, x_3, \ldots, x_n)$ and $U(X_1, X_2, X_3, \ldots, X_n)$ is a real-valued function, then

$$E\left[U(X_1,X_2,X_3,\ldots,X_n)\right]=$$
$$\int_{-\infty}^{\infty}\int_{-\infty}^{\infty}\int_{-\infty}^{\infty}\cdots\int_{-\infty}^{\infty}U(x_1,x_1,\ldots,x_n)f(x_1,x_2,\ldots,x_n)\,\mathrm{d}x_1\mathrm{d}x_2\ldots\mathrm{d}x_n \tag{A2.29}$$

for a continuous distribution, with an analogous definition for a discrete distribution.

A sequence of random variables of great importance for our purposes is a sequence of iid variables $X_1, X_2, X_3, \ldots, X_n$ from the same population, i.e. a random sample from that population. Since these variables are independent, the joint pdf of $X_1, X_2, X_3, \ldots, X_n$ is just the product of the separate pdfs, that is,

$$f(x_1,x_2,x_3,\ldots,x_n)=f(x_1)f(x_2)f(x_3)\ldots f(x_n).$$

We are now able to demonstrate very simply that the expected value of the sample mean from a normal population with mean μ and variance σ^2 can be calculated from

$$E\left[\hat{X}\right]=E\left[\frac{1}{n}\sum_i X_i\right]=\frac{1}{n}\sum_i E\left[X_i\right]=\frac{1}{n}n\mu=\mu \tag{A2.30}$$

Thus, the sample mean is indeed an unbiased estimator of the population mean. Now consider the expected value of the maximum likelihood estimate of the variance.

$$E\left[\hat{\sigma}^2\right]=E\left[\frac{1}{n}\sum_i(X_i-\bar{X})^2\right]=\frac{1}{n}\left[\sum_i(X_i-\bar{X})^2\right]=\frac{n-1}{n}\sigma^2.$$

Details are left to the reader as an exercise. So, the maximum likelihood estimator (mle) of the variance from a normal population is biased. For that reason, we prefer to use the unbiased estimator given by (A2.31):

$$\hat{\sigma}^2=s^2=\frac{1}{n-1}\sum_i(X_i-\bar{X})^2 \tag{A2.31}$$

It can also be shown that the variance of the sample mean is given by

$$\mathrm{Var}(\bar{X})=\frac{\sigma^2}{n}, \tag{A2.32}$$

and the variance of the sample variance is

$$\mathrm{Var}(\hat{\sigma}^2)=\frac{\sigma^4}{2(n-1)}. \tag{A2.33}$$

There are other estimation techniques besides maximum likelihood that will be introduced as needed, but the *mle* is the one of the most important we consider.

A2.4 BAYESIAN METHODOLOGY

The discussion above of probability distributions and density functions treats the parameters as fixed quantities that are unknown and to be estimated from the data. The data are described by distribution functions dependent on these unknown parameters, and the parameters are in turn estimated from the data. In the Bayesian approach to analysis, the parameters are not fixed but are themselves described by probability distributions or density functions. These distributions are conditioned on the data and if θ is the (possibly vector valued) parameter and y is the underlying random variable defining the data, the density function of θ has the form $p(\theta|y)$. This density function may itself depend on other parameters. For example, θ may have a normal distribution with mean μ and variance σ^2, in which case, $p(\theta)=N(\theta|\mu,\sigma^2)$. We can describe the joint pdf of θ and y by the product of the separate density functions

$$p(\theta,y)=p(\theta)p(y|\theta). \tag{A2.34}$$

In the formulation (A2.34), $p(\theta)$ is called the prior density function and $p(y|\theta)$ is referred to as the sampling density function. From Bayes' rule, we can obtain the conditional density of θ, given y, as

$$p(\theta|y) = \frac{p(\theta,y)}{p(y)} = \frac{p(\theta)p(y|\theta)}{p(y)}, \qquad \text{(A2.35)}$$

where

$$p(y) = \sum_{\theta} p(\theta)p(y|\theta), \qquad \text{for a discrete density, and}$$
$$p(y) = \int_{\theta} p(\theta)p(y|\theta)\mathrm{d}(\theta), \qquad \text{for a continuous density}$$

(A2.36)

The density function $p(\theta|y)$ is called the posterior density function of θ and will be recognized as the likelihood function. We can think of this as evaluating our prior understanding of the parameter in light of the data. A natural question to ask is how do we obtain the prior density function $p(\theta)$. There are several basic approaches to answering this question. Essentially every scientist uses his or her expert opinion in deciding how to design and interpret experiments. One of the rationales for Bayesian methodology is that expert opinion can be incorporated into the decision making process in a formal way. One way of doing this is called eliciting of expert opinion. In the words of Choy et al. (2009), "This use of the Bayesian framework reflects the scientific 'learning cycle,' where prior or initial estimates are updated when new data become available." Choy et al. (2009) is an excellent introduction to eliciting expert opinion and has an extensive list of references. Other references for this approach include Warren-Hicks and Moore (1998), Hora and Iman (1989), Ortiz et al. (1991), Cooke (1991), Meyer and Booker (2001), Morgan and Henrion (1992), and Hora (1993). In practice, it can be very challenging to build prior density functions from elicited expert judgment.

Either as an alternative to eliciting expert judgment or as a supplement, prior information can be obtained from experiments or other studies. Indeed, some authors think the only objective prior information comes from previous studies (Clyde, 1999; Hobbs and Hilborn, 2006). In the absence of other information, sometimes a class of "conjugate" priors can be identified. If the posterior distribution, $p(\theta|x)$, is in the same family as the prior probability distribution, $p(\theta)$, the prior and posterior are then called conjugate distributions, and the prior is called a conjugate prior for the likelihood function. It can be seen from Eqs. (A2.35) and (A2.36) that the integration required can be daunting. If the prior and posterior are conjugates, the integration will often lead to a simple algebraic, closed form. Thus, the use of conjugate priors is often a mathematical convenience rather than the result of insight. Gelman et al. (2004) contains a list of conjugate priors for many familiar (and some unfamiliar) likelihood functions.

Finally, so-called uninformative priors can be used. An uninformative prior is one that is not subjectively elicited. The uniform prior is a commonly used example. The use of an uninformative prior often yields results similar to non-Bayesian statistical analysis, since the likelihood function often yields more information than the uninformative prior. One uses whatever prior information is available to understand the data. Gelman et al. (2004) is among several comprehensive texts describing Bayesian methods in much greater detail than is possible here.

The Bayesian approach will be illustrated here using the beta-binomial model introduced by Williams (1975) and refined by Williams et al. (1988). This model is useful in describing incidence rates where the probability of an event or incident is unknown or random. For example, the incidence of some fetal abnormality in a rodent study may vary according to the litter size even though all dams are exposed to the same environmental conditions, such as the level of some chemical in a toxicity experiment. A simple binomial density function with a fixed incidence rate may not describe the data well. Other situations where the beta binomial might be useful are in a traditional aquatic toxicity experiment where multiple organisms (e.g. fish) are placed in each tank and there are multiple tanks at the same nominal test concentration. There often is more variation among incidence rates in tanks than a simple binomial density function would predict. Such "overdispersion" is common in many ecotoxicity experiments, as discussed by Oris et al. (2012) for fish early life stage studies.

The binomial density function (or likelihood) with parameters n and θ is given by Eq. (A2.9), which is reproduced here with a change of notation to facilitate the current discussion:

$$p(x|n,\theta) = \binom{n}{x}\theta^{x}(1-\theta)^{n-x} \qquad \text{if } 0 \le x \le n \text{ and } x \text{ is an integer}$$
$$p(x|n,\theta) = 0 \qquad \text{otherwise.}$$

(A2.9*)

A conjugate prior density function is the beta density function with hyper-parameters α and β, given by

$$p(\theta|\alpha,\beta) = \mathrm{Beta}(\alpha,\beta) = \frac{\theta^{\alpha-1}(1-\theta)^{\beta-1}}{B(\alpha,\beta)} = \frac{\Gamma(\alpha+\beta)}{\Gamma(\alpha)\Gamma(\beta)}\theta^{\alpha-1}(1-\theta)^{\beta-1},$$

(A2.37)

where $B(\alpha, \beta)$ is the beta function defined by

$$B(\alpha,\beta) = \int_{0}^{1} t^{\alpha-1}(1-t)^{\beta-1}\,\mathrm{d}t, \qquad \text{(A2.38)}$$

and

$$\Gamma(\alpha)=\int_0^\infty t^{\alpha-1}e^{-t}dt. \qquad (A2.39)$$

For positive integer values of α, the gamma function simplifies to

$$\Gamma(\alpha)=(\alpha-1)!. \qquad (A2.40)$$

While tables of the beta function (e.g. Askey and Roy, 2010) and gamma function (e.g. Hazewinkel, 2002) exist, fortunately software is readily available (e.g. the *PDF* and *gamma* functions in SAS and the *beta* and *gamma* functions in *R*) from which to obtain the beta density function.

It follows that the posterior distribution of the beta-binomial is given by

$$\begin{aligned} &p(x|\alpha,\beta)\\ &=\int_0^1 p(x|n,\theta)p(\theta|\alpha,\beta)d\theta\\ &=\binom{n}{x}\frac{\Gamma(\alpha)\Gamma(\beta)}{\Gamma(\alpha+\beta)}\int_0^1\theta^{x+\alpha-1}(1-\theta)^{n-x+\beta-1}\,d\theta\\ &=\binom{n}{x}\frac{B(x+\alpha,\, n-x+\beta)}{B(\alpha,\beta)}\\ &=\frac{\Gamma(n+1)}{\Gamma(x+1)\Gamma(n-x+1)}\,\frac{\Gamma(x+\alpha)\Gamma(n+\beta-x)}{\Gamma(\alpha+\beta+n)}\,\frac{\Gamma(\alpha+\beta)}{\Gamma(\alpha)\Gamma(\beta)} \end{aligned} \qquad (A2.41)$$

It is simple to verify that the mean and variance of a binomial are given by $n\theta$ and $n\theta(1-\theta)$ and the mean and variance of a beta distribution are given by

$$\frac{\alpha}{\alpha+\beta} \qquad (A2.42)$$

and

$$\frac{\alpha\beta}{(\alpha+\beta)^2(\alpha+\beta+1)}, \qquad (A2.43)$$

respectively. The mean and variance of the beta binomial can be shown to be

$$E[X]=\frac{n\alpha}{\alpha+\beta}, \qquad (A2.44)$$

and

$$\mathrm{Var}[X]=\frac{n\alpha\beta(\alpha+\beta+n)}{(\alpha+\beta)^2(\alpha+\beta+1)}. \qquad (A2.45)$$

Since the mean of the binomial is $n\theta$ and the mean of the beta-binomial is $n\alpha/(\alpha+\beta)$, if we let $\theta=\alpha/(\alpha+\beta)$ be the "realized" or observed binomial probability, then the variance of the beta-binomial can be written as

$$\mathrm{Var}[X]=n\theta(1-\theta)\left[1+(n-1)\rho\right], \qquad (A2.46)$$

where $p=1/(\alpha+\beta-1)$. The term ρ is called the overdispersion parameter. Values of ρ less than 0 or near 0 correspond to distributions that are essentially binomial, while large values of ρ correspond to distributions with notable overdispersion. In Chapter 9, these concepts will be extended to multinomial distributions.

A2.5 ANALYSIS OF TOXICITY EXPERIMENTS

With these basic concepts from probability and statistics sketched out, we now turn to the analysis of toxicity experiments. We will start with what is called the *simple means model.*

The basic population of interest will be some species, such as fathead minnow or sugar beet or female Wistar rat. Random samples will be drawn from the population at large and randomly assigned to treatment groups. The sample assigned to a particular treatment then represents, for example, the population of all fathead minnows that might be given that treatment. We want to explore whether one of more of these treatment populations is different in some biologically meaningful way from the control or undisturbed population from which we drew all of our subjects. Once the subjects have been treated, there will be separate, not necessarily equal, distributions for each treatment population. Designate the distribution of the ith treatment population as F_i.

So our data consists of sequences $Y_{i1}, Y_{i2}, Y_{i3}, \ldots, Y_{ini}$, of random variables with distribution F_i, i=0 to k. We want to explore whether the distributions F_i are all still the same. Note that F is also F_0. Usually, but by no means always, these distributions will differ only in their means. For example, if we assume $F_i(y) = \Phi((y-\mu_i)/\sigma)$, then our problem reduces to one of comparing the means, μ_i. To do that, we will need to estimate those means and the common variance, σ^2.

In the simplest situation, there are k treatment groups plus a control group, with n_i subjects in the ith group individually housed in separate containers, one subject per container. The statistical model can be expressed as follows:

$$Y_{ij}=\mu_i+e_{ij} \qquad (A2.47)$$

where Y_{ij} is the observation on the jth subject in group i and e_{ij} are iid random errors with mean 0 and variance σ^2.

Here μ_i is the mean response of the population from which the subjects in group i have been drawn and σ^2 is the common variance of those $k+1$ populations. If we place no restrictions on the population means, so model 1.1 applies, the mle for the μ_i and σ^2 and are given by

$$\bar{y}_{i.} = \hat{\mu}_i = \frac{1}{n_i}\sum_j y_{ij}, \tag{A2.48}$$

and

$$\hat{\sigma}^2 = s_p^2 = \frac{1}{n - k - 1}\sum_i (n_i - 1) s_i^2. \tag{A2.49}$$

Equation (A2.49) is the *pooled sample variance estimate*. Here $n = n_0 + n_1 + n_2 + \cdots + n_k$, the biased *mle* estimator of σ^2 has been obtained by replacing $n.$ in the denominator of the leading factor by what is shown, and

$$s_i^2 = \frac{1}{n_i - 1}\sum_j \left(y_{ij} - \bar{y}_{i.}\right)^2, \tag{A2.50}$$

is the sample variance of the ith treatment group. Once μ_i and σ^2 have been estimated, the predicted value of Y in treatment group i is the treatment mean response, $\hat{\mu}_i$, and the variance of this predicted value is obtained from Eq. (A2.50) to be σ^2/n_i. The sample variance of the estimator $\hat{\mu}_i$ of μ_i is s_p^2/n_i. The standard deviation of $\hat{\mu}_i$ is referred to as the standard error of the estimate and is given by $s_p/\sqrt{n_i}$.

A general test statistic that can be used to test the hypothesis that $\mu_i = \mu_0$ has the form

$$\begin{aligned} T &= \frac{\hat{\mu}_i - \hat{\mu}_0}{\sqrt{\operatorname{Var}(\hat{\mu}_i - \hat{\mu}_0)}} = \frac{\hat{\mu}_i - \hat{\mu}_0}{\sqrt{\operatorname{Var}(\hat{\mu}_i) + \operatorname{Var}(\hat{\mu}_0)}} \\ &= \frac{\hat{\mu}_i - \hat{\mu}_0}{\sqrt{\left(s_i^2/n_i\right) + \left(s_0^2/n_0\right)}} = \frac{\hat{\mu}_i - \hat{\mu}_0}{s_p\sqrt{\left(1/n_i\right) + \left(1/n_0\right)}}. \end{aligned} \tag{A2.51}$$

We will explore the distribution and use of this test statistic and alternatives to test the indicated hypothesis in Section 3.2.1.

If Eq. (A2.49) is applied to the complete sample of $n.$ observations, one gets

$$s_{\text{Total}}^2 = \frac{1}{n. - 1}\sum_{i,j}\left(Y_{ij} - \bar{Y}_{..}\right)^2.$$

Discarding the division by $n.-1$, one obtains the total sum of squares

$$\text{SS}_{\text{Total}} = \sum_{i,j}\left(Y_{ij} - \bar{Y}_{..}\right)^2 \tag{A2.52}$$

In the current context, we can decompose the total sum of squares into two parts, a sum of squares indicating what is "explained" by the simple model and a sum of squares indicated what is "unexplained" or error:

$$\begin{aligned} \text{SS}_{\text{Total}} &= \sum_{i,j}\left(Y_{ij} - \bar{Y}_{..}\right)^2 = \sum_{i,j}\left(Y_{ij} - \bar{Y}_{i.}\right)^2 + \sum_{i,j}\left(\bar{Y}_{i.} - \bar{Y}_{..}\right)^2 \\ &= \text{SS}_{\text{Error}} + \text{SS}_{\text{Model}}. \end{aligned} \tag{A2.53}$$

It is left as an exercise to demonstrate that Eq. (A2.53) is true for both balanced and unbalanced designs.

It is customary to summarize the data in what is called an ANOVA table (see Table A2.1).

The divisors for SS_{Error} and SS_{Total} come from the variance formulas (A2.49) and (A2.54). The model sum of squares can be thought of as a summary statistics developed from the estimates of the treatment means and the divisor simply makes the first two terms under the heading degrees of freedom (*df*) add to the third term. A more complete explanation is possible based on a matrix representation of the model. It is not the intention of this text to cover all the matrix methods applicable to statistical analysis. A thorough treatment of such methods can be found in Gruber (2013). We will review only a few key ideas that will facilitate the current presentation. The vector $\boldsymbol{Y}$ of $n.$ observations first expressed in Eq. (A2.47) can be written as

$$\boldsymbol{Y} = \begin{bmatrix} y_{01} \\ y_{02} \\ \vdots \\ y_{0n_0} \\ y_{11} \\ y_{12} \\ \vdots \\ y_{1n_1} \\ \vdots \\ y_{k1} \\ y_{k2} \\ \vdots \\ y_{kn_k} \end{bmatrix} = \begin{bmatrix} \boldsymbol{Y}_0 \\ \boldsymbol{Y}_1 \\ \vdots \\ \boldsymbol{Y}_k \end{bmatrix}, \tag{A2.54}$$

where $\boldsymbol{Y}_i$ is the n_i-by-1 column vector of observations in treatment group i. If $J(a,b,c)$ is an a-by-b matrix (a rows

Table A2.1 ANOVA Table for Simple Means Model

Source	SS	df	Mean square
Model	SS_{Model}	k	$\text{SS}_{\text{Model}}/k$
Error	SS_{Error}	$n..-k-1$	$\text{SS}_{\text{Err}}/(n..-k-1)$
Total	SS_{Total}	$n..-1$	$\text{SS}_{\text{Tot}}/(n..-1)$

and b columns) whose entrees are all equal to c, then the grand mean can be written as

$$\bar{Y}_{..} = Y'*\frac{1}{n_.}J(n_.,1,1) = \frac{1}{n_.}J(1,n_.,1)*Y, \quad \text{(A2.55)}$$

and the ith treatment mean is

$$\bar{Y}_{i.} = Y_i'*\frac{1}{n_i}J(n_i,1,1) = \frac{1}{n_i}J(1,n_i,1)*Y_i, \quad \text{(A2.56)}$$

so the vector of treatment means is given by

$$\langle \bar{Y}_{i.} \rangle_{k+1} = \text{Col}_{k+1}(\bar{Y}_i) = \text{Diag}_{k+1}\left[\frac{1}{n_i}J(1,n_i,1)\right]*Y. \quad \text{(A2.57)}$$

If there is no treatment effect, then all the observations come from a population with the same mean, best estimated by the grand mean. Thus differences of treatment means from the grand mean indicate possible treatment effects. The greater the differences, the stronger the evidence. A "treatment effects" vector can be written as

$$\langle \bar{Y}_{i.} - \bar{Y}_{..} \rangle_{k+1} = \text{Col}_{k+1}(\bar{Y}_{i.} - \bar{Y}_{..}) = \left[\text{Diag}_{k+1}\left(\frac{1}{n_i}J(1,n_i,1)\right) - \frac{1}{n_.}J(k+1,n_.,1)\right]*Y. \quad \text{(A2.58)}$$

The matrix inside the square brackets of Eq. (A2.58) has rank k, and every term of the resulting column $k+1$ element vector on the far left of that equation is a linear combination of the last k terms of that vector. An equivalent way to express this idea is that the last k terms of that vector are linearly independent, or, finally, the vector has k degrees of freedom.

It follows that the total, error, and model sums of squares can be written as

$$\text{SS}_{\text{Total}} = Y'*\left[I_{n_.} - \frac{1}{n_.}J(n_.,n_.,1)\right]*Y, \quad \text{(A2.59)}$$

$$\text{SS}_{\text{Error}} = Y'*\left[I_{n_.} - \text{Diag}_{k+1}\left(\frac{1}{n_i}J(n_i,n_i,1)\right)\right]*Y, \quad \text{(A2.60)}$$

$$\text{SS}_{\text{Model}} = Y'*\left[\text{Diag}_{k+1}\left(\frac{1}{n_i}J(n_i,n_i,1)\right) - \frac{1}{n_.}J(n_.,n_.,1)\right]*Y, \quad \text{(A2.61)}$$

where I_n denotes an n-by-n identity matrix, that is, an n-by-n matrix with value 1 at each diagonal element and value 0 for all off-diagonal elements, $\text{Diag}_{k+1}(B_i)$ denotes a block diagonal matrix with $k+1$ diagonal blocks with ith diagonal block equal to B_i, and for any matrix X, X' is obtained by transposing the rows and columns of X. It is left to the exercises to show that the above matrix expressions do lead to the algebraic expressions in formulas (A2.49), (A2.52), and (A2.54) and the ranks of the matrices inside the square brackets of Eqs. (A2.59)–(A2.61) are $n_.-1$, $n_.-k-1$, and k, respectively. The larger the model sum of squares, the stronger is the evidence of a treatment effect. We will develop formal test statistics that will allow us to quantify the strength of the evidence that treatment effects exist. One simple test assesses the model sum of squares against the error sum of squares. As we shall see, there are much better tests for treatment effects.

It will be helpful to express model (A2.47) in matrix form. Consider first a special case of two treatments plus a control with means μ_0, μ_1, μ_2. Suppose further that there are three observation in each treatment group. The model can be expressed as

$$Y = X\mu + e, \quad \text{(A2.62)}$$

where

$$\mu = \begin{bmatrix} \mu_0 \\ \mu_1 \\ \mu_2 \end{bmatrix},$$

e is a 9-by-1 vector of random errors distributed as iid $N(0, \sigma^2)$, and X is the design matrix

$$X = \begin{bmatrix} 1 & 0 & 0 \\ 1 & 0 & 0 \\ 1 & 0 & 0 \\ 0 & 1 & 0 \\ 0 & 1 & 0 \\ 0 & 1 & 0 \\ 0 & 0 & 1 \\ 0 & 0 & 1 \\ 0 & 0 & 1 \end{bmatrix}.$$

This design matrix can be described as block diagonal, with each diagonal block a column of 1's and each off-diagonal block a column of zeros. To emphasize this, it may be convenient to write the matrix X as $I_3 \otimes J(3,1,1)$, where $\otimes$ indicates the Kronecker product. In general, if A and B are matrices, then $A \otimes B$ is obtained by replacing each element a_{ij} of A by $a_{ij}B$. So if A is n-by-m and B is p-by-q, then $A \otimes B$ is pn-by-qm.

The same pattern applies when $k>2$. Consider the simple case of Eq. (A2.52) where $\Sigma = \sigma^2 I_{n_.}$, where $I_{n_.}$ is the

n-by-n identity matrix, i.e. a square matrix of n rows and n columns with value 1 along the diagonal (from upper left to lower right) and value 0 at all off-diagonal positions. So this is the case of independent observations.

As an exercise, it can be shown that Eqs. (A2.48) and (A2.49) are equivalent to

$$\hat{\mu} = \left(X' * X\right)^{-1} * X' * Y, \tag{A2.63}$$

and

$$\hat{\sigma}^2 = \frac{1}{n - k} Y' * \left[I_n - X\left(X' * X\right)^{-1} X'\right] * Y. \tag{A2.64}$$

Similar expressions are developed for Eqs. (A2.52)–(A2.53) in the exercises.

A2.5.1 Estimation when There are Restrictions on the Treatment Means

Consider the same toxicity experiment with the restriction that the treatment means satisfy a linear model:

$$\mu_i = \mu + \beta d_i$$

where d_i is the dose or concentration of a single test chemical defining treatment group i. The problem now is to obtain the *mle* of the parameters σ^2, μ and β for normal populations with common within-treatment variance σ^2. Equation (A2.20) is modified to obtain

$$\mathrm{LL}(y; \mu, \beta, \sigma) = \sum_{i=0}^{k} \sum_{j=1}^{n_i} \mathrm{Log}\left(\frac{1}{\sigma} \mathrm{e}^{-\left(y_{ij} - \mu - \beta d_i\right)^2 / 2\sigma^2}\right) \tag{A2.65}$$

where now $y' = (y_{01}, y_{02}, y_{03}, \ldots, y_{0n0}, y_{11}, y_{12}, y_{13}, \ldots, y_{1n1}, y_{21}, y_{22}, \ldots, y_{2n2}, \ldots, y_{knk})$. There are now three partial derivatives to compute to find where the maximum value of the likelihood function occurs.

$$\frac{\partial \mathrm{LL}}{\partial \sigma^2} = -\frac{n}{2\sigma^2} + \frac{1}{2\sigma^4} \sum_i \sum_j \left(y_{ij} - \mu - \beta d_i\right)^2, \tag{A2.66}$$

$$\text{which is 0 iff } \sigma^2 = \frac{1}{n_{..}} \sum_i \sum_j \left(y_{ij} - \mu - \beta d_i\right)^2. \tag{A2.67}$$

The *mle* estimate of variance can be understood as the variation about the fitted line, rather than the pooled variation within treatment groups that was obtained for the unconstrained case, found in Eq. (A2.64). The remaining partial derivatives are presented next.

$$\frac{\partial \mathrm{LL}}{\partial \mu} = \frac{1}{\sigma^2} \sum_i \sum_j \left(y_{ij} - \mu - \beta d_i\right), \tag{A2.68}$$

from which it follows that

$$\mu = \bar{y}_{..} - \beta \sum_i n_i d_i = \bar{y}_{..} - n \beta \bar{d}, \tag{A2.69}$$

where

$$\bar{d} = \frac{1}{n} \sum_i n_i d_i. \tag{A2.70}$$

It should be noted that if there is no dose, that is, there is only one population, then Eq. (A2.70) reduces to Eq. (A2.23). The remaining partial derivative is

$$\frac{\partial \mathrm{LL}}{\partial \beta} = \frac{1}{\sigma^2} \sum_i \sum_j d_i \left(y_{ij} - \mu - \beta d_i\right), \tag{A2.71}$$

Which after substituting Eq. (A2.70) for μ, setting the partial derivative equal to zero, and doing some algebraic manipulation, yields

$$\beta = \frac{\sum_i n_i d_i \left(\bar{y}_{i.} - \bar{y}_{..}\right)}{(1/n) \sum_i n_i \left(d_i - \bar{d}\right)^2}. \tag{A2.72}$$

As in the previous cases, the *mle* of σ^2 is biased and n in Eq. (A2.67) must be replaced by $n - 2$ to obtain an unbiased estimate. This can be extended to apply to a very large class of models that are linear in the parameters. For example, a model of the form

$$\mu_d = \beta_0 + \beta_1 d + \beta_2 d^2 + \beta_3 d^{0.5} + \beta_4 \log(d) + \beta_5 10^d$$

is linear in the parameters β_0, β_1, β_2, β_3, β_4, β_5, and closed form expressions analogous to Eqs. (A2.67), (A2.69), and (A2.72) can be derived. In fact, if the model is of the form

$$\mu_d = \beta_0 g_0(d) + \beta_1 g_1(d) + \beta_2 g_1(d) + \cdots + \beta_m g_m(d),$$

where d is the dose defining the treatment, the functions $g_i(d)$ do not depend on any parameters and are linearly independent, and then closed form expressions for those parameters can be derived from the log-likelihood equations.

Unfortunately, most models of interest in toxicity studies are not linear in the parameters. To see why this matters, consider a 3-parameter exponential model:

$$\mu(d) = \alpha \mathrm{e}^{-\beta d^\gamma},$$

where α, β, and γ are the parameters to be estimated. It will be left as an exercise to derive the four log-likelihood equations and show the following: (i) A straightforward biased estimate for σ^2, analogous to Eq. (A2.67), is obtained from setting the partial derivative with respect to σ^2 equal to zero.

This estimate can be realized as the sum of squared deviations of the observations about the fitted model, divided by a constant, and a simple adjustment of the constant (replacing n by $n-4$) gives an unbiased estimate of σ^2 (once estimates for the other parameters are inserted). (ii) Setting the partial derivative with respect to α to zero leads to a simple expression for α in terms of β and γ. (iii) Setting the remaining two partial derivatives to zero leads to highly nonlinear equations that defy simple solution. None of the three log-likelihood equations apart from that for σ^2 require knowledge or estimates for σ^2. Thus, the issue of estimating the variance is separated from the issue of estimating the other parameters, and this is true of all the models so far considered. This realization is behind the concept of *restricted maximum likelihood (REML) estimation*, which is a general method of separating the estimation of the variance(s) from that of estimating the other model parameters. While REML, as well as ML, estimators will be used in this book, their development is highly technical and not central to our purpose. Hocking (1985, pp. 244ff) has a though treatment of REML estimation as well as relevant literature.

While in theory, there are solutions to the log-likelihood equations, they must be obtained by numeric methods. Since this is typical of nonlinear models and nonlinear models are the rule in modeling toxicity studies, methods for obtaining solutions to the log-likelihood equations will be developed next.

A2.6 NEWTON'S OPTIMIZATION METHOD

Consider first the case of a distribution with a single parameter to estimate. Indeed, to simplify matters, consider a more general problem of solving an equation of the form $g(\alpha)=0$ for α, where g is a function with continuous first and second derivatives. If α^* is a solution of the equation $g(\alpha)=0$, i.e. $g(\alpha^*)=0$, it follows from the mean value theorem of calculus, that for any value of α,

$$\frac{g(\alpha)-g(\alpha^*)}{\alpha-\alpha^*}=g'(t) \tag{A2.73}$$

or

$$g(\alpha)=g(\alpha^*)+g'(t)(\alpha-\alpha^*) \tag{A2.74}$$

for some number t between α^* and α. This can be rearranged as

$$\alpha=\alpha^*+\frac{g(\alpha)}{g'(t)}, \tag{A2.75}$$

since $g(\alpha^*)=0$. Newton's method develops an iterative process to estimate α^* from the above by starting with an initial "guess" or starting point α^0 of α^* and updates the estimate at step $m+1$ from the equation

$$\alpha^{m+1}=\alpha^m+\frac{g(\alpha^m)}{g'(\alpha^m)} \tag{A2.76}$$

If the sequence $\alpha^0, \alpha^1, \alpha^2, \ldots$ converges to some number α^* and $g'(\alpha^*)$ is not zero and not infinite, then $g(\alpha^*)=0$, i.e. α^* is a solution of the equation $g(\alpha)=0$.

A2.6.1 Extension to Multiple Parameters

Newton's method in the form of Eq. (A2.76) can be extended to a vector value function $\mathbf{g}$ of several variables, as follows:

$$\mathbf{g}(\boldsymbol{\alpha})=\mathbf{g}(\boldsymbol{\alpha}^*)+\left.\frac{\partial \mathbf{g}}{\partial \boldsymbol{\alpha}}\right|_{\boldsymbol{\alpha}=\mathbf{t}}(\boldsymbol{\alpha}-\boldsymbol{\alpha}^*), \tag{A2.77}$$

for some vector $\mathbf{t}$ between $\boldsymbol{\alpha}$ and $\boldsymbol{\alpha}^*$. The bold type indicates a vector, $'$ indicates the transpose of the indicated column vector, and the indicated partial derivative is a matrix composed of the partial derivatives of each component of $\mathbf{g}$ with respect to every component of $\boldsymbol{\alpha}$. To be clear, if $\mathbf{g}=(g_1, g_2, g_3, \ldots, g_n)'$ and $\boldsymbol{\alpha}=(\alpha_1, \alpha_2, \ldots, \alpha_m)'$, then

$$\mathbf{G}=\frac{\partial \mathbf{g}}{\partial \boldsymbol{\alpha}}=\begin{bmatrix} \frac{\partial g_1}{\partial \alpha_1} & \cdots & \frac{\partial g_1}{\partial \alpha_m} \\ \vdots & \ddots & \vdots \\ \frac{\partial g_n}{\partial \alpha_1} & \cdots & \frac{\partial g_n}{\partial \alpha_m} \end{bmatrix}, \tag{A2.78}$$

which is an $n \times m$ matrix. If $n=m$ and the matrix $\mathbf{G}$ is full rank, i.e. has an inverse, then the iterative process is defined by the equation

$$\boldsymbol{\alpha}^{k+1}=\boldsymbol{\alpha}^k-\mathbf{G}^{-1}\mathbf{g}(\boldsymbol{\alpha}^k), \tag{A2.79}$$

where the matrix G is evaluated at $\boldsymbol{\alpha}^k$ and $\boldsymbol{\alpha}^0$ is an initial guess. In the present context, the equation $\mathbf{g}(\boldsymbol{\alpha})=0$ being solved is the derivative of the log-likelihood function:

$$\mathbf{g}(\mathbf{x};\boldsymbol{\alpha})=\frac{\partial \mathrm{LL}(\mathbf{x};\boldsymbol{\alpha})}{\partial \boldsymbol{\alpha}}=\begin{bmatrix} \frac{\partial \mathrm{LL}}{\partial \alpha_0} \\ \frac{\partial \mathrm{LL}}{\partial \alpha_1} \\ \vdots \\ \frac{\partial \mathrm{LL}}{\partial \alpha_m} \end{bmatrix}, \tag{A2.80}$$

where there are $m+1$ parameters to be estimated. The matrix G in Eq. (A2.78) becomes the *Hessian matrix*,

$$G(\boldsymbol{x};\boldsymbol{\alpha})=\begin{bmatrix} \frac{\partial^2 \mathrm{LL}}{\partial\alpha_0\partial\alpha_0} & \frac{\partial^2 \mathrm{LL}}{\partial\alpha_0\partial\alpha_1} & \frac{\partial^2 \mathrm{LL}}{\partial\alpha_0\partial\alpha_2} & \cdots & \frac{\partial^2 \mathrm{LL}}{\partial\alpha_0\partial\alpha_m} \\ \frac{\partial^2 \mathrm{LL}}{\partial\alpha_1\partial\alpha_0} & \frac{\partial^2 \mathrm{LL}}{\partial\alpha_1\partial\alpha_1} & \frac{\partial^2 \mathrm{LL}}{\partial\alpha_1\partial\alpha_2} & \cdots & \frac{\partial^2 \mathrm{LL}}{\partial\alpha_1\partial\alpha_m} \\ \vdots & \vdots & \ddots & & \vdots \\ \frac{\partial^2 \mathrm{LL}}{\partial\alpha_m\partial\alpha_0} & \frac{\partial^2 \mathrm{LL}}{\partial\alpha_m\partial\alpha_1} & \frac{\partial^2 \mathrm{LL}}{\partial\alpha_m\partial\alpha_2} & \cdots & \frac{\partial^2 \mathrm{LL}}{\partial\alpha_m\partial\alpha_m} \end{bmatrix}. \tag{A2.81}$$

There are numerous modifications of Newton's method, and it is not our purpose to give a thorough discussion of methods for solving systems of equations. However, we will have occasion to use one of these methods and it will be helpful to understand the basic concept. These and other numerical methods are discussed in numerous references, including Kennedy and Gentle (1980) and Lange (2012) and mathematical development is found in Apostol (1974) and almost any advanced calculus text (e.g. Beyer, 2010).

The Hessian is important in part because it is the inverse of the variance–covariance matrix of the estimated parameter vector $\hat{\boldsymbol{\alpha}}$. Often, the estimation of the parameter σ^2 is separated from the estimation of the other parameters and that parameter estimate is independent of the others, so that

$$\mathrm{Var}(\boldsymbol{\alpha})=\hat{\sigma}^2\boldsymbol{H}^{-1},$$

where $H= G(\boldsymbol{x};\boldsymbol{\alpha})$ evaluated at $\hat{\boldsymbol{\alpha}}$, $\hat{\sigma}^2$=SSERR/$(n.-m-1)$, and SSERR is the sum of squared residuals from the model. Many improvements on the basic Newton–Raphson algorithm indicated above to implement Newton's method have been published. One of the most useful is the Levenberg–Marquardt algorithm (Levenberg, 1944; Marquardt, 1963; see also Moré and Sorensen, 1983).

A2.7 THE DELTA METHOD

Related to Newton's method is a useful technique for approximating the distribution of a function of a random variable from the distribution of the variable. Consider first the simple case of a univariate distribution depending on one location parameter, θ, and one variance parameter, σ^2. Suppose Y_n is an asymptotically normal random variable with asymptotic variance σ^2. That is,

$$\sqrt{n}\left(Y_n-\theta\right)\xrightarrow{D}N\left(0,\sigma^2\right), \tag{A2.82}$$

where $\xrightarrow{D}$ denotes converges in distribution, which in turn means that if F_n is the distribution of Y_n, then $\lim_{n\to\infty} F_n(t)=F(t)$ for each number t at which $F(t)$ is continuous.

If g is a differentiable function and $g'(\theta)\neq 0$, then

$$\sqrt{n}\left(g(Y_n)-g(\theta)\right)\xrightarrow{D}N\left(0,\sigma^2\left(g'(\theta)\right)^2\right). \tag{A2.83}$$

This can be interpreted to mean that $g(Y_n)$ is approximately normally distributed with mean $g(\theta)$ and variance $\sigma^2(g'(\theta))^2/n$.

It should be understood that $g(Y_n)$ may not be a good approximation of $g(\theta)$ for small n. Consider the example of the sample mean, $\bar{X}_n$, of n observations from a normal distribution $N(\mu, \sigma^2)$ and consider $g(\mu) = \mu^2$. Now

$$\sqrt{n}\left(\bar{X}_n-\mu\right)\xrightarrow{D}N\left(0,\sigma^2\right).$$

If $g(\mu) = \mu^2$ we need to be careful about concluding $\bar{X}_n^2$ is approximately normal with mean μ^2 and variance $4\mu^2\sigma^2/n$. While that is true asymptotically, the approximation is not very good for small n. The expected value of $\bar{X}_n^2$ is $\mu^2 + \sigma^2/n$. A better small sample estimator of μ^2 is $\bar{X}_n^2 - S_n^2/n$, where S_n^2 is the sample variance.

It will be important to extend the delta method to the multivariate case, which has many applications of interest. Suppose Y_n is an asymptotically multivariate normal random variable with asymptotic variance–covariance matrix Σ, that is,

$$\sqrt{n}\left(Y_n-\theta\right)\xrightarrow{D}N\left(0,\Sigma\right), \tag{A2.84}$$

where now $\boldsymbol{\theta}$ is a parameter vector. If g is a scalar-valued function of $\boldsymbol{\theta}$ with continuous first partial derivatives that are nonzero at $\boldsymbol{\theta}$ and $\nabla=\nabla g(\boldsymbol{\theta})$ is the column vector of first partial derivatives of g evaluated at $\boldsymbol{\theta}$, then

$$\sqrt{n}\left(g(Y_n)-g(\theta)\right)\xrightarrow{D}N\left(0,\nabla^{\mathrm{T}}\Sigma\nabla\right), \tag{A2.85}$$

where ∇^{T} is the transpose of ∇.

Equation (A2.85) is interpreted as meaning that $g(Y_n)$ is approximately normally distributed with mean $g(\boldsymbol{\theta})$ and variance–covariance matrix $\nabla^{\mathrm{T}}\Sigma\nabla/n$. Often Σ is replaced by the estimated variance–covariance matrix. Again, this is an asymptotic result and the small sample behavior of $g(Y_n)$ may not be ideal. The delta method is very useful when there is no small sample alternative. One important application of the delta method is to the Cedergreen et al. hormetic model, HORM2, discussed in Chapter 4.

Application to HORM2 ECx Estimation

The HORM3 model was presented in Chapter 4 for data exhibiting hormesis or low dose stimulation. This model fit to such a dataset has the form in Eq. (4.3):

$$y = \frac{y_0 + fe^{-U^{-a}}}{1+(U/d)^b},$$

where the five parameters y_0, a, b, d, and f are estimated from the data and U is the test concentration. If $x=100p$, then ECx is estimated by solving the equation

$$\frac{y_0 + fe^{-U^{-a}}}{1+(U/d)^b} - y_0(1-p) = 0, \tag{A2.86}$$

for U, which will be ECx. Equation (A2.86) implicitly defines ECx as a function $g(y_0, a, b, d, f)$. That there is no explicit formula for ECx means a simple application of the delta method is not appropriate. According to Mi (2010), when the function of the parameters is only implicitly defined, the variance formula has to be modified. If T is defined as the left-hand side of the above equation, then $T=0$ implicitly defines ECx as a function $g(y_0, a, b, d, f)$ of the parameters estimated by fitting the model HORM3 to the data. T is thus a function of those same parameters, and we use the delta method to obtain the approximate variance of T evaluated at the point $g(y_0, a, b, d, f)$ that makes $T=0$, which is the approximate variance of ECx. From Mi (2010), the variance term is

$$\frac{\nabla' \hat{\Sigma} \nabla}{(\partial T/\partial U)^2}, \tag{A2.87}$$

where $\hat{\Sigma}$ is the sample variance–covariance matrix of the parameter vector $\hat{\theta}$, ∇ is the vector of first partial derivatives of T, and $\hat{\theta}'=(y_0, a, b, d, f)$.

The variance–covariance matrix $\hat{\Sigma}$ can be obtained from the model fitting program. The vector ∇ of first partial derivatives can be worked out from simple applications of calculus. Some key operations from calculus are

1. $$k^a = e^{a\,\mathrm{Ln}(k)}. \tag{A2.88}$$

2–4. If $f(x)$ and $g(x)$ are differentiable functions, then

$$\frac{d}{dx}\left[f(x)g(x)\right] = f'(x)g(x) + f(x)g'(x)$$

$$\frac{d}{dx}\frac{f(x)}{g(x)} = \frac{g(x)f'(x) - g'(x)f(x)}{g^2(x)}, \tag{A2.89}$$

$$\frac{d}{dx}f(g(x)) = f'(g(x))g'(x)$$

where

$$\frac{d}{dx}h(x) \equiv h'(x)$$

is the derivative of the function h evaluated at x.

Applying the above calculus formulas to y in Eq. (4.3), it can be shown that

$$\begin{aligned}
\frac{\partial T}{\partial y_0} &= \frac{1}{1+(U/d)^b} - \left(1 - \frac{x}{100}\right)\\
\frac{\partial T}{\partial a} &= \frac{afe^{-2U^{-a}}U^{-a-1}}{1+(U/d)^b}\\
\frac{\partial T}{\partial b} &= \frac{-\left(y_0 + fe^{-U^{-a}}\right)(U/d)^b\,\mathrm{Ln}(U/d)}{\left[1+(U/d)^b\right]^2}.\\
\frac{\partial T}{\partial d} &= \frac{\left(y_0 + fe^{-U^{-a}}\right)bU^b(d)^{-b-1}}{\left[1+(U/d)^b\right]^2}\\
\frac{\partial T}{\partial f} &= \frac{e^{-U^{-a}}}{1+(U/d)^b}
\end{aligned} \tag{A2.90}$$

To this must be added:

$$\frac{\partial T}{\partial U} = \frac{af\left[1+(U/d)^b\right]e^{-U^{-a}}U^{-a-1} - \left[y_0 + fe^{-U^{-a}}\right]bU^{b-1}/d^b}{\left[1+(U/d)^b\right]^2}. \tag{A2.91}$$

A program to implement the delta rule to approximate the resulting confidence bounds on ECx from the fit of model HORM2 is given in Appendix 1.

A2.8 VARIANCE COMPONENTS

In addition to most models of interest in toxicology being nonlinear, it is also true that many toxicity experiments are designed in a more complex manner. It is very common for subjects to be housed together in small groups in replicate vessels, with multiple vessels in each treatment group. For example, there may be 5 plants per pot and 10 pots per treatment group or 20 tadpoles per tank and 4 tanks per treatment group, or 2 birds per cage and 8 cages per treatment, and so on. In these designs, the subjects in the same vessel generally cannot be considered to be independent. (Recall the discussion in Section 1.3.2.) There are then two sources of random variation: subjects within replicate and replicate within treatment. The model for the mean response is not changed (usually), but the error structure is more complicated, as are the likelihood equations. For this type of study, it is necessary to consider multivariate density functions that are not simply the product of univariate densities. For example, if $y = (y_1, y_2, \ldots, y_n)'$ is an n-variate random variable, the multivariate normal density function for y with mean μ and variance Σ is defined by

$$f(y;\mu,\Sigma)=\frac{1}{\sqrt{(2\pi)^n|\Sigma|}}\mathrm{e}^{-(1/2)(y-\mu)'\Sigma^{-1}(y-\mu)} \qquad \text{(A2.92)}$$

where $|\Sigma|$ denotes the determinant of the square matrix Σ. If $\boldsymbol{y}_{ij}$ is an n_{ij}-variate random variable designating the observations on all n_{ij} subjects in the jth replicate vessel receiving the ith treatment, then

$$\begin{aligned}\operatorname{Cov}(y_{ijk}, y_{i^*j^*k^*}) &= 0 \text{ if } i \neq i^*,\\ &= 0 \text{ if } i = i^* \text{ and } j \neq j^*,\\ &= \phi_1 \text{ if } (i,j) = (i^*, j^*) \text{ and } k \neq k^*,\\ &= \phi_0 + \phi_1 \text{ if } (i,j,k) = (i^*, j^*, k^*).\end{aligned} \qquad \text{(A2.93)}$$

Thus ϕ_1 is the covariance between different observations in the same replicate, and ϕ_0 is the inherent variance of individual observations. There are other ways to parameterize the variance–covariance matrix Σ. The indicated parameterization follows Hocking (1985), which also contains additional material on estimation of these parameters. Searle (1987) and Searle et al. (2007) are other valuable references for these unbalanced models. The resulting variance–covariance matrix is given by

$$\Sigma_{ij} = \phi_0 \boldsymbol{I}_{n_{ij}} + \phi_1 \boldsymbol{J}(\boldsymbol{n}_{ij}, \boldsymbol{n}_{ij}, \mathbf{1}), \qquad \text{(A2.94)}$$

where $\boldsymbol{I}_n$ is an n-by-n identity matrix and $J(a,b,c)$ is an a-by-b matrix (a rows and b columns) whose entrees are all equal to c. If there are r_i replicates in treatment i with n_{ij} subjects in the jth replicate, $k+1$ treatments in all (including a control), the log-likelihood equation for the $n_{..}$–variate random variable

$$Y = \begin{bmatrix} y_{01}\\ y_{02}\\ \vdots\\ y_{0r_0}\\ y_{11}\\ y_{12}\\ \vdots\\ y_{1r_1}\\ \vdots\\ y_{k1}\\ y_{k2}\\ \vdots\\ y_{kr_k}\end{bmatrix} = \begin{bmatrix} Y_0\\ Y_1\\ \vdots\\ Y_k\end{bmatrix} \qquad \text{(A2.95)}$$

is the product of $r_.$ density functions of the form in Eq. (A2.92), where $\boldsymbol{y}$ and $\boldsymbol{\mu}$ are replaced by $\boldsymbol{y}_{ij}$ and $J(n_{ij}, n_{ij}, \mu_i)$, respectively, and $\boldsymbol{Y}_i$ is an $n_{i.}$-by-1 column vector. The equations to solve for the parameters resulting from the various partial derivatives are messy and generally can only be solved by numerical methods except in the completely balanced case (equal numbers of replicates and subjects per replicate). The likelihood equations are easily extended to include unequal variances in different treatment groups or unequal correlations among subjects in different replicates. Additional complexity can be introduced for experiments in which each subject is measured at multiple times during the study, not just once at the end.

The total, model, and error sums of squares can be written as

$$\mathrm{SS}_{\mathrm{Total}} = Y'*\left(I_{n_{..}} - \frac{1}{n_{..}}J(n_{..}, n_{..}, 1)\right)*Y, \qquad \text{(A2.96)}$$

$$\mathrm{SS}_{\mathrm{Error}} = Y'*\left(I_{n_{..}} - \mathrm{Diag}_{k+1}\left(\frac{1}{r_i}J(r_i, r_i, 1)\right)\right)*Y. \qquad \text{(A2.97)}$$

A2.8.1 Design Matrix and Estimation of Variance Components

It will sometimes be convenient to express the model in matrix terms and write

$$Y = X\mu + e, \qquad \text{(A2.98)}$$

where $\boldsymbol{Y}$ is a vector of $n(k+1)$ observations, $\boldsymbol{\mu}$ is a vector of $k+1$ population means, $\boldsymbol{e}$ is a vector of $n(k+1)$ *iid* random errors with density f, and $\boldsymbol{X}$ is the $n(k+1)$ by $(k+1)$ "design" matrix. X is a block diagonal matrix with $k+1$ blocks. In the case $k=2$ and $n=3$,

$$\mu = \begin{bmatrix}\mu_0\\ \mu_1\\ \mu_2\end{bmatrix} \qquad \text{(A2.99)}$$

and

$$X = \begin{bmatrix}1&0&0\\1&0&0\\1&0&0\\0&1&0\\0&1&0\\0&1&0\\0&0&1\\0&0&1\\0&0&1\end{bmatrix}. \qquad \text{(A2.100)}$$

The design matrix (A2.100) can be described as block diagonal, with each diagonal block a column of 1's and each off-diagonal block a column of zeros. To emphasize this, it may be convenient to write the matrix X as $I_3 \otimes J(3,1,1)$, where $\otimes$ indicates the Kronecker product. In general, if A and B are matrices, then $A \otimes B$ is obtained by replacing each element a_{ij} of A by $a_{ij}B$. So if A is n-by-m and B is p-by-q, then $A \otimes B$ is pn-by-qm.

The same pattern applies when $k>2$. Consider the simple case of Eq. (A2.52) where $\Sigma = \sigma^2 I_n$, where I_n is the n-by-n identity matrix, i.e. a square matrix of n rows and n columns with value 1 along the diagonal (from upper left to lower right) and value 0 at all off-diagonal positions. So this is the case of independent observations.

Appendix 3

Tables

Statistical Analysis of Ecotoxicity Studies, First Edition. John W. Green, Timothy A. Springer, and Henrik Holbech.

Companion website: www.wiley.com/go/Green/StatAnalysEcotoxicStudy

Table A3.1 Studentized Maximum Distribution

df/k	2	3	4	5	6	7	8	9	10	11	12	13	14	15	16	17	18	19	20
2	4.075	4.834	5.397	5.842	6.208	6.516	6.781	7.014	7.22	7.406	7.573	7.727	7.867	7.998	8.118	8.231	8.337	8.436	8.53
3	3.09	3.551	3.888	4.154	4.372	4.557	4.717	4.858	4.983	5.096	5.199	5.293	5.38	5.46	5.535	5.605	5.67	5.732	5.791
4	2.722	3.08	3.34	3.544	3.711	3.852	3.974	4.082	4.179	4.266	4.345	4.417	4.484	4.546	4.604	4.658	4.709	4.757	4.803
5	2.532	2.84	3.062	3.234	3.376	3.495	3.599	3.69	3.772	3.845	3.912	3.974	4.03	4.083	4.132	4.178	4.222	4.263	4.301
6	2.417	2.696	2.894	3.049	3.175	3.282	3.374	3.455	3.528	3.593	3.653	3.707	3.758	3.805	3.849	3.89	3.928	3.965	3.999
7	2.34	2.599	2.783	2.925	3.041	3.139	3.224	3.299	3.365	3.425	3.48	3.53	3.577	3.62	3.66	3.698	3.733	3.767	3.798
8	2.285	2.53	2.703	2.837	2.946	3.038	3.117	3.187	3.25	3.306	3.357	3.404	3.447	3.488	3.525	3.56	3.594	3.625	3.655
9	2.243	2.479	2.644	2.772	2.875	2.962	3.038	3.104	3.163	3.216	3.265	3.309	3.351	3.389	3.424	3.458	3.489	3.519	3.547
10	2.211	2.439	2.598	2.72	2.82	2.904	2.976	3.039	3.096	3.147	3.193	3.236	3.275	3.312	3.346	3.378	3.408	3.436	3.463
11	2.186	2.407	2.561	2.68	2.776	2.857	2.927	2.988	3.042	3.092	3.136	3.177	3.215	3.251	3.283	3.314	3.343	3.371	3.396
12	2.164	2.38	2.531	2.647	2.74	2.819	2.886	2.946	2.999	3.046	3.09	3.13	3.166	3.201	3.232	3.262	3.29	3.317	3.342
13	2.147	2.359	2.506	2.619	2.71	2.787	2.853	2.911	2.962	3.009	3.051	3.09	3.126	3.159	3.19	3.219	3.246	3.272	3.296
14	2.132	2.34	2.485	2.596	2.685	2.76	2.825	2.881	2.932	2.977	3.018	3.056	3.091	3.124	3.154	3.182	3.209	3.234	3.258
15	2.119	2.324	2.467	2.576	2.663	2.737	2.8	2.856	2.905	2.95	2.99	3.027	3.062	3.093	3.123	3.151	3.177	3.201	3.225
16	2.108	2.311	2.451	2.558	2.645	2.717	2.779	2.834	2.883	2.926	2.966	3.003	3.036	3.067	3.096	3.124	3.149	3.173	3.196
17	2.099	2.299	2.437	2.543	2.628	2.7	2.761	2.815	2.863	2.906	2.945	2.981	3.014	3.044	3.073	3.1	3.125	3.149	3.171
18	2.09	2.288	2.425	2.53	2.614	2.684	2.745	2.798	2.845	2.888	2.926	2.962	2.994	3.024	3.052	3.079	3.104	3.127	3.149
19	2.083	2.279	2.415	2.518	2.601	2.671	2.731	2.783	2.83	2.872	2.91	2.944	2.977	3.006	3.034	3.06	3.085	3.108	3.13
20	2.076	2.271	2.405	2.507	2.59	2.659	2.718	2.77	2.816	2.857	2.895	2.929	2.961	2.99	3.018	3.044	3.068	3.091	3.112
21	2.07	2.263	2.396	2.498	2.58	2.648	2.706	2.758	2.803	2.844	2.881	2.916	2.947	2.976	3.003	3.029	3.053	3.075	3.096
22	2.064	2.256	2.389	2.489	2.57	2.638	2.696	2.747	2.792	2.833	2.869	2.903	2.934	2.963	2.99	3.015	3.039	3.061	3.082
23	2.059	2.25	2.382	2.481	2.562	2.629	2.687	2.737	2.782	2.822	2.858	2.892	2.923	2.951	2.978	3.003	3.026	3.048	3.069
24	2.055	2.245	2.375	2.474	2.554	2.621	2.678	2.728	2.772	2.812	2.848	2.882	2.912	2.94	2.967	2.992	3.015	3.037	3.057
25	2.051	2.239	2.369	2.468	2.547	2.613	2.67	2.72	2.764	2.803	2.839	2.872	2.902	2.931	2.957	2.981	3.004	3.026	3.047
26	2.047	2.235	2.364	2.462	2.541	2.606	2.663	2.712	2.756	2.795	2.831	2.864	2.894	2.922	2.948	2.972	2.995	3.016	3.037
27	2.043	2.23	2.359	2.456	2.535	2.6	2.656	2.705	2.749	2.788	2.823	2.856	2.885	2.913	2.939	2.963	2.986	3.007	3.027
28	2.04	2.226	2.354	2.451	2.529	2.594	2.65	2.699	2.742	2.781	2.816	2.848	2.878	2.905	2.931	2.955	2.978	2.999	3.019
29	2.037	2.222	2.35	2.446	2.524	2.589	2.644	2.693	2.736	2.774	2.809	2.841	2.871	2.898	2.924	2.948	2.97	2.991	3.011
30	2.034	2.219	2.346	2.442	2.519	2.584	2.639	2.687	2.73	2.768	2.803	2.835	2.864	2.891	2.917	2.941	2.963	2.984	3.004
35	2.022	2.204	2.329	2.424	2.499	2.563	2.617	2.664	2.706	2.743	2.777	2.809	2.837	2.864	2.889	2.912	2.934	2.954	2.973
40	2.014	2.194	2.317	2.41	2.485	2.547	2.6	2.647	2.688	2.725	2.758	2.789	2.817	2.843	2.868	2.891	2.912	2.932	2.951
45	2.007	2.186	2.308	2.4	2.474	2.535	2.588	2.634	2.674	2.711	2.744	2.774	2.802	2.828	2.852	2.874	2.895	2.915	2.934
50	2.001	2.179	2.3	2.391	2.465	2.526	2.578	2.623	2.663	2.699	2.732	2.762	2.79	2.815	2.839	2.861	2.882	2.901	2.92
60	1.993	2.169	2.289	2.379	2.451	2.511	2.563	2.607	2.647	2.683	2.715	2.744	2.771	2.796	2.82	2.842	2.862	2.881	2.899
80	1.984	2.157	2.275	2.364	2.435	2.494	2.544	2.588	2.627	2.662	2.693	2.722	2.749	2.773	2.796	2.817	2.837	2.856	2.874
100	1.978	2.15	2.267	2.355	2.425	2.483	2.533	2.577	2.615	2.649	2.681	2.709	2.735	2.759	2.782	2.803	2.823	2.841	2.859
120	1.974	2.145	2.261	2.349	2.418	2.476	2.526	2.569	2.607	2.641	2.672	2.7	2.726	2.75	2.773	2.793	2.813	2.831	2.849
200	1.966	2.135	2.25	2.337	2.405	2.463	2.511	2.554	2.591	2.625	2.655	2.683	2.709	2.732	2.754	2.775	2.794	2.812	2.829
∞	1.957	2.124	2.237	2.322	2.39	2.446	2.494	2.536	2.573	2.605	2.635	2.662	2.687	2.711	2.732	2.752	2.771	2.788	2.805

Columns are for k=2–20 treatment levels, and rows are degrees of freedom 2–30, 35, 40, 45, 50, 60, 80, 100, 120, 200, infinity, and α =0.05.

Table A3.2 Studentized Maximum Modulus Distribution

df/*k*	2	3	4	5	6	7	8	9	10	11	12	13	14	15	16	17	18	19	20
2	5.571	6.34	6.886	7.306	7.645	7.929	8.172	8.385	8.573	8.741	8.894	9.033	9.162	9.28	9.39	9.493	9.589	9.68	9.766
3	3.96	4.43	4.764	5.023	5.233	5.41	5.562	5.694	5.812	5.918	6.015	6.103	6.184	6.259	6.328	6.394	6.455	6.512	6.567
4	3.382	3.745	4.003	4.203	4.366	4.503	4.621	4.725	4.817	4.9	4.975	5.044	5.107	5.166	5.221	5.272	5.32	5.366	5.409
5	3.091	3.399	3.619	3.789	3.928	4.044	4.145	4.233	4.312	4.382	4.447	4.506	4.56	4.61	4.657	4.701	4.743	4.782	4.818
6	2.916	3.193	3.389	3.541	3.664	3.769	3.858	3.937	4.008	4.071	4.128	4.181	4.23	4.275	4.317	4.357	4.394	4.429	4.462
7	2.8	3.055	3.236	3.376	3.489	3.585	3.668	3.74	3.805	3.863	3.916	3.965	4.009	4.051	4.09	4.126	4.16	4.192	4.223
8	2.718	2.958	3.127	3.258	3.365	3.454	3.532	3.6	3.66	3.715	3.764	3.81	3.852	3.891	3.927	3.961	3.993	4.023	4.052
9	2.657	2.885	3.046	3.171	3.272	3.357	3.43	3.494	3.552	3.603	3.65	3.694	3.733	3.77	3.805	3.837	3.868	3.896	3.923
10	2.609	2.829	2.983	3.103	3.199	3.281	3.351	3.412	3.467	3.517	3.562	3.603	3.641	3.677	3.71	3.741	3.77	3.797	3.823
11	2.571	2.784	2.933	3.048	3.142	3.22	3.288	3.347	3.4	3.448	3.491	3.531	3.568	3.602	3.634	3.663	3.691	3.718	3.743
12	2.54	2.747	2.892	3.004	3.095	3.171	3.236	3.294	3.345	3.391	3.433	3.472	3.507	3.54	3.571	3.6	3.627	3.653	3.677
13	2.514	2.717	2.858	2.967	3.055	3.129	3.193	3.249	3.299	3.344	3.385	3.423	3.457	3.489	3.519	3.547	3.574	3.599	3.622
14	2.493	2.691	2.83	2.936	3.022	3.095	3.157	3.212	3.26	3.304	3.344	3.381	3.415	3.446	3.475	3.503	3.528	3.553	3.576
15	2.474	2.669	2.805	2.909	2.994	3.065	3.126	3.18	3.227	3.27	3.309	3.345	3.378	3.409	3.438	3.464	3.49	3.513	3.536
16	2.458	2.65	2.784	2.886	2.969	3.039	3.099	3.152	3.199	3.241	3.279	3.314	3.347	3.377	3.405	3.431	3.456	3.479	3.501
17	2.444	2.633	2.765	2.866	2.948	3.017	3.076	3.127	3.173	3.215	3.253	3.287	3.319	3.349	3.376	3.402	3.426	3.449	3.471
18	2.432	2.619	2.749	2.849	2.929	2.997	3.055	3.106	3.151	3.192	3.229	3.263	3.295	3.324	3.351	3.376	3.4	3.423	3.444
19	2.421	2.606	2.734	2.833	2.912	2.979	3.037	3.087	3.132	3.172	3.208	3.242	3.273	3.302	3.329	3.354	3.377	3.399	3.421
20	2.411	2.594	2.721	2.819	2.897	2.963	3.02	3.07	3.114	3.154	3.19	3.223	3.254	3.282	3.308	3.333	3.357	3.378	3.399
21	2.402	2.584	2.71	2.806	2.884	2.949	3.005	3.054	3.098	3.137	3.173	3.206	3.236	3.264	3.29	3.315	3.338	3.36	3.38
22	2.394	2.574	2.699	2.795	2.872	2.936	2.992	3.041	3.084	3.123	3.158	3.191	3.22	3.248	3.274	3.298	3.321	3.342	3.363
23	2.387	2.566	2.69	2.784	2.861	2.925	2.98	3.028	3.071	3.109	3.144	3.176	3.206	3.234	3.259	3.283	3.306	3.327	3.347
24	2.38	2.558	2.681	2.775	2.851	2.914	2.969	3.016	3.059	3.097	3.132	3.164	3.193	3.22	3.246	3.269	3.292	3.313	3.333
25	2.374	2.551	2.673	2.766	2.841	2.904	2.959	3.006	3.048	3.086	3.12	3.152	3.181	3.208	3.233	3.257	3.279	3.3	3.319
26	2.369	2.544	2.666	2.758	2.833	2.895	2.949	2.996	3.038	3.076	3.11	3.141	3.17	3.197	3.222	3.245	3.267	3.288	3.307
27	2.364	2.538	2.659	2.751	2.825	2.887	2.941	2.987	3.029	3.066	3.1	3.131	3.16	3.186	3.211	3.234	3.256	3.277	3.296
28	2.359	2.532	2.653	2.744	2.818	2.88	2.933	2.979	3.02	3.057	3.091	3.122	3.15	3.177	3.201	3.224	3.246	3.266	3.286
29	2.354	2.527	2.647	2.738	2.811	2.873	2.925	2.971	3.012	3.049	3.083	3.113	3.142	3.168	3.192	3.215	3.237	3.257	3.276
30	2.35	2.522	2.641	2.732	2.805	2.866	2.918	2.964	3.005	3.042	3.075	3.105	3.133	3.16	3.184	3.207	3.228	3.248	3.267
35	2.333	2.502	2.619	2.708	2.779	2.839	2.89	2.935	2.975	3.01	3.043	3.073	3.1	3.125	3.149	3.171	3.192	3.212	3.23
40	2.321	2.488	2.602	2.69	2.76	2.819	2.869	2.913	2.952	2.987	3.019	3.048	3.075	3.1	3.123	3.145	3.165	3.184	3.203
45	2.311	2.476	2.59	2.676	2.745	2.803	2.853	2.896	2.935	2.969	3.001	3.029	3.056	3.08	3.103	3.124	3.145	3.163	3.181
50	2.304	2.467	2.58	2.665	2.734	2.791	2.84	2.883	2.921	2.955	2.986	3.014	3.04	3.065	3.087	3.108	3.128	3.147	3.164
60	2.292	2.454	2.564	2.649	2.716	2.772	2.821	2.863	2.9	2.934	2.964	2.992	3.018	3.041	3.064	3.084	3.104	3.122	3.139
80	2.278	2.437	2.546	2.628	2.694	2.75	2.797	2.838	2.875	2.907	2.937	2.964	2.989	3.013	3.034	3.054	3.073	3.091	3.108
100	2.27	2.427	2.535	2.616	2.682	2.736	2.783	2.823	2.859	2.892	2.921	2.948	2.973	2.995	3.017	3.037	3.055	3.073	3.089
120	2.264	2.42	2.527	2.608	2.673	2.727	2.773	2.814	2.849	2.881	2.91	2.937	2.961	2.984	3.005	3.025	3.043	3.061	3.077
200	2.253	2.407	2.513	2.592	2.656	2.709	2.755	2.794	2.829	2.861	2.889	2.915	2.939	2.961	2.982	3.001	3.019	3.036	3.052
2.24	2.392	2.495	2.573	2.636	2.688	2.733	2.771	2.806	2.836	2.864	2.89	2.913	2.934	2.955	2.973	2.991	3.008	3.023	2.24

Columns are for k=2–20 treatment levels, and rows are degrees of freedom 2–30, 35, 40, 45, 50, 60, 80, 100, 120, 200, infinity, and α =0.05.

Table A3.3 Linear and Quadratic Contrast Coefficients

Group	1	2	3	4	5	6	7	8	9	10	11	12	13	14	15	16	17	18	19	20
TREND20	19	17	15	13	11	9	7	5	3	1	−1	−3	−5	−7	−9	−11	−13	−15	−17	−19
QUAD20	228	156	92	36	−12	−52	−84	−108	−124	−132	−132	−124	−108	−84	−52	−12	36	92	156	228
TREND19	9	8	7	6	5	4	3	2	1	0	−1	−2	−3	−4	−5	−6	−7	−8	−9	
QUAD19	51	34	19	6	−5	−14	−21	−26	−29	−30	−29	−26	−21	−14	−5	6	19	34	51	
TREND18	17	15	13	11	9	7	5	3	1	−1	−3	−5	−7	−9	−11	−13	−15	−17		
QUAD18	68	44	23	5	−10	−22	−31	−37	−40	−40	−37	−31	−22	−10	5	23	44	68		
TREND17	8	7	6	5	4	3	2	1	0	−1	−2	−3	−4	−5	−6	−7	−8			
QUAD17	40	25	12	1	−8	−15	−20	−23	−24	−23	−20	−15	−8	1	12	25	40			
TREND16	15	13	11	9	7	5	3	1	−1	−3	−5	−7	−9	−11	−13	−15				
QUAD16	35	21	9	−1	−9	−15	−19	−21	−21	−19	−15	−9	−1	9	21	35				
TREND15	7	6	5	4	3	2	1	0	−1	−2	−3	−4	−5	−6	−7					
QUAD15	91	52	19	−8	−29	−44	−53	−56	−53	−44	−29	−8	19	52	91					
TREND14	13	11	9	7	5	3	1	−1	−3	−5	−7	−9	−11	−13						
QUAD14	13	7	2	−2	−5	−7	−8	−8	−7	−5	−2	2	7	13						
TREND13	6	5	4	3	2	1	0	−1	−2	−3	−4	−5	−6							
QUAD13	22	11	2	−5	−10	−13	−14	−13	−10	−5	2	11	22							
TREND12	−11	−9	−7	−5	−3	−1	1	3	5	7	9	11								
QUAD12	55	25	1	−17	−29	−35	−35	−29	−17	1	25	55								
TREND11	−5	−4	−3	−2	−1	0	1	2	3	4	5									
QUAD11	15	6	−1	−6	−9	−10	−9	−6	−1	6	15									
TREND10	−9	−7	−5	−3	−1	1	3	5	7	9										
QUAD10	6	2	−1	−3	−4	−4	−3	−1	2	6										
TREND9	−4	−3	−2	−1	0	1	2	3	4											
QUAD9	28	7	−8	−17	−20	−17	−8	7	28											
TREND8	−7	−5	−3	−1	1	3	5	7												
QUAD8	7	1	−3	−5	−5	−3	1	7												
TREND7	−3	−2	−1	0	1	2	3													
QUAD7	5	0	−3	−4	−3	0	5													
TREND6	−5	−3	−1	1	3	5														
QUAD6	5	−1	−4	−4	−1	5														
TREND5	−2	−1	0	1	2															
QUAD5	2	−1	−2	−1	2															
TREND4	−3	−1	1	3																
QUAD4	1	−1	−1	1																
TREND3	−1	0	1																	
QUAD3	1	−2	1																	
TREND2	−1	1																		

TRENDx and QUADx are the coefficients for linear and quadratic contrasts for x group means. Column headers indicate the group, listed from smallest to largest, starting with 1 corresponding to the control.

Table A3.4 Williams' Test $\bar{t}_{\alpha,k}$ for $\alpha = 0.05$

df\k	2		3		4		5		6		7		8		9		10	
	Q	*B*	*Q*	*B*	*Q*	*B*	*Q*	*B*	*Q*	*B*	*Q*	*B*	*Q*	*B*	*Q*	*B*	*Q*	*B*
5	2.142	2	2.186	4	2.209	5	2.223	5	2.232	6	2.238	6	2.243	6	2.247	6	2.250	6
6	2.058	2	2.098	4	2.119	5	2.131	5	2.139	6	2.144	6	2.149	6	2.152	6	2.154	6
7	2.002	2	2.039	4	2.058	5	2.069	5	2.076	6	2.081	6	2.085	6	2.088	7	2.091	7
8	1.962	2	1.997	4	2.014	5	2.024	5	2.031	6	2.036	6	2.040	6	2.042	7	2.045	7
9	1.931	2	1.965	4	1.981	5	1.991	5	1.998	6	2.002	6	2.006	6	2.008	7	2.010	7
10	1.908	3	1.940	4	1.956	5	1.965	5	1.971	6	1.976	6	1.979	6	1.981	7	1.983	7
11	1.889	3	1.920	4	1.935	5	1.944	5	1.950	6	1.954	6	1.958	6	1.960	7	1.962	7
12	1.873	3	1.903	4	1.918	5	1.927	5	1.933	6	1.937	6	1.940	6	1.942	7	1.944	7
13	1.860	3	1.890	4	1.904	5	1.913	5	1.919	6	1.923	6	1.926	6	1.928	7	1.930	7
14	1.849	3	1.878	4	1.892	5	1.901	5	1.906	6	1.910	6	1.913	6	1.915	7	1.917	7
15	1.839	3	1.868	4	1.882	5	1.891	5	1.896	6	1.900	6	1.903	6	1.905	7	1.907	7
16	1.831	3	1.860	4	1.873	5	1.882	5	1.887	6	1.891	6	1.893	6	1.896	7	1.897	7
17	1.824	3	1.852	4	1.866	5	1.874	5	1.879	6	1.883	6	1.885	6	1.888	7	1.889	7
18	1.818	3	1.845	4	1.859	5	1.867	5	1.872	6	1.876	6	1.878	6	1.880	7	1.882	7
19	1.812	3	1.840	4	1.853	5	1.861	5	1.866	6	1.869	6	1.872	6	1.874	7	1.876	7
20	1.807	3	1.834	4	1.847	5	1.855	5	1.860	6	1.864	6	1.866	6	1.868	7	1.870	7
21	1.803	3	1.829	4	1.843	5	1.850	5	1.855	6	1.859	6	1.861	6	1.863	7	1.865	7
22	1.798	3	1.825	4	1.838	5	1.846	5	1.851	6	1.854	6	1.857	6	1.859	7	1.860	7
23	1.795	3	1.821	4	1.834	5	1.842	5	1.847	6	1.850	6	1.853	6	1.855	7	1.856	7
24	1.791	3	1.818	4	1.830	5	1.838	5	1.843	6	1.846	6	1.849	6	1.851	7	1.852	7
25	1.788	3	1.814	4	1.827	5	1.835	5	1.839	6	1.843	6	1.845	6	1.847	7	1.849	7
26	1.785	3	1.811	4	1.824	5	1.831	5	1.836	6	1.840	6	1.842	6	1.844	7	1.846	7
27	1.783	3	1.809	4	1.821	5	1.828	5	1.833	6	1.837	6	1.839	6	1.841	7	1.843	7
28	1.780	3	1.806	4	1.819	5	1.826	5	1.831	6	1.834	6	1.836	6	1.838	7	1.840	7
29	1.778	3	1.804	4	1.816	5	1.823	5	1.828	6	1.831	6	1.834	6	1.836	7	1.837	7
30	1.776	3	1.801	4	1.814	5	1.821	5	1.826	6	1.829	6	1.831	6	1.833	7	1.835	7
31	1.774	3	1.799	4	1.812	5	1.819	5	1.824	6	1.827	6	1.829	6	1.831	7	1.833	7
32	1.772	3	1.797	4	1.810	5	1.817	5	1.821	6	1.825	6	1.827	6	1.829	7	1.831	7
33	1.770	3	1.796	4	1.808	5	1.815	5	1.820	6	1.823	6	1.825	6	1.827	7	1.829	7
34	1.769	3	1.794	4	1.806	5	1.813	5	1.818	6	1.821	6	1.823	6	1.825	7	1.827	7
35	1.767	3	1.792	4	1.804	5	1.811	5	1.816	6	1.819	6	1.822	6	1.824	7	1.825	7
34	1.769	3	1.794	4	1.806	5	1.813	5	1.818	6	1.821	6	1.823	6	1.825	7	1.827	7
35	1.767	3	1.792	4	1.804	5	1.811	5	1.816	6	1.819	6	1.822	6	1.824	7	1.825	7
36	1.766	3	1.791	4	1.803	5	1.810	5	1.814	6	1.818	6	1.820	6	1.822	7	1.823	7
37	1.764	3	1.789	4	1.801	5	1.808	5	1.813	6	1.816	6	1.819	6	1.820	7	1.822	7
38	1.763	3	1.788	4	1.800	5	1.807	5	1.812	6	1.815	6	1.817	6	1.819	7	1.820	7
39	1.762	3	1.787	4	1.799	5	1.806	5	1.810	6	1.813	6	1.816	6	1.818	7	1.819	7

(*Continued*)

Table A3.4 (Continued)

df\k	2		3		4		5		6		7		8		9		10	
	Q	*B*	*Q*	*B*	*Q*	*B*	*Q*	*B*	*Q*	*B*	*Q*	*B*	*Q*	*B*	*Q*	*B*	*Q*	*B*
40	1.761	3	1.785	4	1.797	5	1.804	5	1.809	6	1.812	6	1.814	6	1.816	7	1.818	7
41	1.759	3	1.784	4	1.796	5	1.803	5	1.808	6	1.811	6	1.813	6	1.815	7	1.816	7
42	1.758	3	1.783	4	1.795	5	1.802	5	1.807	6	1.810	6	1.812	6	1.814	7	1.815	7
43	1.757	3	1.782	4	1.794	5	1.801	5	1.805	6	1.809	6	1.811	6	1.813	7	1.814	7
44	1.756	3	1.781	4	1.793	5	1.800	5	1.804	6	1.808	6	1.810	6	1.812	7	1.813	7
45	1.755	3	1.780	4	1.792	5	1.799	5	1.803	6	1.807	6	1.809	6	1.811	7	1.812	7
46	1.755	3	1.779	4	1.791	5	1.798	5	1.802	6	1.806	6	1.808	6	1.810	7	1.811	7
47	1.754	3	1.778	4	1.790	5	1.797	5	1.802	6	1.805	6	1.807	6	1.809	7	1.810	7
48	1.753	3	1.777	4	1.789	5	1.796	5	1.801	6	1.804	6	1.806	6	1.808	7	1.809	7
49	1.752	3	1.777	4	1.788	5	1.795	5	1.800	6	1.803	6	1.805	6	1.807	7	1.808	7
50	1.751	3	1.776	4	1.788	5	1.795	5	1.799	6	1.802	6	1.804	6	1.806	7	1.808	7
51	1.751	3	1.775	4	1.787	5	1.794	5	1.798	6	1.801	6	1.804	6	1.805	7	1.807	7
52	1.750	3	1.774	4	1.786	5	1.793	5	1.797	6	1.801	6	1.803	6	1.805	7	1.806	7
53	1.749	3	1.774	4	1.785	5	1.792	5	1.797	6	1.800	6	1.802	6	1.804	7	1.805	7
54	1.749	3	1.773	4	1.785	5	1.792	5	1.796	6	1.799	6	1.801	6	1.803	7	1.805	7
55	1.748	3	1.772	4	1.784	5	1.791	5	1.795	6	1.798	6	1.801	6	1.803	7	1.804	7
56	1.748	3	1.772	4	1.784	5	1.790	5	1.795	6	1.798	6	1.800	6	1.802	7	1.803	7
57	1.747	3	1.771	4	1.783	5	1.790	5	1.794	6	1.797	6	1.800	6	1.801	7	1.803	7
58	1.747	3	1.771	4	1.782	5	1.789	5	1.794	6	1.797	6	1.799	6	1.801	7	1.802	7
59	1.746	3	1.770	4	1.782	5	1.789	5	1.793	6	1.796	6	1.798	6	1.800	7	1.801	7
60	1.745	3	1.770	4	1.781	5	1.788	5	1.792	6	1.795	6	1.798	6	1.799	7	1.801	7
61	1.745	3	1.769	4	1.781	5	1.787	5	1.792	6	1.795	6	1.797	6	1.799	7	1.800	7
62	1.745	3	1.769	4	1.780	5	1.787	5	1.791	6	1.794	6	1.797	6	1.798	7	1.800	7
63	1.744	3	1.768	4	1.780	5	1.786	5	1.791	6	1.794	6	1.796	6	1.798	7	1.799	7
64	1.744	3	1.768	4	1.779	5	1.786	5	1.790	6	1.793	6	1.796	6	1.797	7	1.799	7
65	1.743	3	1.767	4	1.779	5	1.786	5	1.790	6	1.793	6	1.795	6	1.797	7	1.798	7
66	1.743	3	1.767	4	1.778	5	1.785	5	1.789	6	1.792	6	1.795	6	1.796	7	1.798	7
67	1.742	3	1.766	4	1.778	5	1.785	5	1.789	6	1.792	6	1.794	6	1.796	7	1.797	7
68	1.742	3	1.766	4	1.778	5	1.784	5	1.789	6	1.792	6	1.794	6	1.796	7	1.797	7
69	1.742	3	1.766	4	1.777	5	1.784	5	1.788	6	1.791	6	1.793	6	1.795	7	1.797	7
70	1.741	3	1.765	4	1.777	5	1.783	5	1.788	6	1.791	6	1.793	6	1.795	7	1.796	7
71	1.741	3	1.765	4	1.776	5	1.783	5	1.787	6	1.790	6	1.793	6	1.794	7	1.796	7
72	1.741	3	1.764	4	1.776	5	1.783	5	1.787	6	1.790	6	1.792	6	1.794	7	1.795	7
73	1.740	3	1.764	4	1.776	5	1.782	5	1.787	6	1.790	6	1.792	6	1.794	7	1.795	7
74	1.740	3	1.764	4	1.775	5	1.782	5	1.786	6	1.789	6	1.791	6	1.793	7	1.795	7
75	1.740	3	1.763	4	1.775	5	1.782	5	1.786	6	1.789	6	1.791	6	1.793	7	1.794	7

76	1.739	3	1.763	4	1.775	5	1.781	5	1.786	6	1.789	6	1.791	6	1.793	7	1.794	7
77	1.739	3	1.763	4	1.774	5	1.781	5	1.785	6	1.788	6	1.790	6	1.792	7	1.794	7
78	1.739	3	1.762	4	1.774	5	1.781	5	1.785	6	1.788	6	1.790	6	1.792	7	1.793	7
79	1.738	3	1.762	4	1.774	5	1.780	5	1.785	6	1.788	6	1.790	6	1.792	7	1.793	7
80	1.738	3	1.762	4	1.773	5	1.780	5	1.784	6	1.787	6	1.789	6	1.791	7	1.793	7
81	1.738	3	1.762	4	1.773	5	1.780	5	1.784	6	1.787	6	1.789	6	1.791	7	1.792	7
82	1.738	3	1.761	4	1.773	5	1.779	5	1.784	6	1.787	6	1.789	6	1.791	7	1.792	7
83	1.737	3	1.761	4	1.772	5	1.779	5	1.783	6	1.786	6	1.789	6	1.790	7	1.792	7
84	1.737	3	1.761	4	1.772	5	1.779	5	1.783	6	1.786	6	1.788	6	1.790	7	1.791	7
85	1.737	3	1.761	4	1.772	5	1.779	5	1.783	6	1.786	6	1.788	6	1.790	7	1.791	7
86	1.737	3	1.760	4	1.772	5	1.778	5	1.783	6	1.786	6	1.788	6	1.789	7	1.791	7
87	1.736	3	1.760	4	1.771	5	1.778	5	1.782	6	1.785	6	1.787	6	1.789	7	1.791	7
88	1.736	3	1.760	4	1.771	5	1.778	5	1.782	6	1.785	6	1.787	6	1.789	7	1.790	7

If c = number of reps in controls, r = number of reps in each treatment, $w = c/r$, then $\bar{t}_{\alpha,k} = Q - 10^{-2}B(1 - 1/w)$.

References

Abbott, W.S. (1925). A method of computing the effectiveness of an insecticide. *Journal of Economic Entomology 18*: 265–267.

Agresti, A. (2002). *Categorical Data Analysis*, 2e. Hoboken, NJ: Wiley.

Agresti, A. (2007). *An Introduction to Categorical Data Analysis*, 2e. Hoboken, NJ: Wiley.

Agresti, A. (2012). *Categorical Data Analysis*, 3e. Hoboken, NJ: Wiley.

Agresti, A. (2015). *Foundations of Linear and Generalized Linear Models*. Hoboken, NJ: Wiley.

Akritas, M.G. (1990). The rank transform method in some two factor designs. *Journal of the American Statistical Association 85*: 73–78.

Akritas, M.G., Arnold, S.F., and Brunner, E. (1997). Nonparametric hypotheses and rank statistics for unbalanced factorial designs. *Journal of the American Statistical Association 92*: 258–265.

Aldenberg, T. (2015). Some species sensitivity distribution statistics revisited – a governmental perspective. *Environmental Toxicology and Chemistry 34*: 2442–2444.

Aldenberg,T.(2016).OECD-TG305R-PackagebcmfRUserGuide(v0.3-2). https://www.oecd.org/chemicalsafety/testing/OECD-TG305%20R-Package%20bcmfR%20User%20Guide%20(v0.3-2)_2016-12-09.pdf (accessed 13 July 2017).

Aldenberg, T. and Jaworska, J.S. (2000). Uncertainty of the hazardous concentration and fraction affected for normal species sensitivity distributions. *Ecotoxicology and Environmental Safety 46*: 1–18.

Aldenberg, T., Jaworska, J.S., and Traas, T.P. (2002). Normal species sensitivity distributions and probabilistic ecological risk assessment. In: *Species Sensitivity Distributions in Ecotoxicology* (ed. L. Posthuma, G.W. Suter and T.P. Traas). Boca Raton, FL: Lewis Publishers.

Aldenberg, T. and Slob, W. (1993). Confidence limits for hazardous concentrations based on logistically distributed NOEC toxicity data. *Ecotoxicology and Environmental Safety 25*: 48–63.

Aldenberg, T., van Vlaardingen, P.L.A., Traas, T.P., and Wintersen, A.M. (2017). ETX 2.2 normal distribution based hazardous concentration and fraction affected. http://www.rivm.nl/Documenten_en_publicaties/Professioneel_Praktisch/Software/ETX_2_2/Download/ETX_2_2 (accessed 6 February 2018).

Almstrup, K., Fernández, M.F., Petersen, J.H. et al. (2002). Dual effects of phytoestrogens result in U-shaped dose-response curves. *Environmental Health Perspectives 110*: 743–748.

Alvord, W.G. and Rossio, J.L. (1993). Determining confidence limits for drug potency in immunoassay. *Journal of Immunological Methods 157*: 155–163.

Anderson, D.R. and Burnham, K.P. (2010). *Model Selection and Multi-Model Inference: A Practical Information-Theoretic Approach*. New York: Springer-Verlag.

Anderson, M.J. and Walsh, D.C.I. (2013). PERMANOVA, ANOSIM, and the Mantel test in the face of heterogeneous dispersions: what null hypothesis are you testing? *Ecological Monographs 83*: 557–574.

Apostol, T.M. (1974). *Mathematical Analysis*, 2e. Reading, MA: Addison-Wesley.

Armitage, P. (1955). Tests for linear trends in proportions and frequencies. *Biometrics 11*: 375–386.

Armitage, P. and Allen, I. (1950). Methods of estimating the LD50 in general quantal response data. *Journal of Hygiene 48*: 298–322.

ASA (2016). *American Statistical Association Releases Statement on Statistical Significance and P-Values: Provides Principles to Improve the Conduct and Interpretation of Quantitative Science*. (7 March 2016). http://www.amstat.org/asa/files/pdfs/P-ValueStatement.pdf (accessed 14 August 2017).

Ashauer, R., Albert, C., Augustine, S. et al. (2016). Modelling survival: exposure pattern, species sensitivity and uncertainty. *Scientific Reports 6*: 29178. doi: 10.1038/srep29178.

Ashauer, R., Annika Agatz, A., Albert, C. et al. (2011). Toxicokinetic-toxicodynamic modelling of quantal and graded sublethal endpoints – a brief discussion of concepts. *Environmental Toxicology and Chemistry 30*: 2519–2524.

Ashauer, R., Thorbek, P., Warinton, J.S. et al. (2013). A method to predict and understand fish survival under dynamic chemical stress using standard ecotoxicity data. *Environmental Toxicology and Chemistry 32*: 954–965.

Askey, R.A. and Roy, R. (2010). Beta function. In: *NIST Handbook of Mathematical Functions* (ed. F.W.J. Olver, D.M. Lozier, R.F. Boisvert and C.W. Clark). Cambridge: Cambridge University Press.

Ayer, M., Brunk, H.D., Ewing, G.M. et al. (1955). An empirical distribution function for sampling with incomplete information. *Annals of Mathematical Statistics 26*: 641–647.

Statistical Analysis of Ecotoxicity Studies, First Edition. John W. Green, Timothy A. Springer, and Henrik Holbech.

Companion website: www.wiley.com/go/Green/StatAnalysEcotoxicStudy

Backhaus, T. and Faust, M. (2012). Predictive environmental risk assessment of chemical mixtures: a conceptual framework. *Environmental Science & Technology 46*: 2564–2573.

Backhaus, T. and Karlsson, M. (2014). Screening level mixture risk assessment of pharmaceuticals in STP effluents. *Water Research 49*: 157–165.

Bailer, A.J. and Portier, C.J. (1988). Effects of treatment-induced mortality and tumor-induced mortality on tests for carcinogenicity in small samples. *Biometrics 44*: 417–431.

Barlow, R.E., Bartholomew, D.J., Bremmer, J.M., and Brunk, H.D. (1972). *Statistical Inference under Order Restrictions*. London: Wiley.

Bartholomew, D.J. (1961). Ordered tests in the analysis of variance. *Biometrika 48*: 325–332.

Bates, D.M. and Watts, D.G. (2007). *Nonlinear Regression Analysis and Its Applications*. New York: Wiley.

Bathke, A. and Lankowski, D. (2005). Rank procedures for a large number of treatments. *Journal of Statistical Planning and Inference 133*: 223–238.

Bauer, P. (1997). A note on multiple testing procedures in dose finding. *Biometrics 53*: 1125–1128.

Baumans, V. (2005). Environmental enrichment for laboratory rodents and rabbits: requirements of rodents, rabbits, and research. *Institute for Laboratory Animal Research Journal 46*: 162–170.

Beasley, T.M., Erickson, S., and Allison, D.B. (2009). Rank-based inverse normal transformations are increasingly used, but are they merited? *Behavior Genetics 39*: 580–595.

Bedaux, J.J.M. and Kooijman, S.A.L.M. (1994). Statistical analysis of bioassays, based on hazard modelling. *Environmental and Ecological Statistics 1*: 303–314.

Belden, J., Gillom, R., and Lydy, M. (2007). How well can we predict the toxicity of pesticide mixtures to aquatic life? *Integrated Environmental Assessment and Management 3*: 364–372.

Bennett, B.M. (1952). Estimation of LC50 by moving averages. *Journal of Hygiene 50*: 157–164.

Bennett, B.M. (1963). Optimum moving averages for the estimation of median effective dose in bioassay. *Journal of Hygiene 61*: 401–406.

Bennett, B.M. (1968). Rank-order tests of linear hypotheses. *Journal of the Royal Statistical Society. Series B 30*: 483–489.

Benoit, D.A. (1982). *User's Guide for Conducting Life-Cycle Chronic Toxicity Tests with Fathead Minnows (Pimephales promelas)*, EPA 600/8-81-011. Environmental Research Laboratory. Duluth, MN.

Berenbaum, M.C. (1989). What is synergy. *Pharmacological Reviews 41*: 93–141.

Berry, D.A. and Stangl, D. (1996). *Bayesian Biostatistics*. New York: CRC Press.

Beyer, H.R. (2010). *Calculus and Analysis: A Combined Approach*. Hoboken, NJ: Wiley.

Bland, J.M. and Altman, D.G. (2004). The log-rank test. *British Journal of Medicine 328*: 1072.

Bliss C. 1967. *Statistics in Biology*. New York: McGraw-Hill. ASIN: B007ZKQ322.

Blom, G. (1958). *Statistical Estimates and Transformed Beta-Variables*. New York: Wiley.

Boesten, J., Aden, K., Beigel, C. et al. (2006). Guidance Document on Estimating Persistence and Degradation Kinetics from Environmental Fate Studies on Pesticides in EU Registration. Final Report. FOCUS Work Group on Degradation Kinetics, EC Document Reference Sanco/10058/2005 version 2.0. Brussels, Belgium.

Bolch, B.W. (1968). More on unbiased estimation of the standard deviation. *The American Statistician 22*: 27.

Bolker, B.M., Brooks, M.E., Clark, C.J. et al. (2008). Generalized linear mixed models: a practical guide for ecology and evolution. *Trends in Ecology & Evolution 24*: 127–135.

Box, G.E.P. (1953). Non-normality and tests on variances. *Biometrika 40*: 318–335.

Box, G.E.P. (1954). Some theorems on quadratic forms applied in the study of analysis of variance problems, I. Effect of inequality of variance in the one-way classification. *Annals of Mathematical Statistics 25*: 290–302.

Box, G.E.P. and Cox, D.R. (1964). An analysis of transformations. *Journal of the Royal Statistical Society. Series B B-26*: 211–252.

Box, G.E.P. and Hill, W.J. (1974). Correcting inhomogeneity of variance with power transformation weighting. *Technometrics 16*: 385–389.

Box, G.E.P. and Tidwell, P.W. (1962). Transformations of the independent variables. *Technometrics 4*: 531–550.

Brain, P. and Cousens, R. (1989). An equation to describe dose responses where there is stimulation of growth at low doses. *Weed Research 29*: 93–96.

Brenner, D.J. and Hall, E.J. (2007). Computed tomography — an increasing source of radiation exposure. *New England Journal of Medicine 357*: 2277–2284.

Brenner, D.J. and Sachs, R.K. (2006). Estimating radiation-induced cancer risks at very low doses: rationale for using a linear no-threshold approach. *Radiation and Environmental Biophysics 44*: 253–256.

Bretz, F. (1999). Powerful modifications of Williams' test on trends. Dissertation. University of Hanover.

Bretz, F. and Hothorn, L.A. (2000). A powerful alternative to Williams' test with application to toxicological dose-response relationships of normally distributed data. *Environmental and Ecological Statistics 7*: 135–254.

Brock, T.C.M., Arts, G.H.P., Ten Hulscher, T.E.M. et al. (2011). Aquatic Effect Assessment for Plant Protection Products: A Dutch Proposal That Addresses the Requirements of the Plant Protection Product Regulation and Water Framework Directive, 140 pp. Wageningen, Alterra, Alterra Report 2235.

Brock, T.C.M., Hammers-Wirtz, M., Hommen, U. et al. (2015). The minimum detectable difference (MDD) and the interpretation of treatment-related effects of pesticides in experimental ecosystems. *Environmental Science and Pollution Research 22*: 1160–1174.

Brock, W.J., Mounho, B., and Fu, L. (2014). *The Role of the Study Director in Nonclinical Studies Pharmaceuticals, Chemicals, Medical Devices, and Pesticides*. Hoboken, NJ: Wiley.

Brown, C.C. (1978). The statistical analysis of dose-response relationships. In: *Principals of Ecotoxicology, Scope 12* (ed. G.C. Butler). Chichester, New York: Wiley. https://dge.carnegiescience.edu/SCOPE/SCOPE_12/SCOPE_12_2.1_chapter6_115-148.pdf (accessed 31 July 2017).

Brown, M.B. and Forsythe, A.B. (1974). Robust tests for the equality of variances. *Journal of the American Statistical Association 69*: 364–367.

Bruce, R.D. and Versteeg, D.J. (1992). A statistical procedure for modeling continuous toxicity data. *Environmental Toxicology and Chemistry 11*: 1485–1494. doi: 10.1897/1552-8618(1992)11[1485:aspfmc]2.0.co;2.

Brünning, H. and Kössler, W. (1999). The asymptotic power of Jonckheere-type tests for ordered alternatives. *Australian & New Zealand Journal of Statistics 41*: 67–77.

Bunke, O., Droge, B., and Polzehl, J. (1999). Model selection, transformations and variance estimation in nonlinear regression. *Statistics 33*: 197–240.

Burnham, K.P. and Anderson, D.R. (2004). Multimodel inference understanding AIC and BIC in model selection. *Sociological Methods & Research 33*: 261–304.

Calabrese, E.J. (2005a). Cancer biology and hormesis: human tumor cell lines commonly display hormetic (biphasic) dose-responses. *Critical Reviews in Toxicology 35*: 463–582.

Calabrese, E.J. (2005b). Paradigm lost, paradigm found: the re-emergence of hormesis as a fundamental dose response model in the toxicological sciences. *Environmental Pollution 138*: 378–411.

Calabrese, E.J. and Baldwin, L.A. (2001). U-shaped dose-responses in biology, toxicology, and public health. *Annual Review of Public Health 22*: 15–33.

Calabrese, E.J. and Baldwin, L.A. (2002). Defining hormesis. *Human & Experimental Toxicology 21*: 91–97.

Calabrese, E.J. and Blain, R. (2005). The occurrence of hormetic dose-responses in the toxicological literature, the hormesis database: an overview. *Toxicology and Applied Pharmacology 202*: 289–301.

Cameron, A.C. and Trivedi, P.K. (2013). *Regression Analysis of Count Data*, 2e. Cambridge: Cambridge University Press.

Carakostos, M.C. and Green, J.W. (1992). Using the correct regression analysis technique in method comparison studies. *Veterinary Clinical Pathology 20*: 91–94.

Carelli, G. and Iavicoli, I. (2002). Defining hormesis: the necessary tool to clarify experimentally the low dose-response relationship. *Human & Experimental Toxicology 21*: 103.

Carroll, R.J., David, R., Stefanski, L.A., and Crainiceanu, C. (2006). *Measurement Error in Nonlinear Models: A Modern Perspective*, 2e. Boca Raton, FL: CRC Press.

Cavanaugh, J.E. (2012). *The Bayesian Information Criterion*. The University of Iowa. http://myweb.uiowa.edu/cavaaugh/doc/ms/ms_lec_5_ho.pdf (accessed 7 January 2016).

Cedergreen, N. (2014). Quantifying synergy: a systematic review of mixture toxicity studies within environmental toxicology. *PLoS One 9* (5): e96580.

Cedergreen, N., Ritz, C., and Streibig, J.C. (2005). Improved empirical models describing hormesis. *Environmental Toxicology and Chemistry 24*: 3166–3172.

Cedergreen, N., Streibig, J.C., Kudsk, P. et al. (2006). The occurrence of hormesis in plants and algae. *Dose Response 5*: 150–162.

CETIS (2016). Tidepool Scientific Software. https://tidepool-scientific.com/Cetis/Cetis.html (accessed 25 July 2017).

Chambers, J.M., Cleveland, W.S., Kleiner, B., and Tukey, P.A. (1983). *Graphical Methods for Data Analysis*, The Wadsworth Statistics/Probability Series. Boston: Duxbury Press.

Chapman P.F., Reed M., Hart A., Roelofs W., Aldenberg T., Soloman K., Tarazona J., Liess M., Byrne P., Powley W., Green J., Ferson S., Galicia H. 2007. Methods of uncertainty analysis. In: *EUFRAM: Concerted Action to Develop a European Framework for Probabilistic Risk Assessment of the Environmental Impacts of Pesticides, Volume 2, Detailed Reports on Role, Methods, Reporting & Validation*, http://myweb.uiowa.edu/cavaaugh/doc/ms/ms_lec_5_ho.pdf (18 February 2018).

Charles, S., Ducrot, V., Azam, D. et al. (2016). Optimizing the design of a reproduction toxicity test with the pond snail *Lymnaea stagnalis*. *Regulatory Toxicology and Pharmacology 81*: 47–56.

Chatfield, C. (1995). Model uncertainty, data mining and statistical inference. *Journal of the Royal Statistical Society. Series A 158* (Part 3): 419–466.

Chen, D.G. (2007). Dose-time-response cumulative multinomial generalized linear model. *Journal of Biopharmaceutical Statistics 17*: 173–185.

Chen, C.W. and Selleck, R.E. (1969). A kinetic model of fish toxicity threshold. *Journal – Water Pollution Control Federation 41*: 294–308.

Chernick, M.R. (1999). *Bootstrap Methods: A Practitioner's Guide*. New York: Wiley.

Chiu, W.A. and Slob, W. (2015). A unified probabilistic framework for dose-response assessment of human health effects. *Environmental Health Perspectives 123*: 1241–1254. doi: http://dx.doi.org/10.1289/ehp.1409385.

CHMP (2013). Guideline on adjustment for baseline. Committee for Medicinal Products for Human Use. http://www.ema.europa.eu/docs/en_GB/document_library/ (last accessed 6 August 2009).

Choy, S.L., O'Leary, R., and Mengersen, K. (2009). Elicitation by design in ecology: using expert opinion to inform priors for Bayesian statistical models. *Ecology 90*: 265–277.

Christensen, E.R., Kusk, K.O., and Nyholm, N. (2009). Dose-response regressions for algal growth and similar continuous endpoints: calculation of effective concentrations. *Environmental Toxicology and Chemistry 28*: 826–835.

Christiansen, S., Scholze, M., Dalgaard, M. et al. (2009). Synergistic disruption of external male sex organ development by a mixture of four antiandrogens. *Environmental Health Perspectives 117*: 1839–1846.

Clayton, C.A., Starr, T.B., Sielken, R.L. Jr. et al. (2003). Using a nonlinear mixed effects model to characterize cholinesterase activity in rats exposed to Aldicarb. *Journal of Agricultural, Biological, and Environmental Statistics 8*: 420–437.

Clyde, M.A. (1999). Comment on Bayesian model averaging: a tutorial by Höeting *et al. Statistical Science 14*: 382–417.

Coady, K., Marino, T., Thomas, J. et al. (2013). An evaluation of 2,4-dichlorophenoxyacetic acid in the amphibian metamorphosis assay and the fish short-term reproduction assay. *Ecotoxicology and Environmental Safety 90*: 143–150.

Cochran, W.G. (1954). Some methods for strengthening the common χ^2-tests. *Biometrics 10*: 417–451.

Cohen, A.C. (1966). A note on certain discrete mixed distributions. *Biometrics 22*: 566–572.

Collett, D. (2002). *Modelling Binary Data*, 2e. London: Chapman and Hall.

Collett, D. (2014). *Modelling Survival Data in Medical Research*, 3e. London: Chapman and Hall.

Conover, W.J. and Iman, R.L. (1981). Rank transformations as a bridge between parametric and nonparametric statistics. *American Statistician 35*: 124–128.

Conover, W.J. and Iman, R.L. (1982). Analysis of variance using the rank transformation. *Biometrics 38*: 715–724.

Cook, J.D. (2009). Notes on the negative binomial distribution. https://www.johndcook.com//negative_binomial.pdf. (accessed 29 January 2017).

Cook D., Duckworth W.M., Kaiser M.S., Meeker W.Q., and Stephenson W.R. 1999. Principles of maximum likelihood estimation and the analysis of censored data. In: *Beyond Traditional Statistical Methods*, Chapter 2. http://www.public.iastate.edu/~stat415/meeker/ml_estimation_chapter.pdf (accessed 2 August 2017).

Cooke, R.M. (1991). *Experts in Uncertainty, Opinions and Subjective Probability in Science*. Oxford: Oxford Universiy Press.

Cox, C. (1987). Threshold dose-response models in toxicology. *Biometrics 43*: 511–523.

Cran (2016). Package "Multcomp" Developed by L. Hothorn. https://cran.r-project.org/web/packages/multcomp/index.html (accessed 25 July 2017).

Crawley, M.J. (2013). *The R Book*, 2e. Singapore: Wiley.

CSIRO (2010). BurrliOZ 2.0. https://research.csiro.au/software/burrlioz/ (accessed 7 February 2018).

Cuppen, J.G.M., Van den Brink, P.J., Uil, K.F. et al. (2000). Impact of the fungicide carbendazim in freshwater microcosms. II. Water quality, breakdown of particulate organic matter and responses of macro-invertebrates. *Aquatic Toxicology 48*: 233–250.

D'Agostino, R.B. and Stephens, M.A. (1986). *Goodness-of-Fit Techniques*. New York: Marcel Dekker.

Dagenais, M.G. and Dagenais, D.L. (1997). Higher moment estimators for linear regression models with errors in the variables. *Journal of Econometrics 76*: 193–221.

Danbaba, A. (2012). Comparison of a class of rank-score tests in two-factor designs. *Nigerian Journal of Basic and Applied Science 20* (4): 305–314.

Davis, J.M. and Svendsgaard, D.J. (1990). U-shaped dose-response curves: their occurrence and implications for risk assessment. *Journal of Toxicology and Environmental Health 30*: 71–83.

Delignette-Muller, M.L., Lopes, C., Veber, P., and Charles, C. (2013). Statistical handling of reproduction data for exposure-response modeling. *Environmental Science & Technology 48*: 7544–7551.

Deneer, J. (2000). Toxicity of mixtures of pesticides in aquatic systems. *Pest Management Science 65*: 516–551.

Dennis, J.E. and Schnabel, R.B. (1996). *Numerical Methods for Unconstrained Optimization and Nonlinear Equations*. Philadelphia: SIAM.

Dixon, W.J. and Mood, A.M. (1948). A method for obtaining and analyzing sensitivity data. *Journal of the American Statistical Association 43*: 109–126.

Dixon, D.G. and Sprague, J.B. (1981). Acclimation to copper by rainbow trout (*Salmo gairdneri*) – a modifying factor in toxicity. *Canadian Journal of Fisheries and Aquatic Sciences 38*: 880–888.

DK-EPA (2013). *Information/Testing Strategy for Identification of Substances with Endocrine Disrupting Properties*. Danish Center on Endocrine Disrupters (CeHoS). http://findresearcher.sdu.dk/portal/en/publishers/cehos-danish-centre-on-endocrine-disrupters(53b7b958-0673-43ab-b6bb-03633994825b).html (accessed 29 August 2017).

Dowse, R., Tang, D., Palmer, C.G., and Kefford, B.J. (2013). Risk assessment using the species sensitivity distribution method: data quality versus data quantity. *Environmental Toxicology and Chemistry 32*: 1360–1369.

Draper, N.R. and Cox, D.R. (1969). On distributions and their transformations to normality. *Journal of the Royal Statistical Society. Series B B-31*: 472–476.

Draper, N.R. and Smith, H. (2012). *Applied Regression Analysis*, 3e. New York: Wiley.

DuBoudin, C., Ciffroy, P., and Magaud, H. (2004). Effects of data manipulation and statistical methods on species sensitivity distributions. *Environmental Toxicology and Chemistry 23*: 489–499.

Ducrot, V., Ashauer, R., Bednarska, A.J. et al. (2015). Using toxicokinetic-toxicodynamic modeling as an acute risk assessment refinement approach in vertebrate ecological risk assessment. *Integrated Environmental Assessment and Management 12*: 32–45.

Duke, S.O., Cedergreen, N., Velini, E.D., and Belz, R.G. (2014). Hormesis: is it an important factor in herbicide use and allelopathy? *Outlooks on Pest Management 17*: 29–33.

Dunn, O.J. (1964). Multiple comparisons using rank sums. *Technometrics 6*: 241–252.

Dunnett, C.W. (1955). A multiple comparison procedure for comparing several treatments with a control. *Journal of the American Statistical Association 50*: 1096–1121.

Dunnett, C.W. (1964). New tables for multiple comparisons with a control. *Biometrics 20*: 482–491.

Dunnett, C.W. (1980). Pairwise multiple comparisons in the unequal variance case. *Journal of the American Statistical Association 75*: 796–800.

Dunnett, C.W. (1989). Algorithm AS 251: multivariate normal probability integrals with product correlation structure. *Journal of the Royal Statistical Society. Series C, Applied Statistics 38*: 564–579.

Dunnett, C.W., Horn, M., and Vollandt, R. (2001). Sample size determination in step-down and step-up multiple tests for comparing treatments with a control. *Journal of Statistical Planning and Inference 97*: 367–384.

Dunnett, C.W. and Tamhane, A.C. (1991). Step-down multiple tests for comparing treatments with a control in unbalanced one-way layout. *Statistics in Medicine 10*: 939–947.

Dunnett, C.W. and Tamhane, A.C. (1992). A step-up multiple test procedure. *Journal of the American Statistical Association 87*: 162–170.

Dunnett, C.W. and Tamhane, A.C. (1995). Step-up multiple testing of parameters with unequally correlated estimates. *Biometrics 51*: 217–227.

Dunnett, C.W. and Tamhane, A.C. (1998). New multiple test procedures for dose finding. *Journal of Biopharmaceutical Statistics 8*: 353–366.

ECOFRAM (1999). Aquatic draft report Ecological Committee on Federal, Insecticide, Fungicide and Rodenticide Act, Risk Assessment. https://archive.epa.gov/oppefed1/web/pdf/aquareport.pdf (accessed 14 February 2017).

Efron, B.E., Hastie, T., Johnstone, I., and Tibshirani, R. (2004). Least angle regression (with discussion). *Annals of Statistics 32*: 407–499.

EFSA (2005). Opinion of the Scientific Panel on Plant Health, Plant Protection Products and their Residues on a request from EFSA related to the assessment of the acute and chronic risk to aquatic organisms with regard to the possibility of lowering the uncertainty factor if additional species were tested. *EFSA Journal 301*: 1–45.

EFSA (2013). Guidance on tiered risk assessment for plant protection products for aquatic organisms in edge-of-field surface waters. EFSA panel on plant protection products and their residues (PPR). Parma, Italy. *EFSA Journal 11* (7): 3290 [268 pp.].

EFSA (2014). Scientific opinion on good modelling practice in the context of mechanistic effect models for risk assessment of plant protection products. *EFSA Journal 12* (3): 3589. doi: 10.2903/j.efsa.2014.3589.

EFSA (2017). Update: use of the benchmark dose approach in risk assessment. EFSA Scientific Committee. *EFSA Journal 15*: 4658.

Ema, M., Aoyama, H., Arima, A. et al. (2012). Historical control data on prenatal developmental toxicity studies in rabbits. *Congenit Anom (Kyoto) 52*: 155–161.

Engleman, R.M., Otis, D.L., and Dusenberry, J.A. (1986). Small sample comparison of Thompson's estimator to some common bioassay estimators. *Journal of Statistical Computation and Simulation 25*: 237–250.

Epstein, S.S. (1973). The Delaney amendment. *Ecologist 3*: 424–430.

Erickson, T. and Whited, T.M. (2002). Two-step GMM estimation of the errors-in-variables model using high-order moments. *Econometric Theory 18*: 776–799.

EU (2006). Regulation (EC) No 1907/2006 of the European Parliament and of the Council of 18 December 2006 concerning the Registration, Evaluation, Authorisation and Restriction of Chemicals (REACH), establishing a European Chemicals Agency, amending Directive 1999/45/EC and repealing Council Regulation (EEC) No 793/93 and Commission Regulation (EC) No 1488/94 as well as Council Directive 76/769/EEC and Commission Directives 91/155/EEC, 93/67/EEC, 93/105/EC and 2000/21/EC.

EU (2012). Commision Regulation (EU) No 528/2012 of the European parliament and of the Council of 22 May 2012 concerning the making available on the market and use of biocidal products.

EU (2013). Commission Regulation (EU) No 283/2013 of 1 March 2013 setting out the data requirements for active substances, in accordance with Regulation (EC) No 1107/2009 of the European Parliament and of the Council concerning the placing of plant protection products on the market Text with EEA relevance.

Farewell, D.M. and Farewell, V.T. (2013). Dirichlet negative multinomial regression for overdispersed correlated count data. *Biostatistics 14*: 395–404.

Faust, M., Altenburger, R., Backhaus, T. et al. (2001). Predicting the joint algal toxicity of multi-component s-triazine mixtures at low-effect concentrations of individual toxicants. *Aquatic Toxicology 56*: 13–32.

Fears, T.R., Tarone, R.E., and Chu, K.C. (1977). False-positive and false-negative rates for carcinogenicity screens. *Cancer Research 37*: 1941–1945.

Field, A. (2016). Discovering Statistics *Analysis of Covariance*. http://www.statisticshell.com/docs/ancova.pdf (accessed 14 December 2016).

Field, A. and Miles, J. (2010). *Discovering Statistics Using SAS*. London: Sage Publications.

Field, A., Miles, J., and Field, Z. (2012). *Discovering Statistics Using R*. London: Sage Publications There is also a version of this book using SPSS.

Filliben, J.J. (1975). The probability plot correlation coefficient test for normality. *Technometrics 17*: 111–117.

Finney, D.J. (1948). The Fisher-Yates test of significance in 2×2 contingency tables. *Biometrika 35*: 145–156.

Fisher, R.A. (1934). *Statistical Methods for Research Workers*, 4e. London: Oliver and Boyd.

Fisk, P. (2014). *Chemical Risk Assessment: A Manual for REACH*. Chichester, UK: Wiley.

Fleming, R. and Retnakaran, A. (1985). Evaluating single treatment data using Abbott's formula with reference to insecticides. *Journal of Economic Entomology 78*: 1179–1181.

Fox, D. (2008). NECS, NOECS and the ECX. *Australasian Journal of Ecotoxicology 14*: 7–9.

Frey, H.C. and Zhao, Y. (2002). Quantification of uncertainty and variability for censored data sets in air toxics. *Presented at 11th International Emission Inventory Conference – "Emission Inventories – Partnering for the Future."* http://www.epa.gov/ttn/chief/conference/ei11/toxics/frey.pdf (accessed 6 February 2018).

Fuller, W.A. (1987). *Measurement Error Models*. Hoboken, NJ: Wiley.

Fung, K.Y.D., Krewski, D., Rao, J.N.K., and Scott, A.J. (1994). Tests for trend in developmental toxicity experiments with correlated binary data. *Risk Analysis 14*: 639–648.

Furuhashi, T. (2015). *Basics of Multivariate Analysis II (Principal Component Analysis): Theory and Exercise Using R*. ASIN: B013HNBQS. Amazon Digital Services LLC

Gad S.C. 2006. *Statistics and Experimental Design for Toxicologists and Pharmacologists*, 4, esp, pp 98ff. Taylor & Francis. Boca Raton, FL.

Gao, Z., Green, J.W., Vanderborght, J., and Schmitt, W. (2011). Improving uncertainty analysis in kinetic evaluations using iteratively reweighted least squares. *Environmental Toxicology and Chemistry 30*: 2363–2371.

Gart, J.J., Chu, K.C., and Tarone, R.E. (1979). Statistical issues in interpretation of chronic bioassay tests for carcinogenicity. *Journal of the National Cancer Institute 62*: 957–974.

Gart, J.J., Krewski, D., Lee, P.N. et al. (1986). Statistical methods in cancer research, volume III – the design and analysis of long-term animal experiments. *IARC Scientific Publications* (79): 1–219.

Gaylor, D.W., Sheehan, D.M., Young, J.F., and Matteson, D.R. (1988). The threshold dose question in teratogenesis. *Teratology 38*: 389–391.

Gelman, A., Carlin, J.B., Stern, H.S., and Rubin, D.B. (2004). *Bayesian Data Analysis*, 3e. Boca Raton, FL: Chapman & Hall/CRC.

Gelman, A., Hill, J., and Yajima, M. (2012). Why we (usually) don't have to worry about multiple comparisons. *Journal of Research on Educational Effectiveness 5*: 189–211.

Gelman, A. and Rubin, D.B. (1995). Avoiding model selection in Bayesian social research. *Sociological Methodology 25*: 165–173.

Gergs, A., Gabsi, F., Zenker, A., and Preuss, T.G. (2016). Demographic toxicokinetic-toxicodynamic modeling of lethal effects. *Environmental Science & Technology 50*: 6017–6024.

Gessner, P. (1995). Isobolographic analysis of interactions: an update on applications and utility. *Toxicology 105*: 161–179.

Gillies, K., Krone, S.M., Nagler, J.J., and Schultz, I.R. (2016). A computational model of the rainbow trout hypothalamus-pituitary-ovary-liver axis. *PLoS Computational Biology 12*: doi: 10.1371/journal.pcbi.1004874.

Gilliom, R.J., Barbash, J.E., Kolpin, D.W., and Larson, S.J. (1999). Testing water quality for pesticide pollution. *Environmental Science & Technology 33*: 164A–169A.

Gilmour, S.G. (1996). The interpretation of Mallows's Cp-statistic. *Journal of the Royal Statistical Society, Series D 45* (1): 49–56.

Giraud, C. (2015). *Introduction to High-Dimensional Statistics*. Boca Raton, FL: Chapman & Hall/CRC.

Goerlitz, L., Gao, Z., and Schmitt, W. (2011). Statistical analysis of chemical transformation kinetics using Markov-chain Monte-Carlo methods. *Environmental Science & Technology 45*: 4429–4437.

Good, P.I. (2005). *Permutation, Parametric, and Bootstrap Tests of Hypotheses*, Springer Series in Statistics. New York: Springer-Verlag.

Google Search 2015. https://www.google.com/search?hl=en&site=imghp&tbm=isch&source=hp&biw=1120&bih=584&q=sarcomeres+in+muscle+myofibrils+&oq=sarcomeres+in+muscle+myofibrils+&gs_l=img.3...2503.2503.0.4527.1.1.0.0.0.0.56.56.1.1.0....0...1ac.1.64.img..1.0.0.lH0iQKWSAJw#tbm=isch&tbs=rimg%3ACZTOElgPd-B5IjiaGYIBKn37ASTnL_1nf-q-YlOrvLl2PoFUElMpAiin6EhfeYfKl-3rIMMe4yBuVn58UsOi3eZGQ6SoSCZoZggEqffsBEbXX_1hXXIdXOKhIJJOcv-d_16r5gRiEpR0S4vSJgqEgmU6u8uXY-gVRFcFKGPchtAoioSCQSUykCKKfoSEZp_1ss4Sa70qKhIJF95h8qX7esgRNDXTmvR46AIqEgkwx7jIG5WfnxFWE2ppNV-MVyoSCRSw6Ld5kZDpEccA6qcKMECE&q=sarcomeres%20in%20muscle%20myofibrils%20&hl=en&imgrc=lM4QiA934HlqSM%3A

Goutelle, S., Maurin, M., Rougier, F. et al. (2008). The Hill equation: a review of its capabilities in pharmacological modelling. *Fundamental & Clinical Pharmacology 22* (2008): 633–648.

Greco, W.R., Bravo, G., and Parsons, J.C. (1995). The search for synergy: a critical review from a response surface perspective. *Pharmacy Review 47*: 331–385.

Green, J.W. (1985). Variance components: estimates and diagnostics. Dissertation. Texas A & M University.

Green, J.W. (2013). Statistical analysis of fish early lifestage experiments. SETAC Glasgow. Platform presentation 316.

Green, J.W. (2014a). Power and control choice in aquatic experiments with solvents. *Ecotoxicology and Environmental Safety 102C*: 142–146.

Green, J.W. (2014b). Biostatistics and data interpretation. In: *The Study Director in Nonclinical Studies for Drugs, Chemicals, Pesticides, and Devices* (ed. B. Brock, B. Mounho and L. Fu). Hoboken, NJ: Wiley.

Green, J.W. (2015a). Issues with using only regression models for ecotoxicology studies. *Integrated Environmental Assessment and Management 15*: doi: 10.1002/ieam.1723.

Green, J.W. (2015b). Statistical challenges in ecotoxicology. *Environmental Toxicology and Chemistry 34*: 2437.

Green, J.W. (2017a). Comparison of statistical approaches for quantal data from ecotoxicity studies. In preparation. Presented to SETAC Brussels 2017.

Green, J.W. (2017b). Species sensitivity distributions with censored data. Manuscript for submission.

Green, J.W. (2017c). Comparison of statistical approaches for count data from ecotoxicity studies. In preparation. Presented to SETAC Brussels 2017.

Green, J.W. and Hocking, R.R. (1988). Model based diagnostics for variance components in a general mixed linear model. *Proceedings of the Thirty Third Conference on the Design of Experiments in Army Research Development and Testing*, ARO Report 88-2, 91–121.

Green, J.W., Hocking, R.R., and Bremer, R.H. (1987). Estimation of variance components in mixed factorial models including model-based diagnostics. *Proceedings of 12-th Annual Conference of SAS Users Group International*, 1162–1167.

Green, J.W., Springer, T.A., Saulnier, A.L., and Swintek, J. (2014). Statistical analysis of histopathology endpoints. *Environmental Toxicology and Chemistry 33*: 1108–1116. doi: 10.1002/etc.2530.

Green, J.W., Springer, T.A., and Staveley, J.P. (2012). The drive to ban the NOEC/LOEC in favor of EC*x* is misguided and misinformed. *Integrated Environmental Assessment and Management 9*: 12–16.

Green, J.W. and Wheeler, J. (2013). The use of carrier solvents in regulatory aquatic toxicology testing: practical, statistical and regulatory considerations. *Aquatic Toxicology 144-145*: 242–249.

Greenland, S., Senn, S.J., Rothman, K.J. et al. (2016). Statistical tests, P values, confidence intervals, and power: a guide to misinterpretations. *European Journal of Epidemiology 31*: 337–350.

Grimm, V., Augusiak, J., Focks, A. et al. (2014). Towards better modelling and decision support: documenting model development, testing, and analysis using TRACE. *Ecological Modelling 280*: 129–139.

Gruber, M.H.J. (2013). *Matrix Algebra for Linear Models*. Hoboken, NJ: Wiley.

Grynovicki, J.O. and Green, J.W. (1988). Estimation of variance components and model-based diagnostics in a repeated measures design. *Proceedings of the Thirty-third Conference on the Design of Experiments in Army Research Development and Testing*, ARO, Report 88-2.

Guimarães, P. and Lindrooth, R. (2005). Dirichlet-multinomial regression. https://ideas.repec.org/p/wpa/wuwpem/0509001.html. (accessed 2 August 2017).

Gutenbrunner, C. and Jureckova, J. (1992). Regression rank scores and regression quantiles. *Annals of Statistics 20*: 305–330.

Hackett, P.K., Silkov, M.R., Mast, T.J. et al. (1987). *Inhalation Developmental Toxicology Studies: Teratology Study of 1,3-Butadiene in Mice.* Prepared for the National Institute of Environmental Health Sciences, National Toxicology Program, Pacific Northwest Laboratory, Battelle Memorial Institute, Final Report No. NIH-401-ES-40131, Richland, WA.

Hadrup, N., Taxvig, C., Pedersen, M. et al. (2013). Concentration addition, independent action and generalized concentration addition models for mixture effect prediction of sex hormone synthesis in vitro. *PLoS One 8* (8): e70490.

Hamilton, M.A., Russo, R.C., and Thurston, R.V. (1977). Trimmed Spearman–Karber method for estimating median lethal concentrations in toxicity bioassays. *Environmental Science & Technology 11*: 714–719.

Harrell, F.E. (2015). *Regression Modeling Strategies: With Applications to Linear Models, Logistic Regression, and Survival Analysis*, 2e. New York: Springer-Verlag.

Harris, E.K. (1959). Confidence limits for the LD50 using the moving average-angle method. *Biometrics 15*: 424–432.

Hart, A. (2001). *Probabilistic Risk Assessment for Pesticides in Europe: Implementation & Research Needs*, A report from the European Workshop on Probabilistic Risk Assessment for the Environmental Impacts of Plant Protection Products (EUPRA), The Netherlands, June 2001, published by Central Science Laboratory, Sand Hutton, UK.

Hartley, H.O. (1950). The use of range in analysis of variance. *Biometrika 37*: 271–280.

Haseman, J.K. (1983). Patterns of tumor incidence in two-year cancer bioassay feeding studies in Fischer 344 rats. *Fundamental and Applied Toxicology 3*: 1–9.

Hayes, A.W. (2007). *Principles and Methods of Toxicology*, 5e, 410. New York: Informa HealthCare.

Hazewinkel, M. ed. (2002). *Encyclopaedia of Mathematics*. Berlin, Heidelberg, New York: Springer-Verlag.

Headrick, T.C. and Sawilowsky, S.S. (2000). Properties of the rank transformation in factorial analysis of covariance. *Communication in Statistics – Simulation and Computation 29*: 1059–1087.

Helsel, D.R. (2005). *Nondetects and Data Analysis Statistics for Censored Environmental Data*. Hoboken, NJ: Wiley.

Helsel, D.R. (2009). Summing nondetects: incorporating low-level contaminants in risk assessment. *Integrated Environmental Assessment and Management 6*: 361–366.

Helsel, D.R. (2012). *Statistics for Censored Environmental Data Using Minitab and R*, 2e. Hoboken, NJ: Wiley.

Helsel, D.R. (2017). Practical *Stats Statistics Down to Earth*. http://www.practicalstats.com/nada (accessed 24 April 2017).

Helsel, D.R. and Gilliom, R.J. (1986). Estimation of distributional parameters for censored trace level water quality data. 2. Verification and applications. *Water Resources Research 22*: 147–155.

Helsel D.R. and Hirsch R.M. 2002. *Statistical Methods in Water Research*. United States Geological Survey. Available for free download at http://www.practicalstats.com/nada (accessed 12 February 2018).

Hilbe, J.M. (2007). *Negative Binomial Regression*, 2e. Cambridge: Cambridge University Press.

Hilbe, J.M. (2013). Beta binomial regression. https://works.bepress.com/joseph_hilbe/43/ (accessed 31 August 2017).

Hirji, K.F. and Tang, M.-L. (1998). A comparison of tests for trend. *Communications in Statistics – Theory and Methods 27*: 943–963.

Hoaglin, D., Mosteller, F., and Tukey, J.W. (2000). *Understanding Robust and Exploratory Data Analysis*. New York: Wiley Interscience.

Hobbs, N.T. and Hilborn, R. (2006). Alternatives to statistical hypothesis testing in ecology: a guide to self teaching. *Ecological Applications 16*: 519.

Hochberg, Y. and Tamhane, A.C. (2009). *Multiple Comparison Procedures*. Hoboken, NJ: Wiley.

Hocking, R.R. (1985). *The Analysis of Linear Models*. Monterey, CA: Brooks/Cole.

Hocking, R.R., Bremer, R.H., and Green, J.W. (1989). Variance component estimation with model based diagnostics. *Technometrics 31*: 227–240.

Hoekstra, J.A. (1989). Estimation of the ED50. *Biometrics 45*: 337–338.

Hoekstra, J.A. (1991). Estimation of the LC50: a review. *Environmetrics 2*: 139–152.

Hogg, R.V. and Tanis, E.A. (1988). *Probability and Statistical Inference*, 3e, 424. New York: Macmillan.

Hollander, M. and Wolfe, D.A. (1973). *Nonparametric statistical methods*. New York, Sydney, Tokyo, Mexico City: Wiley.

Holm, S. (1979). A simple sequentially rejective multiple test procedure. *Scandinavian Journal of Statistics 6*: 65–70.

Hooda, B.K., Mishra, K., and Singh, K.P. (2008). A procedure for identification of principal variables by least generalized dependence. *Communications in Statistics – Simulation and Computation 37*: 167–177.

van der Hoop, J. (2013). *Bioamplification, Bioaccumulation and Bioconcentration*. http://mercurypolicy.scripts.mit.edu/blog/?p=499 (accessed 23 April 2017).

Hoppe, F.M. (1993). *Multiple Comparisons, Selection, and Applications in Biometry*. New York: Marcel Dekker.

Hora, S.C. (1993). Acquisition of expert judgment: examples from risk assessment. *Journal of Energy Engineering 118*: 136–148.

Hora, S.C. and Iman, R.L. (1989). Expert opinion in risk analysis: the NEUREG-1150 methodology. *Nuclear Science and Engineering 102*: 323–331.

Horn, M. and Dunnett, C.W. (2004). Power and sample size comparisons of stepwise FWE and FDR controlling test procedures in the normal many-one case. In: *Recent Developments in Multiple Comparison Procedures*, Institute of Mathematical Statistics Lecture Notes – Monograph Series 47, 48–64. Beachwood, OH: Institute of Mathematical Statistics.

Hothorn, L.A. (2014). Statistical evaluation of toxicological bioassays – a review. *Toxicology Research 3*: 418–432.

Hothorn, L.A. (2016). *Statistics in Toxicology Using R*. Boca Raton, FL: CRC Press.

Hothorn, T., Bretz, F., and Westfall, P. (2016). Simultaneous inference in general parametric models. *Biometrical Journal 50*: 346–363.

Hothorn, L. and Hauschke, D. (2000). Identifying the maximum safe dose: a multiple testing approach. *Journal of Biopharmaceutical Statistics 10*: 15–30.

Hougaard, P. (2000). *Analysis of Multivariate Survival Data*. New York: Springer-Verlag.

Howard, G.J. and Webster, T.F. (2009). Generalized concentration addition: a method for examining mixtures containing partial agonists. *Journal of Theoretical Biology 259*: 469–477.

Hsu, J.C. (1996). *Multiple Comparisons Theory and Methods*. Boca Raton, FL: Chapman and Hall/CRC.

Hubert, J.J. (1996). *Environmental Risk Assessment*. Guelph, Canada: Department of Mathematics and Statistics, University of Guelph.

Huet, S., Bouvier, A., Poursat, M.-A., and Jolivet, E. (2004). *Statistical Tools for Nonlinear Regression: A Practical Guide with S-PLUS and R Examples*, 2e. New York: Springer-Verlag.

Hurlbert, S.H. (1984). Pseudoreplication and the design of ecological field experiments. *Ecological Monographs 54*: 187–211.

Hurlbert, S.H. (2004). On misinterpretations of pseudoreplication and related issues: a reply to Oksanen. *Oikos 104*: 591–597.

Hutchinson, T., Shillabeer, N., Winter, M., and Pickford, D. (2006). Acute and chronic effects of carrier solvents in aquatic organisms: a critical review. *Aquatic Toxicology 76* (1): 69–92.

Ibrahim, J.G., Ryan, C., and Chen, M.H. (1998). Using historical controls to adjust for covariates in trend tests for binary data. *Journal of the American Statistical Association 93*: 1282–1293.

Irwin, J.G. (1935). Tests of significance for differences between percentages based on small numbers. *Metron 12*: 83–94.

Isnard, P., Flammarion, P., Roman, G. et al. (2001). Statistical analysis of regulatory ecotoxicity tests. *Chemosphere 45*: 659–669.

ISO 22030:2005 (2005). *Soil Quality – Biological Methods – Chronic Toxicity in Higher Plants*. http://www.iso.org/iso/home/store/catalogue_tc/catalogue_detail.htm?csnumber=36065 (accessed 6 February 2018).

ISO/TS 20281-2006 (2006). *International Standard. Water Quality – Guidance on Statistical Interpretation of Ecotoxicity Data*. www.iso.org (accessed 6 February 2018).

Iwasaki, Y. and Hanson, N. (2014). Using population level consequences as a basis for determining the "x" in ECx for toxicity testing. *Integrated Environmental Assessment and Management 9*: 344–345.

Iwasaki, Y., Kotani, K., Kashiwada, S., and Masunaga, S. (2015). Does the choice of NOEC or EC10 affect the hazardous concentration for 5% of the species? *Environmental Science & Technology 49*: 9326–9330.

Jacobs, M.N., Laws, S.C., Willett, K. et al. (2013). In vitro metabolism and bioavailability tests for endocrine active substances: what is needed next for regulatory purposes? *ALTEX 30* (3): 331–351.

Jager, T., Albert, C., Preuss, T.G., and Ashauer, R. (2011). General unified threshold model of survival – a toxicokinetic-toxicodynamic framework for ecotoxicology. *Environmnetal Science & Technology 45*: 2529–2540.

Jagoe, R.H. and Newman, M.C. (1997). Bootstrap estimation of community NOEC values. *Ecotoxicology 6*: 293–306.

James, W.H. (1998). Re: the use of offspring sex ratios in the search for endocrine disruptors. *Environmental Health Perspectives 106*: A472–A473.

Jennrich, R.L. (1969). Asymptotic properties of nonlinear least squares estimators. *Annals of Mathematical Statistics 40*: 633–643.

Johnson, N.L., Kotz, S., and Balakrishnan, N. (1994). *Continuous Univariate Distributions*, 2e, vol. *1*. Hoboken, NJ: Wiley.

Johnson, N.L., Kotz, S., and Kemp, A.W. (2005). *Univariate Discrete Distributions*, 3e. Hoboken, NJ: Wiley.

Jolliffe, I.T. (2002). *Principal Component Analysis*, 2e. New York: Springer-Verlag.

Jonckheere, A.R. (1954). A distribution-free k-sample test against ordered alternatives. *Biometrika 41*: 133.

Jones, B.S. (2017). *Complications in Event History I: Frailty Models*. http://psfaculty.ucdavis.edu/bsjjones/frailtyslides.pdf (accessed 19 June 2017).

Judson, R.S., Magpantay, F.M., Chickarmane, V. et al. (2015). Integrated model of chemical perturbations of a biological pathway using 18 in vitro high-throughput screening assays for the estrogen receptor. *Toxicological Sciences 148* (1): 137–154. doi: 10.1093/toxsci/kfv168.

Kabacoff, R.I. (2015). *R in Action Data Analysis and Graphics with R*, 2e. Shelter Island, NY: Manning Publications Co.

Kass, R.E. and Raftery, A.E. (1995). Bayes factors. *Journal of the American Statistical Association 90*: 773–795.

Kass, R.E. and Wasserman, L. (1995). A reference Bayesian test for nested hypotheses and its relationship to the Schwarz criterion. *Journal of the American Statistical Association 90*: 928–934.

Kavlock, R.J. and Setzer, R.W. (1996). The road to embryologically based dose: response models. *Environmental Health Perspectives 104*: 107–121.

Kendall, M. and Stuart, A. (1979). *The Advanced Theory of Statistics*, 4e, vol. 2. London: Charles Griffin & Company.

Kennedy, W.J. and Gentle, J.E. (1980). *Statistical Computing*. New York: Dekker.

Kerr, D.R. and Meador, J.P. (1996). Modeling dose response using generalized linear models. *Environmental Toxicology and Chemistry 15*: 395–401.

Khan, A. and Rayner, G.D. (2003). Robustness to non-normality of common tests for the many-sample location problem. *Journal of Applied Mathematics and Decision Sciences 7*: 187–206.

Kienzler, A., Bopp, S.K., van der Linden, S. et al. (2016). Regulatory assessment of chemical mixtures: requirements, current approaches and future perspectives. *Regulatory Toxicology and Pharmacology 80*: 321–334.

Klein, J.P. and Moeschberger, M.L. (2003). *Survival Analysis Techniques for Censored and Truncated Data*, 2e. New York: Springer.

Kleinman, K. and Horton, N.J. (2014). *SAS and R Data Management, Statistical Analysis, and Graphics*, 2e. Boca Raton, FL: CRC Press, Chapman & Hall.

Klepper, O. and Bedaux, J.J.M. (1997). Nonlinear parameter estimation for toxicological threshold models. *Ecological Modelling 102*: 315–324.

Klimisch, H.J., Andreae, M., and Tillmann, U. (1997). A systematic approach for evaluating the quality of experimental toxicological and ecotoxicological data. *Regulatory Toxicology and Pharmacology 25* (1): 1–5. doi: 10.1006/rtph.1996.1076.

Klockars, A.J. and Moses, T.P. (2002). Type I error rates for rank-based tests of homogeneity of slopes. *Journal of Modern Applied Statistical Methods 1*: 452–460.

Koenker, R. (1994). Confidence intervals for regression quantiles. In: *Asymptotic Statistics: Proceedings of the Fifth Prague Symposium, Held from 4–9 September 1993* (ed. P. Mandl and M. Huskova), 349–359. Berlin: Springer-Verlag.

Koenker, R. and Bassett, G.W. (1978). Regression quantiles. *Econometrica 46*: 33–50.

Koenker, R. and d'Orey, V. (1993). Computing regression quantiles. *Applied Statistics 43*: 410–414.

Koenker, R. and Hallock, K. (2001). Quantile regression: an introduction. *Journal of Economic Perspectives 15*: 143–156.

Koenker, R. and Machado, A.F. (1999). Goodness of fit and related inference processes for quantile regression. *Journal of the American Statistical Association 94*: 1296–1310.

Kon Kam King G., Veber P., Charles S., Delignette-Muller M.L.. 2014. MOSAIC_SSD: a new web-tool for species sensitivity distribution to include censored data by maximum likelihood. *Environmental Toxicology and Chemistry 33*: 2133–2139. The software is located at http://pbil.univ-lyon1.fr/software/mosaic/ssd (accessed 12 February 2018).

Kong, M. and Lee, J.J. (2006). A generalized response surface model with varying relative potency for assessing drug interaction. *Biometrics 62*: 986–995.

Kooijman, S.A.L.M. (1987). A safety factor for LC50 values allowing for differences in sensitivity among species. *Water Research 21*: 269–276.

Kooijman, S.A.L.M. and Bedaux, J.J.M. (1996). *The Analysis of Aquatic Toxicity Data*. Amsterdam: VU University Press.

Korn, E.L. (1982). Confidence bands for isotonic dose-response curves. *Applied Statistics 31*: 59–63.

Kortenkamp, A., Backhaus, T., and Faust, M. (2009). *State of the Art on Mixture Toxicity*. Final report prepared on behalf of the European Union.

Lammer, E., Carr, G.J., Wendler, K. et al. (2009). Is the fish embryo toxicity test (FET) with the zebrafish (*Danio rerio*) a potential alternative for the fish acute toxicity test? *Comparative Biochemistry and Physiology, Part C 149*: 196–209.

Landis, W.G. and Chapman, P.M. (2011). Well past time to stop using NOELs and LOELs. *Integrated Environmental Assessment and Management 7*: vi–viii.

Lange, K. (2012). *Numerical Analysis for Statisticians*, 2e. New York: Springer–Verlag.

Latscha, R. (1955). Tests of significance in a 2 × 2 contingency table: extension of Finney's table. *Biometrika 40*: 74–86.

Lazic, S.E. (2017). *Experimental Design for Laboratory Biologists*. Cambridge: Cambridge University Press.

Lee, E.T. and Wang, J.W. (2013). *Statistical Methods for Survival Data Analysis*, 4e. Hoboken, NJ: Wiley.

Lee E.W., Wei L.J., and Amato D.A. 1992. Cox-type regression analysis for large numbers of small groups of correlated failure time observations. In *Survival Analysis: State of the Art*. Dordrecht, The Netherlands: Kluwer Academic Publishers.

Lehmann, E.L. (2006). *Nonparametrics: Statistical Methods Based on Ranks*, revised edition. New York: Springer-Verlag.

Lehmann, R., Bachmann, J., Maletzki, D. et al. (2016). A new approach to overcome shortcomings with multiple testing of reproduction data in ecotoxicology. *Stochastic Environmental Research and Risk Assessment 30*: 871–882.

Levenberg, K. (1944). A method for the solution of certain non-linear problems in least squares. *Quarterly of Applied Mathematics 2*: 164–168.

Li, J., Nordheim, E.V., Zhang, C., and Lehner, C.E. (2008). Estimation and confidence regions for multi-dimensional effective dose. *Biometrical Journal 50*: 110–122.

Lloyd, R. (1987). Special tests in aquatic toxicity for chemical mixtures: interactions and modifications of response by variation of physiochemical conditions. In: *Methods for Assessing the Effects of Mixtures of Chemicals*, SCOPE Report 30. WHO IRIS (ed. V.B. Vouk, G.C. Butler, A.C. Upton, et al.). New York: Wiley.

Lorenzen, T.J. and Anderson, V.L. (1993). *Design of Experiments: A No-Name Approach*. New York: Marcel Dekker.

Lyles, R.H., Poindexter, C., Evans, A. et al. (2008). Nonlinear model-based estimates of IC_{50} for studies involving continuous therapeutic dose-response data. *Contemporary Clinical Trials 29*: 878–886.

Mallows, C.L. (1973). Some comments on CP. *Technometrics 15* (4): 661–675.

Mansouri, H. and Chang, G.-H. (1995). A comparative study of some rank tests for interaction. *Computational Statistics and Data Analysis 19*: 85–96.

Mantel, N. (1963). Part IV. The concept of threshold in carcinogenesis. *Clinical Pharmacology and Therapeutics 4*: 104–109.

Marcus, R. (1976). The powers of some tests of the equality of normal means against an ordered alternative. *Biometrics 63*: 177–183.

Marcus, R., Peritz, E., and Gabriel, K.R. (1976). On closed testing procedures with special reference to ordered analysis of variance. *Biometrika 63*: 655–660.

Marquardt, D. (1963). An algorithm for least-squares estimation of nonlinear parameters. *SIAM Journal on Applied Mathematics 11*: 431–441.

Martin, B.T., Zimmer, E.I., Grimm, V., and Jager, T. (2012). Dynamic energy budget theory meets individual-based modelling: a generic and accessible implementation. *Methods in Ecology and Evolution 3*: 445–449.

Martinez E.Z., Achcar J.A., and Arago D.C. 2015. Parameter estimation of the beta-binomial distribution: an application using the SAS software. *Ciência e Natura, Santa Maria 37*: 12–19. (Used in SAS program of beta-binomial).

McArdle, B.H. (1988). The structural relationship: regression in biology. *Canadian Journal of Zoology 66* (11): 2329–2339.

McArdle, B.H. (2003). Lines, models and errors: regression in the field. *Limnology and Oceanography 48* (3): 1363–1366.

McCullagh, P. and Nelder, J.A. (1983). *Generalized Linear Models*. London: Chapman and Hall.

McCulloch, C.E., Searle, S.R., and Neuhaus, J.M. (2008). *Generalized, Linear, and Mixed Models*, 2e. Hoboken, NJ: Wiley.

Mead, R., Curnow, R.N., and Hasted, A.M. (2002). *Statistical Methods in Agriculture and Experimental Biology*, 3e. London: Chapman and Hall.

Mebane, C. (2012). Biological arguments for selecting ECx values in chronic aquatic toxicity testing. Presentation to Society of Environmental Toxicology and Chemistry North America. 33rd Annual Conference, Long Beach, CA (15 November 2012).

Mebane, C.A. (2015). biological arguments for selecting effect sizes in ecotoxicological testing – A governmental perspective. *Environmental Toxicology and Chemistry 34*: 2440–2442.

Merrington, G. and Schoeters, I. (2010). *Soil Quality Standards for Trace Elements: Derivation, Implementation, and Interpretation*. Boca Raton, FL: Taylor & Francis, CRC Press.

Meyer, M.A. and Booker, J.M. (2001). *Eliciting and Analyzing Expert Judgment: A Practical Guide*. Philadelphia: Society for Industrial and Applied Mathematics.

Mi, J. (2010). Estimation of quantities determined as implicit functions of unknown parameters. *Metrika 71*: 353–359.

Miller, R.J. (1981). *Simultaneous Statistical Inference*, 2e. New York: Springer-Verlag.

Miller, G.A. and Chapman, J.P. (2001). Misunderstanding analysis of covariance. *Journal of Abnormal Psychology 110*: 40–48.

Miller, R.G. Jr. (1968). Jackknifing variances. *Annals of Mathematical Statistics 39*: 567–582.

MITI (2012). Website of the Ministry of Economy, Trade and Industry of Japan http://www.meti.go.jp/english/other/terms_of_use.html (accessed 6 February 2018).

Moermond, C.T.A., Kase, R., Korkaric, M., and Agerstrandk, M. (2016). CRED: criteria for reporting and evaluating ecotoxicity data. *Environmental Toxicology and Chemistry 35* (5): 1297–1309. doi: 10.1002/etc.3259.

Monte-Silva, K., Kuo, M.-F., Thirugnanasambandam, N. et al. (2009). Dose-dependent inverted U-shaped effect of dopamine (D2-like) receptor activation on focal and nonfocal plasticity in humans. *The Journal of Neuroscience 29*: 6124–6131.

Montgomery, D.C., Peck, E.A., and Vining, G.G. (2013). *Introduction to Linear Regression Analysis*, 5e. Hoboken, NJ: Wiley.

Moon, H., Ahn, H., and Kodell, R.L. (2006). A computational tool for testing dose-related trend using an age-adjusted bootstrap-based poly-k test. *Journal of Statistical Software 16*: doi: 10.18637/jss.v016.i07. https://www.jstatsoft.org/article/view/v016i07 (accessed 12 February 2018).

Moore, D.R.J. and Caux, P.-Y. (1997). Estimating low toxic effects. *Environmental Toxicology and Chemistry 16*: 794–801.

Moré, J.J. and Sorensen, D.C. (1983). Computing a trust-region step. *SIAM Journal on Scientific and Statistical Computing 4*: 553–572.

Morel, J.G. and Neerchal, N. (2012). *Overdispersion Models in SAS*. Cary, NC: SAS Institute, Inc.

Morgan, B.J.T. (1992). *Analysis of Quantal Response Data*. London: Chapman and Hall.

Morgan, M.G. and Henrion, M. (1992). *Uncertainty: A Guide to Dealing with Uncertainty in Quantitative Risk and Policy Analysis*. Cambridge: Cambridge University Press.

Morrison, D.F. (1983). *Applied Linear Statistical Methods*. Englewood Cliffs, NJ: Prentice Hall.

Morrison, D.F. (2004). *Multivariate Statistical Methods*, 4e. Belmont, CA: Duxbury Press.

Mosimann, J.E. (1962). On the compound multinomial distribution, the multivariate β-distribution, and correlations among proportions. *Biometrika 49*: 65–78.

Motulsky, H. and Christopoulos, A. (2004). *Fitting Models to Biological Data Using Linear and Nonlinear Regression: A Practical Guide to Curve Fitting*. Oxford: Oxford University Press.

Murie, C., Barette, C., Button, J. et al. (2014). Improving detection of rare biological events in high-throughput screens. *Journal of Biomolecular Screening 20*: 230–241.

Mylchreest, E. and Harris, S.B. (2013). Historical control data in reproductive and developmental toxicity studies. *Methods in Molecular Biology 947*: 275–294.

Neath, A.A. and Cavanaugh, J.E. (2012, 2012). The Bayesian information criterion: background, derivation, and applications. *WIREs Computational Statistics 4*: 199–203. doi: 10.1002/wics.199.

Nelder, J.A. and Wedderburn, R.W.M. (1972). Generalized linear models. *Journal of the Royal Statistical Society, Series A 135*: 370–384.

Nelson K.P., Lipsitz S.R., Fitzmaurice G.M., Ibrahim J., Parzen M., Strawderman R. 2006. Use of the probability integral transformation to fit nonlinear mixed-effects models with nonnormal random effects, *Journal of Computational and Graphical Statistics 15*: 39–57. (Used in SAS program of beta-binomial).

Newman, M.C. (1995). *Quantitative Methods in Aquatic Ecotoxicology*. Boca Raton, FL: CRC Press, Lewis Publishers. ASIN: B008T1D9C8.

Newman, M.C. (2008). "What exactly are you inferring?" A closer look at hypothesis testing. *Environmental Toxicology and Chemistry 27*: 1013–1019.

Newman, M.C. and McCloskey, J.T. (2000). The individual tolerance concept is not the sole explanation for the probit dose-effect model. *Environmental Toxicology and Chemistry 19*: 520–526.

Newman, M.C., Ownby, D.R., Mézin, L.C.A. et al. (2000). Applying species sensitivity distributions in ecological risk assessment, assumptions of distribution type and sufficient number of species. *Environmental Toxicology and Chemistry 19*: 508–515.

Nielsen, O.K., Ritz, C., and Streibig, J.C. (2004). Nonlinear mixed-model regression to analyze herbicide dose-response relationships. *Weed Technology 18*: 30–37.

Nieuwkoop, P.D. and Faber, J. (1994). *Normal Table of Xenopus laevis (Daudin)*. New York: Garland Publishing, Inc.

Nowicki, S. (2004). *Biology: The Science of Life*. Chantilly, VA: The Teaching Company.

Nyholm, N., Sorensen, P.S., Kusk, K.O., and Christensen, E.R. (1992). Statistical treatment of data from microbial toxicity tests. *Environmental Toxicology and Chemistry 11* (2): 157–167. doi: 10.1897/1552-8618(1992)11[157:stodfm]2.0.co;2.

Nyman, A.M., Schirmer, K., and Ashauer, R. (2012). Toxicokinetic-toxicodynamic modelling of survival of *Gammarus pulex* in multiple pulse exposures to propiconazole: model assumptions, calibration data requirements and predictive power. *Ecotoxicology 21*: 1828–1840.

O'Brien, R.G. (1981). A simple test for variance effects in experimental designs. *Psychological Bulletin 89*: 570–574.

OCSPP (2016). OCSPP Harmonized Test Guidelines – Master List. https://www.epa.gov/sites/production/files/2016-12/documents/ocspp-testguidelines_masterlist-2016-12-28.pdf (accessed 6 February 2018).

Odeh, R.E. (1971). On Jonckheere's *k*-sample test against ordered alternatives. *Technometrics 13*: 912–918.

Odeh, R.E. (1972). On the power of Jonckheere' *k*-sample test against ordered alternatives. *Biometrika 59*: 467–471.

OECD (1984a). *Test No. 207: Earthworm, Acute Toxicity Tests*. Paris: OECD Publishing. http://dx.doi.org/10.1787/9789264070042-en (accessed 6 February 2018).

OECD (1984b). *TG 204 Fish, Prolonged Toxicity Test: 14-Day Study*. OECD Guideline for Testing of Chemicals. http://www.oecd-ilibrary.org/environment/test-no-204-fish-prolonged-toxicity-test-14-day-study_9789264069985-en (accessed 12 February 2018).

OECD (1984c). *TG 206 Avian Reproduction*. OECD Guideline for Testing of Chemicals. http://www.oecd-ilibrary.org/environment/test-no-206-avian-reproduction-test_9789264070028-en (accessed 12 February 2018).

OECD (1992). *Test No. 203: Fish, Acute Toxicity Test*. Paris: OECD Publishing. http://dx.doi.org/10.1787/9789264069961-en (accessed 12 February 2018).

OECD (1997). *Test No. 476: In Vitro Mammalian Cell Gene Mutation Test*. Paris: OECD Publishing.

OECD (2000). *Guidance Document on Aquatic Toxicity Testing of Difficult Substances and Mixtures*, OECD Environmental Health & Safety Publication, Series on Testing & Assessment No. 23, 53. Paris: Organisation for Economic Cooperation & Development (OECD).

OECD (2004). *Test No. 202: Daphnia sp. Acute Immobilisation Test*: OECD Publishing, Paris. http://dx.doi.org/10.1787/9789264069947-en (accessed 12 February 2018).

OECD (2006a). *Current Approaches in the Statistical Analysis of Ecotoxicity Data: A Guidance to Application*, OECD Series on Testing and Assessment, Number 54, ENV/JM/MONO(2006)18. Paris: Environment Directorate, Organisation for Economic Co-Operation and Development.

OECD (2006b). Current Approaches in the Statistical Analysis of Ecotoxicity Data: A Guidance to Application – Annexes. http://www.oecd.org/officialdocuments/publicdisplaydocumentpdf/?cote=ENV/JM/MONO(2006)18/ANN&docLanguage=En (accessed 6 February 2018).

OECD (2006c). *TG 208 Terrestrial Plant Test: Seedling Emergence and Seedling Growth Test*. OECD Guideline for Testing of Chemicals. http://www.oecd-ilibrary.org/content/book/9789264070066-en (accessed 19 February 2018).

OECD (2010a). Report of the OECD Workshop on Statistical Analysis of Aquatic Toxicity Data OECD Series on Testing and Assessment Number 10. http://www.oecd.org/officialdocuments/publicdisplaydocumentpdf/?doclanguage=en&cote=env/mc/chem(98)18 (accessed 6 February 2018).

OECD (2010b). *Cutting Costs in Chemicals Management: How OECD Helps Governments and Industry*. http://www.oecd.org/env/ehs/47813784.pdf (accessed 6 September 2017).

OECD (2011). *Test No. 201: Freshwater Alga and Cyanobacteria, Growth Inhibition Test*. Paris: OECD Publishing. http://dx.doi.org/10.1787/9789264069923-en (accessed 6 September 2017).

OECD (2012a). *Guidance Document 150 on Standardised Test Guidelines for Evaluating Chemicals for Endocrine Disruption*, Series on Testing and Assessment No. 150ENV/JM/MONO(2012)22. Paris: OECD Publishing.

OECD (2012b). *Test No. 229: Fish Short Term Reproduction Assay*. Paris: OECD Publishing. http://dx.doi.org/10.1787/9789264185265-en (accessed 12 February 2018).

OECD (2013). *Test No. 210: Fish, Early-Life Stage Toxicity Test*. Paris: OECD Publishing. http://dx.doi.org/10.1787/9789264203785-en (accessed 12 February 2018).

OECD (2014a). *Current Approaches in the Statistical Analysis of Ecotoxicity Data: A Guidance to application* (annexes to this publication exist as a separate document). OECD Publishing, Paris. http://dx.doi.org/10.1787/9789264085275-en (accessed 12 February 2018).

OECD (2014b). *Guidance Document 116 on the Conduct and Design of Chronic Toxicity and Carcinogenicity Studies, Supporting Test Guidelines 451, 452 and 453*, 2e. Paris: OECD Publishing. http://dx.doi.org/10.1787/9789264221475-en (accessed 12 February 2018).

OECD (2014c). *Fish Toxicity Testing Framework*. Paris: OECD Publishing. http://dx.doi.org/10.1787/9789264221437-en (accessed 12 February 2018).

OECD (2015). *Test No. 240: Medaka Extended One Generation Reproduction Test (MEOGRT)*. Paris: OECD Publishing. http://dx.doi.org/10.1787/9789264242258-en (accessed 12 February 2018).

OECD (2016a). *Test No. 431: In Vitro Skin Corrosion: Reconstructed Human Epidermis (RHE) Test Method*. Paris: OECD Publishing. http://dx.doi.org/10.1787/9789264264618-en (accessed 12 February 2018).

OECD (2016b). *Test No. 476: In Vitro Mammalian Cell Gene Mutation Tests Using the Hprt and Xprt Genes*. Paris: OECD Publishing. http://dx.doi.org/10.1787/9789264264809-en (accessed 12 February 2018).

OECD (2016c). *Test No. 487: In Vitro Mammalian Cell Micronucleus Test*. Paris: OECD Publishing. http://dx.doi.org/10.1787/9789264264861-en (accessed 12 February 2018).

OECD (2017). Guidance Document on Aspects of OECD TG 305 on Fish Accumulation http://www.oecd.org/officialdocuments/publicdisplaydocumentpdf/?cote=ENV/JM/MONO(2017)16&doclanguage=en (accessed 6 February 2018).

Ohara Hines, R.J. (1997). A comparison of tests for overdispersion in generalized linear models. *Journal of Statistical Computation and Simulation 58*: 323–342.

Ohara Hines, R.J. and Lawless, J.F. (1993). Modeling overdispersion in toxicological mortality data grouped over time. *Biometrics 49*: 107–121.

Oksanen, L. (2001). Logic of experiments in ecology: is "pseudoreplication" a pseudoissue? *Oikos 94*: 27–28.

Oksanen, L. (2004). The devil lies in the details: reply to Stuart Hurlbert. *Oikos 104*: 598–605.

Oris, J.T., Belanger, S.E., and Bailer, A.J. (2012). Baseline characteristics and statistical implications for the OECD 210 fish early-life stage chronic toxicity test. *Environmental Toxicology and Chemistry 31*: 370–376.

Ortiz, N.R., Wheeler, M.A., Keeney, R.L. et al. (1991). Use of expert judgment in NEUREG-1150. *Nuclear Engineering and Design 126*: 313–331.

Ott, R.L. and Longnecker, M.T. (2015). *An Introduction to Statistical Methods and Data Analysis*, 7e. Belmont, CA: Brooks-Cole.

Pal, M. (1980). Consistent moment estimators of regression coefficients in the presence of errors in variables. *Journal of Econometrics 14*: 349–364.

Parzen, E. (1992). *Modern Probability Theory and Its Applications*. New York: Wiley Classics Library.

Parzen, E. and Karlin, S. (2013). *Stochastic Processes: Holden-Day*. San Francisco: Literary Licensing, LLC.

Peddada, S.D. and Kissling, E. (2006). A survival-adjusted quantal-response test for analysis of tumor incidence rates in animal carcinogenicity studies. *Environmental Health Perspectives 114*: 537–541.

Peltier, W.H. and Weber, C.I. (1985). *Methods for Measuring the Acute Toxicity of Effluents to Freshwater and Marine Organisms* (Appendix E), 170–181 (March). Cincinnati, OH: U.S. Environmental Monitoring and Support Laboratory.

Petkau, A.J. and Sitter, R. (1989). Models for quantal response experiments over time. *Biometrics 45*: 1299–1308.

Peto, R., Pike, M.C., Day, N.E. et al. (1980). Guidelines for simple, sensitive significance tests for carcinogenic effects in long-term animal experiments. In: *Suppl. 2: Long-Term and Short-Term Screening Assays for Carcinogens – A Critical Appraisal*, IARC Monographs on the Evaluation of Carcinogenic Risks to Humans. Lyon: International Agency for Research on Cancer.

Petterino, C. and Argentino-Storino, A. (2006). Clinical chemistry and haematology historical data in control Sprague–Dawley rats from pre-clinical toxicity studies. *Experimental and Toxicologic Pathology 57*: 213–219.

Pinheiro, J.C. and Bates, D.M. (1995). Approximations to the log-likelihood function in the nonlinear mixed-effects model. *Journal of Computational and Graphical Statistics 4*: 12–35.

Plummer, J.L. and Short, T.G. (1990). Statistical modeling of the effects of drug combinations. *Journal of Pharmacological Methods 23*: 297–309.

Poon, A.H. (1980). A Monte-Carlo study of the power of some k-sample tests for ordered binomial alternatives. *Journal of Statistical Computation and Simulation 11*: 251–259.

Portier, C.J. and Bailer, A.J. (1989). Testing for increased carcinogenicity using a survival-adjusted quantal response test. *Fundamental and Applied Toxicology 12*: 731–737.

Portier, C., Hedges, J., and Hoel, D. (1986). Age-specific models of mortality and tumor onset for historical control animals in the National Toxicology Program's carcinogenicity experiments. *Cancer Research 46*: 4372–4378.

Posada, D. and Buckley, T.R. (2004). Model selection and model averaging in phylogenetics: advantages of Akaike information criterion and Bayesian approaches over likelihood ratio tests. *Systematic Biology 53*: 793–808.

Posthuma, L., Dyer, S.D., de Zwart, D. et al. (2016). Eco-epidemiology of aquatic ecosystems: separating chemicals from multiple stressors. *Science of the Total Environment 573*: 1303–1319.

Posthuma, L., Sutter, G.W. II, and Traas, T.P. ed. (2002). *Species Sensitivity Distributions in Ecotoxicology*. Boca Raton, FL: Lewis Publishers.

Potter, R.W. and Sturm, G.W. (1981). The power of Jonckheere's test. *The American Statistician 35*: 249–250.

Pratheeshkumar, P., Budhraja, A., Son, Y.-O. et al. (2012). Quercetin inhibits angiogenesis mediated human prostate tumor growth by targeting VEGFR-2 regulated AKT/mTOR/P70S6K signaling pathways. *PLoS One 7* (10): e47516. doi: 10.1371/journal.pone.0047516.

Purchase, I.F.H. and Auton, T.R. (1995). Thresholds in chemical carcinogenesis. *Regulatory Toxicology and Pharmacology 22*: 199–205.

Puri, M.L. (1965). Some distribution-free k-sample rank tests of homogeneity against ordered alternatives. *Communications on Pure and Applied Mathematics 18*: 51–63.

Rao, J.N.K. and Scott, A.J. (1992). A simple method for the analysis of clustered binary data. *Biometrics 48*: 577–585.

Rao, J.N.K. and Scott, A.J. (1999). A simple method for analyzing overdispersion in clustered Poisson data. *Statistics in Medicine 18*: 1373–1385.

Rasch, D., Kubinger, K.D., and Moder, K. (2011). The two-sample t test: pretesting its assumptions does not pay. *Statistical Papers 52*: 219–231.

Ratkowsky, D.A. (1993). Principles of nonlinear regression. *Journal of Industrial Microbiology 12*: 195–199.

R-Blogger (2017a). predictNLS (Part 1, Monte Carlo simulation): confidence intervals for "nls". https://www.r-bloggers.com/predictnls-part-1-monte-carlo-simulation-confidence-intervals-for-nls-models/ (accessed 12 February 2018).

R-Blogger (2017b). predictNLS (Part 2, Taylor approximation): confidence intervals for "nls". https://www.r-bloggers.com/predictnls-part-2-taylor-approximation-confidence-intervals-for-nls-models/ (accessed 12 February 2018).

Rencher, A.C. and Christensen, W.F. (2012). *Methods of Multivariate Analysis*, 3e. Hoboken, NJ: Wiley.

Ritz, C., Baty, F., Streibig, J.C., and Gerhard, D. (2015). Dose-response analysis using R. *PLoS One 10* (12): e0146021. https://doi.org/10.1371/journal.pone.0146021 (accessed 12 February 2018).

Robertson, T., Wright, F.T., and Dykstra, R.L. (1988). *Order Restricted Statistical Inference*. Chichester: Wiley.

Rochon, J., Gondan, M., and Kieser, M. (2012). To test or not to test: preliminary assessment of normality when comparing two independent samples. *BMC Medical Research Methodology 12*: 81.

Rodney, S.I., Moore, D., and Teed, R.S. (2008). *SSD Master version 2*. Intrinsik Environmental Sciences. www.intrinsik.com (accessed 12 February 2018).

Rudin, W. (1974). *Principles of Mathematical Analysis*, 3e. New York: McGraw-Hill Science/Engineering/Math.

Rufli, H. and Springer, T.A. (2011). Can we reduce the number of fish in the OECD acute toxicity test? *Environmental Toxicology and Chemistry 30*: 1006–1011.

Ryan, L. (1993). Using historical controls in the analysis of developmental toxicity data. *Biometrics 49*: 1126–1135.

SAICM (2017). International Chemicals Management, SAICM overview. http://www.saicm.org/About/SAICMOverview/tabid/5522/language/en-US/Default.aspx (accessed 6 February 2018).

SAS Institute Inc. (2011). *SAS® 9.2 Language Reference: Dictionary*, 4e. Cary, NC: SAS Institute Inc.

SAS Institute Inc. (2015). *SAS/STAT® 14.1 User's Guide*. Cary, NC: SAS Institute Inc.

Schabenberger, O., Tharp, B.E., Kells, J.J., and Penner, D. (1999). Statistical tests for hormesis and effects dosages in herbicide dose response. *Agronomy Journal 91*: 713–721.

Scheffe, H. (1959). *The Analysis of Variance*. New York: Wiley.

Schervish, M.J. (1984). Algorithm AS 195: multivariate normal probabilities with error bound. *Journal of the Royal Statistical Society. Series C, Applied Statistics 33*: 81–94.

Schlain, B., Jethwa, H., Subramanyam, M. et al. (2001). Designs for bioassays with plate location effects. *BioPharm 14*: 40–44.

Schmidt, K., Schmidtke, J., Kohl, C. et al. (2016). Enhancing the interpretation of statistical *P* values in toxicology studies: implementation of linear mixed models (LMMs) and standardized effect sizes (SESs). *Archives of Toxicology 90*: 731–751.

Schmolke, A., Thorbek, P., DeAngelis, D.L., and Grimm, V. (2010). Ecological models supporting environmental decision making: a strategy for the future. *Trends in Ecology & Evolution 25*: 479–486.

Schneider K., Schwarz M., Burkholder I., Kopp-Schneider A., Edler L., Kinsner-Ovaskainen A., Hartung T., Hoffmann S. 2009. "ToxRTool," a new tool to assess the reliability of toxicological data. *Toxicology Letters*, *189*(2), 138–144. doi:10.1016/j.toxlet.2009.05.013.

Schoenfeld, D. (1986). Confidence intervals for normal means under order restrictions, with applications to dose-response curves, toxicology experiments and low dose extrapolation. *Journal of the American Statistical Association 81*: 186–195.

Schwarz, C.J. (2011). An assessment of the effect of hardness on the dose-response curves to sulphates through the use of model averaging. http://www.env.gov.bc.ca/wat/wq/homesubs/appendices/appendix_e.pdf (accessed 9 October 2017).

Schwarz, G. (1978). Estimating the dimension of a model. *The Annals of Statistics 6*: 461–464.

Scott, B.R. (2008). Low-dose radiation risk extrapolation fallacy associated with the linear-no-threshold model. *Human & Experimental Toxicology 27*: 163–168.

Searle, S.R. (1987). *Linear Models for Unbalanced Data*. New York: Wiley.

Searle, S.R., Casella, G., and McCulloch, C.E. (2007). *Variance Components*. New York: Wiley.

Seber, G.A.F. and Wild, C.J. (2003). *Nonlinear Regression*. Hoboken, NJ: Wiley.

SEDEC (2012). SEquential DEsign Calculator: software for avian acute oral toxicity test. https://urldefense.proofpoint.com/v2/url?u=http-3A__www.oecd.org_env_ehs_testing_softwaretobeusedwithtg223.htm&d=DgIGaQ&c=zRqMG_fghhK--2M6Q5UUdA&r=U3Vbkkld qivnj1oR4teCMmGVqxksRpB1JLTA8uhjRhE&m=RGGXXsQQG_-kIAcg5QQ_IxA1ixSOB_M1oXnpg2DP9yM&s=xosdDRs3FZycTxa RfHvB6aR9Jr5or8DUtJKMr-sZ1eQ&e (accessed 12 February 2018).

Seefeldt, S.S., Jensen, J.E., and Fuerst, E.P. (1995). Log-logistic analysis of herbicide dose-response relationships. *Weed Technology 9*: 218–227.

Selwyn, M.R. (1988). Preclinical safety assessment. In: *Biopharmaceutical Statistics for Drug Development* (ed. K.E. Peace), 231–272. New York: Marcel Dekker.

Serfling, R.J. (2008). *Approximation Theorems of Mathematical Statistics*. New York: Wiley.

Shao, Q. (2000). Estimation for hazardous concentrations based on NOEC toxicity data: an alternative approach. *Environmetrics 11*: 583–595.

Sheskin, D.J. (2011). *Handbook of Parametric and Nonparametric Statistical Procedures*, 5e. Boca Raton, FL: CRC Press.

Shirley, E.A. (1979). The comparison of treatment to control group means in toxicology studies. *Applied Statistics 28*: 144–151.

Shirley, E.A.C. (1981). A distribution-free method for analysis of covariance based on ranked data. *Journal of the Royal Statistical Society. Series C, Applied Statistics 30*: 158–162.

Shirley, E.A.C. and Newnham, P. (1984). The choice between analysis of variance and analysis of covariance with special reference to the analysis of organ weights in toxicology studies. *Statistics in Medicine 3*: 85–91.

Slob, W. (2002). Dose-response modelling of continuous endpoints. *Toxicological Sciences 66*: 298–312.

Slob, W. (2003). *PROAST. A General Software Tool for Dose-Response Modelling*. Bilthoven: RIVM.

Slob, W. and Setzer, R.W. (2014). Shape and steepness of toxicological dose-response relationships of continuous endpoints. *Critical Reviews in Toxicology 44*: 270–297.

Šmilauer, P. and Lepš, J. (2015). *Multivariate Analysis of Ecological Data Using CANOCO 5*, 2e. Cambridge: Cambridge University Press.

Snedecor, G.W. and Cochran, W.G. (1989). *Statistical Methods*, 8e. Ames, IA: Iowa State University Press.

Sroka C., Lee J., Kapat P., Zhang X. 2006. A brief overview of Bayesian model averaging. http://www.stat.osu.edu/~yklee/882/882group4.pdf (accessed 6 April 2017).

StatCHARRMS 2017. https://cran.r-project.org/web/packages/StatCharrms/StatCharrms.pdf (accessed 12 February 2018).

Staveley, J.P., Green, J.W., Nusz. J., Edwards, D., Henry, K., Kern, M., Deines, A.M., Brain, R., Glenn, B., Ehresman, N., Kung, T., Ralston-Hooper, K., Kee, F., and McMaster, S. (2018). Variability in Non-Target Terrestrial Plant Studies Should Inform Endpoint Selection. *IEAM* (to appear).

Stebbing, A.R.D. (1982). Hormesis – the stimulation of growth by low levels of inhibitors. *Science of the Total Environment 22*: 213–234.

Stephan, C.E. (1972). Methods for calculating an LC50. In: *Aquatic Toxicology and Hazard Evaluation. First Annual Symposium on Aquatic Toxicology* (ed. F.L. Mayer and J.L. Hammerlink), 65–84. Philadelphia: ASTM.

Stephan C., Mount D., Hansen D.J., Gentile J.H., Chapman G.A., Brungs W.A. 1985. *Guidelines for Deriving Numerical National Water Quality Criteria for the Protection of Aquatic Organisms and Their Uses*, PB85-227049. Springfield, VA: U.S. Environmental Protection Agency, National Technical Information Service.

Stephenson, W.R. and Jacobson, D. (1988). A comparison of non-parametric analysis of covariance techniques. *Communications in Statistics – Simulations 17*: 451–461.

Stevens, M.T. (1976). The value of relative organ weights. *Toxicology 5*: 311–318.

Stoline, M.R. and Ury, H.K. (1979). Tables of the studentized maximum modulus distribution and an application to multiple comparisons among means. *Technometrics 21*: 87–93.

Straetemans, R., O'Brien, T., Wouters, L. et al. (2005). Design and analysis of drug combination experiments. *Biometrical Journal 47*: 299–308.

Swintek, J. (2016). RSCABS for R. https://cran.r-project.org/web/packages/RSCABS/RSCABS.pdf (accessed 27 July 2017).

Takizawa, T. (1978). An unbiased comparison of organ weights when an inequality in body weight exists. *Toxicology 9*: 353–360.

Tamhane, A.C. (1979). A comparison of procedures for multiple comparison of means with unequal variances. *Journal of the American Statistical Association 74*: 471–480.

Tamhane, A.C. (2009). *Statistical Analysis of Designed Experiments: Theory and Applications*. New York: Wiley.

Tamhane, A.C. and Dunlop, D.D. (1999). *Statistics and Data Analysis: From Elementary to Intermediate*. Upper Saddle River, N J: Prentice Hall.

Tamhane, A.C. and Dunnett, C.W. (1996). Multiple test procedures for dose finding. *Biometrics 52*: 21–37.

Tamhane, A.C., Dunnett, C.W., Green, J.W., and Weatherington, J.D. (2001). Multiple test procedures for identifying the maximum safe dose. *Journal of the American Statistical Association 96*: 835–843.

Tamhane, A.C. and Logan, B.R. (2004). Finding the maximum safe dose level for heteroscedastic data. *Journal of Biopharmaceutical Statistics 14*: 843–856.

Tarone, R.E. (1979). Testing the goodness of fit of the binomial distribution. *Biometrika 66*: 585–590.

Tarone, R.E. (1982). The use of historical control information in testing for a trend in proportions. *Biometrics 38*: 215–220.

Tarone, R.E., Chu, K.C., and Ward, J.M. (1981). Variability in the rates of some common naturally occurring tumors in Fischer 334 rats and (C57BL/6M X C3H/HeN)F1 (B6C3 F1) mice. *Journal of the National Cancer Institute 66*: 1175–1181.

Tarone, R.E. and Gart, J.J. (1980). On the robustness of combined tests for trends in proportions. *Journal of the American Statistical Association 75*: 110–116.

Ter Braak C.J.F. 1988. *CANOCO – a FORTRAN Program for Canonical Community Ordination by [Partial] [Etrended] [Canonical] Correspondence Analysis, Principal Components Analysis and Redundancy Analysis (Version 2.1)*. http://library.wur.nl/WebQuery/wurpubs/436745 (accessed 29 August 2017).

Ter Braak C.J.F. and Smilauer P. 2002. *CANOCO Reference Manual and CanoDraw for Windows User's Guide: Software for Canonical Community Ordination (Version 4.5)*. http://agris.fao.org/agris-search/search.do?recordID=NL2012058438 (accessed 29 August 2017).

Therneau, T.M. and Grambsch, P.M. (2000). *Modeling Survival Data: Extending the Cox Model*. New York: Springer-Verlag.

Thiébaut, R. and Jacqmin-Gadda, H. (2004). Mixed models for longitudinal left-censored repeated measures. *Computer Methods and Programs in Biomedicine 74*: 255–260.

Thisted, R.A. (1988). *Elements of Statistical Computing: Numerical Computation*. Boca Raton, FL: Chapman & Hall/CRC.

Thompson, W.R. (1947). Use of moving averages and interpolation to estimate median effective dose I: fundamental formulas, estimation of error and relation to other methods. *Bacteriological Reviews 11*: 115–145.

Thompson, W.R. and Weil, C.S. (1952). On the construction of tables for moving average interpolation. *Biometrics 8*: 51–54.

Thorsén, E. (2014). Multinomial and Dirichlet-multinomial modeling of categorical time series. Thesis. Matematiska institutionen, Stockholms universitet. http://www2.math.su.se/matstat/reports/seriec/2014/rep6/report.pdf (accessed 6 February 2018).

Tibshirani, R. (1996). Regression shrinkage and selection via the lasso. *Journal of the Royal Statistical Society, Series B 58*: 267–288.

ToxRat (2016). https://www.toxrat.com (accessed 25 July 2017).

Tukey, J.W. (1977). *Exploratory Data Analysis*. Reading, MA: Addison-Wesley.

Ulm, K.W. (1991). A statistical method for assessing a threshold in epidemiological studies. *Statistics in Medicine 10*: 341–349.

UN (2013). *Globally Harmonized System of Classification and Labelling of Chemicals (GHS) Annex 1. Classification and Labeling Summary Tables*. New York and Geneva: United Nations.

UN (2015). *UN Globally Harmonized System of Classification and Labelling of Chemicals (GHS)*, 6 revised edition. New York and Geneva: United Nations.

UNEP (2007). *Strategic Approach to International Chemicals Management. SAICM Texts and Resolutions of the International Conference on Chemicals Management*. Geneva: UNEP.

USEPA (1985). Guidelines for Deriving Numerical National Water Quality Criteria for the Protection of Aquatic Organisms and Their Uses. *PB85-227049*. Springfield, VA: U.S. Environmental Protection Agency, National Technical Information Service.

USEPA (1996). Federal Insecticide, Fungicide, and Rodenticide Act (FIFRA). 7 U.S.C. §136 et seq..

USEPA (1998a). Federal Register/Vol. 63, No. 154/Tuesday, 11 August 1998/Notices. EPA [OPPTS-42206; FRL-6021-3].

USEPA (1998b). Federal Register/Vol. 63, No. 248/Monday, 28 December 1998/Notices. EPA [OPPTS-42208; FRL-6052-9].

USEPA (2015). Endocrine Disrupter Screening Program Test Guidelines OCSPP 890.2300: Larval Amphibian Metamorphosis Growth and Development (LAGDA), EPA Pub No. 740-C-15-001 (July). https://www.regulations.gov/document?D=EPA-HQ-OPPT-2009-0576-0018 (accessed 23 August 2017).

USEPA (2016a). The Frank R. Lautenberg Chemical Safety for the 21st Century Act. H. R. 2576.

USEPA (2016b). OCSPP 850.1000 Background and Special Considerations-Tests with Aquatic and Sediment-Dwelling Fauna and Aquatic Microcosms EPA 712-C-16-014 (October).

USEPA (2016c). Test Guidelines: OCSPP Series 850 Group A: Ecological Effects Test Guidelines. https://www.regulations.gov/document?D=EPA-HQ-OPPT-2009-0154-0043 (accessed 6 February 2018).

USEPA (2016d). Toxicity ForeCaster (ToxCast) data. https://www.epa.gov/chemical-research/toxicity-forecaster-toxcasttm-data (accessed 29 August 2017).

USFDA (2001). *Guidance for Industry: Statistical Aspects of the Design, Analysis, and Interpretation of Chronic Rodent Carcinogenicity Studies of Pharmaceuticals*. Draft Guidance. Silver Spring, MD: FDA/CDER.

USGPO (2017). Title 40: Protection of Environment. Chapter I – Environmental Protection Agency? Subchapter E – Pesticide Programs? Part 158 – Data Requirements for Pesticides. https://www.gpo.gov/fdsys/granule/CFR-2012-title40-vol25/CFR-2012-title40-vol25-part158 (accessed 19 February 2018).

Van den Brink, P.J., Hattink, J., Bransen, F. et al. (2000). Impact of the fungicide carbendazim in freshwater microcosms. II. Zooplankton, primary producers and final conclusions. *Aquatic Toxicology 48*: 251–264.

Van den Brink, P.J. and Ter Braak, C.J.F. (1999). Principal response curves: analysis of time-dependent multivariate responses of a biological community to stress. *Environmental Toxicology and Chemistry 18*: 138–148.

Van den Brink, P.J., Van Donk, E., Gylstra, R. et al. (1995). Effects of chronic low concentrations of the pesticides chlorpyrifos and atrazine in indoor freshwater microcosms. *Chemosphere 31*: 3181–3200.

Van den Brink, P.J., Van Wijngaarden, R.P.A., Lucassen, W.G.H. et al. (1996). Effects of the insecticide Dursbant 4E (active ingredient chlorpyrifos) in outdoor experimental ditches: II. Invertebrate community responses. *Environmental Toxicology and Chemistry 15*: 1143–1153.

Van der Hoeven, N. (2001). Estimating the 5-percentile of the species sensitivity distributions without any assumptions about the distribution. *Ecotoxiciology 10*: 25–34.

Van der Hoeven, N. (2010). Is it safe to pool the blank control data with the solvent control data? *Ecotoxicology and Environmental Safety 73*: 1480–1483.

Van Leeuwen, C.J. and Vermeire, T.G. (2007). *Risk Assessment of Chemicals, an Introduction*, 2e. Dordrecht, The Netherlands: Springer.

Van Wijngaarden, R.P.A., Van den Brink, P.J., Crum, S.J.H. et al. (1996). Effects of the insecticide Dursbant 4E (active ingredient chlorpyrifos) in outdoor experimental ditches: I. Comparison of short-term toxicity between laboratory and field. *Environmental Toxicology and Chemistry 15*: 1133–1142.

Van Wijngaarden, R.P.A., van den Brink, P.J., Oude Voshaar, J.H., and Leeuwangh, P. (1995). Ordination techniques for analyzing response of biological communities to toxic stress in experimental ecosystems. *Ecotoxicology 4*: 61–77.

Veith, G.D. and Konasewich, D.E. ed. (1975). *Structure-Activity Correlations in Studies of Toxicity and Bioconcentration with Aquatic Organisms*. Windsor, Ontario: Great Lakes Research Advisory Board.

van Vlaardingen, P.L.A., Traas, T.P., Wintersen, A.M., and Aldenberg, T. (2005). ETX 2.0. A program to calculate hazardous concentrations and fraction affected, based on normally distributed toxicity data. http://www.rivm.nl/en/Documents_and_publications/Scientific/Reports/2005/februari/ETX_2_0_A_Program_to_Calculate_Hazardous_Concentrations_and_Fraction_Affected_Based_on_Normally_Distributed_Toxicity_Data (accessed 6 February 2018).

Vock, D.M., Davidian, M., and Tsiatis, A.A. (2012). Mixed model analysis of censored longitudinal data with flexible random-effects density. *Biostatistics 13* (1): 61–73.

Vose, D. (2000). *Risk analysis*, 2e. Chichester, England: Wiley.

van der Waerden, B.L. (1952). Order tests for the two-sample problem and their power. *Proc Koninklijke Nederlandse Akademie van Wetenschappen Ser A 55*: 453–458.

Wang, Q., Denton, D.L., and Shukla, R. (2000). Applications and statistical properties of minimum significant difference-based criterion testing in a toxicity testing program. *Environmental Toxicology and Chemistry 19*: 113–117.

Wang, M., Hahne, J., and Ebeling, M. (2016). Reliability of BMD and ECx calculations required by Regulation EU 283/2013 for bird and mammal reproduction studies. Poster presented at SETAC Europe 26th Annual Meeting Nantes 2016.

Warren-Hicks, W.J. and Moore, D.R.J. (1998). Uncertainty analysis in Ecological risk assessment. *Proceedings from the Pellston Workshop on Uncertainty in Ecological Risk Assessment* (23–28 August 1995). Society for Environmental Toxicology and Chemistry (SETAC).

Weil, C.S. (1952). Tables for convenient calculation of median effective dose (LD50 or ED50) and instructions in their use. *Biometrics 8*: 249–263.

Weill, J.W. (1968). Testing for lack of fit in nonlinear models. *The Annals of Statistics 16*: 733–740.

Welling, P.G. (1997). *Pharmacokinetics Processes, Mathematics, and Applications*, 2e. Washington, DC: American Chemical Society.

Westfall, P.H., Tobias, R.D., Rom, D. et al. (2011). *Multiple Comparisons and Multiple Tests*, 2e. Cary, NC: SAS Institute.

Wheeler, M.W. and Bailer, A.J. (2009). Comparing model averaging with other model selection strategies for benchmark dose estimation. *Environmental and Ecological Statistics 16*: 37–51.

Wheeler, J.R., Grist, E.P.M., Leung, K.M.Y. et al. (2002). Species sensitivity distributions: data and model choice. *Marine Pollution Bulletin 45*: 192–202.

Widianarko, B. and Van Straalen, N. (1996). Toxicokinetics-based survival analysis in bioassays using nonpersistent chemicals. *Environmental Toxicology and Chemistry 15*: 402–406.

Wiki (2017). Survival analysis. http://en.wikipedia.org/wiki/Survival_analysis (accessed 1 February 2017).

Williams, D.A. (1971). A test for differences between treatment means when several dose levels are compared with a zero dose control. *Biometrics 27*: 103–117.

Williams, D.A. (1972). The comparison of several dose levels with a zero dose control. *Biometrics 28*: 519–531.

Williams, D.A. (1975). The analysis of binary responses from toxicological experiments involving reproduction and teratotlogy. *Biometrics 31*: 949–952.

Williams, D.A. (1977). Some inference procedures for monotonically ordered normal means. *Biometrika 64*: 9–14.

Williams, R. (2010). Fitting heterogeneous choice models with oglm. *The Stata Journal 10*: 540–567.

Williams, D.A., Kupper, L.L., Portier, C., and Hogan, M.D. (1988). Estimation bias using the beta-binomial distribution in teratology. *Biometrics 44*: 305–309.

van der Woude, H., Alink, G.M., and Rietjens, I.M. (2005). The definition of hormesis and its implications for in vitro to in vivo extrapolation and risk assessment. *Critical Reviews in Toxicology 35*: 603–607.

Wright, S.P. (1992). Adjusted P-values for simultaneous inference. *Biometrics 48*: 1005–1013.

Xenbase (2017). Developmental Stage Ontology. http://www.xenbase.org/anatomy/static/Xenopus_stages/NFstages.jsp (accessed 19 February 2018).

Yanagimoto, T. and Yamamoto, E. (1979). Estimation of safe doses: critical review of the hockey stick regression method. *Environmental Health Perspectives 32*: 193–199.

Yates, F. (1934). Contingency tables involving small numbers and the χ^2 test. *Supplement to the Journal of the Royal Statistical Society 1*: 217–235.

Yee, T.W. (2015). *Vector Generalized Linear and Additive Models with an Implementation in R*. New York: Springer.

Young, L.J. and Morfeld, P. (2016). Statistical considerations for a chronic bioassay study: exposure to decamethylcyclopentasiloxane (D5) and incidence of uterine endometrial adenocarcinomas in a 2-year inhalation study with Fischer rats. *Regulatory Toxicology and Pharmacology 74*: S14–S24.

Zajdlik, B.A. et al. (2005). Statistical Analysis of the SSD Approach for Development of Canadian Water Quality Guidelines. Project # 354-2005 prepared for Canadian Council of Ministers of the Environment (CCME). https://www.ccme.ca/files/Resources/supporting_scientific_documents/pn_1414_e.pdf (19 February 2018).

Zhang, J., Bailer, A.J., and Oris, J.T. (2012). Bayesian approach to estimating reproductive inhibition potency in aquatic toxicity testing. *Environmental Toxicology and Chemistry 31*: 916–927.

Zhang, G., Truong, L., Tanguay, R.L., and Reif, D.M. (2017). A new statistical approach to characterize chemical-elicited behavioral effects in high-throughput studies using zebrafish. *PLoS One 12* (1): e0169408. doi: 10.1371/journal.pone.0169408.

Zhao, Y. and Newman, M.C. (2007). The theory underlying dose-response models influences predictions for intermittent exposures. *Environmental Toxicology and Chemistry 26*: 543–547.

Zimmermann, H., Gerhard, D., Dingermann, T., and Hothorn, L.A. (2009). Statistical aspects of design and validation of microtitre-plate-based linear and non-linear parallel in vitro bioassays. *Biotechnology Journal 5* (1): 62–74.

Author Index

Abbott, W. S., 140, 159, 190, 223, 245
Akritas, M. G., 132
Aldenberg, T., 294, 300, 302, 339–341
Almstrup, K., 47
Alvord, W. G., 114
Anderson, D. R., 137
Anderson, M. J., 29
Apostol, T. M., 357
Armitage, P., 206
Ashauer, R., 331, 341, 342
Askey, R. A., 352
Ayer, M., 68

Backhaus, T., 323
Bailer, A. J., 175, 176
Barlow, R. E., 68
Bartholomew, D. J., 74
Bates, D. M., 32
Bathke, A., 132
Bauer, P., 13, 24
Baumans, V., 249
Beasley, T. M., 132
Bedaux, J. J. M., 342
Belden, J., 324
Bennett, B. M., 135, 206, 208
Benoit, D. A., 290
Berenbaum, M. C., 323
Berry, D. A., 15
Beyer, H. R., 357
Bland, J. M., 262
Bliss, C., 132
Blom, G., 132
Boesten, J., 143
Bolch, B. W., 300
Bolker, B. M., 59
Box, G. E. P., 29, 58
Brain, P., 92
Brenner, D. J., 89
Bretz, F., 71
Brock, T. C. M., 81, 302, 310, 311
Brock, W. J., 12
Brown, C. C., 94
Brown, M. B., 58, 68
Bruce, R. D., 92, 93
Brünning, H., 74
Bunke, O., 114
Burnham, K. P., 103

Calabrese, E. J., 13, 47
Cameron, A. C., 29, 37, 140, 223
Carakostos, M. C., 323
Carelli, G., 13
Carroll, R. J., 322
Cavanaugh, J. E., 104
Cedergreen, N., 47, 92, 93, 104, 109, 323
Chambers, J. M., 125, 132
Chapman, P. F., 294, 295
Charles, S., 234, 235
Chatfield, C., 137
Chen, C. W., 33
Chen, D. G., 199
Chernick, M. R., 302
Chiu, W. A., 327
Choy, S. L., 30
Christensen, E. R., 190
Christiansen, S., 323
Clayton, C. A., 331, 332
Clyde, M. A., 30
Coady, K., 254
Cohen, A. C., 173
Collett, D., 37, 182, 191, 208, 209, 229, 261, 266, 268
Conover, W. J., 132
Cook, D., 229
Cook, J. D., 235
Cooke, R. M., 30
Cox, C., 89, 92, 95
Crawley, M. J., 31
Cuppen, J. G. M., 310, 311

Dagenais, M. G., 322
D'Agostino, R. B., 56, 58, 295
Danbaba, A., 62
Davis, J. M., 13, 24, 47
Delignette-Muller, M. L., 234, 235
Deneer, J., 324
Dennis, J. E., 98
Dixon, D. G., 13
Dixon, W. J., 204
Dowse, R., 294, 302
Draper, N. R., 32, 58, 112, 114, 137
DuBoudin, C., 293
Ducrot, V., 331, 342
Duke, S. O., 13
Dunn, O. J., 53, 65, 80
Dunnett, C. W., 10, 29, 50, 53, 63, 79, 80, 330

Efron, B. E., 137
Ema, M., 15
Engleman, R. M., 206
Epstein, S. S., 33
Erickson, T., 322

Farewell, D. M., 249
Faust, M., 323, 324
Fears, T. R., 15
Field, A., 125
Filliben, J. J., 132
Finney, D. J., 159
Fisher, R. A., 134
Fisk, P., 275
Fleming, R., 191
Fox, D., 53
Frey, H. C., 229

Statistical Analysis of Ecotoxicity Studies, First Edition. John W. Green, Timothy A. Springer, and Henrik Holbech.

Companion website: www.wiley.com/go/Green/StatAnalysEcotoxicStudy

Fuller, W. A., 321, 322
Fung, K. Y. D., 167, 249
Furuhashi, T., 313

Gad, S. C., 33, 124
Gao, Z., 144, 332
Gart, J. J., 15, 177
Gaylor, D. W., 89
Gelman, A., 30, 327, 351
Gergs, A., 331, 342, 343
Gessner, P., 323
Gillies, K., 343
Gilliom, R. J., 323, 338
Gilmour, S. G., 136
Giraud, C., 136, 137
Goerlitz, L., 144
Good, P. I., 60
Goutelle, S., 94
Greco, W. R., 323
Green, J. W., 19, 29, 38–41, 51, 53, 74, 152, 169, 186, 193, 198, 224, 228, 245, 246, 248, 249, 256, 304, 305
Greenland, S., 15
Grimm, V., 343
Gruber, M. H. J., 353
Grynovicki, J. O., 151
Guimarães, P., 249
Gutenbrunner, C., 251

Hackett, P. K., 3
Hadrup, N., 324
Hamilton, M. A., 210, 211
Harrell, F. E., 137
Harris, E. K., 206, 208, 209
Hart, A., 32
Hartley, H. O., 58
Haseman, J. K., 15
Hayes, A. W., 124
Hazewinkel, M., 352
Headrick, T. C., 132
Helsel, D. R., 229, 302, 336–338
Hilbe, J. M., 29, 37, 140, 165, 175, 223, 235
Hirji, K. F., 166
Hoaglin, D., 55
Hobbs, N. T., 30, 351
Hockberg, Y., 29, 48, 49, 51–53, 63
Hocking, R. R., 76, 138, 151, 356, 359
Hoekstra, J. A., 206
Hogg, R. V., 310
Hollander, M., 73
Holm, S., 48
Hooda, B. K., 137
Hoppe, F. M., 49
Hora, S. C., 30, 351
Horn, M., 79, 80
Hothorn, L., 15, 31, 68, 102, 173, 176, 223, 247, 256, 330
Hothorn, T., 223
Hougaard, P., 269
Howard, G. J., 324
Hsu, J. C., 48, 49, 51, 52
Hubert, J. J., 12
Huet, S., 98, 146, 299
Hurlbert, S. H., 9, 10

Hutchinson, T., 38

Ibrahim, J. G., 15
Irwin, J. G., 158
Isnard, P., 14
Iwasaki, Y., 14, 293

Jacobs, M. N., 276
Jager, T., 331, 341, 342
Jagoe, R. H., 295
James, W. H., 234
Jennrich, R. L., 97
Johnson, N. L., 158, 229, 235, 304, 311
Jolliffe, I. T., 313
Jonckheere, A. R., 53
Jones, B. S., 269

Kabacoff, R. I., 31
Kass, R. E., 104
Kavlock, R. J., 95
Kendall, M., 229
Kennedy, W. J., 32, 357
Kerr, D. R., 163, 223
Khan, A., 29
Kienzler, A., 324
Klein, J. P., 266, 268
Kleinman, K., 31
Klepper, O., 89
Klimisch, H. J., 291
Klockars, A. J., 132
Koenker, R., 251
Kon Kam King, G., 294, 302
Kong, M., 324–326
Kooijman, S. A. L. M., 13, 33, 35, 89, 92, 294, 342
Korn, E. L., 72
Kortenkamp, A., 324

Lammer, E., 321, 322
Landis, W. G., 53
Lange, K., 357
Latscha, R., 159
Lazic, S. E., 316, 317
Lee, E. T., 37, 261, 268, 302
Lee, E. W., 268
Lehman, R., 223
Lehmann, E. L., 66, 73
Levenberg, K., 357
Li, J., 199
Lloyd, R., 13
Lorenzen, T. J., 29
Lyles, R. H., 113, 114

Mallows, C. L., 136
Mansouri, H., 132
Mantel, N., 33
Marcus, R., 71
Marquardt, D., 357
Martin, B. T., 343
Martinez, E. Z., 163
McArdle, B. H., 322
McCullagh, P., 223
McCulloch, C. E., 208
Mead, R., 124, 125

Mebane, C., 14
Mebane, C. A., 74
Merrington, G., 293
Meyer, M. A., 30, 351
Mi, J., 93, 358
Miller, G. A., 124, 125
Miller, R. J., 48, 58
Moermond, C. T. A., 291
Monte-Silva, K., 47
Montgomery, D. C., 32
Moon, H., 176
Moore, D. R. J., 190, 227
Moré, J. J., 357
Morel, J. G., 173, 174
Morgan, B. J. T., 30, 206
Morgan, M. G., 351
Morrison, D. F., 137, 313
Mosimann, J. E., 248
Motulsky, H., 32, 98, 103, 112–114, 146, 299
Murie, C., 316
Mylchreest, E., 15

Neath, A. A., 104
Nelder, J. A., 223
Nelson, K. P., 163
Newman, M. C., 15, 294, 295, 336, 337, 342
Nielsen, O. K., 223
Nieuwkoop, P. D., 4, 233, 249
Nowicki, S., 27
Nyholm, N., 190
Nyman, A. M., 331

O'Brien, R. G., 58
Odeh, R. E., 74
Ohara Hines, R. J., 199
Oksanen, L., 9, 10
Oris, J. T., 160, 163, 186, 351
Ortiz, N. R., 30, 351
Ott, R. L., 114

Pal, M., 322
Parzen, E., 346, 350
Peddada, S. D., 175, 176
Peltier, W. H., 206
Petkau, A. J., 199
Peto, R., 177
Petterino, C., 15
Pinheiro, J. C., 331
Plummer, J. L., 325
Poon, A. H., 71
Portier, C., 175
Portier, C. J., 175
Posada, D., 104
Posthuma, L., 294, 300, 323
Potter, R. W., 71
Pratheeshkumar, P., 316
Proschan, M., 124
Purchase, L. F. H., 89
Puri, M. L., 71

Rao, J. N. K., 167, 249
Rasch, D., 29
Ratkowsky, D. A., 113
Rencher, A. C., 313

Ritz, C., 94
Robertson, T., 53, 68
Rochon, J., 29
Rodney, S. I., 300
Rudin, W., 349
Rufli, H., 206
Ryan, L., 15

Schabenberger, O., 92, 93
Scheffe, H., 52
Schervish, M. J., 330
Schlain, B., 316
Schmidt, K., 15
Schmolke, A., 343
Schneider, K., 291
Schoenfeld, D., 72
Schwartz, G., 104
Schwarz, C. J., 104
Scott, B. R., 89
Searle, S. R., 76, 131, 359
Seber, G. A. F., 114
Seefeldt, S. S., 94
Selwyn, M. R., 177
Serfling, R. J., 174, 300, 349
Shao, Q., 294, 296, 300
Sheskin, D. J., 66, 73
Shirley, E. A. C., 71, 124, 125, 132, 135
Slob, W., 16, 92–94
Šmilauer, P., 316
Snedecor, G. W., 58
Sroka, C., 327
Staline, M. R., 63
Staveley J. P., 81
Stebbing, A. R. D., 13
Stephan, C. E., 206, 293
Stephenson, W. R., 132
Stevens, M. T., 124
Straetemans, R., 316
Swintek, J., 249

Takizawa, T., 124
Tamhane, A. C., 29, 63, 110, 329, 330
Tarone, R. E., 15, 161, 166
ter Braak, C. J. F., 313–316
Therneau, T. M., 269
Thiébaut, R., 229
Thisted, R. A., 55
Thompson, W. R., 206
Thorsén, E., 248
Tibshirani, R., 137
Tukey, J. W., 60

Ulm, K. W., 89

Van den Brink, P. J., 310, 311, 313–316
Van der Hoeven, N., 295
van der Hoop, J., 335
van der Waerden, B. L., 132
Van der Woude, H., 13
Van Leeuwen, C. J., 275
Van Wijngaarden, R. P. A., 314, 315
Veith, G. D., 275
Vock, D. M., 229
Vose, D., 12

Wang, M., 327
Wang, Q., 114
Warren-Hicks, W. J., 12, 30, 351
Weil, C. S., 206
Weill, J. W., 97
Welling, P. G., 331, 332, 335, 337
Westfall, P. H., 80
Wheeler, J. R., 293, 294
Wheeler, M. W., 326
Widianarko, B., 342
Williams, D. A., 10, 53, 68, 69, 71, 72, 163, 351
Williams, R., 29
Wright, S. P., 48, 49

Yanagimoto, T., 92, 95
Yates, F., 158
Yee, T. W., 173
Young, L. J., 176

Zajdlik, B. A., 293
Zhang, G., 15, 316
Zhang, J., 15
Zhao, Y., 342
Zimmermann, H., 316, 317

Subject Index

Abbott's formula, 190, 191
Abiotic test, 275, 290
Absorption, distribution, metabolism and excretion (ADME), 275
Abundance, 310–312, 314, 315, 343
Adjustment for multiple comparisons
  Bonferroni, 48
  Bonferroni–Holm, 48–49
  comparison wise error (CWE), 48
  family wise error (FWE), 48, 49
AIC (*see* Akaiki information criteria)
Akaiki information criteria, 103
  with benchmark dose, 326
  with small sample correction (AICc), 326
  weighted models and AIC, 103
Amphibian metamorphosis assay (AMA) TG 231, 5
ANCOVA, 123ff
  interaction, 132
  multiple factors, 135
  nonparametric, 131ff
  nonparametric with transformed covariate, 135
  parametric, 127–131
  tests for equal slopes, 125
ANOVA table, 353
  error sum of squares, 353
  model sum of squares, 353
  total sum of squares, 353–354
Area under the growth curve (AUC), 332–333, 343

Back transformation, 95
Bayes
  Bayesian methodology, 19, 25, 30
  expert judgment, 30
  posterior density, 30
  prior density, 30
  uninformative prior, 30
Bayesian information criterion (BIC), 104, 326
Bayes' rule, 30, 346
Benchmark dose
  BMDL, 326
  model, 326–327
  software (*see* Benchmark dose software)
Benchmark dose software
  BMD, 326
  PROAST, 326
Bioaccumulation, 335–336
Bioamplification, 335
Biocides in the Biocidal Product Regulation (BPR), 287, 290, 291
Bioconcentration, 335–341
  dietary exposure, 341
Biomagnification, 335
Biotic systems, 276, 281
Blom formulation, 132
BMD (*see* Benchmark dose, model)
Bonferroni–Holm adjustment, 159, 167
Bootstrap for SSD, 302
Box–Cox transform, 340, 341

C(α) (*see* Tarone's C(α))
CANOCO, 314
Censored data, 228
  likelihood function for, 229
Censoring, 269
  relation to sacrificial deaths, 268
Central limit theorem, 27, 35
Chemical Substances Control Law (CSCL), 280, 287
CMAX, 332
Cochran–Armitage
  comparison with GLMM, 166
  example, 162, 166, 167, 178
  independence of dose-response shape, 166
  power, 166
  statement, 165–166
Coefficient of variation, 96
  in mesocosm studies, 310
Comparison of Models, 102–104
Conditional probability, 345–346
Confidence bounds
  for ECx estimates, 90, 92, 93, 96, 99
Conjugate prior, 163, 351
Controls
  historical, 15
  multiple, 8, 38–41
  need for, 8, 9
  negative, 38
  positive, 41
  solvent, 39, 40
  statistical handling, 38–41
Covariate, 123
  fetal weight, 124
  liter size, 124
  pup weight, 124
Cox proportional hazard regression, 266ff

Data assessment, 29–30
  influential observation, 29, 31, 43
  outlier, 29, 31, 43
  scatter plot, 31, 36
  visual, 29–31
Degradation rate in soil, 143ff
Delta method, 93, 94, 108, 357
Depuration, 336
Developmental stage, 6, 229
  Nieuwkoop and Faber scale for, 233
Deviance, 163
Distribution for survival data
  negative binomial, 264
  normal, 265, 266
  Poisson, 262–266, 272
Distribution function, 346

Statistical Analysis of Ecotoxicity Studies, First Edition. John W. Green, Timothy A. Springer, and Henrik Holbech.

Companion website: www.wiley.com/go/Green/StatAnalysEcotoxicStudy

Distributions
beta, 163
beta-binomial, 248
binomial, 25, 140
for developmental stage, 254, 255
Dirichlet, 248, 249
gamma-Poisson, 36
maximum modulus, 63
multinomial, 162, 164, 245, 247, 248
M variate central t, 330
negative binomial, 36, 140
normal, 26, 150
Poisson, 25, 140, 150, 164
prior, 30
probability, 24, 25
probability density function, 24
studentized maximum, 63, 72
Distributions for SSDs
Burr III, 294, 296, 301, 302
exponential, 294–297
Fisher–Tippett, 294, 296, 298
loglogistic, 294, 296
lognormal, 294–296
reciprocal Pareto, 294, 296
reciprocal Weibull, 294, 296
Dose response, decreasing, 93
Dose response, increasing, 93, 94
Dunn (*see* Statistical tests)
Dunnett (*see* Statistical tests)

Effect concentration (EC*x*), 89
confidence intervals, 33, 34
effects concentration, 19, 32
EFSA guidance, 326
Endocrine Disruptor Screening Program (EDSP), 280, 290
Errors-in-variables regression, 321ff
Error structure (*see* Distributions)
EUFRAM, 294, 299, 301, 302
European Chemical Agency (ECHA), 275
European Union Reference Laboratory for alternatives to animal testing (EURL-ECVAM), 276
Event, 345
Examples
algae cell density, 65
algal density, 112
collembola reproduction, 64, 110, 119
cucumber dry weight, 112
Daphnia chronic length, 120
Daphnia chronic length, weight, and reproduction, 120
Daphnia chronic reproduction, 83, 220
Daphnia chronic size, 86, 87
Daphnia mortality, 214, 216
Daphnia reproduction, 77, 82
Daphnia survival, 204
Daphnid mortality, 217
degradation dataset, 1, 2, 145, 155
Duckweed growth, 82
earthworm mortality, 217
earthworm reproduction, 82, 83, 120
earthworm weight, 119
ELS first day of swimup, 238, 242
fish acute mortality, 217
fish screens, 82
hockey stick model fit to Selenastrum cell counts, 104
Lemna gibba frond count and biomass, 117
medaka secondary sex characteristics, 239
medaka sex ratio, 157
medaka VTG, 87
MEOGRT egg production, 134ff
Midge emergence, 179
mite survival, 183
mollusc (snail) reproduction, 161, 236
mouse reproduction, 126, 128
mysid mortality, 216
neurotoxicity study, 147ff
non-target terrestrial plant (NTTP) emergence, 187
NTTP vegetative vigor, 167
phenotypic sex of zebrafish, 195
pond mesocosm myriophyllum, 82–84
rainbow trout early life stage length and weight, 120
rainbow trout first day of swimup, 230
rainbow trout hatching, 86
rainbow trout last day of hatch, 241, 241rainbow trout length, 50, 51, 54, 69, 82, 90
rainbow trout mortality, 217
rainbow trout with hormesis, 104–105, 110
Rao–Scott Cochran–Armitage, 167–168
rodent motion data, 152ff
Shirley antidote data, 132
shortened cervical ribs, 173
soybean shoot height with hormesis, 105
stickleback mortality, 217
sugarbeet height and weight of shoot and root, 120
survival analysis of Daphnia first day of reproduction, 261
survival analysis of LAGDA time to metamorphosis, 269ff
tamoxifen-citrate, 160, 193ff
4-tert octylphenol, 139, 158
wheat root weight, 120
Xenopus laevis, 70, 87
zebrafish intersex, 162
zebrafish sex ratio, 164
Examples for SSD
SSD for algal data, 308
SSD for aquatic plants, 297, 298
SSD for aquatic species with censored data, 308
SSD for non-target plants, 308
SSD with heavy tail, 297
Expected value, 349
Experimental design, 8ff, 146
Extrapolation, 89ff

False positive rate (FPR), 21, 131
The Federal Food, Drug and Cosmetic Act (FFDCA), 280
Federal Insecticide, Fungicide, and Rodenticide Act (FIFRA), 280, 291
Fiducial bounds, 96
Fieller's theorem, 208–210
Filiben formula, 132
Fish Embryo Acute Toxicity (FET) test, 317
Frog Embryonic Toxicology Assay Xenopus (FETAX), 317, 318
F-test (*see* Statistical tests)
F-test before effects tests, 51, 52

Gamma-Poisson (Bayes), 234
gPBVP, 235
gPHorm, 235
gPLLog, 235
gPOE2, 235
gPOE3, 235
Geometric mean, 96
Geometric standard deviation, 96
Goodness of Fit, 90, 97, 98, 102, 112
about fitted model, 97
F test for, 97, 100
F value for, 97
lack of fit test, 97
likelihood ratio test for, 97
pure error, 97
R2 98
regression, 97
residuals, 97–101
Greenwood's formula, 261

Hazard concentration, 293
Hazard function
cumulative, 259
definition, 259
relation to survival function, 259
HC5 definition, 293
dependence on distribution, 294, 295ff
HC5LB, 301, 303
Helmert contrast, 330
Hessian 357
Heterogeneity, 29
Histogram, 54, 55, 71, 81, 99
Histopathology, 4–6, 243ff
Homogeneity, 29, 31, 35
Hormesis, 92, 104–107, 110, 120
Hypothesis test, 41, 48, 52, 55, 56, 58, 79–81

Individual tolerance, 342
Influential observation, 60
In silico, 275, 276, 278
Interagency Coordinating Committee on the Validation of Alternative Methods (ICCVAM), 276
Intergovernmental Forum on Chemical Safety (IFCS), 279
International Labour Organization (ILO), 279
Inter-Organization Programme for the Sound Management of Chemicals (IOMC), 279
Interpolation, 89ff, 98, 99
In vitro, 275–278, 291
In vivo, 275, 276, 278
Isobologram, 323, 324
Isotonic means, 68–70, 72
Iteratively reweighted least squares, 146

Japan chemical regulation and testing
Chemical Substances Control Law (CSCL), 287
Japanese Center for the Validation of Alternative methods (JaCVAM), 276
Ministry of Agriculture, Forestry and Fisheries, 287
Ministry of Economy, Trade and Industry (MITI), 287
Ministry of Health, Labor and Welfare, 287
Ministry of the Environment, 287

Kaplan–Meier product limit estimator, 261ff
Kronecker product, 354, 360
Kruskal–Wallis (*see* Statistical tests)

LC50, estimation
binary method, 210
as function of time (*see* Models)
moving average angle, 206ff
probit (*see* Probit model)
sequential estimation, 206
Spearman–Karber (*see* Spearman–Karber)
use in SSDs, 305
Levene (*see* Tests for variance homogeneity)
Limit test, 327
LOEC, 19
Logit, 189
Log likelihood function, 348
Log transformation, 95

Mallow's Cp (*see* Model selection)
Mann–Whitney (*see* Statistical tests)
MATC, 20
Maximum likelihood, 92, 93
estimates (MLEs), 96
method, 95, 348ff
Maximum unsafe dose, 329–330
MEOGRT, 137ff (*see also* Examples; OECD Test Guidelines; Test guidelines with ordinal responses; Time-to-Event data; USEPA Test Guidelines)
Mesocosm, 310ff
multivariate analysis, 313ff
Metabolite, 144
Methods of least squares, 97
Microcosm, 309–311
Microplate experiment, 316ff
regression models for, 318, 319, 321
titerplate, 316
Williams' test for, 318, 343
Minimum detectible difference (MDD), 81, 343
classes of minimum detectible difference, 312
MDDEFF, 310ff
MDD%, %MDD, 311, 312
Minimum safe dose, 329–330
Mixtures, 323ff
additivity, 323
analysis of, 324–326
antagonism, 324
concentration addition, 323, 324
synergism, 323

MLE (maximum likelihood estimates), 350
Model assessment for continuous models, 99, 104, 112–114
Models
ANCOVA, 16
ANOVA, 9, 16
biologically based, 13, 34–35
Bruce–Versteeg (BVP), 193, 198
compartment model for absorption and elimination, 331–332, 342
DebTox, 199
gamma-Poisson (*see* Gamma-Poisson (Bayes))
GNLM, 194–197 (*see also* Models, NLME)
hierarchy, 12–13
Hill model for benchmark dose, 326
hormesis, 12, 13
LC50 as function of time, 199
linearized multistage model (LMS), 326
logistic, 188
NLME, 254
normal vs. Poisson errors, 225ff
OE2 196, 199
OE3 193
OE3 for benchmark dose, 326
OE5 for benchmark dose, 326
pairwise, 21
Petkau–Sitter, 201
probit (*see* Probit model)
probit for benchmark dose, 326
regression, 13
requirements, 15
requirements, assessing, 21, 24, 29–31
SFO (OE2), 332, 334–336
simple, 12
TKTD, 199
trend, 12–13, 21
Weibull, 188, 198
Models, biology based, 89, 92
Models, continuous nonlinear, 92, 95, 98, 100
Brain and Cousens, 92
Bruce–Versteeg (BVP), 92, 93
Cedergreen, 93
Exponential with shape parameter (OE3), 92, 93
Exponential with shape parameter and lower bound (OE5) 92, 93
Hill, 92, 94
Hormetic (HORM1, HORM2), 92–93
Log logistic (3PL, 4PL), 93, 94
Michaelis–Menten (MM), 92, 93
Schabenberger, 93
Simple Exponential (OE2), 93
Simple Exponential with Lower Bound (OE4), 93, 97, 115
Threshold (Hockey stick, HS3PL, HS4PL, HS3PE, and HS4PE), 94–95, 104
Model selection, 136, 228 (*see also* Akaiki information criteria)
Mallow's *C*p, 136
stepwise regression, 136

Models for time to event data, 262, 264, 266, 269, 271, 272
frailty, 269, 271
GLMM, 269, 271
sandwich, 268, 269, 271
Model uncertainty, 326
Monte Carlo, 100
Simulation for SSD modelling, 300–302
Mutual Acceptance of Data (MAD), 276

NEC, 33, 34
Nieuwkoop and Faber (*see* Nieuwkoop and Faber severity scores)
Nieuwkoop and Faber severity scores, 4
NOEC
biological meaning, 20
confidence intervals, 20
relation to experimental design, 20, 21, 23, 30, 31, 37
statistical meaning, 19
No Effect Concentration (NEC), 89
Nondetects, 337, 338
Normality, 95, 97 (*see also* Tests for normality)
Normalization to control, 190ff
comparison with untransformed data, 192, 193, 197, 198
Normalized rank, 132, 150

Occam's razor, 104, 136
OECD Guidance Documents
Fish Toxicity Framework (GD, 171), 290
Framework for Testing and Assessment of Endocrine Disrupters (GD, 150), 278, 290
TG, 201 Algal growth inhibition test, 281, 287
TG, 202 Daphnia acute toxicity test, 290
TG, 202, part, 1 Daphnia acute toxicity test, 287
TG, 202, part, 2 Daphnia reproduction test, 287
TG, 203 Fish acute toxicity test 281, 283, 287
TG, 207 Earthworm acute toxicity test, 281, 287, 290
TG, 229 Fish Short Term Reproduction Assay (FSTRA), 278, 280, 284, 290
TG, 231 (*see* Amphibian metamorphosis assay (AMA) TG 231)
TG, 236 (*see* Fish Embryo Acute Toxicity (FET) test)
TG, 240 Medaka Extended One Generation Reproduction Test (MEOGRT), 278, 280, 285
TG, 431 *In vitro* skin corrosion, 276
TG, 476 *In vitro* Mammalian Cell Gene Mutation Test, 276
TG, 487 *In vitro* mammalian cell micronucleus test, 276
OECD test guidelines, 38, 40, 198, 276, 281
Office of Chemical Safety and Pollution Prevention (OCSPP), 280
Optimization methods
Newton 356
Newton–Raphson, 357

Ordinal scale, 243–245, 254
Organisation for Economic Co-operation and Development (OECD), 275
Outlier, 61
  Tukey rule, 60
Overdispersion, 53, 67, 76, 140, 144, 160, 162, 169, 187, 351, 352
  in multinomial response, 245, 247, 248

Parent compound, 144
Pathology severity score, 243–245, 247–250
Percent effect, 33
  absolute, 34
  additional, 34
  relative, 34
Percentile, definition, 60
Phenotypic sex, 193
Plant Protection Product Regulation (PPPR), 287, 291
Pool-the-adjacent-violators (PAVA) algorithm, 68, 69
Power, 77–81
  calculation, 79, 80
  definition, 77
  examples, 78
Power Of tests for quantal responses, 168ff
Power To detect non-normality, 56–58, 294–296
predicted environmental concentrations (PEC), 287
Principal component analysis (PCA), 313
Principal response curve, 313, 315
Prior density function, 351
Probability density function, 346
  beta-binomial density, 351–352
  binomial density function, 347
  multivariate normal density, 357, 359
  normal density, 348
  Poisson density function, 347
  standard normal, 348
Probability function, 345
Probit model
  with background, 181ff
  with ECx as parameter, 182
  for mortality, 181, 184–185
  for survival, 181, 184–185

QQ-plot, 99
Quality of ECx estimates, 113
  confidence limits, 114
  in relation to control variability, 110
  in relation to test concentrations, 98, 100, 110, 114
Quantile, definition, 60
  equivalence to percentile, 61
Quantitative structure–activity relationship (QSAR), 275

Random effects, 75–76, 352
  continuous, 346
  count, 346
  ordinal, 346
  quantal, 346, 347, 349
Randomization, 11

Registration, Evaluation, Authorization and Restriction of Chemicals (REACH), 279, 280, 287, 290, 291
Regression
  effects concentration (ECx), 32
  model choice, 32
  parameter, 32, 34, 36
  parameter estimate, 33
Reparameterization, 93, 94
Repeated measures, 137
  model with multiple repeated factors, 150
  two repeated factors, 147ff
Replication, 9–11
  need, 9
  pseudo, 9
Residuals
  definition, 54, 58, 60
  deviance, 264, 271
  Pearson, 263–265
  pure-error, 54, 55
  raw, 263
  regression, 56
  studentized, 264, 265
  use, 61, 71
Response
  continuous, 2
  count, 3
  discrete, 29, 30, 35–37
  ordinal, 4–6
  quantal, 2

Safety data Sheets (SDS), 279
Sample description space, 345
Sample space (*see* Sample description space
  Satterthwaite (*see also individual tests*)
  degrees of freedom, 63, 67
  formula, 328
Scatterplot, 99
Schwartz information criterion (SIC), 104
Severity score, 4–6 (*see also* Nieuwkoop and Faber)
Shapiro–Wilk (*see* Tests for normality
Significance
  biological, 1, 5, 8, 13–15
  statistical, 1, 8, 13–15
Simple first order (SFO) model, 145
  Software for SSD (*see also* Nieuwkoop and Faber)
  BurrliOZ 294, 296, 299, 301, 302, 308
  ETX, 2.0 296, 299, 308
  MOSAIC, 294, 302, 303, 308
  SAS, 299, 302, 303, 308
  SSD Master, 294, 296, 298, 299
Spaghetti plot, 301
Spearman–Karber, 210
  to daphnia survival, 212, 213
  to mite survival, 212
SSD
  Aldenberg–Jaworska factors, 301
  with censored data, 302ff
  definition, 293
  treatment of multiple studies with same endpoint, 293, 303
  weighted model, 299

Starting values, 97
Statistical Software (*see also* Software for SSD)
  CETIS, 50, 68
  FORTRAN, 208
  PROAST, 95
  SAS quantreg procedure, 251, 252
  SEDEC, 205
  S-plus, 95
  SPSS, 95
  StatCHARRMS, 68, 73, 140, 247, 249, 256, 257
  ToxRat, 50, 68
Statistical tests
  adjustments for multiple comparisons, 22, 41
  Anderson–Darling, 147
  beta-binomial, 163, 164
  Bonferroni, 22, 23
  Bonferroni–Holm, 23
  CABS, 247, 248
  chi-square, 159
  Cochran–Armitage, 184
  Dunn, 64–65
  Dunnett, 22, 23, 29, 49–53, 62
  Fisher' exact, 158, 167
  F-test, 51–53
  GLMM, 59, 65, 76–77, 140, 162, 167, 250, 254–256
  Holm, 23
  Jonckheere–Terpstra, 73, 75, 77, 78, 81
  JTABS, 249
  Kruskal–Wallis, 53, 159
  LD50-only, 205
  LD50-slope, 205
  Levene, 35, 44
  Limit dose, 206
  Mann–Whitney, 66–67
  multi-quantal Jonckheere–Terpstra (MQJT), 249, 250ff
  nonparametric, 47
  1-sided or one-sided, 22, 24
  with ordinal data, 244, 245, 249–251
  pairwise, 22–24, 47–51, 62, 63, 65, 78
  parametric, 47
  Peto cause of death, 177–178
  with Poisson errors, 255, 256
  poly-3 test, 175–176
  Rao–Scott Cochran–Armitage, 167
  Rao–Scott Cochran–Armitage by Slices (RSCABS), 245ff
  Shapiro–Wilk, 147
  step-down, 22–24
  Tamhane–Dunnett (T3), 29, 51, 63–64
  Tarone's C(α)183, 186, 194, 215
  trend, 22–24, 29, 31, 35, 67–71, 73–75, 77, 78
  t-test, 63, 66
  Tukey outlier rule, 162
  2-sided or two-sided, 21, 22
  Wilcoxon (*see* Mann–Whitney)
  Williams, 68–72, 211
Zero-inflated binomial, 172ff

Step-function, 346
Stochastic death, 342
Strategic Approach to International Chemicals Management (SAICM), 279
Survival analysis with grouped data, 268ff
Survival function
  definition, 259
  empirical, 260

Tamhane–Dunnett (*see* Statistical tests)
Tarone's C(α), 161, 162, 164
  example, 162, 164
  statement, 161
Test guidelines
  ISO, 20281, 191
  ISO, 22030, 190
  OCSPP, 850.2000, 205
  OECD, 202, 199
  OECD, 203, 199
  OECD, 204, 199
  OECD, 208, 187, 191
  OECD, 210, 104, 120, 167, 186
  OECD, 211, 121, 199, 220
  OECD, 221, 190
  OECD, 223, 205
  OECD, 227, 105, 112, 120, 167, 191
  OECD, 231 AMA, 233
  OECD, 232, 119
  OECD, 234 FSDT, 157, 160, 193
  OECD, 240 MEOGRT, 137
  OECD, 241 LAGDA 228, 233, 270ff
  OECD, 243, 161, 234
  OECD, 307, 143
  OPPTS, 850.1075, 206
  USEPA890_2100JQTT, 137
Test guidelines with ordinal responses
  OECD TG, 229 (FSTRA), 253
  OECD TG, 231 (AMA), 249, 250, 253, 254
  OECD TG, 240 (MEOGRT), 240
  OECD TG, 241 (LAGDA), 243, 249, 254
  USEPA TG, 890.2100 (JQTT), 243
Tests for normality
  Anderson–Darling, 56–58
  application to SSDs, 297, 299, 313
  Cramer–von Mises, 56–58
  Kolmogorov–Smirnov, 56–58
  Shapiro–Wilk, 50, 56–58, 71, 81
Tests for survival data
  with censored data, 261
  Dunn, 262, 266, 272
  Dunnett–Hsu, 262, 263, 271, 272
  Jonckheere–Terpstra, 262, 266, 269, 271, 272
  Logrank, 266, 268
  Mann–Whitney, 262, 266
  Peto, 265
  Wilcoxon, 265, 266
Tests for variance homogeneity
  Bartlett, 58
  goodness-of-fit, 55, 58
  Levene, 58, 71, 81
Threshold concentration, 89, 91–93, 94ff, 104, 342
Time-to-event data
  LAGDA, 37
  MEOGRT, 37
Titerplate, 316 (*see also* Microplate)
TMAX, 332
Tox21, 276, 277
Toxicokinetic–toxicodynamic (TKTD) models, 331ff
  toxicodynamics, 331
  toxicokinetics, 331
Toxic Substances Control Act (TSCA), 279, 280, 290
Transformations, 59–60, 62, 131, 132, 135, 138, 150
  arc-sine square-root, 59
  Box–Cox, 59
  for continuous, 59
  for counts, 76
  effect on NOEC, 59
  effect on regression, 59
  Freeman–Tukey, 59
  Log, 58, 71
  for proportions, 59
  quantal data as proportions, 197
  rank order, 62, 67
  square-root, 64, 65, 76, 96, 117
  square-root of count data, 219
Treatment group
  number, 11
  spacing, 11
t-test (*see* Statistical tests)

UN Environment Programme (UNEP), 279
Uninformative prior, 351
United Nations (UN) Globally Harmonized System of Classification and Labelling of Chemicals (GHS), 275, 279
United States Environmental Protection Agency (USEPA), 280, 288
Unit of analysis, 47, 50, 75
Uptake, 337
USEPA Test Guidelines
  OCSPP, 850.1010: Aquatic Invertebrate Acute Toxicity Test, Freshwater Daphnids, 280, 287, 288
  OCSPP, 850.1025 Oyster Acute Toxicity Test (Shell Deposition), 280, 287, 288
  OCSPP, 850.1045: Penaeid Acute Toxicity Test, 280, 287, 288
  OCSPP, 850.1055 Bivalve Acute Toxicity Test (Embryo-Larval), 280, 287, 288
  OCSPP, 890.2200 Medaka Extended One Generation Reproduction Test (MEOGRT), 280
USEPA Toxicity Forecaster named (ToxCast), 276, 277

Variance, 349
  conditional, 164
  posterior, 164
  pooled sample variance, 353
Variance component, 358ff
Variance–covariance matrix, 352, 357
Variance heterogeneity, 99, 110
Variation
  sources, 15
  uncertainty, 15
Vegan package, 314
Visual assessment of data (*see also* Histogram)
  quantile plot (QQ-plot), 55
  scatter plot, 55
  stem-and-leaf, 55

Wilcoxon (*see* Statistical tests)
Williams' type contrasts, 176

Xenopus Embryonic Thyroid signaling Assay (XETA), 317ff

ZEOGRT, 143